# RNA Therapeutics in Human Diseases

Phei Er Saw • Erwei Song

# RNA Therapeutics in Human Diseases

Springer

Phei Er Saw
Sun Yat-Sen Memorial Hospital
Sun Yat-Sen University
Guangzhou, China

Erwei Song
Sun Yat-Sen Memorial Hospital
Sun Yat-Sen University
Guangzhou, China

ISBN 978-981-96-3043-1 ISBN 978-981-96-3041-7 (eBook)
https://doi.org/10.1007/978-981-96-3041-7



This Springer imprint is published by the registered company Springer Nature Singapore Pte Ltd.
The registered company address is: 152 Beach Road, #21-01/04 Gateway East, Singapore 189721, Singapore

# Foreword

First of all, I am very honored to provide a foreword for a book on such a groundbreaking and rapidly evolving field of RNA therapeutics in human diseases. My heartiest congratulation to Erwei and his team for their relentless effort in pioneering the field of immunotherapy and RNA-based therapeutics. Erwei has been outstanding since he was a post-doctoral fellow in my lab. He is now a distinguished Chinese oncologist, known for his significant contributions to medical science, particularly in the field of oncology. His research has been recognized internationally, earning him prestigious awards and recognition.

One of the prominent findings from his team includes the discovery related to neutrophil extracellular traps (NETs) and their role in promoting cancer metastasis. Specifically, his research identified that the DNA of NETs acts as a chemotactic factor attracting cancer cells and that targeting CCDC25 could be a promising strategy to prevent cancer metastasis. This study, published in *Nature*, has significant implications for the development of new therapeutic strategies for cancer treatment. Most recently, Erwei's team unraveled a unique role of circRNA in coding peptides as tumor-specific antigen (TSA), the first proof-of-principle study that opens a new horizon in the field of RNA therapeutics.

Moreover, Erwei has been honored with the 2022 TWAS award in Medical Sciences for his research on the influence of non-coding RNAs in regulating tumor plasticity. This award further highlights his contributions to understanding complex biological mechanisms that underpin cancer progression and metastasis, offering potential pathways for innovative treatments (Asian Scientist Magazine). Throughout his career, Song has received several honors, including the State Natural Science Award (Second Class), the Science and Technology Award from the Ho Leung Ho Lee Foundation, and he was elected as a Member of the Chinese Academy of Sciences in November 2019.

In the annals of medical research and therapeutics, the discovery and application of RNA technologies stand as a beacon of innovation, holding the promise to redefine our approach to treating a myriad of human diseases. As we stand on the precipice of a new era in medicine, *RNA Therapeutics in Human Diseases* aims to serve as both a comprehensive guide and a visionary outlook on the role of RNA-based therapies in combating a wide range of illnesses, from infectious diseases to cancer.

The journey of RNA therapeutics from theoretical possibility to clinical reality is a testament to the relentless pursuit of knowledge and the collaborative spirit that defines the scientific community. It was not so long ago that the idea of manipulating RNA to treat diseases was relegated to the realm of science fiction. However, through the dedicated efforts of countless researchers and clinicians, including those at the forefront of immunology and molecular biology, RNA therapeutics have become one of the most promising and dynamic areas of medical research.

As a professor at Harvard, deeply involved in studying both the innate and adaptive immune responses to infection and cancer, I have witnessed first-hand the transformative potential of RNA technologies. My work, focusing on how the immune system can be harnessed to fight disease, intersects with the principles of RNA therapeutics in ways that continually inspire and challenge me. The ability to design RNA molecules that can, for example, silence genes, edit genomes, or stimulate immune responses against pathogens and tumors, offers an unprecedented opportunity to treat diseases at their molecular roots.

This book is structured to provide a foundational understanding of RNA biology, followed by an exploration of the various types of RNA molecules (including messenger RNA, microRNA, and small interfering RNA) and their mechanisms of action. It delves into the historical development of RNA therapeutics, highlighting key milestones and the scientific challenges overcome along the way. Further, it provides an up-to-date overview of the current state of RNA therapy research, detailing the most promising approaches and their applications in treating diseases, such as cancer, genetic disorders, and viral infections.

A significant portion of this book is dedicated to the revolutionary impact of mRNA vaccines, a topic that has captivated the global community's attention in the wake of the COVID-19 pandemic. The swift development and deployment of these vaccines not only underscored the potential of RNA therapeutics to address immediate public health crises but also highlighted the scalability and adaptability of RNA technology. This achievement is perhaps one of the clearest demonstrations of how theoretical research can quickly transition into life-saving clinical applications.

Moreover, *RNA Therapeutics in Human Diseases* does not shy away from addressing the challenges and limitations of current RNA therapies. Issues such as delivery mechanisms, off-target effects, and immune responses are critically examined, offering insights into the ongoing research aimed at overcoming these hurdles. This book also casts a forward-looking gaze, discussing the future directions of RNA therapy research, including the exploration of novel delivery systems, the development of more targeted therapies, and the potential for personalized medicine.

As we look to the future, the possibilities of RNA therapeutics are boundless. With continued investment in research and development, coupled with a collaborative approach that transcends disciplinary and institutional boundaries, there is hope for groundbreaking treatments that can significantly improve patient outcomes and potentially cure diseases that were once thought to be untreatable.

*RNA Therapeutics in Human Diseases* is not just a book for scientists and students in the field of medical research; it is a beacon for anyone interested in the future of medicine. It serves as a reminder of the power of scientific innovation and collaboration to overcome the greatest challenges in health and disease. As we move forward, it is imperative that we continue to foster a culture of curiosity, perseverance, and openness, ensuring that the potential of RNA therapeutics is fully realized for the benefit of all humanity.

In closing, I am both humbled and excited to contribute to this monumental work. The journey of RNA therapeutics is far from complete, and this book represents both a reflection on our past achievements and a roadmap for the future. It is my sincere hope that this will inspire a new generation of scientists, clinicians, and thinkers to explore the untapped potential of RNA, pushing the boundaries of what is possible in the quest to cure disease and improve human health.

Boston, MA, USA Judy Lierberman

# Preface

Welcome to *RNA Therapeutics in Human Diseases*, a comprehensive exploration of one of the most exciting and rapidly advancing fields in biomedical research. This book aims to elucidate the transformative role of RNA-based technologies in developing novel therapeutic strategies for a wide array of human diseases. The discovery and subsequent understanding of RNA as a central molecule in cellular function and regulation have ushered in a new era of medical possibilities. RNA therapeutics, encompassing a variety of approaches such as small interfering RNA (siRNA), microRNA (miRNA) mimics and inhibitors, antisense oligonucleotides, and messenger RNA (mRNA) based therapies, represent a frontier in the quest to manipulate cellular processes at the genetic level.

This book begins with a historical perspective, tracing the journey from the early discoveries of RNA to the latest advances in RNA technology that have made therapeutic applications possible. The initial chapters introduce the fundamental concepts of RNA science, providing the necessary background to appreciate the mechanisms by which RNA can be harnessed for therapy. Subsequent chapters delve deeply into specific diseases where RNA therapeutics have shown promise. These include genetic disorders, cancers, infectious diseases, and chronic conditions such as heart disease and diabetes. Each chapter is authored by leading experts who provide insights into the current status, challenges, and future prospects of RNA therapies in these areas. A significant focus of this book is on the technological advancements that have propelled RNA therapeutics from theory to clinical reality. These include the development of stable, safe, and efficient delivery systems, and overcoming immunogenicity and off-target effects, which have historically impeded the progress of RNA-based drugs.

As we stand on the cusp of potentially revolutionary changes in the treatment of diseases, this book offers both a deep dive for researchers and clinicians already familiar with the field and a clear introduction for those new to the world of RNA therapeutics. Our discussion extends beyond the science and into the ethical, regulatory, and commercial landscapes that will shape the future of RNA medicine. It is my hope that *RNA Therapeutics in Human Diseases* will serve as a valuable resource, inspiring ongoing innovation and dedicated research in this dynamic field of medicine. Together, we are moving toward a future where RNA therapeutics can provide new hope for patients suffering from previously untreatable conditions.

Guangzhou, China Phei Er Saw
Guangzhou, China Erwei Song

# Contents

# About the Authors

**Phei Er Saw** Originally from Malaysia, Phei Er Saw earned her Bachelor's degree in Biotechnology from the University of Malaysia Sabah (UMS) and completed her PhD studies under the supervision of Dr. Sangyong Jon at KAIST, Korea, in 2014. She acquired exceptional expertise in the field of bio-nanomedicine, with emphasis on developing novel lipid nanoparticles as well as understanding the uptake mechanisms of lipid-based nanoparticles for cancer therapy. Her prior research combined the development of novel lipid-based nanoparticles, micelles, bicelles, and liposomes as well as drug and siRNA encapsulation and targeted delivery in vitro and in vivo. Her works have led to the publication of numerous papers and the filing of an international patent and two Korean domestic patents. During her postdoctoral fellowship at Harvard Medical School, she focused on stimuli-responsive polymeric NPs for the treatment of various cancers, focusing on HER2+ breast cancer and metastatic breast cancer (MBC). Her research experience to date has already had a strong impact on the ongoing design and synthesis of nanoparticle (NP) systems encapsulating small biologic drug conjugates (SBDCs) to overcome the bottleneck of current clinical failures of antibody–drug conjugate (i.e. herceptin–mertansine conjugate; Kadcyla) and small-ligand drug conjugate (i.e. folic-acid conjugated desacetylvinblastine hydrazide; Vintafolide). In her new position at Sun Yat-sen Memorial Hospital, Sun Yat-sen University, she is now focusing on using various bioresponsive nano-delivery system to target and modulate tumor microenvironment for immunotherapy. She is also developing fully natural, bio-active nanomedicine for medical applications and is currently developing a lipid-based cancer vaccine. To date, she has received three grants from the China National Science Foundation and is a recipient of Guangdong Province Excellent Young Researcher Award. She has so far authored approximately 100 papers that have received more than 5000 citations. Recently, she authored multiple review papers with conceptual advances, including the introduction of tumor ecosystem and introducing cancer stem cell as "mimicry."

**Erwei Song** is Professor of Breast Surgery at Sun Yat-sen University (SYSU), a Member of the Chinese Academy of Sciences, a "Cheung Kong Scholar" Chair Professor, a CMB (China Medical Board) Distinguished Professor, and a Fellow of the Royal Society of Biology (UK). He currently serves as President of Sun Yat-sen Memorial Hospital, SYSU.

Erwei obtained his M.D. and Ph.D. from Zhongshan Medical School and was trained as a surgeon at Sun Yat-sen Memorial Hospital, SYSU. He received postdoctoral training at Harvard Medical School in 2002 and became an instructor at the CBR Institute for Biomedical Research at Harvard in 2004. His early publication, "RNA Interference Targeting Fas Protects Mice from Fulminant Hepatitis," in *Nature Medicine* was the first to report that siRNA could be used therapeutically in whole-animal disease model. This finding was selected as one of the "Top Ten breakthroughs of 2003" by Science. In 2005, he returned to China and set up his own lab at Sun Yat-sen University.

As a clinician-scientist, he is keen to address scientific questions derived from clinical practice. He is among the first doctors in China to perform breast-conserving surgery for breast cancer patients, which preserves the microenvironment tissues for activation of anti-tumor immunity. His study has proposed and solidified "the Theory of Tumor Ecosystem," which sheds new lights on immunotherapy. He has identified various new subtypes of cancer-associated immune cells and mesenchymal cells, new receptors for crucial cytokines, and a new archetype of long non-coding RNAs that function as signaling molecules to regulate intracellular or intercellular signal transduction. These findings provide new therapeutic strategies for targeting tumor microenvironment as well as immunotherapies for solid tumors. By comparing patient samples before and after treatment, he has also demonstrated the critical role of miRNAs in regulating the biology of cancer stem cells (CSC).

Prof. Song has published 156 SCI research articles, including papers in journals such as *Nature* (1 paper), Cell (3 papers), *Cancer Cell* (3 papers), *Nature Immunology* (2 papers), *Nature Cell Biology*, *Nature Cancer*, *Science Translational Medicine*, *Nature Communications* (4 papers), among others. Owing to his academic achievements, Prof. Song was invited to contribute a review article on the tumor microenvironment for *Nature Reviews Drug Discovery* and to chair an international Cell Symposium for Cell Press on "Functional RNAs." As a surgeon, Dr. Song compiled the first Chinese Expert Guidelines for Breast-Conserving Surgeries in Early Breast Cancer. To date, his research papers have been cited 12,012 times. As the corresponding author, he has published 73 research articles, with 1 of them having been cited 1458 times and 16 of them having been cited more than 100 times.

Prof. Song serves as the associate editor for *BMC Cancer*, *Cancer Science*, *Science China-Life Sciences*, and the *Journal of Bio-X Research* and a reviewing editor for the *Journal of Biological Chemistry*, as well as an editorial board member of *Signal Transduction and Targeted Therapy*. His research work was selected as one of the Top Ten Scientific and Technological Breakthroughs by the Ministry of Education in China in 2008 and 2018, respectively. He has won numerous awards, including the "Ho Leung Ho Lee Foundation Award" and the "National Natural Science Award," conferred by the State Council of China, among others. In 2020, he was granted the National Innovation Award by the Ministry of Science and Technology, the JCA International Award by the Japanese Cancer Association, and the TWAS Award in Medical Sciences.

# Chapter 1
# Historical Overview of RNA Therapeutics

**Abstract** Recent progress in producing, purifying, and delivering RNA to cells has paved the way for the creation of RNA-based treatments across a wide range of uses. These RNA-based drugs represent a quickly growing sector of medicine poised to revolutionize treatment standards for numerous conditions and bring personalized healthcare to fruition. These therapies are economical, relatively straightforward to produce, and capable of targeting pathways that were previously considered untreatable. This innovative therapeutic approach allows both small biotechnological firms and academic researchers to swiftly create novel and customized RNA molecules. In this chapter, we delve into the essential principles underlying several categories of RNA-based therapies. We cover a range of treatments, including antisense oligonucleotides (ASOs), which are designed to interfere with gene expression at the RNA level; aptamers, short RNA or DNA molecules with unique 3-dimentional structure that can bind to specific target molecules; small interfering RNAs (siRNAs), which play a role in the RNA interference pathway to regulate gene expression; microRNAs (miRNAs) which is involved in gene regulation and expression at the post-transcriptional level; and messenger RNA (mRNA), which is used for replacement therapy for loss-of-function diseases. Additionally, we present a historical summary of RNA-based treatments currently in clinical use. We also examine the benefits and challenges of RNA-based medications as well as different strategies for delivering RNA.

## 1.1 Introduction

Similar to deoxyribonucleic acid (DNA), ribonucleic acid (RNA) consists of nucleic acids, which are crucial components found in the nuclei of both plant and animal cells. These nucleic acids are large, complex molecules composed of thousands of smaller units known as nucleotides linked together. They serve as the repository and conveyance mechanism for genetic characteristics, essentially acting as the blueprint for the organism's protein synthesis machinery. Differing from DNA, RNA is not confined to the nucleus but is also prevalent in other cellular regions,

 
P. E. Saw, E. Song, *RNA Therapeutics in Human Diseases*,
https://doi.org/10.1007/978-981-96-3041-7_1

predominantly in the cytoplasm in various forms. RNA molecules are typically single-stranded, setting them apart from the double-stranded structure of DNA, and they possess a lower molecular weight. Each nucleotide within an RNA strand comprises three components: a ribose sugar, a phosphate group, and one of four nitrogenous bases. A critical difference between RNA and DNA is found in their sugar components. RNA is composed of ribose, a type of pentose sugar, which includes an oxygen atom that DNA's sugar, deoxyribose, does not have. This absence of an oxygen atom in DNA's sugar component contributes to differences in their respective stability and functionality. RNA is synthesized using the same nucleotides as DNA, except for the bases, where RNA uses adenine, cytosine, guanine, and uracil (A, C, G, U) instead of thymine (T) found in DNA. Structurally, RNA is characterized by a backbone consisting of alternating sugar and phosphate groups, from which nitrogenous bases project outward. This configuration resembles a coiled slinky toy. If this structure were to be stretched out, it would extend lengthwise, revealing a linear sequence of bases. This flexible, yet stable backbone allows RNA to adopt various shapes and forms, which are critical for its role in biological processes such as protein synthesis and gene regulation.

## 1.2 History of RNA Biology

The study of RNA biology begins in 1958 when Francis Crick laid down a cornerstone of molecular biology with his formulation of the central dogma, describing the flow of genetic information from DNA to RNA to proteins, which is crucial in the regulation of normal biological functions. This principle has guided much of the research in genetics for decades. However, recent advancements in technology, particularly through next-generation sequencing, have led to revelations that expand our understanding of genetic transcription. Initiatives such as the Functional Annotation of the Mammalian Genome (FANTOM) and the Encyclopedia of DNA Elements (ENCODE) have demonstrated that transcription is more extensive and complex than previously thought. These efforts have uncovered a vast array of non-coding RNAs and other transcriptional phenomena that play critical roles in gene regulation and expression, challenging and enriching our understanding of genomic architecture. Investigations revealed that although around 80% of DNA is transcribed into RNA, only approximately 1.5% of this RNA is actively involved in protein synthesis [1, 2]. Since methods such as deep sequencing have revealed that a substantial part of our genome is transcribed into RNA, the big question is—what is the fate of these RNAs? We now know that RNAs are generally classified into two: the coding RNAs and the non-coding RNAs—although the line between coding and non-coding RNAs is getting thinner by the day (discussed in detail in Chap. 14 of this book). Interestingly, only a small portion of the genome is responsible for coding proteins, as highlighted by findings from The ENCODE Project Consortium in 2012. While much research has focused on the coding RNA known as the messenger RNAs (mRNAs); ncRNAs, previously dismissed as evolutionary leftovers, are now recognized for their critical roles in various molecular functions.

As mentioned earlier, the "RNA world" hypothesis suggests RNA was the first form of life, predating DNA and proteins. Recent findings highlight RNA's versatility in disease mechanisms, epigenetics, and regulatory functions, thanks to its catalytic potential and capacity to carry genetic information [3–5]. ncRNAs are pivotal in biological processes such as development and maintaining homeostasis. ncRNAs are generally categorized by size with a cut-off value of 200 nucleotides. Those smaller than 200 nt are called small ncRNAs (including the well-studied miRNAs); while those larger than 200 nt are long ncRNAs [6]. The rise in complexity from simple organisms to humans is mirrored in the increasing number of introns and intergenic regions, which are finely regulated by mechanisms like alternative splicing, reducing proteome size [7]. Eukaryotes exhibit intricate RNA processing and regulation systems, including RNA interference (RNAi) and chromatin modification, highlighting a shift from protein-based to RNA-based regulation for controlling gene expression and phenotypic traits [8, 9]. This shift underscores the importance of ncRNAs in understanding genetic regulation and their impact on health and disease.

## 1.3 Classification of ncRNA

To date, there are more than 40 classifications of ncRNA, indicating their vast role in cellular regulations and functions. For instance, they are involved in the maturation and processing of RNA, participate in signaling pathways, regulate gene expression, and influence protein synthesis. This classification highlights the diverse functions and complexities of RNA molecules in cellular biology, underscoring their importance beyond their size and type [10, 11]. The presence and conservation level of ncRNA across species are closely linked, with estimates suggesting that each cell contains around 107 molecules of ncRNA, predominantly consisting of snRNA, snoRNA, miRNA, rRNA, and lncRNA. Despite the identification of approximately 53,000 unique human lncRNAs, only around 1000 exist in sufficient quantities to confirm their functional relevance [12]. This study also discusses various RNA types and their unique characteristics [3]. The distinctions among RNA types are depicted in Fig. 1.1.

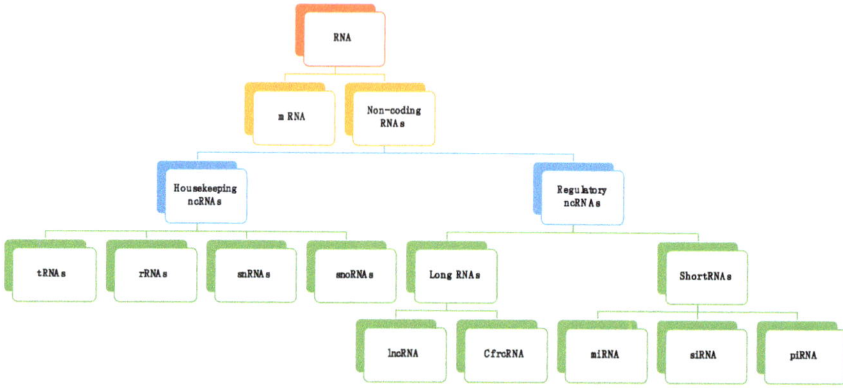


**Fig. 1.1** This is an overview of the classification of RNA, delineating its primary types and subtypes. RNA is broadly divided into messenger RNA (mRNA) and non-coding RNAs (ncRNAs). The non-coding RNAs are further categorized into housekeeping ncRNAs and regulatory ncRNAs. Housekeeping ncRNAs include transfer RNAs (tRNAs), which are involved in protein synthesis; ribosomal RNAs (rRNAs), which form the core of ribosome's structure and function; small nuclear RNAs (snRNAs), which are involved in RNA splicing; and small nucleolar RNAs (snoRNAs), which guide chemical modifications of other RNAs. Regulatory ncRNAs are split into long RNAs and short RNAs. Long RNAs include long non-coding RNAs (lncRNAs) and circular RNAs (circRNAs), which play roles in gene regulation and expression. Short RNAs encompass microRNAs (miRNAs), which regulate gene expression by targeting mRNAs for degradation or translation repression; small interfering RNAs (siRNAs), which are involved in RNA interference and gene silencing; and Piwi-interacting RNAs (piRNAs), which protect the genome integrity by silencing transposable elements in germ cells. This classification highlights the diverse functional roles of RNA in cellular processes and gene regulation

## 1.4 ncRNAs: Biogenesis and Functions

RNA molecules transcend their classical function as mere intermediaries in protein synthesis. It is now understood that non-coding transcripts may act in ways akin to proteins, affecting a broad spectrum of cellular mechanisms. Not only can these RNAs interact with other RNA molecules, they also have the capacity to engage with DNAs and proteins, sometimes forming a new and complex 3-dimentional structure. Regulatory RNAs, encompassing both small and long ncRNAs, are chiefly distinguished by their length, underscoring the diversity in their structure and function within the cell. Over the last decade, the diversity and number of functional ncRNAs, ranging from 20 nucleotides to several thousand nucleotides in length, have significantly expanded. A multitude of genes throughout the entire human genome are responsible for ncRNA production, suggesting the existence of unique transcriptional units with autonomous functions. The creation of functional RNA involves several stages: it begins with transcription, followed by maturation within the nucleus. After maturation, the RNA is exported to the cytoplasm where it undergoes further processing. The culmination of these steps results in the formation of functional RNA that is ready to perform its biological roles [3]. ncRNA is crucial in the biomedical field, especially in genomics, proteomics, and post-translational epigenetics. RNA might have played a fundamental role in the origin of life, with the ability to carry and replicate genetic information. This suggests that RNA's functions extend well beyond mere participation in cellular processes to potentially being instrumental in the early development of life itself [13]. The overview of this process is depicted in Fig. 1.2.

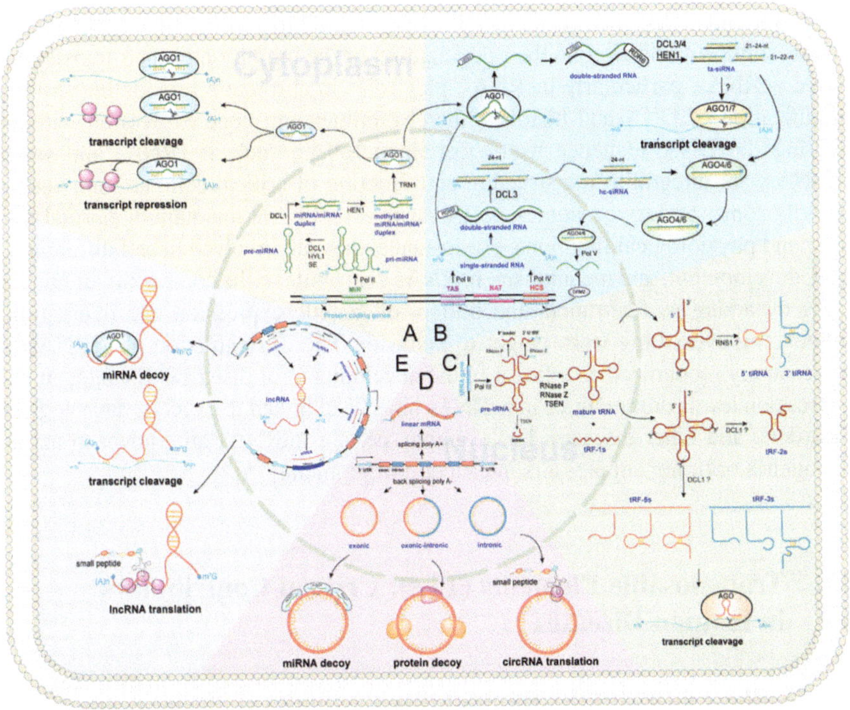


**Fig. 1.2** Detailed overview of RNA processing and function in eukaryotic cells. This diagram illustrates the complex pathways and mechanisms of RNA processing within the nucleus and cytoplasm, emphasizing the diverse roles and fates of various RNA species, including microRNAs (miRNAs), long non-coding RNAs (lncRNAs), and circular RNAs (circRNAs). (**a**) Depicts the genesis of miRNA from primary transcripts through cleavage by Drosha (DC3L complex) and export to the cytoplasm where Dicer completes the maturation. (**b**) Shows the pathway of linear mRNA processing from transcription, splicing, and polyadenylation. (**c**) Outlines the formation and function of circRNAs, highlighting their roles in sequestering proteins and acting as miRNA decoys. (**d**) Illustrates the production of transfer RNAs (tRNAs) from pre-tRNA transcripts, including cleavage and modification processes leading to mature tRNAs ready for amino acid loading. E: Focuses on lncRNAs, detailing their diverse functions from acting as scaffolds for chromatin modification complexes to being translated into small peptides. (Figure is reproduced from [14] under Creative Commons CC-BY 4.0 license)

## 1.5 ncRNAs: Role in Major Human Diseases

The traditional view of RNA was mainly limited to mRNA, rRNA, and tRNA, which play direct roles in protein synthesis. However, advances in genomics and bioinformatics have led to the discovery of numerous other ncRNA species, each with unique functions. Over 40 types of ncRNAs have been identified, including miRNAs, lncRNAs, siRNAs, and piRNAs, among others. Many ncRNAs are involved in the regulation of gene expression at both the transcriptional and post-transcriptional levels. For example, miRNAs can bind to complementary sequences on target mRNAs, leading to their degradation or the inhibition of their translation. Some ncRNAs, particularly lncRNAs, play roles in chromatin remodeling and the modification of DNA and histones, thereby influencing gene expression without altering the DNA sequence itself. Certain ncRNAs, such as rRNAs and small snRNAs, are integral to the structure and function of critical cellular components like ribosomes and spliceosomes. ncRNAs are essential for maintaining normal cellular and physiological functions. For instance, they are involved in cell differentiation, development, and metabolism. ncRNAs contribute to the complexity of human traits by acting as fine-tuners and buffers of genetic expression and interaction, influencing everything from height to behavior. The dysregulation of ncRNAs is increasingly recognized as a factor in the development of diseases. Changes in the expression levels or mutations in ncRNAs have been linked to cancers, neurological disorders, and other diseases. For example, certain miRNAs are known to act as oncogenes or tumor suppressors in cancer development [15].

## 1.6 Transposable Elements (TEs): Critical Contributors in Human Diseases

Transposable elements (TEs), also known as transposons or "'jumping genes," are sequences of DNA that can change their position within the genome. This can occur via a "cut and paste" mechanism (for DNA transposons) or a "copy and paste" mechanism (for retrotransposons). TEs can impact genomic stability and are a source of genetic diversity; they also have the potential to alter gene expression when they insert near or within genes. Many small ncRNAs may have originated from transposable elements. This can occur when sequences derived from TEs are transcribed into RNA, which is then processed into functional small ncRNAs. This process can turn sequences originally meant for TE movement into regulatory elements that influence gene expression. Besides potentially originating from TEs, small ncRNAs often have a role in regulating TEs. For example, piRNAs are crucial in silencing TEs in the germ line, thereby preventing their transposition and protecting genomic integrity. This regulatory relationship is vital for controlling the potentially harmful effects of uncontrolled TE movement, such as mutations or chromosomal rearrangements. There are ample of evidence that points toward the fact that many small ncRNAs either originated from TEs or having a close association with TEs. When linked with TE sequences, ncRNAs are known to participate in a large scope of cellular regulatory functions [16]. TEs are acknowledged as key contributors to the adaptability and evolutionary progression of genomes, especially through TEs are involved in numerous molecular processes such as insertional mutations, DNA recombination, and chromosomal rearrangements. Additionally, they influence changes in gene expression and adjustments in epigenetic regulation, illustrating their multifaceted role in genomic and cellular dynamics [17].

TEs play a significant role in the development of human cancers, notably through the integration of Alu elements and Long Interspersed Nuclear Element-1 (LINE-1) repeats, marks them as potential molecular triggers for various diseases, especially in cancers. Changes in the expression level of LINE-1 in the genome altered the expression of TEs, which in turn could drive mutations in tumorigenesis, and therefore could be considered an important hallmark of cancer. These integrations can lead to chromosomal alterations and epigenetic changes, highlighting the critical impact TEs can have on the genomic stability and regulatory frameworks within cells. A wide range of clinical conditions, including hemoglobinopathies, metabolic, neurological, and joint disorders, might be linked to structural alterations, or change in expression level, or resulting in abnormal chromosomal recombination—all caused by TEs. TEs are also implicated in the aging process. It is believed that environmental stressors and toxins significantly influence the epigenetic activation and heightened mobility of TEs, which can potentially trigger the onset of disease. These factors may alter the normal regulatory mechanisms of TEs, leading to increased activity that poses health risks. Enhancing our understanding of TEs could therefore help in discovering new markers for disease diagnosis [18].

## 1.7 Inherent Vs. Exosomal ncRNAs in Human Diseases

The transport of ncRNAs via exosomes presents a fascinating avenue for intercellular communication. To acquire more stability in extracellular ncRNAs, these ncRNAs are usually packaged or encapsulated within a vesicle, known as extracellular vesicles (EVs), such as cell-derived exosomes, microvesicles from micro-and macro-pinocytosis, and apoptotic bodies (lesser known). This encapsulation allows ncRNAs to be securely transferred between cells, where they can influence cellular behavior and function. Once released from the originating cells, these ncRNAs can have regulatory impacts on a variety of diseases, depending on the cells that receive them [19]. It is widely recognized that ncRNAs are not confined to intracellular environments but are also present in extracellular areas in biofluids, such as blood plasma, lymph fluids, urine, and saliva. The resilience of extracellular ncRNAs in these biofluids, despite the presence of significant extracellular RNase activity, implies that circulating ncRNAs are protected from degradation by adverse conditions [19]. This widespread distribution highlights their potential as biomarkers for diagnosing and monitoring health conditions. The biogenesis and biological function of exosomal ncRNAs will be highlighted in Chaps. 12 and 13 of this book.

## 1.8 Historical Overview of RNA Therapy

A multitude of breakthroughs has cemented the role of RNA therapy as a critical innovation in the treatment of human diseases. RNA's pivotal role in genetic information flow was first highlighted by Crick in his seminal "Central Dogma of Molecular Biology," a role later underscored by the identification of mRNA, demonstrating its crucial function as a mediator in translating genetic codes [20–22]. Yet, there was another discovery of equal importance but less frequently mentioned, pivotal to RNA science: the realization that two RNA strands could base-pair with each other. This concept, now widely accepted, was initially met with scepticism, as early researchers doubted RNA's capability to form a double helical structure (a feature specific to DNA). However, in 1956, this doubt was diminished by Rich and Davies, as they proved the hybridization of RNA. This significantly advanced our understanding of RNA by demonstrating the first nucleic acid hybridization reactions. Their groundbreaking work proved that RNA could form structures similar to DNA by engaging in complementary base-pairing. This discovery highlighted RNA's versatility and its capacity to assume complex configurations, crucial for its various biological functions beyond mere genetic information transfer [23]. This pivotal breakthrough laid the groundwork for significant discoveries in the field of RNA biology, notably the identification of miRNAs in 1993 and RNA interference (RNAi) in 1998; a finding that leads to the Nobel Prize. These findings underscored the essential function of RNA duplex formation in RNA silencing mechanisms, demonstrating how RNA can regulate gene expression through complex interactions. These developments further illustrated RNA's critical roles in cellular regulation and provided a foundation for future research in gene therapy and genetic engineering [24–26].

The pioneering use of RNA base-pairing for therapeutic purposes was first documented in 1978 [27], through the creation of an antisense oligonucleotide (ASO) aimed at the 35S RNA sequence of the Rous sarcoma virus (RSV), to curb its replication. About two decades thereafter, the inaugural drug employing an ASO received the approval from the United States Food and Drug Administration (US FDA) to combat cytomegalovirus retinitis [28]. In 1977, the first description of RNA splicing, a crucial post-transcriptional process that connects exons and excises introns from the initial RNA transcripts was documented [29, 30] and has since been recognized as a significant factor in diseases, with splicing anomalies associated with various human conditions [31, 32]. Traditional small molecule drugs are nearly ineffective at rectifying or influencing these splicing errors, a gap that RNA-based medicines, particularly ASOs, can fill. Remarkably, a 1993 study marked the first instance of ASOs being used to manipulate alternative splicing [33]. In this study, ASOs were used to target the splice sites and branch points of thalassaemic pre-mRNA, correcting its abnormal splicing and reducing the patient's symptoms. This study has since become foundational in the quest for new treatments for numerous challenging neurological disorders.

Unlike the extensive timeline involved in developing ASO-based therapies, the progression from identifying siRNAs to utilizing them as therapeutic agents was remarkably rapid. RNA interference (RNAi) was initially defined in a pivotal 1998 study, demonstrating that by exposing *C. elegans* embryos to a mix of sense and antisense RNAs, the embryos could effectively took up both RNAs, and selectively inhibit specific endogenous mRNAs [26]. Due to its simplicity and efficacy, RNAi quickly became a staple in research across the scientific community, leading to its broad application in a short span of time. An early application was in 2002 when RNAi was used to curb hepatitis C virus replication in mice [34]. Inspired by this, in 2003, we made a groundbreaking discovery by demonstrating that siRNAs, which are double-stranded RNAs about 20–22 nucleotides in length, could be used in vivo to knock down disease-causing genes with complementary sequences. In this study, we showed that silencing the Fas death receptor could protect mice from rapidly fatal autoimmune hepatitis. Noticing that RNAs are not efficiently taken up by cells, we quickly identified that the primary challenge for RNA-based therapeutics would be delivering them into the cytosol of target cells, where the RNA interference machinery operates [35]. To solve this problem, our second pioneering paper complexed siRNAs to a fusion protein composed of an antibody fragment or single-chain antibody fused to an RNA-binding peptide from protamine, the protein that nucleates genomic DNA in sperm, efficiently deliver siRNA into target organ [36], opening vast avenues for the clinical translation of siRNAs in the biomedical field; marking a significant step toward exploring RNAi for therapeutic uses. This progress spurred the launch of the first RNAi-based clinical trials in 2010. In a trial for patients with advanced melanoma, the treatment was done using an siRNA that could specifically target the M2 subunit of ribonucleotide reductase to inhibit tumor growth. This marked a momentous milestone the clinical application of RNAi technology, exploring its potential to combat complex diseases by manipulating genetic expression [37]. This successful trial showcased precise cleavage of the targeted mRNA through the use of a siRNA, which was administered via a specialized nanoparticle delivery system. Following this achievement, a diverse range of siRNA-based drugs were tested for various diseases. This series of evaluations culminated in the approval of the first siRNA therapy in 2018, specifically for hereditary transthyretin-mediated amyloidosis (hATTR) [38].

Although mRNA's role as a genetic translation messenger was first recognized in 1961, it was not until nearly three decades later that researchers began to harness mRNA for translational biomedical uses [39] (Fig. 1.3). The identification of RNA-related enzymes several decades ago was pivotal in propelling the field of RNA research forward. Specifically, in the 1960s, researchers discovered RNA replicase, also termed as RNA-dependent RNA polymerase (RdRp), through their studies on viruses including mengovirus and poliovirus. This enzyme's ability to replicate RNA from an RNA template was a critical finding, revealing mechanisms of viral replication and RNA synthesis. This discovery not only deepened our

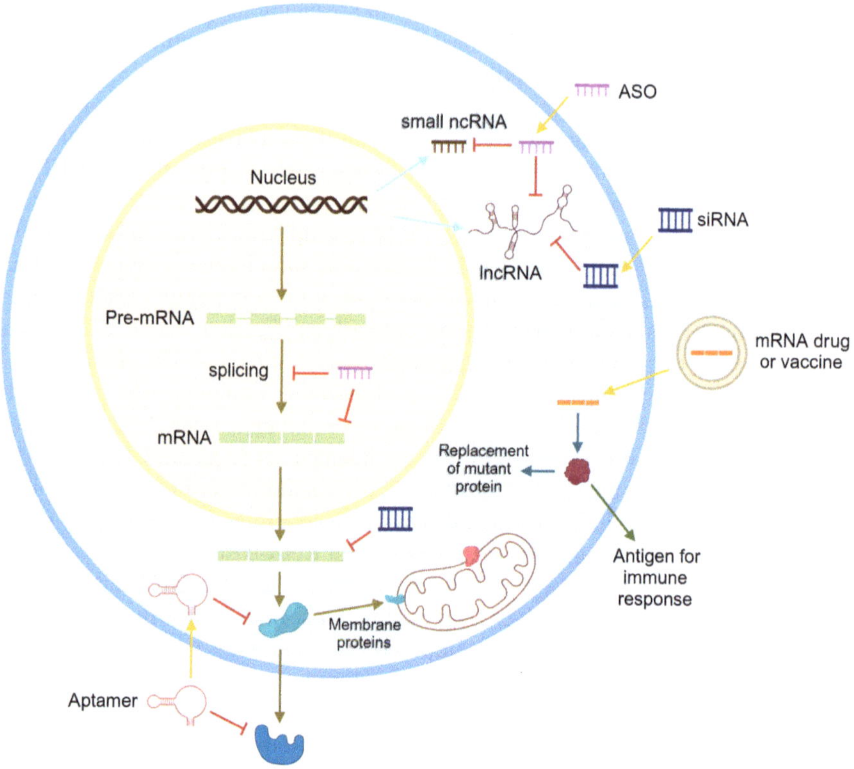


**Fig. 1.3** RNA therapies are versatile, capable of targeting a wide array of cellular molecules across different stages of gene expression for both non-coding and protein-coding genes. Antisense oligonucleotides (ASOs) have the ability to modify splicing mechanisms, influencing the maturation process of messenger RNAs (mRNAs). These mRNAs can also be targeted for intervention using either ASOs or small interfering RNAs (siRNAs), which can selectively inhibit their function. Similarly, non-coding RNAs (ncRNAs), which include both small ncRNAs and long ncRNAs (lncRNAs), can be effectively regulated or silenced through the use of ASOs or siRNAs. Additionally, the activity of specific proteins can be manipulated through the binding of aptamers, which are specially designed to specifically bind with high affinity to their target protein or receptors, thereby influencing protein function and the biological pathways in which they are involved. Moreover, the introduction of exogenous mRNAs into cells enables the production of specific proteins, either to supplement an absent enzyme or to serve as antigens, thereby stimulating a precise immune response. (Adapted from [39], under Creative Commons Attribution 4.0 International License)

understanding of viral biology but also laid essential groundwork for developing antiviral strategies and RNA-based technologies [40, 41], where it was noted that both viruses did not respond to actinomycin D, a known inhibitor of DNA-dependent RNA polymerase. Presently, RdRp is crucial for generating self-amplifying vaccine for RNA molecules [42]. Similarly, reverse transcriptase (RT), also initially identified in viruses, is used by researchers to clone complementary DNA templates for mRNA production [43, 44].

Another popular method of therapy is the "protein replacement therapy"—a medical term that is used where the treatment is aimed to replenish or substitute certain proteins that are absent or mutated as a result of a disease [45]. Since mRNA could be transcribed into protein, this has opened the gateway for protein replacement therapy. Instead of delivering a bulk protein, researchers can choose to deliver shorter mRNA into the cell cytoplasm, in hope that these mRNA could be transcribed into the protein of desire. The effort to synthesize specific proteins by introducing exogenous mRNA into cells gained significant momentum in the 1990s. A notable experiment by Wolff and colleagues showed that directly injecting mRNA coding for various reporter genes into mouse skeletal muscle led to prolonged gene expression. Remarkably, this expression was achieved even in the absence of a specialized delivery system, such as lipid nanoparticles, polymeric nanoparticle platforms or biological viral-like particles (detailed in Chap. 25). This breakthrough demonstrated the potential of mRNA as a tool for genetic engineering and protein production, setting the stage for future developments in gene therapy and vaccine technology [46]. This research paved the way for using mRNA transcripts as vaccines, a concept first tested in 1993 with synthesized mRNAs for the influenza nucleoprotein encapsulated in liposomes, showing that the virus could successfully induce the activation of cytotoxic T lymphocytes in mice [47]. The first cancer treatment mRNA vaccine was developed by Conry et al. in 1995, using a vector expressing the human carcinoembryonic antigen (CEA) in mouse models [48]. In 2008, Weide and colleagues published the initial results from clinical trials using mRNA vaccines to treat metastatic melanoma. These findings indicated that mRNA vaccines could potentially amplify the vaccine-directed T-cell responses, providing a promising avenue for cancer immunotherapy. This early research highlighted the capability of mRNA vaccines not just to trigger an immune response, but to specifically enhance the body's T-cell mediated attack on cancer cells, suggesting a new method for targeting and treating complex malignancies like melanoma [49, 50].

## 1.9 Overview of Each Type of RNA Therapeutics

***Antisense Oligonucleotide (ASO).*** Antisense oligonucleotides (ASOs) are concise, single-stranded molecules that can be synthesized from a variety of materials including DNA, phosphorothioate DNA, RNA analogs, locked nucleic acids (LNA), or morpholino phosphorodiamidate oligonucleotides. They are meticulously crafted to be complementary to specific RNA sequences they intend to target, enabling them to bind precisely to these sequences. The structural modifications to their backbone and sugar components enhance the ASOs' affinity toward their targets and increase their stability, making them more effective in their function [51]. This specific binding allows ASOs to effectively regulate gene expression by interfering with the function of their target RNA, making them powerful tools for genetic research and therapeutic applications. ASOs are categorized into two main types: RNase H-dependent ASOs and RNase H-independent (steric block) ASOs.

**The RNase H-dependent class ASO:** It is more prevalently utilized, leverages the cell's native RNase H enzyme to cleave the RNA strand within an RNA/DNA duplex, leading to gene expression knockdown. This mechanism is noted for its efficiency in reducing gene expression compared to the RNase H-independent ASOs, which operate through physical blockade mechanisms to interfere with translation or splicing without inducing RNA degradation [52–54]. The foundational evidence of RNase H-dependent antisense oligonucleotide (ASO) activity was demonstrated by Minshull and Hunt in 1986 using a reticulocyte lysate system. In their experiment, the introduction of DNA-mRNA hybrid complexes treated with RNase H led to the disappearance of full-length translation products, underscoring the capability for targeted gene silencing [55]. Giles et al. enhanced our understanding of the RNase H-mediated antisense effect in human leukemia cells. They achieved this by identifying mRNA fragments that were cleaved by RNase H, using a reverse ligation-mediated PCR technique. This method allowed them to precisely detect and analyze the RNA fragments resulting from RNase H activity, providing stronger proofs to which and why the ASOs could effectively inhibit gene expressions in cancer cells [56].

**Steric block ASOs:** It operates by directly interfering with the translation or splicing of mRNA, without the need for enzymatic degradation. These ASOs can be specifically designed to have a steric hindrance toward the target, which could achieve a range of regulatory results, such as blocking polyadenylation, either inhibiting or boosting translation, or altering splicing pathways [57–60]. Baker et al. demonstrated the use of this approach by developing RNase H-independent 2′-O-methoxyethyl ASOs that target the 5′ cap region of the mRNA encoding the intercellular adhesion molecule 1 (ICAM-1) in human umbilical vein endothelial cells. These ASOs successfully reduced protein expression by blocking the formation of the 80S translation initiation complex, thereby interfering with the protein synthesis process [58]. The versatility of steric block ASOs in modulating gene expression highlights their potential utility in a wide array of therapeutic applications, offering targeted solutions for gene regulation.

As of now, the FDA has approved three ASO drugs: (1) Inotersen (Ionis Pharmaceuticals and Akcea Therapeutics), (2) Eteplirsen (Sarepta Therapeutics), and (3) Nusinersen, developed by Ionis Pharmaceuticals and approved in December 2016. Nusinersen is a 2′-O-methoxyethyl phosphorothioate-modified antisense oligonucleotide that alters the splicing of the SMN2 gene. It promotes the inclusion of exon 7 in the transcript, thereby enhancing the production of full-length SMN protein and addressing the protein deficiency caused by the defective SMN1 gene [61, 62]. Eteplirsen, approved by the FDA in September 2016, is a treatment for Duchenne muscular dystrophy (DMD), a condition caused by mutations in the DMD gene that disrupt the production of dystrophin protein. Eteplirsen is a 30-mer phosphomorpholidate oligonucleotide designed to target exon 51 of the DMD gene. It facilitates exon skipping during pre-mRNA splicing, leading to the production of a shortened yet functional form of the dystrophin protein [63]. Approved in October 2018, Inotersen is used to treat familial amyloid polyneuropathy, a condition caused by mutations in the transthyretin (TTR) gene that result in the abnormal buildup of

TTR protein in various organs. Inotersen, a 20-mer chemically modified RNA molecule, works by hybridizing with TTR mRNA, thereby inhibiting the production of the TTR protein and addressing the underlying cause of the disease [64]. Additionally, volanesorsen (Waylivra), approved in Europe in May 2019 for familial chylomicronemia syndrome, is an antisense oligonucleotide that targets apolipoprotein CIII mRNA in the 3′-UTR, promoting its degradation and addressing the lipid metabolism disorder [65]. These approvals highlight the versatility and potential of ASOs in targeting a range of genetic diseases.

***Small interfering RNAs.*** siRNAs are compact, non-coding RNA duplexes, which can be either naturally transcribed intracellularly or synthetically introduced exogenously, varying in size (30–100 bp). These precursor molecules undergo processing by the cell's inherent Dicer enzyme, resulting in siRNAs that are 20–30 base-pairs in length, featuring two-nucleotide overhangs at their 3′ ends. These siRNAs then interact with the cell's RNA-induced silencing complex (RISC), initiating the RNAi mechanism. Within RISC, the argonaute 2 (AGO2) enzyme selectively cleaves the sense strand, preserving the antisense strand. This antisense strand directs the activated RISC to its corresponding mRNA target. Upon localization, AGO2 proceeds to cleave the mRNA's phosphodiester backbone. The antisense strand is usually complementary to the target mRNA's coding sequence, allowing siRNA to specifically downregulate a single gene, demonstrating its precision in gene-silencing applications [66, 67]. In 1998, Fire et al. marked a significant milestone in molecular biology by being the first to demonstrate RNAi through the blockade of gene expression for unc-22, unc-54, fem1, and hlh-1 in *C. elegans*. Their groundbreaking work established that double-stranded RNA (dsRNA) is more potent than single-stranded RNA (ssRNA) in artificially inducing RNAi to target and degrade specific mRNA molecules [26, 68]. This foundational research received the Nobel Prize and also paved the way for the development of clinical siRNA-based therapeutics. The first siRNA drug to be commercialized was Patisiran (Onpattro, developed by Alnylam Pharmaceuticals), which received its FDA approval in 2018. Patisiran is used to treat polyneuropathy with hereditary transthyretin (TTR)-mediated amyloidosis, utilizing dsRNA machinery to trigger RNAi and promote the degradation of mRNA that encodes for TTR protein [69]. After the approval of Patisiran, the FDA approved Givosiran (Givlaari, developed by Alnylam Pharmaceuticals) 2019 for the treatment of acute hepatic porphyria. Givosiran functions by targeting the mRNA of aminolevulinate synthase 1 in the liver, thereby decreasing the levels of neurotoxic intermediates such as aminolevulinic acid and porphobilinogen, which are implicated in the disease [70, 71]. These advancements underscore the therapeutic potential of siRNA technology in addressing a variety of genetic diseases.

***MicroRNAs.*** MicroRNAs (miRNAs) are essential regulators of gene expression, influencing a variety of cellular processes. miRNAs exert their regulatory effects by either (1) blocking the translation of proteins or (2) promoting the degradation of target mRNA molecules. By binding to complementary sequences on their target mRNAs, miRNAs can effectively silence gene expression, playing a pivotal role in maintaining cellular homeostasis and contributing to the fine-tuning of various biological pathways. First identified in *C. elegans* [24], miRNAs are first transcribed from DNA as primary miRNAs (pri-miRNAs), which form a characteristic hairpin loop structure. The pri-miRNAs are then processed by the enzyme Drosha into shorter precursor miRNAs (pre-miRNAs). These pre-miRNAs are exported to the cytoplasm, where the enzyme Dicer further processes them into 18–25 base-pair RNA duplexes. These duplexes mature into functional miRNAs, which are then incorporated into the RNA-induced silencing complex (RISC). Once part of RISC, the mature miRNAs guide the complex to specific target mRNAs, resulting in gene silencing through mechanisms such as inhibition of protein translation or promotion of mRNA degradation [68, 72, 73]. miRNA-based therapeutics can be divided into several categories, each designed to manipulate miRNA activity for therapeutic benefit. The primary categories include miRNA mimics and miRNA inhibitors.

**miRNA mimics.** These are double-stranded RNA molecules engineered to resemble endogenous miRNAs. They are introduced into cells to restore or enhance the function of specific miRNAs that may be under-expressed or lost due to disease. By emulating natural miRNAs, mimics can help regulate target gene expression, thereby potentially correcting dysregulated pathways involved in various conditions.

**miRNA inhibitors.** These are single-stranded oligonucleotides specifically designed to bind to and inhibit the function of endogenous miRNAs. By blocking the activity of overexpressed or aberrantly functioning miRNAs, inhibitors can prevent these miRNAs from downregulating their target mRNAs. This can restore the normal expression levels of genes that are otherwise suppressed, offering therapeutic potential in diseases where miRNA overactivity is a problem. Although no miRNA-based drugs have been commercially released, several promising candidates are undergoing clinical trials.

***Aptamers.*** Aptamers are short, single-stranded nucleic acids that can be either RNA or DNA-based molecule that binds to target using its 3-dimentional loop structure. These molecules are uniquely capable of binding to a varied array of targets, such as proteins, peptides, and carbohydrates. This selective binding is a key feature of aptamers, enabling them to interact with high specificity and affinity. Aptamers are engineered using a process known as Systematic Evolution of Ligands by Exponential Enrichment (SELEX). This technique involves iterative rounds of selection and amplification to identify nucleic acid sequences that fold into shapes capable of binding tightly to the target molecules. Through SELEX, aptamers are optimized to achieve the desired binding characteristics, making them powerful tools for therapeutic and diagnostic applications due to their precision and effectiveness in targeting specific biomolecules [74–81].

## 1.10 mRNA as Therapeutics

mRNAs act as crucial intermediaries, translating the genetic code from DNA into functional proteins. Essentially serving as temporary gene blueprints, mRNA transcripts carry genetic instructions to the protein synthesis machinery, enabling the production of proteins encoded by genomic DNA [46, 48, 82]. Early experiments by Wolff et al. demonstrated the potential of exogenous mRNA to express proteins in vivo, where synthetic mRNA encoding various proteins was injected into mouse skeletal muscle, achieving protein expression levels comparable to those from DNA vector injections. The therapeutic potential of mRNA was demonstrated by successfully reversing diabetes insipidus in Brattleboro rats, which suffer from a genetic defect that prevents vasopressin production. They achieved this by injecting mRNA encoding vasopressin directly into the hypothalamus of the rats. The synthetic mRNA was engineered to closely resemble natural mRNA, incorporating untranslated regions, a 5′ cap to initiate translation, and a 3′ poly(A) tail to enhance the mRNA stability. This meticulous design ensures that upon reaching the intracellular environment, the mRNA could be efficiently translated into the desired protein within the cytoplasm and exerts its desired effect. Another important point is to ensure that this mRNA could be subsequently degraded without being integrated into the host genome [48, 83–85]. mRNA-based therapies span several approaches: (1) protein replacement therapy to supplement defective genes or proteins, (2) in situ production of peptides for vaccination purpose to induce protective immunity, (3) cell therapy, by using ex vivo mRNA for cell transfection to modify cell functions before reintroducing them into the patient.

## 1.11 Advantages of RNA Therapy

### *1.11.1 Targeting Undruggable Targets*

RNA-based drugs offer a substantial benefit due to their unlimited ability to interact with almost any desired gene in a cellular compartment. This capability allows them to target numerous areas that might be unreachable for conventional therapeutic approaches such as small molecules and antibodies. The versatility of RNA-based drugs stems from their molecular design, which can be engineered to specifically bind with and modulate the function of various genetic targets, including those deeply embedded in cellular processes. This broad targeting potential makes RNA-based therapies particularly promising for addressing complex diseases with genetic foundations that have been challenging to treat with traditional drugs. Particularly for non-coding RNAs, especially small RNAs, their RNA sequence is often the only distinguishing feature. Drugs such as ASOs and siRNA leverage sequence-specific binding to their targets, making them exceptionally potent in silencing the desired gene, without having the side effects of a small molecule drug. Given the abundance of ncRNAs relative to proteins within the human genome [86], the applications of ncRNAs in the clinical setting is enormous, and the growing body of evidence underscoring their pivotal roles in the onset and progression of numerous diseases [87–89]. Furthermore, it has been posited that a minor fraction of human proteins are amenable to modulation by small molecules, a bottleneck that could be easily be done by RNAi technology [90, 91]. The challenge of selectively targeting individual proteins, due to structural similarities among many proteins, complicates direct intervention. The challenge intensifies when targeting membrane-integrated proteins, which are notoriously difficult for small molecules or antibodies to reach due to their restricted interactions with the cytoplasm. These proteins, embedded within cellular membranes, present a unique obstacle as traditional therapeutic agents often lack the necessary access or capability to effectively interact with these components. RNA-based therapeutics, by inhibiting protein biogenesis, offer a more direct and potentially more effective means of curtailing the production of problematic proteins, thus enhancing therapeutic outcomes.

### *1.11.2 Fast Production*

The journey to create a new antibody-based drugs or new small molecule (inhibitors) typically spans several years or decades. Conversely, once the chemical structure and delivery method for RNAi are determined, RNA-based therapeutics can be rapidly designed and synthesized for clinical testing. For instance, modifying the sequence of an siRNA developed for a disease characterized by gene overexpression in a particular organ can potentially extend its therapeutic application to other diseases affecting the same organ. This adaptability underpins the extensive research into siRNA drugs linked with liver-specific conjugates, targeting a spectrum of diseases related to liver metabolism [92]. The expedited development of COVID-19 vaccines illustrates the agility of RNA-based technologies. After the novel coronavirus genetic sequence was revealed, both Moderna and BioNTech/Pfizer quickly developed mRNA sequences as their lead candidate in their vaccines. This rapid response enabled them to move from animal studies to clinical trials in a remarkably short period of time. The swift progression demonstrated the agility of mRNA technology in addressing emerging global health threats, allowing these companies to initiate testing and subsequently respond to the pandemic with unprecedented speed [93, 94]. This rapid development cycle, facilitated by a standardized production process, positions RNA technology as a pivotal asset for pandemic preparedness and response.

### *1.11.3 Long-Term Effect*

While natural RNAs are prone to rapid degradation by endogenous nucleases, their stability can be significantly enhanced through various chemical modifications introduced during their synthesis [95]. Moreover, encapsulating these modified RNAs within protective carriers like liposomes shields them from nuclease degradation after systemic administration [96], thereby extending their persistence in the host. The stability of RNA-based therapeutics is well demonstrated by inclisiran, a synthetic siRNA specifically designed to target the mRNA of proprotein convertase subtilisin/kexin type 9 (PCSK9). Remarkably, a single dose of inclisiran can maintain its gene-silencing effect for over 6 months [97], facilitating longer intervals between administrations and potentially lowering toxicity risks in comparison to small molecule drugs, which often see a significant reduction in effectiveness within days.

### *1.11.4 Usefulness for Rare (Orphan) Disease*

The pharmaceutical industry often hesitates to develop new treatments for very rare diseases due to the disproportionate investment required relative to the anticipated profit. However, RNA-based therapeutics present a unique advantage in this regard. The first establishment of RNAi platform might be hard. However, once the fundamental chemistry, modification, and delivery systems are established, the subsequent costs that are required to develop variants from this basis is significantly decreased. For example, for a rare genetic disorder that might include several different genes to silence in different individual, the inclusion of different siRNA/ASO cocktail could be tailored individually for these patients [98]. Such personalized treatment would be unfeasible with other technologies. The prior establishment of antisense drug development technologies enabled the production of this groundbreaking therapy, illustrating how advancements in RNA-based drug development can reduce both the complexity and expense of generating novel treatments for rare conditions.

### *1.11.5 No Risk of Genotoxicity*

RNA therapy offers a distinct safety advantage over DNA-based treatments, primarily due to its minimal genotoxic effects. In DNA therapies, delivery into cells typically involves a viral vector, carrying the inherent risk of genomic integration and potential mutation induction. This concern is notably mitigated when employing RNA as the therapeutic agent, bypassing the risk associated with DNA integration into the host genome. This fundamental difference underscores the relative safety of RNA therapy in avoiding genetic alterations that could lead to mutations.

## 1.12 Conclusion

Since gene therapy was first proposed as a treatment approach 50 years ago, considerable scientific and technological progress has pioneered the creation of numerous safe and effective platform technologies for nucleic acid therapeutics. These advancements have provided hope for previously undruggable diseases, opening treatment options for patients who previously had limited or no alternatives. Researchers are now focused on refining the RNAi therapeutics and ensuring better approaches to ensure that therapeutic agents more precisely target affected tissues, improving efficacy and reducing potential side effects. Additionally, there is a push to broaden the scope of applications, particularly in the use of biocompatible lipid nanoparticle (LNP) or highly transfectable adeno-associated virus (AAV) vector systems. While the development of treatments includes ex vivo drug products like modified dendritic cells, this discussion focuses on in vivo approaches and their socioeconomic implications. Bioconjugation techniques have improved the pharmacokinetics and receptor-specific interactions of ASO and siRNA therapeutics, enhancing their gene-silencing efficiency in targeted cells. Gene editing-based treatments offer precise interventions for disorders caused by dominant negative mutations, with in vivo and ex vivo gene editing therapies undergoing clinical translation for various diseases. Despite the tolerability of approved platform technologies, managing infusion or immune-related reactions often requires pre-or co-medications. The socioeconomic implications of nucleic acid therapeutics, especially in terms of treatment pricing, continue to be a contentious issue. The advancement of nucleic acid therapeutics marks a significant transition from theoretical frameworks to clinical application, offering innovative treatments for orphan diseases and facilitating rapid vaccine development during pandemics. Despite ongoing challenges related to manufacturing, toxicity, and socioeconomic factors, the potential of nucleic acid therapeutics to transform the treatment landscape for conditions that were previously untreatable is undeniable.

## References

1. Carninci P, Kasukawa T, Katayama S, Gough J, Frith MC, Maeda N, et al. The transcriptional landscape of the mammalian genome. Science (New York, NY). 2005;309(5740):1559–63.
2. Hangauer MJ, Vaughn IW, McManus MT. Pervasive transcription of the human genome produces thousands of previously unidentified long intergenic noncoding RNAs. PLoS Genet. 2013;9(6):e1003569.
3. Bhatti GK, Khullar N, Sidhu IS, Navik US, Reddy AP, Reddy PH, et al. Emerging role of non-coding RNA in health and disease. Metab Brain Dis. 2021;36(6):1119–34.
4. Robertson MP, Joyce GF. The origins of the RNA world. Cold Spring Harb Perspect Biol. 2012;4(5):a003608.
5. Amaral PP, Dinger ME, Mattick JS. Non-coding RNAs in homeostasis, disease and stress responses: an evolutionary perspective. Brief Funct Genomics. 2013;12(3):254–78.
6. Kim VN, Han J, Siomi MC. Biogenesis of small RNAs in animals. Nat Rev Mol Cell Biol. 2009;10(2):126–39.
7. Mattick JS. Non-coding RNAs: the architects of eukaryotic complexity. EMBO Rep. 2001;2(11):986–91.
8. Mattick JS. RNA regulation: a new genetics? Nat Rev Genet. 2004;5(4):316–23.
9. Guttman M, Donaghey J, Carey BW, Garber M, Grenier JK, Munson G, et al. lincRNAs act in the circuitry controlling pluripotency and differentiation. Nature. 2011;477(7364):295–300.

10. Kung JT, Colognori D, Lee JT. Long noncoding RNAs: past, present, and future. Genetics. 2013;193(3):651–69.
11. Morris KV, Mattick JS. The rise of regulatory RNA. Nat Rev Genet. 2014;15(6):423–37.
12. Djebali S, Davis CA, Merkel A, Dobin A, Lassmann T, Mortazavi A, et al. Landscape of transcription in human cells. Nature. 2012;489(7414):101–8.
13. Joyce GF. RNA evolution and the origins of life. Nature. 1989;338(6212):217–24.
14. Chao H, Hu Y, Zhao L, Xin S, Ni Q, Zhang P, et al. Biogenesis, functions, interactions, and resources of non-coding RNAs in plants. Int J Mol Sci. 2022;23(7):3695.
15. Li X, Yang L, Chen LL. The biogenesis, functions, and challenges of circular RNAs. Mol Cell. 2018;71(3):428–42.
16. Hadjiargyrou M, Delihas N. The intertwining of transposable elements and non-coding RNAs. Int J Mol Sci. 2013;14(7):13307–28.
17. Chénais B. Transposable elements and human diseases: mechanisms and implication in the response to environmental pollutants. Int J Mol Sci. 2022;23(5):2551.
18. Pradhan RK, Ramakrishna W. Transposons: unexpected players in cancer. Gene. 2022;808:145975.
19. Li C, Ni YQ, Xu H, Xiang QY, Zhao Y, Zhan JK, et al. Roles and mechanisms of exosomal non-coding RNAs in human health and diseases. Signal Transduct Target Ther. 2021;6(1):383.
20. Crick FH. On protein synthesis. Symp Soc Exp Biol. 1958;12:138–63.
21. Brenner S, Jacob F, Meselson M. An unstable intermediate carrying information from genes to ribosomes for protein synthesis. Nature. 1961;190:576–81.
22. Gros F, Hiatt H, Gilbert W, Kurland CG, Risebrough RW, Watson JD. Unstable ribonucleic acid revealed by pulse labelling of Escherichia coli. Nature. 1961;190:581–5.
23. Rich A, Davies DR. A new two stranded helical structure: polyadenylic acid and polyuridylic acid. J Am Chem Soc. 1956;78(14):3548–9.
24. Lee RC, Feinbaum RL, Ambros V. The C. elegans heterochronic gene lin-4 encodes small RNAs with antisense complementarity to lin-14. Cell. 1993;75(5):843–54.
25. Wightman B, Ha I, Ruvkun G. Posttranscriptional regulation of the heterochronic gene lin-14 by lin-4 mediates temporal pattern formation in C. elegans. Cell. 1993;75(5):855–62.
26. Fire A, Xu S, Montgomery MK, Kostas SA, Driver SE, Mello CC. Potent and specific genetic interference by double-stranded RNA in Caenorhabditis elegans. Nature. 1998;391(6669):806–11.
27. Zamecnik PC, Stephenson ML. Inhibition of Rous sarcoma virus replication and cell transformation by a specific oligodeoxynucleotide. Proc Natl Acad Sci USA. 1978;75(1):280–4.
28. Roehr B. Fomivirsen approved for CMV retinitis. J Int Assoc Phys AIDS Care. 1998;4(10):14–6.
29. Berget SM, Moore C, Sharp PA. Spliced segments at the 5′terminus of adenovirus 2 late mRNA. Proc Natl Acad Sci USA. 1977;74(8):3171–5.
30. Chow LT, Gelinas RE, Broker TR, Roberts RJ. An amazing sequence arrangement at the 5′ends of adenovirus 2 messenger RNA. Cell. 1977;12(1):1–8.
31. Faustino NA, Cooper TA. Pre-mRNA splicing and human disease. Genes Dev. 2003;17(4):419–37.
32. Scotti MM, Swanson MS. RNA mis-splicing in disease. Nat Rev Genet. 2016;17(1):19–32.
33. Dominski Z, Kole R. Restoration of correct splicing in thalassemic pre-mRNA by antisense oligonucleotides. Proc Natl Acad Sci USA. 1993;90(18):8673–7.
34. McCaffrey AP, Meuse L, Pham TT, Conklin DS, Hannon GJ, Kay MA. RNA interference in adult mice. Nature. 2002;418(6893):38–9.
35. Song EW, Lee SK, Wang J, Ince N, Ouyang N, Min J, et al. RNA interference targeting Fas protects mice from fulminant hepatitis. Nat Med. 2003;9(3):347–51.
36. Song E, Zhu P, Lee SK, Chowdhury D, Kussman S, Dykxhoorn DM, et al. Antibody mediated in vivo delivery of small interfering RNAs via cell-surface receptors. Nat Biotechnol. 2005;23(6):709–17.

37. Davis ME, Zuckerman JE, Choi CH, Seligson D, Tolcher A, Alabi CA, et al. Evidence of RNAi in humans from systemically administered siRNA via targeted nanoparticles. Nature. 2010;464(7291):1067–70.
38. Adams D, Gonzalez-Duarte A, O'Riordan WD, Yang CC, Ueda M, Kristen AV, et al. Patisiran, an RNAi therapeutic, for hereditary transthyretin amyloidosis. N Engl J Med. 2018;379(1):11–21.
39. Kim Y-K. RNA therapy: rich history, various applications and unlimited future prospects. Exp Mol Med. 2022;54(4):455–65.
40. Baltimore D, Franklin RM. Preliminary data on a virus-specific enzyme system responsible for the synthesis of viral RNA. Biochem Biophys Res Commun. 1962;9:388–92.
41. Reich E, Franklin RM, Shatkin AJ. Tatumel. Action of actinomycin D on animal cells and viruses. Proc Natl Acad Sci USA. 1962;48(7):1238–45.
42. Bloom K, van den Berg F, Arbuthnot P. Self-amplifying RNA vaccines for infectious diseases. Gene Ther. 2021;28(3–4):117–29.
43. Baltimore D. RNA-dependent DNA polymerase in virions of RNA tumour viruses. Nature. 1970;226(5252):1209–11.
44. Temin HM, Mizutani S. RNA-dependent DNA polymerase in virions of Rous sarcoma virus. Nature. 1970;226(5252):1211–3.
45. Vavilis T, Stamoula E, Ainatzoglou A, Sachinidis A, Lamprinou M, Dardalas I, et al. mRNA in the context of protein replacement therapy. Pharmaceutics. 2023;15(1):166.
46. Wolff JA, Malone RW, Williams P, Chong W, Acsadi G, Jani A, et al. Direct gene transfer into mouse muscle in vivo. Science (New York, NY). 1990;247(4949 Pt 1):1465–8.
47. Martinon F, Krishnan S, Lenzen G, Magné R, Gomard E, Guillet JG, et al. Induction of virus-specific cytotoxic T lymphocytes in vivo by liposome-entrapped mRNA. Eur J Immunol. 1993;23(7):1719–22.
48. Conry RM, LoBuglio AF, Wright M, Sumerel L, Pike MJ, Johanning F, et al. Characterization of a messenger RNA polynucleotide vaccine vector. Cancer Res. 1995;55(7):1397–400.
49. Weide B, Carralot JP, Reese A, Scheel B, Eigentler TK, Hoerr I, et al. Results of the first phase I/II clinical vaccination trial with direct injection of mRNA. J Immunother (Hagerstown, Md: 1997). 2008;31(2):180–8.
50. Weide B, Pascolo S, Scheel B, Derhovanessian E, Pflugfelder A, Eigentler TK, et al. Direct injection of protamine-protected mRNA: results of a phase 1/2 vaccination trial in metastatic melanoma patients. J Immunother (Hagerstown, Md: 1997). 2009;32(5):498–507.
51. Bennett CF, Swayze EE. RNA targeting therapeutics: molecular mechanisms of antisense oligonucleotides as a therapeutic platform. Annu Rev Pharmacol Toxicol. 2010;50:259–93.
52. Larrouy B, Blonski C, Boiziau C, Stuer M, Moreau S, Shire D, et al. RNase H-mediated inhibition of translation by antisense oligodeoxyribonucleotides: use of backbone modification to improve specificity. Gene. 1992;121(2):189–94.
53. Dean NM, McKay R, Condon TP, Bennett CF. Inhibition of protein kinase C-alpha expression in human A549 cells by antisense oligonucleotides inhibits induction of intercellular adhesion molecule 1 (ICAM-1) mRNA by phorbol esters. J Biol Chem. 1994;269(23):16416–24.
54. Dias N, Stein CA. Antisense oligonucleotides: basic concepts and mechanisms. Mol Cancer Ther. 2002;1(5):347–55.
55. Minshull J, Hunt T. The use of single-stranded DNA and RNase H to promote quantitative 'hybrid arrest of translation' of mRNA/DNA hybrids in reticulocyte lysate cell-free translations. Nucleic Acids Res. 1986;14(16):6433–51.
56. Giles RV, Spiller DG, Tidd DM. Detection of ribonuclease H-generated mRNA fragments in human leukemia cells following reversible membrane permeabilization in the presence of antisense oligodeoxynucleotides. Antisense Res Dev. 1995;5(1):23–31.
57. Vickers TA, Wyatt JR, Burckin T, Bennett CF, Freier SM. Fully modified 2' MOE oligonucleotides redirect polyadenylation. Nucleic Acids Res. 2001;29(6):1293–9.
58. Baker BF, Lot SS, Condon TP, Cheng-Flournoy S, Lesnik EA, Sasmor HM, et al. 2'-O-(2-Methoxy)ethyl-modified anti-intercellular adhesion molecule 1 (ICAM-1) oligonucleotides selectively increase the ICAM-1 mRNA level and inhibit formation of the ICAM-1

translation initiation complex in human umbilical vein endothelial cells. J Biol Chem. 1997;272(18):11994–2000.
59. Liang XH, Shen W, Sun H, Migawa MT, Vickers TA, Crooke ST. Translation efficiency of mRNAs is increased by antisense oligonucleotides targeting upstream open reading frames. Nat Biotechnol. 2016;34(8):875–80.
60. Hua Y, Vickers TA, Baker BF, Bennett CF, Krainer AR. Enhancement of SMN2 exon 7 inclusion by antisense oligonucleotides targeting the exon. PLoS Biol. 2007;5(4):e73.
61. Rigo F, Hua Y, Krainer AR, Bennett CF. Antisense-based therapy for the treatment of spinal muscular atrophy. J Cell Biol. 2012;199(1):21–5.
62. Wurster CD, Ludolph AC. Nusinersen for spinal muscular atrophy. Ther Adv Neurol Disord. 2018;11:1756285618754459.
63. Lim KR, Maruyama R, Yokota T. Eteplirsen in the treatment of Duchenne muscular dystrophy. Drug Des Devel Ther. 2017;11:533–45.
64. Mathew V, Wang AK. Inotersen: new promise for the treatment of hereditary transthyretin amyloidosis. Drug Des Devel Ther. 2019;13:1515–25.
65. Paik J, Duggan S. Volanesorsen: first global approval. Drugs. 2019;79(12):1349–54.
66. Wittrup A, Lieberman J. Knocking down disease: a progress report on siRNA therapeutics. Nat Rev Genet. 2015;16(9):543–52.
67. Dana H, Chalbatani GM, Mahmoodzadeh H, Karimloo R, Rezaiean O, Moradzadeh A, et al. Molecular mechanisms and biological functions of siRNA. Int J Biomed Sci. 2017;13(2):48–57.
68. Lam JK, Chow MY, Zhang Y, Leung SW. siRNA versus miRNA as therapeutics for gene silencing. Mol Ther Nucleic Acids. 2015;4(9):e252.
69. Kristen AV, Ajroud-Driss S, Conceição I, Gorevic P, Kyriakides T, Obici L. Patisiran, an RNAi therapeutic for the treatment of hereditary transthyretin-mediated amyloidosis. Neurodegen Dis Manag. 2019;9(1):5–23.
70. A. Alnylam Announces Approval of GIVLAARITM (givosiran) by the U.S. Food and Drug Administration (FDA) Alnylam Pharm. Inc. 2020b; http://investors.alnylam.com/news-releases/news-releasedetails/alnylam-announces-approval-givlaaritm-givosiran-us-food-and-drug. Accessed 3 Jan 2020.
71. G. Now Approved: GIVLAARITM (givosiran) GIVLAARI. 2020; https://www.givlaari.com. Accessed 3 Jan 2020.
72. Lou S, Sun T, Li H, Hu Z. Mechanisms of microRNA-mediated gene regulation in unicellular model alga Chlamydomonas reinhardtii. Biotechnol Biofuels. 2018;11:244.
73. O'Brien J, Hayder H, Zayed Y, Peng C. Overview of MicroRNA biogenesis, mechanisms of actions, and circulation. Front Endocrinol. 2018;9:402.
74. Dollins CM, Nair S, Boczkowski D, Lee J, Layzer JM, Gilboa E, et al. Assembling OX40 aptamers on a molecular scaffold to create a receptor-activating aptamer. Chem Biol. 2008;15(7):675–82.
75. McNamara JO, Kolonias D, Pastor F, Mittler RS, Chen L, Giangrande PH, et al. Multivalent 4-1BB binding aptamers costimulate CD8+ T cells and inhibit tumor growth in mice. J Clin Invest. 2008;118(1):376–86.
76. Santulli-Marotto S, Nair SK, Rusconi C, Sullenger B, Gilboa E. Multivalent RNA aptamers that inhibit CTLA-4 and enhance tumor immunity. Cancer Res. 2003;63(21):7483–9.
77. Berezhnoy A, Stewart CA, McNamara JO 2nd, Thiel W, Giangrande P, Trinchieri G, et al. Isolation and optimization of murine IL-10 receptor blocking oligonucleotide aptamers using high-throughput sequencing. Mol Ther. 2012;20(6):1242–50.
78. Schrand B, Berezhnoy A, Brenneman R, Williams A, Levay A, Kong LY, et al. Targeting 4-1BB costimulation to the tumor stroma with bispecific aptamer conjugates enhances the therapeutic index of tumor immunotherapy. Cancer Immunol Res. 2014;2(9):867–77.
79. Schrand B, Verma B, Levay A, Patel S, Castro I, Benaduce AP, et al. Radiation-induced enhancement of antitumor T-cell immunity by VEGF-targeted 4-1BB costimulation. Cancer Res. 2017;77(6):1310–21.

80. McNamara JO 2nd, Andrechek ER, Wang Y, Viles KD, Rempel RE, Gilboa E, et al. Cell type-specific delivery of siRNAs with aptamer-siRNA chimeras. Nat Biotechnol. 2006;24(8):1005–15.
81. Lozano T, Soldevilla MM, Casares N, Villanueva H, Bendandi M, Lasarte JJ, et al. Targeting inhibition of Foxp3 by a CD28 2'-Fluro oligonucleotide aptamer conjugated to P60-peptide enhances active cancer immunotherapy. Biomaterials. 2016;91:73–80.
82. Gilbert WJn. Origin of life: The RNA world. Nature. 1986;319(6055):618.
83. Jirikowski GF, Sanna PP, Maciejewski-Lenoir D, Bloom FE. Reversal of diabetes insipidus in Brattleboro rats: intrahypothalamic injection of vasopressin mRNA. Science (New York, NY). 1992;255(5047):996–8.
84. Whisenand JM, Azizian KT, Henderson JM, Shore S, Shin D, Lebedev A, et al. Considerations for the design and cGMP manufacturing of mRNA therapeutics, vol. 26. San Diego, CA: TriLink BioTechnologies; 2017.
85. Shin H, Park SJ, Yim Y, Kim J, Choi C, Won C, et al. Recent advances in RNA therapeutics and RNA delivery systems based on nanoparticles. Adv Therapeutics. 2018;1(7):1800065.
86. Frankish A, Diekhans M, Ferreira AM, Johnson R, Jungreis I, Loveland J, et al. GENCODE reference annotation for the human and mouse genomes. Nucleic Acids Res. 2019;47(D1):D766–d73.
87. Esteller M. Non-coding RNAs in human disease. Nat Rev Genet. 2011;12(12):861–74.
88. Beermann J, Piccoli MT, Viereck J, Thum T. Non-coding RNAs in development and disease: background, mechanisms, and therapeutic approaches. Physiol Rev. 2016;96(4):1297–325.
89. Song J, Kim YK. Targeting non-coding RNAs for the treatment of retinal diseases. Mol Therapy Nucleic Acids. 2021;24:284–93.
90. Fauman EB, Rai BK, Huang ES. Structure-based druggability assessment—identifying suitable targets for small molecule therapeutics. Curr Opin Chem Biol. 2011;15(4):463–8.
91. Finan C, Gaulton A, Kruger FA, Lumbers RT, Shah T, Engmann J, et al. The druggable genome and support for target identification and validation in drug development. Sci Transl Med. 2017;9(383):eaag1166.
92. Springer AD, Dowdy SF. GalNAc-siRNA conjugates: leading the way for delivery of RNAi therapeutics. Nucleic Acid Therapeutics. 2018;28(3):109–18.
93. Jackson LA, Anderson EJ, Rouphael NG, Roberts PC, Makhene M, Coler RN, et al. An mRNA vaccine against SARS-CoV-2—preliminary report. N Engl J Med. 2020;383(20):1920–31.
94. Mulligan MJ, Lyke KE, Kitchin N, Absalon J, Gurtman A, Lockhart S, et al. Phase I/II study of COVID-19 RNA vaccine BNT162b1 in adults. Nature. 2020;586(7830):589–93.
95. Roberts TC, Langer R, Wood MJA. Advances in oligonucleotide drug delivery. Nat Rev Drug Discov. 2020;19(10):673–94.
96. Thi TTH, Suys EJA, Lee JS, Nguyen DH, Park KD, Truong NP. Lipid-based nanoparticles in the clinic and clinical trials: from cancer nanomedicine to COVID-19 vaccines. Vaccine. 2021;9(4):359.
97. Ray KK, Landmesser U, Leiter LA, Kallend D, Dufour R, Karakas M, et al. Inclisiran in patients at high cardiovascular risk with elevated LDL cholesterol. N Engl J Med. 2017;376(15):1430–40.
98. Kim J, Hu C, Moufawad El Achkar C, Black LE, Douville J, Larson A, et al. Patient-customized oligonucleotide therapy for a rare genetic disease. N Engl J Med. 2019;381(17):1644–52.

# Chapter 2
# RNA: An Introduction

**Abstract** This chapter delves into the intriguing world of RNA, emphasizing its discovery, structural components, and multifaceted functions. The narrative begins with the historical milestones marking RNA's discovery in the late nineteenth century, tracing its evolution from being overshadowed by DNA to recognition as a central molecule in genetic regulation and expression. This chapter proceeds to explain the building blocks of RNA—nucleotides, which consist of a ribose sugar, a phosphate group, and one of four nitrogenous bases: adenine, guanine, cytosine, or uracil. These components are outlined in detail, highlighting how their specific arrangements define the structure and function of various RNA types. Focusing on RNA function, this chapter provides an overview of the diverse roles RNA plays in cellular machinery and genetic regulation. It discusses RNA's role in protein synthesis through messenger RNA (mRNA), its structural contributions via ribosomal RNA (rRNA), and its involvement in amino acid transport through transfer RNA (tRNA). Additionally, this chapter explores the regulatory functions of non-coding RNAs, such as microRNAs and long non-coding RNAs, which play critical roles in gene expression and epigenetic regulation. By integrating historical context with detailed descriptions of RNA's molecular structure and function, this chapter aims to provide readers with a comprehensive understanding of RNA's essential role in biology. It sets the stage for further exploration of RNA's potential in advancing genetic research and medical therapies, highlighting its importance in the past, present, and future of biological sciences.

## 2.1 RNA Building Blocks

### *2.1.1 RNA Structure*

RNA is generally composed of a single strand, constructed from ribonucleotides connected via phosphodiester bonds. Each ribonucleotide within the RNA sequence comprises a ribose sugar, one of four nitrogen bases (A, U, G, and C), and a



P. E. Saw, E. Song, *RNA Therapeutics in Human Diseases*,
https://doi.org/10.1007/978-981-96-3041-7_2

phosphate group [1, 2]. The minor structural variance between the sugar components renders DNA more stable, positioning it as the preferable medium for long-term genetic data storage [3]. In contrast, RNA's relative instability aligns it with more transient roles. RNA uniquely incorporates the pyrimidine uracil, which pairs complementarily with adenine, in lieu of thymine found in DNA [4]. Despite RNA's single-stranded nature, many RNA varieties exhibit significant intramolecular base pairing among complementary sequences within the strand itself, forming a three-dimensional structure that is crucial for their functional roles (Fig. 2.1).

Ribonucleotides serve not only as fundamental components in the synthesis of RNA but also play vital roles within the cell, fulfilling various functions [5]. These nucleotides may contain varying numbers of phosphate groups attached to their 5′ carbon. For example, adenosine monophosphate (AMP) features a single phosphate group. This molecule consists of a ribose sugar, with an adenine base linked to the 1′ carbon of the ribose, and a phosphate group attached to the 5′ carbon. In contrast, adenosine diphosphate (ADP) includes two phosphate groups, and adenosine triphosphate (ATP) has three phosphate groups, each also bonded to the 5′ carbon of the ribose (Fig. 2.2), [6]. The number of phosphate groups is crucial as it determines the molecule's role and energy-carrying capacity within the cell. The hydrolysis of ATP to ADP and the release of a phosphate group (known as orthophosphate, pi) liberates energy that is essential for the function of numerous enzymes across virtually all cellular processes [7]. Furthermore, GTP serves as a structural element in a wide array of enzymes and proteins, which are often called small GTPases, reflecting their typical size and enzymatic functions [8, 9] (Fig. 2.3).

## 2.2 RNA Folding

Ribonucleotides and deoxyribonucleotides leverage their 5′-phosphate groups to create phosphodiester bonds, which are crucial for constructing the long chains of RNA or DNA molecule. In these nucleic acid structures, the sequence of nucleotides is joined together such that the phosphate group of one nucleotide links to the sugar component of the next [10]. This results in polymers with distinct ends: an unbound 5′ end, where the terminal 5′ carbon is not connected to another nucleotide

**Fig. 2.1** (continued) sugar-phosphates on the exterior and base pairs located internally. Conversely, RNA is depicted with a single strand of sugar-phosphates, with nitrogenous bases extending along the helix's length. (**b**) An illustration depicting RNA's ability to fold back onto itself demonstrates that bases linked to the sugar-phosphate backbone can establish hydrogen bonds when there are sequences of complementary bases positioned some distance apart on the lengthy strand. In contrast, certain areas lack these hydrogen bonds. (**c**) DNA is generally characterized by its double-stranded formation, in contrast to RNA, which is predominantly single-stranded. (**d**) Despite its single-stranded nature, RNA has the capability to fold back onto itself. This folding is stabilized by brief segments of complementary base pairing found within the RNA molecule itself, resulting in a three-dimensional conformation. (Figure is drawn in biorender.com)

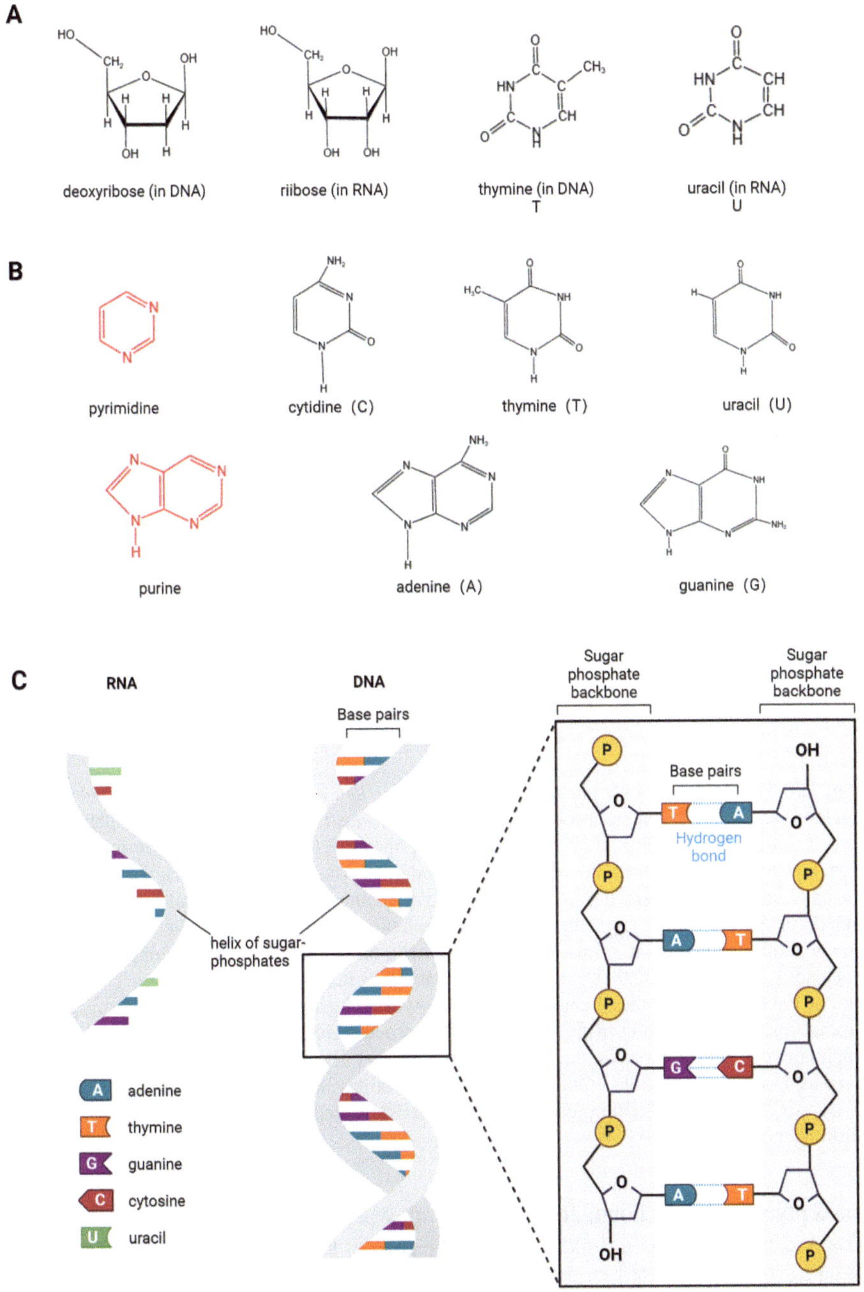


**Fig. 2.1** (**a**) Ribonucleotides are characterized by the presence of the pentose sugar ribose, as opposed to deoxyribose found in deoxyribonucleotides. (**b**) RNA is distinguished by its inclusion of the pyrimidine uracil, replacing the thymine present in DNA. (**a**) An illustration of DNA and RNA structures reveals that DNA adopts a double helix configuration with the helices of

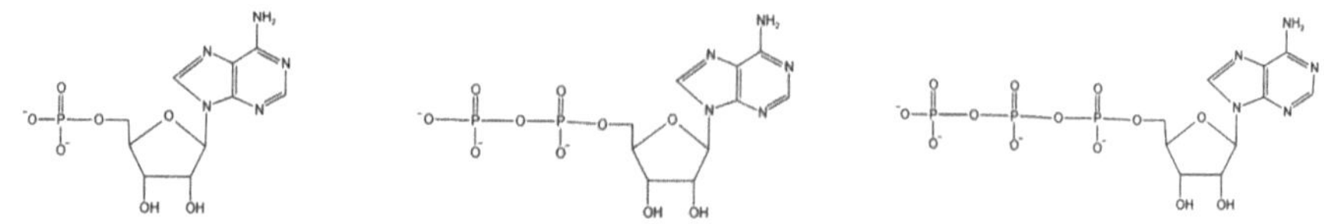


**Fig. 2.2** Nucleosides form nucleotides through the covalent attachment to phosphate groups at their 5′-hydroxyl group. The categorization of these molecules is based on the number of phosphate groups they contain. Specifically, nucleotides are classified as nucleotide monophosphates, nucleotide diphosphates, or nucleotide triphosphates, depending on whether they have one, two, or three phosphate groups attached, respectively. For instance, nucleotide monophosphates include AMP (adenosine monophosphate), GMP (guanosine monophosphate), UMP (uridine monophosphate), and CMP (cytidine monophosphate), each featuring a single phosphate group. Nucleotide diphosphates, such as ADP (adenosine diphosphate), GDP (guanosine diphosphate), UDP (uridine diphosphate), and CDP (cytidine diphosphate), have two phosphate groups. Lastly, nucleotide triphosphates like ATP (adenosine triphosphate), GTP (guanosine triphosphate), UTP (uridine triphosphate), and CTP (cytidine triphosphate) contain three phosphate groups. This classification is crucial as it reflects the molecules' potential for energy storage and transfer, which is fundamental to their biological roles. (Figure is drawn in Biorender.com)

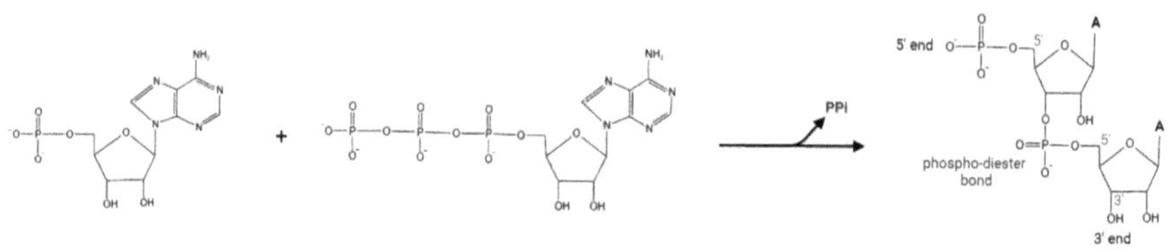


**Fig. 2.3** The formation of a phosphodiester bond is a key chemical reaction that involves two nucleotides coming together. Specifically, this bond is formed when adenosine triphosphate (ATP) interacts with the 3′-carbon of an adenosine monophosphate (AMP) molecule. During this reaction, a phosphate group from the ATP is transferred and forms a bridge between the 5′-phosphate group of one nucleotide and the 3′-hydroxyl group of another. This linkage results in the formation of a dinucleotide, a small segment of what will eventually become a longer nucleic acid chain. The newly formed dinucleotide is characterized by having an unlinked 5′ end, meaning that this end does not participate in another phosphodiester linkage at this stage, leaving it free. The opposite end of the dinucleotide, the 3′ end, retains a free hydroxyl group. This structural arrangement is crucial for the continuation of the polymerization process, as it allows for the addition of more nucleotides to extend the chain, facilitating the synthesis of RNA or DNA molecules essential for various biological functions. Other triphosphonucleotides are capable of reacting with these 3′-hydroxyl groups, facilitating the extension of the RNA chain at its 3′ end. (Figure is drawn in Biorender.com)

via a phosphodiester bond, and a free 3′ end, where the ribose or deoxyribose sugar's 3′ carbon remains unlinked [10]. In the case of DNA, these molecules typically exist as double-stranded structures. This arrangement involves two complementary strands of single-stranded DNA that interact through hydrogen bonds between their respective nucleotides, forming the characteristic double helix [11]. Each strand runs in opposite directions, which is essential for the processes of replication and transcription [11]. The interaction between the two DNA strands adds to the stability and protective properties of the genetic material, ensuring fidelity in the genetic information carried from one generation to the next. For the formation of double-stranded DNA, base pairs engage in hydrogen bonding: adenine pairs with thymine, and guanine pairs with cytosine [12].

The pairing between guanine and cytosine is established through three hydrogen bonds, whereas adenine and thymine are linked by only two hydrogen bonds, making their connection thermodynamically less stable [13]. The structure of DNA was deciphered in the 1950s by James Watson and Francis Crick, who introduced the iconic DNA double helix model [14]. According to the Watson-Crick model, two DNA strands wrap around a common axis to form a double helix. These strands run in opposite directions (antiparallel) and feature the sugar-phosphate backbone on the exterior, while the bases face inward, creating the hydrogen bonds previously mentioned. Notably, the DNA double helix is distinguished by the presence of a minor and a major groove, which serves as key interaction sites for proteins that engage with the DNA. DNA is capable of adopting different structural conformations, which include A-DNA, B-DNA, and Z-DNA [15]. These forms vary based on factors such as the nucleotide sequence, salt concentrations, and other environmental conditions. Each conformation has distinct characteristics, such as variations in the helix diameter, the size of the grooves in the DNA, and the number of nucleotides that make up each turn of the helix. A-DNA is a right-handed helix, similar to B-DNA, but with a shorter, more compact helical structure with approximately 11 nucleotides per turn [16, 17]. The major groove in A-DNA is narrower and deeper, while the minor groove is wider and shallower compared to B-DNA [18]. B-DNA is the most common DNA conformation under physiological conditions and is characterized by a right-handed helix with about 10.5 nucleotides per turn [19]. B-DNA features a larger helix diameter and provides well-defined major and minor grooves, which are critical for protein-DNA interactions [19]. Unlike A-DNA and B-DNA, Z-DNA forms a left-handed helix and has a more irregular structure with 12 nucleotides per turn [15, 20]. The major groove is almost flat, making it distinctly different from the other forms, and the minor groove is very narrow and deep. These different DNA conformations are not just structurally unique but are also functionally significant, influencing how DNA interacts with proteins, how it is expressed, and its overall stability in varying cellular environments. Understanding these conformations and their triggers is crucial for comprehending the dynamic nature of DNA in different biological contexts.

In the process of RNA synthesis, DNA acts as the template harboring genetic instructions. Typically, RNA molecules are produced as single strands by cellular enzymes known as RNA polymerases [21]. Nevertheless, RNA can also adopt secondary structures. Although double-stranded RNA formations are relatively uncommon in higher eukaryotes, two separate RNA strands can occasionally pair up to form double-stranded RNAs [22]. These RNA double helices, influenced by RNA's unique chemical properties, significantly differ from DNA helices. RNA helices usually adopt structures akin to DNA's A form, thus these double-stranded RNA configurations are referred to as A-RNA [23]. Unlike DNA, which predominantly sees guanine (G) pairing with cytosine (C) and adenine (A) with thymine (T), double-stranded RNAs often feature what are known as wobble base pairs [22]. The G–U wobble base pair, a common non-traditional interaction, involves guanine forming two hydrogen bonds with uracil (U), diverging from the canonical base pairing observed in DNA [24].

## 2.3 Category of RNA Structures

Just as proteins exhibit complexity through their primary, secondary, tertiary, and quaternary structural levels, RNA molecules similarly adopt intricate spatial configurations across analogous structural categories [44, 45]:

**Box 2.1 Z-RNA**

Z-RNA represents a high-energy, left-handed form of RNA, with its function still largely unknown. Increasing evidence suggests that Z-RNA may have regulatory roles in the immune response. In this review, we explore how the presence of Z-RNA within cellular RNAs—particularly those with retroelements—could be identified by a family of winged helix proteins, influencing host defense mechanisms. Over time, observations have gradually reduced the controversy surrounding the existence of Z-RNA in cells. Among the many proteins that recognize nucleic acids, several specifically identify Z-conformations of DNA and RNA through a similar winged helix Zα domain [25–27]. These Z-binding proteins are notably involved in viral infections and the innate immune response [28]. RNA that interacts with antibodies against Z-RNA has been detected in the cytoplasm, and Z-binding proteins have been found in stress-related cytoplasmic membraneless condensates [29, 30]. Recent studies have shown that altering the ability of these proteins to recognize Z-RNA in cancer cell lines and mouse models is associated with hyperinflammation phenotypes [31–33]. Collectively, these findings suggest a regulatory role for this transient RNA conformation, particularly in cellular defense against invaders.

One of the key mechanisms of the innate immune system is the detection of viral infections through the recognition of foreign dsRNA in the cytoplasm [34]. In mammals, this process is primarily conducted by dsRNA sensors. Notably, the retinoic acid-inducible gene-like receptors (RLR) family, which includes RIG-I and MDA5, activates the interferon (IFN) pathway upon detecting dsRNA via their helicase domains [34]. Additionally, protein kinase R (PKR) detects dsRNA to trigger stress granule formation and translational

(continued)

**Box 2.1** (continued)
shutdown [34–36]. Another dsRNA sensor, ZBP1, uniquely senses dsRNA through two Zα domains [34, 37, 38]. ZBP1 initiates apoptosis and necroptosis pathways via its RHIM domains [39]. To deactivate the innate immune response, ADAR1 inhibits these pathways by shielding dsRNA regions, thereby preventing ZBP1 and PKR activation [33, 40–43]. There is a consensus that ADAR1 also mitigates MDA5 activation by converting adenosines to inosines in dsRNA regions, which destabilizes their helical structure.

*When does Z-RNA play a role in the innate immune response?*

Host Zα-containing proteins are induced by interferon, while viral Zα-containing proteins are highly expressed during infection. Therefore, the question arises whether Z-RNA accumulates during the innate immune response due to the increase in Zα domains or if innate immune response proteins possess Zα domains because Z-RNA becomes prevalent during the interferon response through other mechanisms. Further research is necessary to clarify the interplay between the editing-related and editing-unrelated functions of Zα domains.

### 2.3.1 *Primary Structure*

The primary structure of RNA is defined by its sequence of nucleotides, which are the building blocks of RNA, much like amino acids are for proteins. This sequence dictates all higher levels of RNA structure and ultimately determines the molecule's function and interactions within the cell.

### 2.3.2 *Secondary Structure*

Secondary structures in RNA arise from the folding of the single-stranded RNA molecule back upon itself. Within the molecule, regions that have partially complementary sequences can pair up to form double-stranded sections, while unpaired regions form loops. Common structures include hairpins or stem-loops, where the paired region forms the "stem" and the loop consists of unpaired nucleotides. Hairpins can range dramatically in size, from minimal loops in small RNAs to extensive and complex looping structures in larger RNA molecules. Additionally, interruptions in the stem, known as bulges or mismatches, can influence the stability and function of these secondary structures.

### 2.3.3 Tertiary Structure

Tertiary structures involve more complex and distant interactions beyond the local secondary structures, allowing the RNA molecule to fold into a compact, three-dimensional shape. These interactions are often mediated by hydrogen bonding, metal ion coordination, and hydrophobic effects. The precise folding pattern is crucial for the molecule's biological activity, particularly in catalytic RNAs (ribozymes) and in RNAs that bind small molecules or interact with proteins.

### 2.3.4 Quaternary Structure

In quaternary structures, multiple RNA molecules, or RNA and protein molecules, come together to form larger complexes. An exemplary quaternary structure is the ribosome, essential for protein synthesis in all living cells. Here, various rRNA molecules interact with each other and with numerous proteins to form a complex capable of translating mRNA into protein. These interactions facilitate the ribosome's catalytic activity and ensure efficient protein synthesis. Table 2.1 provides a structured overview of the roles RNA plays within cells, highlighting its versatility and essential functions in various biological processes.

**Table 2.1** Cellular functions of RNAs

| Function of RNA | Description | Ref |
|---|---|---|
| mRNA | Carries genetic information from DNA to ribosomes, where it guides the synthesis of proteins according to the genetic code | [46] |
| Base pairing capability | Forms base pairs with complementary nucleic acids (either DNA or RNA), directing proteins to specific RNA targets to facilitate modification or degradation of these targets | [47] |
| Genetic material storage | Serves as genetic material in retroviruses. RNA in these viruses must be reverse-transcribed into DNA to integrate into the host's genome, a crucial step for virus replication and persistence | [48] |
| Structural role | Acts as scaffolds in the assembly of complex RNA-protein structures, essential for various cellular functions such as the formation of ribosomes and spliceosomes | [48] |
| Cellular sensing mechanisms | Detects and reacts to changes in the cellular environment, including fluctuations in temperature, ion concentrations, or the presence of specific metabolites, contributing to cellular adaptation and survival | [48, 49] |
| Regulation and expression of genetic information | Integral to the regulation and expression of genes, playing a critical role in the control of gene expression at both transcriptional and post-transcriptional levels | [50] |

## 2.4 Essential Protein Classes in RNA Functionality and Regulation

Within cells, nucleic acids interact closely with proteins, performing a range of functions that are explored throughout this text. DNA is organized into complex chromatin structures, where it is coiled around proteins known as histones. In a similar manner, RNAs form part of RNA–protein complexes. These intricate assemblies are commonly known as ribonucleoprotein particles or RNPs. For instance, mRNAs are integrated into mRNA–protein complexes (mRNPs) at various stages of their lifecycle.

***RNA-Binding Proteins:*** In any given RNP assembly, at least one protein component is required to interact directly with the RNA, highlighting the crucial role of specialized RNA-binding proteins in both the formation of RNPs and the subsequent functionality of the RNA [51]. Based on the RNA structure to which they attach, RNA-binding proteins are categorized into those that bind to single-stranded RNA and those that prefer double-stranded RNA [46]. This section will delve into these two protein classes and their unique RNA-binding motifs, providing insight into their specific interactions and functions.

***Proteins Involved in Single-Stranded RNA Interactions:*** Proteins capable of binding RNA are notably prevalent across various organisms. The RNA recognition motif (RRM) emerges as the most ubiquitous structural motif for binding single-stranded RNA. For instance, in humans, the RRM motif is found in approximately 0.5–1.0% of all proteins. RNA recognition motifs (RRMs), also known as RNA-binding domains (RBDs), are characterized by their unique structural features, primarily the ribonucleoprotein domains 1 (RNP1) and RNP2. These domains are located on two antiparallel beta-sheets that form the core of the RRM structure. These RNP motifs are crucial for the interaction with RNA, enabling the RRM to bind to nucleic acids. The functionality of RRMs varies: some are highly specific and interact with single-stranded RNA molecules in a sequence-dependent manner, recognizing and binding to precise RNA sequences. Others have a broader functionality, capable of interacting with a variety of RNA molecules without sequence specificity. This flexibility allows RRMs to play diverse roles in cellular processes, adapting to different RNA targets as needed for RNA processing, transport, stability, and regulation. Proteins containing RRM domains are pivotal in virtually all RNA-related processes. Frequently, proteins with RRM domains consist of multiple RRMs, which can affect each other's RNA-binding capabilities. Beyond RNA interaction, RRM domains can also function as platforms for protein interactions. The positioning of the alpha helix a1 relative to the RNP-containing beta-sheets dictates whether an RRM domain will interact with RNA or proteins [46, 52, 53].

***Importance of KH domains in RNA.*** KH domains are specialized protein domains renowned for their ability to interact with single-stranded RNAs, and they derive their name from the heterogeneous nuclear ribonucleoprotein K (hnRNP K), which is why they are referred to as hnRNP K-homology (KH) domains [54, 55]. These domains are critical components in RNA-binding proteins and play pivotal roles in various cellular processes involving RNA metabolism. KH domains are named after their prototype, hnRNP K, a protein first identified for its association with heterogeneous nuclear ribonucleoproteins involved in RNA processing. The term "KH domain" reflects the homology to this original protein, emphasizing a common structural motif that characterizes this family. KH domains are ubiquitous, found in a wide range of organisms across all domains of life: bacteria, archaea, and eukaryotes. This wide distribution underscores their fundamental role in cellular biology, particularly in processes involving RNA. Typically, KH domains consist of approximately 70 amino acids. They are structured to include several alpha-helices and beta-sheets. This arrangement allows the KH domain to maintain a stable yet flexible configuration conducive to nucleic acid binding [55]. A specific sequence within the KH domain facilitates its binding to nucleic acids, primarily through hydrogen bonds. Unlike many other RNA-binding motifs that require specific nucleotide sequences, KH domains are notable for their lack of sequence specificity. This feature allows them to interact with a broad range of single-stranded nucleic acids, including both RNA and DNA [56]. KH domains often do not occur in isolation within proteins; rather, they frequently appear in multiple copies within a single RNA-binding protein. This redundancy can enhance the binding strength and specificity to RNA targets, similar to RNA recognition motifs (RRMs). The presence of multiple KH domains can also allow the protein to engage with complex or structured RNAs by providing multiple points of contact, facilitating robust and versatile interactions. The ability of KH domains to bind RNA without sequence specificity is particularly significant. It enables these domains to participate in a variety of RNA-related functions, such as splicing, mRNA stability, and the regulation of gene expression. The flexibility and ubiquity of KH domains highlight their evolutionary importance as versatile tools in RNA biology [46, 52, 57].

## 2.5 Proteins that Interact with dsRNAs

dsRNA can form in several ways: through the folding of a single RNA molecule back onto itself, creating regions of complementary base pairing, or through the pairing of two RNA molecules with complementary sequences. These structures are not only common in the coding sequences of mRNA but also in non-coding RNAs, which play various regulatory roles within the cell. In response to the prevalence and functional importance of dsRNA, evolutionary processes have led to the development of specialized proteins that recognize and bind to these structures [58]. These proteins are crucial in managing the functions of dsRNAs, which include roles in gene expression, RNA silencing, and the defense against viral infections. Proteins that bind dsRNA typically feature one or more double-stranded RNA-binding domains (DRBDs) [46]. A DRBD usually consists of about 65–68 amino acids and adopts a characteristic α-β-β-β-α fold, which facilitates stable binding to dsRNA [59]. This structure allows for the recognition and binding to the shape and form of dsRNA, rather than to specific nucleotide sequences, enabling these proteins to interact with a wide range of RNA molecules. DRBD-containing proteins are found in a wide array of organisms, including bacteria, archaea, eukaryotes, and viruses, reflecting the fundamental nature of RNA processes across life forms. In eukaryotic cells, proteins with DRBDs can be involved in processes such as editing RNA transcripts (e.g., ADAR proteins), modulating RNA stability, and regulating translation [60]. In viruses, DRBDs are often involved in the suppression of the host's RNA interference machinery, crucial for viral replication and evasion of host defenses. While some proteins may possess multiple DRBDs, each potentially binding to different or the same dsRNA molecule(s) to regulate complex RNA structures or pathways, viral proteins often contain just a single DRBD. The presence of multiple DRBDs can enhance the binding affinity and specificity, allowing for more robust interactions with RNA, which is essential for proteins involved in critical regulatory pathways. The interaction of DRBD-containing proteins with dsRNA is critical in many biological processes. For instance, these proteins are essential in RNAi pathway, where they participate in the processing of miRNAs and siRNAs, which are key regulatory molecules that affect gene silencing. They also play roles in antiviral responses, recognizing viral dsRNA as a pathogen-associated molecular pattern (PAMP) and triggering immune responses [61].

## 2.6 RNA Helicases

RNAs have the ability to adopt complex secondary and tertiary formations. However, for certain functionalities, like the translation of mRNAs into proteins, these formations need to be unwound. Additionally, RNPs undergo continuous remodeling, with proteins binding to or releasing from RNPs [62]. To facilitate these processes, a specialized class of proteins known as RNA helicases has evolved. These helicases are distinguished by their capacity to use ATP hydrolysis energy to separate RNA strands, earning them the occasional nickname of unwindases [63]. Helicases, whether they act on RNA or DNA, are identified by several unique domains and motifs, leading to their classification into five superfamilies, SF1 through SF5, with the majority of RNA helicases falling into SF2 and a smaller number into SF1 [64]. RNA helicases are marked by seven to nine conserved motifs forming the core of the helicase [64]. Typically, these include motifs found in either the DEAD box, DEAH box, or DExD/H box, thus many RNA helicases are referred to by these box names [65].

The conserved motifs crucial for the functionality of RNA helicases are organized into two distinct domains. Specifically, motifs I, Ia, Ib, II, and III are grouped within domain 1, while motifs IV, V, and VI are categorized under domain 2. Motifs I and II, also referred to as Walker A and B motifs (named after Nobel laureate John E. Walker), are common not only in RNA helicases but also in various other ATPase proteins [66, 67]. These motifs play a fundamental role in the binding and hydrolysis of ATP, a process essential for the energy-driven functions of these proteins. Motif III directly connects ATP hydrolysis with the helicase activity that unwinds RNA, making it a critical component in the mechanism of RNA processing. Similarly, motif VI is vital for both RNA binding and the hydrolysis of ATP, integral to the helicase's functionality [68]. The additional motifs, Ia, Ib, IV, and V, primarily engage in binding RNA substrates, demonstrating their roles in recognizing and stabilizing RNA structures during unwinding [69]. A particularly notable element among these motifs is the Q motif, characterized by a conserved glutamine (Q) residue. This motif is unique to DEAD box helicases and serves a regulatory function in ATP binding and hydrolysis [70]. The presence and distinct function of the Q motif highlight the specialized adaptations within DEAD box helicases that tailor their enzymatic activities to specific cellular requirements, particularly in RNA metabolism. RNA helicases play a crucial role in resolving secondary structures within single-stranded RNA molecules, typically facilitating the separation of complementary RNA strands. Beyond their capacity to dissociate RNA duplexes, RNA helicases also serve a vital role in ribonucleoprotein (RNP) complex remodeling. It has been shown that these enzymes actively displace RNA-binding proteins from RNA molecules. Consequently, RNA helicases are integral to RNA metabolism, performing two primary functions: (i) unwinding double-stranded RNAs and (ii) detaching proteins from RNA molecules [71].

## 2.7 RNA from a Historical Perspective

As the complexity of the RNA world continues to expand, with new types of RNA being identified nearly every month, it prompts us to revisit the fundamental question: What defines or characterizes RNA? The journey to understand RNA began nearly a century ago when scientists first investigated a substance then known as yeast nucleic acid or zymonucleic acid, which was later renamed pentose nucleic acid (PNA) before ultimately being identified as RNA. Pioneering researchers like Phoebus A.T. Levene and Jean L.A. Brachet were instrumental in these early discoveries, setting the foundation for our current understanding of RNA. By the late 1950s and early 1960s, it had become clear that most cellular RNA is synthesized from DNA templates via enzymatic processes. During this period, the essential role of RNA in protein synthesis also came to light, underscored by the identification of various forms of RNA, such as microsomal RNA (rRNA), soluble RNA (tRNA), and mRNA. These discoveries provided a deeper insight into the complex mechanisms of gene expression.

In the subsequent decades, research continued to advance, leading to the discovery and characterization of additional types of RNA, including small nuclear RNAs (snRNAs) and small nucleolar RNAs (snoRNAs). These findings expanded our understanding of RNA's diverse functions in the cell, from structural components and catalysts to regulators of gene expression. The evolution of our knowledge about RNA illustrates its central role in biology and its complexity as a molecular entity. Initially, RNAs, with the exception of mRNAs, were categorized as structural RNAs, a term likely influenced by the structural functions attributed to them, as seen with rRNA, which was considered merely a scaffold for ribosomal proteins [72]. This classification may also have stemmed from the belief that these RNA species were capable of forming complex structures, similar to tRNA, a concept that was thought to contrast with the structure of mRNAs. Interestingly, in the early 1960s, a few researchers foresaw the significance of RNA in both the evolution of life and its functional and regulatory roles in current organisms. RNA has a rich history of discovery and research leading to the development of current RNA therapeutics. Below is a detailed timeline highlighting key milestones from the initial discovery of RNA to the recent advancements in RNA-based treatments (Fig. 2.4).

## 2.8 RNA Molecule: The Controversy

Recent advancements in experimental methodologies, including microchip assays and in-depth sequencing of cellular RNA components, have uncovered an extensive array of transcripts, some of which are present in very low quantities. There is ongoing debate about the functional significance of these transcripts. One viewpoint suggests that nearly all identified transcripts, including those that appear to be degradation by-products, serve a functional role [73–79]. Conversely, some argue that while it is plausible that the genomes of many organisms encode, beyond mRNAs, a vast number of functional RNAs, a significant portion, especially those transcripts found in small amounts, might simply be the by-product of random transcription initiation in areas between genes [79–86], previously unrecognized UTRs that are alternatively spliced [87], extensions beyond established gene termination points [88], and relatively stable remnants from the processing of initial transcripts. Noteworthy among these are discarded introns, internal transcribed spacers, and sequences at the beginning and end of genes, etc., and in some cases, sequences that include spliced exons, particularly if the genes act as vehicles for miRNAs or snoRNAs without (or having lost) their protein-coding potential [89]. Additionally, this category includes molecules produced during the breakdown and recycling of mature RNAs [90–96]. Importantly, the phenomenon of transcriptional interference, increasingly recognized as a critical aspect of gene regulation, generates RNA molecules with sequences that have no direct relevance; the mere process of

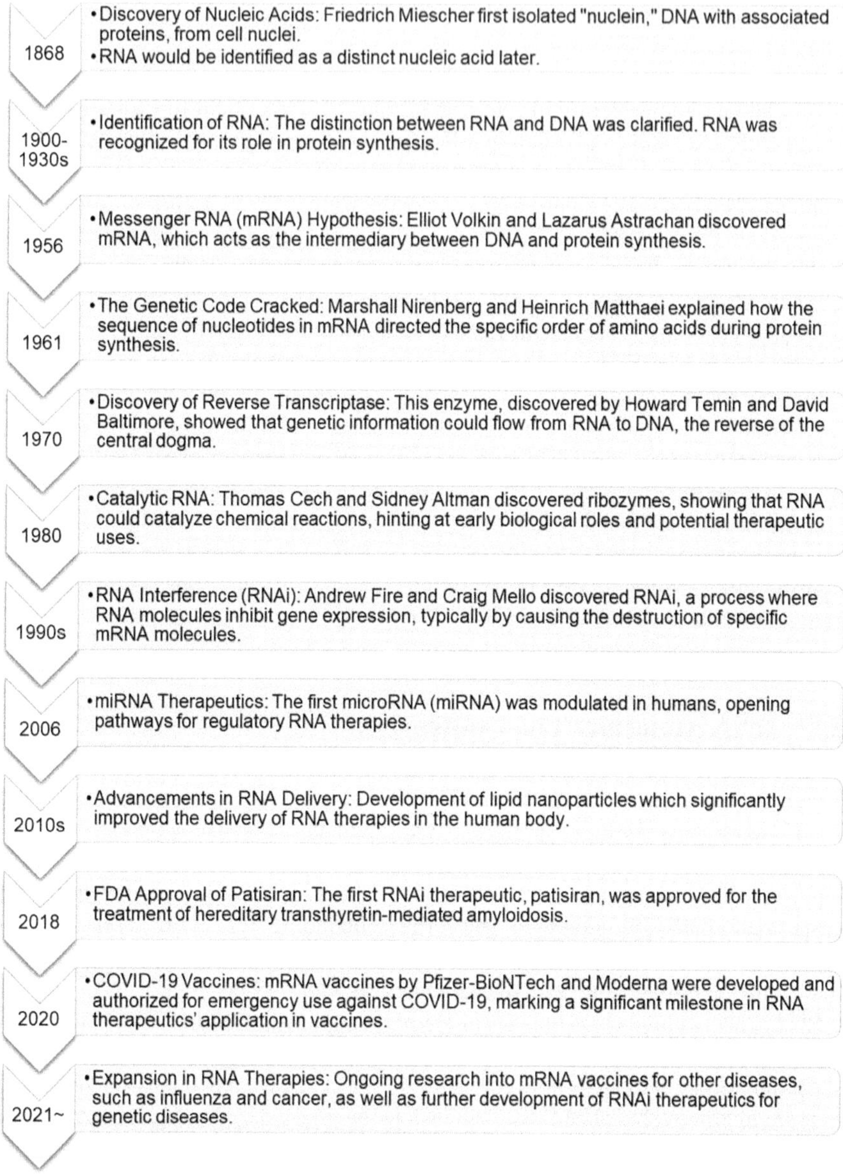


**Fig. 2.4** This timeline underscores the transformative journey of RNA from a fundamental biological molecule to a powerful tool in medicine. Each step has built on the previous discoveries, leading to innovative RNA-based therapies that are now impacting lives globally. Despite the recognition of a few RNA types and the speculated remnants from the RNA world, the broader implications of RNA in cellular functionality and evolutionary history were not fully appreciated until decades later. It is only in recent years that RNA has gained the recognition it deserves, although this has occasionally led to overstated claims about its functions. While all RNAs transcribed from DNA are technically transcripts, the debate remains whether each transcript, despite being chemically a ribonucleic acid, should be classified as RNA in a biological sense. This calls for a more discerning approach in the categorization of RNAs

transcription influences other genetic elements, such as downstream promoters, through mechanisms like blocking transcription factor attachment sites [97–100] or modifying the structure of chromatin [101, 102].

## 2.9 RNA Vs. Transcript: Different Classification, Same Purpose

Considering an RNA chain, it is often linked with specific functional (regulatory or structural) roles, similar to proteins. However, not every random transcript, degradation fragment, or other types of "non-functional" RNAs should be automatically assigned such significance. Following the somewhat paradoxical designation of ncRNAs for those non-messenger RNAs (mRNAs) that are functional, there is a temptation to label RNAs lacking function as "non-RNAs." Nevertheless, some advocate for the broad term "transcripts" for these RNA types, noting that certain ones were previously recognized as either stable untranslated transcripts or transcripts with unidentified functions [103, 104]. In a nutshell, Stable Untranslated Transcripts (SUTs) are RNA molecules that do not code for proteins and are not classified under known categories of ncRNAs (such as rRNA, tRNA, and miRNA). SUTs are typically characterized by their stability in the cell and may be involved in regulatory functions, although their precise roles are often not well understood. They are distinguished from other non-coding RNAs by their longer lengths and their apparent lack of translation into proteins. Transcripts with Unidentified Functions (TUFs) are used for RNA transcripts whose functions have not been identified. These transcripts are detected as part of transcriptional output from genomes but do not match known functional RNA categories, and no clear function or product has been attributed to them. Like SUTs, TUFs may include both coding and non-coding RNAs, and their designation as "unidentified" reflects a gap in understanding rather than a definitive functional characteristic. Given the challenge in determining whether such transcripts lack translation or functionality altogether, and the fact that ultra-deep RNA sequencing reveals even minimal, potentially incidental or background transcripts, some researchers favor the term "transcripts" [105].

The term "transcripts" refer to a broad category of RNA molecules, which includes random transcripts that are not part of a structured gene expression process, fragments left from the initial processing of transcripts, and degradation remnants from the turnover of mature RNA. Additionally, this category encompasses molecules involved in transcriptional regulation. Given the diverse origins and functions of these RNAs, they are classified into a distinct group known as class 0d. This classification is designed to recognize these transcripts as a separate entity from other more defined classes of RNA. Class I encompasses functional RNAs, which includes structural RNAs that play a critical role in maintaining the cellular infrastructure and facilitating biochemical processes. Class II, on the other hand, consists of mRNAs that carry the genetic information necessary for the synthesis of peptides and proteins. By categorizing "transcripts" as class 0d, the classification system acknowledges the complexity and variety of RNA types and functions, highlighting

**Table 2.2** Classification of each class of RNA based on transcripts

| Class | Description | Includes | Function |
|---|---|---|---|
| Class 0d | Miscellaneous and regulatory transcripts that do not fit into more defined RNA classes | • Random transcripts<br>• Transcriptional fragments<br>• Degradation remnants<br>• Regulatory RNAs | Varied, ranging from potential regulatory functions to byproducts of cellular processes |
| Class I | Functional RNAs crucial for cellular infrastructure and biochemical processes | • Ribosomal RNAs (rRNAs)<br>• Transfer RNAs (tRNAs)<br>• Small nuclear RNAs (snRNAs) | Maintain cellular structure and facilitate essential biochemical processes such as protein synthesis and mRNA processing |
| Class II | Messenger RNAs (mRNAs) that carry genetic information for protein synthesis | • Messenger RNAs (mRNAs) | Convey genetic information necessary for synthesizing peptides and proteins, pivotal in cellular functions |

the intricate nature of gene expression and regulation. This structure provides a framework for researchers to better understand and study the various roles that RNA plays in cellular biology [106]. Table 2.2 classifies the RNA transcripts into the "proposed" Class 0d, Class I, and Class II.

## 2.10 Gray Areas in RNA Classification

In biological systems, the concept of distinct boundaries or states often becomes blurred, presenting us with wide, indistinct interfaces or chimeric entities. It is clear that most messenger RNAs (mRNAs, class II) possess not just an ORF but also carry additional coding sequences within UTRs or even inside ORFs themselves. These sequences act as regulatory elements that affect mRNA stability or their ability to be translated, making mRNAs a hybrid of class I and II RNAs [107–110]. The delineation between RNA and mRNA is equally ambiguous, as certain authentic class I RNAs can encode ORFs that might be sporadically translated [111] (thus being dual-functional RNAs) [112, 113]. Similarly, an RNA undergoing exaptation to become an mRNA is another instance of such fluidity [114]. A notable chimera is the bacterial tmRNA, a merger of transfer RNA and mRNA functions, previously identified as 10Sa RNA. This RNA rescues ribosomes stuck on incomplete mRNAs lacking a stop codon [115]. Additionally, some ribosomal RNAs (rRNAs) have been found to encode short peptides that provide antibiotic resistance [116]. The transitions from transcripts to RNAs or mRNAs are also areas of significant interest, spanning a wide spectrum. This spectrum ranges from transcripts evolving toward functional roles to those on the path to being forgotten [80, 117–119]. In contrast, the transition from transcripts to mRNAs is likely narrower due to the requirement for mRNAs to temporarily associate with ribosomes, a process that is experimentally verifiable [120]. Although this association does not guarantee translatability, its absence does not preclude it either, as such interactions may not have been studied in all relevant cellular or developmental contexts.

RNAs that function by binding one or more proteins, facilitating their joint function or targeting them to specific locations within a cell, perform a clear role and are thus considered class I RNAs. However, transcripts or degradation products associated with proteins do not inherently possess a function, classifying them as transcripts (class 0). RNAs that function as decoys or molecular sinks, capturing other RNAs, proteins, or molecules, fall under the category of functional RNAs, designated as class I. This category includes specific types of circRNAs and non-translatable transcripts derived from duplicated genes, such as retroposed pseudogenes. These RNAs play specialized roles within the cellular environment, interacting with and regulating other molecular entities. Despite this, the vast majority of circRNAs, which are often products of aberrant splicing events, are presumed to lack functional roles and are therefore classified as transcripts (class 0). This distinction reflects a broader understanding that not all RNA molecules produced within a cell contribute directly to its biochemical activities. Additionally, RNA molecules that result from transcriptional regulatory processes are not considered true RNAs in the strictest biological sense. This perspective, however, does not preclude the potential for a non-functional class 0 transcript to acquire a functional role as an RNA or mRNA in the future. This idea aligns with the notion first proposed by Henry Harris over fifty years ago, which suggested that such extra transcripts could serve as evolutionary material, potentially gaining new functions as the organism adapts over time. This concept underscores the dynamic and evolutionary nature of RNA within biological systems [121]. However, with the fast-growing identification of new RNAs, the classification of a coding vs. ncRNA, more researches are pointing toward the dual-functionality of RNAs, making it difficult to fully assign an RNA into a specific category. Perhaps we could look forward to a more comprehensive classification in the near future.

## 2.11 Conclusion

The burgeoning field of ncRNA research unveils a richer spectrum of nature's diversity, underscoring the intricate role of ncRNAs in cellular biology far beyond what was initially anticipated. Despite the relatively recent uncovering of ncRNAs, there has been a notable surge in their clinical and diagnostic application breakthroughs. Such research is set to forge new paths in the development of more effective, targeted drugs and bespoke medical strategies, thereby revolutionizing patient care standards. The swift advancements in bioinformatics, sequencing technologies, proteomics, and microarrays have led to the discovery of a multitude of ncRNA, key regulators of most cellular processes, predominantly associated with the complexity in eukaryotes. Understanding the specific roles of these non-coding RNAs becomes increasingly challenging as they form intricate networks and pathways within biological systems. The integration of ncRNA-based therapies into conventional drug development is anticipated to grow, necessitating further research into their pharmacokinetics and dynamics, alongside comprehensive toxicological evaluations. When analyzing the vast array of ncRNAs, machine learning (ML) proves superior by swiftly pinpointing the core issue. The current ML-based analysis of ncRNAs, through the categorization of healthy and diseased samples, shows promising accuracy, suggesting clear differentiation patterns in such cases. Future investigations should focus on refining ML models to better distinguish each disease's unique pattern. Moreover, the forthcoming years are likely to witness a significant uptick in the utilization of ncRNAs, propelling the advancement of precision medicine and the customization of treatment options.

## References


1. Leslie E, Grover N. RNA: composition and base pairing. In: Grover N, editor. Fundamentals of RNA structure and function. Cham: Springer International Publishing; 2022. p. 1–19.
2. Kallenbach NR, Berman HM. RNA structure. Q Rev Biophys. 1977;10(2):138–236.
3. Tolstorukov MY, Jernigan RL, Zhurkin VB. Protein-DNA hydrophobic recognition in the minor groove is facilitated by sugar switching. J Mol Biol. 2004;337(1):65–76.
4. Sire J, Quérat G, Esnault C, Priet S. Uracil within DNA: an actor of antiviral immunity. Retrovirology. 2008;5:45.
5. Kruger K, Grabowski PJ, Zaug AJ, Sands J, Gottschling DE, Cech TR. Self-splicing RNA: autoexcision and autocyclization of the ribosomal RNA intervening sequence of Tetrahymena. Cell. 1982;31(1):147–57.
6. Dunn J, Grider MH. Physiology, adenosine triphosphate. In: StatPearls. Treasure Island (FL) ineligible companies. Disclosure: Michael Grider declares no relevant financial relationships with ineligible companies. StatPearls Publishing. Copyright © 2024.; 2024.
7. Rajendran M, Dane E, Conley J, Tantama M. Imaging adenosine triphosphate (ATP). Biol Bull. 2016;231(1):73–84.
8. Sharma A, Kumar G, Sharma S, Walia K, Chouhan P, Mandal B, et al. Methods for binding analysis of small GTP-binding proteins with their effectors. Methods Cell Biol. 2021;166:235–50.
9. Zhao S, Xu W, Jiang W, Yu W, Lin Y, Zhang T, et al. Regulation of cellular metabolism by protein lysine acetylation. Science (New York, NY). 2010;327(5968):1000–4.
10. Minchin S, Lodge J. Understanding biochemistry: structure and function of nucleic acids. Essays Biochem. 2019;63(4):433–56.
11. Johansson E, Dixon N. Replicative DNA polymerases. Cold Spring Harb Perspect Biol. 2013;5(6):a012799.
12. Slocombe L, Winokan M, Al-Khalili J, Sacchi M. Proton transfer during DNA strand separation as a source of mutagenic guanine-cytosine tautomers. Commun Chem. 2022;5(1):144.
13. Brela MZ, Klimas O, Boczar M, Nakajima T, Wójcik MJ. A comparison of the hydrogen bond interaction dynamics in the adenine and thymine crystals: BOMD and spectroscopic study. Spectrochim Acta A Mol Biomol Spectrosc. 2020;237:118398.
14. Watson JD, Crick FH. Molecular structure of nucleic acids; a structure for deoxyribose nucleic acid. Nature. 1953;171(4356):737–8.

15. Bukowiecka-Matusiak M, Woźniak LA. DNA structure from A to Z—biological implications of structural diversity of DNA. Postepy Biochem. 2006;52(3):229–38.
16. Cheatham TE 3rd, Kollman PA. Insight into the stabilization of A-DNA by specific ion association: spontaneous B-DNA to A-DNA transitions observed in molecular dynamics simulations of d[ACCCGCGGGT]2 in the presence of hexaamminecobalt(III). Structure (London, England: 1993). 1997;5(10):1297–311.
17. Nishimura Y, Torigoe C, Tsuboi M. Salt induced B—a transition of poly(dG).poly(dC) and the stabilization of A form by its methylation. Nucleic Acids Res. 1986;14(6):2737–48.
18. Timsit Y. DNA structure and polymerase fidelity. J Mol Biol. 1999;293(4):835–53.
19. Baker ES, Bowers MT. B-DNA helix stability in a solvent-free environment. J Am Soc Mass Spectrom. 2007;18(7):1188–95.
20. Wang G, Vasquez KM. Z-DNA, an active element in the genome. Front Biosci. 2007;12:4424–38.
21. Uzman A. Molecular biology of the cell (4th ed.): Alberts, B., Johnson, A., Lewis, J., Raff, M., Roberts, K., and Walter, P. Biochem Mol Biol Educ. 2003;31(4):212–4.
22. Gantier MP, Williams BR. The response of mammalian cells to double-stranded RNA. Cytokine Growth Factor Rev. 2007;18(5–6):363–71.
23. Masliah G, Barraud P, Allain FH. RNA recognition by double-stranded RNA binding domains: a matter of shape and sequence. Cell Mol Life Sci. 2013;70(11):1875–95.
24. Varani G, McClain WH. The G x U wobble base pair. A fundamental building block of RNA structure crucial to RNA function in diverse biological systems. EMBO Rep. 2000;1(1):18–23.
25. Gajiwala KS, Burley SK. Winged helix proteins. Curr Opin Struct Biol. 2000;10(1):110–6.
26. Placido D, Brown BA 2nd, Lowenhaupt K, Rich A, Athanasiadis A. A left-handed RNA double helix bound by the Z alpha domain of the RNA-editing enzyme ADAR1. Structure (London, England: 1993). 2007;15(4):395–404.
27. Zhang T, Yin C, Boyd DF, Quarato G, Ingram JP, Shubina M, et al. Influenza virus Z-RNAs induce ZBP1-mediated necroptosis. Cell. 2020;180(6):1115–29.e13.
28. Athanasiadis A. Zalpha-domains: at the intersection between RNA editing and innate immunity. Semin Cell Dev Biol. 2012;23(3):275–80.
29. Zarling DA, Calhoun CJ, Hardin CC, Zarling AH. Cytoplasmic Z-RNA. Proc Natl Acad Sci USA. 1987;84(17):6117–21.
30. Ng SK, Weissbach R, Ronson GE, Scadden AD. Proteins that contain a functional Z-DNA-binding domain localize to cytoplasmic stress granules. Nucleic Acids Res. 2013;41(21):9786–99.
31. de Reuver R, Verdonck S, Dierick E, Nemegeer J, Hessmann E, Ahmad S, et al. ADAR1 prevents autoinflammation by suppressing spontaneous ZBP1 activation. Nature. 2022;607(7920):784–9.
32. Hubbard NW, Ames JM, Maurano M, Chu LH, Somfleth KY, Gokhale NS, et al. ADAR1 mutation causes ZBP1-dependent immunopathology. Nature. 2022;607(7920):769–75.
33. Zhang T, Yin C, Fedorov A, Qiao L, Bao H, Beknazarov N, et al. ADAR1 masks the cancer immunotherapeutic promise of ZBP1-driven necroptosis. Nature. 2022;606(7914):594–602.
34. Hur S, Double-Stranded RNA. Sensors and modulators in innate immunity. Annu Rev Immunol. 2019;37:349–75.
35. Anderson P, Kedersha N. RNA granules. J Cell Biol. 2006;172(6):803–8.
36. Protter DSW, Parker R. Principles and properties of stress granules. Trends Cell Biol. 2016;26(9):668–79.
37. Jiao H, Wachsmuth L, Kumari S, Schwarzer R, Lin J, Eren RO, et al. Z-nucleic-acid sensing triggers ZBP1-dependent necroptosis and inflammation. Nature. 2020;580(7803):391–5.
38. Balachandran S, Mocarski ES. Viral Z-RNA triggers ZBP1-dependent cell death. Curr Opin Virol. 2021;51:134–40.
39. Snyder AG, Oberst A. The antisocial network: cross talk between cell death programs in host defense. Annu Rev Immunol. 2021;39:77–101.

40. Okonski KM, Samuel CE. Stress granule formation induced by measles virus is protein kinase PKR dependent and impaired by RNA adenosine deaminase ADAR1. J Virol. 2013;87(2):756–66.
41. Chung H, Calis JJA, Wu X, Sun T, Yu Y, Sarbanes SL, et al. Human ADAR1 prevents endogenous RNA from triggering translational shutdown. Cell. 2018;172(4):811–24.e14.
42. de Reuver R, Dierick E, Wiernicki B, Staes K, Seys L, De Meester E, et al. ADAR1 interaction with Z-RNA promotes editing of endogenous double-stranded RNA and prevents MDA5-dependent immune activation. Cell Rep. 2021;36(6):109500.
43. Karki R, Sundaram B, Sharma BR, Lee S, Malireddi RKS, Nguyen LN, et al. ADAR1 restricts ZBP1-mediated immune response and PANoptosis to promote tumorigenesis. Cell Rep. 2021;37(3):109858.
44. Tanner NK. Ribozymes: the characteristics and properties of catalytic RNAs. FEMS Microbiol Rev. 1999;23(3):257–75.
45. Kirchner S, Ignatova Z. Emerging roles of tRNA in adaptive translation, signalling dynamics and disease. Nat Rev Genet. 2015;16(2):98–112.
46. Corley M, Burns MC, Yeo GW. How RNA-binding proteins interact with RNA: molecules and mechanisms. Mol Cell. 2020;78(1):9–29.
47. Wang X-W, Liu C-X, Chen L-L, Zhang QC. RNA structure probing uncovers RNA structure-dependent biological functions. Nat Chem Biol. 2021;17(7):755–66.
48. Hiller DA, Strobel SA. The chemical versatility of RNA. Philos Trans R Soc Lond Ser B Biol Sci. 2011;366(1580):2929–35.
49. Cai C, Tang YD, Xu G, Zheng C. The crosstalk between viral RNA-and DNA-sensing mechanisms. Cell Mol Life Sci. 2021;78(23):7427–34.
50. Licatalosi DD, Darnell RB. RNA processing and its regulation: global insights into biological networks. Nat Rev Genet. 2010;11(1):75–87.
51. Tauber D, Tauber G, Parker R. Mechanisms and regulation of RNA condensation in RNP granule formation. Trends Biochem Sci. 2020;45(9):764–78.
52. Maris C, Dominguez C, Allain FH. The RNA recognition motif, a plastic RNA-binding platform to regulate post-transcriptional gene expression. FEBS J. 2005;272(9):2118–31.
53. Barraud P, Allain FH. Solution structure of the two RNA recognition motifs of hnRNP A1 using segmental isotope labeling: how the relative orientation between RRMs influences the nucleic acid binding topology. J Biomol NMR. 2013;55(1):119–38.
54. Hasan MK, Jeannine BL. Nucleic acid-binding KH domain proteins influence a spectrum of biological pathways including as part of membrane-localized complexes. J Struct Biol. 2024;10:100106.
55. Valverde R, Edwards L, Regan L. Structure and function of KH domains. FEBS J. 2008;275(11):2712–26.
56. Olejniczak M, Jiang X, Basczok MM, Storz G. KH domain proteins: another family of bacterial RNA matchmakers? Mol Microbiol. 2022;117(1):10–9.
57. Hollingworth D, Candel AM, Nicastro G, Martin SR, Briata P, Gherzi R, et al. KH domains with impaired nucleic acid binding as a tool for functional analysis. Nucleic Acids Res. 2012;40(14):6873–86.
58. Ryter JM, Schultz SC. Molecular basis of double-stranded RNA-protein interactions: structure of a dsRNA-binding domain complexed with dsRNA. EMBO J. 1998;17(24):7505–13-13.
59. Tian B, Bevilacqua PC, Diegelman-Parente A, Mathews MB. The double-stranded-RNA-binding motif: interference and much more. Nat Rev Mol Cell Biol. 2004;5(12):1013–23.
60. Deng P, Khan A, Jacobson D, Sambrani N, McGurk L, Li X, et al. Adar RNA editing-dependent and -independent effects are required for brain and innate immune functions in Drosophila. Nat Commun. 2020;11(1):1580.
61. Saunders LR, Barber GN. The dsRNA binding protein family: critical roles, diverse cellular functions. FASEB J. 2003;17(9):961–83.
62. Faoro C, Ataide SF. Ribonomic approaches to study the RNA-binding proteome. FEBS Lett. 2014;588(20):3649–64.

63. Jarmoskaite I, Helmers AE, Russell R. Measurement of ATP utilization in RNA unwinding and RNA chaperone activities by DEAD-box helicase proteins. Methods Enzymol. 2022;673:53–76.
64. Wu CG, Spies M. Overview: what are helicases? Adv Exp Med Biol. 2013;767:1–16.
65. Ali MAM. DEAD-box RNA helicases: the driving forces behind RNA metabolism at the crossroad of viral replication and antiviral innate immunity. Virus Res. 2021;296:198352.
66. Abou Assi H, Garavís M, González C, Damha MJ. I-Motif DNA: structural features and significance to cell biology. Nucleic Acids Res. 2018;46(16):8038–56.
67. Banroques J, Cordin O, Doère M, Linder P, Tanner NK. A conserved phenylalanine of motif IV in superfamily 2 helicases is required for cooperative, ATP-dependent binding of RNA substrates in DEAD-box proteins. Mol Cell Biol. 2008;28(10):3359–71.
68. Banroques J, Doère M, Dreyfus M, Linder P, Tanner NK. Motif III in superfamily 2 "helicases" helps convert the binding energy of ATP into a high-affinity RNA binding site in the yeast DEAD-box protein Ded1. J Mol Biol. 2010;396(4):949–66.
69. Rocak S, Linder P. DEAD-box proteins: the driving forces behind RNA metabolism. Nat Rev Mol Cell Biol. 2004;5(3):232–41.
70. Cordin O, Tanner NK, Doère M, Linder P, Banroques J. The newly discovered Q motif of DEAD-box RNA helicases regulates RNA-binding and helicase activity. EMBO J. 2004;23(13):2478–87.
71. Sarkar M, Ghosh MK. DEAD box RNA helicases: crucial regulators of gene expression and oncogenesis. Front Biosci (Landmark Ed). 2016;21(2):225–50.
72. Dai Q, Li J, Zhou K, Liang T. Competing endogenous RNA: a novel posttranscriptional regulatory dimension associated with the progression of cancer. Oncol Lett. 2015;10(5):2683–90.
73. Carninci P. RNA dust: where are the genes? DNA Res. 2010;17(2):51–9.
74. Clark MB, Amaral PP, Schlesinger FJ, Dinger ME, Taft RJ, Rinn JL, et al. The reality of pervasive transcription. PLoS Biol. 2011;9(7):e1000625; discussion e1102.
75. Mattick JS. Challenging the dogma: the hidden layer of non-protein-coding RNAs in complex organisms. BioEssays. 2003;25(10):930–9.
76. Mattick JS, Dinger ME. The extent of functionality in the human genome. Hugo J. 2013;7(1):2. https://doi.org/10.1186/1877-6566-7-2. Epub 2013 Jul 15.
77. Kapranov P, St LG. Dark matter RNA: existence, function, and controversy. Front Genet. 2012;3:60.
78. Anon. An integrated encyclopedia of DNA elements in the human genome. Nature. 2012;489(7414):57–74.
79. Kowalczyk MS, Higgs DR, Gingeras TR. Molecular biology: RNA discrimination. Nature. 2012;482(7385):310–1.
80. Brosius J. Waste not, want not—transcript excess in multicellular eukaryotes. Trends Genetics. 2005;21(5):287–8.
81. Struhl K. Transcriptional noise and the fidelity of initiation by RNA polymerase II. Nat Struct Mol Biol. 2007;14(2):103–5.
82. Raj A, van Oudenaarden A. Nature, nurture, or chance: stochastic gene expression and its consequences. Cell. 2008;135(2):216–26.
83. Ulitsky I, Bartel DP. lincRNAs: genomics, evolution, and mechanisms. Cell. 2013;154(1):26–46.
84. Graur D, Zheng Y, Price N, Azevedo RB, Zufall RA, Elhaik E. On the immortality of television sets: "function" in the human genome according to the evolution-free gospel of ENCODE. Genome Biol Evol. 2013;5(3):578–90.
85. Jensen TH, Jacquier A, Libri D. Dealing with pervasive transcription. Mol Cell. 2013;52(4):473–84.
86. Mudge JM, Frankish A, Harrow J. Functional transcriptomics in the post-ENCODE era. Genome Res. 2013;23(12):1961–73.
87. van Bakel H, Nislow C, Blencowe BJ, Hughes TR. Most "dark matter" transcripts are associated with known genes. PLoS Biol. 2010;8(5):e1000371.

88. Richard P, Manley JL. Transcription termination by nuclear RNA polymerases. Genes Dev. 2009;23(11):1247–69.
89. Moore M. Nuclear RNA turnover. Cell. 2002;108(4):431–4.
90. Cole C, Sobala A, Lu C, Thatcher SR, Bowman A, Brown JW, et al. Filtering of deep sequencing data reveals the existence of abundant dicer-dependent small RNAs derived from tRNAs. RNA (New York, NY). 2009;15(12):2147–60.
91. Lee YS, Shibata Y, Malhotra A, Dutta A. A novel class of small RNAs: tRNA-derived RNA fragments (tRFs). Genes Dev. 2009;23(22):2639–49.
92. Keene JD. Minireview: global regulation and dynamics of ribonucleic acid. Endocrinology. 2010;151(4):1391–7.
93. Khayrullina GA, Raabe CA, Hoe CH, Becker K, Reinhardt R, Tang TH, et al. Transcription analysis and small non-protein coding RNAs associated with bacterial ribosomal protein operons. Curr Med Chem. 2012;19(30):5187–98.
94. Kishore S, Khanna A, Zhang Z, Hui J, Balwierz PJ, Stefan M, et al. The snoRNA MBII-52 (SNORD 115) is processed into smaller RNAs and regulates alternative splicing. Hum Mol Genet. 2010;19(7):1153–64.
95. Gebetsberger J, Polacek N. Slicing tRNAs to boost functional ncRNA diversity. RNA Biol. 2013;10(12):1798–806.
96. Martens JA, Laprade L, Winston F. Intergenic transcription is required to repress the Saccharomyces cerevisiae SER3 gene. Nature. 2004;429(6991):571–4.
97. Shearwin KE, Callen BP, Egan JB. Transcriptional interference—a crash course. Trends Genetics. 2005;21(6):339–45.
98. Palmer AC, Ahlgren-Berg A, Egan JB, Dodd IB, Shearwin KE. Potent transcriptional interference by pausing of RNA polymerases over a downstream promoter. Mol Cell. 2009;34(5):545–55.
99. Kornienko AE, Guenzl PM, Barlow DP, Pauler FM. Gene regulation by the act of long non-coding RNA transcription. BMC Biol. 2013;11:59.
100. Petruk S, Sedkov Y, Riley KM, Hodgson J, Schweisguth F, Hirose S, et al. Transcription of bxd noncoding RNAs promoted by trithorax represses Ubx in cis by transcriptional interference. Cell. 2006;127(6):1209–21.
101. Latos PA, Pauler FM, Koerner MV, Şenergin HB, Hudson QJ, Stocsits RR, et al. Airn transcriptional overlap, but not its lncRNA products, induces imprinted Igf2r silencing. Science (New York, NY). 2012;338(6113):1469–72.
102. Barbieri M. The organic codes: an introduction to semantic biology. New York: Cambridge University Press; 2003.
103. Xu Z, Wei W, Gagneur J, Perocchi F, Clauder-Münster S, Camblong J, et al. Bidirectional promoters generate pervasive transcription in yeast. Nature. 2009;457(7232):1033–7.
104. Bassett AR, Akhtar A, Barlow DP, Bird AP, Brockdorff N, Duboule D, et al. Considerations when investigating lncRNA function in vivo. elife. 2014;3:e03058.
105. Brosius J, Tiedge H. RNomenclature. RNA Biol. 2004;1(2):81–3.
106. Chen H, Blanchette M. Detecting non-coding selective pressure in coding regions. BMC Evol Biol. 2007;7 Suppl 1(Suppl 1):S9.
107. Dinger ME, Pang KC, Mercer TR, Mattick JS. Differentiating protein-coding and noncoding RNA: challenges and ambiguities. PLoS Comput Biol. 2008;4(11):e1000176.
108. Kageyama Y, Kondo T, Hashimoto Y. Coding vs non-coding: translatability of short ORFs found in putative non-coding transcripts. Biochimie. 2011;93(11):1981–6.
109. Mignone F, Gissi C, Liuni S, Pesole G. Untranslated regions of mRNAs. Genome Biol. 2002;3(3):Reviews0004.
110. Ingolia NT, Brar GA, Stern-Ginossar N, Harris MS, Talhouarne GJ, Jackson SE, et al. Ribosome profiling reveals pervasive translation outside of annotated protein-coding genes. Cell Rep. 2014;8(5):1365–79.
111. Baughan T, Shababi M, Coady TH, Dickson AM, Tullis GE, Lorson CL. Stimulating full-length SMN2 expression by delivering bifunctional RNAs via a viral vector. Mol Ther. 2006;14(1):54–62.

112. Wadler CS, Vanderpool CK. A dual function for a bacterial small RNA: SgrS performs base pairing-dependent regulation and encodes a functional polypeptide. Proc Natl Acad Sci USA. 2007;104(51):20454–9.
113. Neme R, Tautz D. Fast turnover of genome transcription across evolutionary time exposes entire non-coding DNA to de novo gene emergence. elife. 2016;5:e09977.
114. Hayes CS, Keiler KC. Beyond ribosome rescue: tmRNA and co-translational processes. FEBS Lett. 2010;584(2):413–9.
115. Tenson T, DeBlasio A, Mankin A. A functional peptide encoded in the Escherichia coli 23S rRNA. Proc Natl Acad Sci USA. 1996;93(11):5641–6.
116. Brosius J, Gould SJ. On "genomenclature": a comprehensive (and respectful) taxonomy for pseudogenes and other "junk DNA". Proc Natl Acad Sci USA. 1992;89(22):10706–10.
117. Polev D. Transcriptional noise as a driver of gene evolution. J Theor Biol. 2012;293:27–33.
118. Wilson BA, Masel J. Putatively noncoding transcripts show extensive association with ribosomes. Genome Biol Evol. 2011;3:1245–52.
119. Schmitz J, Brosius J. Exonization of transposed elements: a challenge and opportunity for evolution. Biochimie. 2011;93(11):1928–34.
120. Babitzke P, Romeo T. CsrB sRNA family: sequestration of RNA-binding regulatory proteins. Curr Opin Microbiol. 2007;10(2):156–63.
121. Harris H. History: non-coding RNA foreseen 48 years ago. Nature. 2013;497(7448):188.

# Chapter 3
# RNA in the Central Dogma: mRNA

**Abstract** Messenger RNA (mRNA) plays a pivotal role in the central dogma of molecular biology, serving as an intermediary between DNA and protein synthesis. Within cells, DNA encodes genetic information that is transcribed into mRNA. This mRNA then carries this genetic code from the nucleus to the cytoplasm, where it directs the synthesis of proteins through a process known as translation. mRNA molecules are thus essential in translating the genetic instructions stored in DNA into functional proteins that perform various vital functions within organisms. Initially, mRNA faced scepticism as a viable therapeutic due to concerns about its stability and challenges in scaling up production. Recent advancements in technology and a deeper understanding of biomolecular processes have largely dispelled these doubts. Current research is heavily focused on overcoming these initial obstacles. Consequently, mRNA is now increasingly recognized for its potential to revolutionize therapeutic strategies such as immunotherapy, regenerative medicine, vaccination, and gene editing. However, realizing the full therapeutic promise of mRNA hinges on improving its production efficiency, stability, and precise delivery to target cells. To address these critical issues, substantial research efforts are underway to explore innovative solutions. These efforts aim to enhance mRNA stability and refine delivery mechanisms, thereby maximizing its therapeutic efficacy.

## 3.1 Introduction

After the detailed elucidation of DNA's structure, function, and biomedical applications, researchers turned their attention to RNA [1, 2]. Unlike DNA, RNA is generally single-stranded but can adopt various secondary structures like hairpins and loops, shaped by specific nucleotide sequences [2, 3] (Table 3.1). These secondary structures play a crucial role as they determine RNA's functional capabilities, influencing its participation in a wide array of cellular processes. The secondary structures of RNA, such as hairpins and loops formed by specific nucleotide sequences, are critical for its functional capabilities and diverse roles in cellular processes [2, 4, 5], detailed below. The secondary structures of RNA are not just

 
P. E. Saw, E. Song, *RNA Therapeutics in Human Diseases*,
https://doi.org/10.1007/978-981-96-3041-7_3

**Table 3.1** General differences between an RNA and DNA molecule

| Feature | RNA | DNA |
|---|---|---|
| Structure | Typically single-stranded | Double-stranded helix |
| Sugar in backbone | Ribose | Deoxyribose |
| Bases | Adenine (A), uracil (u), guanine (g), cytosine (c) | Adenine (A), thymine (t), guanine (g), cytosine (c) |
| Function | Messenger, transfer, ribosomal RNA; also involved in gene regulation and catalysis | Stores genetic information; carries instructions for protein synthesis |
| Stability | Less stable, prone to degradation (due to the single-stranded structure) | More stable, less prone to degradation (due to the double-stranded structure) |
| Location | Found in nucleus and cytoplasm | Found primarily in the nucleus |
| Role in central dogma | Transcribed from DNA; involved in translation into proteins (i.e., mRNA, tRNA, and rRNA) | Serves as the template for RNA synthesis |
| Secondary structures | Can form secondary structures like hairpins and loops | Generally does not form stable secondary structures |
| Genetic information | Transient carrier of genetic information | Permanent repository of genetic information |
| Enzymes | RNA polymerase, RNA-processing enzymes | DNA polymerase, DNA-processing enzymes |
| Examples | mRNA, tRNA, rRNA | Chromosomal DNA, mitochondrial DNA |

passive forms; they actively participate in cellular functions. For instance, RNA hairpin loops can serve as recognition sites for proteins or other RNA molecules, enabling them to interact and form complex regulatory networks within the cell. These interactions are fundamental for processes like gene expression regulation, RNA splicing, and RNA localization. RNA secondary structures often contain regulatory elements that control gene expression. They can act as binding sites for regulatory proteins or microRNAs, influencing whether and how a gene is transcribed or translated into protein. This regulatory role is crucial for maintaining cellular homeostasis and responding to environmental cues. Secondary structures can affect RNA stability by protecting it from degradation or targeting it for degradation, depending on the context. Additionally, RNA folding dynamics play a role in RNA-processing events such as splicing and polyadenylation, which are essential for generating mature RNA transcripts ready for translation. In viruses, RNA secondary structures are vital for their replication and interaction with host cell machinery. Understanding these structures is crucial for developing antiviral therapies. Moreover, in RNA-based therapeutics such as mRNA vaccines, the stability and structure of the mRNA molecule are carefully designed to ensure efficient translation and immune response. RNA molecules can undergo structural changes in response to cellular signals or environmental conditions. These structural switches can alter their functional roles, such as switching between translation and regulatory functions, or responding to stress conditions by adopting protective structures [6]. The general differences between a DNA and RNA molecule is depicted in Table 3.1.

## 3.2 mRNA Structure and Biogenesis

Various types of RNA play indispensable roles in gene expression, a fundamental process where genetic information encoded in DNA is utilized to produce functional molecules such as RNAs and proteins, essential for cellular biology [2]. According to the central dogma of molecular biology, DNA is transcribed into mRNA, which then serves as a template for protein synthesis [1]. mRNA acts as a transient intermediary, faithfully transcribing the genetic code from DNA and specifying the amino acid sequence needed for protein assembly [2, 7]. Additionally, each mRNA molecule includes non-coding or untranslated regions that play critical roles in regulating mRNA processing and translation mechanisms [2, 8]. These regions govern aspects like mRNA stability, localization within the cell, and interactions with regulatory factors, thereby fine-tuning the expression of genes and maintaining cellular functions.

In bacteria, mRNA is synthesized directly from DNA through transcription. The 5′-UTR of bacterial mRNA, often referred to as the Shine-Dalgarno sequence, plays a crucial role in facilitating ribosome binding. This sequence is located upstream of the start codon and serves as a recognition site for the ribosome, enabling efficient translation initiation. The interaction between the Shine-Dalgarno sequence and the ribosome is essential for ensuring proper alignment and initiation of protein synthesis, a fundamental process in bacterial gene expression [2]. Conversely, in eukaryotic organisms, the process of gene expression involves RNA Polymerase II (Pol II) for mRNA synthesis. RNA Polymerase II is responsible for transcribing the DNA sequence into precursor mRNA (pre-mRNA) molecules. This transcription occurs in the nucleus, where RNA Polymerase II binds to specific regions of DNA called promoters and initiates the synthesis of mRNA. Unlike bacteria, eukaryotic mRNA undergoes several processing steps before it becomes mature and functional. This includes capping at the 5′ end, splicing to remove introns, and polyadenylation at the 3′ end. These modifications are crucial for stabilizing the mRNA molecule, enhancing its translation efficiency, and regulating gene expression [9, 10].

Initially, a pre-mRNA, also known as the primary transcript, is synthesized during transcription from DNA [4]. This pre-mRNA undergoes several processing steps within the nucleus to form mature mRNA. These processing steps include capping at the 5′ end, splicing to remove introns, and polyadenylation at the 3′ end, resulting in a mature and functional mRNA molecule ready for translation [4]. Within the mature mRNA, the genetic code for protein synthesis is carried in the form of triplets of nucleotides called codons [2]. Each codon specifies a particular amino acid, ensuring the accurate translation of mRNA into protein by the ribosome. Specific regions within the mRNA molecule play critical roles in translation and mRNA stability. These regions include the UTRs, which flank the coding sequence. The 5′ and 3′ UTRs are involved in regulating mRNA stability, localization within the cell, and interactions with regulatory factors [8]. Proper regulation of these regions is essential for maintaining mRNA integrity and ensuring efficient protein production, thereby contributing to cellular function and homeostasis.

Both prokaryotic and eukaryotic mRNAs share three primary regions, as illustrated in Fig. 3.1 [2, 11, 12]. The 5′-UTR, located at the mRNA's 5′ end, consists of nucleotide sequences that do not encode amino acids [13]. This region serves important regulatory functions, influencing mRNA stability, localization within the cell, and interactions with translation initiation factors and regulatory proteins. During pre-mRNA processing, the addition of a methylated guanine nucleotide at the 5′ end, known as the 5′ cap (Cap-5′), plays multiple critical roles in mRNA functionality [11, 13, 14]. The 5′ cap protects the mRNA from degradation by exonucleases, thereby increasing its stability within the cell. The cap structure is essential for efficient transport of mRNA from the nucleus to the cytoplasm, where protein synthesis occurs. The cap-5′ structure serves as a binding site for the ribosome during translation initiation, ensuring accurate and efficient protein synthesis. Cap-5′ facilitates the export of mature mRNA from the nucleus to the cytoplasm, where it undergoes translation. The presence of the 5′ cap is associated with proper splicing and removal of introns from pre-mRNA molecules, ensuring the production of mature, functional mRNA. The 5′ cap modification reduces the likelihood of mRNA triggering an immune response when used in therapeutic applications, making it more suitable for therapies involving protein replacement. Enhancing the Cap-5′ end to prevent its removal can significantly improve mRNA stabilization and protect it from enzymatic degradation [8, 11, 14].The second region of mRNA is the protein coding sequence, which spans from the start codon, typically AUG encoding methionine, to the stop codon that terminates translation. This sequence consists of a series of codons, each composed of three nucleotides, which dictate the order of amino acids in the resulting protein. The protein coding sequence is fundamental in translating the genetic information encoded in mRNA into a functional polypeptide chain. It begins with initiation at the start codon and continues until a stop codon signals the termination of protein synthesis, ensuring accurate and complete translation of mRNA into protein within the ribosome [2]. Optimizing codons by substituting rare codons with more frequently used synonymous codons can have significant impacts on gene expression levels [8, 15] and reduce the immunogenicity of mRNA [16, 17]. Both coding and non-coding RNAs in eukaryotic cells can undergo specific nucleotide modifications, referred to as epi-transcriptomic changes [18]. These modifications play crucial roles in regulating RNA translation, initiation, stability, localization, and overall function. In 2005, Kariko and Weissmann demonstrated that chemical modifications of mRNA nucleosides can effectively decrease its immunogenicity [19]. This reduction in immune response enhances mRNA translation efficiency and contributes to the effectiveness of mRNA-based vaccines [8]. By modifying nucleosides within the mRNA molecule, researchers can optimize its performance, making it more suitable for therapeutic applications such as protein replacement therapies and vaccine development. Similar to the 5′-UTR, the third

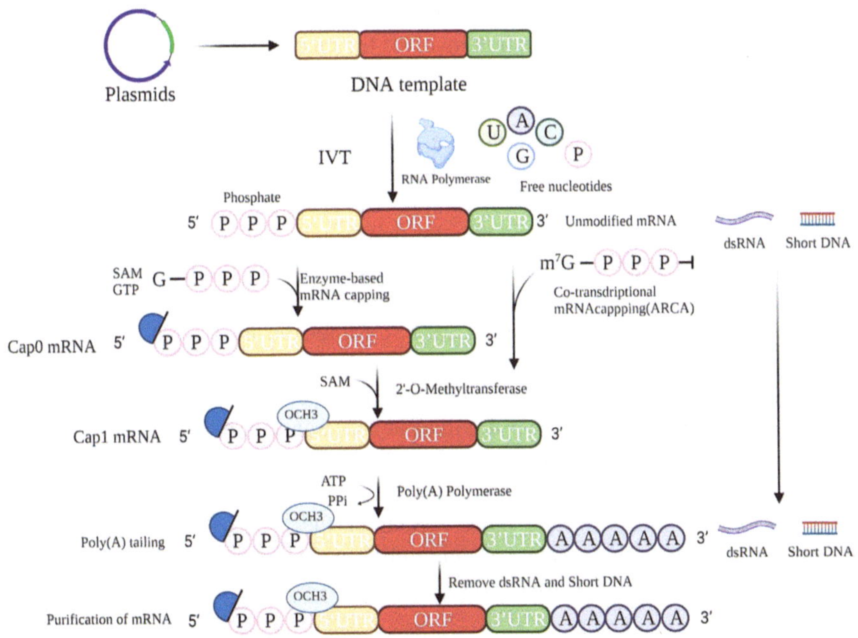


**Fig. 3.1** Overview of the in vitro transcription (IVT) process for mRNA synthesis. This figure provides a detailed schematic of the in vitro transcription (IVT) process used for synthesizing mRNA from a DNA template, which is crucial for various applications, including vaccine development and gene therapy. The process is depicted in several steps: (i) Preparation of Reaction Components—The IVT reaction mixture is prepared by combining a DNA template containing the desired promoter sequence, nucleotide triphosphates (NTPs) (which include ATP, UTP, GTP, and CTP), RNA polymerase enzyme (e.g., T7, T3, SP6), and buffer components optimized for transcriptional activity. (ii) Promoter Recognition—The RNA polymerase enzyme binds specifically to the promoter sequence present on the DNA template. Each bacteriophage RNA polymerase recognizes its respective promoter sequence (e.g., T7 polymerase binds to the T7 promoter, SP6 polymerase to the SP6 promoter, and T3 polymerase to the T3 promoter). (iii) Initiation of Transcription—Once bound to the promoter, the RNA polymerase initiates transcription by synthesizing mRNA in the 5′ to 3′ direction. The mRNA molecule grows in length as nucleotides are sequentially added to the growing RNA strand complementary to the DNA template. (iv) Termination and Processing—Transcription continues until the RNA polymerase reaches a termination sequence or structure, causing it to dissociate from the DNA template. The resulting mRNA transcript may undergo post-transcriptional modifications such as capping at the 5′ end with a modified guanine nucleotide (Cap structure) and addition of a polyadenylated [poly (A/T)] sequence at the 3′ end. (v) Purification and Quality Control—Following transcription, the mRNA product is purified to remove residual DNA template, enzymes, and reaction by-products. Quality control steps ensure the integrity, purity, and functionality of the mRNA product for subsequent applications in research or therapeutic development. (Figure is reproduced from [35] under the Creative Commons Attribution licence (CC-BY))

region of mRNA is the 3′-untranslated region (3′-UTR), a non-coding nucleotide sequence located at its 3′ end [2, 17]. The 3′-UTR plays critical roles in regulating mRNA stability, localization within the cell, and interactions with regulatory factors. One of the key processes involving the 3′-UTR is polyadenylation, where a polyadenylated “tail” (Poly (A)) is added to the mRNA molecule [2, 11, 12], as depicted in Fig. 3.1. Polyadenylation involves the enzymatic addition of a string of adenine nucleotides to the mRNA’s 3′ end. This modification is essential for mRNA stability, protection against degradation, and efficient translation. Additionally, the length of the poly(A) tail can influence mRNA localization, turnover rates, and translational efficiency, thereby regulating gene expression at the post-transcriptional level. The Poly(A) tail serves multiple critical functions in mRNA biology. It prevents degradation by nucleases, thereby significantly enhancing mRNA stability and increasing translation efficiency [20–22]. This stability is crucial for maintaining mRNA integrity and prolonging its half-life within cells, which can lead to higher levels of protein production [2, 13, 15]. Moreover, the Poly(A) tail is extensively utilized in mRNA purification processes due to its distinctive sequence and characteristics. Both the 5′ and 3′ UTRs of mRNA contribute to its stability and regulatory functions [2, 11]. The 5’ UTR influences translation initiation and mRNA stability, while the 3′ UTR plays a role in regulating mRNA degradation rates and localization within the cell. Together, these UTRs orchestrate mRNA dynamics by recruiting RBPs, miRNAs, and other regulatory factors [8, 13]. They influence translation efficiency and dynamics, fine-tuning gene expression to meet cellular demands and environmental cues.

## 3.3 From DNA to mRNA: Therapeutic Implications

During the 1990s, significant advancements in intracellular delivery technology and methods for producing DNA or mRNA reached a level sophisticated enough to support preclinical studies in gene therapy, DNA- or mRNA-based vaccines, and cancer immunotherapy [1, 8, 23]. Both DNA and mRNA serve as templates for protein synthesis, enabling cells, when used therapeutically, to produce deficient or missing proteins in various diseases or deficiencies [15]. These technological breakthroughs paved the way for exploring the therapeutic potential of genetic materials in manipulating cellular processes for medical purposes. DNA-based approaches initially dominated the field of gene therapy, aiming to correct genetic defects by delivering functional genes into target cells. However, the challenges of delivering DNA to the nucleus and ensuring stable integration into the host genome posed significant hurdles, particularly in non-dividing cells. In contrast, mRNA emerged as a promising alternative due to its ability to directly encode therapeutic proteins without integrating into the genome. Advances in mRNA synthesis, modification, and delivery methods have expanded its applications in vaccine development, protein replacement therapies, and cancer treatments. mRNA-based vaccines, for instance, have demonstrated remarkable efficacy and rapid development capabilities, as seen in recent global vaccination efforts against infectious diseases like COVID-19.

**Box 3.1 Advantages of Nucleic Acid-Based Therapies**

Nucleic acid-based therapies provide several advantages over direct protein therapies, including speed, cost-effectiveness, and fewer size constraints. These advantages are particularly pronounced for proteins that are difficult to produce effectively in vitro, such as membrane proteins [24, 25]. Nucleic acid-based therapies can be developed and manufactured more quickly than protein therapies. Once the nucleic acid sequence is known, production can begin rapidly through chemical synthesis or transcription, followed by purification and formulation. Producing nucleic acids, whether DNA or mRNA, is often less expensive than manufacturing proteins. This is partly due to simpler production processes and the potential for scaling up production more efficiently. Nucleic acids can encode large proteins, including those with complex structures such as membrane proteins, without encountering the size limitations that can hinder protein synthesis or purification. Nucleic acid therapies offer flexibility in modifying the sequence or structure of the encoded protein, enabling optimization for therapeutic purposes. This flexibility can be particularly advantageous in addressing specific disease mechanisms or patient needs. They can be engineered to target specific cells or tissues, enhancing therapeutic specificity and reducing off-target effects compared to systemic protein administration. An additional advantage of nucleic acid therapies relates to production issues, particularly the complexities of post-translational modifications crucial for protein functionality, which are difficult to ensure in heterologous protein production systems. Since then, nucleic acid-based therapies represent a versatile and promising approach in biotherapeutics, overcoming many of the limitations associated with traditional protein-based therapies. Their ability to encode complex proteins efficiently, combined with advancements in delivery technologies, positions them as valuable tools in treating a wide range of diseases and medical conditions [8, 15].

## 3.4 Post-Translational Modification (PTM) Vs. Immunogenicity

Challenges in achieving proper post-translational modifications (PTMs) can lead to notable differences between proteins produced endogenously within the body and those produced exogenously for therapeutic purposes. These differences can contribute to increased immunogenicity or the likelihood of triggering an immune response against the therapeutic protein. Many proteins require specific PTMs such as glycosylation, phosphorylation, or disulfide bond formation to achieve their correct structure and function [26]. In vivo, these modifications occur naturally within the cellular environment. However, replicating these PTMs in exogenously produced proteins can be challenging, leading to variations in protein folding, stability, and biological activity. Differences in PTMs between endogenous and exogenous proteins can result in the recognition of therapeutic proteins as foreign by the immune system. This recognition may lead to immune responses such as antibody production or T cell activation, potentially compromising the efficacy and safety of the therapeutic intervention. Ensuring consistent and accurate PTMs in therapeutic proteins requires stringent quality control measures during manufacturing. Variations in PTMs can arise due to differences in cell culture conditions, purification methods, or post-production modifications, all of which can affect the immunogenic potential of the therapeutic protein. Immunogenicity can impact the safety and efficacy of protein-based therapies. It may lead to adverse reactions in patients, including allergic reactions, reduced therapeutic efficacy due to neutralizing antibodies, or even the development of autoimmune responses in some cases [1, 12]. Moreover, delivering protein drugs faces limitations due to their size and instability, which can hinder achieving therapeutic concentrations in vivo, alongside high-production costs [1].

In the manipulation of protein expression, delivering exogenous mRNA into cells has emerged as a promising strategy. Unlike DNA, mRNA does not integrate into the cell's genome but rather undergoes transient activity before natural degradation, a feature that offers distinct advantages for therapeutic applications [1]. Historically, DNA-based therapeutic strategies have been primary in treating inherited diseases caused by deficient or malfunctioning proteins. However, using DNA therapeutically presents challenges. DNA must swiftly reach the cell cytoplasm and effectively penetrate the nucleus for transcription into mRNA, the precursor essential for protein synthesis [13, 15, 27]. This process of nuclear delivery poses a substantial hurdle in post-mitotic cells prevalent in mature tissues, which have limited division capabilities [15, 28]. In contrast, mRNA, upon entering the cytoplasm, can initiate translation immediately without the need to enter the nucleus [29], simplifying its therapeutic application [12, 16]. The first successful delivery of mRNA into cells via liposomes occurred in 1970 [1, 15]. Initially deemed impractical for gene therapy due to instability and immunogenicity concerns [12, 13, 17], mRNA's potential in therapy was not widely recognized because it did not offer a route to permanent gene correction [1]. This perception has evolved recently with deeper insights into mRNA biology, the development of stabilizing modifications, and advancements in nucleic acid delivery technologies [13, 15, 17].

Recent advancements in mRNA technology have significantly enhanced its utility for therapeutic applications. These advancements have led to improvements in mRNA translation efficiency, extended half-life, and enhanced transfection efficiency, thereby reducing its immunogenicity [15, 23, 27]. As a transient carrier of genetic information capable of inducing protein expression, mRNA has emerged as a promising alternative to DNA or recombinant proteins in medicine [8, 15, 23]. Its transient nature offers several advantages. mRNA can be easily modified and synthesized for various therapeutic purposes, allowing rapid adaptation to different targets and conditions [1, 12]. mRNA technology shows promise across a wide range of applications, including vaccination, cell reprogramming, protein replacement therapies, and treatment of genetic disorders [8, 27, 30]. Its ability to transiently express proteins in cells without integrating into the genome offers a safer approach compared to traditional gene therapy strategies [1, 13]. The transient nature of mRNA reduces the risk of long-term genomic alterations or adverse immune responses associated with DNA-based therapies. This combination of enhanced efficiency, safety, and versatility has sparked significant interest in mRNA technologies across various health sectors. Ongoing research continues to explore and optimize mRNA-based therapies, aiming to harness their full potential in revolutionizing treatment modalities for a range of diseases and medical conditions.

Over time, mRNA has evolved from a fragile molecule to a robust therapeutic agent that is currently undergoing clinical studies [15]. Unlike DNA vectors, mRNA exhibits enhanced transfection efficiency even in non-dividing cells [29], and it eliminates the risk of insertional mutagenesis in the genome [13, 28, 31], making it particularly appealing for protein supplementation therapies [17, 27]. Furthermore, mRNA offers several advantages that contribute to its attractiveness in therapeutic applications, including (1) mRNA enables rapid synthesis of proteins within cells, facilitating timely therapeutic effects. (2) The dosage and duration of protein expression can be finely controlled by adjusting mRNA delivery and degradation kinetics [12, 28, 32]. (3) mRNA lacks additional exogenous sequences such as antibiotic resistance genes or viral-derived promoters, which enhances its safety and reduces potential side effects [12, 28, 32]. (4) The structure of mRNA can be optimized to control its pharmacokinetics, allowing for improved temporal regulation and degradation in the body [1, 12, 15]. These attributes make mRNA a versatile tool for developing therapies across various medical fields, including protein replacement therapies, vaccination, and potentially treating genetic disorders. The production of mRNA can be achieved through chemical synthesis or in vitro transcription, each method having its associated limitations. An emerging alternative is in vivo production using recombinant technologies, although scaling this up for large-scale microbial production remains challenging and requires addressing safety concerns to meet regulatory standards [12].

Despite its promising therapeutic potential, mRNA faces challenges such as instability under physiological conditions and strong immunogenicity. These hurdles can limit its effectiveness in clinical applications. To overcome these challenges, several strategies are being explored, including utilizing DDS can help protect mRNA from degradation by nucleases and evade immune detection. DDS can include lipid nanoparticles, polymers, or other carriers designed to efficiently deliver mRNA into cells while maintaining its stability [17, 28]. Exploring methods to reduce mRNA's immunogenicity are under investigation. This includes optimizing mRNA sequences and modifying nucleotides to minimize immune responses without compromising its translational efficiency. Chemical modifications of mRNA molecules can enhance their stability and improve their performance in vivo. These modifications may include modifications to the backbone, base modifications, or incorporation of modified nucleosides to increase resistance to degradation. However, despite these efforts, the transient nature of mRNA can pose challenges in certain therapeutic applications. For diseases requiring prolonged treatment, such as genetic or metabolic disorders, continuous administration of mRNA may be necessary to maintain therapeutic protein levels [8, 15, 27, 33]. This can potentially affect patient compliance and comfort, as frequent dosing may be required. Nevertheless, whether the therapy is transient or prolonged, the primary objective of mRNA therapy remains consistent: efficient delivery to the cytoplasm for effective protein expression with minimal toxicity and immunogenicity [15]. Recent advancements have highlighted mRNA's potential in both prophylactic applications, such as antiviral vaccines, and in treating various diseases including myocardial infarction, HIV infection, and multiple types of cancer [1, 30]. These developments underscore the expanding role of mRNA therapeutics in modern medicine.

## 3.5 Synthesis and Production of mRNA

With the growing recognition of mRNA's therapeutic potential, considerable efforts are underway to overcome challenges related to its production. The primary methods for synthesizing mRNA include chemical synthesis and enzymatic production through in vitro transcription (IVT) from a linearized plasmid DNA (pDNA) template [12, 13, 34]. Alternatively, a PCR-based approach can also be used for mRNA synthesis [12].

### *3.5.1 Enzymatic Strategies for mRNA Synthesis*

**Template Requirements.** A linearized plasmid DNA (pDNA) template serves as the starting material for IVT. This template includes elements such as a bacteriophage promoter (e.g., T7 or SP6 promoter), 3′ and 5′ UTRs, and an open reading frame (ORF) encoding the protein of interest [12, 31, 33].

**Additional Structural Elements.** Optionally, the mRNA sequence may include a 5′ Cap structure, typically a modified guanine nucleotide, which enhances stability and translational efficiency [11, 12, 33]. At the 3′ end, a polyadenylated [poly (A/T)] sequence may be added to stabilize the mRNA molecule and facilitate efficient translation [11, 12, 33].

**RNA polymerases.** RNA polymerases such as T7 or SP6 recognize the bacteriophage promoter and transcribe the mRNA sequence from the pDNA template. Post-transcriptional modifications, such as capping and polyadenylation, may be enzymatically added to enhance mRNA stability and translational efficiency. Incorporating structural elements like the Cap structure at the 5′ end and a polyadenylation sequence (poly (A/T)) at the 3′ end of mRNA plays crucial roles in enhancing its functionality and utility in therapeutic applications. The Cap structure, typically a modified guanine nucleotide added to the 5′ end of mRNA during synthesis, protects the mRNA from degradation by exonucleases. This enhances mRNA stability within the cellular environment, prolonging its half-life and ensuring sustained protein expression. The cap structure also plays a pivotal role in translation initiation. It serves as a recognition site for the ribosome, facilitating the efficient initiation of protein synthesis. This leads to higher translation efficiency, ensuring that the encoded protein is produced in sufficient quantities. The polyadenylation sequence (poly (A/T)) added to the 3′ end of mRNA serves multiple purposes. Firstly, it stabilizes the mRNA molecule by protecting it from enzymatic degradation in the cytoplasm. Secondly, the poly (A/T) sequence facilitates purification of mRNA transcripts from cellular extracts or reaction mixtures. This is achieved through methods such as oligo-dT chromatography, where the poly(A) tail selectively binds to oligo-dT beads, allowing for efficient isolation and purification of mRNA [11–13]. These advancements are crucial for optimizing mRNA characteristics essential for therapeutic applications.

## 3.6 In Vitro Transcription (IVT) of mRNA Production

The process of mRNA production through in vitro transcription (IVT) starts with the assembly of necessary components, including nucleotides and RNA polymerase enzymes derived from bacteriophages such as T7, T3, or SP6. These enzymes are crucial as they specifically recognize and bind to promoter sequences that are integrated into the DNA template [12, 28, 34]. There are 5 basic steps in the preparation of mRNA through IVT production, summarized in Fig. 3.1.

In eukaryotic cells, natural mRNA molecules are characterized by a polyadenylated tail (poly(A)) typically consisting of 50–250 adenosine residues in the 3′ UTR. This poly(A) tail plays a crucial role in mRNA stability, translation efficiency, and cellular localization. In synthetic mRNA synthesis, a poly(A) tail is also incorporated to mimic the structure of endogenous mRNA. Studies have demonstrated that the length of the poly(A) tail can significantly impact protein expression levels. Specifically, increasing the length of the poly(A) tail up to approximately 120 adenosine residues has been shown to enhance protein expression levels. This optimal length promotes mRNA stability and facilitates efficient translation initiation, thereby increasing the yield of translated proteins [11, 12]. Beyond a certain length, usually around 120 adenosine residues, further elongation of the poly(A) tail does not provide additional benefits in terms of protein expression. Longer tails may not be efficiently recognized or utilized by cellular translational machinery, potentially leading to decreased translational efficiency or other undesirable effects [11, 12]. The exact optimal length of the poly(A) tail may vary depending on the specific mRNA sequence, cell type, and experimental conditions. Therefore, experimental optimization is often necessary to determine the ideal poly(A) tail length for maximizing protein expression in a given context.

Chromatography is widely utilized for purifying IVT-derived mRNA because of its effectiveness in removing various impurities. These impurities can include DNA, which is susceptible to degradation by DNases as well as abortive transcripts and double-stranded RNA (dsRNA) [12, 23, 34, 36]. Purification of IVT-derived mRNA is a critical step in ensuring the production of pure and stable mRNA suitable for therapeutic applications. First is the removal of contaminants. Chromatography techniques, such as size-exclusion chromatography or ion-exchange chromatography, are capable of separating mRNA molecules from contaminants like residual DNA, short RNA fragments, and other impurities that could potentially interfere with therapeutic efficacy or safety [12, 23, 34, 36]. Second is to enhance the purity of the synthesized mRNA. By selectively isolating mRNA based on size, charge, or affinity interactions, chromatography achieves high-purity levels that are essential for regulatory compliance and clinical safety standards. This ensures that the mRNA product meets stringent quality criteria for therapeutic use [23, 36]. Third is to maintain the mRNA stability. Purification helps maintain the stability of mRNA by eliminating sources of degradation, such as DNases that can degrade DNA impurities, thereby preserving the integrity and functional competence of the mRNA molecule during storage and administration [12, 23, 37]. Cellular mechanism of how syn-mRNA produces peptides is detailed in Fig. 3.2.

## 3.7 Producing Synthetic mRNAs

Traditional methods for altering cellular expression profiles often involve the use of DNA or RNA-based viral and nonviral vectors [39]. While these approaches have been effective in gene therapy and genetic engineering, they also come with inherent risks, particularly related to genomic integration. DNA-based vectors, whether viral (e.g., adenovirus, lentivirus) or nonviral (e.g., plasmid DNA), have the potential to integrate into the host cell's genome [40]. This integration can lead to permanent genetic alterations, which may pose risks such as oncogenesis (formation of tumors) or disruption of normal cellular functions. The risks associated with genomic integration have raised safety concerns in the context of clinical therapies. Unintended consequences, such as activation of oncogenes or disruption of tumor suppressor genes, could occur if integration events disrupt regulatory elements within the genome. The use of DNA-based vectors in clinical therapies also raises ethical considerations, particularly regarding the long-term effects of genetic modifications in patients. Ensuring patient safety and informed consent becomes crucial when considering therapies that involve permanent alterations to the genome. In contrast, mRNA-based therapies offer a promising alternative. mRNA does not integrate into the genome and remains transient within the cell, minimizing the risks associated with genomic alterations. This transient nature also allows for precise control over protein expression levels without long-term effects on the cell's genetic makeup. Despite these benefits, concerns about RNA degradation have historically

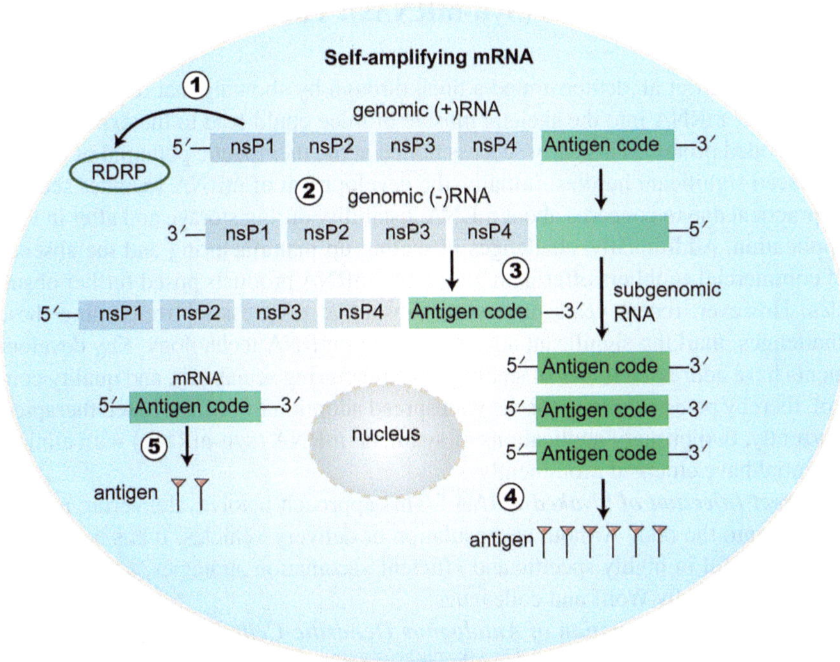


**Fig. 3.2** Mechanism of self-amplifying mRNA for antigen production. This figure illustrates the mechanism by which self-amplifying mRNA (saRNA) functions within a cell to produce antigens, enhancing the efficiency of mRNA-based vaccines. Initial transcription (Step 1): The process begins with the genomic (+)RNA (sense strand) containing the RNA-dependent RNA polymerase (RDRP) and non-structural proteins (nsP1, nsP2, nsP3, nsP4) alongside the antigen-coding sequence. RDRP initiates the transcription of the genomic (+)RNA into genomic (−)RNA (antisense strand). Negative strand RNA (Step 2): The genomic (−)RNA serves as a template for the synthesis of multiple copies of genomic (+)RNA and subgenomic RNA. Subgenomic RNA Production (Step 3): The subgenomic RNA is synthesized, which contains the antigen-coding sequence. This allows for the generation of multiple subgenomic RNA molecules from a single genomic RNA template, amplifying the potential for antigen production. Antigen Expression (Step 4): The subgenomic RNA is translated into the antigen protein by the host cell's ribosomes. Multiple copies of the antigen-coding RNA result in the production of large amounts of antigen protein. Antigen Presentation (Step 5): The produced antigens are then presented on the cell surface or secreted, stimulating an immune response. This self-amplifying mechanism significantly enhances the amount of antigen produced from a small initial amount of mRNA, increasing the efficacy of the vaccine. Overall, this self-amplifying mRNA technology leverages the host cell's machinery to produce large quantities of antigen from a single RNA template, providing a potent and efficient method for generating immune responses in mRNA-based vaccines. (This figure is reproduced from [38] under Creative Commons Attribution (CC-BY) license (https://creativecommons.org/licenses/by/4.0/))

limited the widespread adoption of mRNA-based therapies. Recent advancements in syn-mRNA technology have alleviated many of these concerns, boosting confidence in their application. This overview highlights key features of synthetic mRNAs and explores their potential across various fields of application.

## 3.8 Synthetic mRNAs (Syn-mRNAs): The Advantages

In 1990, Wolff et al. demonstrated a breakthrough by showing that direct injection of "naked" mRNA into the skeletal muscle of mice could lead to the expression of the encoded protein. This early success hinted at the therapeutic potential of mRNA but faced significant hurdles. Initially, the development of mRNA vaccines seemed impractical due to concerns about mRNA instability during storage and after in vivo application. Additionally, challenges in scaling up manufacturing and the absence of commercial suppliers offering high-quality mRNA products posed further obstacles. However, recent years have seen substantial progress in overcoming these challenges, marking significant advancements in mRNA technology. Key developments have addressed issues of stability, manufacturing scalability, and quality control, thereby paving the way for the widespread adoption of mRNA-based therapies. Currently, two primary applications of synthetic mRNA (syn-mRNA) with clinical potential have emerged prominently:

***Direct Injection of "Naked mRNA."*** This approach involves delivering mRNA directly into the body without encapsulation or delivery vehicles. It has been notably successful in highly specific and efficient vaccination strategies, a method initially pioneered by Wolff and colleagues.

**Ex Vivo *Transformation of Autologous Dendritic Cells (DCs)*.** Another application involves isolating dendritic cells from a patient's own immune system, transforming them ex vivo with mRNA encoding antigens, and then reintroducing these modified cells back into the patient. This method offers advantages in terms of targeting specific immune responses and enhancing vaccine effectiveness (as depicted in Fig. 3.3a).

Warren et al. demonstrated successful reprogramming human fibroblasts into induced pluripotent stem cells (iPSCs) using mRNA-mediated techniques. This breakthrough approach, depicted in Figure 3.3b, holds tremendous promise in regenerative medicine by enabling the generation of iPSCs that can be further differentiated into various mature cell types. In the context of mRNA-based therapies, ensuring the purity and integrity of synthetic mRNA molecules is crucial for their efficacy and safety. High-performance liquid chromatography (HPLC) plays a critical role in this regard, particularly in stringent purification procedures. Unlike standard silica-based spin column purification methods, HPLC offers enhanced resolution and separation capabilities, making it effective in removing minor impurities such as short double-stranded RNA fragments. The removal of these fragments is essential for minimizing potential immune responses [36]. Even trace amounts of contaminants could trigger unwanted immune reactions when synthetic mRNA is introduced into the body, underscoring the importance of rigorous purification methods like HPLC in mRNA production for clinical applications.

Future projects leveraging mRNA technologies hold immense promise for enhancing clinical applications of iPSCs. Ongoing research and clinical studies are exploring innovative uses of iPSCs, such as treatments for age-related macular degeneration (AMD), a leading cause of vision loss in older adults. These

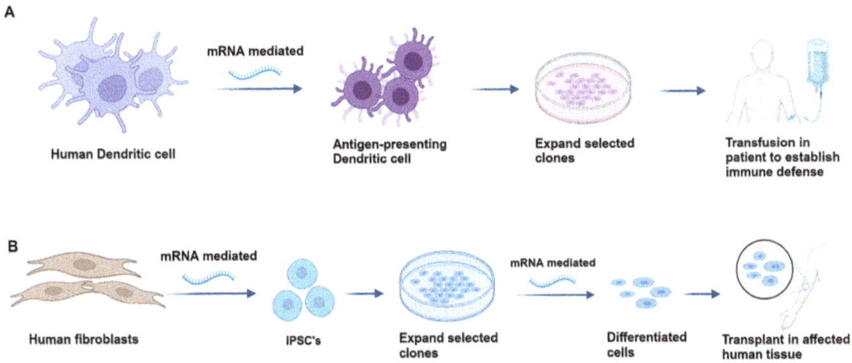


**Fig. 3.3** Development and application of mRNA therapeutics. This figure illustrates two pathways for the development and application of mRNA therapeutics. (**a**) mRNA-based Vaccines. The process begins with the identification of target antigens from pathogens such as viruses or bacteria. mRNA encoding the antigen is synthesized. The synthesized mRNA is encapsulated within lipid nanoparticles to protect it and enhance delivery. The mRNA vaccine is administered to humans, where it instructs cells to produce the antigen, thereby stimulating an immune response. (**b**) mRNA-based protein replacement therapy. This approach starts with identifying the protein that needs replacement due to genetic disorders or diseases. mRNA encoding the therapeutic protein is synthesized. The synthesized mRNA is introduced into cells in culture, leading to the production of the therapeutic protein. The therapeutic protein is purified and then administered to patients, aiming to restore normal function. (Figure is drawn in Biorender.com)

treatments often utilize protein-based approaches derived from iPSCs, aiming to replace or repair damaged retinal cells [41]. In comparison to conventional pharmaceutical drugs, synthetic mRNAs offer several distinct advantages. Firstly, they function similarly to endogenous mRNA molecules, encoding specific proteins within cells without integrating into the genome. This transient nature ensures that mRNA degrades naturally over time, minimizing environmental impact compared to synthetic chemical drugs that may persist in the environment. Furthermore, mRNA-based therapies do not introduce non-biological structures into cells, reducing the risk of adverse reactions and potential long-term effects associated with foreign materials. The inherent characteristics of synthetic mRNAs, such as their ability to transiently induce protein expression and their natural degradation pathway, make them well-suited for a variety of therapeutic applications. These include not only iPSC-based regenerative medicine but also vaccination, gene editing, and personalized medicine approaches. As research progresses and technological advancements continue, the clinical potential of mRNA technologies in transforming medical treatments is expected to expand significantly, offering novel solutions for a wide range of diseases and conditions. Table 3.2 highlights the advantages of synthetic mRNA compared to endogenous mRNA and small molecular compound as a lead molecule in pharmaceutical drugs.

The generation of syn-mRNA presents significant challenges for cell biology laboratories, particularly in the context of research and development phases where multiple constructs are needed in small quantities. These challenges become even more pronounced when larger quantities of high and consistent quality syn-mRNA

**Table 3.2** Detailed comparison between synthetic mRNA and endogenous mRNA, compared to small molecular drugs [12, 34, 42, 43]

| Comparison criteria | Synthetic mRNA | Endogenous mRNA | Small molecule drugs |
|---|---|---|---|
| Delivery and localization | Can be engineered to include sequences that enhance cellular uptake and targeted delivery to specific tissues or cells | Naturally occurs in the cell but lacks specific targeting mechanisms for therapeutic purposes | Can be designed to target specific organs or tissues, but may lack the precision of targeted mRNA therapies |
| Duration of effect | Designed to have a controlled duration of action, which can be tuned by modifying the mRNA sequence or using delivery systems to protect the mRNA | Typically has a short-lived effect due to rapid degradation processes in the cell | Often requires repeated dosing due to metabolism and excretion, which can lead to fluctuating levels of the drug in the body |
| Precision of action | Can be precisely designed to produce a specific protein, providing high specificity in terms of therapeutic action | The production and function are predetermined by the cell's genetic makeup and are not specifically designed for therapeutic intervention | Can be designed to interact with specific molecular targets, but off-target effects and side effects are common |
| Safety and immunogenicity | Reduced risk of integration into the host genome and can be modified to decrease immune responses | Endogenous mRNAs are naturally occurring and typically do not elicit an immune response, but they cannot be easily modified for therapeutic purposes | May cause adverse effects due to interactions with other molecules or cellular components not related to their intended target |
| Scale and speed of development | Rapid synthesis and scalable production allow for quick response to emerging health challenges, such as pandemics | Production is limited to extraction and replication from natural sources, which can be slower and less scalable | Development and scaling can be time-consuming and costly, often taking years to refine and approve |
| Adaptability and versatility | Highly adaptable to different targets by simply changing the sequence that encodes the protein of interest, making it suitable for multiple diseases and conditions | Limited to the existing sequences within the genome, offering less flexibility for adaptation to different therapeutic needs | While versatile in targeting various biochemical pathways, they are generally less adaptable than mRNA for different disease |
| Mechanism of action | Acts through a natural cellular mechanism by directing the synthesis of therapeutic proteins directly within the body's own cells | Functions as part of the cellular machinery but is not typically used for direct therapeutic interventions | Interacts chemically with specific cellular targets to modulate or inhibit their activity, which can be less natural than mRNA action |

are required, especially when adhering to regulatory standards such as current good manufacturing practice (cGMP) principles. Currently, the availability of commercial suppliers capable of meeting these stringent requirements is limited. Established companies like Asuragen in the USA and CureVac in Europe provide GMP-grade mRNA products. These companies ensure compliance with regulatory standards and offer reliability in large-scale manufacturing processes. However, smaller and more flexible companies, such as AmpTec GmbH, play a crucial role in the landscape of mRNA production. They specialize in offering diverse manufacturing options that cater to different needs. For instance, AmpTec GmbH can provide services for producing multiple mRNA sequences in small quantities during the early phases of projects. This flexibility is particularly beneficial for researchers and developers who require varied constructs for initial experimentation and feasibility studies.

## 3.9 Design of Syn-mRNAs

syn-mRNAs offer several advantages in their production and engineering capabilities. Once the technology is established, syn-mRNAs can be scaled up to meet specific needs without significant limitations in size or sequence. RNA sequences can be tailored for stability and translation efficiency, utilizing cis-acting structural elements such as the 5′ cap (available in structural and modified forms), the poly(A) tail (with defined lengths and absence of 3′-terminal 'masking' sequences), untranslated regions, and the coding region sequence, which can be optimized for specific cellular contexts. Additionally, nucleotide modifications can be incorporated if necessary to further enhance desired properties [44].

## 3.10 Syn-mRNAs and Cellular Immune Response: An Advantage or a Problem?

Synthetic mRNAs are recognized as foreign genetic material by the immune system, triggering a responsive activation of the innate immune system [45]. Unlike endogenous mRNA, which originates from the host cell's genome, synthetic mRNAs typically have non-self-origins, resembling RNA from viral invaders or other foreign pathogens. This distinction allows cells to distinguish between self and non-self RNAs, a critical defense mechanism against infections. Cells of the innate immune system are equipped with various receptors specialized in detecting RNA molecules. Toll-like receptors (TLRs), including TLR3, TLR7, and TLR8, as well as RIG-I-like receptors (RLRs), are prominent examples. These receptors are located in specific cellular compartments and have distinct properties for recognizing RNA ligands. Upon binding to synthetic mRNA, these receptors initiate complex signaling pathways that lead to the induction of an antiviral state [46, 47]. The antiviral state triggered by RNA sensing receptors involves several mechanisms, including the activation of innate immune receptors can inhibit the translation of mRNA, thereby reducing protein production, increase the production of interferons (IFNs), which are signaling proteins crucial for coordinating the immune response against viral infections; and in severe cases, activation of innate immune pathways can lead to apoptosis, aiming to eliminate infected cells and prevent viral spread [48]. When introduced into the body as therapeutics, synthetic mRNA molecules can elicit similar immune responses through these sensors [49]. Depending on the therapeutic or research context, this immune reaction can have both beneficial and detrimental effects. Activation of innate immune responses can enhance immune surveillance and stimulate adaptive immune responses, potentially aiding in the clearance of pathogens or promoting therapeutic efficacy in certain contexts. However, excessive or inappropriate activation of immune pathways may lead to inflammation, adverse reactions, or immune-mediated toxicity, limiting the therapeutic application of syn-mRNA.

## 3.11 Innate Immune Response as an Advantage

The immune response elicited by mRNA-based vaccines offers distinct advantages by enhancing the immune reaction triggered by the vaccine. Synthetic mRNA serves a dual role in these vaccines: it encodes the target antigen and acts as an adjuvant, thereby augmenting immunological responses and antigen presentation. Synthetic mRNA carries the genetic blueprint to produce specific antigens that mimic those found on viruses or cancer cells. When the mRNA is delivered into cells, it instructs them to produce the antigen. This process effectively trains the immune system to recognize and mount a response against these antigens. In addition to encoding antigens, synthetic mRNA can stimulate innate immune receptors like TLRs and RIG-I-like receptors (RLRs). Activation of these receptors' triggers signaling pathways that enhance immune activation and antigen presentation. This adjuvant effect is crucial for boosting the immune response and improving the efficacy of the vaccine. Due to these dual functions—antigen encoding and adjuvant activity—syn-mRNA is rapidly gaining traction in vaccination strategies against viral infections and cancer. mRNA vaccines have demonstrated flexibility in encoding different antigens, rapid adaptability to new variants, and a favorable safety profile compared to traditional vaccine platforms.

Clinical investigations are actively exploring the potential of synthetic mRNA vaccination with a focus on DCs. DCs play a crucial role in the immune system by capturing and presenting antigens via MHC class I and II molecules. This presentation is essential for stimulating both cellular and humoral immune responses against pathogens or abnormal cells, such as cancer cells. Loading DCs with antigens can be achieved using various methods, including proteins, complementary DNAs (cDNAs), or syn-mRNAs. The use of nucleic acid vaccinations, particularly mRNA vaccines, offers distinct advantages over traditional methods like whole-cell protein extracts or DNA vaccines. Whole-cell protein extracts contain a wide range of proteins, some of which may be irrelevant to the desired immune response or even potentially tumorigenic. This diversity increases the risk of inducing autoimmune responses or unintended side effects. Thus, their practicality in vaccine development is limited [50].

## 3.12 Syn-mRNA Vaccines

The potential of mRNA vaccines has been recognized since 1993, when Martinon et al. demonstrated their efficacy using a liposome-encapsulated mRNA vaccine against influenza in a mouse model [20]. While the majority of current studies utilizing synthetic mRNA (syn-mRNA) vaccination focus on immunotherapy for various cancers, this early research highlighted the versatility and promise of mRNA as a vaccine platform. Martinon et al.'s study marked a significant milestone in vaccine development by showcasing several key advantages of mRNA vaccines, including being an efficient vaccine delivery system. Liposome encapsulation facilitated the delivery of mRNA into cells, where it could be translated into the target antigen. This approach ensured effective antigen presentation and immune stimulation. The mRNA vaccine-induced robust immune responses against influenza virus antigens in the mouse model, demonstrating its potential to confer protective immunity against infectious diseases. Furthermore, mRNA vaccines offer flexibility in antigen selection and can be rapidly developed to target emerging pathogens or seasonal variants. This adaptability is critical for responding to evolving infectious disease threats.

Since Martinon et al.'s pioneering work, mRNA vaccine technology has advanced significantly. Current research continues to explore mRNA vaccines not only for infectious diseases but also for cancer immunotherapy and other therapeutic applications. The success of mRNA vaccines against COVID-19 has further underscored their potential and accelerated their adoption in public health efforts globally [51]. A comprehensive study by Petsch et al. in 2012 further expanded on this concept [52] and investigations with HIV-1-infected patients have also been initiated [53, 54]. Another promising application involves using mRNA vaccination to safely target specific protection from type I allergies, utilizing a range of allergen-encoding syn-mRNA vaccines that provide enduring protection against sensitization and induce immunity akin to natural responses acquired early in life [55].

Significant advancements in syn-mRNA molecule development have underscored its expanding applications across various fields of biomedical research and therapy. Despite these strides, several challenges remain that require further innovation and refinement to fully harness the potential of syn-mRNA. Enhancing the design and structure of syn-mRNA constructs is crucial for improving their stability, translation efficiency, and immunogenicity profile. This involves optimizing codon usage, modifying UTRs, and incorporating stabilizing elements to prolong mRNA half-life within cells. Developing robust methods for synthesizing and purifying syn-mRNA is essential to ensure consistency, purity, and efficacy. Techniques such as HPLC and stringent purification protocols play a critical role in eliminating impurities and enhancing the stability of mRNA molecules. Formulating syn-mRNA with delivery systems like lipid nanoparticles or other carriers improves cellular uptake and intracellular delivery efficiency. These formulations protect syn-mRNA from degradation and facilitate its transport across cell membranes, optimizing therapeutic outcomes. Engineering syn-mRNA versions with enhanced selectivity for specific cell types can enhance therapeutic efficacy and specificity. This approach is particularly important for targeting therapies to diseased tissues or organs while minimizing off-target effects. Addressing these challenges requires collaborative efforts among researchers, clinicians, and biotechnologists to advance syn-mRNA technology. Overcoming these hurdles will further unlock the potential of syn-mRNA in revolutionizing personalized medicine, genetic therapies, and vaccine development, paving the way for innovative solutions to complex medical conditions.

## 3.13 Conclusion

In conclusion, this chapter has explored the transformative potential of both mRNA and synthetic mRNA (syn-mRNA) in biomedical research and therapeutic applications. Originating from their foundational roles in the central dogma of molecular biology, mRNA molecules have evolved from essential mediators of protein synthesis to versatile tools with profound implications for modern medicine. The journey of mRNA technology began with seminal studies demonstrating its efficacy in vaccine development, notably highlighted by Martinon et al.'s pioneering work on an influenza vaccine in 1993. Subsequent advancements, including the development of synthetic mRNA (syn-mRNA) and sophisticated delivery systems like lipid nanoparticles, have significantly enhanced stability, efficacy, and safety profiles. These innovations have expanded the scope of mRNA applications beyond infectious diseases to include cancer immunotherapy, genetic disorders, and regenerative medicine. Despite these advancements, challenges persist in optimizing syn-mRNA constructs for improved stability, refining manufacturing processes to meet regulatory standards, and enhancing delivery systems to ensure efficient cellular uptake and intracellular activity. Moreover, achieving enhanced selectivity for specific cell types remains a crucial objective to maximize therapeutic efficacy while minimizing off-target effects. Recent breakthroughs, particularly the successful development and global deployment of mRNA-based COVID-19 vaccines, underscore the transformative potential of syn-mRNA in public health emergencies and pandemic preparedness. These achievements highlight the agility and scalability of mRNA technology in rapidly responding to emerging health threats. Looking forward, ongoing research efforts and collaborative initiatives are essential to overcoming existing challenges and fully harnessing the therapeutic potential of both mRNA and syn-mRNA. The convergence of molecular biology, nanotechnology, and immunology promises continued innovation in personalized medicine and precision therapeutics. By addressing these complexities, mRNA and syn-mRNA technologies stand poised to redefine the treatment landscape, offering safer, more effective solutions across a broad spectrum of diseases and medical conditions. Looking forward, while there are challenges to be addressed, the trajectory of mRNA and syn-mRNA technologies represents a paradigm shift in biomedicine, promising unprecedented opportunities to improve human health and well-being in the decades ahead.

## References


1. Hajj KA, Whitehead KA. Tools for translation: non-viral materials for therapeutic mRNA delivery. Nat Rev Mater. 2017;2(10):17056.
2. Pierce B. Genetics: a conceptual approach. Macmillan; 2017.
3. Nguyen TC, Zaleta-Rivera K, Huang X, Dai X, Zhong S. RNA, action through interactions. Trends Genetics. 2018;34(11):867–82.
4. Saldi T, Riemondy K, Erickson B, Bentley DL. Alternative RNA structures formed during transcription depend on elongation rate and modify RNA processing. Mol Cell. 2021;81(8):1789–801.e5.
5. Lehman N. RNA in evolution. Wiley Interdiscip Rev RNA. 2010;1(2):202–13.
6. Minchin S, Lodge J. Understanding biochemistry: structure and function of nucleic acids. Essays Biochem. 2019;63(4):433–56.
7. Bicknell AA, Ricci EP. When mRNA translation meets decay. Biochem Soc Trans. 2017;45(2):339–51.
8. Kowalski PS, Rudra A, Miao L, Anderson DG. Delivering the messenger: advances in Technologies for Therapeutic mRNA delivery. Mol Ther. 2019;27(4):710–28.
9. Kornberg RD. The molecular basis of eukaryotic transcription. Proc Natl Acad Sci USA. 2007;104(32):12955–61.
10. Detke S, Stein JL, Stein GS. Synthesis of histone messenger RNAs by RNA polymerase II in nuclei from S phase HeLa S3 cells. Nucleic Acids Res. 1978;5(5):1515–28.
11. Schlake T, Thess A, Thran M, Jordan I. mRNA as novel technology for passive immunotherapy. Cell Mol Life Sci. 2019;76(2):301–28.
12. Zhong Z, Mc Cafferty S, Combes F, Huysmans H, De Temmerman J, Gitsels A, et al. mRNA therapeutics deliver a hopeful message. Nano Today. 2018;23:16–39.
13. Granot-Matok Y, Kon E, Dammes N, Mechtinger G, Peer D. Therapeutic mRNA delivery to leukocytes. J Control Release. 2019;305:165–75.
14. Jia L, Mao Y, Ji Q, Dersh D, Yewdell JW, Qian SB. Decoding mRNA translatability and stability from the 5' UTR. Nat Struct Mol Biol. 2020;27(9):814–21.
15. Trepotec Z, Lichtenegger E, Plank C, Aneja MK, Rudolph C. Delivery of mRNA therapeutics for the treatment of hepatic diseases. Mol Ther. 2019;27(4):794–802.
16. Magadum A, Kaur K, Zangi L. mRNA-based protein replacement therapy for the heart. Mol Ther. 2019;27(4):785–93.
17. Sahu I, Haque A, Weidensee B, Weinmann P, Kormann MSD. Recent developments in mRNA-based protein supplementation therapy to target lung diseases. Mol Ther. 2019;27(4):803–23.
18. Grozhik AV, Jaffrey SR. Distinguishing RNA modifications from noise in epitranscriptome maps. Nat Chem Biol. 2018;14(3):215–25.
19. Karikó K, Buckstein M, Ni H, Weissman D. Suppression of RNA recognition by toll-like receptors: the impact of nucleoside modification and the evolutionary origin of RNA. Immunity. 2005;23(2):165–75.

20. Eichhorn SW, Subtelny AO, Kronja I, Kwasnieski JC, Orr-Weaver TL, Bartel DP. mRNA poly(A)-tail changes specified by deadenylation broadly reshape translation in Drosophila oocytes and early embryos. elife. 2016;e16955:5.
21. Wiederhold K, Passmore LA. Cytoplasmic deadenylation: regulation of mRNA fate. Biochem Soc Trans. 2010;38(6):1531–6.
22. Dreyfus M, Régnier P. The poly(a) tail of mRNAs: bodyguard in eukaryotes, scavenger in bacteria. Cell. 2002;111(5):611–3.
23. Schlake T, Thran M, Fiedler K, Heidenreich R, Petsch B, Fotin-Mleczek M. mRNA: a novel avenue to antibody therapy? Mol Ther. 2019;27(4):773–84.
24. Corey DR, Damha MJ, Manoharan M. Challenges and opportunities for nucleic acid therapeutics. Nucleic Acid Ther. 2022;32(1):8–13.
25. Kulkarni JA, Witzigmann D, Thomson SB, Chen S, Leavitt BR, Cullis PR, et al. The current landscape of nucleic acid therapeutics. Nat Nanotechnol. 2021;16(6):630–43.
26. Walsh G, Jefferis R. Post-translational modifications in the context of therapeutic proteins. Nat Biotechnol. 2006;24(10):1241–52.
27. Weng Y, Li C, Yang T, Hu B, Zhang M, Guo S, et al. The challenge and prospect of mRNA therapeutics landscape. Biotechnol Adv. 2020;40:107534.
28. Lin CY, Perche F, Ikegami M, Uchida S, Kataoka K, Itaka K. Messenger RNA-based therapeutics for brain diseases: An animal study for augmenting clearance of beta-amyloid by intracerebral administration of neprilysin mRNA loaded in polyplex nanomicelles. J Control Release. 2016;235:268–75.
29. Guan S, Rosenecker J. Nanotechnologies in delivery of mRNA therapeutics using nonviral vector-based delivery systems. Gene Ther. 2017;24(3):133–43.
30. Shin H, Park S-J, Yim Y, Kim J, Choi C, Won C, et al. Recent advances in RNA therapeutics and RNA delivery systems based on nanoparticles. Adv Ther. 2018;1(7):1800065.
31. Cao J, An D, Galduroz M, Zhuo J, Liang S, Eybye M, et al. mRNA therapy improves metabolic and behavioral abnormalities in a murine model of Citrin deficiency. Mol Ther. 2019;27(7):1242–51.
32. Midoux P, Pichon C. Lipid-based mRNA vaccine delivery systems. Expert Rev Vaccines. 2015;14(2):221–34.
33. An D, Frassetto A, Jacquinet E, Eybye M, Milano J, DeAntonis C, et al. Long-term efficacy and safety of mRNA therapy in two murine models of methylmalonic acidemia. EBioMedicine. 2019;45:519–28.
34. Schlake T, Thess A, Fotin-Mleczek M, Kallen KJ. Developing mRNA-vaccine technologies. RNA Biol. 2012;9(11):1319–30.
35. Deng Z, Tian Y, Song J, An G, Yang P. mRNA vaccines: the Dawn of a new era of cancer immunotherapy. Front Immunol. 2022;13:887125.
36. Karikó K, Muramatsu H, Ludwig J, Weissman D. Generating the optimal mRNA for therapy: HPLC purification eliminates immune activation and improves translation of nucleoside-modified, protein-encoding mRNA. Nucleic Acids Res. 2011;39(21):e142.
37. Parkesh R, Childs-Disney JL, Nakamori M, Kumar A, Wang E, Wang T, et al. Design of a bioactive small molecule that targets the myotonic dystrophy type 1 RNA via an RNA motif-ligand database and chemical similarity searching. J Am Chem Soc. 2012;134(10):4731–42.
38. Blakney AK, Ip S, Geall AJ. An update on self-amplifying mRNA vaccine development. Vaccines (Basel). 2021;9(2):97.
39. Nayerossadat N, Maedeh T, Ali PA. Viral and nonviral delivery systems for gene delivery. Adv Biomed Res. 2012;1:27.
40. Zu H, Gao D. Non-viral vectors in gene therapy: recent development, challenges, and prospects. AAPS J. 2021;23(4):78.
41. Cyranoski D. Japan to offer fast-track approval path for stem cell therapies. Nat Med. 2013;19(5):510.
42. Yu AM, Choi YH, Tu MJ. RNA drugs and RNA targets for small molecules: principles, Progress, and challenges. Pharmacol Rev. 2020;72(4):862–98.

43. Qin S, Tang X, Chen Y, Chen K, Fan N, Xiao W, et al. mRNA-based therapeutics: powerful and versatile tools to combat diseases. Signal Transduct Target Ther. 2022;7(1):166.
44. Warren L, Manos PD, Ahfeldt T, Loh YH, Li H, Lau F, et al. Highly efficient reprogramming to pluripotency and directed differentiation of human cells with synthetic modified mRNA. Cell Stem Cell. 2010;7(5):618–30.
45. Akira S, Uematsu S, Takeuchi O. Pathogen recognition and innate immunity. Cell. 2006;124(4):783–801.
46. Kawai T, Akira S. Antiviral signaling through pattern recognition receptors. J Biochem. 2007;141(2):137–45.
47. Yoneyama M, Fujita T. Structural mechanism of RNA recognition by the RIG-I-like receptors. Immunity. 2008;29(2):178–81.
48. Wilkins C, Gale M Jr. Recognition of viruses by cytoplasmic sensors. Curr Opin Immunol. 2010;22(1):41–7.
49. Nallagatla SR, Toroney R, Bevilacqua PC. Regulation of innate immunity through RNA structure and the protein kinase PKR. Curr Opin Struct Biol. 2011;21(1):119–27.
50. Gilboa E, Vieweg J. Cancer immunotherapy with mRNA-transfected dendritic cells. Immunol Rev. 2004;199:251–63.
51. Martinon F, Krishnan S, Lenzen G, Magné R, Gomard E, Guillet JG, et al. Induction of virus-specific cytotoxic T lymphocytes in vivo by liposome-entrapped mRNA. Eur J Immunol. 1993;23(7):1719–22.
52. Petsch B, Schnee M, Vogel AB, Lange E, Hoffmann B, Voss D, et al. Protective efficacy of in vitro synthesized, specific mRNA vaccines against influenza A virus infection. Nat Biotechnol. 2012;30(12):1210–6.
53. Van Lint S, Goyvaerts C, Maenhout S, Goethals L, Disy A, Benteyn D, et al. Preclinical evaluation of TriMix and antigen mRNA-based antitumor therapy. Cancer Res. 2012;72(7):1661–71.
54. Van Gulck E, Vlieghe E, Vekemans M, Van Tendeloo VF, Van De Velde A, Smits E, et al. mRNA-based dendritic cell vaccination induces potent antiviral T-cell responses in HIV-1-infected patients. AIDS (London, England). 2012;26(4):F1–12.
55. Weiss R, Scheiblhofer S, Roesler E, Weinberger E, Thalhamer J. mRNA vaccination as a safe approach for specific protection from type I allergy. Expert Rev Vaccines. 2012;11(1):55–67.

# Chapter 4
# RNA in the Central Dogma: tRNAs

**Abstract** The elucidation of the genetic code and the identification of tRNAs as its translators were transformative events in the field of life sciences. Initially, the spotlight on tRNAs diminished as their basic role in protein synthesis was established, prompting researchers to explore other aspects of RNA biology. Recently, there has been a revival in tRNA research, demonstrating their involvement in various biological pathways beyond simple translation. This renewed interest underscores the adaptability of tRNAs to modify protein synthesis in accordance with environmental shifts. The emerging roles of tRNAs are closely connected to the existence of various tRNA forms, including isoacceptors and isodecoders, as well as numerous base modifications, a wide range of protein interactions, and tRNA fragmentation. These factors introduce a remarkable complexity to the tRNA milieu, providing cells with an elaborate selection of tRNA species. These species are crucial in preserving cellular equilibrium and modulating cellular activities in response to environmental variability, thereby ensuring the health and stability of the overall ecosystem. This enhanced understanding of tRNA functions not only deepens our insights into cellular biology but also unveils new possibilities for investigating how these primordial molecules influence intricate cellular mechanisms and adapt to environmental challenges. The continued exploration of tRNA biology is expected to further elucidate their significant impacts on health and disease.

## 4.1 Introduction

More than 50 years ago, the unveiling of the universal genetic code and tRNAs fundamentally transformed life sciences by providing a systematic approach to understanding biological mechanisms (Fig. 4.1). During this time, the critical function of tRNAs was exemplified by their involvement in aminoacylation reactions, mediated by aminoacyl-tRNA synthetases (aaRSs). These enzymes are crucial for translating mRNA sequences, as they are responsible for the catalysis process that attach specific amino acids to their corresponding tRNAs, resulting in aminoacylated tRNAs. This is a process that requires energy and therefore ATP is involved.



P. E. Saw, E. Song, *RNA Therapeutics in Human Diseases*,
https://doi.org/10.1007/978-981-96-3041-7_4

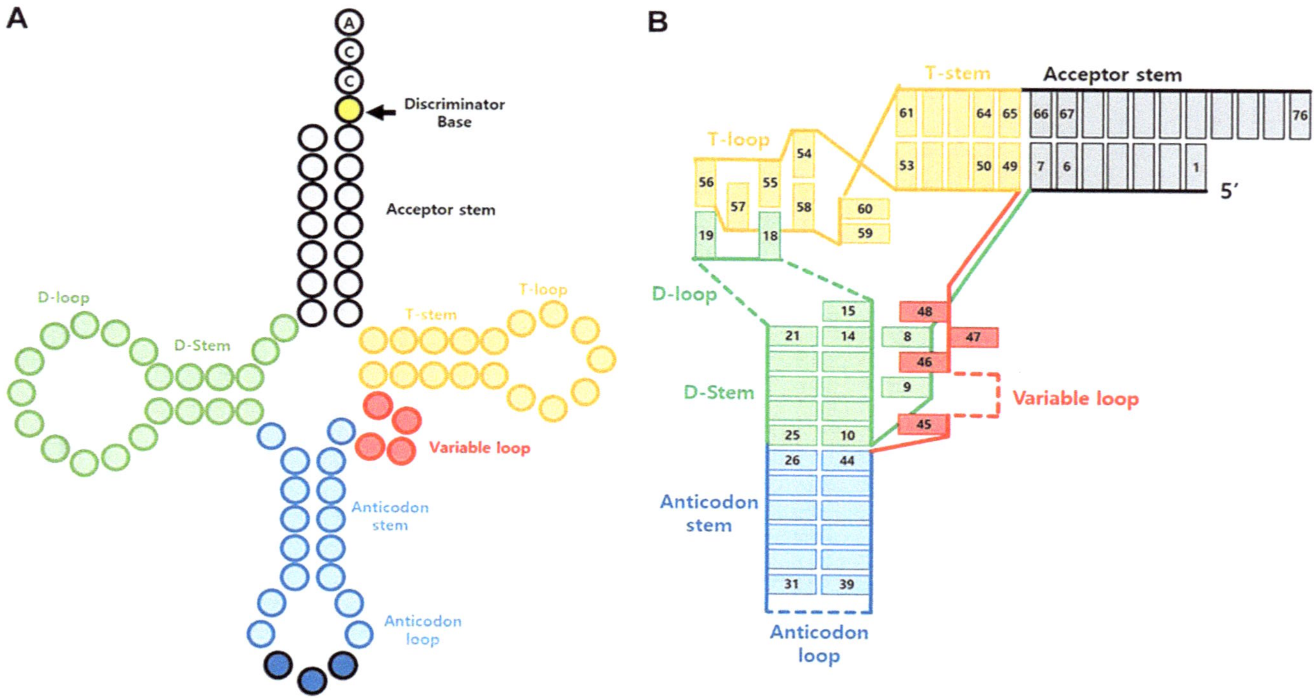


**Fig. 4.1** Representation of tRNA structure and numbering. (**a**) Schematic diagram of a tRNA molecule, illustrating its cloverleaf structure. The diagram identifies key components including the D-stem (green), anticodon stem (blue), T-stem (yellow), and variable loop (red). The discriminator base is highlighted at the top of the acceptor stem (black). (**b**) Linear representation of the tRNA showing the base sequence from 5′ to 3′ end. The secondary structure elements are color-coded consistent with (**a**), emphasizing the connectivity between the stems and loops. Numbers represent the position of bases within the tRNA sequence. The dashed lines indicate the boundaries of the variable loop, distinguishing its specific location and sequence within the overall structure. (Figure is reproduced from [5] under Creative Common CC-BY license)

The translation elongation factors then transported these aminoacylated tRNAs to the ribosomes, where they are made into a peptide chain, playing a key role in protein synthesis. Extensive research, including detailed structural analyses of tRNA and its associated complexes, led to the belief that our understanding of tRNA biology was quite comprehensive.

However, recent discoveries have revealed additional layers of complexity in tRNA function and regulation, indicating that there is still much to learn about these essential molecules. Recent research has revealed that apart from the traditional role in protein synthesis, tRNAs might have additional roles in other biological processes, and notably, their participation as ncRNAs within expansive regulatory networks. These insights have been driven by advances in RNA sequencing technologies, particularly RNA-Seq, and improved computational methods that facilitate the analysis of large sequence datasets [1–3]. Simultaneously, advancements in proteomics have allowed for the integration of protein and RNA sequencing data, uncovering a previously unrecognized depth in tRNA structure and function.

## 4.2 Structure of tRNAs

The classical tRNA molecule, usually comprised of 76 nucleotides, presents a conserved configuration recognized across all many species [4]. The basic tRNA molecule is shaped like a cloverleaf, usually comprises 76–93 nucleotides (nt). All tRNA molecule ends with a CCA trinucleotide on the 3′-end, and aminoacylation occurs at the terminal adenosine, either at the 2′- or 3′-hydroxyl group (Fig. 4.1).

The cloverleaf structure of tRNA is divided into four clear segments: (1) the acceptor stem where the aminoacylation occurs (ending with -CCA), (2) the dihydrouridine (D) stem-loop, the anticodon stem-loop (matching up with the genetic sequence during translation), and the TψC stem-loop, with ψ representing pseudouridine. When it folds into its tertiary, L-shaped structure, the TψC stem aligns with the acceptor stem, forming a 12-base pair acceptor–TψC minihelix. Simultaneously, the anticodon stem-loop pairs with the D-stem to create a 9–10 base pair anticodon–D-loop dumbbell shape. These segments are connected perpendicularly through between conserved nucleotides in the TψC and D-loops. This connection is essential for stabilizing the L-shaped conformation. tRNAs with nucleotides higher than 76 usually extend into a small variable loop, which protrudes from the back of the L-shaped structure to prevent internal clashes. Importantly, the anticodon triplet, which interfaces with the mRNA codon, is positioned maximally distant from the site of amino acid attachment, with the farthest point within the three-dimensional structure being 76 Å.

## 4.3 Uniqueness of Complex tRNA Structure: Universal Isodecoders and Isoacceptors

Although tRNAs are among the most abundant types of RNA in a cell, the total number of tRNA genes varies between individuals. On average, the human genome contains approximately 500 genes that encode tRNAs and tRNA-like molecules. These genes are responsible for producing the diverse array of tRNAs necessary for decoding mRNA sequences into proteins during translation. Despite this variation, the overall abundance of tRNAs underscores their essential role in maintaining efficient and accurate protein synthesis [6–8]. The human genome encodes approximately 300 distinct cytoplasmic tRNA sequences, while mitochondrial DNA adds another 22 to the mitochondrial tRNA pool. Each amino acid is associated with at least one specific tRNA species. Once an amino acid is aminocylated to its corresponding tRNA, its incorporation into the polypeptide chain is determined by the codon–anticodon matching on mRNA-tRNA, respectively. To reduce the probability of mutation, one amino acid can be represented by multiple tricodons, allowing for redundancy in the coding sequence. This overlap has spurred the development of tRNA isoacceptors—tRNAs that differ in sequence, including their anticodon sequences, but carry the exact same amino acid, elucidating the inherent complexity of the tRNA, which is initially rooted in the extensive variety of tRNA genes and thc redundancy of the genetic code. This is further enhanced by the possible range of base modifications within the tRNAs, their interacting partners like aaRSs, and their fragmentation.

### 4.3.1 *Isoacceptors and Isodecoders*

Isoacceptors are defined as "tRNA genes that possess different anticodons, but are charged with the same amino acids," while isodecoders are defined as "tRNA genes that possess the same anticodons, are different in their general structures and sequences" [9]. For example, in human cytoplasmic tRNAs specific to alanine, known as $tRNA^{Ala}$. The wobble pairing mechanism in codon–anticodon interactions allows for less stable bonds at certain positions (e.g., G: U wobble pairing versus the more stable A:U Watson-Crick pairing). Theoretically, decoding the four GCX codons for alanine (where X can be any base) would require only two $tRNA^{Ala}$ molecules with anticodons UGC or GGC. However, the human genome actually contains 46 genes encoding 39 distinct $tRNA^{Ala}$ sequences. Each of these tRNAs feature a G3–U70 base pair in the acceptor stem, which is specifically recognized by alanine tRNA synthetase (AlaRS) for the attachment of the amino acid. Although this might be seen as redundancy, it highlights the complexity and robustness of the tRNA gene repertoire in humans [10]. Notably, some of these $tRNA^{Alas}$ function as both isoacceptors and isodecoders, as depicted in Fig. 4.2. There are approximately 270 isodecoders corresponding to about 50 isoacceptors among all cytoplasmic tRNAs [7, 11, 12].

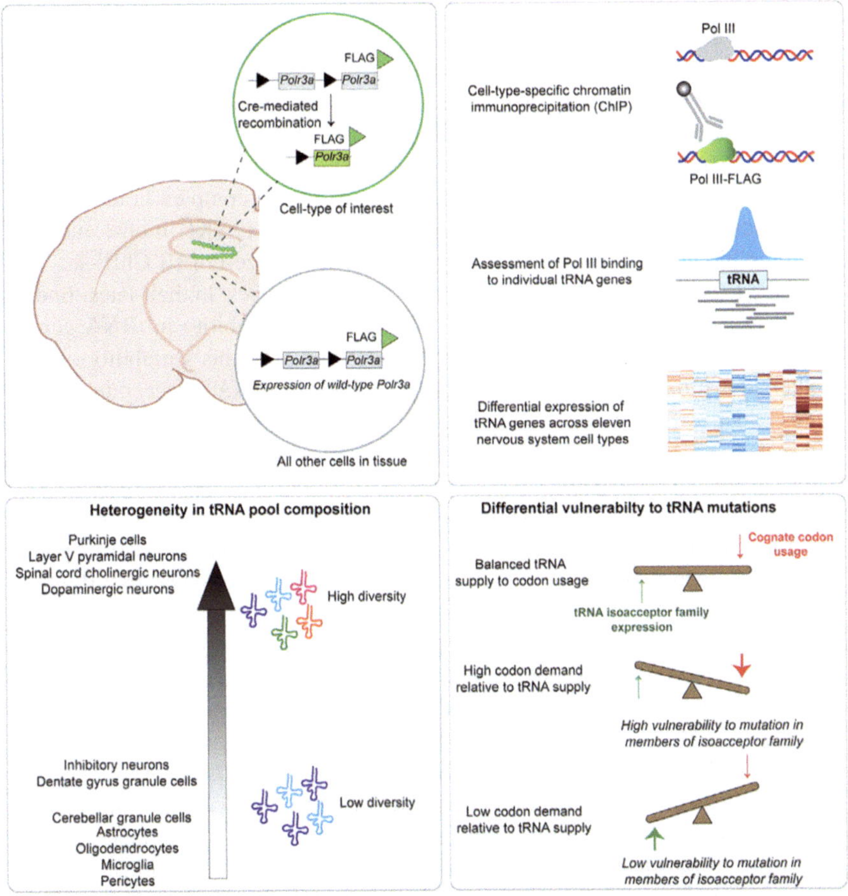


**Fig. 4.2** The graphical abstract from [31] illustrates the cell-type-specific investigation of RNA Polymerase III (Pol III) and transfer RNA (tRNA) dynamics within the nervous system. The upper left panel depicts a Cre-loxP system for the cell-type-specific expression of FLAG-tagged wild-type Pol III subunit (Pol3α), allowing for selective investigation in desired cell types while maintaining wild-type Pol3α expression in other cells. The upper right panels highlight the methodology and outcomes of chromatin immunoprecipitation (ChIP) specific to Pol III, detailing Pol III binding to tRNA genes and subsequent differential tRNA gene expression analysis across various nervous system cell types. The lower left panel shows the heterogeneity in tRNA pool composition among different neuronal and glial cells, pointing to a spectrum of tRNA diversity, from high in Purkinje and dopaminergic neurons to low in microglia and oligodendrocytes. The lower right panel explores the differential vulnerability of cells to tRNA mutations, correlating tRNA isodecoder family expression and codon usage patterns with susceptibility to mutations, thus underlining a balance between tRNA supply and codon demand in neural cell types. (Figure is reproduced from [31], under Creative Commons CC-BY license)

**Box 4.1 Brain-Specific tRNAs**

Defects in tRNA biogenesis are increasingly recognized as key contributors to a variety of neurological disorders [13]. However, understanding how these dysfunctions translate into cell-specific effects within diverse brain cell populations has been challenging, largely due to the difficulty in measuring tRNA expression accurately across distinct cell types within complex brain tissue. To address this, we engineered a conditional knock-in (KI) mouse model expressing epitope-tagged Pol III, enabling us to utilize Pol III ChIP-seq to profile tRNA gene expression in specific neural cell types. In their latest findings in Neuron, Kapur et al. reveal a pronounced variability in tRNA gene expression across nervous system cell types. However, this variability often alters the composition rather than the abundance of tRNA isoacceptor families. Such differential expression of tRNA isoacceptors can influence the translation efficiency and decoding speed of associated codons [14], though the specific biological impact of varying tRNA isodecoder expression remains to be elucidated. Sequence variation within isodecoder regions might affect their translational effectiveness [11], suggesting that distinct isodecoder profiles may help to optimize translation fidelity and responsiveness in different cell types.

Emerging research highlights that tRNAs are precursors to a spectrum of small ncRNAs with diverse cellular roles [15, 16]. A recent study hypothesized that the main function of varied isodecoder expression may be to generate specific tRNA fragments that play critical roles in various cell signaling pathways [17]. For instance, fragments derived from tRNA-Leu-CAG genes are implicated in ribosome biogenesis [18], with this isoacceptor family displaying significant expression differences across neural cell populations. Moreover, we identified that a functional miRNA, miR-1983, is generated from the tRNA-Ile-TAT-2-1 precursor [19–22] and is specifically expressed in certain neuronal subtypes. Notably, this miRNA is exclusively produced from pre-tRNA-Ile-TAT-2-1, not from similar genes like −2-2 or − 2-3, despite these sharing identical mature sequences. The results suggest that differential tRNA isodecoder expression may contribute to the susceptibility of specific neuronal populations to tRNA-deficiency-linked ribosomal stalling and subsequent neurodegeneration. For example, neuron populations susceptible to mutations in tRNA-Arg-TCT-4-1 exhibit a more homogeneous $tRNA^{Arg}$ (TCT) pool compared to resistant cells, such as Purkinje neurons. Similarly, mutations in tRNA-Ile-TAT-2-3 and tRNA-Ile-TAT-1-1 are particularly detrimental to cerebellar granule cells, whereas neuronal populations with a more diverse $tRNA^{Ile}$ (TAT) pool appear to be protected. Nonetheless, the vulnerability of neural cells to tRNA mutations likely depends on more than just individual tRNA gene expression; factors like codon usage preferences or neuron-specific signaling responses, such as mTOR pathway modulation in

(continued)

**Box 4.1** (continued)
reaction to ribosome stalling, might also contribute to these observed susceptibilities [23].

*Gene-specific regulation of tRNA expression*

Research has shown that RNAPol II and histone modifications associated with it are frequently located near tRNA genes that have high occupancy by RNA Pol III. This spatial association implies that chromatin modifications established through Pol II activity might facilitate Pol III access to these regions, essentially creating an open chromatin environment that supports tRNA transcription [24–27]. In fact, the accessibility of chromatin around tRNA genes correlates closely with tRNA abundance as observed in Pol III ChIP-seq experiments [14]. Nonetheless, the control of Pol III-driven transcription at tRNA loci is more intricate than simply modulating chromatin openness. Recent findings indicate that Pol II itself may directly influence tRNA transcription in a gene-specific manner, sometimes silencing particular tRNAs by modulating Pol III activity [28]. Additionally, Pol II-associated transcription factors (TFs) like SOX4 have been found to inhibit Pol III recruitment by blocking TFIIIB binding at certain tRNA genes, such as those in human glioblastoma cells [29]. Numerous binding motifs for Pol II transcription factors are located near tRNA genes, and the presence of these TFs at these sites may alter Pol III complex assembly, providing a potential mechanism for selective, cell-type-specific tRNA regulation [24, 26, 30]. The network of molecular factors and pathways underlying cell-specific expression of tRNAs could operate across groups of tRNA genes, such as those upregulated in cerebellar granule cells compared to other neuronal subtypes, or they may vary widely depending on the tRNA gene in question (Fig. 4.2).

## 4.4 tRNA Modifications

tRNA molecules undergo a variety of modifications that are critical for their structure, stability, and function. These modifications occur at numerous nucleoside residues within the tRNA and can affect the tRNA's ability to accurately and efficiently decode mRNA during protein synthesis. Herein, we summarize the top five modifications in tRNAs.

### *4.4.1 Base Modifications*

Methylation is one of the most common modifications, where methyl groups are added to the bases or the ribose moiety of the nucleosides. Methylation can occur on different bases such as adenine, guanine, and cytosine, leading to modified bases like 1-methyladenosine (m1A), N2-methylguanosine (m2G), and 5-methylcytosine (m5C). Second is pseudouridylation. Second is pseudouridylation, whereby the conversion of uridine to pseudouridine occurs, when the uracil is bonded to the ribose via a carbon-carbon instead of the usual nitrogen-carbon glycosidic bond. This modification provides additional stability to the tRNA structure and is crucial for the proper folding and base pairing properties of tRNA.

### 4.4.2 Ribose Modifications

2'-O-methylation is an addition of a methyl group at the 2′-hydroxyl of the ribose sugar. This modification is important for the correct folding of the tRNA as well as protecting it from enzymatic cleavage.

### 4.4.3 Complex Hypermodifications

Modifications like 4-thiouridine (s4U) where a sulfur atom replaces an oxygen atom in the uridine molecule is called thiolation. These modifications often occur at the wobble position of the anticodon, affecting the tRNA's ability to recognize codons on the mRNA. In isopentenylation, the addition of an isopentenyl group to adenosine (i6A) in certain tRNAs, which is essential for accurate decoding of the mRNA during translation.

### 4.4.4 Modified Wobble Bases

Adenosine in the anticodon can be deaminated to inosine (I), which can pair with U, C, or A in the third position of the mRNA codon, allowing for wobble base pairing and thus greater flexibility in codon recognition. A more complex modification where queuosine (Q) replaces guanine in the wobble position, affecting the tRNA's interaction with the ribosome and influencing translational accuracy and efficiency.

### 4.4.5 Dihydrouridine (D) Modifications

Reduction of uridine to dihydrouridine at various positions in tRNA. Dihydrouridine makes the ribose ring more flexible, facilitating the proper folding and architecture of the tRNA molecule.

## 4.5 Biological Functions of tRNA Modifications

### 4.5.1 Homeostatic Balancing of tRNA Function

Enzymes that catalyze modification reactions can often be complemented by counterparts that catalyze de-modification, maintaining a delicate balance analogous to the yin-yang principle. A recent investigation delved into this balance concerning tRNA methylation [16]. One study focused on the nucleic acid dioxygenase ALKBH1, which removes the modification on N1-methyl group from the adenine at position 58 (A58) in several tRNAs through a non-heme iron-dependent oxidative process. To date, ALKBH1 is the sole tRNA methylation "eraser" thoroughly characterized. Positioned in the TψC-loop, which is at the exterior corner of the L-shaped tertiary structure of tRNAs, A58 plays a crucial role. Experimental results showed that overexpression of ALKBH1 resulted in reduced methylation levels at A58, while depletion of ALKBH1 led to increased methylation. Intriguingly, the study found that in the population of tRNAs, the methylation at A58 was always partial, suggesting that a state of either full modification or complete lack thereof was not attained. This partial modification might be an intrinsic aspect of the cell's homeostatic mechanisms, contributing to a vast diversity of tRNA microspecies. Furthermore, ALKBH1 was identified as a translational repressor. It exerts its repressive effects both at the stage of translation initiation and during elongation. These actions have significant physiological impacts; for instance, an increase in ALKBH1 expression induced by glucose deprivation resulted in a general reduction of the translation initiation process. This highlights the significance of the methylation-demethylation cycle at A58 and their crucial role in regulating this process. This dynamic interplay between methylation and demethylation underscores the complex regulatory capacity of tRNA modifications within cellular translation processes.

### 4.5.2 Allowing Adjustment in Translation

Modifications within the anticodon loop are also important for their role in influencing codon recognition and reading [32, 33]. However, understanding how the intra- and extracellular environment modifies these bases in response to stress or other stimuli was largely unknown [34]. Recent technological developments enable detailed analysis of base modifications across various tRNAs have revolutionized the annotation and study of these modifications, shedding new light on their functional roles and regulation [35–37]. For example, in bacteria, despite hypoxic conditions, both M. bovis and M. tuberculosis can survive in a non-replicative dormant state from which they can later be revived. The central regulator of mycobacterial homeostasis is DosR, a transcription factor activated by kinase. DosR controls the expression of the Dos regulon, which consists of 48 genes [38–41]. These genes are marked by specific codons whose effective translation depends heavily on tRNA modifications, such as 5-oxyacetyl uracil (cmo5U).

In hypoxic conditions, significant fluctuations in tRNA base modifications were observed, including a 350% increase in cmo5U, which correlates with bacterial persistence and potential for revival [38]. Specifically, ACG codons are preferentially recognized by tRNAThr/UGU through G:U wobble base pairing facilitated by the cmo5U modification. Additionally, it was discovered that during early phase of hypoxia, there are significant decrement in the levels of two $tRNA^{Thr}$ isoacceptors—$tRNA^{Thr}$/CGU and $tRNA^{Thr}$/GGU. This reduction led to decreased translation of proteins from genes that are rich in the respective codons for these isoacceptors, particularly the "optimal" ACC codon. These findings highlight a pronounced bias for ACG over ACC codons within the Dos regulon, favoring the persistence of M. bovis under stress conditions [38]. This research underscores how specific modification reactions in tRNAs are crucial for adaptive translation, enhancing cell survival under variable environmental conditions.

**Box 4.2 Dynamic System of Genetic Regulation Using tRNA Modification: Mathematical Possibility**

Within a tRNA molecule, sites may range from unmodified to nearly fully modified. Since modification occurs in a binary manner, it is possible to generate a vast array of microspecies. Considering just the 13 sites associated with disease when unmodified, the potential number of microspecies for each tRNA calculates to $2^{13}$, equating to 8192 [42, 43]. These modifications add a layer of complexity that significantly surpasses the diversity introduced by isoacceptors and isodecoders. In a typical mammalian cell, there are approximately 60 million tRNA molecules. This vast number of tRNAs, each potentially carrying a variety of unique modifications, represents an extensive collection of microspecies. Despite the sheer number and diversity, this intricate network of tRNAs is seamlessly integrated within the overall cellular population, ensuring efficient and accurate protein synthesis while accommodating the cell's diverse functional needs [44]. When we consider the mathematical possibilities of all potential combination of all tRNAs, it will result in an astronomical number, a complexity that far exceeds the capacity of an entire organism to fully encompass. The sheer diversity generated by the possible variations and combinations of tRNA modifications highlights the intricate regulatory mechanisms at play. It underscores the sophisticated level of control required to maintain proper cellular function and protein synthesis, reflecting a highly evolved and dynamic system of genetic regulation and expression.

### 4.5.3 *Variability of tRNA Sequence*

The tRNA isoacceptors is to accommodate the degeneracy of the genetic code, where different codons can carry the same amino acid (redundancy). Additionally, the use of isoacceptors and isodecoders enables gene-to-gene variations in codon usage. This variation provides a mechanism to fine-tune the translation efficiency of different mRNAs, allowing the cell to regulate protein synthesis with greater precision [45]. Although the underlying rationale is still being investigated and has sparked some debate, recent studies have clearly demonstrated the significant role of isoacceptors and isodecoders in regulating various cellular functions through translation. These studies provide strong evidence that these tRNA variants are not merely redundant elements but actively contribute to fine-tuning the efficiency and fidelity of protein synthesis. By influencing how different mRNAs are translated, isoacceptors and isodecoders help coordinate the cell's response to changing conditions and demands, thus playing a crucial role in maintaining cellular homeostasis and adaptability [46].

**Box 4.3 Organelle-Specific tRNAs: Mitochondrial tRNAs**

In addition to cytoplasmic tRNAs, animal cells house a distinct set of mitochondrial tRNAs. These mitochondrial tRNAs are generally shorter than their cytoplasmic counterparts. In humans, mitochondrial tRNAs range from 59–75 nt. These tRNAs are found to have smaller stems and loops, and some may even lack entire domains typically found in cytoplasmic tRNAs. This variability indicates a significant adaptive capacity in tRNAs, suggesting that even truncated tRNAs and their fragments can play crucial roles in cellular functions. Furthermore, mitochondrial tRNAs are notably rich in adenine and uracil, contributing to their reduced stability. Many human diseases linked to mutations in mitochondrial tRNAs are believed to arise largely from base modifications that intensify this instability [47–49].

Mammalian mitochondrial DNA encodes all 22 functional tRNAs required for mitochondrial protein synthesis, corresponding to the 18 amino acids, with two types each for serine and leucine [50]. Since mitochondria is the place to produce energy, these tRNAs are used for the synthesis of 13 essential proteins involved in ATP production and energy generation [51, 52]. Unlike cytoplasmic tRNAs, mitochondrial tRNA for methionine (mt-$tRNA^{Met}{}_{CAU}$, with CAU as the anticodon) is encoded by a single gene and serves a dual function: (1) initiating translation at the start codon and (2) elongating the polypeptide chain during protein synthesis. Notably, mt-$tRNA^{Met}{}_{CAU}$ recognizes both the universal start codon AUG and the normally cytoplasmic isoleucine codon AUA as methionine, highlighting its unique role in mitochondrial translation. The post-translational modifications of tRNAs are depicted in Fig. 4.3.

(continued)

**Box 4.3** (continued)

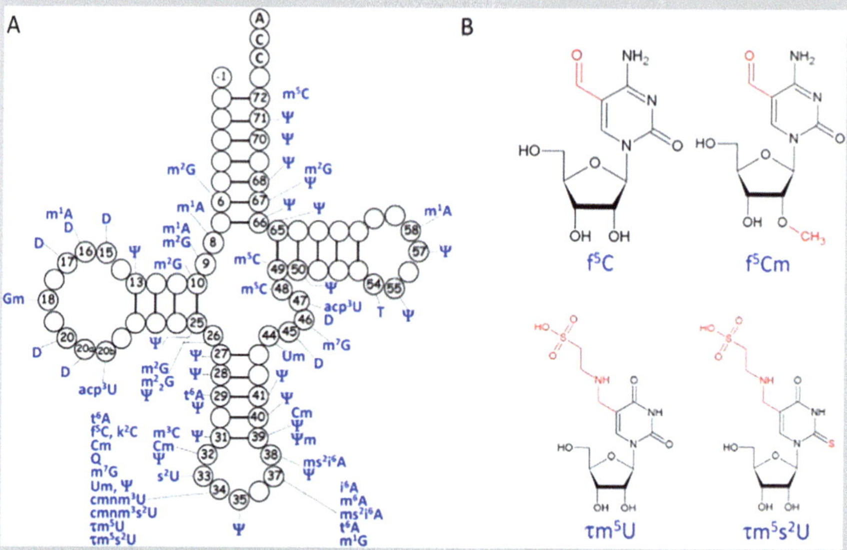


**Fig. 4.3** (**a**) This diagram represents the secondary structure of a typical tRNA molecule, annotated with various post-transcriptional modifications. Each circle represents a nucleotide, and modifications are labeled at their respective positions within the tRNA structure. Modifications include methylations (m), dihydrouridine (**d**), pseudouridine (Ψ), and other complex modifications specific to tRNA molecules, which are crucial for the correct folding, stability, and functionality of the tRNA during protein synthesis. (**b**) The molecular structures of selected modified nucleosides found in tRNA are depicted. These include 5-formylcytosine (f5C), 5-formylcytidine (f5Cm), 5-taurinomethyluridine (tm5U), and 5-taurinomethyl-2-thiouridine (tm5s2U), illustrating the chemical diversity and complexity of tRNA modifications. These modifications can affect tRNA conformation, recognition by aminoacyl-tRNA synthetases, and interaction with the ribosome, thus playing critical roles in the regulation and efficiency of protein synthesis. (Figure is reproduced from [53] under Creative Commons CC-BY international license)

**Mitochondrial tRNA: their role in human diseases**

Mutations in mitochondrial DNA are linked with a broad spectrum of mitochondrial diseases, primarily due to pathogenic alterations in mitochondrial tRNA genes. These mutations affect all mitochondrial genes and are linked to numerous maternally inherited genetic disorders. Remarkably, over 250 mutations in mitochondrial tRNA genes account for a significant proportion of reported disease-causing mutations [54]. The majority of maternally inherited genetic diseases are connected to point mutations in mitochondrial tRNA, which disrupt mitochondrial function [55, 56]. Specifically, three distinct diseases are linked to single point mutations in the human mitochondrial $tRNA^{Met}$, namely the T4499C mutation, A4453G mutation and G4450A mutations.

(continued)

**Box 4.3** (continued)

***T4409C mutation and mitochondrial myopathy***

Mitochondrial myopathies are a group of neuromuscular diseases caused by damage to the mitochondria, the energy-producing organelles within cells. These disorders can vary widely in their presentation but commonly include muscle weakness, exercise intolerance, and a range of other systemic manifestations, depending on the severity and specific tissues affected. The T4409C mutation occurs in the mtDNA, which is inherited maternally and encodes key genes essential for mitochondrial function. Specifically, this mutation affects the mitochondrial tRNA molecule, which plays a crucial role in protein synthesis within the mitochondria. Specifically, this mutation involves a substitution of thymine (T) with cytosine (C) at the nucleotide position 4409. This particular mutation changes a conserved uracil (U) to cytosine (C) at a critical junction between the acceptor stem and the D-stem of the mitochondrial tRNA. The acceptor stem and the D-stem are important structural components of tRNA, responsible for proper tRNA folding and function.

The acceptor stem is where the amino acid is attached, and the D-stem plays a role in the correct positioning of the tRNA on the ribosome during translation. By altering the structure of the tRNA, the T4409C mutation may impair its ability to correctly align with mRNA during protein synthesis, leading to incorrect or incomplete proteins being formed. These dysfunctional proteins can significantly impair the electron transport chain, crucial for energy production in cells. Since mitochondria generate energy through oxidative phosphorylation, and any impairment in this process can reduce the energy available to cells, particularly affecting high-energy-demand tissues like muscles and the nervous system. Clinically, patients with mitochondrial myopathy often exhibit muscle weakness and myalgia, which are the direct result of impaired muscle fiber energy metabolism. The lack of adequate energy production leads to muscle fiber degeneration and atrophy. Exercise intolerance is another hallmark of mitochondrial myopathies, where individuals experience rapid fatigue and discomfort upon physical exertion. This is due to the inability of their muscles to generate sufficient energy under stress. Current diagnosis typically involves a combination of clinical evaluation, muscle biopsy, and genetic testing to identify specific mutations like T4409C. Treatment remains supportive and symptomatic, focusing on managing symptoms and improving quality of life. This may include physical therapy, nutritional support, and in some cases, supplements like coenzyme Q10, which may help improve mitochondrial function [57].

***A4435G mutation and Leber's Hereditary Optic Neuropathy (LHON)***

Leber's Hereditary Optic Neuropathy (LHON) is a mitochondrial genetic disorder primarily affecting the optic nerves, leading to sudden central vision loss. This condition is usually caused by mutations in the mtDNA that affect mitochondrial function in the retinal ganglion cells and other tissues. The

(continued)

**Box 4.3** (continued)
A4435G mutation is one of the less common mutations associated with LHON and offers an illustrative example of how mitochondrial dysfunction can result in specific clinical manifestations. LHON typically presents in young adults, with males being more frequently affected than females. The primary symptom is rapid, painless, and usually bilateral loss of central vision, which can lead to legal blindness. The disease is characterized by degeneration of retinal ganglion cells and subsequent optic nerve atrophy. The A4435G mutation refers to the substitution of adenine (A) with guanine (G) at the nucleotide position 4435 in the mitochondrial genome. This mutation occurs in a gene encoding a mitochondrial tRNA, specifically $tRNA^{Met}$, which is crucial for the initiation and elongation phases of mitochondrial protein synthesis, which are essential for the respiratory chain function. A4435G mutation typically disrupts the normal structure and functionality of the tRNA molecule. This disruption can lead to defective mitochondrial protein synthesis, impacting the components of the respiratory chain. Inefficient respiratory chain function reduces ATP production and increases the production of reactive oxygen species (ROS), which can damage cellular structures, including mitochondrial DNA itself, which then results in the pathophysiology in LHON where the retinal ganglion cells, which are highly energy-dependent, suffer from the decreased efficiency of ATP production due to defective respiratory chain complexes. This energy deficit impairs the ability of these cells to maintain normal function and structural integrity. In serious cases, this A4435G mutation, by compromising mitochondrial function, leads to the loss of retinal ganglion cells through apoptosis, resulting in optic nerve atrophy and the clinical presentation of vision loss. Currently, diagnosis is confirmed through genetic testing, where mtDNA is analyzed for known pathogenic mutations associated with LHON, including the A4435G mutation. However, there is currently no cure for LHON. Management focuses on genetic counseling and supportive measures. Idebenone, a synthetic analog of coenzyme Q10, has shown some promise in treating LHON by potentially enhancing mitochondrial function and reducing oxidative stress. Ongoing research aims to develop gene therapy and other molecular approaches to directly address the underlying mitochondrial dysfunction [58].

***G4450A mutation and splenic lymphoma***

Splenic lymphoma, specifically, is a type of lymphoma that predominantly affects the spleen, often involving other parts of the lymphatic system as well. It is characterized by the presence of malignant lymphocytes in the spleen. The symptoms can include splenomegaly (enlarged spleen), fatigue, weight loss, and night sweats among others. The G4450A mutation involves the substitution of guanine (G) with adenine (A) at the nucleotide position 4450 of the mitochondrial genome. This mutation occurs in the gene encoding

(continued)

**Box 4.3** (continued)
mitochondrial $tRNA^{Gly}$ (glycine), which is crucial for the incorporation of the amino acid glycine during the translation of mitochondrial proteins. Glycine is a significant component in various critical metabolic pathways, including the synthesis of proteins. Similar to the above two case studies, this mutation can disrupt mitochondrial protein synthesis, leading to mitochondrial dysfunction. This dysfunction can manifest as a failure in the respiratory chain, which is critical for ATP production and the regulation of apoptosis, ultimately leading to increased production of ROS, causing further oxidative damage to cellular components, potentially leading to mutagenesis and the promotion of a tumorigenic environment. Although the exact mechanism by which the G4450A mutation contributes to the development of splenic lymphoma is not well understood, it was hypothesized that mitochondrial dysfunction could influence the behavior of lymphocytes, leading to uncontrolled proliferation and the avoidance of apoptosis, typical features of cancer cells. Current diagnosis of splenic lymphoma included imaging studies, blood tests, and possibly a biopsy of the spleen or lymph nodes. One particular test that could be useful is the genetic testing, which includes a screening for G4450A mutations in their mtDNA, particularly in cases where a hereditary predisposition or mitochondrial disease is suspected. Current treatment for splenic lymphoma often involves chemotherapy, radiation therapy, and in some cases, surgical removal of the spleen. Research into targeted therapies, including those that might address specific mitochondrial dysfunctions, is ongoing and may provide new avenues for treatment in the future [59]

Mitochondrial tRNAs are essential for the proper synthesis of mitochondrial proteins, which are crucial components of the oxidative phosphorylation pathway. Disruptions in this pathway due to tRNA mutations can lead to a spectrum of diseases, reflecting the diverse roles of mitochondria in different tissues. For instance, mutations like T4409C disrupt muscle function by impairing energy production, crucial for muscle fiber endurance and repair. Similarly, the A4435G mutation affects the optic nerves, where high energy demands make retinal cells particularly vulnerable to mitochondrial defects. Lastly, the G4450A mutation illustrates the potential for mitochondrial dysfunction to alter immune cell proliferation and survival, contributing to the development of conditions such as splenic lymphoma. These examples collectively demonstrate that mitochondrial tRNA mutations can lead to severe consequences, including compromised muscle function, loss of vision, and aberrant immune cell behavior. They also highlight the need for further research into targeted therapies that can address these specific mitochondrial dysfunctions. Understanding the precise mechanisms by which these mutations alter cellular function will be key to developing effective treatments that can mitigate the severe impacts of these mitochondrial disorders.

### 4.5.4 tRNA Role in Protein Synthesis

The primary role of tRNAs in conventional protein synthesis is well-established. This role involves their initial interaction with cognate aminoacyl-tRNA synthetases (aaRSs), which attach a specific amino acid to its corresponding tRNA [60, 61]. Aminoacyl-tRNA synthetases (aaRSs) are abundant in mammalian cells. Typically, the majority of tRNAs are associated with either these synthetases or are integrated into complexes involving the translation initiation and elongation factors as well as ribosomes. These interactions are crucial for the efficient and accurate synthesis of proteins, ensuring that tRNAs are properly charged with their corresponding amino acids and delivered to the ribosome to be added into the growing polypeptide chain. Furthermore, in humans, in human cells, the accuracy of protein synthesis is crucial and heavily reliant on the correct charging of tRNA molecules by aminoacyl-tRNA synthetases. These enzymes are responsible for attaching the correct amino acid to its corresponding tRNA, a process crucial for the faithful translation of mRNA into protein. However, errors can occur, leading to tRNAs being charged with incorrect amino acids. To mitigate such errors and ensure the fidelity of protein synthesis, human cells utilize additional mechanisms involving trans-acting editing factors like AlaXp [62] and ProXp-Ala [63]. AlaXp and ProXp-Ala are specific proteins that play roles in the error-correction process when the standard editing functions of aminoacyl-tRNA synthetases are insufficient. For instance, if an aminoacyl-tRNA synthetase mistakenly attaches the wrong amino acid to a tRNA, these editing factors can recognize and correct the mischarged tRNA. AlaXp, for example, primarily acts on tRNAs that have been incorrectly charged with alanine, while ProXp-Ala is involved with errors involving proline and alanine. These factors associate with the mischarged tRNAs and help prevent the incorporation of incorrect amino acids into the growing polypeptide chain, thus preventing potential malfunctioning proteins that could disrupt cellular function.

The involvement of human aaRSs significantly contributes to the complexity within the tRNA world, particularly through their roles in non-ribosomal functions such as transcriptional regulation within the nucleus, signaling mediated by extracellular receptors, mTOR signaling, and the regulation of mRNA accessibility through the recruitment of various binding partners [64–70]. Corresponding to these diverse functions, there are numerous splice variants of human aaRSs, exhibiting a broad spectrum of activities [71]. Although many of these variants are catalytically inactive, either partially or completely lacking the active site, several retain domains that are able to interact with tRNA, including anticodon-binding domains [72, 73] and pan-specific tRNA-binding domains. These domains bind to the outer corner of the elbow in the L-shaped tRNA structure, relying on structural interactions rather than sequence-specific ones. They act like tRNA chaperones, stabilizing the L-shaped conformation. These domains are considered functional analogs of the chaperone-like protein Trbp111 [74–76].

## 4.6 Alteration in tRNA Functions Is Linked to Human Diseases

Alterations in tRNA functionality, beyond mere fluctuations in tRNA levels, can also contribute to disease in humans. These functional disruptions include inaccuracies during amino acid charging and deficiencies in post-transcriptional modifications, both of which are critical for accurate recognition of cognate codons.

### *4.6.1 Changes in Amino Acid Charge*

To be functionally active, a tRNA must be correctly charged with the amino acid that aligns with its anticodon sequence, a process mediated by aminoacyl-tRNA synthetase (aaRS) enzymes [6]. Faults in this charging mechanism can disrupt protein synthesis significantly and are linked to a variety of human disorders. Mutations affecting at least 17 different mitochondrial aaRS enzymes have been documented, with most of these genetic alterations causing symptoms that primarily impact the central nervous system (reviewed in [77]).

The first mutation associated with disease in a cytosolic aminoacyl-tRNA synthetase (aaRS) was identified in patients with Charcot–Marie–Tooth disease type 2D. This hereditary neuropathy, which leads to gradual motor and sensory neuron degeneration, manifests as muscle weakness and atrophy in both upper and lower limbs. Mutations in five distinct cytoplasmic glycyl-tRNA synthetases (GARSs) have been linked to this disorder [78], with emerging insights into the molecular mechanisms underlying the pathology. Although the molecular details are still debated, research increasingly suggests that these GARS mutations induce a gain-of-function mechanism that disrupts translation. The mutated GARS appears to abnormally bind and sequester tRNAGly, depleting its availability and causing ribosomes to stall at glycine codons. This stalling trigger persistent activation of the integrated stress response (ISR) through the GCN2 kinase. Evidence from animal models indicates that suppressing the ISR—either by pharmacologically inhibiting GCN2 or by increasing tRNAGly expression—can mitigate the neuropathy symptoms [79, 80]. While it remains uncertain whether this chronic ISR activation is a common feature across disease-related mutations in other aaRSs, dysregulated aaRS function is now recognized as a significant factor in various tRNA-associated diseases.

### *4.6.2 Changes in tRNA Chemical Structure*

tRNA molecules undergo extensive chemical modifications, affecting approximately 10–20% of their nucleotides. These modifications are crucial for maintaining tRNA stability, structure, folding, translational accuracy, and fine-tuning of the translation process [81–83]. In humans, tRNA modifications vary by tissue and cell type, adapting dynamically to changes in cellular conditions and environmental stresses [84, 85]. Cytoplasmic tRNAs carry up to 39 distinct types of modifications, while mitochondrial tRNAs (mt-tRNAs) have around 18 types. The enzymes responsible for these modifications are frequently altered in various diseases, a topic that has been thoroughly reviewed [86]. The field examining RNA chemical modifications is known as epitranscriptomics, while diseases resulting from dysfunctional tRNA modifications are termed tRNA modopathies. Both research areas focus on deciphering the molecular roles of specific tRNA modifications.

Most tRNA modopathies are associated with loss-of-function mechanisms due to mutations or deletions in the modifying enzymes. Although the landscape of tRNA modifications and their related enzymes is not yet fully mapped, disruptions in these modifications affect translation in several ways. These include: modifications in the anticodon loop that either restrict or expand codon recognition; alterations in the tRNA body that impact its folding and stability; and modifications that affect the tRNA's aminoacylation capacity. For instance, abnormal modifications at the anticodon's wobble position (position 34) can alter codon recognition, with the absence of 5-methoxycarbonylmethyluridine (mcm5U) or 5-methoxycarbonylmethyl-2-thiouridine (mcm5s2U) leading to mistranslation, which triggers the unfolded protein and heat shock responses [87, 88]. Conversely, an adenosine-to-inosine modification at position 34 (A34 to I34) enables thc tRNA to read multiple codons with U, C, or A in the third position [89, 90]. The essential nature of I34 for accurate protein synthesis is underscored by the fact that its absence is lethal. Additionally, there is evidence that modification of position 37, such as N6-threonylcarbamoyladenosine (t6A), is critical for proper aminoacylation in human mt-tRNALys, as the lack of t6A at this position drastically reduces aminoacylation efficiency [91].

## 4.7 Regulatory Roles of tRNAs in Human Diseases

Proteins involved in tRNA biogenesis and function present promising therapeutic targets for a variety of human diseases. Among these, aminoacyl-tRNA synthetases (aaRSs) are prominent due to their dual role as potential drug targets, not only in human cells but also against pathogens that rely on distinct aaRSs [92]. Additionally, substantial research is focusing on developing small-molecule inhibitors for RNA-modifying enzymes, including those that act on tRNA modifications [93]. Another exciting therapeutic avenue is mitochondrial genome editing, which is emerging as a potential treatment for mitochondrial disorders [94]. Direct modulation of tRNA molecules themselves offers a unique, yet underexplored, therapeutic pathway. This approach may involve three main strategies: restoring functional tRNA in patients with loss-of-function tRNA-related diseases, downregulating specific tRNAs that are overexpressed in conditions such as cancer (referred to as onco-tRNAs), or expressing engineered tRNAs with modified anticodons to correct premature stop codons. A notable preclinical example of the first strategy is the alleviation of

Charcot–Marie–Tooth disease symptoms in animal models with mutant GARS by overexpressing a functional transgenic tRNAGly [79]. The second strategy targets pathogenic tRNAs, such as the oncogenic tRNA-Arg (anticodon TCT) 4–1, identified as a major player in METTL1-driven tumorigenesis [95]. Therapeutic options for targeting tRNA-Arg (anticodon TCT) 4–1 in tumors include small molecules specifically designed to inhibit this tRNA, RNA interference, or antisense oligonucleotides to disrupt its function [96, 97].

### 4.7.1 Cancer Metastasis

Two isoacceptors, $tRNA^{Glu}$/UUC and $tRNA^{Arg}$/CCG, are observed to be significantly upregulated in breast cancer, and correlates with metastasis [98]. As a result, mRNA transcripts containing codons complementary to these tRNAs are more competitive in securing ribosomes for translation. This competitive advantage shifts the proteomic balance from a state of homeostasis toward increased production of proteins that drive the metastatic pathway. Additionally, the enhanced translation associated with these mRNAs not only boosts protein synthesis but also increases the stability of tumor-promoting mRNAs—since translation efficiency often correlates with reduced mRNA decay [99, 100]. Other research has revealed a link between the upregulation of $tRNA^{Glu}$/UUC and metastasis. This association is particularly evident through the increased expression of the RNA exosome component EXOSC2 and the guanine nucleotide exchange factor GRIPAP1. The elevated levels of $tRNA^{Glu}$/UUC appear to contribute to the enhanced expression of these proteins, which play critical roles in RNA processing and cellular signaling. This connection was not previously recognized in the context of metastatic pathways.

### 4.7.2 Neuronal Homeostasis

A particular isodecoder from the $tRNA^{Arg}$/UCU family, which reads AGA codons, has demonstrated a crucial role in maintaining cellular homeostasis and preventing neuronal death. This isodecoder is uniquely expressed in the CNS, representing the first documented instance of tissue-specific expression of a tRNA species in a vertebrate [101]. In studies with mutant mice lacking this specific isodecoder, with a mutation in the GTP-binding protein 2 (GTPBP2), which is known for rescuing stalled ribosomes, these mice showed translation impairments and significant ribosome pausing at AGA codons, which resulted in neurodegeneration. Remarkably, reintroducing the neuron-specific $tRNA^{Arg}$/UCU isodecoder in these mice alleviated the neurodegenerative conditions. Further investigations revealed that in response to the early signs of neurodegeneration, the activating transcription factor 4 (ATF4)-driven integrated stress response was activated. This activation was facilitated by the phosphorylation of eukaryotic translation initiation factor 2 subunit 1 (eIF2$\alpha$) by the GCN2 kinase [102]. This cascade of molecular events underscores the critical balance within cellular homeostasis, which is highly dependent on the precise levels and functionality of specific tRNA species. The delicate equilibrium maintained by these molecular components illustrates how subtle disruptions in tRNA expression or function can precipitate significant cellular consequences, particularly in sensitive environments such as neuronal tissues.

## 4.8 Conclusion

The remarkable influence of tRNAs on biological processes beyond translation has highlighted numerous unresolved issues. Key questions include the underlying mechanisms through which tiRNAs and tfRNAs contribute to either the suppression or promotion of tumorigenesis. It is also intriguing why sex hormone-sensitive cancers specifically target certain charged tRNAs for cleavage by angiogenin. Given the close relationship between tRNA synthetases, their splice variants, and tRNA biology, it is likely that these proteins may play a crucial role in interactions involving tRNA fragments. The role of tfRNA associations in the pathology of various diseases, from a wide range of cancers to ischemic conditions, remains largely unexplored and ripe for further research. Additionally, the recent discovery of tfRNAs in extracellular vesicles, suggesting their role in intercellular communication, introduces a new layer of complexity. The purpose and vast diversity of tRNA species continue to be a profound mystery. Drawing parallels from modern artificial intelligence systems may provide valuable insights into this complexity. The array of tRNA modifications and fragments might have emerged from a random evolutionary process, creating a vast pool of possibilities for neutral drift along with positive and negative selection to influence, potentially enhancing cellular fitness without necessarily achieving a theoretical "optimum"—a concept akin to strategies in AI gaming. This perspective suggests that the intricate biology of tRNAs began in a simpler state, with complexity accumulating through evolutionary processes. Unlike AI scenarios that start with predefined complexity, the evolution of tRNAs illustrates a gradual build-up of complexity. Ongoing research into the tRNA realm promises to deepen our understanding of life's fundamental principles and could illuminate how the complexity of life as we know it has evolved from simpler origins.

## References

1. Zheng G, Qin Y, Clark WC, Dai Q, Yi C, He C, et al. Efficient and quantitative high-throughput tRNA sequencing. Nat Methods. 2015;12(9):835–7.
2. Cozen AE, Quartley E, Holmes AD, Hrabeta-Robinson E, Phizicky EM, Lowe TM. ARM-seq: AlkB-facilitated RNA methylation sequencing reveals a complex landscape of modified tRNA fragments. Nat Methods. 2015;12(9):879–84.

3. Arimbasseri AG, Blewett NH, Iben JR, Lamichhane TN, Cherkasova V, Hafner M, et al. RNA polymerase III output is functionally linked to tRNA dimethyl-G26 modification. PLoS Genet. 2015;11(12):e1005671.
4. Rich A, RajBhandary UL. Transfer RNA: molecular structure, sequence, and properties. Annu Rev Biochem. 1976;45:805–60.
5. Kim Y, Cho S, Kim JC, Park HS. tRNA engineering strategies for genetic code expansion. Front Genet. 2024;15:1373250.
6. Iben JR, Maraia RJ. tRNA gene copy number variation in humans. Gene. 2014;536(2):376–84.
7. Parisien M, Wang X, Pan T. Diversity of human tRNA genes from the 1000-genomes project. RNA Biol. 2013;10(12):1853–67.
8. Abe T, Inokuchi H, Yamada Y, Muto A, Iwasaki Y, Ikemura T. tRNADB-CE: tRNA gene database well-timed in the era of big sequence data. Front Genet. 2014;5:114.
9. Goodenbour JM, Pan T. Diversity of tRNA genes in eukaryotes. Nucleic Acids Res. 2006;34(21):6137–46.
10. Francklyn C, Schimmel P. Aminoacylation of RNA minihelices with alanine. Nature. 1989;337(6206):478–81.
11. Geslain R, Pan T. Functional analysis of human tRNA isodecoders. J Mol Biol. 2010;396(3):821–31.
12. Dittmar KA, Goodenbour JM, Pan T. Tissue-specific differences in human transfer RNA expression. PLoS Genet. 2006;2(12):e221.
13. Burgess RW, Storkebaum E. tRNA dysregulation in neurodevelopmental and neurodegenerative diseases. Annu Rev Cell Dev Biol. 2023;39:223–52.
14. Gao W, Gallardo-Dodd CJ, Kutter C. Cell type-specific analysis by single-cell profiling identifies a stable mammalian tRNA-mRNA interface and increased translation efficiency in neurons. Genome Res. 2022;32(1):97–110.
15. Avcilar-Kucukgoze I, Kashina A. Hijacking tRNAs from translation: regulatory functions of tRNAs in mammalian cell physiology. Front Mol Biosci. 2020;7:610617.
16. Su Z, Wilson B, Kumar P, Dutta A. Noncanonical roles of tRNAs: tRNA fragments and beyond. Annu Rev Genet. 2020;54:47–69.
17. Torres AG, Reina O, Stephan-Otto Attolini C, Ribas de Pouplana L. Differential expression of human tRNA genes drives the abundance of tRNA-derived fragments. Proc Natl Acad Sci USA. 2019;116(17):8451–6.
18. Kim HK, Fuchs G, Wang S, Wei W, Zhang Y, Park H, et al. A transfer-RNA-derived small RNA regulates ribosome biogenesis. Nature. 2017;552(7683):57–62.
19. Hasler D, Lehmann G, Murakawa Y, Klironomos F, Jakob L, Grässer FA, et al. The lupus autoantigen La prevents Mis-channeling of tRNA fragments into the human MicroRNA pathway. Mol Cell. 2016;63(1):110–24.
20. Babiarz JE, Ruby JG, Wang Y, Bartel DP, Blelloch R. Mouse ES cells express endogenous shRNAs, siRNAs, and other microprocessor-independent, Dicer-dependent small RNAs. Genes Dev. 2008;22(20):2773–85.
21. Shah D, Comba A, Faisal SM, Kadiyala P, Baker GJ, Alghamri MS, et al. A novel miR1983-TLR7-IFNβ circuit licenses NK cells to kill glioma cells, and is under the control of galectin-1. Onco Targets Ther. 2021;10(1):1939601.
22. Chalmers JA, Dalvi PS, Loganathan N, McIlwraith EK, Wellhauser L, Nazarians-Armavil A, et al. Hypothalamic miR-1983 targets insulin receptor β and the insulin-mediated miR-1983 increase is blocked by metformin. Endocrinology. 2022;163(1):bqab241.
23. Schmall A, Al-Tamari HM, Herold S, Kampschulte M, Weigert A, Wietelmann A, et al. Macrophage and cancer cell cross-talk via CCR2 and CX3CR1 is a fundamental mechanism driving lung cancer. Am J Respir Crit Care Med. 2015;191(4):437–47.
24. Oler AJ, Alla RK, Roberts DN, Wong A, Hollenhorst PC, Chandler KJ, et al. Human RNA polymerase III transcriptomes and relationships to pol II promoter chromatin and enhancer-binding factors. Nat Struct Mol Biol. 2010;17(5):620–8.

25. Barski A, Chepelev I, Liko D, Cuddapah S, Fleming AB, Birch J, et al. Pol II and its associated epigenetic marks are present at pol III-transcribed noncoding RNA genes. Nat Struct Mol Biol. 2010;17(5):629–34.
26. Raha D, Wang Z, Moqtaderi Z, Wu L, Zhong G, Gerstein M, et al. Close association of RNA polymerase II and many transcription factors with pol III genes. Proc Natl Acad Sci USA. 2010;107(8):3639–44.
27. Yague-Sanz C, Migeot V, Larochelle M, Bachand F, Wéry M, Morillon A, et al. Chromatin remodeling by pol II primes efficient pol III transcription. Nat Commun. 2023;14(1):3587.
28. Gerber A, Ito K, Chu CS, Roeder RG. Gene-specific control of tRNA expression by RNA polymerase II. Mol Cell. 2020;78(4):765–78.e7.
29. Yang J, Li W, Wang D, Li Y. Single-atom materials: small structures determine macroproperties. Small Struct. 2021;2(2):2000051.
30. Sun Z, Tan J, Zhao M, Peng Q, Zhou M, Zuo S, et al. Integrated genomic analysis reveals regulatory pathways and dynamic landscapes of the tRNA transcriptome. Sci Rep. 2021;11(1):5226.
31. Kapur M, Molumby MJ, Guzman C, Heinz S, Ackerman SL. Cell-type-specific expression of tRNAs in the brain regulates cellular homeostasis. Neuron. 2024;112(9):1397–415 e6.
32. Agris PF, Vendeix FA, Graham WD. tRNA's wobble decoding of the genome: 40 years of modification. J Mol Biol. 2007;366(1):1–13.
33. Maehigashi T, Dunkle JA, Miles SJ, Dunham CM. Structural insights into +1 frameshifting promoted by expanded or modification-deficient anticodon stem loops. Proc Natl Acad Sci USA. 2014;111(35):12740–5.
34. Kirchner S, Ignatova Z. Emerging roles of tRNA in adaptive translation, signalling dynamics and disease. Nat Rev Genet. 2015;16(2):98–112.
35. Suzuki T, Suzuki T. A complete landscape of post-transcriptional modifications in mammalian mitochondrial tRNAs. Nucleic Acids Res. 2014;42(11):7346–57.
36. Linder B, Grozhik AV, Olarerin-George AO, Meydan C, Mason CE, Jaffrey SR. Single-nucleotide-resolution mapping of m6A and m6Am throughout the transcriptome. Nat Methods. 2015;12(8):767–72.
37. Cai WM, Chionh YH, Hia F, Gu C, Kellner S, McBee ME, et al. A platform for discovery and quantification of modified Ribonucleosides in RNA: application to stress-induced reprogramming of tRNA modifications. Methods Enzymol. 2015;560:29–71.
38. Chionh YH, McBee M, Babu IR, Hia F, Lin W, Zhao W, et al. tRNA-mediated codon-biased translation in mycobacterial hypoxic persistence. Nat Commun. 2016;7:13302.
39. Leistikow RL, Morton RA, Bartek IL, Frimpong I, Wagner K, Voskuil MI. The mycobacterium tuberculosis DosR regulon assists in metabolic homeostasis and enables rapid recovery from nonrespiring dormancy. J Bacteriol. 2010;192(6):1662–70.
40. Majumdar SD, Vashist A, Dhingra S, Gupta R, Singh A, Challu VK, et al. Appropriate DevR (DosR)-mediated signaling determines transcriptional response, hypoxic viability and virulence of mycobacterium tuberculosis. PLoS One. 2012;7(4):e35847.
41. Selvaraj S, Sambandam V, Sardar D, Anishetty S. In silico analysis of DosR regulon proteins of mycobacterium tuberculosis. Gene. 2012;506(1):233–41.
42. Torres AG, Batlle E, Ribas de Pouplana L. Role of tRNA modifications in human diseases. Trends Mol Med. 2014;20(6):306–14.
43. Jurga S, Erdmann VA, Barciszewski J. Modified nucleic acids in biology and medicine. Cham: Springer; 2016.
44. Lodish HF, Baltimore D, Rintoul D, Darnell JE. Molecular cell biology: student companion for molecular cell biology. Scientific American Books; 1990.
45. Novoa EM, Ribas de Pouplana L. Speeding with control: codon usage, tRNAs, and ribosomes. Trends Genet. 2012;28(11):574–81.
46. Rudolph KL, Schmitt BM, Villar D, White RJ, Marioni JC, Kutter C, et al. Codon-driven translational efficiency is stable across diverse mammalian cell states. PLoS Genet. 2016;12(5):e1006024.

47. Duechler M, Leszczyńska G, Sochacka E, Nawrot B. Nucleoside modifications in the regulation of gene expression: focus on tRNA. Cell Mol Life Sci. 2016;73(16):3075–95.
48. Schimmel P, Giegé R, Moras D, Yokoyama S. An operational RNA code for amino acids and possible relationship to genetic code. Proc Natl Acad Sci USA. 1993;90(19):8763–8.
49. Agris PF. The importance of being modified: an unrealized code to RNA structure and function. RNA. 2015;21(4):552–4.
50. Anderson S, Bankier AT, Barrell BG, de Bruijn MH, Coulson AR, Drouin J, et al. Sequence and organization of the human mitochondrial genome. Nature. 1981;290(5806):457–65.
51. Attardi G. Animal mitochondrial DNA: an extreme example of genetic economy. Int Rev Cytol. 1985;93:93–145.
52. Anderson S, de Bruijn MH, Coulson AR, Eperon IC, Sanger F, Young IG. Complete sequence of bovine mitochondrial DNA. Conserved features of the mammalian mitochondrial genome. J Mol Biol. 1982;156(4):683–717.
53. Salinas-Giege T, Giege R, Giege P. tRNA biology in mitochondria. Int J Mol Sci. 2015;16(3):4518–59.
54. Lott MT, Leipzig JN, Derbeneva O, Xie HM, Chalkia D, Sarmady M, et al. mtDNA variation and analysis using Mitomap and Mitomaster. Curr Protoc Bioinformatics. 2013;44(123):1.23.1–6.
55. Alston CL, Rocha MC, Lax NZ, Turnbull DM, Taylor RW. The genetics and pathology of mitochondrial disease. J Pathol. 2017;241(2):236–50.
56. de Almeida RA, Fraczek MG, Parker S, Delneri D, O'Keefe RT. Non-coding RNAs and disease: the classical ncRNAs make a comeback. Biochem Soc Trans. 2016;44(4):1073–8.
57. Vissing J, Salamon MB, Arlien-Søborg P, Nørby S, Manta P, DiMauro S, et al. A new mitochondrial tRNA(met) gene mutation in a patient with dystrophic muscle and exercise intolerance. Neurology. 1998;50(6):1875–8.
58. Qu J, Li R, Zhou X, Tong Y, Lu F, Qian Y, et al. The novel A4435G mutation in the mitochondrial tRNAMet may modulate the phenotypic expression of the LHON-associated ND4 G11778A mutation. Invest Ophthalmol Vis Sci. 2006;47(2):475–83.
59. Lombès A, Bories D, Girodon E, Frachon P, Ngo MM, Breton-Gorius J, et al. The first pathogenic mitochondrial methionine tRNA point mutation is discovered in splenic lymphoma. Hum Mutat. 1998;Suppl 1:S175–83.
60. Ibba M, Francklyn C, Cusack S. The aminoacyl-tRNA synthetases. Landes Bioscience; 2005.
61. Gordon R. Encyclopedia of biophysics 2013.
62. Beebe K, Mock M, Merriman E, Schimmel P. Distinct domains of tRNA synthetase recognize the same base pair. Nature. 2008;451(7174):90–3.
63. Ruan LL, Zhou XL, Tan M, Wang ED. Human cytoplasmic ProX edits mischarged tRNAPro with amino acid but not tRNA specificity. Biochem J. 2013;450(1):243–52.
64. Yao P, Fox PL. Aminoacyl-tRNA synthetases in medicine and disease. EMBO Mol Med. 2013;5(3):332–43.
65. Yao P, Poruri K, Martinis SA, Fox PL. Non-catalytic regulation of gene expression by aminoacyl-tRNA synthetases. Top Curr Chem. 2014;344:167–87.
66. Arif A, Terenzi F, Potdar AA, Jia J, Sacks J, China A, et al. EPRS is a critical mTORC1-S6K1 effector that influences adiposity in mice. Nature. 2017;542(7641):357–61.
67. Ofir-Birin Y, Fang P, Bennett SP, Zhang HM, Wang J, Rachmin I, et al. Structural switch of lysyl-tRNA synthetase between translation and transcription. Mol Cell. 2013;49(1):30–42.
68. Ahn YH, Park S, Choi JJ, Park BK, Rhee KH, Kang E, et al. Secreted tryptophanyl-tRNA synthetase as a primary defence system against infection. Nat Microbiol. 2016;2:16191.
69. Motzik A, Nechushtan H, Foo SY, Razin E. Non-canonical roles of lysyl-tRNA synthetase in health and disease. Trends Mol Med. 2013;19(12):726–31.
70. Guo M, Schimmel P. Essential nontranslational functions of tRNA synthetases. Nat Chem Biol. 2013;9(3):145–53.
71. Lo WS, Gardiner E, Xu Z, Lau CF, Wang F, Zhou JJ, et al. Human tRNA synthetase catalytic nulls with diverse functions. Science. 2014;345(6194):328–32.

72. Gale AJ, Schimmel P. Isolated RNA binding domain of a class I tRNA synthetase. Biochemistry. 1995;34(27):8896–903.
73. Wright DJ, Martinis SA, Jahn M, Söll D, Schimmel P. Acceptor stem and anticodon RNA hairpin helix interactions with glutamine tRNA synthetase. Biochimie. 1993;75(12):1041–9.
74. Morales AJ, Swairjo MA, Schimmel P. Structure-specific tRNA-binding protein from the extreme thermophile Aquifex aeolicus. EMBO J. 1999;18(12):3475–83.
75. Swairjo MA, Morales AJ, Wang CC, Ortiz AR, Schimmel P. Crystal structure of trbp111: a structure-specific tRNA-binding protein. EMBO J. 2000;19(23):6287–98.
76. Kushiro T, Schimmel P. Trbp111 selectively binds a noncovalently assembled tRNA-like structure. Proc Natl Acad Sci USA. 2002;99(26):16631–5.
77. Sissler M, González-Serrano LE, Westhof E. Recent advances in mitochondrial aminoacyl-tRNA Synthetases and disease. Trends Mol Med. 2017;23(8):693–708.
78. Antonellis A, Ellsworth RE, Sambuughin N, Puls I, Abel A, Lee-Lin SQ, et al. Glycyl tRNA synthetase mutations in Charcot-Marie-tooth disease type 2D and distal spinal muscular atrophy type V. Am J Hum Genet. 2003;72(5):1293–9.
79. Zuko A, Mallik M, Thompson R, Spaulding EL, Wienand AR, Been M, et al. tRNA overexpression rescues peripheral neuropathy caused by mutations in tRNA synthetase. Science (New York, NY). 2021;373(6559):1161–6.
80. Spaulding EL, Hines TJ, Bais P, Tadenev ALD, Schneider R, Jewett D, et al. The integrated stress response contributes to tRNA synthetase-associated peripheral neuropathy. Science (New York, NY). 2021;373(6559):1156–61.
81. Chou HJ, Donnard E, Gustafsson HT, Garber M, Rando OJ. Transcriptome-wide analysis of roles for tRNA modifications in translational regulation. Mol Cell. 2017;68(5):978–92.e4.
82. de Crécy-Lagard V, Boccaletto P, Mangleburg CG, Sharma P, Lowe TM, Leidel SA, et al. Matching tRNA modifications in humans to their known and predicted enzymes. Nucleic Acids Res. 2019;47(5):2143–59.
83. Alexandrov A, Martzen MR, Phizicky EM. Two proteins that form a complex are required for 7-methylguanosine modification of yeast tRNA. RNA (New York, NY). 2002;8(10):1253–66.
84. Hoffmann A, Fallmann J, Vilardo E, Mörl M, Stadler PF, Amman F. Accurate mapping of tRNA reads. Bioinformatics (Oxford, England). 2018;34(7):1116–24.
85. Pinkard O, McFarland S, Sweet T, Coller J. Quantitative tRNA-sequencing uncovers metazoan tissue-specific tRNA regulation. Nat Commun. 2020;11(1):4104.
86. Suzuki T. The expanding world of tRNA modifications and their disease relevance. Nat Rev Mol Cell Biol. 2021;22(6):375–92.
87. Begley U, Dyavaiah M, Patil A, Rooney JP, DiRenzo D, Young CM, et al. Trm9-catalyzed tRNA modifications link translation to the DNA damage response. Mol Cell. 2007;28(5):860–70.
88. Patil A, Chan CT, Dyavaiah M, Rooney JP, Dedon PC, Begley TJ. Translational infidelity-induced protein stress results from a deficiency in Trm9-catalyzed tRNA modifications. RNA Biol. 2012;9(7):990–1001.
89. Murphy FVT, Ramakrishnan V. Structure of a purine-purine wobble base pair in the decoding center of the ribosome. Nat Struct Mol Biol. 2004;11(12):1251–2.
90. Crick FH. Codon--anticodon pairing: the wobble hypothesis. J Mol Biol. 1966;19(2):548–55.
91. Lin H, Miyauchi K, Harada T, Okita R, Takeshita E, Komaki H, et al. CO(2)-sensitive tRNA modification associated with human mitochondrial disease. Nat Commun. 2018;9(1):1875.
92. Kwon NH, Fox PL, Kim S. Aminoacyl-tRNA synthetases as therapeutic targets. Nat Rev Drug Discov. 2019;18(8):629–50.
93. Yankova E, Blackaby W, Albertella M, Rak J, De Braekeleer E, Tsagkogeorga G, et al. Small-molecule inhibition of METTL3 as a strategy against myeloid leukaemia. Nature. 2021;593(7860):597–601.
94. Mok BY, de Moraes MH, Zeng J, Bosch DE, Kotrys AV, Raguram A, et al. A bacterial cytidine deaminase toxin enables CRISPR-free mitochondrial base editing. Nature. 2020;583(7817):631–7.

95. Orellana EA, Liu Q, Yankova E, Pirouz M, De Braekeleer E, Zhang W, et al. METTL1-mediated m(7)G modification of Arg-TCT tRNA drives oncogenic transformation. Mol Cell. 2021;81(16):3323–38.e14.
96. Lueck JD, Yoon JS, Perales-Puchalt A, Mackey AL, Infield DT, Behlke MA, et al. Engineered transfer RNAs for suppression of premature termination codons. Nat Commun. 2019;10(1):822.
97. Porter JJ, Heil CS, Lueck JD. Therapeutic promise of engineered nonsense suppressor tRNAs. Wiley Interdiscip Rev RNA. 2021;12(4):e1641.
98. Goodarzi H, Nguyen HCB, Zhang S, Dill BD, Molina H, Tavazoie SF. Modulated expression of specific tRNAs drives gene expression and cancer progression. Cell. 2016;165(6):1416–27.
99. Bazzini AA, Del Viso F, Moreno-Mateos MA, Johnstone TG, Vejnar CE, Qin Y, et al. Codon identity regulates mRNA stability and translation efficiency during the maternal-to-zygotic transition. EMBO J. 2016;35(19):2087–103.
100. Presnyak V, Alhusaini N, Chen YH, Martin S, Morris N, Kline N, et al. Codon optimality is a major determinant of mRNA stability. Cell. 2015;160(6):1111–24.
101. Ishimura R, Nagy G, Dotu I, Zhou H, Yang XL, Schimmel P, et al. RNA function. Ribosome stalling induced by mutation of a CNS-specific tRNA causes neurodegeneration. Science. 2014;345(6195):455–9.
102. Ishimura R, Nagy G, Dotu I, Chuang JH, Ackerman SL. Activation of GCN2 kinase by ribosome stalling links translation elongation with translation initiation. elife. 2016;5:5.

# Chapter 5
# Non-coding RNA: Long Non-coding RNA (lncRNA)

**Abstract** This chapter aims to encapsulate recent findings regarding long non-coding RNAs (lncRNAs), highlighting their inherent potential as molecular biomarkers, diagnostics, and therapeutics in the treatment of human diseases. With advancements in sequencing technologies, the exploration of lncRNAs presents a frontier laden with unknowns. Initially, the ongoing refinement and expansion of lncRNAs' definition and categorization reflect their complex and varied nature. Subsequently, diverse methodologies and approaches have been devised to elucidate lncRNA characteristics, encompassing identification of new types, subcellular localization studies, functional analyses, molecular interactions, and bioinformatics assessments. Furthermore, the current basic research has illuminated the operational mechanisms of lncRNAs, which involve intricate networks of interactions with DNA, RNA, and proteins in the cell. Moreover, lncRNAs are indicated to assume critical roles in embryonic development and organ differentiation. Understanding the basics of lncRNA would help us develop more intricate RNA-based medicine for human diseases.

## 5.1 Introduction

### *5.1.1 Definition and Nomenclature of lncRNAs*

Long non-coding RNAs (lncRNAs) are non-coding transcripts with nucleotide counts larger than 200 [1]. This definition distinctly divided ncRNAs into two groups, the larger (>200 nt) and smaller (<200 nt) RNAs, including tiRNAs, siRNAs, miRNAs, snRNAs, snoRNAs, and piRNAs, just to name a few [2]. It also excludes Y RNAs, (~100 nt), which serve as an important scaffolds for RNP complexes [3], vault RNAs (~88–140 nt) which translate extracellular signals into intracellular actions; and some other promoter-associated RNAs along with unconventional small RNAs that result from post-transcriptional processes [4–7]. In the same category as lncRNAs, there are RNAs like 7SK RNA (~330 nt), which plays a role in transcription termination [8, 9] and 7SL RNA (~300 nt), which plays a key

 
P. E. Saw, E. Song, *RNA Therapeutics in Human Diseases*,
https://doi.org/10.1007/978-981-96-3041-7_5

role as a crucial constituent of the signaling pathway that directs proteins from the intracellular region to membranes [10]. Additionally, ancestral elements such as primate Alu RNAs, roughly 280 nucleotides long, also falls into this category [11–13].

Many lncRNAs undergo splicing and polyadenylation, processes typically associated with mRNAs, which has led to them being described as a "mRNA-like" molecule. However, not all lncRNAs possess these features. Some lncRNAs lack polyadenylation or a 7-methylguanosine cap, distinguishing them from traditional mRNAs. Other lncRNAs originate from precursors such as introns and repetitive sequences within the genome. These ncRNAs do not conform to the conventional characteristics of mRNAs and are often referred to as "transcripts of unknown function" (TUFs). This designation underscores the current gap in understanding their roles and functions within cellular processes. The diverse origins and structures of lncRNAs suggest a wide range of potential functions, from regulating gene expression to participating in complex cellular processes. However, the lack of comprehensive functional characterization for many lncRNAs means that their contributions to cellular biology remain an area of active research. Understanding these "transcripts of unknown function" is crucial for uncovering the full spectrum of RNA-mediated regulatory mechanisms in the cell [14–18].

The classification and nomenclature of lncRNAs have been the focus of various initiatives spearheaded by organizations such as the HUGO Gene Nomenclature Committee (HGNC) and the GENCODE consortium. These efforts aim to standardize lncRNA identification and categorization, primarily based on their genomic location and orientation relative to protein-coding genes. By establishing clear criteria and consistent naming conventions, these initiatives facilitate better communication and understanding within the scientific community regarding the diverse roles and characteristics of lncRNAs [19–23]. Associating lncRNAs with adjacent genes has proven insightful, providing context and occasionally hints at their function, such as gene expression regulation often seen with enhancers, though it is important not to automatically attribute enhancer activity to the nearest genes.

## 5.2 Initial Researches on lncRNAs

Initial research predominantly explored long intergenic non-coding RNAs (lincRNAs), which are usually isolated from neighboring protein-coding regions, emphasizing the necessity to differentiate their functions from those of proteins. However, many other lncRNAs intersect with or arise from protein-coding areas or are encoded within introns. Traditional genomic perspectives, which view genomes as linear sequences of distinct protein-coding genes, do not fully account for the complexity revealed in eukaryotic transcription. This complexity is notably illustrated in human and model organisms, where transcription appears as a continuous spectrum [24], characterized by nested genes, interleaved genes, and non-coding sequences that overlap or originate within coding regions [15, 25, 26], complicating genome annotation efforts. In humans and *Drosophila melanogaster*, for example, numerous protein-coding genes have 5′ exons that are utilized during early embryogenesis. These exons are located hundreds of kilobases away from the conventional first exon and span multiple intervening genes [27]. Thus, any nucleotide can represent an exonic, intronic, or intergenic region, contingent on the transcriptional activity within a specific developmental or physiological context [26]. Therefore, unless a lncRNA is antisense to a protein-coding gene, it should be assigned an independent name that reflects its unique characteristics or function, similar to the naming conventions for proteins. Examples of such names include (1) HOX antisense intergenic RNA (HOTAIR), (2) XIST, antisense IGF2R non-protein-coding RNA (AIRN), and (3) auxin-regulated promoter loop (APOLO) [28]. These names should ideally be memorable, with detailed descriptions of exon-intron configurations and genomic locations. In cases lacking biological context, naming should follow the GENCODE system [19]. In light of the difficulty in achieving a more precise classification, it is advisable to continue using the broad term "lncRNA." Interestingly, certain lncRNAs serve dual roles, functioning both as coding and regulatory molecules, and a significant number of cytosolic lncRNAs are known to produce short peptides [29–32]. Moreover, protein-coding regions also give rise to lncRNAs through mechanisms such as alternative splicing [33–35], although we should note that the primary transcript from approximately 17% of human protein-coding genes is actually non-coding [36]. lncRNA genes and mRNA genes alike can generate a variety of functional transcripts, including unspliced, spliced, circular, and intronic RNAs, as well as stable small RNAs derived from these forms [37–39]. Essentially, any lncRNA can have regulatory properties, and any genetic locus has the potential to produce both protein-coding and regulatory lncRNAs (detailed in Table 5.1 below).

## 5.3 LncRNAs Constitute Various RNA Molecules

Classifying lncRNAs presents a significant challenge due to their diverse molecular structures, splicing patterns, and wide range of cellular functions, which arise from their various modes of action. Traditionally, lncRNAs have been categorized based on their genomic locations into five broad types, as summarized in Table 5.1. Nevertheless, as research continues to uncover new varieties of lncRNAs, this traditional framework has proven inadequate for capturing the entire spectrum of these molecules, necessitating a more detailed classification. Researchers are now expanding this classification to detail the spatial relationships between lncRNA genes and adjacent genomic features more comprehensively. However, despite these refinements, the classifications largely adhere to a single criterion based on genomic location. As the discovery of new lncRNA forms progresses, classifications must evolve to incorporate additional criteria such as splicing patterns and functional mechanisms, ensuring a comprehensive framework that captures the full complexity of lncRNA biology (Fig. 5.1).

**Table 5.1** Traditional [40] vs. refined classification [41] of lncRNA

| | Definition, function and mechanism |
|---|---|
| Traditional classification | |
| Sense lncRNAs | A sense lncRNA is a type of long non-coding RNA that is transcribed from the same DNA strand as a nearby coding gene, sharing its transcriptional orientation. This means that the sense lncRNA's sequence overlaps with that of one or more exons of the coding transcript |
| Antisense lncRNAs | Antisense lncRNAs are a class of long non-coding RNAs that are transcribed from the DNA strand opposite to a protein-coding transcript, which means they have sequences complementary to those of a sense or coding mRNA. This antisense transcription allows them to specifically pair with and regulate their corresponding sense RNAs through RNA-RNA interactions. By being transcribed, antisense lncRNAs can interfere with the transcription of the corresponding sense mRNA |
| Bidirectional lncRNAs | Bidirectional lncRNAs are a unique class of lncRNAs that are transcribed from loci where transcription is initiated in both directions, producing two RNA molecules: One on each strand of the DNA. These lncRNAs are typically found near the promoters of protein-coding genes, and their transcription usually occurs concurrently with, or closely related to, the transcription of a neighboring coding gene. Bidirectional lncRNAs are often transcribed from sequences that are very close to the promoters of protein-coding genes |
| Intronic lncRNAs | Intronic lncRNAs are a type of lncRNA transcribed from the intronic regions of a gene—That is, the sequences within a gene that are not translated into protein. These sequences are typically found between exons, which are the coding segments of genes that remain after RNA splicing to form mature mRNA. Intronic lncRNAs are transcribed from the non-coding regions within the introns of a host gene. During the process of mRNA splicing, these intronic sequences are usually excised from the pre-mRNA transcript and degraded |
| Intergenic lncRNAs | Intergenic lncRNAs are transcribed from the regions of the genome located between two genes. These regions, called intergenic spaces, do not contain any overlapping protein-coding sequences, allowing lincRNAs to be distinct entities separate from other types of genes. Intergenic lncRNAs can regulate the expression of neighboring genes. They achieve this through various mechanisms such as modifying the chromatin structure, affecting transcription factor activity, or altering the local epigenetic landscape. This regulation can be in cis, affecting nearby genes, or in trans, influencing the function of distant genes |
| Refined classification | |
| Divergent lncRNAs | Divergent lncRNAs are a category of long non-coding RNAs that are transcribed in the opposite direction to a nearby coding gene. They are often found near the promoters of these genes, sharing the same transcriptional start sites but extending in opposite directions. This bidirectional transcription is a hallmark of divergent lncRNAs. Divergent lncRNAs originate from the same genomic loci as adjacent coding genes but are transcribed in the reverse direction. This setup often results in a tightly coupled regulation with their neighboring genes, influenced by shared regulatory elements like enhancers or repressors |

(continued)

**Table 5.1** (continued)

| | Definition, function and mechanism |
|---|---|
| Convergent lncRNAs | Convergent lncRNAs are a subset of long non-coding RNAs that are transcribed toward each other from opposite strands, usually converging at their 3′ ends near the same genomic location. Unlike divergent lncRNAs, which are oriented head-to-head and share a promoter region, convergent lncRNAs meet tail-to-tail, typically without overlapping promoters. The transcription of convergent lncRNAs involves two RNA polymerases moving toward each other from opposite directions. This setup can lead to complex regulatory scenarios at their point of convergence, possibly affecting the termination of transcription for each transcript |
| Overlapping sense lncRNAs | Overlapping sense lncRNAs share significant sequence overlap with a coding gene, but unlike antisense lncRNAs, they are transcribed in the same direction as the coding transcript. This means that overlapping sense lncRNAs are co-linear with portions of the mRNA transcripts from which proteins are made. These lncRNAs are transcribed from the same DNA strand as an adjacent or overlapping protein-coding gene. Their transcription often occurs simultaneously with the coding gene, which can lead to complex regulatory interactions |
| Overlapping antisense lncRNAs | Overlapping antisense lncRNAs are a specialized class of lncRNAs that are transcribed from the opposite DNA strand relative to a corresponding sense or coding transcript. These lncRNAs overlap in part or entirely with the exons or introns of a protein-coding gene, but because they are transcribed in the reverse direction, they can form complementary base pairs with the mRNA of the overlapping gene. Their transcriptional overlap with coding mRNA transcripts allows for potential RNA-RNA interactions, which can regulate gene expression post-transcriptionally. One of the primary roles of overlapping antisense lncRNAs is the regulation of the expression of their corresponding sense genes |
| Enhancer RNA | Enhancer lncRNAs are transcribed from enhancer regions of the genome. Enhancers are DNA sequences that can increase the transcriptional activity of associated genes, even if they are located far away from the gene's promoter region. Enhancer lncRNAs, therefore, play crucial roles in regulating gene expression by acting through these enhancer elements. Enhancer lncRNAs are produced from the enhancer regions of the DNA, which are key regulatory elements that increase the transcription of specific genes |
| miRNA host gene | LncRNAs can serve as host genes for miRNAs, indicating that the primary transcript of an lncRNA contains sequences that can be processed into miRNAs. This dual functionality adds another layer of complexity to the regulatory roles of lncRNAs in gene expression. These lncRNAs contain embedded miRNA sequences within their primary transcripts. The miRNAs are usually located within the intronic or exonic regions of the lncRNA. Upon transcription, the lncRNA primary transcript can be spliced and processed to release mature miRNAs |

## 5.4 Subcellular Localization of lncRNA

Fluorescence in situ hybridization (FISH) is a powerful technique to determine the subcellular localization of specific lncRNAs. For example, RNA-FISH was employed to study the lncRNA MALAT1, which revealed its abundant presence in the nuclear speckles of interphase cells. Additionally, during mitosis, MALAT1 was

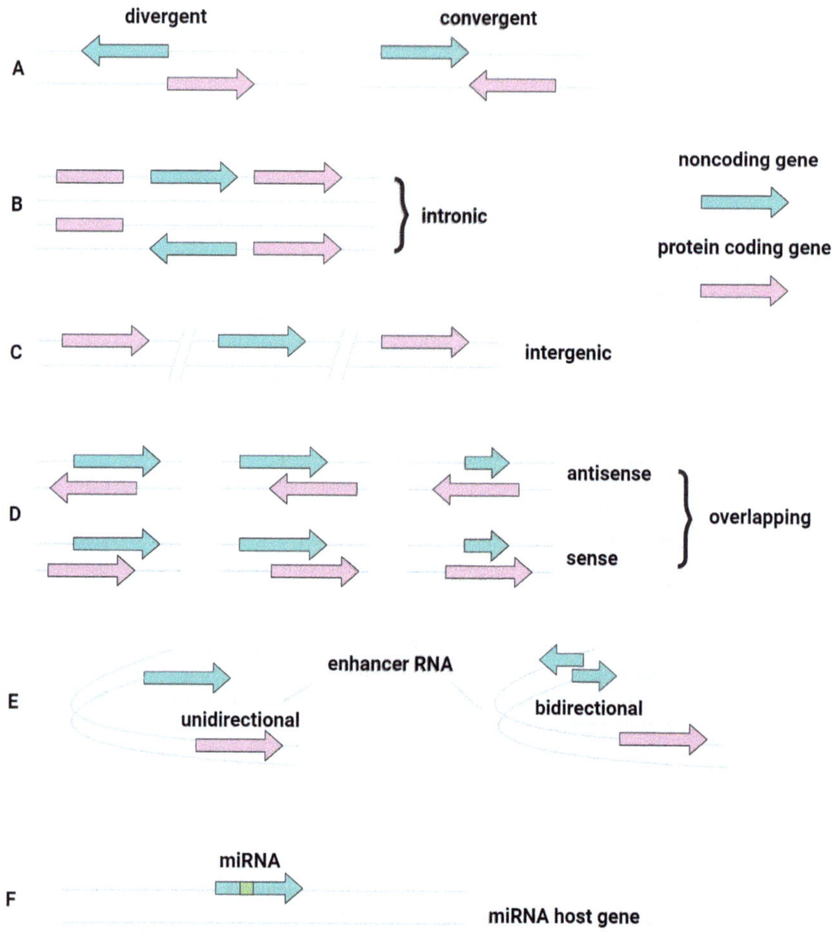


**Fig. 5.1** LncRNAs are categorized based on their location in the genome and their direction of transcription relative to other genes. (**a**) Divergent (pancRNA) and convergent: Divergent lncRNAs, also known as pancRNAs, originate from the opposite strand of a protein-coding gene's promoter region. Convergent lncRNAs are encoded on opposite strands, oriented toward each other. (**b**) Intronic: These lncRNAs are transcribed from within the introns of other genes, functioning independently from the exonic regions that code for proteins. (**c**) Intergenic: Located at least 10 kb away from neighboring genes, these lncRNAs are found in the non-coding regions between the genes of the protein-coding region, often occupying large genomic intervals. (**d**) Overlapping Sense and Overlapping Antisense: Overlapping sense lncRNAs share sequences with other genes on the same strand, while overlapping antisense lncRNAs overlap with genes on the opposite strand, indicating potential regulatory interactions. (**e**) Enhancer RNA: These lncRNAs are transcribed from enhancer regions and can be expressed as either unidirectional or bidirectional transcripts, playing roles in the regulation of gene expression. (**f**) Host of miRNA Gene: These lncRNAs act as primary transcripts for miRNAs, providing RNA sequences from which miRNAs are processed and subsequently involved in post-transcriptional gene regulation [41]. (Figure is adapted from [42] under Creative Commons CC-BY-NC license)

found to concentrate in interchromatin granule clusters. This technique not only provides insights into the spatial distribution of lncRNAs within the cell but also helps in understanding their potential functional roles in various cellular compartments [43]. RNA-FISH has also confirmed the nuclear presence of MIR99HG (MONC) and MIR100HG [44]. Additionally, combining both RNA-FISH and DNA-FISH is useful in confirming the co-localization of an RNA transcript with specific chromatin sequences. This integrated approach allows researchers to simultaneously visualize RNA molecules and their corresponding DNA regions within the cell. By doing so, it provides a more comprehensive understanding of the spatial relationships between lncRNAs and chromatin, which is essential for elucidating their roles in gene regulation and chromatin organization [45]. A more recent technology of fluorescent in situ RNA sequencing (FISSEQ) offers significant advancements over traditional RNA-FISH. FISSEQ involves amplifying cDNA within crosslinked cells and tissue samples, enabling a higher resolution analysis. This technique allows for the identification of a greater number of RNA targets simultaneously, providing a more detailed and comprehensive view of the transcriptome within its native cellular context. By combining the spatial information of RNA-FISH with the sequencing capabilities of RNA-seq, FISSEQ facilitates an in-depth understanding of RNA localization, abundance, and interactions at an unprecedented level of detail [46].

## 5.5 Molecular Interaction of lncRNAs

The functional mechanisms of lncRNAs involve complex molecular interactions among various types of RNA, DNA, and proteins [47–50]. Current technologies used to study these interactions can be broadly categorized into three main groups below.

### *5.5.1 RNA–RNA Interaction (RNA-Centric Methods)*

These techniques focus on identifying and characterizing lncRNAs themselves. Examples include RNA sequencing (RNA-seq) for transcriptome analysis, single-cell RNA sequencing (scRNA-seq) for examining lncRNA expression at the individual cell level, and FISH for visualizing lncRNA localization within cells. RAP-RNA is a variation of the RNA Antisense Purification (RAP) technique specifically designed to detect RNA-RNA interactions. This method allows researchers to differentiate between direct and indirect interactions among RNA transcripts [45]. Cross-linking, ligation, and sequencing of hybrids (CLASH) is a specialized technique developed to capture and analyze direct RNA–RNA interactions within cells. The process begins with UV cross-linking, which induces covalent bonds between interacting RNA molecules, stabilizing their associations and preserving their native interactions within the cellular environment. CLASH is particularly valuable for studying the complex networks of RNA interactions that regulate various cellular processes [51].

### 5.5.2 RNA–DNA Interaction (DNA-Centric Methods)

These approaches investigate how lncRNAs interact with genomic DNA. Chromatin isolation by RNA purification (ChIRP) and capture hybridization analysis of RNA targets (CHART) are used to map the genomic binding sites of lncRNAs [52], thereby elucidating their roles in chromatin organization and gene regulation [45, 53]. RAP uses approximately 120-nucleotide-long antisense RNA probes to identify regions of chromatin that interact with specific RNAs. These antisense probes are specifically designed to be fully complementary to the target RNA. When introduced into a cell extract, they hybridize with their target RNA, forming RNA–RNA duplexes. Once these duplexes are formed, the complexes are isolated using affinity purification methods. The associated chromatin fragments are then identified and analyzed, typically through high-throughput sequencing. RAP enables the precise mapping of RNA-chromatin interactions, revealing how RNAs contribute to the many pathways and regulation of gene expression and chromatin architecture. This method provides valuable insights into the spatial organization of the genome and the role of RNAs in modulating chromatin states and gene activity [54].

### 5.5.3 RNA–Protein Interactions (Protein-Centric Methods)

These methods explore the interactions between lncRNAs and proteins. Two popular methods are RNA immunoprecipitation (RIP) and cross-linking immunoprecipitation (CLIP), which are commonly employed to identify protein partners of lncRNAs and to understand how these interactions influence cellular processes [45]. High-throughput sequencing cross-linking immunoprecipitation (HITS-CLIP/CLIP-Seq) is a powerful technique for studying RNA-protein interactions. This method begins by cross-linking RNA and proteins within cells using ultraviolet (UV) light, which creates covalent bonds between RNA molecules and their associated proteins. After cross-linking, the cells are lysed, and the immunoprecipitation was carried out to obtain RNA-protein complexes using specific antibodies that target the protein of interest. The RNA molecules bound to these proteins are then isolated and sequenced. HITS-CLIP/CLIP-Seq technic provides high-resolution spatiotemporal maps of RNA-protein interaction sites, allowing researchers to identify specific binding sites of RNA-binding proteins across the transcriptome. This information is crucial for understanding the regulatory roles of proteins in RNA processing, stability, transport, and translation. By revealing the precise locations where proteins interact with RNA, this technique offers insights into the mechanisms of post-transcriptional regulation and the functional consequences of these interactions in various cellular contexts [45].

## 5.6 Gain or Loss of Function of lncRNA

To elucidate the function of a novel molecule, traditional methods often involve altering the expression levels of the target through overexpression or knockdown techniques. One study utilized an overexpression plasmid and siRNA/shRNA to, respectively, increase and decrease the levels of the lncRNA HULC. Their experiments demonstrated that overexpressing HULC enhanced lipogenesis in HepG2 and Huh7 cells, whereas knocking down HULC with siRNA reversed the phenomenon [55]. Similarly, Nie et al. employed these methods to demonstrate that reducing ANRIL expression inhibited cell proliferation and induced apoptosis in both cellular and animal models [56]. Furthermore, to achieve stable gene modulation, lentiviral technology has been widely utilized for many years. Recently, CRISPR/Cas9 and NgAgo, more advanced and precise gene-editing technologies have emerged. These innovative tools enable researchers to make targeted modifications to the genome with high accuracy. CRISPR/Cas9, in particular, uses a guide RNA to direct the Cas9 enzyme to specific DNA sequences, allowing for precise cutting and subsequent gene-editing. NgAgo, another promising technology, offers an alternative approach for genome editing, potentially expanding the range of applications and improving the precision of genetic modifications. These advanced methods hold the promise of achieving more accurate gene edits than ever before, despite ongoing debates regarding their application and efficacy. Compared to traditional techniques, these modern tools may offer enhanced precision in gene-editing.

## 5.7 Bioinformatics Analysis

The landscape of lncRNA research databases has expanded considerably alongside advances in lncRNA research. Notable among these resources is lncRNAdb, a prominent database offering comprehensive annotations on functional lncRNAs, now updated to version 2.0 [57]. Another critical repository, NONCODE, provides detailed annotations of lncRNAs, including expressions and functions as predicted by its computational tool, ncFANs, now at version v4 [58]. LNCipedia focuses on the sequences and structural annotations of human lncRNAs [59, 60]. Additionally, the LncRNA disease database catalogues lncRNAs implicated in various diseases [61], while NRED offers expression data for human and mouse lncRNAs derived from microarray experiments [62]. ChIPBase provides a comprehensive platform for mapping lncRNAs and annotating transcriptional regulation, integrating data from high-throughput RNA-seq and ChIP-seq experiments to identify lncRNAs, their expression profiles, and transcription factor binding sites [63]. The fRNAdb aggregates non-coding transcripts from sources like H-invitation, NONCODE, and RNAdb, offering search functionality across various categories of ncRNAs [64]. Starbase provides the most extensive experimental support through CLIP-Seq for regulatory networks involving miRNAs and lncRNAs, RNA-binding proteins and lncRNAs, and competing endogenous RNA (ceRNA) networks, along with predictive tools for lncRNA functions [65]. The bioinformatics analysis segment in lncRNA research is integral, not just as a standalone section but as a pervasive element across various research facets. It offers vital information for investigating new functions of known lncRNAs or enriching the understanding of newly discovered lncRNAs.

## 5.8 Working Mechanisms of lncRNAs

lncRNAs play a pivotal role in the regulation of gene expression through various mechanisms. These mechanisms include chromatin remodeling, histone modification, DNA methylation, and the regulation of transcription either as factors or enhancers. LncRNAs can alter the architecture of chromatin by recruiting chromatin-modifying enzymes to specific genomic loci, leading to changes in histone marks and subsequently affecting gene accessibility and expression. Additionally, some lncRNAs directly interact with DNA to influence methylation patterns, which can either suppress or enhance gene expression depending on the context. Moreover, lncRNAs can also act as scaffolds or decoys for transcription factors and other proteins that regulate transcription, thereby influencing the transcriptional landscape of cells in diverse ways. These multifaceted roles make lncRNAs crucial components in cellular processes and developmental pathways, underscoring their potential as targets in therapeutic strategies and as biomarkers for various diseases.

### *5.8.1 lncRNA as Competitive Endogenous RNAs (ceRNAs)*

The concept that interactions between lncRNAs and other RNAs are largely interconnected suggests that the identification of ceRNA functions for lncRNAs is not unexpected, given their significant biological implications. lncRNAs contribute to the ceRNA network by interacting with other RNAs through shared miRNAs [66]. These externally expressed transcripts often possess multiple miRNA response elements (MREs) which enable them to bind selectively to specific miRNAs or combinations thereof [67]. By sequestering these miRNAs, lncRNAs prevent them from binding to their target mRNAs, thereby effectively lifting repression on these targets [68, 69]. The ceRNA function of lncRNAs in human cancers has been investigated. For example, research has demonstrated that the MALAT1 lncRNA acts as a sponge for miR-145, which affects the radiosensitivity of HPV+ cervical cancer [70]. MALAT1 has also been shown to interact with miR-124, thereby upregulating RBG2 expression and supporting the proliferation and invasion of HPV(+) cervical cancer cells. This interaction exemplifies how MALAT1 acts as a competitive endogenous RNA, sequestering miR-124 to prevent it from suppressing RBG2, thereby influencing cancer cell behavior [71]. The HOTAIR lncRNA has been identified as interacting with miR-331-3p, which plays a significant role in regulating the expression of HER2 in gastric cancer. This interaction is a key example of how HOTAIR can influence oncogenic pathways. By binding to miR-331-3p, HOTAIR reduces the miRNA's availability to target and suppress HER2 mRNA, thereby potentially increasing HER2 protein levels. HER2, known for its involvement in cell growth and survival pathways, can contribute to the progression and aggressiveness of gastric cancer when overexpressed [72]. HOTAIR lncRNA has also been found to suppress the activity of miR-568, which leads to an increase in the levels of S100 calcium-binding protein A1 (S100A1) in breast cancer cells. This action of HOTAIR contributes to the progression and metastatic potential of breast cancer. By inhibiting miR-568, HOTAIR prevents the microRNA from targeting and reducing the expression of S100A1, a protein known to regulate several cellular functions that include cell cycle progression and motility. Elevated levels of S100A1, as a result of this inhibition, are associated with enhanced metastatic capabilities of cancer cells, highlighting a critical molecular pathway by which HOTAIR can influence tumor behavior and metastasis [73]. A recent discovery has involved the identification of B-Raf proto-oncogene, serine/threonine kinase (BRAF) pseudogene 1, which plays a crucial role as a competing endogenous RNA (ceRNA). This pseudogene functions by sponging or sequestering specific microRNAs, namely miR-30a, miR-182, and miR-876, thereby regulating the activity of the BRAF gene. By absorbing these microRNAs, the pseudogene prevents them from binding to and inhibiting the mRNA of the BRAF gene, potentially influencing the expression and function of BRAF. This interaction can have significant implications in the modulation of cellular pathways controlled by BRAF, including those involved in cell growth and proliferation, and might impact the development or progression of certain cancers by altering the signaling dynamics within cells [74]. This variety of interactions underscores the diverse and critical roles that lncRNAs play in gene regulation and disease pathogenesis.

### *5.8.2 lncRNA Structural Interactions*

LncRNAs are capable of forming intricate secondary and tertiary structures, which facilitate specific interactions with a variety of proteins. These interactions are vital for the regulation of gene expression and play a significant role in affecting cellular functions. For instance, ANRIL assists in guiding the polycomb repressive complexes PRC1 and PRC2 to the INK4b-ARF-INK4a gene cluster, resulting in the creation of heterochromatin and consequently inhibiting gene expression [75]. H19, a transcript expressed maternally and subject to genomic imprinting, engages with a protein complex made up of heterogeneous nuclear ribonucleoprotein U (hnRNP U), lysine acetyltransferase 2B (PCAF), and RNA polymerase II (RNA Pol II). This interaction enhances the activation of the miR-200 family by increasing histone acetylation, which in turn impacts the miR-200 pathway. This process contributes to the mesenchymal-to-epithelial transition (MET) and aids in reducing tumor metastasis in hepatocellular carcinoma (HCC) [76]. MALAT1 has been shown to promote cell proliferation, migration, and invasion in gastrointestinal cancer through its interaction with PRKA kinase anchor protein 9 (AKAP-9). Additionally, it interacts with the splicing factor proline/glutamine-rich (SFPQ), causing the release of polypyrimidine tract binding protein 2 (PTBP2) from the SFPQ/PTBP2 complex. This mechanism further enhances growth and metastasis in colon cancer [77, 78]. LincRNA-p21 interacts with heterogeneous nuclear ribonucleoprotein K (hnRNP-K) to enhance the binding efficiency of the tumor suppressor protein p53 to the promoter of the p21 gene. As a result, this interaction initiates the transcription of the p21 gene, effectively activating its expression. The upregulation of p21 is crucial as it contributes to cell cycle regulation and can have implications in tumor suppression and cellular aging [79]. These examples illustrate not only the diversity of lncRNA interactions with proteins but also their ability to influence critical cellular pathways. It is important to note that these interactions do not usually occur in isolation but rather in concert with other molecular interactions within the cell. This interconnectedness highlights the complexity and integrated nature of regulatory mechanisms mediated by lncRNAs in cellular contexts.

### *5.8.3 LncRNAs as Signaling Molecules*

Studies have shown that lncRNAs are selectively transcribed in response to diverse stimuli, thereby influencing specific signaling pathways. These lncRNAs can operate autonomously or synergize with particular proteins, including transcription factors, to modulate the transcriptional activity of genes. This interaction often leads to either the activation or repression of gene expression; an action that depends on the nature of the stimulus and the specific signaling pathways involved. Such mechanisms highlight the versatility and critical role of lncRNAs in cellular regulation and response to environmental changes.

As mentioned previously, the lncRNA-p21 plays a significant role in gene regulation within the p53 signaling pathway by interacting with nuclear heterogeneous ribonucleoprotein K (hnRNP-K). This interaction is critical for the modulation of gene expression, particularly in response to cellular stress and DNA damage, which are central to the functions mediated by the tumor suppressor protein p53 [80]. Similarly, another lncRNA known as PANDA is activated through the synergy between the p53 and CDKN1A, particularly under conditions of DNA damage. These interactions underscore the intricate network of lncRNAs in mediating cellular responses through specific and targeted gene regulation mechanisms [81]. The functionality of lncRNAs in transcriptional regulation is particularly advantageous homeostatically because it enables a swift response to external stimuli. This rapid reaction is facilitated through the modulation of protein activity, showcasing the dynamic capabilities of RNA in cellular regulation.

## 5.9 LncRNAs as Decoy Molecules

LncRNAs often function as decoy molecules, impeding specific molecular pathways by directly interacting with proteins (i.e., regulators of transcription or chromosome folding protein) to inhibit their functions. This interaction typically disrupts the activity of transcription factors, leading to suppressed transcription of downstream genes. For example, the lncRNA PANDA is activated during DNA damage, which can directly bind to the nuclear transcription factor NF-YA, thereby disrupting its function. This binding interaction blocks the expression of genes related to apoptosis, effectively inhibiting the pathways that rely on NF-YA to trigger cell death processes [81]. Additionally, lncRNAs can interfere with proteins' ability to regulate mRNA expression by binding to them. For instance, one of the critical mechanisms by which MALAT1 influences cellular processes is through the modulation of the activity of serine/arginine-rich (SR) splicing factors.

## 5.10 LncRNAs as Guide Molecules

lncRNAs function as guide molecules in several crucial biological processes by directing the activity and location of chromatin-modifying enzymes to specific genomic locations [47, 49, 82]. This guiding role allows them to play a significant part in regulating gene expression through epigenetic mechanisms.

***Recruiting Chromatin-Modifying Complexes***. lncRNAs can specifically bind to various chromatin-modifying complexes, such as PRC2, or involved in DNA methylation or histone modification. Once bound, lncRNAs guide these complexes to specific sites on the DNA. This targeted recruitment can lead to the silencing or activation of genes depending on the nature of the modifications induced. For example, the lncRNA HOTAIR is known to interact with PRC2, guiding it to the HOXD locus to establish repressive chromatin marks that silence gene expression [83].

***Interaction with TFs and Other Regulatory Proteins***. Some lncRNAs function by binding to transcription factors and guiding them to their target genes, thereby influencing transcription initiation and progression. By modulating the activity and specificity of transcription factors, lncRNAs can exert profound effects on gene regulatory networks.

***Influence on RNA Polymerase II***. Certain lncRNAs guide RNA Pol II to specific promoters or regulate its pausing and release at gene loci, thus controlling the transcription of mRNA. This interaction can either enhance or repress the transcriptional output, impacting cellular phenotype and function.

***Modulation of Chromosomal Architecture.*** *l*ncRNAs can also guide structural proteins to establish loops or interactions between distant regions of chromatin. These interactions can bring enhancers into close proximity with promoters, enabling or inhibiting gene expression. For instance, lncRNA-mediated looping can influence the level of gene expression of the various genes involved in development and differentiation.

***Spatial Organization within the Nucleus.*** Guide lncRNAs are involved in organizing the architecture of the nucleus by directing certain genes or gene regions to specific nuclear compartments. This positioning can affect gene expression levels and the phase separation within the nucleus, which in turn influences cellular metabolism and response to environmental stimuli. The guiding function of lncRNAs demonstrates their versatility and importance in cellular regulation. By acting as molecular scaffolds that bring together various components of the transcriptional machinery, lncRNAs help orchestrate complex gene expression patterns that are essential for normal development and cellular function, and also in the onset of various diseases states such as cancer. This functional versatility makes them potential targets for therapeutic intervention, providing opportunities for the development of novel treatments based on modulating lncRNA activity.

***Cis Regulation by lncRNAs.*** cis regulation is the mechanism of affecting the transcription of proximal genes that are on the same chromosome. A notable example illustrates how SWI/SNF chromatin remodeling complex was recruited by lncTCF7 to the promoter of the TCF7 gene. This recruitment significantly enhances the expression of TCF7. The upregulation of the TCF7 gene plays a pivotal role in activating the Wnt signaling pathway, which promotes the proliferation of liver cancer cells [84].

***Trans Regulation by lncRNAs.*** Conversely, lncRNAs also facilitate trans-regulation, affecting the transcription of genes located far away or even on different chromosomes. A significant example of this is the interaction between the lncRNA HOTAIR and the polycomb repressive complex 2 (PRC2), which results in a changed rearrangement of PRC2 in the genome. This interaction facilitates the methylation of at multiple gene loci, affecting gene expression across various regions [83]. Further detailing this mechanism, Tsai et al. made a significant discovery regarding the structure and function of the lncRNA HOTAIR. They found that HOTAIR's different domains interact with distinct histone modification complexes to regulate gene expression. The 5′-domain of HOTAIR binds to the PRC2, influencing the methylation of histones, which typically acts to suppress gene expression, while the 3′-domain of HOTAIR binds to the LSD1/CoREST/REST tricomplex, which is involved in the demethylation of histones, a process that generally facilitates gene activation. This bifunctional role of HOTAIR in modulating chromatin state highlights its complex and critical involvement in gene regulatory networks. These interactions facilitate the assembly of diverse histone modification enzymes, which then target various gene regions to modify histone methylation patterns, ultimately influencing the transcription of multiple genes [85]. These guiding roles of lncRNAs underscore their significance in cellular processes, where they orchestrate complex gene regulation networks through direct interactions with complexes that could modify the chromatin, enabling precise genetic control crucial for both normal cellular function and in disease states like cancer.

**Box 5.1 Cis-and Trans-Regulatory lncRNAs and Their Association with Immune-Related Pathways**

In a very recent study, Liu et al. identified 27 cis-regulatory and 34 trans-regulatory lncRNAs involved in various immune-related pathways in non-small cell lung cancer (NSCLC), including primary immunodeficiency, focal adhesion, and extracellular matrix (ECM)-receptor interactions. It is well-documented that primary immunodeficiency can increase susceptibility to malignancies [86, 87], and it may also influence the success of chemotherapy treatments [88]. A comprehensive transcriptomic analysis of NSCLC revealed the presence of diverse B-cell subtypes, suggesting their multifunctional roles within the tumor microenvironment [89]. Our findings indicate that cis-regulatory lncRNAs are highly concentrated in pathways associated with primary immunodeficiency and B-cell receptor signaling regulation, potentially opening new avenues for immunotherapy research. Furthermore, focal adhesion signaling, which is vital for tumor cell motility, has been closely linked with cancer resistance to radiotherapy and chemotherapy [90–92]. Prior research has shown that CX3CL1 enhances lung cancer cell migration and invasion via focal adhesion signaling activation [92], and G9A has been implicated in advancing NSCLC cell metastasis by activating focal adhesion kinase [93]. Additionally, the ECM-receptor interaction pathway is implicated in the progression of various cancers [94–97]. We observed that trans-regulatory lncRNA-mRNA interactions are predominantly enriched in immune-related biological processes. In a study utilizing single-cell sequencing, researchers noted that certain lncRNAs were associated with immune cell infiltration and found to be highly expressed in immune cells, such as B cells and T cells, in lung cancer [89]. Notably, lncRNA NRON has been identified as a regulator of the resting state in T cells [98]. Collectively, these findings provide a theoretical foundation for understanding the role of lncRNAs in immune regulation within the tumor microenvironment.

## 5.11 LncRNAs as Scaffold Molecules

This critical scaffolding function enables lncRNAs to play an essential role in the formation of diverse macromolecular complexes in the intracellular environment. lncRNA serves as an important bridge for the integration of multiple signaling pathways, thereby exerting a substantial influence on cellular functions and regulatory mechanisms. This integration helps coordinate complex biological processes, from gene expression to cellular response to environmental cues, highlighting the complex roles of lncRNAs in cellular biology [82, 99] in Table 5.2 and Fig. 5.2.

## 5.12 lncRNAs in Human Diseases: Biomedical Applications

lncRNAs are involved in various biological processes, including gene regulation, cellular structure, and intracellular signaling. Recent advances have highlighted their roles in numerous diseases, such as cancer, cardiovascular, and

**Table 5.2** Examples of Scaffold lncRNAs

| Scaffold lncRNAs | Function and mechanism |
|---|---|
| Xist (X-inactive specific transcript) | Plays a crucial role in dosage compensation in mammals. This process is essential for balancing the expression of X chromosome-derived genes between males (who have one X chromosome) and females (who have two). Xist RNA achieves this by recruiting the polycomb repressive complexes PRC1 and PRC2 to one of the X chromosomes in female cells. These complexes facilitate the suppression of gene expression on that chromosome, effectively ensuring equal gene dosage between the sexes [101, 102] |
| HOTAIR (HOX transcript antisense RNA) | HOTAIR plays a pivotal role in modulating gene expression by facilitating the recruitment and interaction of various regulatory proteins. Specifically, it enhances the interaction between lysine-specific demethylase 1A (LSD1) and hepatitis B X-interacting protein (HBXIP). This interaction is instrumental in activating the transcription factor c-Myc, a key regulator of cell growth and proliferation. Once activated, c-Myc promotes the expression of several important genes, including cyclin A, eIF4E, and LDHA. These genes are crucial for cell cycle progression, protein synthesis, and metabolism, respectively, illustrating how HOTAIR can influence a broad spectrum of cellular functions through its regulatory actions [103] |
| INK4 (lncRNA INK4, ANRIL) An antisense transcription product of the CDKN2B gene (p15 INK4b) | Serves as a structural scaffold for PRC1 and PRC2. This interaction is fundamental in the epigenetic regulation within cells, specifically leading to the silencing of the tumor suppressor gene CDKN2A (p16 INK4a). The suppression of p16 INK4a is a significant event in cancer biology, as this gene plays a key role in regulating the cell cycle and preventing uncontrolled cell proliferation. By facilitating the silencing of p16 INK4a, INK4 contributes to a crucial regulatory mechanism that, when disrupted, can lead to the development and progression of various cancers [104] |
| CCAT1 (colon cancer-associated Transcript-1): | Its ability to interact with two distinct epigenetic modification complexes. Specifically, the 5′-domain of CCAT1 binds to the PRC2, while the 3′-domain associates with the histone methyltransferase SUV39H1. These interactions play a crucial role in altering the histone methylation patterns at the SPRY4 gene promoter. By modulating these epigenetic marks, CCAT1 enhances the transcriptional activity of the SPRY4 gene, which is known to promote cell proliferation and metastasis. This regulatory mechanism is particularly significant in the context of esophageal squamous cell carcinoma (ESCC), where such changes in gene expression contribute to the aggressive nature of the cancer, aiding in its proliferation and the spread of tumor cells to other parts of the body [105] |

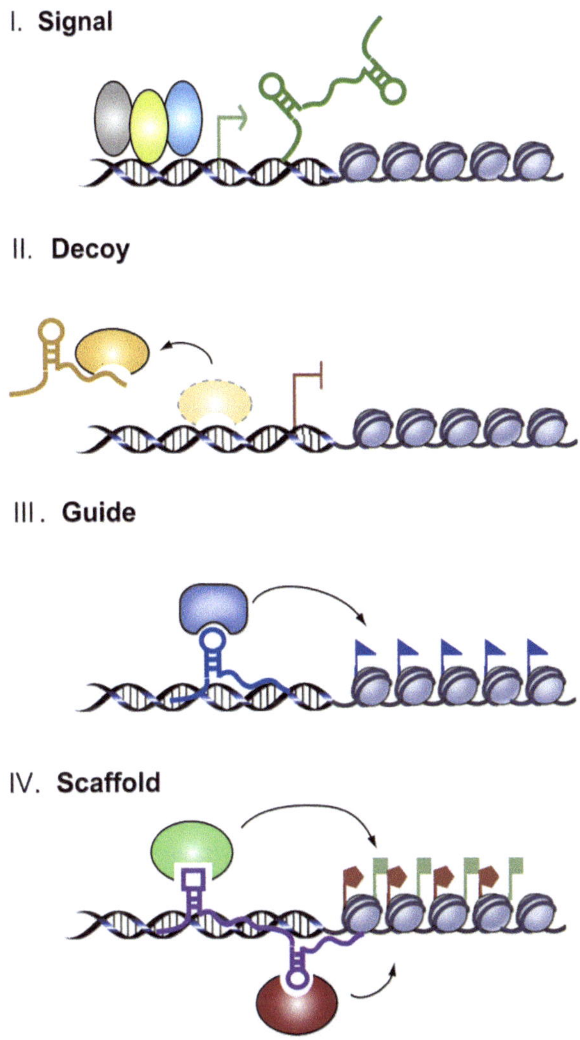


**Fig. 5.2** Functional diversity of lncRNAs in gene regulation. This illustration categorizes four primary functions of lncRNAs within cellular processes. (i) lncRNA as signaling molecule: This panel depicts lncRNAs acting as signaling molecules that are conditionally expressed and activated in response to various stimuli. The lncRNA is shown facilitating the interaction between transcription factors (represented by colored orbs) and specific gene loci to initiate transcription. (ii) lncRNA as decoy: Demonstrates lncRNAs functioning as decoys by binding and sequestering proteins away from their usual targets. Here, the lncRNA captures a transcriptional repressor (golden orb), preventing it from binding to DNA and thereby allowing gene expression to proceed. (iii) lncRNA as guide molecule: Shows lncRNAs guiding chromatin-modifying enzymes to specific locations in the genome. The lncRNA is illustrated bringing a histone modification enzyme (blue box) to a particular gene, directing targeted epigenetic alterations that affect gene expression. (4) lncRNA as scaffold molecule: Illustrates the scaffolding role of lncRNAs, where they serve as central platforms assembling multiple proteins to form ribonucleoprotein complexes. In this panel, the lncRNA connects different enzymatic and regulatory proteins (various colored shapes), facilitating complex formation and function at specific genomic sites. (Figure is reproduced from [100] with permission from Elsevier (License number: 5910500322933))

neurodegenerative disorders, marking them as promising targets for novel interventions. LncRNAs originating from tumor cells or circulating in the bloodstream present a cost-effective, accessible, and stable method for improving the detection of cancers and their various subtypes. This potential makes lncRNAs highly valuable in advancing precision medicine and developing more targeted, effective treatments [106].

Eight types of lncRNAs were investigated in 240 HCC patients and found that all these lncRNA were upregulated in liver cancer with vascular invasion (HULC). Vascular invasion has a positive correlation with both overall survival and disease-free survival in HCC [107]. Compared to traditional HCC biomarkers like AFP, three lncRNAs—RP11-160H22.5, XLOC_014172, and LOC149086—may serve as potential indicators for predicting tumorigenesis. Additionally, XLOC_014172 and LOC149086 are also predictive of metastasis [108]. Furthermore, lncRNA SPRY4 intronic transcript 1 (SPRY4-IT1) has been identified as an independent prognostic marker for overall survival in patients with urothelial carcinoma of the bladder (UCB) [109]. Sørensen and colleagues conducted groundbreaking research on the prognostic value of lncRNA profiles in BC patients. Their study successfully differentiated between metastatic and non-metastatic patients, achieving over 90% sensitivity and 64–65% specificity [110].

Due to the diverse and complex nature of human cancer, personalized medicine has been a major focus for many years. Precision medicine has become a crucial area of research and treatment. This approach aims to develop targeted therapies that consider genetic, environmental, and lifestyle factors, thereby improving the efficacy and outcomes of cancer treatments. Currently, several therapeutic strategies have been developed. Among them, the use of RNA interference to target growth-promoting lncRNAs using antisense technologies represents a significant advancement [106]. Additionally, lncRNAs can act as sponges for mRNA or miRNA, obstructing pro-tumorigenic signaling pathways, thereby enabling the potential use of synthetically engineered lncRNAs in replacement therapy to combat tumor growth [111].

Another innovative approach involves the use of dextran nanoparticles to deliver chemotherapeutic agents directly to target nucleus-localized lncRNAs [106]. First, they used DTA-H19 plasmid to selectively target cancer cells by leveraging the H19 promoter to control the expression of a diphtheria toxin subunit. When this plasmid is injected into tumors, the H19 promoter—active primarily in cancer cells—triggers the expression of the diphtheria toxin. This targeted activation leads to the production of the toxin specifically within H19-expressing tumor cells, causing their destruction while sparing healthy cells. This innovative approach exemplifies how gene therapy can be used to create highly specific cancer treatments based on the unique genetic characteristics of tumor cells [112, 113]. Emerging strategies also include the usage of ASOs to suppress the effect of lncRNAs [106]. These molecules can be administered subcutaneously, penetrate cellular and endosomal membranes without a delivery vehicle, and reach the nucleus to disrupt interactions between PRC2 and lncRNAs [114]. Remarkably, these compounds are also capable of crossing the blood-brain barrier (BBB) without requiring the help of a carrier [106], offering promising avenues for targeting brain tumors and metastases.

## 5.13 Conclusion

The field of transcriptomics is rapidly evolving, leading to a greater appreciation of ncRNAs as critical elements of genomic expression. Far from serving merely as messengers, ncRNAs exhibit sophisticated structural and informational capacities that evolution seems to have leveraged for diverse biological functions. Despite these advances, significant gaps remain in our knowledge, particularly regarding the percentage of function long ncRNAs (not limited to lncRNAs), defining their localization and their modus operandi in the cellular environment. To address these gaps, genome-wide methodologies is essential to uncover the full range of ncRNA expression. Advances in sequencing technology such as NGS offer unbiased approaches that transcend the limitations of protein-focused annotations. This approach is expected to gradually establish a comprehensive database of ncRNAs, delineating shared characteristics that facilitate the prediction and identification of their functional aspects. This will be further enriched through rigorous experimental studies on individual ncRNAs to elucidate the specific mechanisms they employ. The integration of methodologies from various disciplines, including genomics, proteomics, live cell RNA imaging, AI-mediated deciphering of RNA structural biology, etc., will definitely enhance our understanding of long ncRNAs. These RNAs may potentially match the functional diversity of proteins, albeit within a different context. It is clear that ncRNAs, irrespective of their protein-coding potential, possess unique intrinsic functions, pointing to a genomic landscape where RNA-based information plays a more complex role than previously thought. If many of ncRNAs are found to be functional, their detailed characterization will greatly enhance our understanding of genetic programming in complex organisms. This in-depth study will offer fresh insights into evolutionary biology, shedding light on how these molecules have influenced genetic diversity and adaptation. Additionally, it will elucidate developmental processes by revealing how ncRNAs regulate gene expression during growth and differentiation. Furthermore, understanding the roles of ncRNAs in disease pathology will open new avenues for diagnosing, treating, and potentially preventing various conditions, significantly impacting biomedical research and clinical practice.

## References

1. Mercer TR, Dinger ME, Mattick JS, Mercer TR, Dinger ME, Mattick JS. Long non-coding RNAs: insights into functions. Nat Rev Genet. 2009;10:155–9. Nature Reviews Genetics. 2009;10(3):155–9.
2. Parrott AM, Tsai M, Batchu P, Ryan K, Ozer HL, Tian B, et al. The evolution and expression of the snaR family of small non-coding RNAs. Nucleic Acids Res. 2011;39(4):1485–500.
3. Täuber H, Hüttelmaier S, Köhn M. POLIII-derived non-coding RNAs acting as scaffolds and decoys. J Mol Cell Biol. 2019;11(10):880–5.
4. Anon. Post-transcriptional processing generates a diversity of 5′-modified long and short RNAs. Nature. 2009;457(7232):1028–32.

5. Hahne JC, Lampis A, Valeri N. Vault RNAs: hidden gems in RNA and protein regulation. Cell Mol Life Sci. 2021;78(4):1487–99.
6. Kapranov P, Cheng J, Dike S, Nix DA, Duttagupta R, Willingham AT, et al. RNA maps reveal new RNA classes and a possible function for pervasive transcription. Science. 2007;316(5830):1484–8.
7. Preker P, Almvig K, Christensen MS, Valen E, Mapendano CK, Sandelin A, et al. PROMoter uPstream transcripts share characteristics with mRNAs and are produced upstream of all three major types of mammalian promoters. Nucleic Acids Res. 2011;39(16):7179–93.
8. Castelo-Branco G, Amaral PP, Engström PG, Robson SC, Marques SC, Bertone P, et al. The non-coding snRNA 7SK controls transcriptional termination, poising, and bidirectionality in embryonic stem cells. Genome Biol. 2013;14(9):R98.
9. Flynn RA, Do BT, Rubin AJ, Calo E, Lee B, Kuchelmeister H, et al. 7SK-BAF axis controls pervasive transcription at enhancers. Nat Struct Mol Biol. 2016;23(3):231–8.
10. Gussakovsky D, McKenna SA. Alu RNA and their roles in human disease states. RNA Biol. 2021;18(sup2):574–85.
11. Tsirigos A, Rigoutsos I. Alu and b1 repeats have been selectively retained in the upstream and intronic regions of genes of specific functional classes. PLoS Comput Biol. 2009;5(12):e1000610.
12. Ullu E, Tschudi C. Alu sequences are processed 7SL RNA genes. Nature. 1984;312(5990):171–2.
13. Zhang XO, Gingeras TR, Weng Z. Genome-wide analysis of polymerase III–transcribed Alu elements suggests cell-type–specific enhancer function. Genome Res. 2019;29(9):1402.
14. Cheng J, Kapranov P, Drenkow J, Dike S, Brubaker S, Patel S, et al. Transcriptional maps of 10 human chromosomes at 5-nucleotide resolution. Science. 2005;308(5725):1149–54.
15. Gingeras TR. Origin of phenotypes: genes and transcripts. Genome Res. 2007;17(6):682–90.
16. Wilusz JE, Freier SM, Spector DL. 3′ end processing of a long nuclear-retained noncoding RNA yields a tRNA-like cytoplasmic RNA. Cell. 2008;135(5):919–32.
17. Wu H, Yin QF, Luo Z, Yao RW, Zheng CC, Zhang J, et al. Unusual processing generates SPA LncRNAs that sequester multiple RNA binding proteins. Mol Cell. 2016;64(3):534–48.
18. Yin QF, Yang L, Zhang Y, Xiang JF, Wu YW, Carmichael GG, et al. Long noncoding RNAs with snoRNA ends. Mol Cell. 2012;48(2):219–30.
19. Frankish A, Diekhans M, Jungreis I, Lagarde J, Loveland JE, Mudge JM, et al. GENCODE 2021. Nucleic Acids Res. 2021;49(D1):D916–d23.
20. Mattick JS, Rinn JL. Discovery and annotation of long noncoding RNAs. Nat Struct Mol Biol. 2015;22(1):5–7.
21. Seal RL, Chen LL, Griffiths-Jones S, Lowe TM, Mathews MB, O'Reilly D, et al. A guide to naming human non-coding RNA genes. EMBO J. 2020;39(6):e103777.
22. Uszczynska-Ratajczak B, Lagarde J, Frankish A, Guigó R, Johnson R. Towards a complete map of the human long non-coding RNA transcriptome. Nat Rev Genet. 2018;19(9):535–48.
23. Wright MW. A short guide to long non-coding RNA gene nomenclature. Hum Genomics. 2014;8(1):7.
24. Mattick JS. Challenging the dogma: the hidden layer of non-protein-coding RNAs in complex organisms. BioEssays. 2003;25(10):930–9.
25. Carninci P, Kasukawa T, Katayama S, Gough J, Frith MC, Maeda N, et al. The transcriptional landscape of the mammalian genome. Science. 2005;309(5740):1559–63.
26. Kapranov P, Willingham AT, Gingeras TR. Genome-wide transcription and the implications for genomic organization. Nat Rev Genet. 2007;8(6):413–23.
27. Willingham AT, Dike S, Cheng J, Manak JR, Bell I, Cheung E, et al. Transcriptional landscape of the human and fly genomes: nonlinear and multifunctional modular model of transcriptomes. Cold Spring Harb Symp Quant Biol. 2006;71:101–10.
28. Ariel F, Lucero L, Christ A, Mammarella MF, Jegu T, Veluchamy A, et al. R-loop mediated trans action of the APOLO long noncoding RNA. Mol Cell. 2020;77(5):1055–65.e4.

29. Wright BW, Yi Z, Weissman JS, Chen J. The dark proteome: translation from noncanonical open reading frames. Trends Cell Biol. 2022;32(3):243–58.
30. Wu P, Mo Y, Peng M, Tang T, Zhong Y, Deng X, et al. Emerging role of tumor-related functional peptides encoded by lncRNA and circRNA. Mol Cancer. 2020;19(1):22.
31. Dinger ME, Gascoigne DK, Mattick JS. The evolution of RNAs with multiple functions. Biochimie. 2011;93(11):2013–8.
32. Makarewich CA, Olson EN. Mining for micropeptides. Vaccine. 2017;35 Pt.B:35.
33. Hube F, Guo J, Chooniedass-Kothari S, Cooper C, Hamedani MK, Dibrov AA, et al. Alternative splicing of the first intron of the steroid receptor RNA activator (SRA) participates in the generation of coding and noncoding RNA isoforms in breast cancer cell lines. DNA Cell Biol. 2006;25(7):418–28.
34. Williamson L, Saponaro M, Boeing S, East P, Mitter R, Kantidakis T, et al. UV irradiation induces a non-coding RNA that functionally opposes the protein encoded by the same gene. Cell. 2017;168(5):843–55.e13.
35. Grelet S, Link LA, Howley B, Obellianne C, Palanisamy V, Gangaraju VK, et al. A regulated PNUTS mRNA to lncRNA splice switch mediates EMT and tumour progression. Nat Cell Biol. 2017;19:1105.
36. Gonzàlez-Porta M, Frankish A, Rung J, Harrow J, Brazma A. Transcriptome analysis of human tissues and cell lines reveals one dominant transcript per gene. Genome Biol. 2013;14(7):R70.
37. Tuck AC, Tollervey D. RNA in pieces. Trends Genet. 2011;27(10):422–32.
38. Chan SN, Pek JW. Stable Intronic sequence RNAs (sisRNAs): an expanding universe. Trends Biochem Sci. 2019;44(3):258–72.
39. Patop IL, Wüst S, Kadener S. Past, present, and future of circRNAs. EMBO J. 2019;38(16):e100836.
40. Ponting CP, Oliver PL, Reik W. Evolution and functions of long noncoding RNAs. Cell. 2009;136(4):629–41.
41. Schmitz SU, Grote P, Herrmann BG. Mechanisms of long noncoding RNA function in development and disease. Cell Mol Life Sci. 2016;73(13):2491–509.
42. Wu T, Du Y. LncRNAs: from basic research to medical application. Int J Biol Sci. 2017;13(3):295–307.
43. Tripathi V, Ellis JD, Shen Z, Song DY, Pan Q, Watt AT, et al. The nuclear-retained noncoding RNA MALAT1 regulates alternative splicing by modulating SR splicing factor phosphorylation. Mol Cell. 2010;39(6):925–38.
44. Emmrich S, Streltsov A, Schmidt F, Thangapandi VR, Reinhardt D, Klusmann JH. LincRNAs MONC and MIR100HG act as oncogenes in acute megakaryoblastic leukemia. Mol Cancer. 2014;13:171.
45. Kashi K, Henderson L, Bonetti A, Carninci P. Discovery and functional analysis of lncRNAs: methodologies to investigate an uncharacterized transcriptome. Biochim Biophys Acta. 2016;1859(1):3–15.
46. Lee JH, Daugharthy ER, Scheiman J, Kalhor R, Ferrante TC, Terry R, et al. Fluorescent in situ sequencing (FISSEQ) of RNA for gene expression profiling in intact cells and tissues. Nat Protoc. 2015;10(3):442–58.
47. Oo JA, Brandes RP, Leisegang MS. Long non-coding RNAs: novel regulators of cellular physiology and function. Pflugers Arch—Eur J Physiol. 2022;474(2):191–204.
48. Camilleri-Robles C, Amador R, Klein CC, Guigó R, Corominas M, Ruiz-Romero M. Genomic and functional conservation of lncRNAs: lessons from flies. Mamm Genome. 2022;33(2):328–42.
49. Ferrer J, Dimitrova N. Transcription regulation by long non-coding RNAs: mechanisms and disease relevance. Nat Rev Mol Cell Biol. 2024;25(5):396–415.
50. Statello L, Guo C-J, Chen L-L, Huarte M. Gene regulation by long non-coding RNAs and its biological functions. Nat Rev Mol Cell Biol. 2021;22(2):96–118.

51. Kudla G, Granneman S, Hahn D, Beggs JD, Tollervey D. Cross-linking, ligation, and sequencing of hybrids reveals RNA-RNA interactions in yeast. Proc Natl Acad Sci USA. 2011;108(24):10010–5.
52. Chu C, Qu K, Zhong FL, Artandi SE, Chang HY. Genomic maps of long noncoding RNA occupancy reveal principles of RNA-chromatin interactions. Mol Cell. 2011;44(4):667–78.
53. Simon MD, Pinter SF, Fang R, Sarma K, Rutenberg-Schoenberg M, Bowman SK, et al. High-resolution Xist binding maps reveal two-step spreading during X-chromosome inactivation. Nature. 2013;504(7480):465–9.
54. Engreitz JM, Pandya-Jones A, McDonel P, Shishkin A, Sirokman K, Surka C, et al. The Xist lncRNA exploits three-dimensional genome architecture to spread across the X chromosome. Science. 2013;341(6147):1237973.
55. Cui M, Xiao Z, Wang Y, Zheng M, Song T, Cai X, et al. Long noncoding RNA HULC modulates abnormal lipid metabolism in hepatoma cells through an miR-9-mediated RXRA signaling pathway. Cancer Res. 2015;75(5):846–57.
56. Nie FQ, Sun M, Yang JS, Xie M, Xu TP, Xia R, et al. Long noncoding RNA ANRIL promotes non-small cell lung cancer cell proliferation and inhibits apoptosis by silencing KLF2 and P21 expression. Mol Cancer Ther. 2015;14(1):268–77.
57. Quek XC, Thomson DW, Maag JL, Bartonicek N, Signal B, Clark MB, et al. lncRNAdb v2.0: expanding the reference database for functional long noncoding RNAs. Nucleic Acids Res. 2015;43(Database issue):D168–73.
58. Zhao Y, Yuan J, Chen R. NONCODEv4: annotation of noncoding RNAs with emphasis on long noncoding RNAs. Methods Mol Biol. 2016;1402:243–54.
59. Volders PJ, Helsens K, Wang X, Menten B, Martens L, Gevaert K, et al. LNCipedia: a database for annotated human lncRNA transcript sequences and structures. Nucleic Acids Res. 2013;41(Database issue):D246–51.
60. Volders PJ, Verheggen K, Menschaert G, Vandepoele K, Martens L, Vandesompele J, et al. An update on LNCipedia: a database for annotated human lncRNA sequences. Nucleic Acids Res. 2015;43(8):4363–4.
61. Chen G, Wang Z, Wang D, Qiu C, Liu M, Chen X, et al. LncRNADisease: a database for long-non-coding RNA-associated diseases. Nucleic Acids Res. 2013;41(Database issue):D983–6.
62. Dinger ME, Pang KC, Mercer TR, Crowe ML, Grimmond SM, Mattick JS. NRED: a database of long noncoding RNA expression. Nucleic Acids Res. 2009;37(Database issue):D122–6.
63. Yang JH, Li JH, Jiang S, Zhou H, Qu LH. ChIPBase: a database for decoding the transcriptional regulation of long non-coding RNA and microRNA genes from ChIP-Seq data. Nucleic Acids Res. 2013;41(Database issue):D177–87.
64. Kin T, Yamada K, Terai G, Okida H, Yoshinari Y, Ono Y, et al. fRNAdb: a platform for mining/annotating functional RNA candidates from non-coding RNA sequences. Nucleic Acids Res. 2007;35(Database issue):D145–8.
65. Yang JH, Li JH, Shao P, Zhou H, Chen YQ, Qu LH. starBase: a database for exploring microRNA-mRNA interaction maps from Argonaute CLIP-Seq and Degradome-Seq data. Nucleic Acids Res. 2011;39(Database issue):D202-9:D202.
66. Li T, Mo X, Fu L, Xiao B, Guo J. Molecular mechanisms of long noncoding RNAs on gastric cancer. Oncotarget. 2016;7(8):8601–12.
67. Sanchez-Mejias A, Tay Y. Competing endogenous RNA networks: tying the essential knots for cancer biology and therapeutics. J Hematol Oncol. 2015;8:30.
68. Ebert MS, Neilson JR, Sharp PA. MicroRNA sponges: competitive inhibitors of small RNAs in mammalian cells. Nat Methods. 2007;4(9):721–6.
69. Ebert MS, Sharp PA. MicroRNA sponges: progress and possibilities. RNA. 2010;16(11):2043–50.
70. Lu H, He Y, Lin L, Qi Z, Ma L, Li L, et al. Long non-coding RNA MALAT1 modulates radiosensitivity of HR-HPV+ cervical cancer via sponging miR-145. Tumour Biol. 2016;37(2):1683–91.

71. Liu S, Song L, Zeng S, Zhang L. MALAT1-miR-124-RBG2 axis is involved in growth and invasion of HR-HPV-positive cervical cancer cells. Tumour Biol. 2016;37(1):633–40.
72. Liu XH, Sun M, Nie FQ, Ge YB, Zhang EB, Yin DD, et al. Lnc RNA HOTAIR functions as a competing endogenous RNA to regulate HER2 expression by sponging miR-331-3p in gastric cancer. Mol Cancer. 2014;13:92.
73. Li JT, Wang LF, Zhao YL, Yang T, Li W, Zhao J, et al. Nuclear factor of activated T cells 5 maintained by Hotair suppression of miR-568 upregulates S100 calcium binding protein A4 to promote breast cancer metastasis. Breast Cancer Res. 2014;16(5):454.
74. Karreth FA, Reschke M, Ruocco A, Ng C, Chapuy B, Léopold V, et al. The BRAF pseudogene functions as a competitive endogenous RNA and induces lymphoma in vivo. Cell. 2015;161(2):319–32.
75. Yap KL, Li S, Muñoz-Cabello AM, Raguz S, Zeng L, Mujtaba S, et al. Molecular interplay of the noncoding RNA ANRIL and methylated histone H3 lysine 27 by polycomb CBX7 in transcriptional silencing of INK4a. Mol Cell. 2010;38(5):662–74.
76. Zhang L, Yang F, Yuan JH, Yuan SX, Zhou WP, Huo XS, et al. Epigenetic activation of the MiR-200 family contributes to H19-mediated metastasis suppression in hepatocellular carcinoma. Carcinogenesis. 2013;34(3):577–86.
77. Ji Q, Zhang L, Liu X, Zhou L, Wang W, Han Z, et al. Long non-coding RNA MALAT1 promotes tumour growth and metastasis in colorectal cancer through binding to SFPQ and releasing oncogene PTBP2 from SFPQ/PTBP2 complex. Br J Cancer. 2014;111(4):736–48.
78. Yang MH, Hu ZY, Xu C, Xie LY, Wang XY, Chen SY, et al. MALAT1 promotes colorectal cancer cell proliferation/migration/invasion via PRKA kinase anchor protein 9. Biochim Biophys Acta. 2015;1852(1):166–74.
79. Dimitrova N, Zamudio JR, Jong RM, Soukup D, Resnick R, Sarma K, et al. LincRNA-p21 activates p21 in cis to promote Polycomb target gene expression and to enforce the G1/S checkpoint. Mol Cell. 2014;54(5):777–90.
80. Huarte M, Guttman M, Feldser D, Garber M, Koziol MJ, Kenzelmann-Broz D, et al. A large intergenic noncoding RNA induced by p53 mediates global gene repression in the p53 response. Cell. 2010;142(3):409–19.
81. Hung T, Wang Y, Lin MF, Koegel AK, Kotake Y, Grant GD, et al. Extensive and coordinated transcription of noncoding RNAs within cell-cycle promoters. Nat Genet. 2011;43(7):621–9.
82. Wu R, Su Y, Wu H, Dai Y, Zhao M, Lu Q. Characters, functions and clinical perspectives of long non-coding RNAs. Mol Gen Genomics. 2016;291(3):1013–33.
83. Gupta RA, Shah N, Wang KC, Kim J, Horlings HM, Wong DJ, et al. Long non-coding RNA HOTAIR reprograms chromatin state to promote cancer metastasis. Nature. 2010;464(7291):1071–6.
84. Wang Y, He L, Du Y, Zhu P, Huang G, Luo J, et al. The long noncoding RNA lncTCF7 promotes self-renewal of human liver cancer stem cells through activation of Wnt signaling. Cell Stem Cell. 2015;16(4):413–25.
85. Tsai MC, Manor O, Wan Y, Mosammaparast N, Wang JK, Lan F, et al. Long noncoding RNA as modular scaffold of histone modification complexes. Science. 2010;329(5992):689–93.
86. Mayor PC, Eng KH, Singel KL, Abrams SI, Odunsi K, Moysich KB, et al. Cancer in primary immunodeficiency diseases: cancer incidence in the United States immune deficiency network registry. J Allergy Clin Immunol. 2018;141(3):1028–35.
87. Kebudi R, Kiykim A, Sahin MK. Primary immunodeficiency and cancer in children; a review of the literature. Curr Pediatr Rev. 2019;15(4):245–50.
88. Sun J, Bai H, Wang Z, Duan J, Li J, Guo R, et al. Pegylated recombinant human granulocyte colony-stimulating factor regulates the immune status of patients with small cell lung cancer. Thoracic Cancer. 2020;11(3):713–22.
89. Chen J, Tan Y, Sun F, Hou L, Zhang C, Ge T, et al. Single-cell transcriptome and antigen-immunoglobin analysis reveals the diversity of B cells in non-small cell lung cancer. Genome Biol. 2020;21(1):152.

90. Zhao X, Guan JL. Focal adhesion kinase and its signaling pathways in cell migration and angiogenesis. Adv Drug Deliv Rev. 2011;63(8):610–5.
91. Eke I, Cordes N. Focal adhesion signaling and therapy resistance in cancer. Semin Cancer Biol. 2015;31:65–75.
92. Paluch EK, Aspalter IM, Sixt M. Focal adhesion-independent cell migration. Annu Rev Cell Dev Biol. 2016;32:469–90.
93. Sun T, Zhang K, Pangeni RP, Wu J, Li W, Du Y, et al. G9a promotes invasion and metastasis of non-small cell lung cancer through enhancing focal adhesion kinase activation via NF-κB signaling pathway. Mol Cancer Res. 2021;19(3):429–40.
94. Yang X, Chen L, Mao Y, Hu Z, He M. Progressive and prognostic performance of an extracellular matrix-receptor interaction signature in gastric cancer. Dis Markers. 2020;2020:8816070.
95. Bao Y, Wang L, Shi L, Yun F, Liu X, Chen Y, et al. Transcriptome profiling revealed multiple genes and ECM-receptor interaction pathways that may be associated with breast cancer. Cell Mol Biol Lett. 2019;24:38.
96. Machackova T, Vychytilova-Faltejskova P, Souckova K, Trachtova K, Brchnelova D, Svoboda M, et al. MiR-215-5p reduces liver metastasis in an experimental model of colorectal cancer through regulation of ECM-receptor interactions and focal adhesion. Cancers. 2020;12(12):3518.
97. Wei Z, Zhongqiu T, Lu S, Zhang F, Xie W, Wang Y. Gene coexpression analysis offers important modules and pathway of human lung adenocarcinomas. J Cell Physiol. 2020;235(1):454–64.
98. Sharma S, Findlay GM, Bandukwala HS, Oberdoerffer S, Baust B, Li Z, et al. Dephosphorylation of the nuclear factor of activated T cells (NFAT) transcription factor is regulated by an RNA-protein scaffold complex. Proc Natl Acad Sci USA. 2011;108(28):11381–6.
99. Spitale RC, Tsai MC, Chang HY. RNA templating the epigenome: long noncoding RNAs as molecular scaffolds. Epigenetics. 2011;6(5):539–43.
100. Wang Kevin C, Chang HY. Molecular mechanisms of long noncoding RNAs. Mol Cell. 2011;43(6):904–14.
101. Bousard A, Raposo AC, Żylicz JJ, Picard C, Pires VB, Qi Y, et al. The role of Xist-mediated Polycomb recruitment in the initiation of X-chromosome inactivation. EMBO Rep. 2019;20(10):e48019.
102. Colognori D, Sunwoo H, Kriz AJ, Wang CY, Lee JT. Xist Deletional analysis reveals an interdependency between Xist RNA and Polycomb complexes for spreading along the inactive X. Mol Cell. 2019;74(1):101–17.e10.
103. Li J, Chen JN, Zeng TT, He F, Chen SP, Ma S, et al. CD133+ liver cancer stem cells resist interferon-gamma-induced autophagy. BMC Cancer. 2016;16:15.
104. Meseure D, Vacher S, Alsibai KD, Nicolas A, Chemlali W, Caly M, et al. Expression of ANRIL-Polycomb complexes-CDKN2A/B/ARF genes in breast tumors: identification of a two-gene (EZH2/CBX7) signature with independent prognostic value. Mol Cancer Res. 2016;14(7):623–33.
105. Zhang E, Han L, Yin D, He X, Hong L, Si X, et al. H3K27 acetylation activated-long non-coding RNA CCAT1 affects cell proliferation and migration by regulating SPRY4 and HOXB13 expression in esophageal squamous cell carcinoma. Nucleic Acids Res. 2017;45(6):3086–101.
106. Qi X, Zhang DH, Wu N, Xiao JH, Wang X, Ma W. ceRNA in cancer: possible functions and clinical implications. J Med Genet. 2015;52(10):710–8.
107. Yang C, Wu D, Gao L, Liu X, Jin Y, Wang D, et al. Competing endogenous RNA networks in human cancer: hypothesis, validation, and perspectives. Oncotarget. 2016;7(12):13479–90.
108. Qu L, Ding J, Chen C, Wu ZJ, Liu B, Gao Y, et al. Exosome-transmitted lncARSR promotes Sunitinib resistance in renal cancer by acting as a competing endogenous RNA. Cancer Cell. 2016;29(5):653–68.

109. Jia P, Cai H, Liu X, Chen J, Ma J, Wang P, et al. Long non-coding RNA H19 regulates glioma angiogenesis and the biological behavior of glioma-associated endothelial cells by inhibiting microRNA-29a. Cancer Lett. 2016;381(2):359–69.
110. Hardwick SA, Joglekar A, Flicek P, Frankish A, Tilgner HU. Getting the entire message: Progress in isoform sequencing. Front Genet. 2019;10:709.
111. Lim B, Lin Y, Navin N. Advancing cancer research and medicine with single-cell genomics. Cancer Cell. 2020;37(4):456–70.
112. Tomczak K, Czerwińska P, Wiznerowicz M. The cancer genome atlas (TCGA): an immeasurable source of knowledge. Contemp Oncol (Pozn). 2015;19(1a):A68–77.
113. Xiyuan L, Dechao B, Liang S, Yang W, Shuangsang F, Hui L, et al. Using the NONCODE database resource. Curr Protoc Bioinformatics. 2017;58(1):2.6.1–9.
114. Morlando M, Fatica A. Alteration of epigenetic regulation by long noncoding RNAs in cancer. Int J Mol Sci. 2018;19(2):570.

# Chapter 6
# Non-coding RNAs: MicroRNAs (miRNAs)

**Abstract** microRNAs (miRNAs) serve as post-translational modulators for numerous genes, acting not only as indicators of cancer onset and progression but also as markers for the efficacy of anticancer treatments. Their ability to adjust cellular responses to treatments positions them as valuable targets in the battle against various diseases. With the advancement of technologies, researchers have developed numerous strategies to explore the optimal pathways for miRNA delivery to diseased cells, including the use of synthetic miRNA analogs, miRNA-encoding DNA plasmids, and compounds that modify miRNA expression through epigenetic mechanisms. Evidence from animal studies and preclinical research in both solid tumors and blood cancers has supported the efficacy of microRNA-based treatment approaches. However, challenges persist in the clinical application of miRNAs, such as creating stable nanoformulations, selecting appropriate delivery methods, and managing biodistribution. This chapter aims to outline the process of miRNA biogenesis and their use in treating human diseases.

## 6.1 MicroRNA Biogenesis

MicroRNAs (miRNAs) are abundant across various species and can be generated by numerous genes. The transcription process of miRNAs typically yields a primary miRNA transcript (pri-miRNA). These pri-miRNAs feature a cap at their 5′-end and are polyadenylated at their 3′-end, similar to mRNA processing. This structure is crucial for the subsequent processing and maturation of miRNAs into their functional forms, which play key roles in regulating gene expression by targeting messenger RNAs for degradation or translational repression [1]. A single transcript might carry more than one miRNA. This is especially true when the miRNA sequence embedded within an intron. It is believed that many new miRNAs in animals originate through the gradual modification of nucleotide sequences rather than through gene duplication [2]. When a novel miRNA sequence develops in an existing transcription unit, it can be expressed right away, eliminating the need for new enhancers and promoters. This process facilitates the emergence of new miRNA



P. E. Saw, E. Song, *RNA Therapeutics in Human Diseases*,
https://doi.org/10.1007/978-981-96-3041-7_6

genes without the necessity for gene duplication, possibly explaining why miRNA genes frequently produce multiple distinct products (i.e., pri-miRNA, miRNA, and proteins).

Further processing of miRNAs heavily depends on the ability of the miRNA sequence to form a stem-loop 3D structure. Typically, a primary miRNA (pri-miRNA) features an imperfectly-paired stem approximately 33 base pairs in length, topped by a terminal loop, and flanked by additional sequences. This specific structure is crucial as it is recognized and processed by the microprocessor complex, which includes the Drosha enzyme and its partner DGCR8. The complex cleaves the pri-miRNA near the base of the stem, removing the flanking sequences and the terminal loop, and leaving behind a precursor miRNA (pre-miRNA). This pre-miRNA is then further processed to form a mature miRNA, which is capable of regulating gene expression. The structural requirements of pri-miRNAs ensure the accuracy and efficiency of these cleavage events, critical for the generation of functional miRNAs [3]. For the processing steps, the initial phase occurs in the nucleus, where the stem-loop structure of the primary miRNA (pri-miRNA) is separated from the rest of the transcript, resulting in a precursor miRNA (pre-miRNA). This cleavage is typically carried out by a nucleas enzyme belonging to the RNase III family. In animals, this enzyme is termed Drosha, while in plants, it is known as Dcl1. This process is critical as it ensures that the pri-miRNA is trimmed to a shorter, more manageable pre-miRNA, which is then exported to the cytoplasm for further processing [1]. Although Drosha is crucial in the processing of pri-miRNA, its function depends on a protein cofactor that improves the precision and efficiency of the cleavage. This cofactor, which contains two ds RNA-binding domains (dsRBDs), forms a stable association with Drosha, creating what is known as the Microprocessor complex [4, 5]. Nevertheless, in animals, there is an alternative pathway for generating pre-miRNAs that circumvents the microprocessor. This route involves splicing pri-miRNA transcripts to release introns, known as mirtrons, which have a structure similar to pre-miRNAs [6, 7]. These mirtrons subsequently integrate into the miRNA processing pathway without relying on microprocessor activity. While not very common, mirtrons are found in different animal species. In the next processing stage, the terminal loop form the pre-miRNA has to be removed to produce a mature miRNA duplex that is ~22 bp [3].

One notable distinction between miRNAs and siRNAs lies in the accuracy of their terminal structures. miRNAs, resembling traditional gene products, typically exhibit highly specific ends with minimal variation. Conversely, siRNAs display a greater heterogeneity at their ends. This characteristic of miRNAs likely contributes to their ability to bind target mRNAs with high specificity, despite not requiring strict complement or extensive overlap in their target mRNA. Therefore, the enzymatic process is specially designed to produce miRNA duplexes with accurately defined 5′- and 3′-ends, making the initial cleavage essential. In mammals, this vital incision is performed by Drosha, aided by its partner protein DGCR8, which contains dsRBD domains. DGCR8 could have a direct interaction with both the stem area of the pri-miRNA, and also and single-stranded region of the neighbors [8]. In this intricate system, Drosha performs the cleavage of primary miRNA (pri-miRNA), but its functionality hinges significantly on DGCR8. Drosha itself does not directly bind to the pri-miRNA. Instead, its interaction and positioning are mediated by DGCR8, which acts as a molecular guide. DGCR8 identifies the junction between the stem of the pri-miRNA and the flanking unstructured sequences. This identification is critical as Drosha's catalytic activity requires precise placement relative to this stem-flank junction. The enzymatic cleavage by Drosha is specifically situated one helical turn, or 11 base pairs, away from this junction. This strategic cleavage defines one end of the mature miRNA, emphasizing the importance of the stem's end in determining the mature miRNA's final structure [1, 9]. One key component is the PAZ domain within Dicer, which plays a critical role in binding to the 2-nucleotide 3′ overhangs typically present at the ends of dsRNA substrates. This binding effectively positions the RNA within the Dicer enzyme in a manner akin to using a ruler, ensuring that cleavage occurs at a consistent and specific distance from the RNA end. The catalytic mechanism of Dicer involves its RNase III domains, which are responsible for the enzymatic cleavage of RNA. These domains are situated such that the cleavage site is precisely two helical turns, or about 22 base pairs, away from the RNA terminus where the PAZ domain binds. This architectural feature of Dicer allows it to produce RNA duplexes with remarkably consistent lengths, essential for the functionality of miRNAs in gene regulation [10].

Regulating the biogenesis of miRNAs is critical, yet it remains underexplored. Interestingly, numerous miRNA genes are regulated by the very targets they modulate. In Drosophila, the regulatory relationship between the miR-7 gene and the transcription factor Yan exemplifies a sophisticated feedback mechanism within cellular signaling pathways. Yan, which contains an ETS domain, actively represses the transcription of the miR-7 gene. This regulatory interaction ensures that when Yan is active, the levels of miR-7 are kept low, maintaining a controlled balance of gene expression. In a reciprocal manner, miR-7 targets and inhibits the translation of the Yan protein. This forms a double-negative feedback loop, where each component suppresses the other to maintain cellular homeostasis. Such a feedback mechanism is crucial for fine-tuning gene expression and cellular responses, particularly in developmental processes and cell differentiation where precise control of transcription factors like Yan is required [11]. In *C. elegans,* a similar feedback mechanism involves the let-7 miRNA and the protein Lin28, showcasing an intricate regulatory loop that plays a crucial role in timing developmental transitions and maintaining cellular fate decisions. The let-7 miRNA directly targets and suppresses the translation of Lin28, a regulatory protein that has significant implications in developmental timing and stem cell renewal. Conversely, Lin28 inhibits the maturation of let-7 miRNA by binding to the pre-let-7 (precursor let-7), preventing its processing into mature miRNA. This suppression of let-7 by Lin28 serves to maintain higher levels of Lin28 during earlier developmental stages, which is critical for the proper timing of developmental events [12]. Lin28, which shows minimal nuclear presence but is predominantly cytoplasmic, impacts let-7 processing in both locations. It hinders Drosha-mediated cleavage and also blocks Dicer's action [13–16]. Additionally, Lin28 induces polyuridylation at the 3′ terminus of pre-let-7

[14], which makes it resistant to Dicer, suggesting a mechanism by which Lin28 could inhibit miRNA maturation. Alternatively, polyuridylation may target residual pre-let-7 RNAs that Dicer cannot cleave due to Lin28 interference. The importance of these double-negative feedback loops lies in their ability to tightly regulate miRNA biogenesis. This stringent control prevents miRNA misexpression, which can mimic loss-of-function phenotypes of their target genes. Such regulation likely serves to minimize unintended effects from errant miRNAs, maintaining specificity and efficacy in gene regulation (Fig. 6.1).

## 6.2 Post-Transcriptional Repression by miRNAs

miRNAs act as adaptors within the miRISC complex, allowing it to selectively target and regulate specific mRNAs. When miRISC is associated with an exogenous RNA recognition factor, it can suppress mRNAs that lack miRNA binding sites [18]. In animals, miRNA binding sites are primarily located within the 3′-UTRs of mRNAs, often appearing in clusters. Unlike their plant counterparts, animal miRNAs typically engage their targets through partial matches that may include mismatches and bulges. Crucially, the recognition between miRNA and its target mRNA hinges on Watson-Crick base pairing within the miRNA's "seed region," which spans 2–8 nt. This specific region is critical for ensuring the correct targeting and binding of the miRNA to its complementary mRNA sequence. In animals, the partial complementarity often leads to translational repression or mRNA destabilization, whereas in plants, the high complementarity can trigger direct mRNA cleavage and degradation. This difference underlines the diverse evolutionary adaptations of miRNA-mediated gene regulation across different kingdoms. High complementarity permits Ago to catalyze mRNA strand cleavage, while mismatches in central regions generally lead to repression of mRNA translation without cleavage. This observation has supported the view that plant and animal miRNAs function differently, with cleavage excluding translational repression. Nevertheless, recent genetic studies have challenged this perspective.

The debate around how the miRISC affects translation primarily centers on whether it impacts the initiation phase or the post-initiation stages of this process. Understanding the basics of translation initiation helps clarify the potential points of interference by miRISC. Translation initiation in eukaryotic cells begins with the eIF4E subunit of the eIF4F complex recognizing and binding to the 5′ cap of the mRNA. eIF4G enhances the interaction between the mRNA and the ribosomal machinery by linking with eIF3, which aids in recruiting the 40S ribosomal subunit. Once the 40S preinitiation complex is formed, it scans the mRNA to locate the AUG start codon, where it joins with the 60S ribosomal subunit, thus forming the complete 80S ribosome and initiating elongation. eIF4G also connects with the polyA-binding protein (PABP1) at the mRNA's 3′ end, facilitating the circularization of the mRNA. This dual interaction allows eIF4G to effectively circularize the mRNA, a configuration that significantly boosts translation efficiency, especially when mRNA

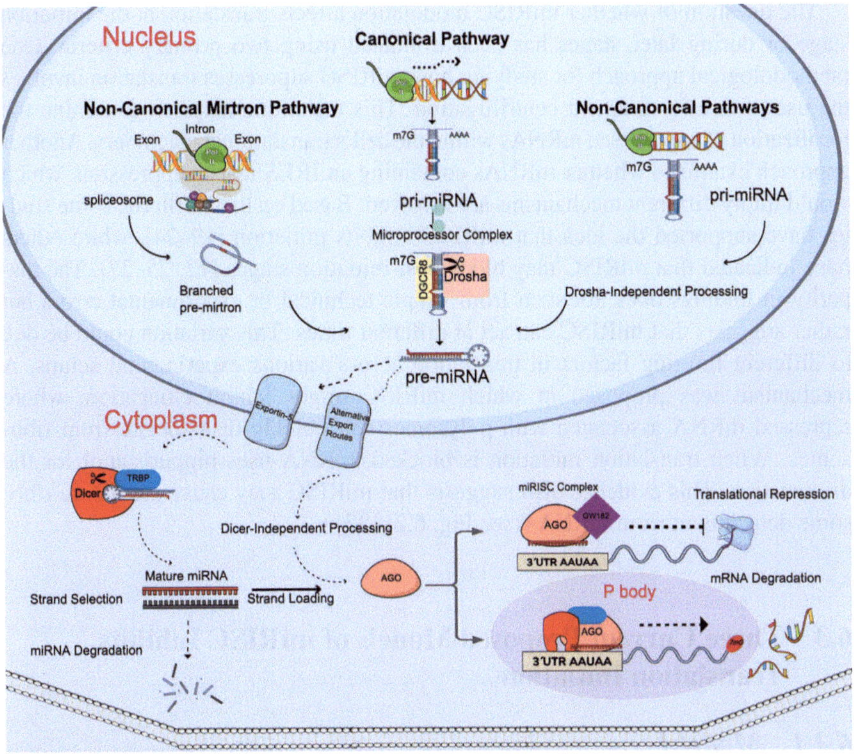


**Fig. 6.1** Schematic overview of miRNA biogenesis pathways and their function in gene regulation. This diagram illustrates the complex processes involved in the generation and function of microRNAs (miRNAs) within the cell, distinguishing between canonical and non-canonical pathways. (i) Canonical pathway begins in the nucleus where the primary miRNA (pri-miRNA) is transcribed. The pri-miRNA undergoes processing by the microprocessor complex, which includes Drosha and DGCR8, resulting in the precursor miRNA (pre-miRNA). The pre-miRNA is then exported to the cytoplasm via Exportin-5. In the cytoplasm, Dicer, along with the TAR RNA-binding protein (TRBP), further processes the pre-miRNA into mature miRNA. (ii) Non-canonical pathways are divided into two distinct pathways, namely (**a**) Mirtron Pathway: This non-canonical pathway involves splicing of introns (mirtrons) which bypass Drosha processing. The resulting branched pre-mirtrons are then debranched and mimic pre-miRNA, entering the miRNA processing pathway directly in the cytoplasm. (**b**) Drosha-Independent Processing: Some pri-miRNAs are processed without Drosha, relying on alternative splicing and other mechanisms to generate pre-miRNAs. Once mature miRNAs are generated, they are incorporated into the RNA-induced silencing complex (miRISC), which includes the Argonaute (AGO) protein. The miRISC complex facilitates the miRNA-mediated gene regulation through translational repression or mRNA degradation. This regulation occurs primarily in the cytoplasm, often in structures known as P-bodies, where mRNA is stored and degraded. (Figure is reproduced from [17] under Creative Commons Attribution 4.0 International License)

levels are high. Certain viral mRNAs can begin translation independently of traditional initiation factors or might only need a limited set since there are internal ribosome entry site (IRES) present.

The question of whether miRISC modulation affects translation at the initiation stage or during later stages has been evaluated using two primary criteria. One methodological approach for studying how miRISC suppresses translation involves the use of density gradient centrifugation. This technique helps to determine the localization of suppressed mRNAs within the cell's translational machinery. Another approach examines whether mRNAs containing an IRES resist suppression, which would imply different mechanisms are involved. Based on these criteria, some studies have supported the idea that miRISC inhibits initiation [19–24], while others have indicated that miRISC may block post-initiation stages [12, 25–27]. The disparity in findings does not stem from simple technical or experimental errors but rather suggests that miRISC can act at different times. This variation could be due to different limiting factors in translation across various experimental setups. A mechanism was proposed in which miRISC might inhibit elongation, where repressed mRNA associated with polysomes can quickly disassociate from ribosomes. When translation initiation is blocked, mRNA uses hippuristanol for the dissociation. This evidence also suggests that miRISC may cause premature ribosome detachment from mRNAs (see Fig. 6.2) [27].

## 6.3 Three Current Proposed Models of miRISC Inhibits Translation Initiation

### *6.3.1 Model 1*

Cell-free systems have played a crucial role in uncovering the mechanisms through which the miRISC inhibits translation initiation, although these studies have sparked some debate. Presently, there are three main theoretical models proposed to explain how miRISC might suppress translation initiation. One prevailing theory posits that the miRISC competes with the initiation factor eIF4E for binding to the 5′ cap of mRNA. If miRISC competes with eIF4E at this critical site, it could interfere with the initiation of translation. To test this interaction, experiments have demonstrated that adding an excess of the eIF4F complex, which includes eIF4E, which will then counteract the repressive effect of miRISC. Specifically, in an ascites-derived cell-free system, the introduction of purified eIF4F successfully alleviated miRISC-mediated repression. This finding not only supports the competitive interaction theory but also highlights a potential intervention strategy where modulating the components of the translation initiation complex could relieve negative regulation by miRISC, potentially enhancing protein synthesis in conditions where it is undesirably suppressed [22]. It has been posited that the mid-domain of human Ago2, containing two phenylalanine residues, may mimic the eIF4E structure and thereby compete for cap binding. Mutating the phenylalanine residues in Ago2 significantly diminishes its ability to repress translation and its capacity to bind to m7GTP-coupled beads in vitro. This suggests that Ago2 may directly compete with eIF4E, the eukaryotic initiation factor that normally binds to the cap structure of mRNA,

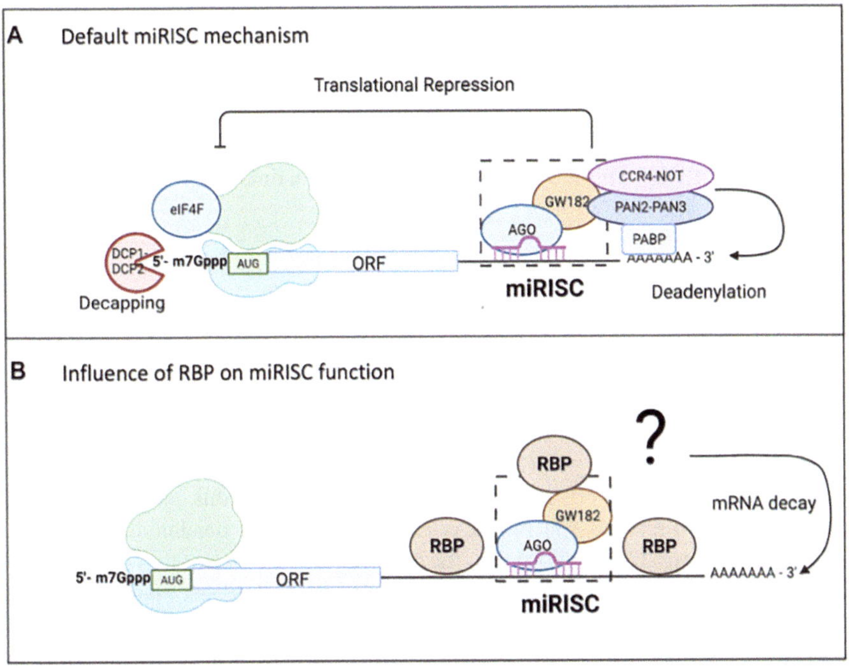


**Fig. 6.2** Mechanisms of miRNA-induced gene silencing and the modulatory role of RNA-binding proteins (RBPs) on miRISC activity. (**a**) Default miRISC mechanism: In this panel, the core mechanism of miRNA-induced gene silencing by the microRNA-induced silencing complex (miRISC) is illustrated. miRISC is composed of Argonaute (AGO) proteins bound to a miRNA strand, which guides the complex to complementary sequences in the 3′ untranslated region (UTR) of target mRNAs. Binding of miRISC to mRNA leads to translational repression and initiates two primary mRNA degradation processes: deadenylation and decapping. For deadenylation, miRISC interacts with GW182, which recruits the CCR4-NOT and PAN2-PAN3 deadenylase complexes, promoting the gradual removal of the poly(A) tail by these deadenylases, with poly(A)-binding protein (PABP) also shown associating with the mRNA. The loss of the poly(A) tail destabilizes the mRNA, marking it for degradation. Concurrently, decapping of the mRNA 5′ cap structure (m7Gppp) is catalyzed by the DCP1-DCP2 complex, leading to further mRNA destabilization and degradation. Both deadenylation and decapping promote mRNA decay, reducing the target gene expression at the post-transcriptional level. (**b**) Influence of RNA-binding proteins (RBPs) on miRISC function: This panel illustrates how RBPs can modulate the silencing function of miRISC. RBPs, which are shown binding at various locations on the mRNA, can interact with or influence miRISC activity, although the exact effects of these interactions remain incompletely understood (indicated by the question mark). RBPs may either facilitate or inhibit miRISC-mediated repression and mRNA decay, depending on the specific RBP and its binding context. This complex interplay between RBPs and miRISC can modulate mRNA stability and turnover, ultimately impacting the efficiency of miRNA-induced gene silencing. (Figure is reproduced from [28] under Creative Commons CC-BY license)

for cap binding. This finding is crucial as it hints at a direct interference mechanism by which Ago2 could disrupt the normal initiation process of translation, diverting the mRNA from being efficiently translated [21]. However, subsequent research challenges this hypothesis, particularly studies on Drosophila. In Drosophila, the

Ago1 Mid domain also contains phenylalanine residues; their mutation disrupts Ago1's translational repression ability but not its binding to m7GTP beads. Instead, these mutations impair Ago1's interaction with GW182. Furthermore, evidence shows that GW182 alone can repress translation when bound to mRNAs, suggesting that GW182 or another intermediary, rather than Ago1 itself, might be the actual competitor of eIF4E. This shifts the focus away from a direct cap-binding competition involving Ago1.

### 6.3.2 Model 2

A second hypothesis posits that the miRISC promotes the deadenylation of the mRNA tail, leading to translational repression. This occurs because the cap structure of the mRNA and the poly(A)-binding protein 1 (PABP1)-free tail of the deadenylated mRNA cannot effectively circularize. Supporting this model, numerous studies have documented that miRNAs can trigger mRNA deadenylation in vivo [29–31] and in vitro [23]. Importantly, the occurrence of deadenylation is not merely a secondary result of inhibited initiation, as evidenced by IRES-containing mRNAs that also undergo deadenylation despite their resistance to miRISC-mediated translational repression [23]. The mechanism driving this deadenylation involves GW182, which has been demonstrated to initiate both deadenylation and translational repression in Drosophila cells [29]. However, this model is not without its challenges, as evidence are pointing toward the fact that miRNAs can silence translation in mRNAs that is lacking a poly(A) tail [20, 31, 32]. This repression is contingent on the presence of Ago and GW182 [32].

### 6.3.3 Model 3

A third proposed model for miRISC-mediated translation repression suggests that miRISC blocks the joining of the 60S ribosomal subunit to the 40S preinitiation complex. This inhibition mechanism is crucial for understanding how gene expression is controlled at the translational level. Supporting this model, there is evidence showing that human Ago2 physically interacts with eIF6 and the 60S ribosomal subunits in vitro. This interaction suggests a direct role of Ago2 in preventing the assembly of a functional 80S ribosome, thereby halting translation initiation. This model highlights the intricate nature of cellular control mechanisms and underscores the potential of targeting these interactions in therapeutic applications, particularly in diseases where translational control is disrupted [33]. Notably, experimental depletion of eIF6 in human cells or in the model organism *Caenorhabditis elegans* can counteract miRNA-mediated inhibition. This suggests that disrupting the interaction between eIF6 and ribosomal subunits can relieve the miRNA-imposed repression of translation. This insight reveals a potential therapeutic avenue where modulating eIF6 levels could serve to control miRNA activity, thereby adjusting the translation process in diseases where miRNA regulation is out of balance [32]. However, similar experiments in Drosophila cells show little to no change in silencing, indicating species-specific mechanisms or differences in miRISC function across organisms. Additional support for this model is derived from studies using a reticulocyte cell-free system [34]. In this context, when targeted mRNAs are introduced to a miRNA-programmed lysate, they predominantly associate with 40S ribosomal subunits rather than 60S subunits.

## 6.4 mRNA Degradation Via miRNAs

Initial research suggested that miRNA did have the ability to repress translation did not necessarily involve mRNA destabilization. Nonetheless, subsequent studies have shown that certain miRNA-target interactions lead to a marked decrease in mRNA levels, primarily due to enhanced mRNA degradation rather than Ago-catalyzed cleavage [29–31, 35, 36]. This degradation process begins with deadenylation, followed by decapping, and concludes with exonucleolytic digestion of the mRNA [29–31]. Evidence indicates that degradation can occur independently of translation. For example, mRNAs whose translation has been inhibited still undergo miRNA-dependent deadenylation [23, 31], suggesting that degradation could serve as a separate mechanism of miRNA-mediated repression for specific targets. The reasons why only certain targets undergo degradation remain unclear [37].

**Box 6.1 Mechanism of miRNA Silencing: Blind Men and the Elephant?**
Reconciling the varied accounts of miRNA mechanisms is challenging, as the diversity in observations might be attributed to differences in experimental methods, yet other factors are likely more influential. One possible explanation could be that miRNA silencing operates via a single overarching mechanism, with the observed variations being downstream secondary events. Another explanation posits the existence of multiple miRNA mechanisms, each predominating under specific but not yet fully understood conditions. This situation resembles the 'blind men and the elephant' fable, where a complete understanding remains elusive. This complexity underlines the importance of considering these phenomena within the broader context of cellular activity and post-transcriptional gene regulation. A notable example of context-dependent miRNA function is observed in the cell cycle. Research

(continued)

**Box 6.1** (continued)
has shown that the microRNA let-7 and a synthetic miRNA targeting CXCR4 can inhibit protein synthesis in rapidly dividing human cells. Intriguingly, these miRNAs exhibit a functional shift under certain conditions; when cellular division is paused at the G1 checkpoint due to lack of serum, they begin to act as enhancers of translation instead [38]. This change in function does not depend on whether cells are in a static or dynamic state but rather on the specific phase of the cell cycle. Similarly, arrest induced by aphidicolin at the G1 phase leads to translational activation, while arrest caused by nocodazole at the G2/M phase results in translational repression. This pattern, however, is not universally observed as studies in Drosophila eye cells indicate that G1 arrest still results in miRNA-mediated repression [11]. For instance, under normal homeostasis condition, miR-369-3p acts as a repressor. However, upon stress (i.e., lymphocyte growth arrest) the same miRNA activates TNF-$\alpha$, a process crucial for macrophage maturation [38].

## 6.5 miRNA Interaction with Other Organelles

The organization and modulation of small RNA regulation within cells is a crucial area of study. Proteins from the Ago family are frequently associated with cellular compartments involved in membrane trafficking, such as the ER and the Golgi apparatus [39]. Evidence from cytosolic extraction studies supports this association [20]. It is therefore plausible to suggest that miRISC components may be tethered within specific subcellular locations. Notably, distinct cytoplasmic structures such as P bodies or GW bodies where translationally inactive mRNAs tend to gather, also accumulate components of miRISC such as miRNAs, Ago, and GW182 along with their silenced targets [20, 40, 41]. However, the presence of GW bodies is not critical for the repression activity of miRNAs, as evidenced by experiments showing that depletion of certain GW body proteins leads to their dispersal but does not impact miRNA function [42]. Instead, the formation of GW bodies requires a functional miRNA pathway, indicating that the initiation of silencing takes place somewhere else in the cytoplasm and the aggregation into GW bodies is a subsequent event. Fluorescence recovery after photobleaching studies have shown that the exchange of Ago proteins within GW bodies with other cytoplasmic Ago molecules is remarkably slow [43]. This raises the question of whether these bodies are the final destinations for miRISC machinery bound to their targets, or if mechanisms exist for recycling miRISC to engage with additional targets. Interestingly, the protein Imp8, belonging to the Importin family, has been shown to enhance miRISC-mediated silencing by aiding the binding of Ago2 complexes to target mRNAs [44].

## 6.6 miRNA in Diseases

As mentioned above, the regulation of gene expression by miRNAs is a fundamental aspect of cellular homeostasis and normal physiological functions. These small, non-coding RNA molecules influence gene activity by binding to target mRNAs and regulating their stability or translation. This regulatory role is crucial for fine-tuning cellular processes such as differentiation, proliferation, and apoptosis—key elements in maintaining healthy cellular function. When the normal function of miRNAs becomes dysregulated—whether through alterations in their expression levels, mutations affecting their structure, or changes in their target specificity—it can lead to significant disruptions in cellular signaling networks [45]. Such disruptions are a common pathway toward the development and progression of various diseases. For example, in cancer, certain miRNAs known as "oncomiRs" may become overexpressed and suppress the function of tumor suppressor genes, promoting cancer cell growth and survival [46]. Conversely, miRNAs that normally inhibit cancerous behaviors might be downregulated in tumor cells, further contributing to malignant progression.

For example, in cardiovascular diseases, miRNAs have been shown to influence heart muscle cell function, vascular development, and the response to myocardial infarction (heart attack). Dysregulated miRNA expression can lead to arrhythmias, hypertrophy, or heart failure by affecting how heart cells grow, die, or respond to stress. Similarly, in neurodegenerative diseases such as Alzheimer's or Parkinson's, abnormal miRNA levels can affect the expression of genes involved in nerve cell function and survival, potentially leading to progressive loss of neuronal function and disease symptoms [47]. The pervasive influence of miRNAs in such a broad range of diseases highlights their potential as targets for therapeutic intervention. By modulating the levels or activity of specific miRNAs, researchers hope to correct the underlying genetic and molecular imbalances that drive disease. Additionally, because miRNA expression patterns are often uniquely altered in different diseases, these molecules also hold promise as biomarkers for more precise diagnosis and disease monitoring. Their distinct role and function in human diseases are summarized in Table 6.1.

## 6.7 Circulating miRNAs and Their Implication in Diseases

Extensive research has examined the concentrations of circulating miRNAs across a spectrum of human diseases. For example, specific alterations in miRNA levels have been associated with tumor progression in cancer patients, damage to heart tissue in those with cardiovascular diseases, and brain injuries in stroke survivors. Understanding these correlations could pave the way for the development of miRNA-based diagnostics and therapeutics, offering a more nuanced and effective approach to managing these serious health conditions [63, 64]. The connection

**Table 6.1** Specific miRNAs and their role in various human diseases

| miRNA | Disease subtype | Function | Ref |
|---|---|---|---|
| In cancer | | | |
| miR-15 and miR-16 | Chronic lympocytic leukemia (CLL) | Were reduced or absent in approximately 68% in CLL patients | [48] |
| miR-205 | Lung cancer | Differentiate between the two primary forms of non-small-cell lung cancer (NSCLS) with over 90% specificity | [49] |
| Autoimmune diseases | | | |
| miR-146 and miR-155 | Rheumatoid Arthritis (RA) | Increment of miR-146 and miR-155 in RA patients' synovial fluid and PBMC | [50] |
| miR-203, mi-346, miR-124a | Rheumatoid Arthritis (RA) | Increment of miR-203 and miR-346 Decrease in miR-124a | [51] |
| miR-21, miR-155, and miR-146a | Crohn's disease and ulcerative colitis | miR-21, miR-155, and miR-146a are upregulated in the inflamed intestinal tissues of IBD patients | |
| miR-146a, miR-203, miR-21, miR-125b, miR-31 | Psoriasis | miR-146a, miR-203, and miR-21 are upregulated, while miR-125b and miR-31 are downregulated in psoriatic skin lesion | |
| miR-21, miR-155, and miR-146a | Asthma | Regulating the production of inflammatory cytokines, the proliferation of airway smooth muscle cells, and mucus secretion | |
| miR-21, miR-155, miR-146a, miR-29 family | Type 1 diabetes (T1D) | Dysregulated in the pancreatic islets or immune cells of T1D patients. These miRNAs influence immune cell function, cytokine production, and beta cell apoptosis | |
| miR-146a, miR-155, miR-21, and miR-221 | Autoimmune thyroid diseases | Dysregulation of these miRNAs play roles in regulating immune responses, thyroid hormone synthesis, and thyroid cell proliferation | |
| Neurological diseases | | | |
| miR-29, miR-107 | Alzheimer's disease (AD) | Changes in miRNA levels within various brain regions, including cerebrospinal fluid (CSF) and PBMC, with significant reductions in miR-29 and miR-107, both of which target BACE | [52] |
| miR-133b, miR-34b/c | Parkinson disease (PD) | Elevated levels of miR-133b in the brain tissues of PD patients. This elevation is part of a regulatory feedback loop involving miR-133b and the Pitx3 transcription factor, which miR-133b targets | [53] |
| | | miR-34b/c levels are found to be reduced in PD cases | [52] |

(continued)

**Table 6.1** (continued)

| miRNA | Disease subtype | Function | Ref |
|---|---|---|---|
| miR-181 | Stroke | Distinct regional variations in miR-181 levels have been observed, with increased levels in ischemic areas and decreased levels in the surrounding penumbra | [54] |
| miR-124, miR-497, miR-12, miR-298, miR-50, and miR-672 | | Upregulated in ischemic region | [55, 56] |
| miR-155, miR-362-3p | | Downregulated in ischemic region | [55, 56] |
| miR-210 | | Emerging as a potential novel blood biomarker for the diagnosis and prognosis of acute cerebral ischemia | [57] |
| miR-24, miR-26b, miR-29c, and miR-7 | Schizophrenia and bipolar disorder | Identified alterations in the levels of miR-24, miR-26b, miR-29c, and miR-7, as well as an increase in dicer activity | [52, 58] |
| miR-30a, miR-195 | Schizophrenia | The expression of BDNF is influenced by miR-30a and miR-195 | [58] |
| miR-34a, miR-152, miR-155, and miR-221 | Bipolar disorder | Modifications in miRNA expression have been observed in both brain tissue and plasma samples | [59] |
| Cardiovascular diseases | | | |
| miR-1 | Heart | Regulates processes such as morphogenesis, electrical conduction, and cell cycle control | [60, 61] |
| miRs-208a, miR-208b, and miR-499 (myomiRs) | | Altered levels in cardiac conditions | [60] |
| miR-15 | Cardiac ischemic injury | Blocking miR-15 has been shown to protect mice from cardiac ischemic injury by reducing the size of the infarct and improving heart function | [62] |
| miR-21, miR-133, miR-208a/b | Cardiac hypertrophy and heart failure | It has been associated with fibrosis, apoptosis, and oxidative stress in the heart | |
| miR-29 | Cardiac fibrosis | miR-29 targets genes involved in extracellular matrix remodeling | |
| miR-126 | Atherosclerosis and ischemic heart disease | This miRNA is involved in angiogenesis and vascular integrity | |
| miR-34a | Cardiac aging and heart failure | It promotes cardiomyocyte senescence and apoptosis | |

between circulating miRNAs and cancer was first established in 2008. Researchers observed a significant increase in the levels of miR-155, miR-21, and miR-210 in the serum of patients with diffuse large B-cell lymphoma (DLBCL). Specifically, the levels of miR-21 were found to correlate with the duration of relapse-free survival. This finding highlighted the potential of circulating miRNAs as biomarkers for cancer prognosis and disease monitoring [65]. In a study involving lung cancer patients, circulating miRNAs were used as markers to observe the patients' length of survival. Interestingly, it was found that in patients with shorter survival times, the levels of miR-486 and miR-30d were elevated, whereas the levels of miR-1 and miR-499 were decreased [66]. In another study, researchers isolated exosomes from the plasma of patients with non-small cell lung cancer (NSCLC) and healthy controls. They found that 12 miRNAs were detected at similar levels in both the tumor tissues and plasma of NSCLC patients, yet these miRNAs were undetectable in the healthy control group [67]. This observation suggests that specific miRNAs could serve as potential prognostic indicators, offering insights into the aggressiveness of the disease and possibly guiding treatment decisions.

In solid tumors, circulating miRNAs have shown potential as biomarkers in several types of cancer, including colorectal, ovarian, and breast cancers. Studies focusing on these cancers have identified a dysregulated expression profile of several circulating miRNAs, specifically let-7a, miR-10b, miR-21, miR-34a, miR-155, and miR-195. The altered levels of these miRNAs in the bloodstream reflect changes occurring in the tumor environment, suggesting that they could serve as accessible and non-invasive markers for cancer diagnosis, prognosis, and possibly for monitoring treatment responses [68–71]. For example, elevated level of miR-195 is a biomarker to determine malignancy of breast cancer [72]. Interestingly, miR-34a, miR-10b, and miR-155 could be used to indicate different clinical stages of breast cancer [70]. Additionally, in the context of breast cancer (BC) treatment, levels of circulating miR-195 and let-7a were observed to decrease post-operation. This change suggests that these specific miRNAs could be useful in monitoring therapeutic efficacy, providing a potential tool for clinicians to assess the success of surgical interventions and possibly other treatments in real-time [69]. In ovarian cancer, studies have identified specific patterns in the levels of circulating miRNAs, which could serve as potential biomarkers. Specifically, miRNAs such as miR-21, miR-92, miR-93, miR-126, and miR-29a were found to be elevated in patients, while levels of miR-155, miR-127, and miR-99b were reduced [73]. In colorectal cancer, research is actively exploring the use of various circulating miRNAs as diagnostic biomarkers. Key miRNAs such as miR-17-3p, miR-29a, miR-92a, miR-141, and miR-221 are under investigation for their potential to help in early detection and diagnosis. These miRNAs have been identified because of their altered levels in the blood of colorectal cancer patients compared to healthy individuals [74]. Elevated serum miR-29a levels have been associated with liver metastasis in CRC [63].

Levels of circulating miRNA have been assessed across various cardiac conditions. In cases of acute myocardial infarction, significant changes were observed in the levels of specific miRNAs, including miR-1, miR-133, miR-208a, miR-328, and miR-499. These miRNAs are potentially valuable as biomarkers for early detection and assessment of cardiac injury. Their altered levels can provide critical insights into the extent of myocardial damage, helping clinicians in the diagnosis and management of heart attack patients [60]. Additionally, the levels of miR-208b and miR-499 were observed to correlate with the serum concentrations of troponin T and creatine phosphokinase (CPK), which are established biomarkers indicative of cardiac-related injury. This correlation suggests that miR-208b and miR-499 could serve as complementary biomarkers for assessing the severity and extent of myocardial damage [60]. In cases of coronary artery disease (CAD), significant alterations were observed in the levels of six specific miRNAs: miR-17, miR-92a, miR-126, miR-130a, miR-221, and miR-222. These miRNAs have shown potential as biomarkers for detecting and monitoring CAD, reflecting changes in vascular function and atherosclerotic processes. Meanwhile, miR-423-5p was primarily impacted in cases of congestive heart failure, suggesting its unique role in reflecting the pathophysiological changes associated with heart failure [50, 63]. Circulating miRNAs were also observed in autoimmune diseases including SLE, sclerosis, and psoriasis [75, 76]. Notably, serum levels of miR-200a, miR-200b, miR-200c, miR-429, miR-205, and miR-192 were found to be diminished in patients with systemic lupus erythematosus (SLE) [76]. Additionally, circulating miRNAs have been linked to inflammatory conditions. Studies indicate that in cases of sepsis and systemic inflammatory response syndrome (SIRS), levels of miR-126 and miR-146a are diminished in SIRS patients. Similarly, significant reductions in miR-126, miR-146a, and miR-223 have been observed in those suffering from sepsis. These findings suggest that specific miRNAs could serve as crucial indicators of inflammation severity and could potentially guide therapeutic strategies. Targeting these miRNAs may offer a way to modulate the inflammatory response and improve treatment outcomes for patients with these severe conditions. This area of research holds significant promise for the development of RNA-based therapies that could precisely adjust miRNA levels to manage or mitigate inflammatory diseases [63]. Table 6.2 highlights the primary differences in location, function, stability, source, diagnostic

**Table 6.2** Major differences between cellular miRNAs and circulating miRNAs

| Aspect | Cellular miRNA | Circulating miRNA |
|---|---|---|
| Location | Found within cells, particularly in the cytoplasm | Found in bodily fluids such as blood, saliva, urine, and cerebrospinal fluid |
| Function | Regulate gene expression within the cell by targeting mRNAs for degradation or inhibiting their translation | Act as biomarkers for various diseases; potential role in cell-cell communication |
| Stability | Relatively stable within the cellular environment, protected by cellular structures | Highly stable in extracellular environment, often protected within exosomes or bound to proteins |
| Source | Produced by the cell itself | Derived from various cells and can be secreted into fluids via mechanisms like exosome release |
| Diagnostic relevance | Studied primarily for their role in cellular processes and pathologies | Increasingly used as non-invasive biomarkers for diagnosing and monitoring diseases |
| Extraction and analysis | Requires cell lysis and is often studied in tissue samples | Can be extracted from easily accessible bodily fluids, allowing for non-invasive sampling |

relevance, and methods of extraction and analysis between cellular and circulating miRNAs. Cellular miRNAs are integral to internal cellular regulation, whereas circulating miRNAs have crucial implications for non-invasive medical diagnostics and potentially for intercellular communication.

## 6.8 Conclusion

A significant amount of research highlights the essential role of miRNAs in many biological processes, and their role in human disease is vital, including their implications in disease pathways. Alterations in the genes that encode miRNAs can lead to changes in miRNA expression. These changes have been consistently linked with the onset of various human diseases. Moreover, specific alterations in miRNA profiles have been detected in connection with different diseases. As a result, the analysis of miRNA levels in affected tissues or the examination of circulating miRNAs in the bloodstream, and these miRNAs has been proved to be promising for their use as diagnostic or prognostic biomarkers in managing diseases in the biomedical field. This emerging understanding presents a fertile ground for RNA therapeutics, offering new avenues for targeted treatments that can adjust miRNA levels to treat or potentially even prevent disease.

## References

1. Kim VN. MicroRNA biogenesis: coordinated cropping and dicing. Nat Rev Mol Cell Biol. 2005;6(5):376–85.
2. Lu J, Shen Y, Wu Q, Kumar S, He B, Shi S, et al. The birth and death of microRNA genes in drosophila. Nat Genet. 2008;40(3):351–5.
3. Bartel DP. MicroRNAs: genomics, biogenesis, mechanism, and function. Cell. 2004;116(2):281–97.
4. Lee Y, Ahn C, Han J, Choi H, Kim J, Yim J, et al. The nuclear RNase III Drosha initiates microRNA processing. Nature. 2003;425(6956):415–9.
5. Denli AM, Tops BB, Plasterk RH, Ketting RF, Hannon GJ. Processing of primary microRNAs by the microprocessor complex. Nature. 2004;432(7014):231–5.
6. Okamura K, Hagen JW, Duan H, Tyler DM, Lai EC. The mirtron pathway generates microRNA-class regulatory RNAs in Drosophila. Cell. 2007;130(1):89–100.
7. Ruby JG, Jan CH, Bartel DP. Intronic microRNA precursors that bypass Drosha processing. Nature. 2007;448(7149):83–6.
8. Han J, Lee Y, Yeom KH, Nam JW, Heo I, Rhee JK, et al. Molecular basis for the recognition of primary microRNAs by the Drosha-DGCR8 complex. Cell. 2006;125(5):887–901.
9. Vermeulen A, Behlen L, Reynolds A, Wolfson A, Marshall WS, Karpilow J, et al. The contributions of dsRNA structure to dicer specificity and efficiency. RNA (New York, NY). 2005;11(5):674–82.
10. Ender C, Krek A, Friedländer MR, Beitzinger M, Weinmann L, Chen W, et al. A human snoRNA with microRNA-like functions. Mol Cell. 2008;32(4):519–28.

11. Li X, Carthew RW. A microRNA mediates EGF receptor signaling and promotes photoreceptor differentiation in the drosophila eye. Cell. 2005;123(7):1267–77.
12. Seggerson K, Tang L, Moss EG. Two genetic circuits repress the Caenorhabditis elegans heterochronic gene lin-28 after translation initiation. Dev Biol. 2002;243(2):215–25.
13. Viswanathan SR, Daley GQ, Gregory RI. Selective blockade of microRNA processing by Lin28. Science. 2008;320(5872):97–100.
14. Heo I, Joo C, Cho J, Ha M, Han J, Kim VN. Lin28 mediates the terminal uridylation of let-7 precursor MicroRNA. Mol Cell. 2008;32(2):276–84.
15. Newman MA, Thomson JM, Hammond SM. Lin-28 interaction with the Let-7 precursor loop mediates regulated microRNA processing. RNA (New York, NY). 2008;14(8):1539–49.
16. Rybak A, Fuchs H, Smirnova L, Brandt C, Pohl EE, Nitsch R, et al. A feedback loop comprising lin-28 and let-7 controls pre-let-7 maturation during neural stem-cell commitment. Nat Cell Biol. 2008;10(8):987–93.
17. Nalbant E, Akkaya-Ulum YZ. Exploring regulatory mechanisms on miRNAs and their implications in inflammation-related diseases. Clin Exp Med. 2024;24(1):142.
18. Pillai RS, Artus CG, Filipowicz W. Tethering of human Ago proteins to mRNA mimics the miRNA-mediated repression of protein synthesis. RNA (New York, NY). 2004;10(10):1518–25.
19. Humphreys DT, Westman BJ, Martin DI, Preiss T. MicroRNAs control translation initiation by inhibiting eukaryotic initiation factor 4E/cap and poly(a) tail function. Proc Natl Acad Sci USA. 2005;102(47):16961–6.
20. Pillai RS, Bhattacharyya SN, Artus CG, Zoller T, Cougot N, Basyuk E, et al. Inhibition of translational initiation by Let-7 MicroRNA in human cells. Science. 2005;309(5740):1573–6.
21. Kiriakidou M, Tan GS, Lamprinaki S, De Planell-Saguer M, Nelson PT, Mourelatos Z. An mRNA m7G cap binding-like motif within human Ago2 represses translation. Cell. 2007;129(6):1141–51.
22. Mathonnet G, Fabian MR, Svitkin YV, Parsyan A, Huck L, Murata T, et al. MicroRNA inhibition of translation initiation in vitro by targeting the cap-binding complex eIF4F. Science. 2007;317(5845):1764–7.
23. Wakiyama M, Takimoto K, Ohara O, Yokoyama S. Let-7 microRNA-mediated mRNA deadenylation and translational repression in a mammalian cell-free system. Genes Dev. 2007;21(15):1857–62.
24. Ding XC, Grosshans H, Repression of C. Elegans microRNA targets at the initiation level of translation requires GW182 proteins. EMBO J. 2009;28(3):213–22.
25. Maroney PA, Yu Y, Fisher J, Nilsen TW. Evidence that microRNAs are associated with translating messenger RNAs in human cells. Nat Struct Mol Biol. 2006;13(12):1102–7.
26. Nottrott S, Simard MJ, Richter JD. Human let-7a miRNA blocks protein production on actively translating polyribosomes. Nat Struct Mol Biol. 2006;13(12):1108–14.
27. Petersen CP, Bordeleau ME, Pelletier J, Sharp PA. Short RNAs repress translation after initiation in mammalian cells. Mol Cell. 2006;21(4):533–42.
28. Jungers CF, Djuranovic S. Modulation of miRISC-mediated gene silencing in eukaryotes. Front Mol Biosci. 2022;9:832916.
29. Behm-Ansmant I, Rehwinkel J, Doerks T, Stark A, Bork P, Izaurralde E. mRNA degradation by miRNAs and GW182 requires both CCR4:NOT deadenylase and DCP1:DCP2 decapping complexes. Genes Dev. 2006;20(14):1885–98.
30. Giraldez AJ, Mishima Y, Rihel J, Grocock RJ, Van Dongen S, Inoue K, et al. Zebrafish MiR-430 promotes deadenylation and clearance of maternal mRNAs. Science. 2006;312(5770):75–9.
31. Wu L, Fan J, Belasco JG. MicroRNAs direct rapid deadenylation of mRNA. Proc Natl Acad Sci USA. 2006;103(11):4034–9.
32. Eulalio A, Huntzinger E, Izaurralde E. GW182 interaction with Argonaute is essential for miRNA-mediated translational repression and mRNA decay. Nat Struct Mol Biol. 2008;15(4):346–53.
33. Chendrimada TP, Finn KJ, Ji X, Baillat D, Gregory RI, Liebhaber SA, et al. MicroRNA silencing through RISC recruitment of eIF6. Nature. 2007;447(7146):823–8.

34. Wang B, Yanez A, Novina CD. MicroRNA-repressed mRNAs contain 40S but not 60S components. Proc Natl Acad Sci USA. 2008;105(14):5343–8.
35. Bagga S, Bracht J, Hunter S, Massirer K, Holtz J, Eachus R, et al. Regulation by let-7 and lin-4 miRNAs results in target mRNA degradation. Cell. 2005;122(4):553–63.
36. Lim LP, Lau NC, Garrett-Engele P, Grimson A, Schelter JM, Castle J, et al. Microarray analysis shows that some microRNAs downregulate large numbers of target mRNAs. Nature. 2005;433(7027):769–73.
37. Alemán LM, Doench J, Sharp PA. Comparison of siRNA-induced off-target RNA and protein effects. RNA (New York, NY). 2007;13(3):385–95.
38. Vasudevan S, Tong Y, Steitz JA. Switching from repression to activation: microRNAs can up-regulate translation. Science. 2007;318(5858):1931–4.
39. Cikaluk DE, Tahbaz N, Hendricks LC, DiMattia GE, Hansen D, Pilgrim D, et al. GERp95, a membrane-associated protein that belongs to a family of proteins involved in stem cell differentiation. Mol Biol Cell. 1999;10(10):3357–72.
40. Jakymiw A, Lian S, Eystathioy T, Li S, Satoh M, Hamel JC, et al. Disruption of GW bodies impairs mammalian RNA interference. Nat Cell Biol. 2005;7(12):1267–74.
41. Liu J, Valencia-Sanchez MA, Hannon GJ, Parker R. MicroRNA-dependent localization of targeted mRNAs to mammalian P-bodies. Nat Cell Biol. 2005;7(7):719–23.
42. Eulalio A, Behm-Ansmant I, Schweizer D, Izaurralde E. P-body formation is a consequence, not the cause, of RNA-mediated gene silencing. Mol Cell Biol. 2007;27(11):3970–81.
43. Leung AK, Calabrese JM, Sharp PA. Quantitative analysis of Argonaute protein reveals microRNA-dependent localization to stress granules. Proc Natl Acad Sci USA. 2006;103(48):18125–30.
44. Weinmann L, Höck J, Ivacevic T, Ohrt T, Mütze J, Schwille P, et al. Importin 8 is a gene silencing factor that targets argonaute proteins to distinct mRNAs. Cell. 2009;136(3):496–507.
45. Vaghf A, Khansarinejad B, Ghaznavi-Rad E, Mondanizadeh M. The role of microRNAs in diseases and related signaling pathways. Mol Biol Rep. 2022;49(7):6789–801.
46. Chen D, Yang X, Liu M, Zhang Z, Xing E. Roles of miRNA dysregulation in the pathogenesis of multiple myeloma. Cancer Gene Ther. 2021;28(12):1256–68.
47. Roy B, Lee E, Li T, Rampersaud M. Role of miRNAs in neurodegeneration: from disease cause to tools of biomarker discovery and therapeutics. Genes. 2022;13(3):425.
48. Calin GA, Dumitru CD, Shimizu M, Bichi R, Zupo S, Noch E, et al. Frequent deletions and down-regulation of micro-RNA genes miR15 and miR16 at 13q14 in chronic lymphocytic leukemia. Proc Natl Acad Sci USA. 2002;99(24):15524–9.
49. Lebanony D, Benjamin H, Gilad S, Ezagouri M, Dov A, Ashkenazi K, et al. Diagnostic assay based on hsa-miR-205 expression distinguishes squamous from nonsquamous non-small-cell lung carcinoma. J Clin Oncol. 2009;27(12):2030–7.
50. Fichtlscherer S, De Rosa S, Fox H, Schwietz T, Fischer A, Liebetrau C, et al. Circulating microRNAs in patients with coronary artery disease. Circ Res. 2010;107(5):677–84.
51. Stanczyk J, Ospelt C, Karouzakis E, Filer A, Raza K, Kolling C, et al. Altered expression of microRNA-203 in rheumatoid arthritis synovial fibroblasts and its role in fibroblast activation. Arthritis Rheum. 2011;63(2):373–81.
52. De Smaele E, Ferretti E, Gulino A. MicroRNAs as biomarkers for CNS cancer and other disorders. Brain Res. 2010;1338:100–11.
53. Tufekci KU, Genc S, Genc K. The endotoxin-induced neuroinflammation model of Parkinson's disease. Parkinsons Dis. 2011;2011:487450.
54. Ouyang YB, Lu Y, Yue S, Xu LJ, Xiong XX, White RE, et al. miR-181 regulates GRP78 and influences outcome from cerebral ischemia in vitro and in vivo. Neurobiol Dis. 2012;45(1):555–63.
55. Jeyaseelan K, Lim KY, Armugam A. MicroRNA expression in the blood and brain of rats subjected to transient focal ischemia by middle cerebral artery occlusion. Stroke. 2008;39(3):959–66.

56. Liu DZ, Tian Y, Ander BP, Xu H, Stamova BS, Zhan X, et al. Brain and blood microRNA expression profiling of ischemic stroke, intracerebral hemorrhage, and kainate seizures. J Cereb Blood Flow Metab. 2010;30(1):92–101.
57. Zeng L, Liu J, Wang Y, Wang L, Weng S, Tang Y, et al. MicroRNA-210 as a novel blood biomarker in acute cerebral ischemia. Front Biosci (Elite Ed). 2011;3(4):1265–72.
58. Forero DA, van der Ven K, Callaerts P, Del-Favero J. miRNA genes and the brain: implications for psychiatric disorders. Hum Mutat. 2010;31(11):1195–204.
59. Chen H, Wang N, Burmeister M, McInnis MG. MicroRNA expression changes in lymphoblastoid cell lines in response to lithium treatment. Int J Neuropsychopharmacol. 2009;12(7):975–81.
60. Dorn GW 2nd. MicroRNAs in cardiac disease. Transl Res. 2011;157(4):226–35.
61. van Rooij E, Sutherland LB, Liu N, Williams AH, McAnally J, Gerard RD, et al. A signature pattern of stress-responsive microRNAs that can evoke cardiac hypertrophy and heart failure. Proc Natl Acad Sci USA. 2006;103(48):18255–60.
62. Hullinger TG, Montgomery RL, Seto AG, Dickinson BA, Semus HM, Lynch JM, et al. Inhibition of miR-15 protects against cardiac ischemic injury. Circ Res. 2012;110(1):71–81.
63. Reid G, Kirschner MB, van Zandwijk N. Circulating microRNAs: association with disease and potential use as biomarkers. Crit Rev Oncol Hematol. 2011;80(2):193–208.
64. Chen X, Ba Y, Ma L, Cai X, Yin Y, Wang K, et al. Characterization of microRNAs in serum: a novel class of biomarkers for diagnosis of cancer and other diseases. Cell Res. 2008;18(10):997–1006.
65. Lawrie CH, Gal S, Dunlop HM, Pushkaran B, Liggins AP, Pulford K, et al. Detection of elevated levels of tumour-associated microRNAs in serum of patients with diffuse large B-cell lymphoma. Br J Haematol. 2008;141(5):672–5.
66. Hu Z, Chen X, Zhao Y, Tian T, Jin G, Shu Y, et al. Serum microRNA signatures identified in a genome-wide serum microRNA expression profiling predict survival of non-small-cell lung cancer. J Clin Oncol. 2010;28(10):1721–6.
67. Rabinowits G, Gerçel-Taylor C, Day JM, Taylor DD, Kloecker GH. Exosomal microRNA: a diagnostic marker for lung cancer. Clin Lung Cancer. 2009;10(1):42–6.
68. Asaga S, Kuo C, Nguyen T, Terpenning M, Giuliano AE, Hoon DS. Direct serum assay for microRNA-21 concentrations in early and advanced breast cancer. Clin Chem. 2011;57(1):84–91.
69. Heneghan HM, Miller N, Lowery AJ, Sweeney KJ, Newell J, Kerin MJ. Circulating microRNAs as novel minimally invasive biomarkers for breast cancer. Ann Surg. 2010;251(3):499–505.
70. Roth C, Rack B, Müller V, Janni W, Pantel K, Schwarzenbach H. Circulating microRNAs as blood-based markers for patients with primary and metastatic breast cancer. Breast Cancer Res. 2010;12(6):R90.
71. Zhu W, Qin W, Atasoy U, Sauter ER. Circulating microRNAs in breast cancer and healthy subjects. BMC Res Notes. 2009;2:89.
72. Heneghan HM, Miller N, Kelly R, Newell J, Kerin MJ. Systemic miRNA-195 differentiates breast cancer from other malignancies and is a potential biomarker for detecting noninvasive and early stage disease. Oncologist. 2010;15(7):673–82.
73. Resnick KE, Alder H, Hagan JP, Richardson DL, Croce CM, Cohn DE. The detection of differentially expressed microRNAs from the serum of ovarian cancer patients using a novel real-time PCR platform. Gynecol Oncol. 2009;112(1):55–9.
74. Nugent M, Miller N, Kerin MJ. MicroRNAs in colorectal cancer: function, dysregulation and potential as novel biomarkers. Eur J Surg Oncol. 2011;37(8):649–54.
75. Makino K, Jinnin M, Kajihara I, Honda N, Sakai K, Masuguchi S, et al. Circulating miR-142-3p levels in patients with systemic sclerosis. Clin Exp Dermatol. 2012;37(1):34–9.
76. Wang G, Tam LS, Li EK, Kwan BC, Chow KM, Luk CC, et al. Serum and urinary free microRNA level in patients with systemic lupus erythematosus. Lupus. 2011;20(5):493–500.

# Chapter 7
# Non-coding RNAs: Circular RNAs (circRNAs)

**Abstract** Circular RNAs (circRNAs) are a unique class of endogenous RNAs that are characterized by their covalently closed loop structures, with no 5′ end cap or 3′ polyadenylated tail. They are formed through a process called reverse splicing, where the downstream splice donor of precursor mRNA (pre-mRNA) is joined to an upstream splice acceptor. This process can occur across numerous genes within eukaryotic organisms, giving rise to a diverse array of circRNA molecules. Although circRNAs are generally expressed at lower levels compared to linear RNAs, recent studies have shed light on their expression patterns, which are notably cell type-specific and tissue-specific. This specificity suggests that circRNAs play crucial, context-dependent roles within the biological processes of different cells and tissues. Further investigations into the regulatory mechanisms of circRNA biogenesis have revealed a complex network of influences that govern their formation. These mechanisms ensure that circRNAs are produced in a controlled manner, contributing to their functional roles in the cells. One of the most intriguing aspects of circRNA function is their capacity to influence gene expression. CircRNAs can act as molecular sponges for miRNAs, a process by which they bind and sequester these small, non-coding RNAs, thereby modulating their ability to regulate target mRNAs. This interaction can have profound effects on gene expression patterns and cellular function. In addition to their role in microRNA sequestration, circRNAs can also affect gene expression directly by interacting with transcription machinery or indirectly by interfering with splicing processes. By disrupting or modifying splicing, circRNAs can alter the splicing patterns of other pre-mRNAs, leading to diverse outcomes in gene expression profiles. These multifaceted roles of circRNAs add a significant layer of complexity and diversity to the transcriptomes of eukaryotic organisms, indicating that circRNAs are more than mere by-products of splicing—they are active participants in the regulation of gene expression and have potential implications in development, differentiation, and disease. The ongoing exploration of circRNAs promises to further unravel their functions and mechanisms, potentially opening new avenues for therapeutic intervention.



P. E. Saw, E. Song, *RNA Therapeutics in Human Diseases*,
https://doi.org/10.1007/978-981-96-3041-7_7

## 7.1 Introduction

Circular RNAs (circRNAs) represent a class of closed, single-stranded RNA molecules that encompass a diverse range of RNA species. Originally, circRNAs were considered to be accidental products of splicing errors, thought to possess minimal functional importance. This perception stemmed from their unusual formation process and lack of polyadenylation, features that typically signify aberrant or non-functional RNA molecules in eukaryotic cells [1–5]. The application of RNA-seq has revealed that circRNAs are not merely sporadic anomalies but are instead broadly expressed across various non-polyadenylated transcriptomes. This widespread expression suggests a more substantial and possibly regulatory role in cellular processes. As RNA-seq provides deep insights into RNA profiles with unprecedented resolution and sensitivity, it has allowed researchers to detect and quantify circRNAs in numerous biological contexts, highlighting their potential relevance in gene regulation, cellular differentiation, and disease pathology. The enhanced understanding of circRNAs through next-generation sequencing has transformed them into a significant focus of current genomic and transcriptomic research. This shift underscores the evolving nature of our understanding of RNA as a central element in cellular biology, extending beyond its conventional roles to include diverse and complex regulatory functions [6–13].

## 7.2 circRNA Biogenesis: The Basics

The formation of circular RNAs (circRNAs) originates from canonical splice sites, with their biogenesis largely dependent on the standard splicing machinery. Experimental studies involving circRNA expression vectors and the splicing inhibitor isoginkgetin in HeLa cells demonstrate that disruption of spliceosome assembly halts circRNA production, confirming its reliance on canonical splicing pathways [14, 15]. Conversely, research in Drosophila melanogaster has shown that when the spliceosome is inhibited, for instance, by depleting U2 snRNP components, there is a marked increase in circular RNA formation compared to linear RNA. This suggests that slowing conventional pre-mRNA processing may reroute nascent transcripts toward alternative back-splicing pathways [16]. Further support for this idea comes from studies in *D. melanogaster* showing that reducing multiple splicing factors cumulatively enhances circRNA levels, indicating that individual splicing factors likely play unique roles in circRNA biogenesis [17]. The prevailing model of back-splicing proposes that the looping of intronic sequences, specifically those flanking downstream splice donor and upstream splice-acceptor sites, brings these sites into proximity, facilitating circularization (Fig. 7.1 (Upper)). This intron looping can be driven either by base pairing between inverted repeat elements (such as Alu elements) within these introns [9, 12, 19] or by dimerization of RNA-binding proteins (RBPs) like quaking (QKI) or FUS, which bind to motifs in the flanking introns [20, 21]. In addition to these mechanisms, studies in *D. melanogaster*

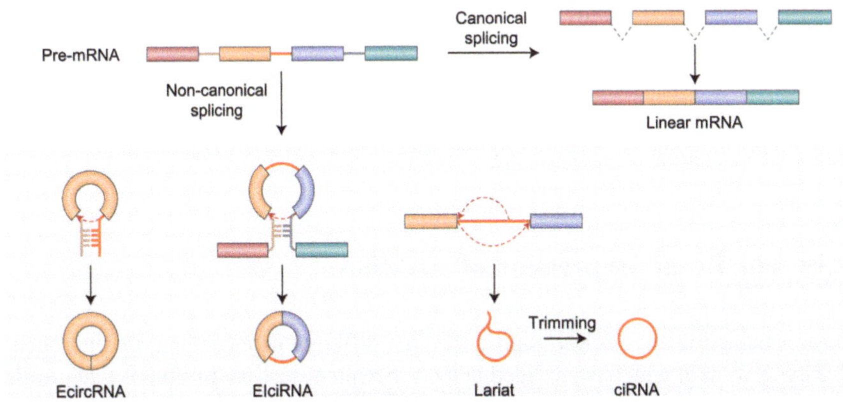


**Fig. 7.1** Schematic representation of mRNA splicing and circular RNA formation. This diagram illustrates the different pathways of mRNA processing from a pre-mRNA transcript, delineating both canonical and non-canonical splicing events. The leftmost segment of the figure depicts the pre-mRNA undergoing non-canonical splicing, resulting in the formation of various types of circular RNAs (circRNAs). These include Exonic circRNA (EcircRNA) and ElciRNA, both formed by back-splicing where the downstream splice donor joins to an upstream splice acceptor. The process and intermediates are illustrated, from the looping of the transcript to the final circular structure. The central section shows the transformation of a lariat structure into a circular intronic RNA (ciRNA) through a trimming process that removes the lariat tail, leaving a circularized exon. Meanwhile, the rightmost segment presents canonical splicing of pre-mRNA leading to the formation of linear mRNA. This figure encapsulates the complexity and diversity of RNA processing mechanisms in the cell, highlighting the formation of functionally diverse RNA molecules from a single pre-mRNA sequence. Figure is reproduced from [18] under Creative Commons CC BY International License

indicate that both cis-acting elements and various trans-acting splicing factors—including hnRNPs and SR proteins (rich in serine and arginine residues)—can influence circRNA formation [17]. The biogenesis of circRNAs is also regulated by factors that suppress circRNA formation. Double-stranded RNA (dsRNA)-specific adenosine deaminase (ADAR) enzymes, which convert adenosine to inosine in dsRNA to prevent immune activation [22], and ATP-dependent RNA helicase A (DHX9) [23] inhibit circRNAs that rely on intronic base pairing. ADAR's RNA editing and DHX9's helicase activity both disrupt the intronic loops necessary for circRNA formation by unwinding dsRNA structures [12, 24]. Contrarily, proteins NF90 and NF110—products of the interleukin enhancer-binding factor 3 (ILF3) gene—promote circRNA formation by stabilizing RNA intronic pairs and are implicated in antiviral responses [25, 26].

Another source of circRNAs arises during exon skipping events, where exons are removed from the mature mRNA transcript and contained within lariats. These lariats can undergo back-splicing internally, leading to circRNA formation [19, 27]. Similarly, when intronic lariats evade debranching, they can become circular intronic RNAs (ciRNAs) [28] (Fig. 7.1 (Lower)). Highly expressed circRNAs often stem from genes with extended flanking introns around backspliced exons and are typically derived from genes with active promoters [7, 29–31]. Epigenetic factors, including histone modifications and gene body methylation, can influence alternative splicing and potentially impact circRNA formation. Although the exact effects of histone modifications on circRNA biogenesis have not been fully explored, recent evidence suggests that DNA methyltransferase DNMT3B, which primarily methylates gene bodies [32], can alter circRNA expression independently of linear gene expression, suggesting a gene-context-dependent role for DNA methylation in circRNA regulation [31].

## 7.3 The Broad Expression of circRNAs

Various categories of circRNAs arise through different mechanisms, a comprehensive exposition of which is provided in the subsequent section. In studies that profile polyadenylated RNA, circRNAs formed through pre-mRNA back-splicing of exons are seldom identified due to their absence of poly(A) tails. However, recent advancements in the study of circRNAs have enabled researchers to detect them by utilizing non-polyadenylated transcriptomes or selectively enriching circRNAs through RNase R treatment. The RNase R enzyme catalyzes the digestion of RNAs that are linear while preserving circRNAs. By utilizing RNA sequencing and specialized computational pipelines, researchers have identified a vast number of circRNAs spanning back-splicing junctions in a reversed orientation.

Despite often being expressed at modest levels, suggesting the potential inertness of many circRNAs as splicing by-products, around 50 genes exhibit higher expression levels of circRNAs compared to their isoforms that are linear in structure, in all cell lines examined [33]. Furthermore, Furthermore, experimental validation confirms the heightened expression of several circRNAs in a tissue-specific or cell type-specific way [15, 34]. Notably, the brain harbors the accumulation of thousands of circRNAs, and hundreds of them are increased in expression during EMT. Intriguingly, no discernable correlation was identified between circRNA and mRNA levels [10, 20, 33–36], indicating that increased circRNA expression in certain tissues or cells implies functionality instead of merely manifesting as a by-product of elevated transcription. Nevertheless, a genome-wide analysis underscores that the cell type specificity of high-confidence circRNAs is not markedly distinct from their linear counterparts [10]. The extent to which circRNAs varies from that of linear transcripts across different tissues and cell types remains incompletely elucidated. Additionally, given the stability of circRNAs [7, 8, 28] and their potential differential accumulation in cells with varying rates of division, it is crucial to investigate whether detected steady-state circRNA levels accurately reflect circRNA processing kinetics, and the analysis of various circRNAs are summarized in Fig. 7.2.

**Fig. 7.2** (continued) analyses of circRNAs using techniques like CRISPR-Cas13 and RNA interference (RNAi) are presented. (**j**) Gain-of-function (GoF) analyses of circRNAs utilizing expression vectors that often produce linear concurrent RNAs. (**k**) Additional GoF analyses involving circRNAs synthesized in vitro through an engineering process called direct ligation using T4 RNA ligase into a permuted autocatalytic intron are explained. Figure is adapted with permission from Elsevier (license number: 590167097653)

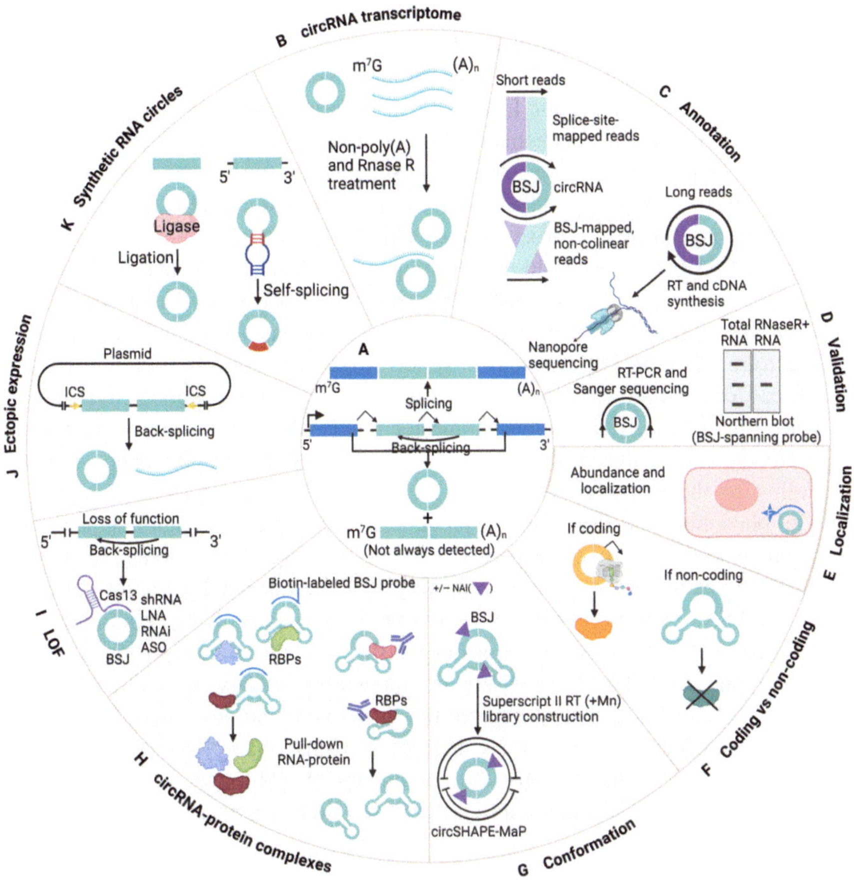


**Fig. 7.2** Detailed exploration and functional analysis of circRNAs. (**a**) The biogenesis of circRNAs, where the spliceosome performs back-splicing on exons within pre-RNA, leading to a circular RNA structure. This back-splicing forms a 3′-5′ phosphodiester bond at the back-splicing junction (BSJ) site. It is important to note that back-splicing occurs with less efficiency compared to canonical splicing, resulting in lower overall expression of circRNAs. (**b**) The biochemical methods used to enrich transcriptomes for circRNAs prior to conducting RNA-seq analyses, enhancing the detection of circRNAs. (**c**) An overview of RNA-seq techniques and algorithms used to annotate circRNAs is presented. All sequencing approaches require mapping of RNA-seq reads to rearranged sequences at the BSJ site for accurate circRNA detection. (**d**) Experimental validation strategies for confirming circRNA candidates are depicted. These include Sanger sequencing via divergent PCR that spans the proposed BSJ site and Northern blotting (NB) with probes that target exons involved in forming the circRNAs at the BSJ site. (**e**) Investigations into the abundance and subcellular localization of circRNAs are highlighted as key factors in understanding their cellular roles, demonstrated through RNA fluorescence in situ hybridization (FISH) using antisense probes targeting the BSJ site of circRNA. (**f**) The importance of distinguishing between non-coding and coding circRNAs to clarify their regulatory roles is discussed. (**g**) An illustration of circRNA structural analysis through chemical probing techniques, specifically demonstrated by circSHAPE-MaP, is shown. (**h**) This panel displays the use of antisense oligonucleotides that bind to circRNA sequences at the BSJ site, and in vitro synthesized RNA circles were used as baits to fish out and identify proteins that interact with specific circRNAs. (**i**) Loss-of-function (LoF)

**Box 7.1 Varieties of Circular RNAs**

Besides circRNAs that are generated through exon back-splicing circularization at numerous eukaryotic loci [6–14, 19], various other types exist, originating from different mechanisms. Firstly, viral circRNA genomes [37, 38] undergo ligation by enzymes found in host cells, resulting in the formation of 2′,5′- [39] or 3′,5′-phosphodiester bonds [40]. Secondly, circRNA intermediates emerge during the processing of rRNA [41, 42]. In both Achaea and algae, tRNA is also involved in their biogenesis [43]. Thirdly, certain housekeeping non-coding RNAs, such as the ribozyme RNase P and small nucleolar RNAs, have been identified in Archaea as circular forms [42]. Lastly, a spectrum of circRNAs arises from spliced exons and introns. Notably, self-spliced, excised groups I and II introns undergo ligation to generate circRNAs with either 2′,5′- or 3′,5′-phosphodiester bonds [44–47]. In the context of tRNA molecules, introns are removed by specialized tRNA splicing enzymes. Following this excision, the ends of the tRNA are joined together through a 3′,5′-phosphodiester bond ligation process. This splicing and ligation sequence occurs in both metazoan species and archaea, ensuring the functional maturity of tRNA necessary for protein synthesis [48, 49]. Additionally, the formation of circRNAs from lariat introns, which are by-products of the splicing process, has been observed in various organisms, including Xenopus tropicalis and human cells. These circRNAs are noteworthy because they represent a unique type of circular RNA formed not from exonic but intronic sequences, highlighting the diverse sources and mechanisms of circRNA biogenesis. This phenomenon underscores the complexity of RNA processing and the variety of RNA molecules that can have functional significance in cellular biology [28, 50, 51].

## 7.4 The Formation of circRNA via Cis-elements Facilitation

The majority of circRNAs typically encompass two or three exons [9], and they do not require a specific sequence except for the splice site [15]. Alternative splicing can produce multiple circRNAs from the same gene, including or excluding internal introns [9, 10, 33, 52]. The length of the exon may affect the back-splicing, with highly expressed single-exon circRNAs having longer exons compared to those comprising multiple exons [9, 53]. Regulatory elements (Res) are usually located in the introns in between the exons destined for circularization are critical in promoting back-splicing. It is common for circRNAs to be processed from internal exons flanked by long introns that have reverse complementary sequences. Since they are fully complementary, it is common for these sequences to pair up to form RNA duplexes, a factor that greatly enhances the efficiency of back-splicing. This arrangement underscores the importance of specific intronic sequences in facilitating the circularization process in RNA biology [7, 9, 17, 34]. These complementary sequences may originate from REs (namely Alu elements) [9, 17, 34] or non-repetitive complementary sequences [9]. Short sequences, ranging around 30–40 nt, can promote the biogenesis of circRNA [34], even though longer sequences result in more robust circRNA production [9]. The efficiency of back-splicing is influenced by competition within individual introns among RNA pairings across flanking introns. RNA pairings within individual introns are associated with canonical splicing, whereas those spanning flanking introns facilitate circRNA production [9]. The specific function of cis-elements in intronic sequences during circRNA formation has yet to be fully determined. It is hypothesized that short cis-regulatory elements within these introns, which can be targeted by RBPs, might play a crucial role in facilitating exon circularization. This suggests that interactions between these cis-elements and RBPs could be key drivers in the process, although more research is needed to clarify their exact contributions [14, 20]. Moreover, splicing silencers or enhancers within introns that affect back-splicing efficiency may also play a role [54]. It is important to note that the specific combinations of RBPS with their splicing regulatory elements for conventional splicing and back-splicing may vary. This distinction underscores the need for additional research to fully understand how these cis-elements influence the formation of circRNAs. Such studies are essential to determine the regulatory mechanisms that govern circRNA biogenesis.

## 7.5 Regulation by RNA-Binding Proteins (RBPs)

Regulatory proteins binding to RNA, known as RBPs, extend their role beyond cis-elements and encompasses circRNA biogenesis [12, 14, 17, 20, 35]. Notably, Muscleblind (Mbl), a splicing regulator in the model organism *D. melanogaster*, orchestrates the generation of circRNA derived from its own precursor mRNA through interaction with multiple Mbl-binding sites situated within the introns adjacent the exon undergoing circularization [14]. Enhanced circularization is observed with either the overexpression of Mbl or the incorporation of additional Mbl-binding sites within the introns bordering the exon targeted for circularization in a vector. This observation underscores the promotive role of Mbl in circRNA biosynthesis [14]. Similarly, another significant discovery pertains to the RBP quaking (QKI), which markedly increases circRNA formation through its binding to precursor mRNA, particularly evident during EMT in humans, where an upregulation of circRNAs mediated by QKI suggests a deliberate, cell type-specific generation of circRNAs [20]. Introduction of synthetic QKI-binding sites into the introns has been demonstrated to be adequate for circRNA induction, suggesting a dimeric binding mechanism wherein QKI potentially facilitate the proximity of circularized exons, thereby fostering the generation of circRNA [20]. Conversely, Adenosine deaminase 1 acting on RNA (ADAR1) exerts suppressive effects on circRNA expression, primarily attributed to its adenosine-to-inosine (A-to-I) editing activity on RNA duplexes formed by sequences surrounding circularized exons [12, 35]. The A-to-I editing disrupts RNA pairing, thereby attenuating back-splicing. Nonetheless, the impact of ADAR1-mediated editing seems restricted, as only a subset of genes demonstrates altered circularization upon ADAR1 knockdown [35]. Additionally, recent findings also suggest a dual role for ADAR1, not solely confined to editing but extending to double-stranded RNA-binding, hinting at a potential editing-independent regulation of circRNA formation [55]. Given the multitude of RBPs, uncovering additional players in circRNA biogenesis represents a critical future endeavor, further compounded by the necessity to elucidate the complex interplay of RBPs in circRNA production. One study has shed light on the regulation of back-splicing in the Drosophila melanogaster Laccase 2 gene, revealing that this process is influenced by a combination of intronic repeats along with the cooperative activities of various heterogeneous nuclear ribonucleoproteins (hnRNPs) and Serine/Arginine-rich (SR) proteins [17]. This finding emphasizes that the regulatory mechanism governing circRNA production in cellular contexts are considerably more intricate than previously understood.

## 7.6 Gene Regulation by circRNAs

circRNAs are historically viewed as minor and incidental products within eukaryotic transcriptomes, were often overlooked due to their low expression levels and presumed lack of significant biological function. Their sparse presence in cells suggested that they might not play critical roles in cellular processes. However, recent advancements in RNA research have begun to change this perspective significantly. Emerging studies have highlighted the potential importance of certain circRNAs, especially in complex physiological contexts such as neural functions. These studies indicate that circRNAs could have multifaceted roles in regulating gene expression at various levels. For instance, in neural tissue, circRNAs have been implicated in the fine-tuning of neuronal development, synaptic functioning, and even in the neuroplasticity that underpins learning and memory.

The ability of circRNAs to modulate gene expression can occur through several mechanisms. One key method is through acting as "molecular sponges," where circRNAs bind to and sequester microRNAs, preventing these miRNAs from suppressing their target mRNAs. This sequestration can lead to increased levels of certain proteins crucial for cellular functions. Additionally, some circRNAs can interact directly with RNA-binding proteins, influencing the stability and translation of mRNAs or altering the splicing of pre-mRNAs. Furthermore, circRNA possess increased stability of circRNAs, due to their circular conformation which resists degradation by exonucleases, makes them suitable candidates for regulatory roles that require persistence and resilience within the cellular environment. This stability, coupled with their ability to modulate complex gene regulatory networks, underscores the emerging view of circRNAs as significant players in cellular physiology, beyond being mere by-products of splicing. These insights are propelling a broader recognition of circRNAs' potential roles and are motivating further research to unravel the specific functions and mechanisms through which circRNAs influence gene expression and contribute to physiological and perhaps pathological processes.

## 7.7 CircRNAs can Function as miRNA Sponges

Usually, circRNAs regulate gene expression in the cell cytoplasm by sequestering miRNAs. Predominantly, these circRNAs annotated for back-splicing are found within the cytoplasm [6, 7]. Despite the experimental demonstrations that circRNAs, during integration with ribosome entry sites, can be translated [20, 54, 56]; native circRNAs are yet to be observed in association with ribosomes [7, 10], suggesting their role in post-transcriptional regulation rather than translation. Experimental evidence suggests that stable circRNAs may engage in competition with mRNAs for cytoplasmic attachment of miRNA, thereby influencing gene expression regulation [7]. One of the most extensively studied circRNAs that supports this hypothesis is ciRS-7 (circular RNA sponge for miR-7). ciRS-7 is derived from the antisense transcript of the cerebellar degeneration-related 1 (CDR1) gene in vertebrates (Fig. 7.3a). It is predominantly expressed in the brains of humans and mice and contains more than 60 conserved target sites for miR-7. This high number of binding sites allows ciRS-7 to effectively sequester miR-7, highlighting its potential role in regulating miRNA activity and gene expression within neural tissues [8, 57]. This circRNA is not only abundantly expressed and cytoplasmically localized but also possesses the capability to bind a maximum of 20,000 miR-7 molecules in one cell [8]. This binding capacity suggests a significant role of ciRS-7 in modulating miR-7. A reduction in ciRS-7 levels has been correlated with a decrease in the expression of mRNAs containing miR-7 binding sites, indicating a competitive mechanism for miR-7 binding between ciRS-7 and mRNAs [8, 57]. The functional significance of ciRS-7 is highlighted by its evolutionary conservation; its expression in zebrafish disrupts midbrain development in a manner akin to the effects observed with the silencing of miR-7 [8]. However, the conservation of miRNA target sites in circRNAs across different species is not a prerequisite for their function as miRNA sponges. Interestingly, recent observation indicated that circRNAs in both mice and humans seldom contain miRNA target sites, indicating that most circRNAs may have a limited role in acting as miRNA sponges [10, 36] On the contrary, circRNAs from *D. melanogaster* contain over a thousand conserved miRNA seed matches [11], raising the question of their general function as miRNA sponges. A recent quantitative analysis of miRNA functionality proposes that altering the abundance of miRNA targets generally does not impact gene expression, serving as a note of caution in the field [4].

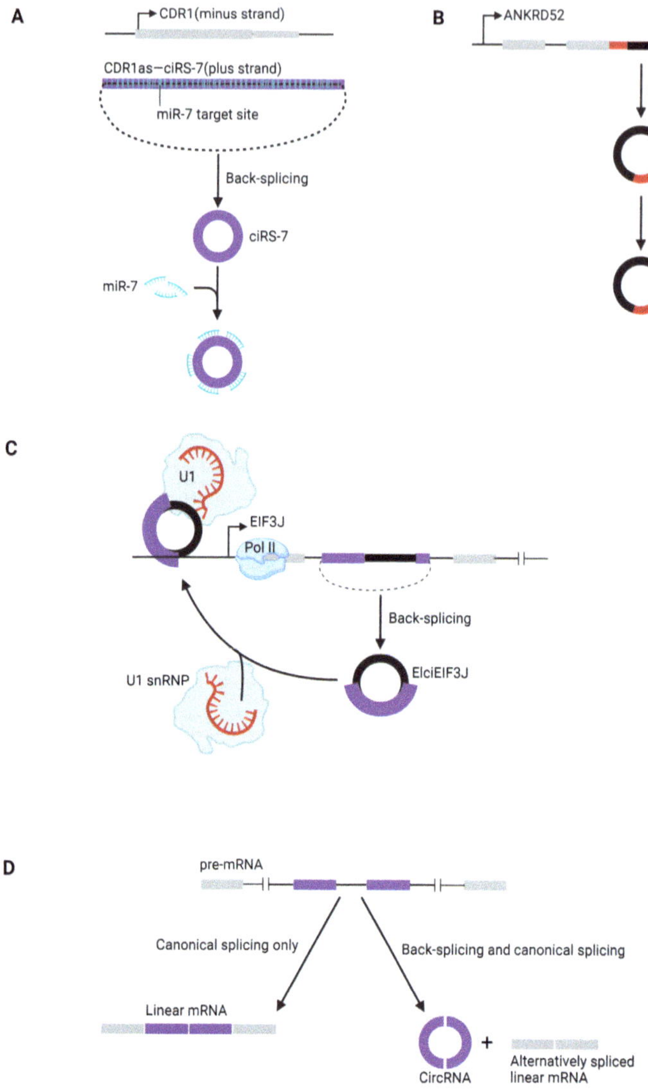


**Fig. 7.3** Potential regulatory roles of circRNAs in gene expression modulation. (**a**) circRNA ciRS-7, also known as the miR-7 sponge, emerges through back-splicing of the CDR1-antisense transcript. This circRNA harbors more than 60 miR-7 binding sites, primarily resides in the cytoplasm, and acts by sequestering miR07. (**b**) ci-ankrd52 originates from the second intron of the ankyrin repeat domain 52 (ANKRD52) gene. Its stability in the nucleus, where it enhances Pol II-mediated transcription of ANKRD52, is attributed to specific RNA sequences that prevent intron lariat debranching. (**c**) EIciEIF3J, a circRNA containing an intron, results from the back-splicing of the eukaryotic translation initiation factor 3J (EIF3J) gene. It facilitates EIF3J transcription by Pol II through interaction with U1 small nuclear ribonucleoprotein (snRNP) and the EIF3J gene promoter. (**d**) The generation of circRNA through back-splicing influences the alternative splicing of pre-mRNAs that give rise to circRNAs, thus affecting gene expression. Figure is adapted with permission from Spinger Nature (License number: 5901681139461)

## 7.8 CircRNAs Regulation of Transcription

circRNAs, predominantly localized in the cytoplasm [6, 7], exhibit a unique subset—those containing introns—that mainly reside within the nucleus of human cells [28, 52]. The mechanisms governing their nuclear retention or export of circRNAs remain to be elucidated. It is posited that the retention of circRNAs with introns in the nucleus may follow a similar paradigm to the nuclear restriction observed for linear RNAs with retained introns [58]. Additionally, a distinct category of nuclear circRNAs known as circular intronic RNAs (ciRNAs) arises from lariat introns. The biogenesis of circRNAs depends on a specific consensus sequence, which features a 7-nucleotide guanine-uracil-rich motif near the 5′ splice site and an 11-nucleotide cytosine-rich motif at the branch point (Fig. 7.3b). This sequence arrangement promotes the formation of a covalently closed loop via a 2′-,5′-phosphodiester bond, effectively excluding the linear segment that extends from the 3′ end of the intron to the branch point [28]. Numerous circRNAs have been catalogued in human cells [1], with a significant subset identified in the oocytes of *Xenopus tropicalis* [50, 51]. These ciRNAs predominantly accumulate in the nucleus, where they play a pivotal role in the regulation of gene transcription in a cis manner. Key examples include circ-ankrd52 and circ-sirt7, which are known to associate with the Pol II elongation complex. The depletion of these ciRNAs has been shown to adversely affect the transcriptional activity of their respective parental genes, namely the sirtuin 7 (SIRT7) and ankyrin repeat domain 52 (ANKRD52) genes, there by underscoring their contributory role in facilitating Pol II-mediated transcription, although the exact mechanism remains to be uncovered [28]. The role of Pol II in the regulation of transcription through nuclear RNA circles has been further elucidated by subsequent research, wherein RNA immunoprecipitation assays were utilized to investigate the association of Pol II with non-coding RNAs [52]. One study revealed the presence of exon–intron circRNAs (EIciRNAs), characterized by back-splicing and the retention of introns, in association with Pol II. The depletion of specific EIciRNAs including EIciPAIP2 and EIciEIF3J, led to a reduction in the transcription levels of their respective genes. Mechanistically, exon-intron ciRNAs (EIciRNAs) have been shown to interact with U1 snRNP (small nuclear ribonucleoprotein) and the promoters of their parent genes. Interestingly, both ciRNA (EIciPAIP2 and EIciEIF3J) contain a potential binding site for U1 snRNA within their retained introns, respectively. Disrupting these RNA-RNA interactions was found to decrease the association between EIciRNAs and RNA Pol II [52]. These findings collectively suggest that certain nuclear RNA circles may play a regulatory role in transcription. However, the exact mechanisms by which they exert this influence are yet to be fully understood. Future investigations are warranted to determine whether circRNAs influence other loci in trans and to explore the potential additional roles of nuclear-retained circRNAs, such as their function as decoys for RBPs [28]. For example, in amyotrophic lateral sclerosis (ALS), TAR DNA-binding protein 43 (TDP43) aggregates and accumulates in degenerating neurons in most patients. In an ALS model, it has been proposed that the accumulation of intronic lariats, due to the absence of the RNA lariat debranching enzyme DBR1, may act as decoys. These intronic lariats sequester TDP43 in the cytoplasm, thereby reducing its toxicity [51, 59] (Fig. 7.3c).

## 7.9 circRNAs Processing Affects Splicing

CircRNAs predominantly originate from the internal exon gene-encoding proteins [9]. Consequently, circRNA biogenesis can influence the alternative splicing mechanisms of these pre-mRNAs, which may result in variations in gene expression profiles. The selection of 5′ and 3′ splice sites during circRNA formation can interfere with the splicing of pre-mRNAs, although back-splicing mechanism is less common than canonical splicing. This interference typically leads to a reduced abundance of linear mRNAs incorporating the exons that undergo circularization (Fig. 7.3d) [9, 14, 19]. Exon skipping is an often-observed alternative splicing mechanism in human pre-mRNAs, is believed to have a direct correlation with circRNA formation, implying that an increase in exon skipping should theoretically enhance circRNA production. This relationship has been validated using circRNA expression vectors and an extensive analysis of RNA-seq data from human endothelial cells activated by either TNF-α or TGF-β [19]. Nevertheless, it is important to note that not every instance of exon skipping results in the formation of circRNAs [19], hinting at the presence of other regulatory factors that may influence exon circularization post-skipping.

## 7.10 Circular RNA-Based Biomedical Applications

The unique structural characteristics, enhanced stability, and reduced immunogenicity of circRNAs have spurred significant efforts in the development of technologies based on RNA circles. These applications include the employment of circRNA as non-coding aptamers to disrupt cellular processes, regulate innate immune mechanisms, capture miRNAs and proteins associated with diseases, and act as antisense

**Box 7.2 Expression Patterns of circRNAs in Tissues and Disease**
circRNAs, products of exon back-splicing circularization, are emerging as significant players in neurological functions. This is evidenced by their preferential origination from neurons-specific genes [14]. Various developmental stages of the mouse cerebrum have shown that numerous circRNAs are highly abundant in the mammalian brain. A notable upregulation of these circRNAs is observed during the neurogenic process, with a pronounced enrichment

(continued)

**Box 7.2** (continued)
seen in the phase of synaptogenesis compared to their linear counterparts [35, 36]. Similarly, in *Drosophila melanogaster*, an age-dependent increase in circRNAs levels in neural tissues has been documented [11]. The expression patterns of certain brain-specific circRNAs are largely conserved across human, mice, and even *D. melanogaster* has been identified [35], underscoring a potential evolutionary conserved role in neuronal functionality. The inherent stability of circRNAs [7, 8], facilitating their accumulation in non-dividing cells [60], complements the low proliferation rate of neurons, possibly leading to their accumulation. Additionally, the regulation of many circRNAs has been observed during EMT in humans, demonstrating their potential role in key EMT processes such as cellular migration, invasion, and metastasis [20]. While the exact roles of circRNAs in neural contexts or EMT are yet to be fully elucidated, certain circRNAs originating from genomic regions linked to development and disease suggest a functional impact in specific physiological and pathological scenarios. For instance, circSRY, derived from the sex-determining region Y, is instrumental in the development of mouse testes [3, 57]. Similarly, the expression of circRNAs linked to the INK4/ARF tumor suppressor locus, mediated by the lncRNA ANRIL, correlates positively with INK4/ARF expression, and is associated with a heightened risk of atherosclerosis [61, 62]. Among the well-characterized circRNAs, circular RNA sponge for miR-7 (CiRS-7), acts as a miR-7 sponge in neuronal tissues [8, 57] and exhibits a marked reduction in sporadic cases of Alzheimer 's disease [63], highlighting a dysregulated ciRS-7-miR-7 axis in affected individuals. Additionally, alterations in circRNA have been linked to cancer-related cell proliferation [60], suggesting a broader significance in disease pathology.

RNAs or templates for prolonged translation. Moreover, circRNAs with relevance to specific diseases are being recognized as potential therapeutic targets, and the presence of circRNAs in circulation is being explored as diagnostic biomarkers (Fig. 7.4).

## 7.11 circRNA as Fluorescent Sensors and Intracellular Modulators

Aptamers, which are short DNA and RNA sequences produced within cells, serve the function of binding to specific molecular targets, thereby influencing internal cellular activities [64]. However, a significant challenge encountered in the deployment of RNA aptamers involves their expression at adequately high levels to ensure

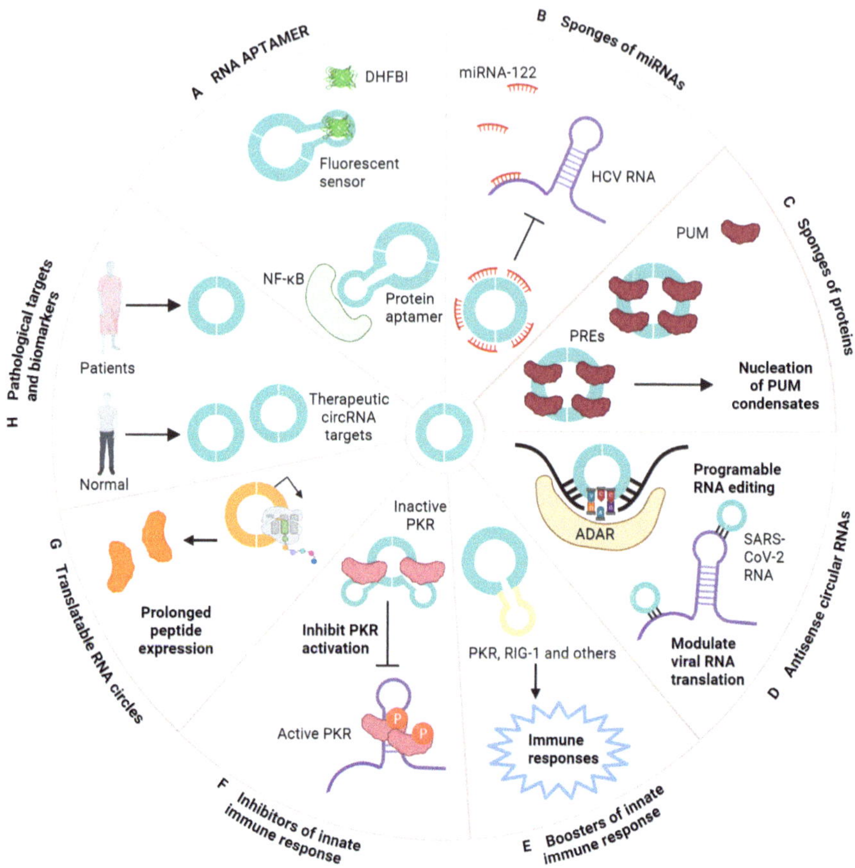


**Fig. 7.4** (**a**) circRNA aptamers possess the capability to attract fluorescent markers as well as potent biosensors for metabolites. (**b**) circRNAs are capable of capturing miRNAs associated with diseases, thereby reducing their availability to target mRNAs. (**c**) circRNAs that contain Pumilio (PUM) response elements (PREs) function as molecular sponges for PUM, thereby influencing the activity of cellular proteins. (**d**) Engineered circRNAs containing antisense RNAs that target conserved regions of the COVID-19 genome can effectively inhibit viral replication. Moreover, targeting mRNAs to recruit ADARs (adenosine deaminases acting on RNA) enables programmable RNA editing. (**e**) Self-splicing-derived RNA circles initiates the expression of RIG-I and activates PKR, leading to the activation of innate immune response pathways. (**f**) RNA circles that feature short dsRNA segments do not elicit an immune response and serve effective suppressors of PKR, which plays a role in innate immunity. (**g**) circRNAs that are capable of being translated facilitate extended expression of proteins. (**h**) Focusing on circRNAs that are linked to diseases or pathological conditions has significant therapeutic potential. Moreover, certain circRNAs may also act as biomarkers for diagnosis. Figure is adapted with permission from Elsevier (License number: 5901670976353)

effective functionality [65]. The advancement of methods for controlled ectopic expression and the in vitro generation of stable RNA circles has made circRNA-based aptamers a more feasible option. In this context, circRNA-encoded fluorescent RNA-based sensors have become particularly promising. Tornado (an RNA sensor) have demonstrated a fluorescence intensity nearly 200 times greater than their linear counterparts [66] when they bind to small molecules within living cells [67]. Furthermore, a circRNA molecule approximately 300 nt in length, incorporating a near 30 nt NF-κB aptamer recognized for its high affinity for the NF-κB p65 subunit [68], outperformed its linear RNA equivalent in inhibiting the NF-κB signaling pathway [67]. Another example includes a circRNA of approximately 150 nucleotides, expressed using the Tornado system, which contains an aptamer targeting S-adenosyl methionine (SAM). This circRNA functioned as an effective metabolite biosensor within cells [67]. These instances underscore the potential wide-ranging utility of circRNA aptamers in modulating diverse intracellular pathways.

## 7.12 Engineered circRNAs as Protein Sponges of Cellular Proteins

Employing synthetic circRNAs as protein sponges or molecular decoys offers a novel approach for modulating protein activity by incorporating specific RBP sites of interest into these circular molecules. These RBP sites can be meticulously selected based on comprehensive libraries such as those generated from crosslinking immunoprecipitation (CLIP) and systematic evolution of ligands by exponential enrichment (SELEX) datasets. A notable example includes the targeting of hnRNP L, a critical splicing regulator that interacts with short CA repeats within pre-mRNAs, that circRNAs containing 20–100 CA dinucleotide repeats, synthesized using T4 RNA ligase or ectopically expressed circRNAs harboring identical CA sequences [69]. These circRNAs exhibited strong binding affinity to hnRNP L due to the sequestration by these cytoplasmic localized CA-repeats-containing circRNAs, leading to alternative splicing patterns akin to those observed with siRNA-mediated depletion of hnRNP L [70].

### *7.12.1 Application to circRNA Therapeutics*

CircRNAs can be engineered to act as "sponges" or decoys for miRNAs or RNA-binding proteins (RBPs), interfering with their function. SELEX can be used to identify circRNA sequences that specifically bind to these miRNAs or RBPs with

**Box 7.3 Systematic Evolution of Ligands by Exponential Enrichment (SELEX) in Producing circRNA Therapeutics**

SELEX is a powerful laboratory technique used to identify oligonucleotides (either DNA or RNA) with high affinity to target molecules [71, 72]. This method has been increasingly applied in the field of circRNA therapeutics [70]. SELEX involves synthesizing a large library of random sequence oligonucleotides, then repeatedly binding these sequences to a target of interest, such as proteins, small molecules, or other RNAs. After each binding round, sequences that do not bind are washed away, and the bound sequences are amplified and subjected to further rounds of selection. This iterative process enriches the pool for sequences with the highest affinity and specificity to the target. Once the selection process is complete, the most effective sequences are identified and can be further characterized and optimized for therapeutic use [73] (Fig. 7.5).

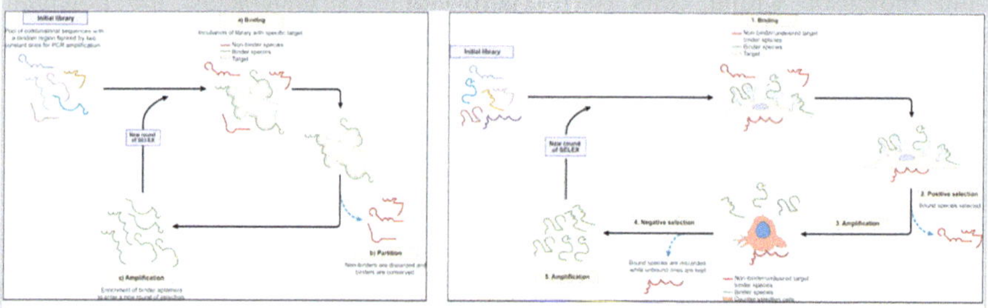


**Fig. 7.5** Overview of SELEX and Cell-SELEX techniques for aptamer selection. Left panel: Conventional SELEX. The SELEX process is used to select specific nucleic acid sequences, known as aptamers, which can bind to a target molecule with high affinity. The process begins with a large initial library of random oligonucleotide sequences, which may contain both RNA and DNA variants. Each sequence has conserved regions flanking random regions to enable PCR amplification. The oligonucleotide library is incubated with a defined target molecule. Target molecules can include proteins, small molecules, or other biomolecules. After incubation, sequences that do not bind to the target are separated and discarded, while those with binding affinity for the target are retained. The retained, target-bound sequences are amplified using PCR to generate an enriched pool of aptamer candidates. The enriched pool is subjected to multiple rounds of SELEX to increase the specificity and affinity of the aptamers for the target molecule. With each round, the selection stringency is increased to improve binding affinity and specificity. Right panel: Cell-SELEX. The Cell-SELEX technique adapts the SELEX process for selecting aptamers that specifically bind to molecular targets on the surface of whole cells, commonly used to identify biomarkers and for therapeutic targeting. Binding: The initial library of random oligonucleotide sequences is incubated with a population of target cells. The target cells may represent a disease state or a specific cell type with distinct surface markers. Aptamers that bind to the surface of target cells are retained, capturing sequences with affinity for surface-expressed biomarkers. The bound sequences are amplified via PCR to enrich the pool of aptamer candidates that specifically recognize the target cells. To improve specificity, the enriched pool is then incubated with non-target (control) cells. Aptamers that bind non-target cells are discarded, while those that remain unbound to non-target cells are retained for the next round. After negative selection, the remaining aptamer candidates are amplified again to create an enriched pool for subsequent rounds. The process is repeated for multiple rounds, alternating positive and negative selection steps, to refine the pool and select aptamers with high specificity and affinity for the target cell population. Figure is reproduced from [74] under Creative Commons CC-BY license

high affinity, thus modulating their activity in a cellular context [70]. Once target sequences are identified, SELEX can further refine these circRNAs to enhance their stability and binding efficacy, crucial for therapeutic effectiveness. SELEX can also help in designing circRNAs or their delivery vectors (i.e., nanoparticles or liposomes) to improve cellular uptake and distribution, ensuring that therapeutic circRNAs reach their intended sites of action within the body. By targeting specific miRNAs or RBPs involved in disease processes, SELEX-derived circRNAs can be used to modulate gene expression patterns extensively, offering potential treatments for diseases like cancer, cardiovascular diseases, or genetic disorders. SELEX enables the tailoring of circRNA therapeutics to individual molecular profiles, enhancing personalized medicine approaches.

While the potential of SELEX in developing circRNA therapeutics is significant, there are challenges. circRNAs can form complex secondary structures that might affect binding and function, requiring advanced computational tools and thorough biochemical characterization during the SELEX process. As with any novel therapeutic, circRNAs face rigorous testing and regulatory approvals, with SELEX-derived candidates needing extensive validation for specificity and off-target effects. In a sum, SELEX offers a systematic and efficient approach to evolve circRNA molecules or their binding partners that can be used in therapeutic contexts, significantly impacting the development of novel RNA-based treatments.

In another instance, cicrRNA NORAD, present at approximately 400 copies per HCT116 cell, plays a vital role in mitigating aberrant mitosis through its inhibition of PUM proteins, which are otherwise engaged with mRNAs encoding crucial cell cycle regulators [75, 76]. NORAD achieves this by containing 18 PUM response elements (PREs), effectively outcompeting a myriad of other PUM-binding transcripts. This results in the aggregation of PUM proteins into phase-separated condensates [77], a process facilitated by NORAD. The expression of circRNAs with up to 8 PREs per molecule, through a method termed Tornado, generated approximately 400 circPREs copies (or about 3200 PREs in total) per cell. This strategy effectively mimicked the NORAD-induced formation of PUM-containing cellular structures, significantly ameliorating chromosome segregation issues in cells deficient in NORAD. This evidence underscores the potent sequestration capacity of NORAD for PUM proteins, thereby ensuring proper mitotic progression [77]. These instances highlight the potential of circRNA-based sponges in the regulation of protein function within cells.

## 7.13 Circular Antisense RNAs

Engineering circular RNAs as stable antisense molecules offers a novel strategy reminiscent of the mechanisms employed by miRNA- and protein-circular RNA sponges. Targeting the 5′ UTR, which maintains structural conservation across the SARS-CoV-2 genome, researchers have successfully developed a cadre of circular antisense RNAs. These engineered RNAs demonstrated a significant inhibition of

SARS-CoV-2 replication in cellular models, evidencing their potential with pronounced durability [78]. Historically, the utilization of ASOs to influence RNA stability, structure, and functionality has been well-established. This is commonly accomplished by incorporating 2′-O-methyl (2′-OMe), 2′-O-methoxyethyl (2′-MOE) nucleotides, or locked nucleic acids (LNAs) to improve RNA base pairing and stability [79, 80]. Integration of antisense sequences into circRNA backbones led to a notable reduction, up to 90%, in viral replication. Notably, unmodified antisense circRNA outperformed their 2′-OMe- and 2′-MOE-modified counterparts in terms of activity, attributed to their enhanced RNA stability [78]. Exploiting the innate mechanism of adenosine (A) to inosine (I) conversion by adenosine deaminases that acts on RNA (ADAR), a novel antisense strategy employing circular guide RNAs (gNRA) was devised. This approach harnesses endogenous ADARs to selectively alter A to I in mRNA transcripts, thereby modulating gene expression without altering the genomic DNA sequence. This method holds promise for therapeutic applications, offering a means to precisely engineer gene expression profiles [81, 82].

## 7.14 Enhancers of Innate Immune Activation

Employing T4 RNA ligase [83, 84] and permuted group I introns [85–88], researchers have successfully engineered RNA circles. The immunogenic qualities of these circRNAs vary significantly based on the production techniques utilized. circRNA synthesize using T4 RNA ligase are characterized by their reduced immunogenicity, whereas those derived from the group I intron of the phage T4 td gene are known to provoke immune reactions by facilitating the recognition by Retinoic acid-inducible gene I (RIG-I, Fig. 7.4e) [85, 89]. A notable factor in this differential response is the lack of m6A modification in the synthesized circRNAs, with m6A-modified circles being processed as "self" by cellular mechanisms, thereby evading the immune response [89]. Beyond the activation of RIG-I, circRNAs originating from the td gene also engage the protein kinase R (PKR), though to a lesser extent compared to longer dsRNAs, suggesting that td-derived circular RNAs might be potent agents for augmenting immune activation [85, 89].

CircRNAs derived from the permuted Anabaena pre-tRNA group I intron were found to increase cellular RIG-I expression and activate the PKR pathway. This activation is believed to result from additional fragments that were unintentionally incorporated into the circRNAs during their synthesis. Although it was initially believed that a permuted Anabaena pre-tRNA group I intron would not provoke cellular RNA sensing mechanisms [88], it has become evident that considerations such as the purity of the RNA, the methods employed for its purification [90], and the specific sequences engineered into the circRNAs should be accounted for. Despite these findings, the mechanism through which RIG-I, a sensor known for its ability to detect 5′-triphosphates for immune surveillance [91], recognizes these additional td and pre-tRNA gene fragments present in circRNAs, thereby initiating an immune response, requires further investigation.

## 7.15 circRNA as Inhibitors of Innate Immune Responses

Significant reductions in circulating endogenous circRNAs have been observed in various autoimmune diseases, including chronic inflammatory skin conditions [92, 93] and systemic lupus erythematosus (SLE) [84]. The introduction of exogenous circRNAs, comprising 16–26 bp of dsRNAs into T cells derived from SLE patients and into peripheral blood mononuclear cells, resulted in the suppression of inappropriate PKR activation and the mitigation of IFN-β-induced expression of genes associated with the SLE phenotype [84]. Notably, circRNAs synthesized by the enzyme ligase T4 RNA were found to exhibit negligible immunogenicity. These RNAs, ranging from 300 to 500 nt and incorporating regions of short dsRNA, demonstrated a capacity to inhibit PKR activation both in cellular environments and in vitro, surpassing the efficacy of conventional small molecule PKR inhibitors by a factor of over a thousand. As circRNAs harboring dsRNA also interact with additional dsRBPs integral to the innate immune response [94], they hold potential as inhibitors of a broader spectrum of dsRBPs engaged in antiviral defenses. These insights pave the way for the potential use of non-immunogenic circRNAs in treating conditions marked by undesirable overactivation of PKR, warranting further investigation through in vivo studies.

## 7.16 Applications of Translatable circRNAs

Owing to their inherent stability, synthesized [88] and AAV-mediated [95] circRNAs with translational capacity have been proposed as adjunctive or alternative therapeutic agents to mRNA-based treatments, such as those developed for the COVID-19 pandemic [96, 97]. The administration of circRNAs, unmodified and containing IRES, through nano-formulation, has shown to sustain protein synthesis over more extended periods compared to their linear RNA counterparts in both cellular models and mouse adipose tissue [88, 98]. Recent advancements have seen the development of circRNA vaccines targeting the receptor-binding domain of SARS-CoV-2 spike protein, which offer prolonged antigen expression, induce the production of neutralizing antibodies, and promote Th1-skewed immune responses [99]. These novel findings underscore the therapeutic promise of circRNAs for sustained gene expression. Nonetheless, several challenges remain, including enhancing the yield and delivery efficiency of circRNAs, refining cap-independent translation through IRES optimization and other strategies [100, 101], and mitigating the cellular immune reactions elicited by the circRNA themselves. Recently, our group have found that tumor circular RNAs could elicit anti-tumor immunity by encoding cryptic peptides.

## 7.17 Targeting Pathological circRNAs

In addition to the emerging technologies leveraging circRNAs, circRNAs with established mechanisms and associations with critical illnesses may be therapeutically targeted, mirroring the clinical approaches taken with lncRNAs [102]. For instance, oncogenic f-circRNAs arise from chromosomal translocations linked to cancer, which reposition the native trans-complementary Alu sequences into cis configurations, thereby facilitating the synthesis of f-circRNAs [102, 103]. f-circPR and f-circM9 are both derived from the common PML/RARa and MLL/AF9 translocations found in acute promyelocytic leukemia (APL) patients. These f-circRNAs are involved in malignant cellular transformation, promoting cell viability, and conferring resistance to therapy, making them valuable targets for therapeutic intervention [104]. Similarly, circMYBL2, which is generated from exons 8 and 9 of the MYBL2 gene, a regulator of cell cycle. It is markedly overexpressed in AML patients harboring internal tandem duplication (ITD) mutations in FMS-like tyrosine kinase-3 (FLT3) gene. The suppression of circMYBL2 hampers the proliferation of AML cell lines. Functionally, circMYBL2 accelerates AML progression by attracting PTBP1 to boost the translation of FLT3 mRNA [105].

## 7.18 Circular RNAs as Biomarkers

Recent studies have increasingly identified the pathological dysregulation of circRNAs in a variety of conditions, including autoimmune disorders [84, 92, 93], prostate cancer [106], Alzheimer's disease (AD) [107], leukemia [104], HCC [108, 109], CRC [60], non-alcoholic fatty liver disease (NAFLD) [110], and non-small cell lung cancer (NSCLC) [103], among others. Notably, the exceptional stability of circRNAs not only within cellular confines but also in extracellular matrices such as extracellular vesicles [111], blood and plasma [112–115], as well as in saliva [116] and urine specimens from individuals with prostate cancer [106], underscores their viability as diagnostic markers. This potential is further enhanced when circRNAs are analyzed through nanopore sequencing technology, which enables the detection of various circRNA isoforms.

## 7.19 Conclusion

A comprehensive understanding at the molecular level is crucial for unraveling the functionality of circRNAs, as they are often co-expressed with their corresponding linear transcripts but possess unique structural characteristics, such as their covalently closed loop formations, which distinguish them from linear RNAs. These structural attributes contribute to their stability and resistance to exonucleases, potentially allowing them to play distinct roles within the cell. To fully elucidate the functions of circRNAs, it is essential to develop and refine methodologies that can specifically investigate these RNA loops without disrupting their associated linear gene counterparts. Such techniques might include advanced RNA sequencing technologies, innovative bioinformatics tools, and sophisticated molecular biology methods designed to isolate and study circRNAs in their native cellular contexts. Moreover, the widespread presence of circRNAs in various tissues and their potential regulatory functions underscore their importance in the complexity of eukaryotic transcriptomes. CircRNAs have been implicated in diverse biological processes, including gene regulation, RNA sponging, and protein interactions. They can modulate gene expression by sequestering miRNAs, interacting with RBPs, and even translating into peptides in some cases. The ability of circRNAs to participate in these varied functions enriches the diversity of the transcriptome and adds another layer of regulatory control in eukaryotic cells. As research progresses, a deeper understanding of circRNA biogenesis, localization, and functional roles will likely reveal new insights into cellular mechanisms and disease pathogenesis, highlighting their potential as biomarkers and therapeutic targets.

## References


1. Nigro JM, Cho KR, Fearon ER, Kern SE, Ruppert JM, Oliner JD, et al. Scrambled exons. Cell. 1991;64(3):607–13.
2. Cocquerelle C, Daubersies P, Majérus MA, Kerckaert JP, Bailleul B. Splicing with inverted order of exons occurs proximal to large introns. EMBO J. 1992;11(3):1095–8.
3. Capel B, Swain A, Nicolis S, Hacker A, Walter M, Koopman P, et al. Circular transcripts of the testis-determining gene Sry in adult mouse testis. Cell. 1993;73(5):1019–30.
4. Cocquerelle C, Mascrez B, Hétuin D, Bailleul B. Mis-splicing yields circular RNA molecules. FASEB J. 1993;7(1):155–60.
5. Pasman Z, Been MD, Garcia-Blanco MA. Exon circularization in mammalian nuclear extracts. RNA. 1996;2(6):603–10.
6. Salzman J, Gawad C, Wang PL, Lacayo N, Brown PO. Circular RNAs are the predominant transcript isoform from hundreds of human genes in diverse cell types. PLoS One. 2012;7(2):e30733.
7. Jeck WR, Sorrentino JA, Wang K, Slevin MK, Burd CE, Liu J, et al. Circular RNAs are abundant, conserved, and associated with ALU repeats. RNA. 2013;19(2):141–57.
8. Memczak S, Jens M, Elefsinioti A, Torti F, Krueger J, Rybak A, et al. Circular RNAs are a large class of animal RNAs with regulatory potency. Nature. 2013;495(7441):333–8.
9. Zhang XO, Wang HB, Zhang Y, Lu X, Chen LL, Yang L. Complementary sequence-mediated exon circularization. Cell. 2014;159(1):134–47.
10. Guo JU, Agarwal V, Guo H, Bartel DP. Expanded identification and characterization of mammalian circular RNAs. Genome Biol. 2014;15(7):409.
11. Westholm JO, Miura P, Olson S, Shenker S, Joseph B, Sanfilippo P, et al. Genome-wide analysis of drosophila circular RNAs reveals their structural and sequence properties and age-dependent neural accumulation. Cell Rep. 2014;9(5):1966–80.
12. Ivanov A, Memczak S, Wyler E, Torti F, Porath HT, Orejuela MR, et al. Analysis of intron sequences reveals hallmarks of circular RNA biogenesis in animals. Cell Rep. 2015;10(2):170–7.

13. Fan X, Zhang X, Wu X, Guo H, Hu Y, Tang F, et al. Single-cell RNA-seq transcriptome analysis of linear and circular RNAs in mouse preimplantation embryos. Genome Biol. 2015;16(1):148.
14. Ashwal-Fluss R, Meyer M, Pamudurti NR, Ivanov A, Bartok O, Hanan M, et al. circRNA biogenesis competes with pre-mRNA splicing. Mol Cell. 2014;56(1):55–66.
15. Starke S, Jost I, Rossbach O, Schneider T, Schreiner S, Hung LH, et al. Exon circularization requires canonical splice signals. Cell Rep. 2015;10(1):103–11.
16. Liang S, Xiao X, Bai L, Liu B, Yuan M, Ma P, et al. Conferring Ti-based MOFs with defects for enhanced sonodynamic cancer therapy. Adv Mater. 2021;33(18):e2100333.
17. Kramer MC, Liang D, Tatomer DC, Gold B, March ZM, Cherry S, et al. Combinatorial control of Drosophila circular RNA expression by intronic repeats, hnRNPs, and SR proteins. Genes Dev. 2015;29(20):2168–82.
18. Hwang HJ, Kim YK. Molecular mechanisms of circular RNA translation. Exp Mol Med. 2024;56(6):1272–80.
19. Kelly S, Greenman C, Cook PR, Papantonis A. Exon skipping is correlated with exon circularization. J Mol Biol. 2015;427(15):2414–7.
20. Conn SJ, Pillman KA, Toubia J, Conn VM, Salmanidis M, Phillips CA, et al. The RNA binding protein quaking regulates formation of circRNAs. Cell. 2015;160(6):1125–34.
21. Errichelli L, Dini Modigliani S, Laneve P, Colantoni A, Legnini I, Capauto D, et al. FUS affects circular RNA expression in murine embryonic stem cell-derived motor neurons. Nat Commun. 2017;8:14741.
22. Eisenberg E, Levanon EY. A-to-I RNA editing - immune protector and transcriptome diversifier. Nat Rev Genet. 2018;19(8):473–90.
23. Koh HR, Xing L, Kleiman L, Myong S. Repetitive RNA unwinding by RNA helicase A facilitates RNA annealing. Nucleic Acids Res. 2014;42(13):8556–64.
24. Aktaş T, Avşar Ilık İ, Maticzka D, Bhardwaj V, Pessoa Rodrigues C, Mittler G, et al. DHX9 suppresses RNA processing defects originating from the Alu invasion of the human genome. Nature. 2017;544(7648):115–9.
25. Wen X, Huang X, Mok BW, Chen Y, Zheng M, Lau SY, et al. NF90 exerts antiviral activity through regulation of PKR phosphorylation and stress granules in infected cells. J Immunol. 2014;192(8):3753–64.
26. Li X, Liu CX, Xue W, Zhang Y, Jiang S, Yin QF, et al. Coordinated circRNA biogenesis and function with NF90/NF110 in viral infection. Mol Cell. 2017;67(2):214–27.e7.
27. Eger N, Schoppe L, Schuster S, Laufs U, Boeckel JN. Circular RNA splicing. Adv Exp Med Biol. 2018;1087:41–52.
28. Zhang Y, Zhang XO, Chen T, Xiang JF, Yin QF, Xing YH, et al. Circular intronic long noncoding RNAs. Mol Cell. 2013;51(6):792–806.
29. Enuka Y, Lauriola M, Feldman ME, Sas-Chen A, Ulitsky I, Yarden Y. Circular RNAs are long-lived and display only minimal early alterations in response to a growth factor. Nucleic Acids Res. 2016;44(3):1370–83.
30. Ferreira HJ, Davalos V, de Moura MC, Soler M, Perez-Salvia M, Bueno-Costa A, et al. Circular RNA CpG island hypermethylation-associated silencing in human cancer. Oncotarget. 2018;9(49):29208–19.
31. Kristensen LS, Okholm TLH, Venø MT, Kjems J. Circular RNAs are abundantly expressed and upregulated during human epidermal stem cell differentiation. RNA Biol. 2018;15(2):280–91.
32. Rinaldi L, Datta D, Serrat J, Morey L, Solanas G, Avgustinova A, et al. Dnmt3a and Dnmt3b associate with enhancers to regulate human epidermal stem cell homeostasis. Cell Stem Cell. 2016;19(4):491–501.
33. Salzman J, Chen RE, Olsen MN, Wang PL, Brown PO. Cell-type specific features of circular RNA expression. PLoS Genet. 2013;9(9):e1003777.
34. Liang D, Wilusz JE. Short intronic repeat sequences facilitate circular RNA production. Genes Dev. 2014;28(20):2233–47.

35. Rybak-Wolf A, Stottmeister C, Glažar P, Jens M, Pino N, Giusti S, et al. Circular RNAs in the mammalian brain are highly abundant, conserved, and dynamically expressed. Mol Cell. 2015;58(5):870–85.
36. You X, Vlatkovic I, Babic A, Will T, Epstein I, Tushev G, et al. Neural circular RNAs are derived from synaptic genes and regulated by development and plasticity. Nat Neurosci. 2015;18(4):603–10.
37. Sanger HL, Klotz G, Riesner D, Gross HJ, Kleinschmidt AK. Viroids are single-stranded covalently closed circular RNA molecules existing as highly base-paired rod-like structures. Proc Natl Acad Sci USA. 1976;73(11):3852–6.
38. Kos A, Dijkema R, Arnberg AC, van der Meide PH, Schellekens H. The hepatitis delta (delta) virus possesses a circular RNA. Nature. 1986;323(6088):558–60.
39. Côté F, Perreault JP. Peach latent mosaic viroid is locked by a 2′,5′-phosphodiester bond produced by in vitro self-ligation. J Mol Biol. 1997;273(3):533–43.
40. Flores R, Grubb D, Elleuch A, Nohales M, Delgado S, Gago S. Rolling-circle replication of viroids, viroid-like satellite RNAs and hepatitis delta virus: variations on a theme. RNA Biol. 2011;8(2):200–6.
41. Tang TH, Rozhdestvensky TS, d'Orval BC, Bortolin ML, Huber H, Charpentier B, et al. RNomics in Archaea reveals a further link between splicing of archaeal introns and rRNA processing. Nucleic Acids Res. 2002;30(4):921–30.
42. Danan M, Schwartz S, Edelheit S, Sorek R. Transcriptome-wide discovery of circular RNAs in Archaea. Nucleic Acids Res. 2012;40(7):3131–42.
43. Soma A, Onodera A, Sugahara J, Kanai A, Yachie N, Tomita M, et al. Permuted tRNA genes expressed via a circular RNA intermediate in Cyanidioschyzon merolae. Science. 2007;318(5849):450–3.
44. Grabowski PJ, Zaug AJ, Cech TR. The intervening sequence of the ribosomal RNA precursor is converted to a circular RNA in isolated nuclei of Tetrahymena. Cell. 1981;23(2):467–76.
45. Cech TR. Self-splicing of group I introns. Annu Rev Biochem. 1990;59:543–68.
46. Li-Pook-Than J, Bonen L. Multiple physical forms of excised group II intron RNAs in wheat mitochondria. Nucleic Acids Res. 2006;34(9):2782–90.
47. Molina-Sánchez MD, Martinez-Abarca F, Toro N. Excision of the Sinorhizobium meliloti group II intron RmInt1 as circles in vivo. J Biol Chem. 2006;281(39):28737–44.
48. Abelson J, Trotta CR, Li H. tRNA splicing. J Biol Chem. 1998;273(21):12685–8.
49. Salgia SR, Singh SK, Gurha P, Gupta R. Two reactions of Haloferax volcanii RNA splicing enzymes: joining of exons and circularization of introns. RNA. 2003;9(3):319–30.
50. Gardner EJ, Nizami ZF, Talbot CC, Gall JG. Stable intronic sequence RNA (sisRNA), a new class of noncoding RNA from the oocyte nucleus of Xenopus tropicalis. Genes Dev. 2012;26(22):2550–9.
51. Talhouarne GJ, Gall JG. Lariat intronic RNAs in the cytoplasm of Xenopus tropicalis oocytes. RNA. 2014;20(9):1476–87.
52. Li Z, Huang C, Bao C, Chen L, Lin M, Wang X, et al. Exon-intron circular RNAs regulate transcription in the nucleus. Nat Struct Mol Biol. 2015;22(3):256–64.
53. Szabo L, Morey R, Palpant NJ, Wang PL, Afari N, Jiang C, et al. Statistically based splicing detection reveals neural enrichment and tissue-specific induction of circular RNA during human fetal development. Genome Biol. 2015;16(1):126.
54. Wang Y, Wang Z. Efficient backsplicing produces translatable circular mRNAs. RNA. 2015;21(2):172–9.
55. Chen T, Xiang JF, Zhu S, Chen S, Yin QF, Zhang XO, et al. ADAR1 is required for differentiation and neural induction by regulating microRNA processing in a catalytically independent manner. Cell Res. 2015;25(4):459–76.
56. Chen CY, Sarnow P. Initiation of protein synthesis by the eukaryotic translational apparatus on circular RNAs. Science. 1995;268(5209):415–7.
57. Hansen TB, Jensen TI, Clausen BH, Bramsen JB, Finsen B, Damgaard CK, et al. Natural RNA circles function as efficient microRNA sponges. Nature. 2013;495(7441):384–8.

58. Braunschweig U, Barbosa-Morais NL, Pan Q, Nachman EN, Alipanahi B, Gonatopoulos-Pournatzis T, et al. Widespread intron retention in mammals functionally tunes transcriptomes. Genome Res. 2014;24(11):1774–86.
59. Armakola M, Higgins MJ, Figley MD, Barmada SJ, Scarborough EA, Diaz Z, et al. Inhibition of RNA lariat debranching enzyme suppresses TDP-43 toxicity in ALS disease models. Nat Genet. 2012;44(12):1302–9.
60. Bachmayr-Heyda A, Reiner AT, Auer K, Sukhbaatar N, Aust S, Bachleitner-Hofmann T, et al. Correlation of circular RNA abundance with proliferation–exemplified with colorectal and ovarian cancer, idiopathic lung fibrosis, and normal human tissues. Sci Rep. 2015;5:8057.
61. Burd CE, Jeck WR, Liu Y, Sanoff HK, Wang Z, Sharpless NE. Expression of linear and novel circular forms of an INK4/ARF-associated non-coding RNA correlates with atherosclerosis risk. PLoS Genet. 2010;6(12):e1001233.
62. Yap KL, Li S, Muñoz-Cabello AM, Raguz S, Zeng L, Mujtaba S, et al. Molecular interplay of the noncoding RNA ANRIL and methylated histone H3 lysine 27 by polycomb CBX7 in transcriptional silencing of INK4a. Mol Cell. 2010;38(5):662–74.
63. Lukiw WJ. Circular RNA (circRNA) in Alzheimer's disease (AD). Front Genet. 2013;4:307.
64. Dunn MR, Jimenez RM, Chaput JC. Analysis of aptamer discovery and technology. Nat Rev Chem. 2017;1(10):0076.
65. Filonov GS, Kam CW, Song W, Jaffrey SR. In-gel imaging of RNA processing using broccoli reveals optimal aptamer expression strategies. Chem Biol. 2015;22(5):649–60.
66. Filonov GS, Moon JD, Svensen N, Jaffrey SR. Broccoli: rapid selection of an RNA mimic of green fluorescent protein by fluorescence-based selection and directed evolution. J Am Chem Soc. 2014;136(46):16299–308.
67. Litke JL, Jaffrey SR. Highly efficient expression of circular RNA aptamers in cells using autocatalytic transcripts. Nat Biotechnol. 2019;37(6):667–75.
68. Wurster SE, Maher LJ. Selection and characterization of anti-NF-kappaB p65 RNA aptamers. RNA. 2008;14(6):1037–47.
69. Hui J, Hung LH, Heiner M, Schreiner S, Neumüller N, Reither G, et al. Intronic CA-repeat and CA-rich elements: a new class of regulators of mammalian alternative splicing. EMBO J. 2005;24(11):1988–98.
70. Schreiner S, Didio A, Hung LH, Bindereif A. Design and application of circular RNAs with protein-sponge function. Nucleic Acids Res. 2020;48(21):12326–35.
71. Chai C, Xie Z, Grotewold E. SELEX (Systematic Evolution of Ligands by EXponential Enrichment), as a powerful tool for deciphering the protein-DNA interaction space. Methods Mol Biol. 2011;754:249–58.
72. Darmostuk M, Rimpelova S, Gbelcova H, Ruml T. Current approaches in SELEX: an update to aptamer selection technology. Biotechnol Adv. 2015;33(Part 2):1141–61.
73. Kohlberger M, Gadermaier G. SELEX: critical factors and optimization strategies for successful aptamer selection. Biotechnol Appl Biochem. 2022;69(5):1771–92.
74. Sola M, Menon AP, Moreno B, Meraviglia-Crivelli D, Soldevilla MM, Cartón-García F, et al. Aptamers against live targets: is in vivo SELEX finally coming to the edge? Mol Ther Nucl Acids. 2020;21:192–204.
75. Lee S, Kopp F, Chang TC, Sataluri A, Chen B, Sivakumar S, et al. Noncoding RNA NORAD regulates genomic stability by sequestering PUMILIO proteins. Cell. 2016;164(1-2):69–80.
76. Tichon A, Gil N, Lubelsky Y, Havkin Solomon T, Lemze D, Itzkovitz S, et al. A conserved abundant cytoplasmic long noncoding RNA modulates repression by Pumilio proteins in human cells. Nat Commun. 2016;7:12209.
77. Elguindy MM, Mendell JT. NORAD-induced Pumilio phase separation is required for genome stability. Nature. 2021;595(7866):303–8.
78. Pfafenrot C, Schneider T, Müller C, Hung LH, Schreiner S, Ziebuhr J, et al. Inhibition of SARS-CoV-2 coronavirus proliferation by designer antisense-circRNAs. Nucleic Acids Res. 2021;49(21):12502–16.

79. Bennett CF, Swayze EE. RNA targeting therapeutics: molecular mechanisms of antisense oligonucleotides as a therapeutic platform. Annu Rev Pharmacol Toxicol. 2010;50:259–93.
80. Crooke ST, Witztum JL, Bennett CF, Baker BF. RNA-targeted therapeutics. Cell Metab. 2018;27(4):714–39.
81. Katrekar D, Yen J, Xiang Y, Saha A, Meluzzi D, Savva Y, et al. Efficient in vitro and in vivo RNA editing via recruitment of endogenous ADARs using circular guide RNAs. Nat Biotechnol. 2022;40(6):938–45.
82. Yi Z, Qu L, Tang H, Liu Z, Liu Y, Tian F, et al. Engineered circular ADAR-recruiting RNAs increase the efficiency and fidelity of RNA editing in vitro and in vivo. Nat Biotechnol. 2022;40(6):946–55.
83. Kaufmann G, Klein T, Littauer UZ. T4 RNA ligase: substrate chain length requirements. FEBS Lett. 1974;46(1):271–5.
84. Liu CX, Li X, Nan F, Jiang S, Gao X, Guo SK, et al. Structure and degradation of circular RNAs regulate PKR activation in innate immunity. Cell. 2019;177(4):865–80.
85. Chen YG, Kim MV, Chen X, Batista PJ, Aoyama S, Wilusz JE, et al. Sensing self and foreign circular RNAs by intron identity. Mol Cell. 2017;67(2):228–38.
86. Ford E, Ares M. Synthesis of circular RNA in bacteria and yeast using RNA cyclase ribozymes derived from a group I intron of phage T4. Proc Natl Acad Sci USA. 1994;91(8):3117–21.
87. Puttaraju M, Been MD. Group I permuted intron-exon (PIE) sequences self-splice to produce circular exons. Nucleic Acids Res. 1992;20(20):5357–64.
88. Wesselhoeft RA, Kowalski PS, Parker-Hale FC, Huang Y, Bisaria N, Anderson DG. RNA circularization diminishes immunogenicity and can extend translation duration in vivo. Mol Cell. 2019;74(3):508–20.
89. Chen YG, Chen R, Ahmad S, Verma R, Kasturi SP, Amaya L, et al. N6-methyladenosine modification controls circular RNA immunity. Mol Cell. 2019;76(1):96–109.
90. Abe BT, Wesselhoeft RA, Chen R, Anderson DG, Chang HY. Circular RNA migration in agarose gel electrophoresis. Mol Cell. 2022;82(9):1768–77.
91. Reikine S, Nguyen JB, Modis Y. Pattern recognition and signaling mechanisms of RIG-I and MDA5. Front Immunol. 2014;5:342.
92. Moldovan LI, Hansen TB, Venø MT, Okholm TLH, Andersen TL, Hager H, et al. High-throughput RNA sequencing from paired lesional- and non-lesional skin reveals major alterations in the psoriasis circRNAome. BMC Med Genet. 2019;12(1):174.
93. Moldovan LI, Tsoi LC, Ranjitha U, Hager H, Weidinger S, Gudjonsson JE, et al. Characterization of circular RNA transcriptomes in psoriasis and atopic dermatitis reveals disease-specific expression profiles. Exp Dermatol. 2021;30(8):1187–96.
94. Xia P, Wang S, Ye B, Du Y, Li C, Xiong Z, et al. A Circular RNA protects dormant hematopoietic stem cells from DNA sensor cGAS-mediated exhaustion. Immunity. 2018;48(4):688–701.
95. Meganck RM, Borchardt EK, Castellanos Rivera RM, Scalabrino ML, Wilusz JE, Marzluff WF, et al. Tissue-dependent expression and translation of circular RNAs with recombinant AAV vectors in vivo. Mol Ther Nucl Acids. 2018;13:89–98.
96. Banerji A, Wickner PG, Saff R, Stone CA, Robinson LB, Long AA, et al. mRNA vaccines to prevent COVID-19 disease and reported allergic reactions: current evidence and suggested approach. J Allergy Clin Immunol Pract. 2021;9(4):1423–37.
97. Polack FP, Thomas SJ, Kitchin N, Absalon J, Gurtman A, Lockhart S, et al. Safety and efficacy of the BNT162b2 mRNA covid-19 vaccine. N Engl J Med. 2020;383(27):2603–15.
98. Wesselhoeft RA, Kowalski PS, Anderson DG. Engineering circular RNA for potent and stable translation in eukaryotic cells. Nat Commun. 2018;9(1):2629.
99. Qu L, Yi Z, Shen Y, Lin L, Chen F, Xu Y, et al. Circular RNA vaccines against SARS-CoV-2 and emerging variants. Cell. 2022;185(10):1728–44.
100. Chen CK, Cheng R, Demeter J, Chen J, Weingarten-Gabbay S, Jiang L, et al. Structured elements drive extensive circular RNA translation. Mol Cell. 2021;81(20):4300–18.

101. Meganck RM, Liu J, Hale AE, Simon KE, Fanous MM, Vincent HA, et al. Engineering highly efficient backsplicing and translation of synthetic circRNAs. Mol Ther Nucl Acids. 2021;23:821–34.
102. Statello L, Guo CJ, Chen LL, Huarte M. Gene regulation by long non-coding RNAs and its biological functions. Nat Rev Mol Cell Biol. 2021;22(2):96–118.
103. Wu K, Liao X, Gong Y, He J, Zhou JK, Tan S, et al. Circular RNA F-circSR derived from SLC34A2-ROS1 fusion gene promotes cell migration in non-small cell lung cancer. Mol Cancer. 2019;18(1):98.
104. Guarnerio J, Bezzi M, Jeong JC, Paffenholz SV, Berry K, Naldini MM, et al. Oncogenic role of fusion-circRNAs derived from cancer-associated chromosomal translocations. Cell. 2016;165(2):289–302.
105. Sun YM, Wang WT, Zeng ZC, Chen TQ, Han C, Pan Q, et al. circMYBL2, a circRNA from MYBL2, regulates FLT3 translation by recruiting PTBP1 to promote FLT3-ITD AML progression. Blood. 2019;134(18):1533–46.
106. Vo JN, Cieslik M, Zhang Y, Shukla S, Xiao L, Zhang Y, et al. The landscape of circular RNA in cancer. Cell. 2019;176(4):869–81.
107. Dube U, Del-Aguila JL, Li Z, Budde JP, Jiang S, Hsu S, et al. An atlas of cortical circular RNA expression in Alzheimer disease brains demonstrates clinical and pathological associations. Nat Neurosci. 2019;22(11):1903–12.
108. Chen S, Cao X, Zhang J, Wu W, Zhang B, Zhao F. circVAMP3 drives CAPRIN1 phase separation and inhibits hepatocellular carcinoma by suppressing c-Myc translation. Adv Sci. 2022;9(8):e2103817.
109. Yu J, Xu QG, Wang ZG, Yang Y, Zhang L, Ma JZ, et al. Circular RNA cSMARCA5 inhibits growth and metastasis in hepatocellular carcinoma. J Hepatol. 2018;68(6):1214–27.
110. Zhao Q, Liu J, Deng H, Ma R, Liao JY, Liang H, et al. Targeting mitochondria-located circRNA SCAR alleviates NASH via reducing mROS output. Cell. 2020;183(1):76–93.
111. Li Y, Zheng Q, Bao C, Li S, Guo W, Zhao J, et al. Circular RNA is enriched and stable in exosomes: a promising biomarker for cancer diagnosis. Cell Res. 2015;25(8):981–4.
112. Li H, Li K, Lai W, Li X, Wang H, Yang J, et al. Comprehensive circular RNA profiles in plasma reveals that circular RNAs can be used as novel biomarkers for systemic lupus erythematosus. Clin Chim Acta. 2018;480:17–25.
113. Li T, Shao Y, Fu L, Xie Y, Zhu L, Sun W, et al. Plasma circular RNA profiling of patients with gastric cancer and their droplet digital RT-PCR detection. J Mol Med. 2018;96(1):85–96.
114. Memczak S, Papavasileiou P, Peters O, Rajewsky N. Identification and characterization of circular RNAs as a new class of putative biomarkers in human blood. PLoS One. 2015;10(10):e0141214.
115. Preußer C, Hung LH, Schneider T, Schreiner S, Hardt M, Moebus A, et al. Selective release of circRNAs in platelet-derived extracellular vesicles. J Extracell Vesicles. 2018;7(1):1424473.
116. Bahn JH, Zhang Q, Li F, Chan TM, Lin X, Kim Y, et al. The landscape of microRNA, Piwi-interacting RNA, and circular RNA in human saliva. Clin Chem. 2015;61(1):221–30.

# Chapter 8
# Non-coding RNAs: siRNAs

**Abstract** RNA interference (RNAi) represents a primeval biological process initially evolved as a defense mechanism against external genetic elements. This pathway can selectively silence any gene associated with disease through a sequence-specific mechanism, positioning small interfering RNAs (siRNAs) as a potent therapeutic tool. Over the past two decades, this concept has transitioned from discovery to clinical application, culminating in the approval of two siRNA-based drugs by Alnylam Pharmaceuticals: ONPATTRO® (Patisiran) and GIVLAARI™ (Givosiran). These approvals mark a significant milestone in the annals of pharmaceutical history, introducing a new paradigm in the treatment and management of diseases. Unlike traditional small molecules and antibodies which needed constant administration, siRNA therapeutics can be administered in a lesser regime (i.e., monthly, quarterly or even bi-annually); while maintaining therapeutic efficacy. The development of siRNA therapies, however, has faced formidable challenges. Nevertheless, one of the key obstacles for siRNA delivery to specific target tissues include developing efficient and safe methods, creating robust nanoplatform for siRNA encapsulation, and optimizing the activity, stability, specificity, all while minimizing potential their off-target effects. This chapter delves into the biogenesis and function of siRNAs, providing a comprehensive overview of the landscape. Additionally, we explore the latest advancements in siRNA delivery platforms that are clinically explored and commercially available, discussing the core principles behind their design.

## 8.1 Introduction

The fundamental activator of RNA interference (RNAi) is typically long, double-stranded RNA (dsRNA). These endogenous or exogenous dsRNAs could be is either directly introduced into the cytoplasm or absorbed from the external environment [1]. In the cytoplasm, these dsRNAs undergo processing by the enzyme Dicer, generating siRNAs that could bind to sense strand of the target gene to orchestrate the gene-silencing process [2, 3]. Initially observed during studies on

 
P. E. Saw, E. Song, *RNA Therapeutics in Human Diseases*,
https://doi.org/10.1007/978-981-96-3041-7_8

transgene- and virus-induced gene silencing in plants, siRNAs have been acknowledged for their innate role in defending the genome [1]. More researches are revealing that endogenous sources of siRNA (endo-siRNAs) could come from transposons, centromeres, and other repetitive elements in the cytoplasm [4], to include convergent mRNA transcripts, natural sense-antisense pairs, duplexes formed between antisense transcripts from pseudogenes, as well as hairpin RNAs (hpRNAs) [5]. In plants, trans-acting siRNAs (ta-siRNAs) have been identified. These ta-siRNAs are first cleaved from specific genomic transcripts and then subsequently regulate distinct groups of genes [6, 7]. It is now evident that siRNAs originate not only from external nucleic acids but also from within the endogenous genomic framework. Unlike many externally derived siRNAs (exo-siRNAs), endogenously arising siRNAs and their precursors typically undergo a critical nuclear phase during their biogenesis. This distinction underscores the complexity and intrinsic nature of siRNA pathways within the cellular environment.

The biogenesis of double-stranded siRNAs by Dicer necessitates further complex assembly pathways for RISC. These assembly mechanisms have been extensively studied and characterized in both Drosophila and humans. First affirmed in Drosophila, the RISC assembly process commences with the formation of a nucleation complex involving a heterodimer called the R2D2/Dicer-2 that could bind to an siRNA duplex [3]. This interaction leads to the recruitment of additional factors (currently unidentified) that contribute to forming the RISC-loading complex (RLC). The RLC facilitates the formation of the pre-RISC, which still contains the siRNA in its duplex form [8]. Here, the Ago2 cleaves the passenger strand of the siRNA [9–11], results in the ejection of the passenger strand and the conversion of the pre-RISC into the active 80S holo-RISC [3]. Interestingly, the assembly of RISC in humans is biochemically different from Drosophila. In human, RISC is made up of a core trimeric complex known as the RISC-loading complex, consisting of (1) Dicer, (2) TAR RNA-binding protein (TRBP), and (3) Ago-2. This trimeric core can form even in the absence of dsRNAs [12, 13]. When bound to dsRNAs, this complex is capable of processing the dsRNAs into siRNA, loading the siRNA into Ago-2, and finally discard the passenger strand to yield a functional RISC [14]. Although other proteins have been found to associate with Ago complexes in human cells [15–17], they are not deemed essential for the RISC loading or target RNA cleavage processes (Fig. 8.1).

Interestingly, experiments in mouse cells have shown that even in the absence of Dicer, siRNA can still be assembled into functional RISCs [19, 20]. This finding suggests that Dicer is not absolutely required for RISC loading in mammals, indicating a divergence in the molecular mechanisms underlying RISC assembly across different species. The products of Dicer's action are initially double-stranded, but it is ultimately a single guide strand that is incorporated into the RISC complex. Interestingly, the selection of this guide strand does not depend on the presence of a matching mRNA target; indicating that the RISC could be pre-programmed to cleave one target, and re-programmed to cleave another target that is later introduced to RISC. Further studies also indicate that the relative thermodynamic stabilities at both ends of siRNA duplex could influence the selection of the guide strand.

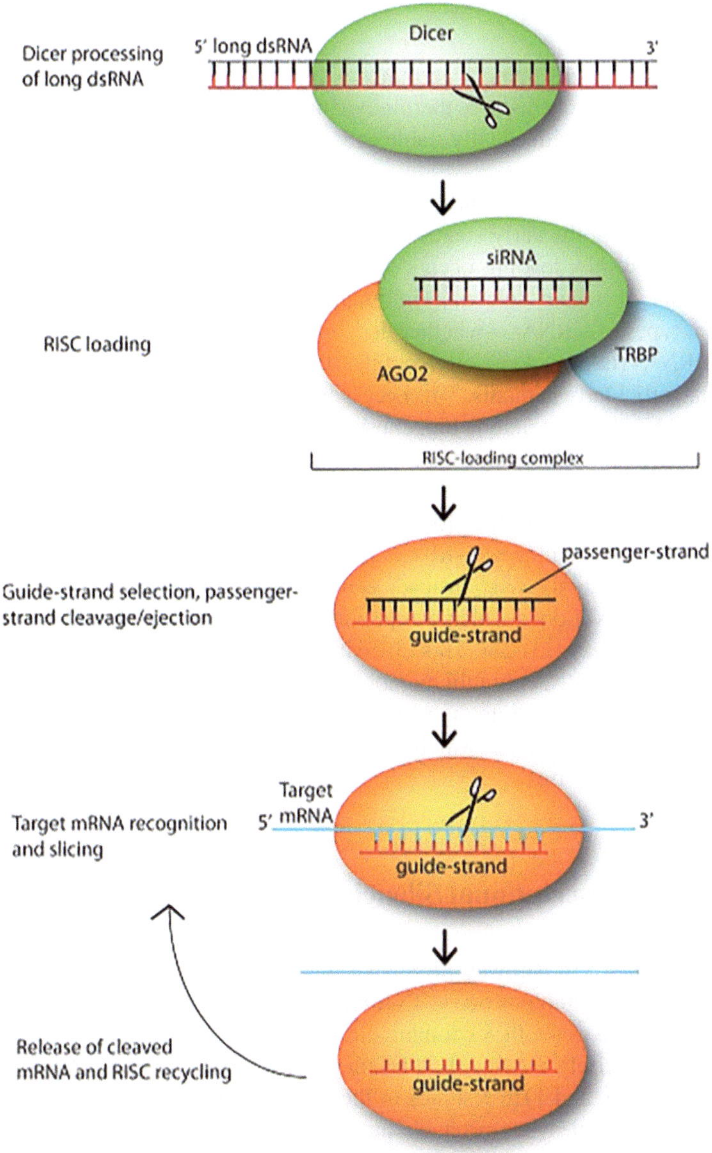


**Fig. 8.1** This figure illustrates the process of RNA interference (RNAi) via the small interfering RNA (siRNA) pathway, leading to target mRNA degradation. The process begins with long double-stranded RNA (dsRNA), which is cleaved by the enzyme Dicer into small siRNA fragments. These siRNAs are then loaded onto the RNA-induced silencing complex (RISC), assisted by Argonaute 2 (AGO2) and the transactivation response RNA-binding protein (TRBP). Within RISC, the siRNA is separated into two strands: the passenger strand is discarded, while the guide strand, which is complementary to the target mRNA, is retained. The guide strand directs RISC to bind to the target mRNA, where AGO2 mediates the cleavage of the mRNA, leading to its degradation and blocking protein translation. Following cleavage, the mRNA fragments are released, and RISC, still bound to the guide strand, is recycled for additional rounds of target mRNA recognition and degradation. This mechanism enables specific gene silencing at the post-transcriptional level and has both natural regulatory functions and therapeutic potential. Figure is reproduced from [18] under Creative Commons Attribution License (http://creativecommons.org/licenses/by/2.5/)

The strand with a less stable 5′ terminus is typically favored to serve as the guide strand [3]. This mechanism is thermodynamically graded, meaning that if RISC encounters siRNAs with equally stable ends, the incorporation of the guide strand is a fully randomized process.

In Drosophila, the R2D2/Dicer-2 heterodimer plays a crucial role in sensing this asymmetry during the early stages of RISC assembly [3]. However, in Dicer-null mouse cells, siRNA loading occurs without apparent issues, leaving the specific mechanisms of strand selection in mammals somewhat ambiguous [19, 20]. Different types of siRNAs—whether derived from viruses, transgenes, or as trans-acting siRNAs (ta-siRNAs)—often require distinct proteins for their processing and function, particularly in plants. This necessity for diverse cofactors underscores that various siRNAs engage unique biogenesis and RISC assembly pathways. This concept extends to animals, evidenced by findings in Drosophila where Dicer-2 is associated with different double-stranded RNA-binding proteins (dsRBDs) for endo-siRNAs and exo-siRNAs [5]. The interaction between Dicer-2 and R2D2 is only well-documented for exo-siRNAs [3], while being less important for the functions of most endo-siRNAs. Instead, endo-siRNAs mostly depend on a protein called Loquacious (Loqs), which was previously identified as a partner of Dicer-1 in the miRNA pathway [21–23]. This diversity in protein requirements among different categories of siRNAs highlights not only the complexity of RNA interference machinery but also illustrates the overlapping and distinct roles of small RNAs within silencing pathways, which continue to blur the functional boundaries between these categories.

## 8.2 Post-transcriptional Silencing by siRNAs

In the canonical RNAi pathway, the siRNA guide strand steers the RISC to precisely bind and degrade complementary RNA targets that is a "perfect fit." This degradation process involves the PIWI domain of the Ago protein, which exhibits a "slicer" activity that precisely cleaves the phosphodiester bonds between the target RNA paired to the 10th and 11th residues of the siRNA guide strand on the 5′-terminus [3]. The resulting cleavage products have a 5′-monophosphate and a 3′-hydroxyl terminus. After this initial incision, cellular exonucleases further degrade these RNA fragments [24]. Additionally, the 3-terminus of the RISC cleavage products may further undergo oligouridylation, which further targets them for exonucleolytic degradation [25]. Once the target RNA is cleaved, it detaches from the siRNA guide, freeing RISC to target additional RNA molecules. However, it has been observed that in some instances, highly purified RISC does not efficiently recycle to cleave new targets without the assistance of extrinsic factors, suggesting that the release of cleavage products and subsequent RISC recycling may be facilitated by ATP hydrolysis [3, 26, 27]. Importantly, mismatches between the siRNA guide strand and the target RNA, particularly near the central region of the duplex, can inhibit endonucleolytic cleavage. Moreover, some Ago proteins, when programmed with siRNA exhibit a decrement in endonuclease activity, even when paired with perfectly complementary target sequences [3]. In other scenarios, targets that are not fully complementary or are bound by endonuclease-inactive siRISC can still be silenced through mechanisms similar to those used by miRNAs, including translational repression or exonucleolytic degradation.

In mammalian systems, specially designed siRNAs that engage targets with partial complementarity function almost identically to miRNAs in their silencing effects. This phenomenon is largely responsible for the "off-target" effects observed with siRNAs, where the imperfectly matched mRNAs could still be silenced in a manner akin to miRNA-mediated repression. Such off-target actions are not only significant for understanding the broader implications of siRNA therapies but also underscore the necessity for precise siRNA design to minimize unintended gene silencing. The degree to which naturally occurring endogenous siRNAs (endo-siRNAs) silence genes by targeting imperfectly matched mRNA sequences is yet to be fully understood, but we could hypothesize that such mechanisms are exploited to some extent within cellular regulatory processes. Typically, the effective phases of post-transcriptional siRNA silencing occur in the cytoplasm. Here, siRNA bound to RISC prompts the re-localization of Ago proteins to P-bodies. These P-bodies are repositories rich in mRNA degradation factors, indicating a focused area for RNA-induced silencing [28]. Despite this, the necessity of P-body localization for effective RNA interference (RNAi) is not absolute [29], and evidence also supports the functionality of RNAi within the nuclear environment [30].

Due to advancement in fluorescence microscopy and cross-correlation spectroscopy, we now know that RISC also exists in the nucleus. However, the nuclear form of RISC is clearly having a smaller size compared to its cytoplasmic counterpart [31]. A genetic screening done in *C. elegans* revealed that Nrde-3, an Ago specific protein is usually located in the cytoplasm. However, upon siRNA binding, Nrde-3 will then be translocated to the nucleus [32]. This is an interesting observation as these insights collectively underscore the dynamic nature of RNAi factor localization and suggest a broad distribution and versatility of the silencing machinery within cells. Such adaptability ensures that RNAi mechanisms are not only confined to the cytoplasm but are also capable of exerting their gene-silencing effects within the nucleus, adapting to where they are most needed for regulating gene expression. This dynamic distribution is crucial for the nuanced control of gene activity across different cellular contexts and conditions.

## 8.3 siRNA Amplification

RNAi is remarkably efficient. With only mere few molecules dsRNA per cell are sufficient to trigger a strong RNAi response [1]. This trait is particularly evident in organisms like *C. elegans*, where when the target mRNA is introduced, the primary dsRNA initiates the synthesis of secondary siRNAs. This synthesis is mediated by

RNA-dependent RNA polymerase (RdRP) enzymes [2], which produce a secondary pool of siRNAs that not only amplifies but also prolongs the RNAi response. In eukaryotic genome, RdRP genes are highly expressed. However, this is not true in insects and vertebrates, where RdRP genes are absent. This absence may contribute to the heightened specificity observed in these groups, allowing for more precise targeting of RNA molecules. For siRNA amplification via RdRP, siRNA could be generated and mapped to mRNA regions beyond those initially targeted by the original dsRNA. This process potentially leads to the silencing of additional transcripts that share conserved sequences or common exons with the original target. When transitive RNAi is absent, the specificity of the RNAi mechanism in these organisms is enhanced, enabling selective targeting of distinct mRNA isoforms from the same gene, even though they were alternatively spliced. This specificity is crucial for research and therapeutic applications where off-target effects can diminish efficacy and increase the risk of side effects.

The extensive cataloguing of secondary siRNAs in nematodes has yielded intriguing findings that have reshaped our understanding of RNAi mechanisms [33, 34]. A surprising aspect of these secondary siRNAs is that they do not correspond to the sense strand of mRNA, rather they are predominantly corresponding to the antisense strand of the mRNA. This observation challenges the conventional understanding that secondary siRNAs are produced from an antisense dsRNA intermediate processed by Dicer. If this were the case, the inherent thermodynamic asymmetry of siRNAs would typically result in a mix of both sense and antisense guide strands. Furthermore, secondary siRNAs are characterized by the presence of di- or triphosphate groups at their 5′ termini. This detail suggests that these siRNAs are likely primary, unprimed products of RdRP activity, indicating that they do not undergo processing by Dicer. This aligns with the hypothesis that secondary siRNAs originate directly from RdRP synthesis rather than from the traditional double-stranded RNA precursors.

## 8.4 siRNAs Can Induce Heterochromatin Formation

siRNAs are known to regulate gene expression, and this regulation mechanism is not limited to post-transcriptional mechanisms. siRNAs was demonstrated to initiate heterochromatin formation in *S. pombe*, aligning with prior observations of transcriptional gene silencing (TGS) in plants [4]. Subsequently, the connection between RNAi mechanism and heterochromatin formation was established across various organisms including animals, plants, and ciliates. In *S. pombe*, the RNA-induced transcriptional silencing (RITS) complex contains Ago1 and targets specific genomic regions like centromeric repeats, guided by its associated siRNAs. There are theories suggesting that siRNAs could identify emerging transcripts [35], a process aided by the interaction of RITS with RNA polymerase II [36, 37]. This interaction leads to the methylation of H3K9 by histone methyltransferases (HMTs), which in turn attracts the chromodomain protein Swi6, resulting in chromatin condensation [4]. RITS also stimulates the RNA-dependent RNA polymerase complex (RDRC), which utilizes its RdRP component Rdp1 to produce secondary siRNAs [38], thereby amplifying and extending the silencing effect. TGS has been thoroughly explored in plants, where it shares many mechanisms with those found in fission yeast, but also differs notably, including the direct methylation of DNA in addition to histone modification [4]. Additionally, specific DNA-dependent RNA polymerases, RNA Pol IV and Pol V, are involved in generating siRNA precursors [39–41]. The field of siRNA-directed transcriptional silencing continues to evolve rapidly and has been comprehensively reviewed recently [42–44].

**Box 8.1 Modification of siRNA and Their Effects in Therapeutic Efficacy**

In the early stages of siRNA therapeutic development, many agents were designed using unmodified or minimally modified siRNAs aimed at achieving targeted gene silencing in specific tissues. These unmodified siRNAs demonstrated the capacity to mediate gene silencing in vivo, particularly in tissues accessible via local administration, such as the eye. However, these approaches often encountered limitations in efficacy and issues with off-target effects. For instance, Kleinman et al. [45] found that the siRNA therapeutics bevasiranib and AGN211745, which were developed for treating age-related macular degeneration [46], activated toll-like receptor 3 (TLR3) and its adapter TRIF, leading to the secretion of pro-inflammatory cytokines like interleukin-12 and interferon-γ. Bevasiranib, targeting VEGFA [47], and AGN211745, targeting VEGFR1 [48], represent unmodified and slightly modified siRNAs, respectively. Additionally, Kleinman's study indicated that siRNA molecules with sequences of 21 nucleotides or longer could suppress choroidal neovascularization (CNV) in mice, regardless of their target gene. To address these limitations, chemically modified siRNAs were developed to enhance therapeutic efficacy and reduce immune activation and toxicity. Modifications such as replacing the 2′-hydroxyl group (2′-OH) with 2′-O-methyl (2′-OMe) [34] or 2′-methoxyethyl (2′-MOE) [49], or incorporating nucleotide analogs like locked nucleic acid (LNA) [50], unlocked nucleic acid (UNA) [51], or glycol nucleic acid (GNA) [52] (illustrated in Fig. 8.2), have been shown to effectively mitigate immunostimulatory responses, increase siRNA activity and specificity, and reduce off-target toxicity. Numerous configurations of these chemical modifications have been tested to optimize siRNA potency and minimize potential side effects [54].

Nucleotide structure allows for chemical modifications at various positions, including the phosphate backbone, ribose sugar, or nucleobase. Typically, multiple modifications are incorporated into siRNAs to enhance their therapeutic properties. For instance, a combination of 2′-O-methyl (2′-OMe) and phosphorothioate (PS) modifications has been shown to improve systemic delivery of cholesterol-conjugated siRNAs, resulting in effective gene silencing in vivo [55]. The dual modification of 2′-OMe and 2′-deoxy-2′-fluoro (2′-F) is utilized in ONPATTRO® [56, 57] (illustrated in Figs. 8.2 and 8.3). In QPI-1007, a naked, unformulated siRNA therapeutic [2], 2′-OMe is selectively incorporated in the antisense strand, while an inverted deoxyabasic moiety and an L-DNA cytidine nucleotide are present in the sense strand [58]. Similarly, modifications with PS, 2′-OMe, 2′-F, and 2′-deoxy have been applied in inclisiran (ALN-PCSsc), an siRNA designed for hypercholesterolemia treatment [54, 59]. These targeted modifications enhance siRNA efficacy, specificity, and stability while reducing toxicity and immunogenicity, making them essential for therapeutic success.

## 8.5 Preclinical Development of siRNA Therapeutics

The selection of therapeutic targets for development typically centers on genes that have been validated through human genetics, the availability of well-defined circulating biomarkers to evaluate therapeutic activity, and a clear pathway toward regulatory approval. These criteria help reduce the risks associated with targeting and facilitate the direct evaluation of how effectively the therapy can be translated into clinical settings.

## 8.6 siRNA Lead Discovery

The process of discovering lead compounds, which spans from the early design to the preclinical setting (i.e., animal experiments) and concludes with the identification of a potential candidate for further preclinical testing, typically requires 9–12 months. This journey begins with an in-silico assessment that analyses potential siRNAs targeting the desired transcript using algorithms designed to predict their potency and specificity. From this analysis, a few hundred siRNAs are synthesized based on established chemical modification frameworks and templates. These are subsequently tested for their activity on and off the target in cultured cells and further assessed for pharmacodynamic properties in rodent models. Only a select few siRNAs that show promise through this extensive optimization, which includes safety evaluations in non-clinical settings, proceed to pharmacological studies targeting specific outcomes in non-human primates [52, 60]. Notably, not all RNA targets are suitable for RNAi approaches. Characteristics such as rapid mRNA turnover can hinder the efficacy of siRNAs [61]. The therapeutic doses are also calibrated based on the necessary extent of knockdown for clinical efficacy. For instance, reducing serum TTR protein levels by approximately 80% is adequate for treating polyneuropathy, whereas a reduction greater than 99% in complement factor 5 (C5) is required for effectively managing paroxysmal nocturnal hemoglobinuria.

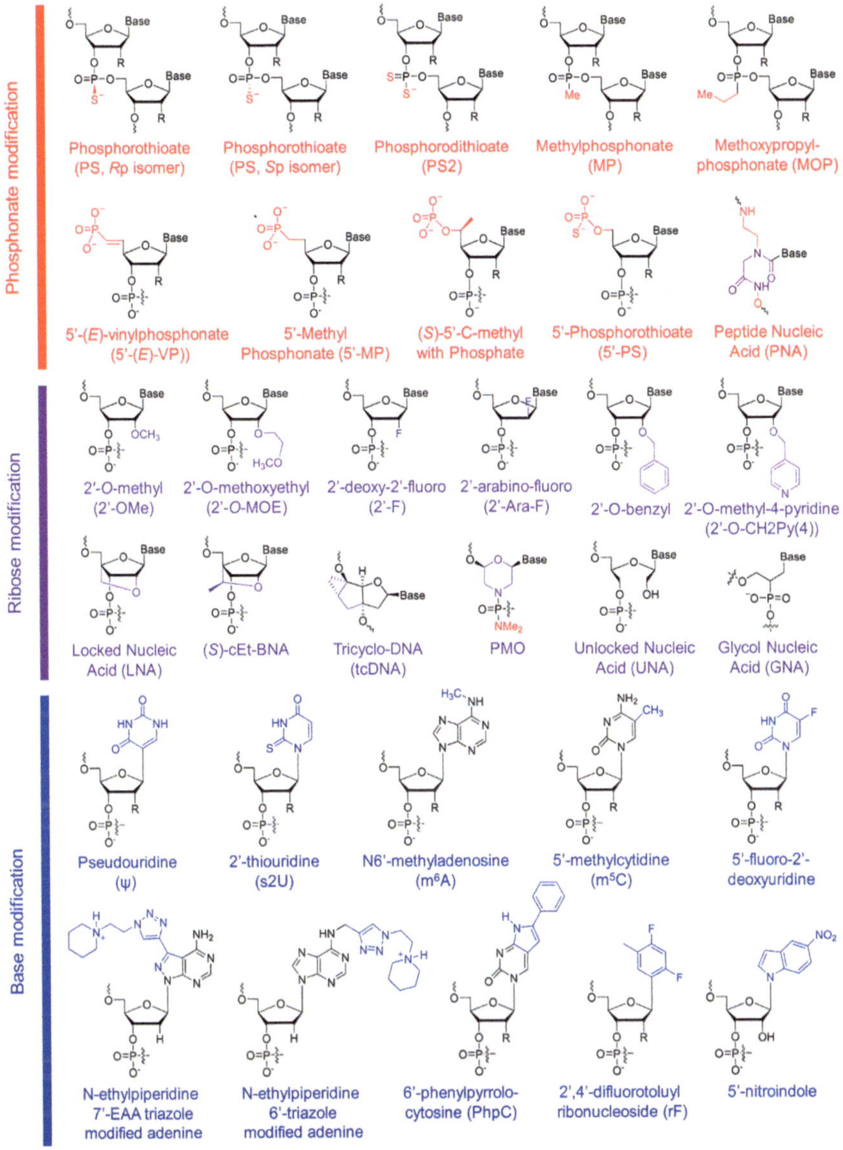


**Fig. 8.2** Overview of nucleic acid modifications. This diagram categorizes and illustrates various chemical modifications of nucleic acids, primarily focusing on RNA analogs. Modifications are divided into three main categories: phosphonate modification, ribose modification, and base modification. Each modification is represented by its chemical structure, highlighting functional groups or atoms that differ from the standard nucleotide structure. Figure is reproduced from [53] under Creative Commons http://creativecommons.org/licenses/by/4.0

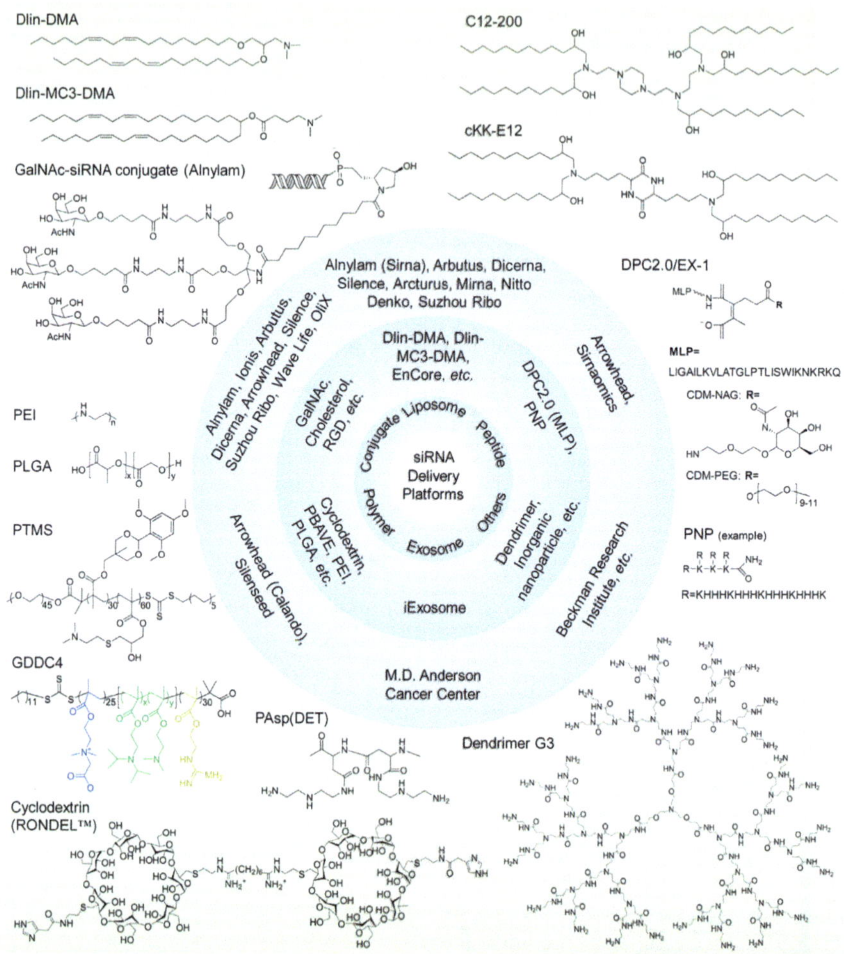


**Fig. 8.3** Comprehensive overview of siRNA delivery platforms. This diagram provides a detailed look at various small interfering RNA (siRNA) delivery systems and their chemical constituents. Central to the figure is a circular diagram highlighting key delivery platforms including conjugates, liposomes, dendrimers, and nanoparticles, with specific examples like Dlin-DMA and GalNAc-siRNA conjugates. Around this central diagram are molecular structures of various compounds used in siRNA delivery, such as PEI (polyethyleneimine), PLGA (polylactic-co-glycolic acid), and cyclodextrin, as well as structures of more specific molecules like Dlin-MC3-DMA and cKK-E12. Figure is reproduced from [53] under Creative Commons http://creativecommons.org/licenses/by/4.0

## 8.7 Pharmacokinetics (PK) and Pharmacodynamics (PD) Prediction of siRNA Therapeutics

Pharmacokinetics (PK) and pharmacodynamics (PD) are critical components in the development of siRNA therapeutics, playing a significant role in their effectiveness and safety profiles. PK studies assess how siRNAs are absorbed and distributed within the body. Understanding the absorption patterns, whether through direct injection or more innovative delivery systems, is vital to ensure that siRNAs reach their intended cellular targets in sufficient concentrations without being prematurely degraded or excreted [62]. siRNAs are susceptible to degradation by nucleases in the bloodstream and within cells. PK studies help in understanding the metabolic pathways and the rate of degradation, which are crucial for optimizing the stability and half-life of siRNA molecules [63]. Additionally, these studies track how siRNAs and their metabolites are excreted, typically via renal or hepatic pathways, which influences dosing schedules and safety profiles. Determining the bioavailability of siRNAs, or the proportion of the administered dose that reaches the systemic circulation and accesses target cells, is fundamental. PK data guide modifications to the siRNA structure or formulation to enhance stability and cellular uptake. PD studies focus on the biological and physiological effects siRNAs have on their target cells and how these effects are achieved [64]. This includes understanding how siRNAs modulate gene expression, either by degrading target mRNA or by inhibiting its translation, leading to a decrease in protein synthesis. PD analysis involves quantifying the relationship between drug concentration (achieved through PK studies) and its therapeutic effect. This includes determining the minimal effective concentration and the maximal response achievable, critical for setting appropriate dosing regimens. PD also evaluates potential side effects caused by off-target effects or immune stimulation. Understanding these responses is essential for assessing the therapeutic window of siRNA molecules—the range between efficacy and toxicity. The integration of PK and PD data allows for the design of siRNA therapeutics that are not only effective but also have an acceptable safety margin. It helps in refining dosing strategies to achieve optimal therapeutic levels that are effective without being toxic. PK/PD studies aid in personalizing treatment plans based on individual variations in metabolism and response, which can be critical due to the genetic specificity of siRNA mechanisms. Comprehensive PK and PD data are crucial for regulatory submissions, helping demonstrate that the siRNA therapeutic is safe, effective, and predictable in its action [62–64].

Upon subcutaneous (*s.c.*) injection, GalNAc-conjugated siRNAs are swiftly absorbed into the bloodstream and predominantly target liver hepatocytes [65]. Their removal through the kidneys represents a minor pathway [65], which is likely due to the administered dose maintaining GalNAc–siRNA concentrations below the saturation level of the ASGPR, facilitating efficient hepatocyte uptake. Extensive preclinical studies of various GalNAc-conjugated siRNAs have been studied. Results confirmed that GalNAc-conjugated siRNAs showed consistent absorption, distribution, metabolism, excretion (ADME), pharmacokinetic (PK) and pharmacodynamic (PD) profiles across different species, the four major points described above [60]. This consistency supports the accurate prediction of human PK and PD metrics, aiding in the translation of dosing regimens for clinical use. The dose-normalized exposure levels for these compounds typically remain within a twofold range of the average for both plasma and liver, underscoring their predictable PK profiles in these matrices.

Distinct from conventional small-molecule drugs or inhibitors, RNAi therapeutics offer a much more flexible approach. Since the active component primarily lies within the siRNA sequence, it is possible to leverage on their universal chemical insights, even among molecules with very different structures. The design parameters, such as the length of siRNA strand, whether or not the siRNA has undergone chemical modifications, and their delivery mechanisms, could be customized independently of the siRNA sequence. All these factors could influence specific RNAi product characteristics to increase potency, duration, specificity, and even specific tissue distribution [53]. Consequently, even when employing the same delivery system, such as a GalNAc ligand, using different siRNAs would typically yields different PK and PD outcomes [53, 60, 66]. Chemical modifications applied to these siRNAs reduce their metabolic breakdown, a property consistently observed across species. In rodents, higher nuclease activity was seen compared to non-human primates and humans [60]. With less than 10% of GalNAc-conjugated siRNAs metabolized in the plasma, the majority undergoes slow hepatic metabolism and is excreted via urine and bile [60]. This low degradation rate contributes to the formation of an intracellular depot [67], prolonging the drug's effect and allowing less frequent dosing schedules in clinical settings.

## 8.8 Safety Concern and Tolerability Effect of RNAi

For RNAi therapeutics, the safety concerns are mainly categorized into three primary areas: (1) the intracellular build-up of siRNAs and their metabolites, (2) disturbances in endo-RNAi mechanisms, and (3) off-target effects resulting from hybridization in non-target ograns. A key focus regarding chemical toxicity relates to the non-natural modifications such as 2′-F and PS, which are employed to enhance siRNA stability. The 2′-fluororibo modification, an effective analog of both natural RNA and DNA, enhances resistance to enzymatic degradation [68]. 2′-F-uridine and cytidine are 2′-fluororibo pyrimidines modification that were utilized in Macugen, an aptamer drug approved for age-related macular degeneration (AMD) [69]. However, due to the severe hepatic toxicity observed with fialuridine, a nucleoside analog containing a 2′-F modification in the arabino configuration and a 5′-iodo group on its uracil base [70], these pyrimidines underwent rigorous safety evaluations before human application [71]. Despite the toxicity seen with fialuridine, no comparable adverse effects were detected with 2′-F pyrimidines in rats or woodchucks. Although high doses of 2′-F nucleotides have been shown to incorporate into DNA and RNA [71, 72], and despite the cytotoxicity noted in vitro with single-stranded ASOs featuring full PS backbones [73], a detailed safety analysis covering all four 2′-fluororibo nucleosides was performed with the first GalNAc–siRNA conjugate, Revusiran [74]. This study revealed that 2′-F-nucleoside triphosphates act as weak inhibitors and are poorly recognized as substrates by mitochondrial DNA and RNA polymerases, do not halt chain elongation, and are efficiently outcompeted by natural nucleotides in enzyme reactions [75]. Despite a mortality imbalance linked to Revusiran, exhaustive investigations did not pinpoint a definitive causative mechanism. While cardiac parameters progressed similarly in both treatment groups, the potential influence of Revusiran or its metabolites, including 2′-fluronucleosides, remains unconfirmed [76].

Current siRNA conjugates frequently employ non-natural chemical modifications, such as phosphorothioate (PS) linkages in the phosphate backbone, to improve metabolic stability [77, 78]. This method was initially introduced by Eckstein [79] and first applied in antisense oligonucleotides (ASOs) [80]. While PS modifications have been linked to adverse effects in extensive applications [81], such issues have not manifested with siRNAs [82, 83]. This discrepancy is likely due to the fewer PS linkages needed to stabilize siRNAs, only 6–8 PS in both strands is sufficient. Furthermore, PS-stabilized siRNA is more structured, double-stranded nature, primarily exposes the hydrophilic sugar–phosphate backbone, thus reducing nonspecific interactions. In addition to potential toxicities from the chemically modified backbone, exo-siRNAs might compete with endogenous miRNAs to activate the RISC, potentially disrupting miRNA functions. This phenomenon has been observed with short hairpin RNAs under strong promoter-driven expression, leading to saturation of the exportin molecule involved in transporting precursor miRNAs (pre-miRNAs) from the nucleus [84, 85]. However, siRNAs directly interact with the RISC and do not follow the pre-miRNA pathway. This interaction was confirmed in studies where hepatotoxic siRNAs were neutralized using Reversir technology, which specifically blocks RISC-loaded siRNA activity [86]. The absence of hepatotoxicity in these cases indicated that it did not stem from RISC competition with endogenous RNAi mechanisms [87]. Furthermore, RISC-engaged siRNAs can induce RNAi-mechanism-based off-target effects resembling miRNA activity, involving seed-region interactions that destabilize and degrade unintended mRNA targets [88–90]. Although such off-target interactions are challenging to prevent due to the prevalence of potential mRNA targets with short seed matches (~7 nucleotides), they can be reduced through careful sequence selection and modifications that thermodynamically weaken seed-mediated binding (ESC+ design) [51, 87, 91]. This approach was validated in preclinical models and later in human studies, demonstrating an enhanced safety profile for GalNAc–siRNA conjugates modified with a single guanidine in the seed region, which had previously been linked to elevated liver enzymes [92, 93]. Overall, with minimal chemical toxicity and reduced risk of disturbing RNAi pathways, the siRNA designs employed in GalNAc conjugates are highly specific and exhibit promising safety profiles.

## 8.9 RNAi Therapeutic in Clinical Development

In recent years, many RNAi-based therapeutics are being developed in the pipelines, with some successful stories of FDA-approval, including the lipid-based nanoparticle (LNP)-encapsulated siRNA, Patisiran [94]. Prior to RNAi, single-stranded ASOs were a hit, as these are the first non-drug molecules being approved by the FDA. Therefore, most RNAi-based therapeutics were aimed to integrate common features of single-stranded ASOs. Similar to ASOs, RNAi drugs are also designed to feature a 19- to 21-bp, sometimes paired with a shorter 11–15 nucleotide sense strand [95]. Study indicated that when the longer antisense strand is strengthened with PS linkages, the siRNA distribution and their efficacy in extra-hepatic tissues will differ [96]. Dicer-substrate siRNA, on the other hand, are on the larger end, with 25–30 bp duplexes. This size is optimal for being recognized and processed by Dicer [97]. Current popular strategies to enhance siRNA stability include 2′-modification with 2′-OMe or 2′-F with the combination of PS linkages on the whole siRNA molecule. Although 25–30 bp is optimal for Dicer, RISC loading actually prefers 19–21 nucleotide duplex structure, with a 2-nucleotide overhand on both strands. Due to its inherent structural bias, the 19–21 siRNA duplex demonstrated excellent silencing efficacy across multiple animal species. This could be attributed to the physical stability, a protection from exonuclease at both siRNA strands, enhancing their durability. A summary of some preclinical development of siRNA delivery systems are summarized in Fig. 8.3.

## 8.10 FDA-Approved Therapies

In 2018, FDA approved the inaugural RNAi therapeutic Patisiran, and it remains the sole LNP-formulated siRNA on the market [57]. Since then, five other RNAi drugs have received approval, all of which are GalNAc–siRNA conjugates, with nedosiran being the latest addition in 2023. Each of these FDA-approved RNAi therapeutics targets gene transcripts that are highly expressed in the liver [57]. The development and application of GalNAc-conjugated siRNA technology have revolutionized nucleic acid-based therapeutics. Despite liposomal nanoparticle systems (LNPs) offering a quicker onset of action [98], the robust and consistent outcomes of subcutaneously administered GalNAc conjugates have led to their dominance as the preferred liver-targeting platform for oligonucleotide treatments, including ASOs [66].

## 8.11 Future Outlook

A variety of RNAi therapeutics are currently utilized or under development for addressing a broad spectrum of liver-associated diseases [53, 99, 100]. Recent enhancements in the GalNAc conjugate platform have led to the creation of molecules with extended durability, capable of providing sustained pharmacological effects potentially suitable for yearly dosing. This prolonged activity, which is unprecedented for non-genome-altering drugs, benefits from the pharmacological strengths of RNAi therapeutics: a predictable, dose-dependent PK–PD profile and a reversible mechanism of action that avoids permanent genetic modifications. A significant breakthrough was seen with the approval of Inclisiran, showcasing the transformative potential of long-acting medications for diseases other than targeting the liver. Inclisiran (Leqvio®) is the first a siRNA drug developed by Novartis to lower systemic lipid level for the treatment of dyslipidemia, hypercholesterolemia or for patients that needs to lower their LDL-C levels in general. The prospect of less frequent dosing intervals could revolutionize the management of widespread chronic conditions like hypertension and gout, where the effectiveness of traditional treatments is often compromised by issues related to tolerance, adherence, and prescribing practices [101–104]. For conditions necessitating profound suppression of a target gene, siRNAs can be strategically employed to substantially halt the majority of target gene expression, while any residual activity may be addressed using additional modalities such as antibodies. Owing to their extended duration of effect, unique action mechanism distinct from other therapies, and minimal off-target impacts.

Moreover, a combination of multiple siRNAs (known as siRNA cocktails) can be utilized concurrently to target different molecules. This capability is particularly advantageous for targeting intracellular proteins, especially the "undruggable" targets, as they could not be modulated by small molecules. For instance, simultaneous targeting of multiple critical pathways involved in tumor growth could decrease the probability of therapeutic resistance in cancer treatment. Similarly, in cardiovascular disease management, where reducing absolute risk is crucial, employing siRNAs to concurrently address multiple risk factors, such as hypertension and hypercholesterolemia, could be highly effective. The safety and efficacy of the GalNAc–siRNA platform for various targets have been established, suggesting their potential to be a robust treatment solution for individuals with metabolic syndrome. This combination strategy could also prove particularly advantageous in combating viral diseases, where targeting multiple distinct genomic sites can significantly reduce the chances of viral resistance [105].

## 8.12 Conclusion

After 20 years of ongoing development, RNAi therapeutics have demonstrated remarkable capabilities such as exceptional specificity, prolonged effect, and a reversible action mechanism. These therapies offer numerous advantages, including their modular nature and reproducibility, a solid track record of human safety backed by successful FDA approvals, and a handful of ongoing clinical trials in the pipeline. The distinct mechanism of action of RNAi also makes it an ideal candidate for combination therapies, giving hope to previously undruggable targets. Since its inception, the translation of RNAi into clinical therapeutics, from in vitro to animal models to human trials are seen to be quite consistent, with relatively high success rates. Although current FDA-approved RNAi therapeutics are focused more on silencing genes in the liver, there is an expanding effort to develop delivery systems that reach other tissues. Such advancements are poised to address a broad spectrum of diseases that currently have substantial unmet medical needs. With six approved therapies and more in the pipeline, RNAi therapeutics have securely positioned themselves as a new and effective modality in the pharmaceutical landscape. Considering the vast potential of this technology and the numerous targets yet to be explored, it is evident that we are only witnessing the beginning stages of this innovative class of bio-medicines.

## References


1. Mello CC, Conte D. Revealing the world of RNA interference. Nature. 2004;431(7006):338–42.
2. Meister G, Tuschl T. Mechanisms of gene silencing by double-stranded RNA. Nature. 2004;431(7006):343–9.
3. Tomari Y, Zamore PD. Perspective: machines for RNAi. Genes Dev. 2005;19(5):517–29.
4. Lippman Z, Martienssen R. The role of RNA interference in heterochromatic silencing. Nature. 2004;431(7006):364–70.
5. Golden DE, Gerbasi VR, Sontheimer EJ. An inside job for siRNAs. Mol Cell. 2008;31(3):309–12.
6. Vazquez F, Vaucheret H, Rajagopalan R, Lepers C, Gasciolli V, Mallory AC, et al. Endogenous trans-acting siRNAs regulate the accumulation of Arabidopsis mRNAs. Mol Cell. 2004;16(1):69–79.
7. Allen E, Xie Z, Gustafson AM, Carrington JC. microRNA-directed phasing during trans-acting siRNA biogenesis in plants. Cell. 2005;121(2):207–21.
8. Kim K, Lee YS, Carthew RW. Conversion of pre-RISC to holo-RISC by Ago2 during assembly of RNAi complexes. RNA. 2007;13(1):22–9.
9. Rand TA, Petersen S, Du F, Wang X. Argonaute2 cleaves the anti-guide strand of siRNA during RISC activation. Cell. 2005;123(4):621–9.
10. Matranga C, Tomari Y, Shin C, Bartel DP, Zamore PD. Passenger-strand cleavage facilitates assembly of siRNA into Ago2-containing RNAi enzyme complexes. Cell. 2005;123(4):607–20.
11. Miyoshi K, Tsukumo H, Nagami T, Siomi H, Siomi MC. Slicer function of Drosophila Argonautes and its involvement in RISC formation. Genes Dev. 2005;19(23):2837–48.
12. Gregory RI, Chendrimada TP, Cooch N, Shiekhattar R. Human RISC couples microRNA biogenesis and posttranscriptional gene silencing. Cell. 2005;123(4):631–40.
13. Maniataki E, Mourelatos Z. A human, ATP-independent, RISC assembly machine fueled by pre-miRNA. Genes Dev. 2005;19(24):2979–90.
14. MacRae IJ, Ma E, Zhou M, Robinson CV, Doudna JA. In vitro reconstitution of the human RISC-loading complex. Proc Natl Acad Sci USA. 2008;105(2):512–7.
15. Meister G, Landthaler M, Peters L, Chen PY, Urlaub H, Lührmann R, et al. Identification of novel argonaute-associated proteins. Curr Biol. 2005;15(23):2149–55.
16. Höck J, Weinmann L, Ender C, Rüdel S, Kremmer E, Raabe M, et al. Proteomic and functional analysis of Argonaute-containing mRNA-protein complexes in human cells. EMBO Rep. 2007;8(11):1052–60.
17. Landthaler M, Gaidatzis D, Rothballer A, Chen PY, Soll SJ, Dinic L, et al. Molecular characterization of human Argonaute-containing ribonucleoprotein complexes and their bound target mRNAs. RNA. 2008;14(12):2580–96.
18. Dana H, Chalbatani GM, Mahmoodzadeh H, Karimloo R, Rezaiean O, Moradzadeh A, et al. Molecular mechanisms and biological functions of siRNA. Int J Biomed Sci. 2017;13(2):48–57.

19. Kanellopoulou C, Muljo SA, Kung AL, Ganesan S, Drapkin R, Jenuwein T, et al. Dicer-deficient mouse embryonic stem cells are defective in differentiation and centromeric silencing. Genes Dev. 2005;19(4):489–501.
20. Murchison EP, Partridge JF, Tam OH, Cheloufi S, Hannon GJ. Characterization of Dicer-deficient murine embryonic stem cells. Proc Natl Acad Sci USA. 2005;102(34):12135–40.
21. Förstemann K, Tomari Y, Du T, Vagin VV, Denli AM, Bratu DP, et al. Normal microRNA maturation and germ-line stem cell maintenance requires Loquacious, a double-stranded RNA-binding domain protein. PLoS Biol. 2005;3(7):e236.
22. Jiang F, Ye X, Liu X, Fincher L, McKearin D, Liu Q. Dicer-1 and R3D1-L catalyze microRNA maturation in Drosophila. Genes Dev. 2005;19(14):1674–9.
23. Saito K, Ishizuka A, Siomi H, Siomi MC. Processing of pre-microRNAs by the dicer-1-loquacious complex in Drosophila cells. PLoS Biol. 2005;3(7):e235.
24. Orban TI, Izaurralde E. Decay of mRNAs targeted by RISC requires XRN1, the Ski complex, and the exosome. RNA. 2005;11(4):459–69.
25. Shen B, Goodman HM. Uridine addition after microRNA-directed cleavage. Science. 2004;306(5698):997.
26. Rivas FV, Tolia NH, Song JJ, Aragon JP, Liu J, Hannon GJ, et al. Purified Argonaute2 and an siRNA form recombinant human RISC. Nat Struct Mol Biol. 2005;12(4):340–9.
27. Förstemann K, Horwich MD, Wee L, Tomari Y, Zamore PD. Drosophila microRNAs are sorted into functionally distinct argonaute complexes after production by dicer-1. Cell. 2007;130(2):287–97.
28. Liu J, Valencia-Sanchez MA, Hannon GJ, Parker R. MicroRNA-dependent localization of targeted mRNAs to mammalian P-bodies. Nat Cell Biol. 2005;7(7):719–23.
29. Chu CY, Rana TM. Translation repression in human cells by microRNA-induced gene silencing requires RCK/p54. PLoS Biol. 2006;4(7):e210.
30. Robb GB, Brown KM, Khurana J, Rana TM. Specific and potent RNAi in the nucleus of human cells. Nat Struct Mol Biol. 2005;12(2):133–7.
31. Ohrt T, Mütze J, Staroske W, Weinmann L, Höck J, Crell K, et al. Fluorescence correlation spectroscopy and fluorescence cross-correlation spectroscopy reveal the cytoplasmic origination of loaded nuclear RISC in vivo in human cells. Nucleic Acids Res. 2008;36(20):6439–49.
32. Guang S, Bochner AF, Pavelec DM, Burkhart KB, Harding S, Lachowiec J, et al. An Argonaute transports siRNAs from the cytoplasm to the nucleus. Science. 2008;321(5888):537–41.
33. Pak J, Fire A. Distinct populations of primary and secondary effectors during RNAi in C. elegans. Science. 2007;315(5809):241–4.
34. Sijen T, Steiner FA, Thijssen KL, Plasterk RH. Retracted: secondary siRNAs result from unprimed RNA synthesis and form a distinct class. Science. 2007;315(5809):244–7.
35. Bühler M, Verdel A, Moazed D. Tethering RITS to a nascent transcript initiates RNAi- and heterochromatin-dependent gene silencing. Cell. 2006;125(5):873–86.
36. Djupedal I, Portoso M, Spåhr H, Bonilla C, Gustafsson CM, Allshire RC, et al. RNA Pol II subunit Rpb7 promotes centromeric transcription and RNAi-directed chromatin silencing. Genes Dev. 2005;19(19):2301–6.
37. Kato H, Goto DB, Martienssen RA, Urano T, Furukawa K, Murakami Y. RNA polymerase II is required for RNAi-dependent heterochromatin assembly. Science. 2005;309(5733):467–9.
38. Sugiyama T, Cam H, Verdel A, Moazed D, Grewal SI. RNA-dependent RNA polymerase is an essential component of a self-enforcing loop coupling heterochromatin assembly to siRNA production. Proc Natl Acad Sci USA. 2005;102(1):152–7.
39. Herr AJ, Jensen MB, Dalmay T, Baulcombe DC. RNA polymerase IV directs silencing of endogenous DNA. Science. 2005;308(5718):118–20.
40. Onodera Y, Haag JR, Ream T, Costa Nunes P, Pontes O, Pikaard CS. Plant nuclear RNA polymerase IV mediates siRNA and DNA methylation-dependent heterochromatin formation. Cell. 2005;120(5):613–22.

41. Wierzbicki AT, Haag JR, Pikaard CS. Noncoding transcription by RNA polymerase Pol IVb/Pol V mediates transcriptional silencing of overlapping and adjacent genes. Cell. 2008;135(4):635–48.
42. Grewal SI, Elgin SC. Transcription and RNA interference in the formation of heterochromatin. Nature. 2007;447(7143):399–406.
43. Henderson IR, Jacobsen SE. Epigenetic inheritance in plants. Nature. 2007;447(7143):418–24.
44. Moazed D. Small RNAs in transcriptional gene silencing and genome defence. Nature. 2009;457(7228):413–20.
45. Kleinman ME, Yamada K, Takeda A, Chandrasekaran V, Nozaki M, Baffi JZ, et al. Sequence- and target-independent angiogenesis suppression by siRNA via TLR3. Nature. 2008;452(7187):591–7.
46. Barakat MR, Kaiser PK. VEGF inhibitors for the treatment of neovascular age-related macular degeneration. Expert Opin Investig Drugs. 2009;18(5):637–46.
47. Reich SJ, Fosnot J, Kuroki A, Tang W, Yang X, Maguire AM, et al. Small interfering RNA (siRNA) targeting VEGF effectively inhibits ocular neovascularization in a mouse model. Mol Vis. 2003;9:210–6.
48. Shen J, Samul R, Silva RL, Akiyama H, Liu H, Saishin Y, et al. Suppression of ocular neovascularization with siRNA targeting VEGF receptor 1. Gene Ther. 2006;13(3):225–34.
49. Song X, Wang X, Ma Y, Liang Z, Yang Z, Cao H. Site-specific modification using the 2′-methoxyethyl group improves the specificity and activity of siRNAs. Mol Ther Nucl Acids. 2017;9:242–50.
50. Fluiter K, Mook OR, Baas F. The therapeutic potential of LNA-modified siRNAs: reduction of off-target effects by chemical modification of the siRNA sequence. Methods Mol Biol. 2009;487:189–203.
51. Bramsen JB, Pakula MM, Hansen TB, Bus C, Langkjær N, Odadzic D, et al. A screen of chemical modifications identifies position-specific modification by UNA to most potently reduce siRNA off-target effects. Nucleic Acids Res. 2010;38(17):5761–73.
52. Janas MM, Harbison CE, Perry VK, Carito B, Sutherland JE, Vaishnaw AK, et al. The nonclinical safety profile of GalNAc-conjugated RNAi therapeutics in subacute studies. Toxicol Pathol. 2018;46(7):735–45.
53. Hu B, Zhong L, Weng Y, Peng L, Huang Y, Zhao Y, et al. Therapeutic siRNA: state of the art. Signal Transduct Target Ther. 2020;5(1):101.
54. Khvorova A, Watts JK. The chemical evolution of oligonucleotide therapies of clinical utility. Nat Biotechnol. 2017;35(3):238–48.
55. Soutschek J, Akinc A, Bramlage B, Charisse K, Constien R, Donoghue M, et al. Therapeutic silencing of an endogenous gene by systemic administration of modified siRNAs. Nature. 2004;432(7014):173–8.
56. Coelho T, Adams D, Silva A, Lozeron P, Hawkins PN, Mant T, et al. Safety and efficacy of RNAi therapy for transthyretin amyloidosis. N Engl J Med. 2013;369(9):819–29.
57. Adams D, Gonzalez-Duarte A, O'Riordan WD, Yang CC, Ueda M, Kristen AV, et al. Patisiran, an RNAi therapeutic, for hereditary transthyretin amyloidosis. N Engl J Med. 2018;379(1):11–21.
58. Titze-de-Almeida R, David C, Titze-de-Almeida SS. The race of 10 synthetic RNAi-based drugs to the pharmaceutical market. Pharm Res. 2017;34(7):1339–63.
59. Fitzgerald K, Kallend D, Simon A. A highly durable RNAi therapeutic inhibitor of PCSK9. N Engl J Med. 2017;376(18):e38.
60. McDougall R, Ramsden D, Agarwal S, Agarwal S, Aluri K, Arciprete M, et al. The nonclinical disposition and pharmacokinetic/pharmacodynamic properties of N-acetylgalactosamine-conjugated small interfering RNA are highly predictable and build confidence in translation to human. Drug Metab Dispos. 2022;50(6):781–97.
61. Larsson E, Sander C, Marks D. mRNA turnover rate limits siRNA and microRNA efficacy. Mol Syst Biol. 2010;6:433.

62. Jeon JY, Ayyar VS, Mitra A. Pharmacokinetic and pharmacodynamic modeling of siRNA therapeutics - a mini review. Pharm Res. 2022;39(8):1749–59.
63. Lumen A, Zhang X, Dutta S, Upreti VV. Predicting clinical pharmacokinetics/pharmacodynamics and impact of organ impairment on siRNA-based therapeutics using a mechanistic physiologically-based pharmacokinetic-pharmacodynamic model. Clin Pharmacol Ther. 2024;115(5):1054–64.
64. Park J, Park J, Pei Y, Xu J, Yeo Y. Pharmacokinetics and biodistribution of recently-developed siRNA nanomedicines. Adv Drug Deliv Rev. 2016;104:93–109.
65. Agarwal S, Simon AR, Goel V, Habtemariam BA, Clausen VA, Kim JB, et al. Pharmacokinetics and pharmacodynamics of the small interfering ribonucleic acid, givosiran, in patients with acute hepatic porphyria. Clin Pharmacol Ther. 2020;108(1):63–72.
66. Springer AD, Dowdy SF. GalNAc-siRNA conjugates: leading the way for delivery of RNAi therapeutics. Nucl Acid Ther. 2018;28(3):109–18.
67. Brown CR, Gupta S, Qin J, Racie T, He G, Lentini S, et al. Investigating the pharmacodynamic durability of GalNAc-siRNA conjugates. Nucleic Acids Res. 2020;48(21):11827–44.
68. Kawasaki AM, Casper MD, Freier SM, Lesnik EA, Zounes MC, Cummins LL, et al. Uniformly modified 2′-deoxy-2′-fluoro phosphorothioate oligonucleotides as nuclease-resistant antisense compounds with high affinity and specificity for RNA targets. J Med Chem. 1993;36(7):831–41.
69. Tobin KA. Macugen treatment for wet age-related macular degeneration. Insight. 2006;31(1):11–4.
70. McKenzie R, Fried MW, Sallie R, Conjeevaram H, Di Bisceglie AM, Park Y, et al. Hepatic failure and lactic acidosis due to fialuridine (FIAU), an investigational nucleoside analogue for chronic hepatitis B. N Engl J Med. 1995;333(17):1099–105.
71. Richardson FC, Tennant BC, Meyer DJ, Richardson KA, Mann PC, McGinty GR, et al. An evaluation of the toxicities of 2′-fluorouridine and 2′-fluorocytidine-HCl in F344 rats and woodchucks (Marmota monax). Toxicol Pathol. 1999;27(6):607–17.
72. Saleh AF, Bachman M, Priestley CC, Gooderham NJ, Andersson P, Henry SP, et al. 2′-O-(2-methoxyethyl) nucleosides are not phosphorylated or incorporated into the genome of human lymphoblastoid TK6 cells. Toxicol Sci. 2018;163(1):70–8.
73. Shen W, Liang XH, Sun H, Crooke ST. 2′-Fluoro-modified phosphorothioate oligonucleotide can cause rapid degradation of P54nrb and PSF. Nucleic Acids Res. 2015;43(9):4569–78.
74. Zimmermann TS, Karsten V, Chan A, Chiesa J, Boyce M, Bettencourt BR, et al. Clinical proof of concept for a novel hepatocyte-targeting GalNAc-siRNA conjugate. Mol Ther. 2017;25(1):71–8.
75. Janas MM, Zlatev I, Liu J, Jiang Y, Barros SA, Sutherland JE, et al. Safety evaluation of 2′-deoxy-2′-fluoro nucleotides in GalNAc-siRNA conjugates. Nucleic Acids Res. 2019;47(7):3306–20.
76. Judge DP, Kristen AV, Grogan M, Maurer MS, Falk RH, Hanna M, et al. Phase 3 multicenter study of revusiran in patients with hereditary transthyretin-mediated (hATTR) amyloidosis with cardiomyopathy (ENDEAVOUR). Cardiovasc Drugs Ther. 2020;34(3):357–70.
77. Vosberg HP, Eckstein F. Effect of deoxynucleoside phosphorothioates incorporated in DNA on cleavage by restriction enzymes. J Biol Chem. 1982;257(11):6595–9.
78. Geary RS, Yu RZ, Levin AA. Pharmacokinetics of phosphorothioate antisense oligodeoxynucleotides. Curr Opin Investig Drugs. 2001;2(4):562–73.
79. Burgers PMJ, Eckstein F. Synthesis of dinucleoside monophosphorothioates via addition of sulphur to phosphite triesters. Tetrahedron Lett. 1978;19(40):3835–8.
80. Crooke ST, Seth PP, Vickers TA, Liang XH. The interaction of phosphorothioate-containing RNA targeted drugs with proteins is a critical determinant of the therapeutic effects of these agents. J Am Chem Soc. 2020;142(35):14754–71.
81. Frazier KS. Antisense oligonucleotide therapies: the promise and the challenges from a toxicologic pathologist's perspective. Toxicol Pathol. 2015;43(1):78–89.

82. Janas MM, Jiang Y, Duncan RG, Hayes AN, Liu J, Kasperkovitz PV, et al. Exposure to siRNA-GalNAc conjugates in systems of the standard test battery for genotoxicity. Nucl Acid Ther. 2016;26(6):363–71.
83. Ray KK, Wright RS, Kallend D, Koenig W, Leiter LA, Raal FJ, et al. Two phase 3 trials of inclisiran in patients with elevated LDL cholesterol. N Engl J Med. 2020;382(16):1507–19.
84. Valdmanis PN, Gu S, Chu K, Jin L, Zhang F, Munding EM, et al. RNA interference-induced hepatotoxicity results from loss of the first synthesized isoform of microRNA-122 in mice. Nat Med. 2016;22(5):557–62.
85. Grimm D, Streetz KL, Jopling CL, Storm TA, Pandey K, Davis CR, et al. Fatality in mice due to oversaturation of cellular microRNA/short hairpin RNA pathways. Nature. 2006;441(7092):537–41.
86. Zlatev I, Castoreno A, Brown CR, Qin J, Waldron S, Schlegel MK, et al. Reversal of siRNA-mediated gene silencing in vivo. Nat Biotechnol. 2018;36(6):509–11.
87. Janas MM, Schlegel MK, Harbison CE, Yilmaz VO, Jiang Y, Parmar R, et al. Selection of GalNAc-conjugated siRNAs with limited off-target-driven rat hepatotoxicity. Nat Commun. 2018;9(1):723.
88. Jackson AL, Bartz SR, Schelter J, Kobayashi SV, Burchard J, Mao M, et al. Expression profiling reveals off-target gene regulation by RNAi. Nat Biotechnol. 2003;21(6):635–7.
89. Jackson AL, Burchard J, Schelter J, Chau BN, Cleary M, Lim L, et al. Widespread siRNA "off-target" transcript silencing mediated by seed region sequence complementarity. RNA. 2006;12(7):1179–87.
90. Birmingham A, Anderson EM, Reynolds A, Ilsley-Tyree D, Leake D, Fedorov Y, et al. 3′ UTR seed matches, but not overall identity, are associated with RNAi off-targets. Nat Methods. 2006;3(3):199–204.
91. Vaish N, Chen F, Seth S, Fosnaugh K, Liu Y, Adami R, et al. Improved specificity of gene silencing by siRNAs containing unlocked nucleobase analogs. Nucleic Acids Res. 2011;39(5):1823–32.
92. Schlegel MK, Janas MM, Jiang Y, Barry JD, Davis W, Agarwal S, et al. From bench to bedside: improving the clinical safety of GalNAc-siRNA conjugates using seed-pairing destabilization. Nucleic Acids Res. 2022;50(12):6656–70.
93. Gane E, Lim YS, Kim JB, Jadhav V, Shen L, Bakardjiev AI, et al. Evaluation of RNAi therapeutics VIR-2218 and ALN-HBV for chronic hepatitis B: results from randomized clinical trials. J Hepatol. 2023;79(4):924–32.
94. Elbashir SM, Harborth J, Lendeckel W, Yalcin A, Weber K, Tuschl T. Duplexes of 21-nucleotide RNAs mediate RNA interference in cultured mammalian cells. Nature. 2001;411(6836):494–8.
95. Grimm D. Asymmetry in siRNA design. Gene Ther. 2009;16(7):827–9.
96. Biscans A, Caiazzi J, Davis S, McHugh N, Sousa J, Khvorova A. The chemical structure and phosphorothioate content of hydrophobically modified siRNAs impact extrahepatic distribution and efficacy. Nucleic Acids Res. 2020;48(14):7665–80.
97. Kim DH, Behlke MA, Rose SD, Chang MS, Choi S, Rossi JJ. Synthetic dsRNA Dicer substrates enhance RNAi potency and efficacy. Nat Biotechnol. 2005;23(2):222–6.
98. Zhang X, Goel V, Robbie GJ. Pharmacokinetics of Patisiran, the first approved RNA interference therapy in patients with hereditary transthyretin-mediated amyloidosis. J Clin Pharmacol. 2020;60(5):573–85.
99. Zhang MM, Bahal R, Rasmussen TP, Manautou JE, Zhong XB. The growth of siRNA-based therapeutics: updated clinical studies. Biochem Pharmacol. 2021;189:114432.
100. Setten RL, Rossi JJ, Han SP. The current state and future directions of RNAi-based therapeutics. Nat Rev Drug Discov. 2019;18(6):421–46.
101. Fields TR. The challenges of approaching and managing gout. Rheum Dis Clin N Am. 2019;45(1):145–57.
102. Flack JM. Epidemiology and unmet needs in hypertension. J Manag Care Pharm. 2007;13(8):2–8.

103. Ceral J, Habrdova V, Vorisek V, Bima M, Pelouch R, Solar M. Difficult-to-control arterial hypertension or uncooperative patients? The assessment of serum antihypertensive drug levels to differentiate non-responsiveness from non-adherence to recommended therapy. Hypertens Res. 2011;34(1):87–90.
104. Tomaszewski M, White C, Patel P, Masca N, Damani R, Hepworth J, et al. High rates of non-adherence to antihypertensive treatment revealed by high-performance liquid chromatography-tandem mass spectrometry (HP LC-MS/MS) urine analysis. Heart. 2014;100(11):855–61.
105. Balakrishnan KN, Abdullah AA, Bala JA, Jesse FFA, Abdullah CAC, Noordin MM, et al. Multiple gene targeting siRNAs for down regulation of Immediate Early-2 (Ie2) and DNA polymerase genes mediated inhibition of novel rat Cytomegalovirus (strain All-03). Virol J. 2020;17(1):164.

# Chapter 9
# Non-coding RNAs: tRNA-Derived Stress-Induced RNAs (tiRNAs)

**Abstract** RNA-derived stress-induced RNAs (tiRNAs) are fragments generated from transfer RNA (tRNA) molecules and are gaining recognition for their pivotal role in cancer progression. These tiRNAs emerge under conditions of cellular stress through the targeted cleavage of mature tRNA by the enzyme angiogenin. As key mediators in the cellular stress response, tiRNAs influence several critical processes. Under stress conditions, angiogenin is activated and cleaves tRNAs at specific sites, leading to the formation of tiRNAs. This cleavage is not random but occurs at precise anticodon loops, producing fragments that are uniquely suited to participate in stress responses. These tiRNAs then engage in a variety of cellular mechanisms primarily aimed at survival and adaptation. They help reprogram protein synthesis by modulating translation, a process critical for reshaping cellular functions under stress. This modulation often involves selective translation of mRNAs that encode proteins crucial for stress mitigation. Moreover, tiRNAs play a significant role in inhibiting apoptosis, the programmed cell death pathway, which is often activated under severe stress conditions. By blocking apoptosis, tiRNAs allow cells additional time to repair and survive. They are also involved in the degradation of mRNAs, selectively removing messages that encode unnecessary or detrimental proteins during stress conditions. Additionally, tiRNAs contribute to the formation of stress granules, which are aggregates of proteins and RNAs that appear in the cytoplasm during stress and help in managing protein synthesis and stabilizing cellular metabolism. This chapter on tiRNAs delves into their genesis, detailing the molecular mechanisms of their formation, their functional roles in cellular stress response, and their significant impact on disease management, particularly in the context of cancer. Understanding the role of tiRNAs in these processes not only illuminates a novel aspect of RNA biology but also opens potential therapeutic avenues where tiRNAs could be targeted to enhance cancer treatment or modulate stress responses more broadly. This growing body of research highlights the complex and nuanced roles that small RNAs play in cellular homeostasis and disease, offering new insights into the intricate dance of molecular interactions within cells.

 
P. E. Saw, E. Song, *RNA Therapeutics in Human Diseases*,
https://doi.org/10.1007/978-981-96-3041-7_9

## 9.1 Introduction

To withstand harsh environmental or internal conditions, cells activate crucial stress response mechanisms. These mechanisms are contingent upon a sophisticated and effective signal transduction system that orchestrates cellular survival strategies. This system is structured around three fundamental stages: (1) Detection of stress signals—the initial step involves the recognition of stressors, which can be varied, ranging from nutrient deprivation and hypoxia to thermal shock or toxic compounds. Specialized cellular sensors, such as membrane receptors or intracellular proteins, are attuned to these stress signals. These sensors are pivotal in identifying specific changes in the cellular or external environment, thereby initiating the stress response; (2) signal relay, amplification, and integration—upon detecting stress signals, cells transmit these cues through a cascade of signaling molecules. This process often involves the phosphorylation of proteins, leading to a sequential activation or deactivation of signal relay molecules. The signal may be amplified as it progresses through the cascade, ensuring a robust response even to a minor initial stimulus. Additionally, signals from different pathways are often integrated to form a coherent response, adapting the cellular machinery to the specific type of stress encountered; (3) coordination of cellular responses—The final stage is the coordination of a comprehensive cellular response. This involves various processes such as altering gene expression, modulating metabolic pathways, and enacting cellular repair mechanisms. The aim is to fortify the cell against the detected stress, mitigate damage, and maintain homeostasis. For example, in response to DNA damage, cells may halt the cell cycle to allow for repair mechanisms to correct the damage before replication proceeds. Alternatively, in response to oxidative stress, antioxidant genes might be upregulated to protect cellular components from oxidative damage [1].

## 9.2 Types of Cellular Stress and Cellular Response

### 9.2.1 *Heat Shock Proteins (HSPs)*

The magnitude of stress that a cell encounter plays a critical role in determining the specificity and intensity of the cellular response. For instance, heat stress induces a highly coordinated reaction involving the activation of heat shock proteins (HSPs), which are crucial for cellular protection and recovery in mammalian cells. When cells are exposed to elevated temperatures, this thermal stress can lead to protein denaturation and aggregation, disrupting cellular functions. In response, cells rapidly increase the production of HSPs. HSPs act as molecular chaperones that bind to partially unfolded proteins, thereby preventing them from aggregating. This is crucial because protein aggregates can be toxic and interfere with cellular processes. Beyond simply preventing aggregation, HSPs actively facilitate the refolding of denatured proteins into their functional conformations. This activity helps in restoring and maintaining the functional proteome of the cell, which is essential for survival under stress conditions. HSPs are integral to the broader network of proteostasis, which encompasses the synthesis, folding, trafficking, and degradation of proteins within the cell. By enhancing the capability to refold or degrade damaged proteins, HSPs help maintain overall protein homeostasis. Apart from their direct chaperone activities, HSPs can modulate signaling pathways that control cell survival, growth, and death. This modulation allows the cell to adapt to prolonged stress conditions and can initiate repair mechanisms or apoptosis if the damage is irreparable [2].

### *9.2.2 DNA Damage Response (DRR) Kinase*

Furthermore, DNA damage elicits a critical cellular response through the activation of DNA damage response (DDR) kinases, such as DNA-dependent protein kinase catalytic subunit (DNA-PKcs), ataxia telangiectasia mutated (ATM), and ATM and Rad3-related (ATR). These kinases are central to the signaling processes that detect and respond to genomic lesions, thereby upholding genomic stability in mammalian cells. Upon detecting DNA damage, such as double-strand breaks (DSBs) or replication stress, ATM, ATR, and DNA-PKcs are rapidly activated. Each of these kinases is specialized for responding to specific types of DNA damage. ATM primarily responds to double-strand breaks, ATR is activated by replication stress and other forms of DNA damage that interfere with replication, and DNA-PKcs is directly involved in the repair of DSBs through the non-homologous end joining (NHEJ) pathway. These kinases phosphorylate a myriad of substrates, including checkpoint proteins, histones, and other kinases. This phosphorylation cascade amplifies the damage signal and recruits various components of the DNA repair machinery to the site of damage. For example, ATM phosphorylates the histone variant H2AX (forming γH2AX), marking an extended region around the DSB and serving as a platform for assembling repair factors. DDR kinases play a crucial role in modulating the cell cycle in response to DNA damage. By activating checkpoint kinases like Chk1 and Chk2, ATM and ATR can induce cell cycle arrest, providing the cell time to repair the damage before proceeding through critical phases like DNA replication and cell division. This arrest helps prevent the propagation of damaged DNA, which could lead to genomic instability and tumorigenesis. In cases of severe or irreparable damage, DDR kinases can also trigger apoptosis or cellular senescence [3]. These responses collectively establish a robust defense system against hostile environments. Nonetheless, extreme stress levels may lead to cellular death [4]. Changes in cellular stress responses are linked to numerous human ailments, particularly cancers. Tumors, for instance, adapt to hypoxia by altering energy metabolism to continue growth [5]. Additionally, the malfunctioning of mitochondrial oxidative stress and the ER stress response plays a significant role in cancer development [6, 7]. Therefore, understanding cellular stress responses is pivotal for cell destiny and could unlock new avenues for treating associated diseases.

## 9.3 RNA Metabolism and Their Complex Network of Cellular Stress Responses

RNA metabolism is central to the intricate web of cellular responses to stress, acting as a fundamental pathway that orchestrates various aspects of cellular function and regulation. This pathway encompasses the processing, modification, and degradation of RNA molecules, a process that is vital for ensuring cellular adaptation and survival under stress conditions [8, 9]. The RNA metabolism pathway indeed represents a complex network involving various types of RNAs that play essential roles in cellular function and regulation. Among these, miRNAs, lncRNAs, tRNAs, and other small ncRNAs each contribute uniquely to the regulation of gene expression and cellular homeostasis, detailed below.

## 9.4 Types of RNAs and Their Roles in RNA Metabolism

The RNA metabolism pathway is not just a collection of isolated RNA-related functions but a deeply interconnected network that responds dynamically to cellular stress. These RNAs interact with each other and with various protein complexes to modulate the cell's transcriptional and translational machinery, enabling it to swiftly adapt to changing conditions. For example, changes in miRNA expression levels can quickly alter protein production rates, impacting cell cycle progression, apoptosis, and other critical cellular functions during stress. The complexity and versatility of RNA types within this network are what allow cells to maintain homeostasis and integrity under stressful conditions. Each type of RNA brings a unique set of mechanisms to this regulation, demonstrating the breadth and depth of RNA metabolism's role in cellular biology.

### *9.4.1 miRNAs and Their Role in RNA Metabolism*

miRNAs play an integral role in RNA metabolism by regulating the stability and translational efficiency of mRNAs in the cell. Their involvement is crucial for maintaining cellular homeostasis, orchestrating developmental timing, and responding to environmental cues. During miRNA processing, miRNA genes are transcribed by RNA Pol II as long primary transcripts known as pri-miRNAs, which are capped and polyadenylated. These transcripts form stem-loop structures that are processed in the nucleus by the Drosha-DGCR8 microprocessor complex to produce pre-miRNAs. These pre-miRNAs are then exported to the cytoplasm where they are further processed by the enzyme Dicer into mature miRNA duplexes. The mature miRNA is loaded into the RISC, where it guides the complex to target mRNAs based on sequence complementarity. In regulatory role, miRNAs primarily function by base-pairing with complementary sequences in the 3′-UTRs of target mRNAs. This interaction can lead to mRNA cleavage if the complementarity is high, but more commonly in mammals, it results in translational repression and mRNA deadenylation. Deadenylation, the shortening of the poly(A) tail, is often a precursor to mRNA decay, reducing mRNA stability and thereby decreasing gene expression. miRNAs can also block the translation of mRNAs by interfering with the initiation and elongation processes of translation. This can occur through mechanisms such as hindering the recruitment of the ribosomal assembly to the mRNA or by disrupting the translation machinery directly. Under stress conditions, the expression levels of specific miRNAs can change rapidly to adjust the protein output of the cell. For example, in response to oxidative stress, cells may upregulate miRNAs that target mRNAs encoding proteins involved in ROS production, thereby helping to mitigate cellular damage. During development, precise spatial and temporal control of gene expression is critical. miRNAs contribute to this regulation by ensuring that certain target mRNAs are downregulated at appropriate times or locations, facilitating the proper progression of developmental processes. Given that a single miRNA can target potentially hundreds of different mRNAs, and a single mRNA can be regulated by multiple miRNAs, miRNAs are pivotal in forming regulatory networks that can influence entire cellular pathways. miRNAs intersect with other RNA silencing pathways, such as those involving siRNAs and piRNAs. These interactions can complicate the regulatory landscape but also provide layers of regulation that enhance the cell's ability to fine-tune gene expression.

### 9.4.2 lncRNAs and Their Role in RNA Metabolism

lncRNAs are a diverse class of RNA molecules, typically longer than 200 nucleotides, which do not code for proteins but have significant roles in regulating gene expression and cellular functions. Their involvement in RNA metabolism is multifaceted, influencing various aspects from transcription to mRNA stability and splicing. LncRNAs are transcribed by RNA Pol II, similar to mRNAs, and they often undergo post-transcriptional modifications such as 5′-capping, polyadenylation, and splicing. However, unlike mRNAs, lncRNAs are generally not translated into proteins. LncRNAs can form complex secondary and tertiary structures, which enable them to interact with a variety of molecular partners including DNA, RNA, and proteins. Some lncRNAs interact with chromatin-modifying enzymes, recruiting them to specific genomic loci to alter chromatin states and thereby regulate transcription. For instance, the lncRNA Xist is essential for X-chromosome inactivation in females, coating the chromosome and recruiting chromatin modifiers that silence gene expression. LncRNAs can also regulate gene expression by hybridizing with complementary DNA sequences, forming RNA-DNA duplexes that block transcription factor binding or the progression of RNA polymerase. LncRNAs can bind to mRNAs, forming complexes that either stabilize the mRNA or lead to its degradation. For example, the lncRNA BACE1-AS regulates the stability of BACE1 mRNA, which is involved in Alzheimer's disease pathology. Some lncRNAs interact with splicing factors and modulate the splicing of pre-mRNAs, affecting the diversity and functionality of proteins encoded by a single gene. LncRNAs can influence translation by interacting directly with mRNAs or with the translation machinery. Some lncRNAs act as competing endogenous RNAs (ceRNAs) or molecular sponges that sequester miRNAs, preventing them from binding to their target mRNAs, thereby indirectly enhancing translation. Certain lncRNAs play a role in the transport of RNA molecules between the nucleus and cytoplasm, influencing which transcripts are available for translation in the cytoplasm or which are retained in the nucleus for further processing. LncRNAs are crucial in the formation and regulation of gene expression networks. They can orchestrate the expression of genes involved in cell differentiation, development, and response to stress. By interacting with structural components within the cell, lncRNAs contribute to the organization of nuclear architecture and the formation of subcellular structures like paraspeckles and nucleoli.

### 9.4.3 tRNAs and Their Role in RNA Metabolism

tRNAs is a critical component of the protein synthesis machinery within cells, playing an essential role in translating the genetic code into proteins. Each tRNA molecule is relatively small, typically about 70–90 nucleotides long, and has a distinctive cloverleaf structure in its secondary form and an L-shaped three-dimensional structure. tRNAs serve as the key interpreters of the genetic code, and their role in RNA metabolism is pivotal to cellular function and protein biosynthesis. tRNA molecules consist of four major arms: the acceptor stem, the TΨC arm, the anticodon arm, and the D-arm. Each arm has a specific role in the stability and function of tRNA. The acceptor stem carries the specific amino acid corresponding to the mRNA codon it reads, and the anticodon arm houses the anticodon triplet that pairs with the codon on the mRNA during translation. tRNA genes are transcribed by RNA polymerase III as precursor tRNA molecules. These precursors undergo extensive processing, including cleavage of extra sequences at both the 5′ and 3′ ends, addition of a CCA sequence at the 3′ end (which serves as the amino acid attachment site), and modification of specific bases and the sugar-phosphate backbone. This processing is critical for the proper functioning and stability of tRNA molecules. tRNAs are central to decoding the mRNA transcript into amino acids, the building blocks of proteins. Each tRNA molecule carries a specific anticodon sequence that recognizes and base-pairs with the complementary codon on the mRNA molecule. Before tRNAs can participate in translation, they must be "charged" with the correct amino acid. This process is catalyzed by a group of enzymes called aminoacyl-tRNA synthetases. Each enzyme recognizes its specific tRNA and the corresponding amino acid, facilitating the attachment of the amino acid to the 3′ end of the tRNA. This charging is crucial for the accuracy of protein synthesis. During protein synthesis, charged tRNAs deliver their amino acids to the growing polypeptide chain in the ribosome. The ribosome moves along the mRNA, matching the mRNA codon with the corresponding tRNA anticodon. Once the match is made, the amino acid is transferred from the tRNA to the nascent polypeptide chain. Beyond their role in translation, tRNAs can regulate gene expression at various levels, including modulating the rate of translation based on the availability of specific tRNAs, thereby influencing cellular response to nutritional and environmental changes. tRNAs are directly involved in maintaining protein homeostasis by ensuring that proteins are synthesized accurately and efficiently, according to the genetic instructions encoded by DNA. Under conditions of amino acid starvation or other stresses, certain tRNAs' levels can fluctuate, impacting the translation rates of specific proteins that are critical for stress response. The availability and modification of tRNAs are linked to cell cycle progression and growth, influencing overall cellular health and proliferation rates.

## *9.4.4 Other Small ncRNAs and Their Role in RNA Metabolism*

Small RNAs, including small nucleolar RNAs (snoRNAs), small interfering RNAs (siRNAs), and Piwi-interacting RNAs (piRNAs), play crucial roles in RNA metabolism, each having distinct functions that contribute to gene regulation, RNA processing, and genome stability.

### 9.4.4.1 Small Nucleolar RNAs (snoRNAs)

snoRNAs are typically encoded within the introns of host genes and are released through the process of pre-mRNA splicing. They are then processed into mature forms within the nucleus. snoRNAs are primarily involved in the chemical modification and processing of rRNA. They fall into two main categories based on their structure and function: C/D box snoRNAs guide the 2′-O-methylation of rRNAs. H/ACA box snoRNAs direct the pseudouridylation of rRNAs. By modifying rRNA molecules, snoRNAs ensure the proper folding, stability, and function of ribosomes, which are essential for protein synthesis. This modulation affects ribosome biogenesis and function, directly impacting cellular capacity for translation.

#### 9.4.4.2 Small Interfering RNAs (siRNAs)

siRNAs originate from long double-stranded RNA (dsRNA) molecules, which can be formed by bidirectional transcription, RNA viruses, or introduced experimentally. These dsRNAs are cleaved by the enzyme Dicer into siRNA duplexes, which are then incorporated into the RISC. The primary role of siRNAs is to mediate RNAi pathway. This process is essential for gene silencing, especially in the context of defense against viruses and the regulation of endogenous gene expression. siRNAs help maintain genome integrity by silencing foreign genetic elements and regulating the expression of genes. They are crucial in developmental timing, pathogen defense, and gene expression adjustments necessary for cellular adaptation to environmental changes.

#### 9.4.4.3 Piwi-interacting RNAs (piRNAs)

piRNAs are generated from single-stranded RNA precursors that come from piRNA clusters, genomic regions rich in transposon-derived sequences. Unlike siRNAs, their processing does not involve Dicer. Instead, piRNAs are processed through a Dicer-independent mechanism that involves Piwi proteins, a subfamily of the Argonaute proteins. piRNAs interact with Piwi proteins to form piRNA complexes, which are involved in the silencing of transposons, particularly in germ cells. This silencing can occur via post-transcriptional gene silencing (PTGS) of transposon RNAs or through transcriptional gene silencing (TGS) mechanisms that modify the chromatin structure. piRNAs are vital for protecting the integrity of the genome in reproductive cells. They prevent transposon mobilization, which if unchecked, could lead to genomic instability and affect gene expression through insertional mutagenesis.

**Box 9.1 tRNA in the Central Dogma Producing ncRNAs?**
RNA-derived stress-induced RNAs (tiRNAs) are considered ncRNAs. This classification stems from their function and genesis rather than their origin from tRNA, which itself is a non-coding RNA involved in the translation process of the central dogma of molecular biology. tRNAs are crucial in decoding the genetic information stored in mRNA during translation, by bringing specific amino acids to the ribosome to build proteins. However, tRNAs do not code for proteins themselves; they play a role in the process of protein synthesis by acting as adaptors. Similarly, tiRNAs, which are derived from the cleavage of tRNAs under stress conditions, do not encode proteins. Instead, their role is primarily regulatory. tiRNAs are involved in various cellular stress response mechanisms, such as inhibiting translation initiation, modulating cell death pathways, and aiding in the formation of stress granules. These functions categorize them clearly as non-coding RNAs, as their purpose is to regulate other molecular processes rather than serve as templates for protein synthesis.

tRNAs, in particular, are central to protein synthesis, translating genetic information into the polypeptide chains that form proteins. Under stress conditions, tRNAs can undergo specific modifications that alter their stability and functionality, thereby adjusting their abundance and activity. tRNA synthetases, which charge tRNAs with their respective amino acids, can be regulated under stress, altering the efficiency of tRNA charging and, consequently, protein synthesis. Under certain stress conditions, tRNAs can be cleaved into smaller fragments, such as tiRNAs. These fragments can inhibit the initiation of protein synthesis, allowing cells to conserve resources under adverse conditions. Chemical modifications of tRNAs, such as methylation or thiolation, can change under stress, affecting their stability and the accuracy of translation. This can lead to the selective translation of proteins that assist in stress responses. These regulatory mechanisms involving tRNAs are crucial for allowing cells to swiftly adapt their protein synthesis machinery in response to environmental or internal stresses. This dynamic response is part of a broader strategy to maintain cellular integrity and functionality under challenging conditions. Additionally, other types of non-coding RNAs in the RNA metabolism pathway, like miRNAs and lncRNAs, also play critical roles. miRNAs generally function by binding to complementary sequences on target mRNAs, leading to mRNA degradation or repression of translation, thus regulating gene expression post-transcriptionally. lncRNAs are involved in diverse cellular processes, including chromatin modification, transcriptional and post-transcriptional regulation, and serving as molecular scaffolds or decoys. These intricate interactions and regulatory mechanisms underscore the complexity and sophistication of RNA metabolism, highlighting the critical role of various RNA types in adapting to and managing cellular stress [10].

Stress triggers ribonucleases to slice mature tRNAs, producing a range of functional fragments. This tRNA cleavage is a ubiquitous mechanism aiding cell survival in hostile environments [11, 12]. Small ncRNAs are integral components of the cell's stress response, showcasing remarkable versatility and specificity in their roles across various types of cellular stress. These molecules are activated or modulated in response to a wide array of environmental and physiological stressors. Each type of stressor can trigger specific ncRNAs, which in turn influence cellular processes to adapt to and mitigate the effects of stress. Table 9.1 listed different stressors impact the activity of various small ncRNAs [15].

The production of tiRNAs primarily involves the cleavage of tRNA molecules by angiogenin, an enzyme that is part of the RNase family. This specific cleavage typically occurs at the anticodon loop of tRNAs, a critical process especially under conditions of cellular stress. Angiogenin is a potent ribonuclease with a key role in RNA metabolism under both normal and stress conditions. During stress, angiogenin's ribonuclease activity is upregulated or activated, allowing it to specifically target and cleave tRNAs. The anticodon loop of a tRNA is a crucial structure that interacts directly with mRNA codons during protein synthesis. Cleavage by angiogenin at this site effectively halts the capacity of the tRNA to participate in translation. This specific cut generates tiRNAs, which are typically 30–40 nucleotides long. Once produced, tiRNAs serve several regulatory functions related to the

**Table 9.1** Consequences of stress toward ncRNAs activities [13, 14]

| Internal/external stress | Consequences |
|---|---|
| Amino acid scarcity | During protein synthesis, a lack of amino acids can signal through specific pathways like the general amino acid control (GAAC) pathway, leading to the upregulation of ncRNAs that adjust metabolic and synthetic pathways to conserve resources and optimize survival |
| Phosphate depletion | Low-phosphate levels can trigger adjustments in the synthesis and activity of ncRNAs involved in phosphate homeostasis. These ncRNAs can regulate the expression of genes involved in phosphate uptake and distribution, helping the cell adapt to phosphate scarcity |
| UV exposure | Ultraviolet radiation can cause direct DNA damage and oxidative stress, prompting the expression of ncRNAs that are involved in DNA repair processes and the modulation of apoptosis. These ncRNAs help mitigate DNA damage and maintain genomic stability |
| Heat shock | Heat stress leads to the production of heat shock proteins, mediated by ncRNAs that regulate the translation and stability of mRNAs encoding these proteins. This response helps in refolding denatured proteins and preventing protein aggregation |
| Hypoxia | Hypoxic conditions stimulate the production of ncRNAs that can reprogram cell metabolism from oxidative phosphorylation to anaerobic glycolysis, reducing oxygen consumption and adjusting energy production to suit the reduced availability of oxygen |
| Oxidative stress | Reactive oxygen species (ROS) generated under oxidative stress can damage cellular components. ncRNAs play roles in activating antioxidant responses and modulating the expression of genes involved in oxidative stress response pathways |
| Viral invasion | Upon viral infection, ncRNAs can be upregulated to suppress viral replication and modulate the immune response. These ncRNAs may target viral RNAs directly or regulate host genes critical for viral entry, replication, and assembly |

cellular stress response. By interfering with the translation initiation process, tiRNAs help conserve energy and resources during cellular stress. tiRNAs can promote the formation of stress granules, which are aggregates of proteins and RNAs that serve as storage centers for translationally stalled mRNAs. Stress granules help cells survive by regulating protein synthesis and stabilizing cell metabolism under stress. Certain tiRNAs have been shown to influence pathways related to apoptosis and cellular survival, thus playing a role in deciding the fate of stressed cells [16–18].

**Box 9.2 Angiogenin and Their Role in Biological Stress**

In normal biological processes, angiogenin is a protein that is secreted as a response to an acute phase of inflammation, usually induced by an inflammatory stimulus [19]. Studies have shown that in situ introduction of a

(continued)

**Box 9.2** (continued)
recombinant angiogenin in a cell line leads to an increased production of tiRNAs. Angiogenin also inhibited the cellular protein synthesis. Given that stress-induced reprogramming of protein translation aids cells in enduring unfavorable environmental conditions, secreted angiogenin might initiate a transmissible stress response mechanism, enabling stressed cells to alert neighboring cells about imminent harmful stimuli. This process could be facilitated at the organismal level through the liver's secretion of angiogenin [19], or at the tissue level via secretion from stressed cells within peripheral tissues. The dual local and systemic effects of angiogenin align with its expression pattern, involving an upstream promoter active in all tissues and a downstream promoter specific to liver cells [20]. Similar to how interferons, triggered by viruses, safeguard nearby or distant cells from viral infections, angiogenin may act as a stress-induced factor that shields nearby or distant cells from the damaging impact of environmental stress. Angiogenin is also involved in the development of ALS, a neurodegenerative condition characterized by the loss of motor neurons. Mutations in the angiogenin gene are present in a minority of both familial and sporadic ALS cases [21, 22]. These ALS-related missense mutations in angiogenin exhibit diminished ribonuclease and angiogenic activities [21, 22]. This suggests that the angiogenin-induced stress response program might be essential for the survival of motor neurons.

## 9.5 Formation of tiRNAs: Biogenesis and Characteristics

Advancements in next-generation sequencing technology have unveiled extensive details of tRNA-derived fragments, arises from the specific cleavage of tRNA by various endonucleases. These fragments are broadly categorized into tRNA halves (tiRNAs) and smaller tRNA segments, distinguished by the site of a specific enzymatic cleavage [18], produced specifically by the action of angiogenin. Experimental evidence shows that recombinant angiogenin can trigger tiRNA production, whereas the absence of angiogenin reduces tiRNA generation in response to arsenite in U2OS cells [23]. Angiogenin is found to accumulate in the cytoplasm under hypoxic conditions [24, 25]. Moreover, one of the ways that angiogenin enhances its RNase activity is by detaching itself from its inhibitor, the ribonuclease/angiogenin inhibitor 1 (RNH1) [11]. Within the cytoplasm, angiogenin processes and cleaves mature tRNAs into two distinct RNAs: (1) the 30–35 nucleotide 5′-tRNA halves (5′-tiRNAs) and the 40–50 nucleotide 3′-tRNA halves (3′-tiRNAs) (Fig. 9.1).

Although tiRNAs are indicated mostly in stressful conditions, yet, instances such as exposure to γ-irradiation, treatment with etoposide, or caffeine in human cells did not result in tRNA cleavage [23, 26]. Furthermore, there are studies which found the existence of tiRNAs under non-stressful conditions, indicating their role in

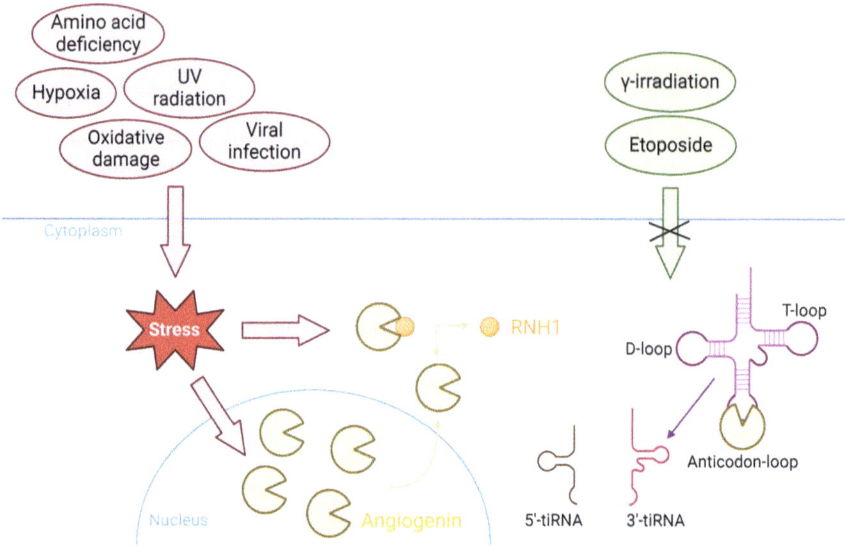


**Fig. 9.1** The biogenesis of tiRNA under various conditions. When specific stress stimuli are present, angiogenin is released from the nucleus, and is translocated into cytoplasm, in the form of angiogenin-NRH1 complex. Once in the cytoplasm, angiogenin detaches itself from RNH1, exposing its active site that could target the anticodon region of the tRNA. This attachment leads to the cleavage of the anticodon loop of the tRNA, resulting in two distinct tiRNAs, namely the 5′-tiRNAs (in brown) and 3′-tiRNAs (in pink). Figure is adapted with an agreement with John Wiley & Sons (License number: 5901690555841)

regulating processes beyond the cellular stress response [27, 28]. Modifications to mature tRNA have been shown to impact the production of tiRNAs. Specifically, the introduction of 5-methylcytidine modifications near the anticodon and the variable loop of tRNAs by DNA methyltransferase 2 and NOP2/Sun RNA methylcytidine 2 can inhibit the production of angiogenin-cleaved tiRNAs [29–31]. Thus, the generation of tiRNAs is contingent upon a multitude of factors including (1) the nature of the stressor, (2) the presence and function of angiogenin, (3) the levels of tRNA substrates, and (4) the overall rate of protein synthesis [32].

## 9.6 tiRNAs Response in Stressful Condition

Adaptation and survival hinge on an organism's capacity to detect environmental shifts and swiftly counteract detrimental stimuli. This capability is crucial for homeostasis maintenance, with its disruption being implicated in various human pathologies [33]. A key aspect of the cellular stress response is translational reprogramming, which alters cellular operations and protein quantity. During stress, tiRNAs halt protein synthesis to decrease cellular functions and conserve energy, thus shielding cells from apoptosis [34]. Under oxidative stress, angiogenin specifically targets the 3′-CCA end of all the tRNAs, rendering the aminoacyl-ends inactive and unfit for translation. Nevertheless, this inhibition is reversible, due to the ATP (CTP): tRNA nucleotidyltransferase enzyme, which permits the swift resumption of translation at minimal metabolic expense [35]. Furthermore, the phosphorylation of eIF2α induced by stress modulates the formation of eIF2α-GTP-tRNAi$^{met}$ ternary complexes, leading to a reduction in overall protein creation [36]. In a study, it was found that the presence of tiRNA could reduce protein production by approximately 20%, without depleting the inherent intracellular tRNA pools [12, 23, 37]. This intriguing finding suggested that the existence of tiRNAs, and not the process of tRNA cleavage, is pivotal in blocking translation.

tiRNAs can be derived from cells treated with angiogenin or synthesized artificially [23]. Recent studies confirm that both natural and synthetically synthesized tiRNAs can directly impede protein translation [38]. Researches also indicated that the length of tiRNA also plays a part in their biological function. For example, 3′-tiRNAs, featuring an extra V-region and 5′-hydroxyl groups, may pose challenges in interacting with other proteins or RNAs, thus impacting their regulatory capabilities. Notably, it has been found that only the 5′-tiRNAs originating from tRNA-Ala (tRNA carrying Alanine) and tRNA-Cys (tRNA carrying Cysteine), and not their 3′-tiRNAs counterpart, are crucial in translation suppression [38]. 5′-tiRNA-Ala/Cys variants feature a terminal oligoguanine motif conducive to the formation of G-quadruplex structures, crucial in regulating various aspects of cellular biology and related to several human diseases [39, 40]. These 5′-tiRNAs interact with Y-box protein 1 (YB-1) to inhibit the eukaryotic initiation factor 4F (eIF4F) complex from starting translation, thereby facilitating stress granule (SG) assembly [38]. Although YB-1's cold shock domain directly engages with tiRNAs to drive SG assembly, research suggests YB-1 might not be essential for the translation suppression effect of 5′-tiRNAs [41]. Stress granules, which emerge in response to various environmental stresses, are dense aggregations of mRNAs that have paused translation. They play a key role in enhancing cell survival by impacting mRNA translation, apoptosis, and nuclear functions [42, 43]. The process by which 5′-tiRNA triggers SG formation is atypical, bypassing the need for eIF2α phosphorylation. In essence, certain 5′-tiRNAs, through G-quadruplex structures, capture the eIF4F complex away from mRNA, obstructing the commencement of translation. Yet, the extent of YB-1's involvement in the translational repression prompted by 5′-tiRNAs and the influence of SGs on cellular processes are areas requiring further investigation.

Research has established that tiRNAs serve various functions in hindering apoptosis [44]. While excessive apoptosis is a hallmark of many pathological states, its suppression can enhance cell survival [45]. Extreme stress prompts cells to initiate apoptosis, suggesting the anti-apoptotic capability of tiRNAs as a potential protective mechanism. It is documented that tiRNAs, in response to hyperosmotic stress, can bind to cytochrome c (Cyt c), forming a ribonucleoprotein complex that either

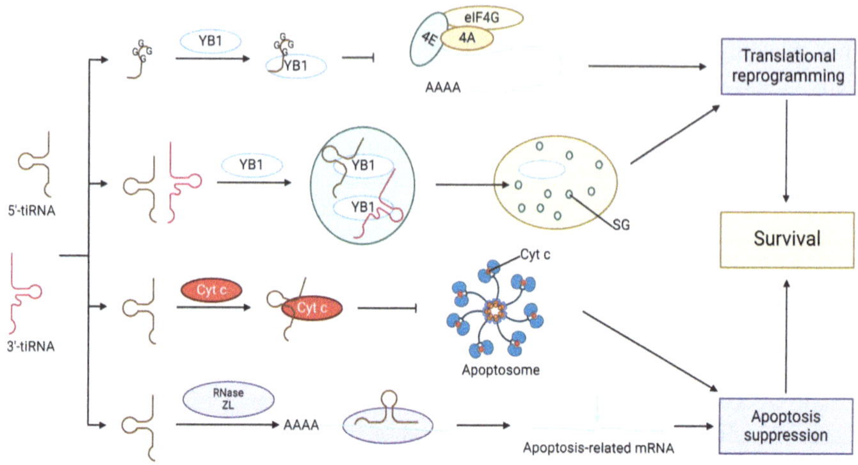


**Fig. 9.2** Stress response mechanisms of tiRNAs after being cleaved from mature tRNAs. As shown in the figure, 5′ and 3′ tiRNAs varies in their responses to different stress stimuli, and each employ multiple strategies to shield cells from multiple types of cellular stress. In the first example, 5′-tiRNAs could form G-quadruplex structures through the partnering of YB-1, to disrupt the initiation of translation by the eIF4F complex, by keeping the complex away from mRNA. In the second example, both 5′- and 3′-tiRNAs can partner with YB-1 to create stress granules, protecting the cells from cell death pathways and enhancing cell survival. In the third example, 5′-tiRNAs can also block apoptosis through interaction with Cyt c. Finally, 5′-tiRNAs may associate with RNase ZL to target and cleave mRNAs related to apoptosis, offering another layer of cellular protection. Figure is adapted with an agreement with John Wiley & Sons (License number: 5901690555841)

curtails the formation of apoptosomes or diminishes their activity, thereby inhibiting apoptosis [44]. Similar to tiRNAs, tRNAs have been shown to prevent apoptosis through interaction with Cyt c [46, 47], though Cyt c exhibits a preference for binding tiRNAs over tRNAs under stress [44]. Furthermore, 5′-tiRNA acts as a small guide RNA (sgRNA), which, along with tRNase ZL, can suppress the expression of certain human genes [48, 49]. Specifically, 5′-tiRNA-Glu acts both in vitro and in vivo as sgRNA, guiding tRNase ZL to cleave targeted mRNAs, with PPM1F mRNA being one such target. Overexpression of PPM1F mRNA triggers apoptosis in Hela cells [50], indicating that tiRNA-mediated mRNA reduction via tRNase ZL cleavage plays a role in apoptosis inhibition (Fig. 9.2). The roles of tiRNAs in stress response represent only a fraction of their potential cellular functions. Beyond their established roles in repressing overall protein synthesis and modulating apoptosis, tiRNAs have additional, yet-to-be-uncovered capabilities. Reports suggest their involvement in the cellular RNAi pathway, influencing the expression of stress-related genes [51], though it remains uncertain if tiRNAs can act similarly to siRNAs or miRNAs.

## 9.7 tiRNA in Cellular Development

Although extensive studies have highlighted the importance of small non-coding RNA (sncRNA) subsets like microRNAs (miRNAs), piwi-interacting RNAs (piRNAs), and small interfering RNAs (siRNAs) in processes such as cell differentiation and developmental pathways, the specific functions of tRNA-derived fragments, or tiRNAs, remain largely obscure [52]. Evidence from previous research suggests that sncRNAs contribute to cell differentiation by displaying unique, tissue- and cell-specific expression patterns, which imply their regulatory influence on protein production through interactions with promoter regions and modulation of gene expression across various cell types, including neurons, muscle, and germ cells [53]. Studies have further shown that differential expression of tRNA genes influences the levels of tRNA-derived fragments without affecting mature tRNAs, underscoring how selective tRNA gene expression may fine-tune noncanonical tRNA functions without depleting the mature tRNA reservoir. This selective expression may play a crucial role in managing tiRNA activity within specific tissues [54]. Notably, certain 5′-tiRNAs, such as those derived from tRNAGlyGCC, tRNAVal-CAC, tRNAGlnCTG, tRNAGluTTC, and tRNALysTTT, have emerged as critical players during developmental processes [52]. These 5′-tiRNAs engage with specific proteins and RNA transcripts during mouse embryonic stem cell (mESC) differentiation, including interactions with Igf2bp1, an RNA-binding protein that modulates c-Myc expression and translation—c-Myc being a key transcription factor involved in cellular differentiation and oncogenic transformation [52, 55].

In hematopoietic stem cells, the enzyme Pseudouridine Synthase 7 (PUS7) has been found essential for protein synthesis and cellular growth by promoting pseudouridylation, an important epigenetic modification [56]. When PUS7 expression is reduced, translation regulation mediated by 5′-tiRNAs—particularly those derived from tRNAs with a 5′ terminal oligoguanine (TOG) motif, such as tRNA-Ala, tRNA-Cys, and tRNA-Val—is disrupted, resulting in increased protein synthesis and abnormal germ layer specification in embryonic stem cells [56]. Additionally, TOG-motif-containing 5′-tiRNAs, such as 5′-tiRNA-Ala and 5′-tiRNA-Cys, inhibit translation by forming RNA G-quadruplexes (RG4s), which replace the eIF4G/eIF4E complex on the mRNA cap (m7GTP) structure, effectively blocking the initiation of translation [57–59].

## 9.8 tiRNA in Aging (Senescence)

tiRNAs appear to hold significant roles in both aging and spermatogenesis. Aging encompasses a range of physiological changes, which have been categorized into eight key hallmarks: proteostasis loss, stem cell depletion, disrupted intercellular signaling, deregulated nutrient sensing, cellular senescence, telomere shortening, mitochondrial dysfunction, genomic instability, and epigenetic modifications [60].

Research in humans and various model organisms indicates that both tiRNAs and tRNA-derived small RNAs (tsRNAs) display distinct expression profiles associated with age-related diseases and natural senescence. During aging, notable shifts occur in circulating levels of specific 5′-tiRNAs. For example, reductions in 5′-tiRNAs derived from $tRNA^{Cys}$GCA and $tRNA^{Lys}$CTT, alongside increases in those from $tRNA^{His}$GTG and $tRNA^{Asp}$GTC, have been documented in aged mice [61]. Additional tiRNAs also show differential expression patterns with age, pointing to a possible regulatory role in the aging process.

## 9.9 tiRNA in Immune-Related Disease

tiRNAs have emerged as promising candidates for systemic immune signaling, with 5′-tiRNAs circulating in the bloodstream and potentially acting like extracellular miRNAs in response to viral challenges and immune stimuli [62, 63]. These tiRNAs are thought to contribute to cellular resilience by promoting the formation of stress granules—aggregates of translationally stalled messenger ribonucleoproteins (mRNPs)—which can transiently suppress mRNA translation, aiding cell survival under stress. Unlike the broadly distributed circulating miRNAs found across peripheral tissues, circulating 5′-tiRNAs appear predominantly within hematological and lymphoid tissues, hinting at a specialized immune function [38].

tiRNAs are also implicated in key immune-related signaling pathways, such as NF-κB, interferon (IFN), and Wnt/β-Catenin, further supporting their role in immune responses [64]. For example, stimulation of Toll-like receptor 4 (TLR-4) with bacterial lipopolysaccharide (LPS) initiates a signaling cascade that activates NF-κB and upregulates ANG exonuclease. This increase in cytosolic angiogenin leads to the cleavage of specific 5′-tiRNAs, which are then encapsulated into extracellular vesicles (EVs) and released as signaling entities. These EV-5′-tiRNA complexes are subsequently delivered to endosomes in recipient cells, where they activate TLR-7, inducing the expression of immune-responsive genes [65]. Although the exact mechanisms by which specific tiRNAs are selected for EV packaging remain to be clarified, selective EV incorporation of both miRNAs and tRNA-derived fragments (tRFs) has been observed, highlighting a potential regulatory process in extracellular RNA communication [66–68]. Interferon expression is tightly regulated through activation of the JAK/STAT pathway. In response to viral infections, interferons are released, leading to a suppression of theANG gene, which is known to influence tiRNA production [69]. When ANG expression is downregulated, tiRNA biogenesis shifts toward generating a distinct set of antiviral-specific tiRNAs. This process is thought to involve alternative RNases, including members of the Schlafen protein family, which have recently been identified for their capacity to degrade the 3′ end and create full 5′ fragments of tiRNAs [70]. For instance, 5′-$tiRNA_{Val}$ can directly interact with the human Frizzled homolog 3 (FZD3) receptor, thereby modulating the Wnt signaling pathway. This interaction affects key immune processes, such as the formation of effector T cells, activation of regulatory

T cells, and maturation of DCs [71, 72]. The genomic organization of the largest tRNA gene cluster in humans and mammals suggests a fundamental link between tiRNAs and immune functions. In humans, this major tRNA gene cluster is located within the MHC, underscoring a possible evolutionary relationship between tRNA-derived fragments and immune system regulation [73].

## 9.10 tiRNA and Their Role in Cancer

The connection between tiRNAs and cancer has emerged as a field of interest in recent studies. Deep sequencing analyses of serum ncRNAs revealed that certain tiRNAs are increased in breast cancer patients compared to healthy controls, showing correlation with clinicopathological features. Similarly, tiRNAs have been identified at higher levels in solid prostate tumors and seminal fluid [74, 75], suggesting their potential as non-invasive biomarkers for cancer detection. Further research underscores the significant role of tiRNAs in oncogenesis. Researchers noted that sex hormone signaling pathways may enhance the angiogenin-mediated cleavage of mature tRNA, with specific 5′-tiRNAs boosting cell growth in hormone-sensitive breast and prostate cancers [76, 77]. In the context of NSCLC, an increase in 5′-tiRNA-Leu-CAG levels was reported in tissues, cell lines, and serum of patients, suggesting this tiRNA is involved in NSCLC development through promotion of cell proliferation and inhibiting cellular apoptosis [78]. In colorectal cancer (CRC), an association between 5′-tiRNA-Val and tumor metastasis was discovered, initiating a proposal that an angiogenin-tiRNAs axis could facilitate CRC metastasis by enhancing cell migration and invasion [79].

While a substantial number of tiRNAs exhibit increased expression in cancerous tissues and cells, certain studies have highlighted that some tiRNAs are diminished in tumors and might act as innovative biomarkers [80] or even hinder cancer advancement [71]. For instance, although previous research indicated that 5′-tiRNA-Val promoted metastasis in CRC, another research proved that this same tiRNA had the opposite effect in breast cancer, by impairing their cell growth, migration and invasion via the obstruction of the FZD3/Wnt/β-Catenin signaling pathway [71]. This interesting finding also suggests that identical tiRNAs could have diverse effects across different cancer types, a finding that warrants deeper studies. Some tiRNAs that has been elucidated in cancer is summarized in Fig. 9.3.

## 9.11 Circulating tRNA Fragments: Significance in Human Diseases

Although circulating tRNA fragments might not be high in quantity, they have demonstrated accuracy and sensitivity to detect many pathological conditions and physiological diseases, namely cancer, inflammatory diseases, aging, organ

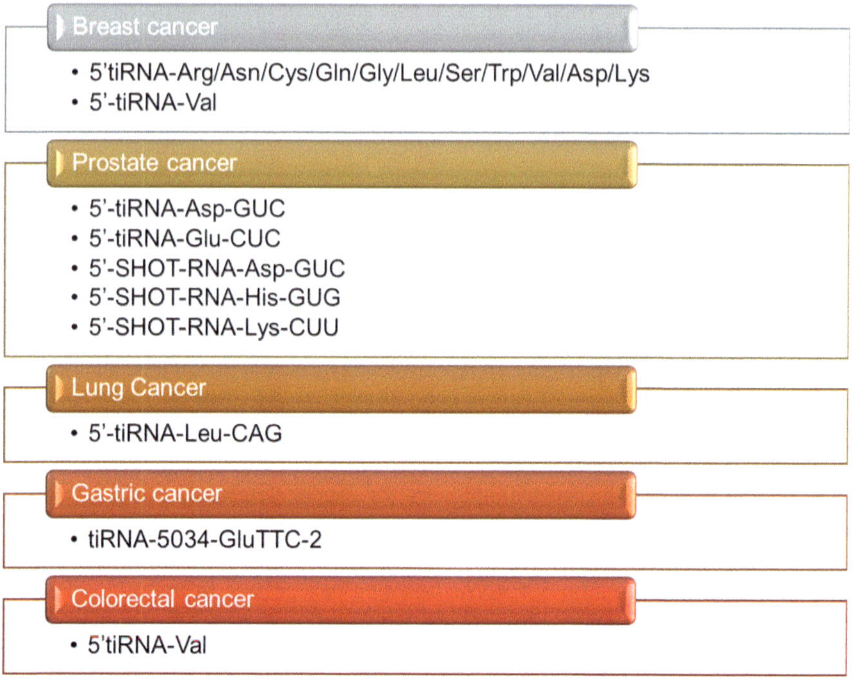


**Fig. 9.3** Novel tiRNAs found in various cancer cell lines or models, reflecting their roles in cancer cell growth and progression

damage (i.e., kidney, liver, lung) [81]. These properties make them promising candidates for non-invasive biomarkers [82–84]. Moreover, these fragments appear to perform specific functions when the body is subjected to different stressors, suggesting a broader role in modulating cellular responses beyond their traditional functions in translation.

### *9.11.1 Hyperosmotic Stress and Apoptosis Modulation*

In a mouse model experiment, under stress condition, the mouse embryonic fibroblast could be less sensitive to apoptosis pathway signaling. During stress, the mouse underwent a hyperosmotic stress-induced apoptosis. Here, the angiogenin first cleaved the tRNA to tiRNAs. The 5′-tiRNA then binds to cytochrome c, a molecule released from mitochondria that is essential for the formation of the apoptosome and the initiation of apoptosis [44]. By binding directly to cytochrome c and forming an RNP complex, these tRNA fragments sequester cytochrome c, thereby inhibiting its apoptotic function. This interaction extends cell survival following hyperosmotic stress.

### 9.11.2 Role in Tumorigenesis

In a specific case, YBX1 was found to bind to the 3′-UTRs of pro-oncogenic transcripts in metastatic invasive ductal carcinoma cell lines in research carried out in MD Anderson Cancer Center. This binding stabilizes and protects the transcripts that promote metastasis. However, under hypoxic conditions, this protective effect is compromised. During hypoxia, YBX1 is sequestered by 5′-tRNA fragments generated by an unknown nuclease. This sequestration prevents YBX1 from binding to the pro-oncogenic transcripts, thereby reducing their stability and potentially hindering the metastatic process. This sequestration diverts YBX1 from stabilizing tumor-promoting transcripts, potentially reducing metastatic capabilities [85]. The role of tRNA fragments in cancer biology is becoming increasingly evident, as seen in a variety of studies that explore their involvement in both the promotion and inhibition of tumorigenesis. One particular investigation into hormone-sensitive cancers highlighted how the aminoacylation of tRNA fragments, a precursor to their angiogenin-mediated fragmentation into tiRNAs, correlates significantly with tumor development [76].

### 9.11.3 Hematopoiesis Regulation

In the process of hematopoiesis (blood cell generation), hematopoietic stem cells (HSCs) are the key players that are responsible in the generation of all types of blood cells, including red blood cells, white blood cells, and platelets. In normal and healthy adults, HSCs are primarily residing in the bone marrow and lymphatic tissues, and these stem cells are ready to differentiate into various blood cell lineages, depending on the cues given by the host. In hematopoiesis regulation, angiogenin plays a vital role by modulating protein synthesis. The regulation is different depending on the cell type involved, whether it be HSCs or lineage-committed myeloid progenitor cells. In HSCs, angiogenin is directed to stress granules, where it generates tiRNAs that suppress protein synthesis. This suppression helps maintain the dormancy of HSCs, ensuring they remain in a quiescent state until they are needed for the production of new blood cells. By controlling protein synthesis in this manner, angiogenin helps balance the proliferation and preservation of stem cells within the bone marrow, contributing to effective and regulated hematopoiesis. Conversely, in myeloid progenitors, angiogenin may also influence tiRNA roles but primarily functions in the nucleus to boost ribosomal RNA production, thus fostering protein synthesis and supporting cell proliferation. Therefore, angiogenin and the tRNA fragments it generates implement a complex, cell-context-dependent, and dual regulatory effect on protein synthesis [86].

### 9.11.4 Regulation of Epigenetic Inheritance

Initial studies on the epigenetic transmission of traits demonstrated that a father's diet can influence the metabolic traits of his offspring, and that changes in sperm epigenetics driven by metabolism can regulate the metabolism of the next generation [87, 88]. It has increasingly been recognized that changes in DNA methylation alone cannot fully account for the paternal contributions to epigenetic inheritance. For instance, there was no observable link between DNA methylation patterns in sperm and the epigenetic inheritance influenced by diet in sperm [89]. Unexpected findings emerged when experiments showed that injecting RNA fragments from sperm into fertilized, normal oocytes could replicate the transmission of acquired male traits [90]. Furthermore, discoveries that tRNA fragments are prevalent in sperm [91] have been supported by two independent studies, establishing the crucial role of tRNA fragments in epigenetic inheritance [27, 92].

In one investigation, the transmission of a metabolic phenotype influenced by the male's diet was proven through the zygotic introduction of sperm-derived small RNAs [27]. Another study revealed that the assortment of small RNAs increases as sperm mature in the epididymis [92]. Predominantly, these small RNAs were derived from tRNAs, with a notable abundance of a 5′-fragment from $tRNA^{Gly/CCC}$. The reasons for this specific prevalence remain unclear. In embryonic stem cells and mouse embryos, this particular fragment inhibited genes regulated by the MERVL retrotransposon, which plays a crucial role in early pre-implantation development [93]. Given that retrotransposons comprise about half of the mammalian genome, tRNA fragments could have extensive interactions with retrotransposon-encoded elements and their functions. This suggests that the epigenetic impact of tRNA fragments could be more substantial than previously thought. The presence of various RNA modifications in sperm tRNA fragments, which respond to environmental shifts, indicates that these fragments might act as carriers for paternal epigenetic information, offering a vast potential reservoir of RNA microspecies [27].

### 9.11.5 Regulation of Pulmonary Diseases

Chronic obstructive pulmonary disease (COPD) stands as the most common lung disorder, with macrophages playing a pivotal role in mediating the inflammatory responses associated with the disease [94]. Recent studies have provided a comprehensive characterization of circulating short non-coding RNAs (sncRNAs) in the plasma of patients suffering from COPD. Researchers have applied a novel approach by pre-treating plasma RNAs with T4 polynucleotide kinase, which modifies RNA ends to make them amenable to RNA-seq analysis. This modification process enables the comprehensive sequencing of a broad spectrum of sncRNAs, including those such as 5′-tRNA halves that contain a 2′,3′-cyclic phosphate. Notably, an accumulation of 5′-half derived from $tRNA^{Val}$CAC was observed in the plasma of

COPD patients, in contrast to the predominance of the 5′-tRNA$^{Gly}$GCC halves in healthy individuals. Moreover, it has been discovered that the 5′-tRNA$^{Val}$CAC half can activate human macrophages via TLR 7, leading to the production of cytokines, thus having an immunoactive signature [95].

### *9.11.6 Rheumatoid Arthritis and Psoriatic Arthritis*

A recent study found a significant reduction, averaging up to threefold, in the levels of tRFs derived from tRNA$_{Gly}$GCC, tRNA$_{Glu}$CTC, and tRNA$_{Val}$CAC/AAC was observed in the plasma of RA patients. Conversely, these same tRFs (tRNA$_{Glu}$CTC and tRNA$_{Val}$CAC/AAC) were found at significantly higher levels, also up to threefold, in PsA patients in comparison to healthy individuals. This trend of altered tRF levels was further substantiated in an experimental model using arthritic IL1Ra-KO mice, where reduced levels of tRNA-Glu-CTC fragments were detected. Moreover, an investigation into the expression of NSUN2, an enzyme responsible for catalyzing tRNA methylation, showed increased activity in the PBMCs of RA patients as opposed to those from PsA patients. However, this finding is not consistently corroborated by available public transcriptomics data, indicating the complexity of tRF regulation and their potential interactions [96]. The distinct patterns of tRF expression in RA and PsA patients suggest these small RNA fragments play significant roles in modulating immune responses within these diseases. The specific increase or decrease in certain tRFs could reflect underlying mechanistic differences between these conditions, influencing how immune cells engage and regulate inflammatory responses. The differential expression of NSUN2 and its impact on tRNA methylation might contribute to these observed variations in tRF profiles, suggesting a methylation-dependent mechanism of tRF modulation in immune-mediated diseases.

## 9.12 Conclusion

Once regarded as mere by-products of tRNA degradation, tiRNAs have emerged as significant players in cellular stress response, particularly in the context of cancer progression. The stress-angiogenin-tiRNAs signaling pathway highlights a sophisticated mechanism whereby cells adapt to environmental and genetic stressors, enhancing their survival and longevity. This pathway's exploration has enriched our understanding of cancer biology and opened promising avenues in RNAi therapy, where tiRNAs can be targeted or mimicked to modulate specific stress responses. Furthermore, the potential of tiRNAs as non-invasive biomarkers and innovative therapeutic targets is being increasingly recognized, offering new possibilities for the early detection and targeted treatment of cancer. This shift in perception underscores the pivotal role of tiRNAs in oncology, heralding a new era of therapeutic strategies that leverage our growing understanding of RNA biology to combat cancer more effectively.

## References


1. Kültz D. Molecular and evolutionary basis of the cellular stress response. Annu Rev Physiol. 2005;67:225–57.
2. Richter K, Haslbeck M, Buchner J. The heat shock response: life on the verge of death. Mol Cell. 2010;40(2):253–66.
3. Sirbu BM, Cortez D. DNA damage response: three levels of DNA repair regulation. Cold Spring Harb Perspect Biol. 2013;5(8):a012724.
4. Fulda S, Gorman AM, Hori O, Samali A. Cellular stress responses: cell survival and cell death. Int J Cell Biol. 2010;2010:214074.
5. Muz B, de la Puente P, Azab F, Azab AK. The role of hypoxia in cancer progression, angiogenesis, metastasis, and resistance to therapy. Hypoxia. 2015;3:83–92.
6. Cubillos-Ruiz JR, Bettigole SE, Glimcher LH. Tumorigenic and immunosuppressive effects of endoplasmic reticulum stress in cancer. Cell. 2017;168(4):692–706.
7. Kim B, Song YS. Mitochondrial dynamics altered by oxidative stress in cancer. Free Radic Res. 2016;50(10):1065–70.
8. Mendell JT, Olson EN. MicroRNAs in stress signaling and human disease. Cell. 2012;148(6):1172–87.
9. Valadkhan S, Valencia-Hipólito A. lncRNAs in stress response. Curr Top Microbiol Immunol. 2016;394:203–36.
10. Torrent M, Chalancon G, de Groot NS, Wuster A, Madan BM. Cells alter their tRNA abundance to selectively regulate protein synthesis during stress conditions. Sci Signal. 2018;11:546.
11. Thompson DM, Parker R. Stressing out over tRNA cleavage. Cell. 2009;138(2):215–9.
12. Thompson DM, Lu C, Green PJ, Parker R. tRNA cleavage is a conserved response to oxidative stress in eukaryotes. RNA. 2008;14(10):2095–103.
13. Miguel V, Lamas S, Espinosa-Diez C. Role of non-coding-RNAs in response to environmental stressors and consequences on human health. Redox Biol. 2020;37:101580.
14. Nemeth K, Bayraktar R, Ferracin M, Calin GA. Non-coding RNAs in disease: from mechanisms to therapeutics. Nat Rev Genet. 2024;25(3):211–32.
15. Li S, Xu Z, Sheng J. tRNA-derived small RNA: a novel regulatory small non-coding RNA. Gene. 2018;9(5):246.
16. Li S, Hu GF. Emerging role of angiogenin in stress response and cell survival under adverse conditions. J Cell Physiol. 2012;227(7):2822–6.
17. Lyons SM, Fay MM, Akiyama Y, Anderson PJ, Ivanov P. RNA biology of angiogenin: current state and perspectives. RNA Biol. 2017;14(2):171–8.
18. Sheng J, Xu Z. Three decades of research on angiogenin: a review and perspective. Acta Biochim Biophys Sin. 2016;48(5):399–410.
19. Olson KA, Verselis SJ, Fett JW. Angiogenin is regulated in vivo as an acute phase protein. Biochem Biophys Res Commun. 1998;242(3):480–3.
20. Dyer KD, Rosenberg HF. The mouse RNase 4 and RNase 5/ang 1 locus utilizes dual promoters for tissue-specific expression. Nucleic Acids Res. 2005;33(3):1077–86.
21. Greenway MJ, Andersen PM, Russ C, Ennis S, Cashman S, Donaghy C, et al. ANG mutations segregate with familial and 'sporadic' amyotrophic lateral sclerosis. Nat Genet. 2006;38(4):411–3.
22. Wu D, Yu W, Kishikawa H, Folkerth RD, Iafrate AJ, Shen Y, et al. Angiogenin loss-of-function mutations in amyotrophic lateral sclerosis. Ann Neurol. 2007;62(6):609–17.

23. Yamasaki S, Ivanov P, Hu GF, Anderson P. Angiogenin cleaves tRNA and promotes stress-induced translational repression. J Cell Biol. 2009;185(1):35–42.
24. Kishimoto K, Yoshida S, Ibaragi S, Yoshioka N, Okui T, Hu GF, et al. Hypoxia-induced up-regulation of angiogenin, besides VEGF, is related to progression of oral cancer. Oral Oncol. 2012;48(11):1120–7.
25. Ramani P, Headford A, Sowa-Avugrah E, Hunt LP. Angiogenin expression in human kidneys and Wilms' tumours: relationship with hypoxia and angiogenic factors. Int J Exp Pathol. 2013;94(2):115–25.
26. Fu H, Feng J, Liu Q, Sun F, Tie Y, Zhu J, et al. Stress induces tRNA cleavage by angiogenin in mammalian cells. FEBS Lett. 2009;583(2):437–42.
27. Chen Q, Yan M, Cao Z, Li X, Zhang Y, Shi J, et al. Sperm tsRNAs contribute to intergenerational inheritance of an acquired metabolic disorder. Science. 2016;351(6271):397–400.
28. Galizi R, Spano F, Giubilei MA, Capuccini B, Magini A, Urbanelli L, et al. Evidence of tRNA cleavage in apicomplexan parasites: Half-tRNAs as new potential regulatory molecules of Toxoplasma gondii and Plasmodium berghei. Mol Biochem Parasitol. 2013;188(2):99–108.
29. Blanco S, Dietmann S, Flores JV, Hussain S, Kutter C, Humphreys P, et al. Aberrant methylation of tRNAs links cellular stress to neuro-developmental disorders. EMBO J. 2014;33(18):2020–39.
30. Frye M, Harada BT, Behm M, He C. RNA modifications modulate gene expression during development. Science. 2018;361(6409):1346–9.
31. Schaefer M, Pollex T, Hanna K, Tuorto F, Meusburger M, Helm M, et al. RNA methylation by Dnmt2 protects transfer RNAs against stress-induced cleavage. Genes Dev. 2010;24(15):1590–5.
32. Saikia M, Krokowski D, Guan BJ, Ivanov P, Parisien M, Hu GF, et al. Genome-wide identification and quantitative analysis of cleaved tRNA fragments induced by cellular stress. J Biol Chem. 2012;287(51):42708–25.
33. Davies KJ. Adaptive homeostasis. Mol Asp Med. 2016;49:1–7.
34. Liu B, Qian SB. Translational reprogramming in cellular stress response. Wiley Interdiscip Rev RNA. 2014;5(3):301–15.
35. Czech A, Wende S, Mörl M, Pan T, Ignatova Z. Reversible and rapid transfer-RNA deactivation as a mechanism of translational repression in stress. PLoS Genet. 2013;9(8):e1003767.
36. Sonenberg N, Hinnebusch AG. Regulation of translation initiation in eukaryotes: mechanisms and biological targets. Cell. 2009;136(4):731–45.
37. Zhang S, Sun L, Kragler F. The phloem-delivered RNA pool contains small noncoding RNAs and interferes with translation. Plant Physiol. 2009;150(1):378–87.
38. Ivanov P, Emara MM, Villen J, Gygi SP, Anderson P. Angiogenin-induced tRNA fragments inhibit translation initiation. Mol Cell. 2011;43(4):613–23.
39. Cammas A, Millevoi S. RNA G-quadruplexes: emerging mechanisms in disease. Nucleic Acids Res. 2017;45(4):1584–95.
40. Rhodes D, Lipps HJ. G-quadruplexes and their regulatory roles in biology. Nucleic Acids Res. 2015;43(18):8627–37.
41. Lyons SM, Achorn C, Kedersha NL, Anderson PJ, Ivanov P. YB-1 regulates tiRNA-induced stress granule formation but not translational repression. Nucleic Acids Res. 2016;44(14):6949–60.
42. Buchan JR, Parker R. Eukaryotic stress granules: the ins and outs of translation. Mol Cell. 2009;36(6):932–41.
43. Kedersha N, Ivanov P, Anderson P. Stress granules and cell signaling: more than just a passing phase? Trends Biochem Sci. 2013;38(10):494–506.
44. Saikia M, Jobava R, Parisien M, Putnam A, Krokowski D, Gao XH, et al. Angiogenin-cleaved tRNA halves interact with cytochrome c, protecting cells from apoptosis during osmotic stress. Mol Cell Biol. 2014;34(13):2450–63.
45. Portt L, Norman G, Clapp C, Greenwood M, Greenwood MT. Anti-apoptosis and cell survival: a review. Biochim Biophys Acta. 2011;1813(1):238–59.

46. Mei Y, Yong J, Liu H, Shi Y, Meinkoth J, Dreyfuss G, et al. tRNA binds to cytochrome c and inhibits caspase activation. Mol Cell. 2010;37(5):668–78.
47. Suryanarayana T, Uppala JK, Garapati UK. Interaction of cytochrome c with tRNA and other polynucleotides. Mol Biol Rep. 2012;39(9):9187–91.
48. Nakashima A, Takaku H, Shibata HS, Negishi Y, Takagi M, Tamura M, et al. Gene silencing by the tRNA maturase tRNase ZL under the direction of small-guide RNA. Gene Ther. 2007;14(1):78–85.
49. Nashimoto M. Specific cleavage of target RNAs from HIV-1 with 5′ half tRNA by mammalian tRNA 3′ processing endoribonuclease. RNA. 1996;2(6):523–4.
50. Tan KM, Chan SL, Tan KO, Yu VC. The Caenorhabditis elegans sex-determining protein FEM-2 and its human homologue, hFEM-2, are $Ca^{2+}$/calmodulin-dependent protein kinase phosphatases that promote apoptosis. J Biol Chem. 2001;276(47):44193–202.
51. Durdevic Z, Mobin MB, Hanna K, Lyko F, Schaefer M. The RNA methyltransferase Dnmt2 is required for efficient Dicer-2-dependent siRNA pathway activity in Drosophila. Cell Rep. 2013;4(5):931–7.
52. Krishna S, Yim DG, Lakshmanan V, Tirumalai V, Koh JL, Park JE, et al. Dynamic expression of tRNA-derived small RNAs define cellular states. EMBO Rep. 2019;20(7):e47789.
53. Stefani G, Slack FJ. Small non-coding RNAs in animal development. Nat Rev Mol Cell Biol. 2008;9(3):219–30.
54. Torres AG, Reina O, Stephan-Otto Attolini C. Ribas de Pouplana L. Differential expression of human tRNA genes drives the abundance of tRNA-derived fragments. Proc Natl Acad Sci USA. 2019;116(17):8451–6.
55. Aoi T, Yae K, Nakagawa M, Ichisaka T, Okita K, Takahashi K, et al. Generation of pluripotent stem cells from adult mouse liver and stomach cells. Science. 2008;321(5889):699–702.
56. Zhang MM, Bahal R, Rasmussen TP, Manautou JE, Zhong XB. The growth of siRNA-based therapeutics: updated clinical studies. Biochem Pharmacol. 2021;189:114432.
57. Liu B, Cao J, Wang X, Guo C, Liu Y, Wang T. Deciphering the tRNA-derived small RNAs: origin, development, and future. Cell Death Dis. 2021;13(1):24.
58. Lyons SM, Gudanis D, Coyne SM, Gdaniec Z, Ivanov P. Identification of functional tetramolecular RNA G-quadruplexes derived from transfer RNAs. Nat Commun. 2017;8(1):1127.
59. Lyons SM, Kharel P, Akiyama Y, Ojha S, Dave D, Tsvetkov V, et al. eIF4G has intrinsic G-quadruplex binding activity that is required for tiRNA function. Nucleic Acids Res. 2020;48(11):6223–33.
60. Yuan Y, Li J, He Z, Fan X, Mao X, Yang M, et al. tRNA-derived fragments as new hallmarks of aging and age-related diseases. Aging Dis. 2021;12(5):1304–22.
61. Victoria B, Dhahbi JM, Nunez Lopez YO, Spinel L, Atamna H, Spindler SR, et al. Circulating microRNA signature of genotype-by-age interactions in the long-lived Ames dwarf mouse. Aging Cell. 2015;14(6):1055–66.
62. Chen Y, Shen J. Mucosal immunity and tRNA, tRF, and tiRNA. J Mol Med. 2021;99(1):47–56.
63. Xie Y, Yao L, Yu X, Ruan Y, Li Z, Guo J. Action mechanisms and research methods of tRNA-derived small RNAs. Signal Transduct Target Ther. 2020;5(1):109.
64. Pandey KK, Madhry D, Ravi Kumar YS, Malvankar S, Sapra L, Srivastava RK, et al. Regulatory roles of tRNA-derived RNA fragments in human pathophysiology. Mol Ther Nucleic Acids. 2021;26:161–73.
65. Pawar K, Shigematsu M, Sharbati S, Kirino Y. Infection-induced 5′-half molecules of tRNA-HisGUG activate toll-like receptor 7. PLoS Biol. 2020;18(12):e3000982.
66. Chiou NT, Kageyama R, Ansel KM. Selective export into extracellular vesicles and function of tRNA fragments during T cell activation. Cell Rep. 2018;25(12):3356–70.
67. Villarroya-Beltri C, Gutiérrez-Vázquez C, Sánchez-Cabo F, Pérez-Hernández D, Vázquez J, Martin-Cofreces N, et al. Sumoylated hnRNPA2B1 controls the sorting of miRNAs into exosomes through binding to specific motifs. Nat Commun. 2013;4:2980.
68. Shurtleff MJ, Temoche-Diaz MM, Karfilis KV, Ri S, Schekman R. Y-box protein 1 is required to sort microRNAs into exosomes in cells and in a cell-free reaction. elife. 2016;5:e19276.

69. McNab FW, Ewbank J, Rajsbaum R, Stavropoulos E, Martirosyan A, Redford PS, et al. TPL-2-ERK1/2 signaling promotes host resistance against intracellular bacterial infection by negative regulation of type I IFN production. J Immunol. 2013;191(4):1732–43.
70. Yang JY, Deng XY, Li YS, Ma XC, Feng JX, Yu B, et al. Structure of Schlafen13 reveals a new class of tRNA/rRNA-targeting RNase engaged in translational control. Nat Commun. 2018;9(1):1165.
71. Mo D, Jiang P, Yang Y, Mao X, Tan X, Tang X, et al. A tRNA fragment, 5′-tiRNA(Val), suppresses the Wnt/β-catenin signaling pathway by targeting FZD3 in breast cancer. Cancer Lett. 2019;457:60–73.
72. Van Lint S, Wilgenhof S, Heirman C, Corthals J, Breckpot K, Bonehill A, et al. Optimized dendritic cell-based immunotherapy for melanoma: the TriMix-formula. Cancer Immunol Immunother. 2014;63(9):959–67.
73. Dhahbi JM. 5′ tRNA Halves: the next generation of immune signaling molecules. Front Immunol. 2015;6:74.
74. Dhahbi JM, Atamna H, Selth LA. Data mining of small RNA-Seq suggests an association between prostate cancer and altered abundance of 5′ transfer RNA halves in seminal fluid and prostatic tissues. Biomark Cancer. 2018;10:1179299x18759545.
75. Zhao C, Tolkach Y, Schmidt D, Muders M, Kristiansen G, Müller SC, et al. tRNA-halves are prognostic biomarkers for patients with prostate cancer. Urol Oncol. 2018;36(11):503.
76. Honda S, Loher P, Shigematsu M, Palazzo JP, Suzuki R, Imoto I, et al. Sex hormone-dependent tRNA halves enhance cell proliferation in breast and prostate cancers. Proc Natl Acad Sci USA. 2015;112(29):E3816–25.
77. Honda S, Kirino Y. SHOT-RNAs: a novel class of tRNA-derived functional RNAs expressed in hormone-dependent cancers. Mol Cell Oncol. 2016;3(2):e1079672.
78. Shao Y, Sun Q, Liu X, Wang P, Wu R, Ma Z. tRF-Leu-CAG promotes cell proliferation and cell cycle in non-small cell lung cancer. Chem Biol Drug Des. 2017;90(5):730–8.
79. Li S, Shi X, Chen M, Xu N, Sun D, Bai R, et al. Angiogenin promotes colorectal cancer metastasis via tiRNA production. Int J Cancer. 2019;145(5):1395–407.
80. Zhu L, Li T, Shen Y, Yu X, Xiao B, Guo J. Using tRNA halves as novel biomarkers for the diagnosis of gastric cancer. Cancer Biomark. 2019;25(2):169–76.
81. Zhang L, Liu J, Hou Y. Classification, function, and advances in tsRNA in non-neoplastic diseases. Cell Death Dis. 2023;14(11):748.
82. Zhang Y, Zhang Y, Shi J, Zhang H, Cao Z, Gao X, et al. Identification and characterization of an ancient class of small RNAs enriched in serum associating with active infection. J Mol Cell Biol. 2014;6(2):172–4.
83. Dhahbi JM, Spindler SR, Atamna H, Yamakawa A, Boffelli D, Mote P, et al. 5′ tRNA halves are present as abundant complexes in serum, concentrated in blood cells, and modulated by aging and calorie restriction. BMC Genomics. 2013;14:298.
84. Mishima E, Inoue C, Saigusa D, Inoue R, Ito K, Suzuki Y, et al. Conformational change in transfer RNA is an early indicator of acute cellular damage. J Am Soc Nephrol. 2014;25(10):2316–26.
85. Goodarzi H, Liu X, Nguyen HC, Zhang S, Fish L, Tavazoie SF. Endogenous tRNA-derived fragments suppress breast cancer progression via YBX1 displacement. Cell. 2015;161(4):790–802.
86. Goncalves KA, Silberstein L, Li S, Severe N, Hu MG, Yang H, et al. Angiogenin promotes hematopoietic regeneration by dichotomously regulating quiescence of stem and progenitor cells. Cell. 2016;166(4):894–906.
87. Carone BR, Fauquier L, Habib N, Shea JM, Hart CE, Li R, et al. Paternally induced transgenerational environmental reprogramming of metabolic gene expression in mammals. Cell. 2010;143(7):1084–96.
88. Ng SF, Lin RC, Laybutt DR, Barres R, Owens JA, Morris MJ. Chronic high-fat diet in fathers programs β-cell dysfunction in female rat offspring. Nature. 2010;467(7318):963–6.
89. Shea JM, Serra RW, Carone BR, Shulha HP, Kucukural A, Ziller MJ, et al. Genetic and epigenetic variation, but not diet, shape the sperm methylome. Dev Cell. 2015;35(6):750–8.

90. Gapp K, Jawaid A, Sarkies P, Bohacek J, Pelczar P, Prados J, et al. Implication of sperm RNAs in transgenerational inheritance of the effects of early trauma in mice. Nat Neurosci. 2014;17(5):667–9.
91. Peng H, Shi J, Zhang Y, Zhang H, Liao S, Li W, et al. A novel class of tRNA-derived small RNAs extremely enriched in mature mouse sperm. Cell Res. 2012;22(11):1609–12.
92. Sharma U, Conine CC, Shea JM, Boskovic A, Derr AG, Bing XY, et al. Biogenesis and function of tRNA fragments during sperm maturation and fertilization in mammals. Science. 2016;351(6271):391–6.
93. Schoorlemmer J, Pérez-Palacios R, Climent M, Guallar D, Muniesa P. Regulation of mouse retroelement MuERV-L/MERVL expression by REX1 and epigenetic control of stem cell potency. Front Oncol. 2014;4:14.
94. Elonheimo HM, Mattila T, Andersen HR, Bocca B, Ruggieri F, Haverinen E, et al. Environmental substances associated with chronic obstructive pulmonary disease-a scoping review. Int J Environ Res Public Health. 2022;19(7):3945.
95. Shigematsu M, Kawamura T, Deshpande DA, Kirino Y. Immunoactive signatures of circulating tRNA- and rRNA-derived RNAs in chronic obstructive pulmonary disease. Mol Ther Nucleic Acids. 2024;35(3):102285.
96. Dunaeva M, Blom J, Thurlings R, van Weijsten M, van de Loo FAJ, Pruijn GJM. Circulating tRNA-derived fragments are decreased in patients with rheumatoid arthritis and increased in patients with psoriatic arthritis. Biomarkers. 2024;29(2):90–9.

# Chapter 10
# Emerging RNAs: Glycan-RNAs

**Abstract** Researchers have discovered a novel class of RNA molecules that are modified by the addition of carbohydrates, a process known as glycosylation. These molecules, aptly termed "glycan-RNAs," are found on the surfaces of cells. To fully appreciate the significance and potential applications of glycan-RNAs, it is essential to start with a foundational understanding of both RNA and the glycosylation process. RNA, or ribonucleic acid, plays crucial roles in the coding, decoding, regulation, and expression of genes. Unlike DNA, RNA is single-stranded and can fold into complex three-dimensional shapes, enabling it to participate in a vast array of cellular functions, from acting as a messenger carrying instructions from DNA for controlling the synthesis of proteins, to regulating gene expression itself. Glycosylation, on the other hand, involves the enzymatic process that attaches glycans (sugar molecules) to proteins, lipids, or other organic molecules. This modification can alter the physical properties and functions of the original molecule, affecting processes such as protein folding, recognition, and signaling. In the context of RNA, glycosylation could similarly impact the molecule's stability, its interactions with other cellular components, and its overall function. Following this introduction to the basic principles, we delve into more specific discussions on the identification and further development of glycoRNAs. This includes techniques used to detect these molecules, their biological implications, and the potential therapeutic applications they might hold. Understanding glycoRNAs could open new avenues in biomedical research, offering insights into novel regulatory mechanisms in cells and potentially leading to breakthroughs in disease treatment and diagnostics.

## 10.1 Introduction

In the expanding field of transcriptomics, researchers have catalogued over 140 types of RNA modifications affecting the four primary nucleobases [1]. Beyond the well-studied 5′ cap and 3′ poly(A) tail, there are five main categories of RNA modifications that have been identified: adenosine methylation, cytosine modifications,

 
P. E. Saw, E. Song, *RNA Therapeutics in Human Diseases*,
https://doi.org/10.1007/978-981-96-3041-7_10

uridine isomerization, ribose modifications, and cytidine acetylation. These modifications are critical not only for the structural integrity and function of RNA molecules but also for their stability and proper biogenesis. Modifications such as adenosine methylation typically occur at specific sites within the RNA molecule, influencing RNA's role in the regulation of gene expression and its ability to perform its cellular functions. Cytosine modifications, for instance, can impact RNA stability and the process of translation, while uridine isomerization alters the structural configuration of RNA, potentially affecting its interaction with other molecules. Modifications in the ribose backbone of RNA molecules, such as 2′-O-methylation, play a pivotal role in protecting RNA from degradation and aiding in the correct folding and processing of ribosomal and transfer RNAs. Cytidine acetylation, on the other hand, can influence the tertiary structure of RNA, affecting its interaction with proteins and other RNAs [2–4]. These RNA modifications are involved in a broad spectrum of biological processes including splicing, translation, and decay, ultimately regulating protein synthesis through a mechanism known as post-transcriptional modification. This intricate system of modifications ensures that proteins are produced accurately and efficiently in response to cellular needs [5–11]. Recently, RNA glycosylation has emerged as a newly identified type of modification, further broadening the spectrum of potential regulatory mechanisms at the RNA level. This discovery extends the traditional boundaries of epigenetics, suggesting a complex layer of regulation that involves the chemical modification of ribose and nucleosides, as well as the attachment of carbohydrate moieties to RNA molecules. This novel class of modification could have profound implications for our understanding of cellular regulation and the genetic encoding of biological complexity [12, 13].

## 10.2 Glycosylation and Glycans

Glycosylation is a highly intricate and critical post-translational modification involving the attachment of carbohydrate moieties, or glycosyl donors, to proteins or lipids [14, 15]. This process results in the formation of glycoconjugates through a series of specific enzymatic reactions, predominantly occurring within the Golgi apparatus and the endoplasmic reticulum. Recognized as one of the most complex forms of post-translational modification, glycosylation significantly alters the structure and function of proteins and lipids. There are two primary types of glycosylation: N-linked and O-linked [16]. N-linked glycosylation attaches a glycan to the nitrogen atom of the asparagine side chain within a specific consensus sequence—Asn-X-Ser/Thr, where "X" represents any amino acid except proline. This type of glycosylation is crucial for protein folding and stability. O-linked glycosylation, in contrast, involves the attachment of a glycan to the oxygen atom of serine (Ser) or threonine (Thr) residues. This modification typically influences protein solubility and cellular localization. Each type of glycosylation plays a vital role in cellular mechanisms, impacting a wide range of biological processes including cell–cell interaction, immune response, and protein clearance, thereby underscoring the complexity and importance of glycosylation in cellular biology [17]. These modifications do not merely add a sugar group; they often drastically alter the properties and functions of the modified molecules. The glycans attached through these processes can vary greatly, resulting in a highly heterogeneous array of glycan structures. This diversity is not random but finely regulated, affecting various biological functions such as protein folding, stability, signaling, and cell–cell interactions [18]. In proteins, these glycan groups can affect how proteins fold, how they are recognized and degraded by cellular machinery, and their movement through and between cells. For lipids, glycosylation can influence membrane composition, the distribution of molecules within the membrane, and the interactions between cells and their environments [19, 20].

Glycoconjugates, essential components of cellular membranes, are complex molecules formed by the covalent bonding of glycans (sugar chains) to proteins or lipids. This bonding results in a diverse group of molecules that play critical roles in cell structure and function. Depending on the nature of their linkage to proteins, lipids, or other glycans, these glycans are classified into various types, each influencing the molecule's function and localization within the cell. There are four major types of glycoconjugates, detailed below.

### 10.2.1 Glycoproteins

These are proteins that have one or several oligosaccharides covalently attached to amino acids within the protein chain. The glycans in glycoproteins are typically linked through the nitrogen atom of asparagine (N-linked) or the oxygen atom of serine or threonine (O-linked). Glycoproteins are versatile and can function in cell adhesion, signaling, and immune recognition [21, 22].

### 10.2.2 Proteoglycans

Distinguished by their heavily glycosylated nature, proteoglycans consist of a core protein to which one or more glycosaminoglycan (GAG) chains are attached. These are typically found in the extracellular matrix where they contribute to the structural integrity of the tissue, mediate interactions between cells and their surroundings, and regulate growth factors [23–25].

### 10.2.3 Glycosphingolipids

These are composed of a glycan moiety attached to a lipid molecule known as sphingosine. Glycosphingolipids are crucial components of the lipid bilayer of cellular membranes and play vital roles in cell recognition, signal transmission, and interactions between cells [26].

### 10.2.4 Glycosylphosphatidylinositol (GPI)-Anchored Glycoproteins

These glycoconjugates are proteins that are anchored to the cell membrane via a GPI linkage. The GPI anchor is a glycan moiety attached to phosphatidylinositol, a type of phospholipid. GPI-anchored glycoproteins are involved in enzymatic processes, signal transduction, and cellular adhesion [18, 27].

## 10.3 Protein Glycosylation

While N-linked and O-linked glycosylations are the most recognized forms, protein glycosylation encompasses a broader range of glycan attachments including glycosaminoglycans (GAGs), phosphorylated glycans, C-mannosylated tryptophan residues, and GPI anchors [28]. Different types of glycosylation are summarized in Table 10.1.

## 10.4 Glycosphingolipids

Glycosphingolipids are complex molecules produced through the process of lipid glycosylation within the secretory pathway, encompassing the endoplasmic reticulum and the Golgi apparatus [38]. These compounds consist of a lipid component, specifically a ceramide (a sphingosine linked to a fatty acid), which is glycosylated—meaning sugar moieties are covalently attached to it. This glycosylation enhances the diversity and functionality of these lipids. Once synthesized, glycosphingolipids are incorporated into the plasma membranes of cells. They play pivotal roles in cell membrane structure and are crucial for the formation and stabilityof lipid rafts. Lipid rafts are specialized, dynamic microdomains within the cell membrane, characterized by their distinct lipid composition and ability to float freely within the more fluid and disordered environment of the membrane. These domains serve as organizing centers for the assembly of signaling molecules, significantly influencing signal transduction and cellular communication [39].

**Table 10.1** Major types of glycosylation [14, 15, 18, 22, 28–30]

| | |
|---|---|
| N-linked glycosylation [31, 32] | N-linked glycosylation, a critical biochemical process, begins with the attachment of a glycan to the nitrogen atom of an asparagine residue within a specific consensus sequence, Asn-X-Ser/Thr, where "X" can be any amino acid except proline. This initial step occurs in the endoplasmic reticulum (ER). After the initial glycosylation, the glycoprotein is transported to the Golgi apparatus, where the glycan undergoes further processing and maturation. This type of glycosylation is crucial for various cellular functions. It plays a significant role in protein folding, ensuring that proteins achieve and maintain their correct three-dimensional structure, which is essential for their functionality. Additionally, N-linked glycans contribute to protein stability, protecting them from degradation, and are involved in cellular signaling processes where they mediate interactions between cells and their environments |
| O-linked glycosylation [33, 34] | O-linked glycosylation is distinct from N-linked glycosylation, primarily occurring within the Golgi apparatus, where sugar molecules are attached to the oxygen atom of serine (Ser) or threonine (Thr) residues. This type of modification plays pivotal roles in the structural and functional aspects of proteins. O-linked glycosylation is instrumental in determining protein conformation, thereby influencing protein stability and extending the functional lifespan of proteins within cellular environments. Furthermore, O-linked glycans are critical for cell–cell recognition, a process fundamental to many biological activities including immune responses, cell signaling, and tissue formation. By modifying the surface properties of proteins, O-linked glycosylation facilitates specific interactions between cells and other molecules, which is crucial for the accurate transmission of signals across cellular membranes and the establishment of cellular networks. The diversity and complexity of the sugar chains in O-linked glycosylation also allow for a wide range of functional outcomes, underscoring its significance in cellular operations and organismal health |
| Glycosaminoglycans (GAGs) [23] | Glycosaminoglycans (GAGs) are unique components of the extracellular matrix and cell surfaces, characterized by their long, unbranched polysaccharide chains composed of repeating disaccharide units. Common types of GAGs include heparan sulfate and chondroitin sulfate. Unlike typical glycosylation processes where sugar chains are directly linked to the peptide backbone of proteins, GAGs are generally found covalently linked to core proteins, forming proteoglycans. This linkage does not occur directly to the peptide sequence but through specific serine residues in the core protein, contributing to the formation of a distinct macromolecular structure<br>Proteoglycans, composed of GAGs, play pivotal roles in cell communication, proliferation, and differentiation. These molecules are essential for constructing a highly regulated extracellular environment that supports a wide range of cellular functions, including the modulation of growth factor activities, the regulation of collagen structure in the extracellular matrix, and the facilitation of cell adhesion and migration. This makes them critical in developmental processes, wound healing, and pathological conditions such as cancer metastasis and inflammation Additionally, the glycan structures of proteoglycans can be further modified by the addition of phosphate groups, a process distinct from traditional glycosylation. Phosphorylation of glycans affects the function and localization of glycoproteins, significantly impacting their biological roles. This modification can modulate signaling pathways, enhancing or inhibiting interactions between cells and their surroundings, and can influence the binding affinity of glycoproteins to other molecules, including growth factors and receptors |

(continued)

**Table 10.1** (continued)

| | |
|---|---|
| Phosphorylated glycans [18] | Phosphorylation of glycan structures is a specialized form of glycosylation that involves the addition of phosphate groups to sugars within the glycan chains of glycoproteins. This modification can profoundly influence the function and localization of these glycoproteins, enabling dynamic responses to cellular cues and environmental changes. Phosphate groups introduce a negative charge to glycan structures, which can affect the three-dimensional conformation of glycoproteins. This alteration may change the protein's overall stability, activity, or localization within the cell, enabling it to interact with new targets or participate in different biological pathways. Phosphorylated glycans can play crucial roles in signaling pathways by modulating the interaction between glycoproteins and other cellular molecules such as receptors, enzymes, or structural proteins. These interactions often trigger downstream signaling events that influence cellular behavior, including growth, division, and apoptosis. In the cellular matrix or among cells, phosphorylated glycans affect adhesion, migration, and communication. The modified glycans can interact differently with molecules like growth factors, cytokines, and extracellular matrix components, influencing processes such as immune responses and tissue repair |
| C-mannosylation [35, 36] | C-mannosylation is a distinct and relatively rare type of glycosylation that involves the attachment of a mannose sugar to the indole C2 carbon of tryptophan residues within proteins. This unique linkage is chemically different from the more typical N-linked or O-linked glycosylation, where sugars are attached to nitrogen or oxygen atoms of amino acids. C-mannosylation significantly contributes to the proper folding and stability of proteins. By attaching mannose to tryptophan, this modification helps stabilize the protein structure, potentially influencing its overall conformation and how it folds. This stability is crucial for maintaining the functional integrity of proteins under physiological conditions. Certain growth factors and receptors are particularly dependent on C-mannosylation for their activity and longevity. These proteins often play critical roles in cell signaling pathways that regulate development, cell growth, and repair processes. The presence of mannose on these proteins can affect their binding affinity, receptor interaction, and signaling efficiency. Proteins modified by C-mannosylation are often involved in significant biological processes, including developmental pathways and immune responses. The modification can influence how proteins interact with other cellular components, impacting signal transduction and cellular communication |
| GPI anchors [37] | Glycosylphosphatidylinositol (GPI) anchoring is a sophisticated post-translational modification that involves attaching a protein to a specific glycan, which is itself linked to phosphatidylinositol, a type of phospholipid. This modification effectively anchors proteins to the cell membrane, integrating them into the cellular architecture with significant implications for their function and interaction. GPI anchoring is a primary mechanism by which proteins are localized to the cell membrane. This localization is crucial for proteins that need to be accessible on the cell surface, such as receptors, enzymes, and adhesion molecules. By anchoring these proteins to the membrane, GPI ensures they are positioned to interact efficiently with external molecules and signals. Many GPI-anchored proteins are involved in signal transduction pathways. Their location on the cell surface allows them to participate directly in signaling events that relay external cues into cellular responses. This is particularly vital for processes like immune activation, hormone response, and growth factor signaling. GPI-anchored proteins often facilitate interactions between cells. This is crucial in diverse biological contexts, from the development and maintenance of tissue architecture to immune system function. GPI anchors allow these proteins to remain mobile within the plane of the membrane, enhancing their ability to engage with ligands on adjacent cells |

The functional significance of glycosphingolipids extends beyond their structural roles in membranes and lipid rafts. They are involved in a variety of biological processes, including cell–cell recognition and adhesion, where glycosphingolipids contribute to the cellular interactions that are essential in many physiological processes, such as immune responses and the development of tissues. By participating in the formation of lipid rafts, glycosphingolipids influence the localization and activities of signaling proteins, thereby modulating various signaling pathways. These molecules are also involved in the regulation of endocytosis and the trafficking of molecules across cell membranes, impacting how cells uptake substances and communicate with their external environment. Alterations in glycosphingolipid metabolism are linked to several diseases, including Gaucher's disease and other lipid storage disorders, highlighting their importance in health and disease [40] (Fig. 10.1).

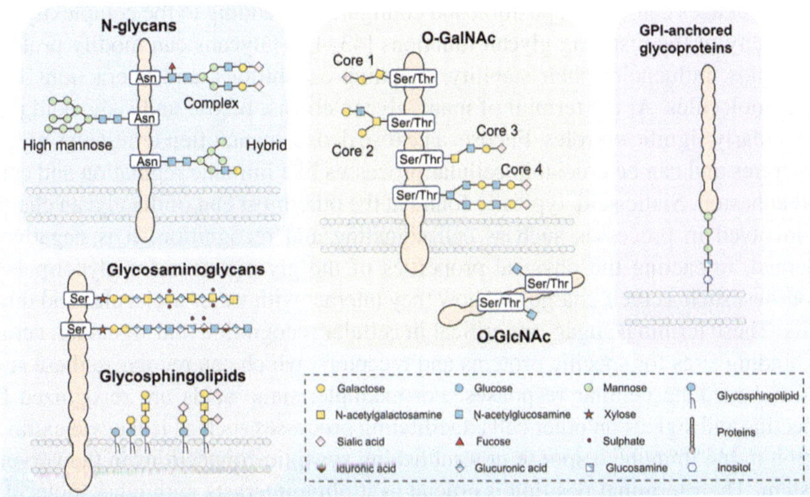


**Fig. 10.1** Schematic representation of glycosylation structures in proteins and lipids. This diagram illustrates various glycosylation patterns found in biomolecules, categorized into three main groups: N-glycans, O-GalNAc (mucin-type), and glycosaminoglycans/glycosphingolipids/GPI-anchored glycoproteins. N-glycans: shows high mannose, hybrid, and complex types, all attached to asparagine (Asn) residues of proteins. O-GalNAc: depicts mucin-type glycosylation with core structures 1–4 attached to serine (Ser) or threonine (Thr) residues. Glycosaminoglycans and glycosphingolipids: illustrates structures like heparan sulfate on serine residues and various glycosphingolipids, respectively. GPI-anchored glycoproteins: outlines the general structure of glycoproteins anchored to the cell membrane via a glycosylphosphatidylinositol (GPI) linkage. Figure is reproduced from [41] under Creative Commons CC-BY license

## 10.5 Glycans

Glycans are multifaceted biomolecules composed of various monosaccharides linked through glycosidic bonds [42]. These structures are notably complex and branched, synthesized from a foundational set of seven basic monosaccharides. Each of these sugars—fucose (Fuc), sialic acid (SA), glucose (Glc), galactose (Gal), mannose (Man), N-acetylglucosamine (GlcNAc), and N-acetyl-galactosamine (GalNAc)—brings unique properties and roles within the larger structure of a glycan. The synthesis of glycans is a meticulous process involving multiple enzymatic steps, each catalyzed by a specific enzyme. The assembly begins in the cytoplasm and endoplasmic reticulum, where nucleotide sugars form and then are transported into the Golgi apparatus. Here, glycosyltransferases, enzymes that facilitate the transfer of sugar moieties from activated donor molecules to specific acceptor molecules, add sugars in a stepwise manner. The sequence and type of sugars added, as well as the linkages formed between them, are tightly regulated, resulting in a diverse array of glycan structures.

The diversity of glycan structures is not merely a consequence of the different monosaccharides used but also of the various linkages that can form between them. These linkages can vary in position and configuration, adding to the complexity and specificity of the resulting glycan functions [43–45]. Glycans can modify proteins and lipids, influencing their stability, signaling capabilities, and interactions with other molecules. At the termini of many glycan chains, fucose and sialic acid play particularly significant roles. Fucose, a hexose deoxy sugar, often terminates glycan structures and can be crucial in cellular processes like immune regulation and cell–cell adhesion. Sialic acid, typically found at the outermost end of the glycan chains, is involved in processes such as cell signaling and recognition. It is negatively charged, impacting the physical properties of the glycoproteins or glycolipids it modifies, such as their charge and how they interact with water molecules and other cells. These terminal sugars are critical in cellular recognition and signaling, acting as binding sites for specific proteins and receptors, which can recognize these sugars and mediate cellular responses. For example, sialic acids are recognized by selectins and siglecs on other cells, facilitating processes such as leukocyte extravasation in the immune response or establishing synaptic connections in the nervous system. Their terminal position is crucial as it often interacts with other molecules and receptors, influencing cellular communication and immune responses [46].

## 10.6 Functions of Glycosylation

### *10.6.1 Promotion of Destabilization of Unfolded Proteins*

Glycosylation contributes to protein stabilization by promoting the destabilization of unfolded proteins, as evidenced by thermodynamic analysis [47]. Specifically, N-glycosylation aids in the proper folding of proteins into their functional three-dimensional structures, which is crucial for biological processes like cell–cell communication and signaling pathways [29, 48]. Furthermore, N-linked glycosylation can inhibit the deamidation process of glycoproteins and glycopeptides, a typical degradation route for proteins [32]. Glycosylation, particularly N-glycosylation, significantly enhances the stability of proteins, primarily through extending their half-lives and reducing susceptibility to enzymatic degradation and denaturation. This stabilizing effect is mediated by several mechanisms, including steric hindrance provided by large N-glycans that prevent chemical modifications and degradation pathways that would otherwise lead to protein instability. Large N-glycans attached to proteins can physically obstruct the access of enzymatic and chemical agents to specific sites on the protein backbone. This obstruction is crucial for protecting sensitive sites against reactions that can lead to protein destabilization and degradation. One of the key aspects of protein degradation involves deamidation, a process where asparagine residues are converted to aspartic acid or isoaspartic acid, forming a cyclic imide intermediate (succinimide) that is prone to hydrolysis. Deamidation can lead to a change in protein structure and function, often resulting in a loss of activity and increased susceptibility to further degradation. Glycosylation, especially near asparagine residues involved in N-linked glycosylation, helps protect these residues from the deamidation process. The attached glycan chains hinder the attack on the side chain amine groups by backbone nitrogen atoms, thus preventing the formation of the succinimide intermediate. This protective action helps maintain the structural integrity and functional activity of the protein over time. Glycosylated proteins generally exhibit improved solubility compared to their non-glycosylated counterparts. Glycans can enhance the hydrophilicity of proteins, making them less prone to aggregation and facilitating better interaction with the surrounding aqueous environment. Furthermore, glycosylation often plays a critical role in proper protein folding. The process of glycosylation can assist in the correct folding of proteins during synthesis in the endoplasmic reticulum, thereby reducing the occurrence of misfolded proteins and enhancing overall cellular protein quality control [49, 50].

### *10.6.2 Complex Interaction with Chaperones*

Nascent glycoproteins within the endoplasmic reticulum (ER) undergo a meticulous process of quality control to ensure that only properly folded and assembled proteins are transported to their functional destinations within or outside the cell. Calnexin and calreticulin, two crucial ER-resident chaperones, play a pivotal role in this process, overseeing the folding and quality control of newly synthesized glycoproteins. Calnexin is a membrane-bound chaperone, while calreticulin is its soluble counterpart within the ER lumen. Both share a similar function: they selectively bind to glycoproteins that exhibit a single glucose residue on their N-linked oligosaccharide chains. This binding is crucial for maintaining the glycoproteins in a state that is conducive to proper folding. The recognition and binding by calnexin and calreticulin are highly specific and glycan-dependent. This specificity is governed by the enzymatic trimming of glycans attached to the nascent glycoproteins. Initially, glucosidases trim the glucose residues on the glycan chains, and if the glycoprotein is not correctly folded, an enzyme called UDP-glucose glucosyltransferase (UGGT) re-glucosylates the glycan, thereby allowing another round of binding and folding assessment by calnexin or calreticulin. The cycle of deglucosylation and reglucosylation can repeat multiple times, with calnexin and calreticulin continuously monitoring and assisting in the folding process. This iterative mechanism helps prevent improperly folded proteins from advancing along the secretory pathway, which could lead to dysfunction or disease. Once a glycoprotein is properly folded, the terminal glucose residue is removed by glucosidase. This removal effectively ends the interaction with calnexin or calreticulin, signaling that the glycoprotein is correctly folded and ready to proceed to the next stage of maturation or to be transported out of the ER [51, 52]. Overall, protein glycosylation plays a vital role in the folding, degradation, and trafficking of proteins, including the targeting of receptors to the cell surface [21].

### *10.6.3 Influencing Ligand–Receptor Interaction*

Glycosylation can influence ligand–receptor interactions, thereby modulating signal transduction processes [18]. Sialylation, a specific type of glycosylation where sialic acid residues are added to the ends of glycan chains on proteins and lipids, plays a multifaceted role in cellular communication and adhesion processes. One intriguing aspect of sialylation's function is its involvement in modulating the activity of cell-surface receptors such as integrins, which are crucial for cell–matrix interactions. Anoikis is a form of programmed cell death that occurs when cells detach from the surrounding extracellular matrix, a crucial mechanism that prevents detached cells from colonizing inappropriate locations, thus inhibiting metastasis. Tumor cells often develop mechanisms to resist anoikis, allowing them to survive after detachment and disseminate to form secondary tumors. In the context of cancer, the sialylation of integrin receptors and subsequent binding to Gal-1 can inhibit the anoikis response. This interaction may allow tumor cells to evade anoikis, facilitating their survival in circulation and enhancing their metastatic potential [53].

The activity of FcγRIIB, an immunoglobulin G (IgG) receptor expressed on endothelial and other immune cells, reveals intricate connections between immune function, metabolic states, and the regulatory impact of glycosylation. Specifically, FcγRIIB has been implicated in obesity-induced insulin resistance, highlighting a significant cross-talk between immune signaling and metabolic pathways. This receptor's activity and function can be notably modulated by the glycosylation state of IgG antibodies, particularly through sialylation. FcγRIIB is a low-affinity receptor for the Fc region of IgG and plays a critical role in modulating immune responses. Its activation typically leads to inhibitory signaling, balancing the activating signals of other Fc receptors and thus maintaining immune homeostasis. In the context of obesity, chronic low-grade inflammation is common, which can enhance the expression or activity of FcγRIIB on endothelial cells. This upregulation is linked to the development of insulin resistance, a hallmark of metabolic syndromes such as type 2 diabetes. The chronic inflammatory state induced by obesity may lead to alterations in the signaling pathways involving FcγRIIB, contributing to endothelial dysfunction and impaired insulin signaling. This process exacerbates insulin resistance, compounding the metabolic challenges associated with obesity [54].

Cell adhesion molecules, particularly those involved in the immune system's trafficking and signaling, are significantly impacted by glycosylation, a modification that is vital for their function. Members of the L-selectin ligand family are prime examples of such molecules. These ligands play crucial roles in lymphocyte homing—a process essential for immune surveillance and the initiation of immune responses. Glycosylation, through modifications such as sialylation and fucosylation, is pivotal in enhancing the functional capabilities of these ligands [55].

**Box 10.1 Evolutionary Adept Blood-Group Antigen: Importance of Glycans in Biological Homeostasis**
Blood-group antigens are a fascinating example of how glycans contribute to cellular identity and immune recognition. These antigens are essentially specific carbohydrate structures located on the surface of red blood cells, and they play a crucial role in transfusion medicine, transplantation, and maternal–fetal interactions. The structure and composition of these glycans determine an individual's blood type, such as A, B, AB, or O, which are differentiated based on the presence or absence of specific sugar residues. Blood-group antigens are composed of glycan structures that are attached to proteins or lipids on the cell surface. These glycans typically consist of a backbone of sugars such as galactose and N-acetylglucosamine, with terminal additions of fucose and sialic acid residues. The presence of specific sugar residues at the terminal positions of these glycan chains—added through enzymatic processes involving glycosyltransferases—defines the blood-group antigen. For instance, the A and B antigens differ by a single sugar residue: A has N-acetylgalactosamine, while B has galactose [56].

## 10.7 Role of Sialic Acid and Fucose

### 10.7.1 Sialic Acid

This negatively charged sugar at the terminal positions of glycan chains influences various biological processes, including cell adhesion and protein recognition. In blood-group antigens, sialic acid can affect the stability and recognition of these molecules [57].

### 10.7.2 Fucose

This sugar is crucial in the H antigen, the precursor to A and B antigens. The presence of fucose is required for the subsequent addition of A or B antigenic sugars. The H antigen itself is significant in the O blood type, which lacks the additional sugars that characterize A and B types [58].

## 10.8 Antigenic Potential of Glycans

### 10.8.1 Immunogenicity

The specific glycan structures of blood-group antigens can elicit immune responses, especially when they are recognized as foreign by the immune system of another individual. This is particularly evident in transfusion and transplantation scenarios, where mismatched blood-group antigens can lead to severe immune reactions [59].

### 10.8.2 Clinical Implications

Understanding the glycosylation pathways that lead to the formation of blood-group antigens has profound clinical implications. It aids in the development of safer blood transfusion practices and improves compatibility assessments for organ transplantation. Beyond their role in transfusion medicine, the antigenic properties of glycans are also being explored for therapeutic and diagnostic applications in areas such as cancer, where altered glycosylation patterns can serve as biomarkers or therapeutic targets. Therefore, blood-group antigens exemplify the critical role of glycans in mediating biological recognition and immunological responses. Their study not only helps in enhancing clinical practices but also broadens our understanding of how glycans can influence health and disease at the molecular level [60, 61].

### 10.8.3 Glycans in the Immune System

As key effectors of the immune system, glycoforms are crucial in the synthesis, stability, signal recognition, functional regulation, and protein–protein interactions of immune proteins [22, 62]. The assembly of MHC I molecules is a critical process in the immune system, involving several key proteins and modifications to ensure that these molecules are correctly formed and functional. This assembly process is essential for presenting endogenous antigens to cytotoxic T cells, a fundamental aspect of immune surveillance, and response to infected or malignant cells. Initially, the unassembled heavy chains of MHC I molecules in the ER associate with calnexin, a membrane-bound chaperone. Calnexin assists in the folding of the MHC I heavy chains and ensures their proper conformation before further assembly. The association with calnexin is dependent on the glycosylation of asparagine (Asn) residues in the heavy chain. These sugar chains are critical for the initial stabilization and correct folding of the heavy chain. Once the heavy chains are sufficiently folded, they disassociate from calnexin. Subsequently, the heavy chains associate with tapasin, a transmembrane glycoprotein that acts as a scaffold. Tapasin is crucial for orchestrating the later stages of MHC I assembly. Tapasin facilitates the integration of the MHC I heavy chains with other key proteins, including calreticulin, ERp57, and the transporter associated with antigen processing (TAP). Calreticulin, another ER chaperone, further stabilizes the heavy chain, while ERp57, an enzyme, assists in forming disulfide bonds that are crucial for the structural integrity of MHC I. The TAP transporter plays a pivotal role by translocating peptides from the cytosol into the ER. These peptides are then loaded onto the MHC I molecule, a crucial step for the presentation of antigens. Once the peptide is successfully loaded onto the MHC I molecule, the complete complex (consisting of the MHC I heavy chain, β2-microglobulin, and the peptide) detaches from the peptide-loading complex. This mature MHC I molecule is then transported to the cell surface for antigen presentation [31, 63]. The integration of antigenic glycopeptides onto peptide moieties and glycans determines the specificity of T cells [64]. Monosaccharide and disaccharide modifications of glycopeptides are recognized by $CD8^+$ and $CD4^+$ T cells through the classical MHC pathway [65, 66].

### 10.8.4 Glycans and Drug Resistance

Protein glycosylation on cell surfaces plays a multifaceted role in cellular functions and has profound implications in disease pathology, including the development of multidrug resistance in cancer [67]. In the context of acute myeloid leukemia (AML), research has illuminated how inflammatory mediators released by AML cells, such as TNF-α and granulocyte colony-stimulating factor (G-CSF), can influence the glycosylation patterns of cell-surface ligands. This modification of glycosylation patterns is particularly significant as it affects how AML cells interact with their environment and respond to therapeutic agents.

TNF-α and G-CSF are cytokines that can have various effects on tumor cells and the tumor microenvironment. In AML, these cytokines are released in response to leukemic activity and cellular stress. These mediators modulate the activity of enzymes such as sialyltransferases and fucosyltransferases, which are responsible for the addition of sialic acid and fucose residues, respectively, to glycan chains on proteins [68, 69]. The alteration of cell-surface glycosylation can affect the stability and availability of drug targets, as well as receptor signaling pathways, leading to reduced efficacy of chemotherapeutic agents. For instance, increased sialylation has been linked to reduced recognition and elimination by immune cells, as well as altered drug uptake. These glycosylation changes may also affect the physical properties of the cell surface, such as permeability and receptor accessibility, which can directly impact the ability of drugs to enter cells and reach their targets [70], potentially causing chemotherapy resistance by modifying the endothelial niche microenvironment and activating the AKT/NF-κB/mTOR pathway [71].

The expression of related glycosylases is influenced by DNA methylation and histone modification [72, 73]. CD63, a member of the tetraspanin family, plays a significant role in various cellular processes including signaling, adhesion, and cellular trafficking. Its function and localization are heavily influenced by its glycosylation status, particularly N-glycosylation, which has profound implications in the context of cancer, notably breast cancer. For CD63, this modification can dictate its cellular localization and functional interactions. Glycosylated CD63 is typically directed toward the plasma membrane, where it interacts with the other membrane proteins and signaling molecules. On the cell surface, glycosylated CD63 can interact with key molecules involved in cancer progression. It links receptor tyrosine kinases and integrins with Src family kinases. These signaling pathways are crucial for cell migration, invasion, and survival. These interactions can enhance the malignancy of cancer cells by promoting invasive behavior and contributing to the development of drug resistance, crucial traits for cancer metastasis, and treatment failure. In contrast, non-glycosylated CD63 tends to remain associated with the endosomal system. This localization restricts its ability to interact with surface receptors and kinases, thereby limiting its potential to promote malignancy. The retention of CD63 in endosomes can prevent it from contributing to the aggressive behaviors typically associated with cancer cells, such as enhanced mobility and evasion of therapeutic agents [74, 75].

N-glycosylation plays a crucial regulatory role in the functionality and stability of key proteins involved in cell signaling and drug resistance, such as the epidermal growth factor receptor (EGFR) and P-glycoprotein. These proteins, when glycosylated, exhibit enhanced stability, altered cellular localization, and increased functionality that contribute significantly to cancer cell survival and multidrug resistance. EGFR, a critical receptor tyrosine kinase involved in regulating cell growth, survival, proliferation, and differentiation, is heavily N-glycosylated in many cancers. These glycan modifications regulate its trafficking, stability, and localization on the cell surface. N-glycans on EGFR can affect receptor conformation and accessibility, impacting the binding of ligands and efficacy of EGFR-targeted therapies, such as tyrosine kinase inhibitors (TKIs). Glycosylation can shield specific epitopes on EGFR from being recognized by therapeutic antibodies or inhibit the proper binding of TKIs, contributing to a drug-resistant phenotype in cancer cells. P-glycoprotein, a well-known efflux transporter associated with multidrug resistance, pumps a variety of anticancer drugs out of cells, reducing their efficacy. The stability and function of P-glycoprotein are significantly enhanced by N-glycosylation. Glycosylated P-glycoprotein is more stable than its non-glycosylated form, predominantly due to its robust anchoring in the cell membrane. This stability enhances its drug efflux capacity, thereby contributing to the multidrug resistance phenotype observed in many tumors [76–79].

## 10.9 GlycoRNA: Direct Linking of Glycans to RNA

The recent discovery of glycosylated ncRNAs, termed glycoRNAs, marks a significant advancement in the fields of RNA biology and glycobiology. This novel class of ncRNAs is characterized by the presence of N-glycans, particularly those that are highly enriched in sialic acid and fucose. Found on the cell surface, glycoRNAs represent an emerging paradigm where RNA molecules are directly involved in cell-surface biochemistry and signaling, a role traditionally attributed to proteins and lipids. The synthesis of glycoRNAs utilizes the conventional N-glycan biosynthetic pathway, which is primarily located in the endoplasmic reticulum and Golgi apparatus. This pathway, typically associated with protein glycosylation, involves a series of enzymatic reactions that assemble complex glycan structures. In the case of glycoRNAs, this machinery facilitates the addition of sialic acid and fucose, sugars known for their roles in cellular recognition and signaling due to their terminal positions in glycan chains.

On the cell surface, glycoRNAs interact with specific biological molecules, including anti-dsRNA antibodies and members of the Siglec family, which are sialic acid-binding immunoglobulin-type lectins. These interactions suggest potential roles in immune modulation, cellular communication, and perhaps even viral defense mechanisms, given the involvement of dsRNA and Siglec pathways in these processes. The glycoRNAs identified span several classes of small ncRNAs, including YRNA, snRNA, rRNA, snoRNA, and tRNA, with Y5 RNA being notably prevalent among the glycoRNAs analyzed. This suggests a selective process in which certain RNA transcripts are more likely to be glycosylated, hinting at a functional specificity that remains to be fully understood. This breakthrough expands our understanding of glycosylation, traditionally viewed as a modification for proteins and lipids, to include RNA—introducing an entirely new layer of complexity and functionality to ncRNAs. It opens new avenues for research into how glycosylation affects RNA stability, localization, and interactions with other cellular components, which could have profound implications for understanding cellular processes and developing new therapeutic approaches targeting RNA molecules directly [11].

Y5 RNA, a key member of the human Y ncRNA family, is distinguished by a highly conserved stem loop structure [80, 81]. This RNA is synthesized by RNA polymerase III [82], and it has a propensity to associate with specific proteins such as the Ro60 protein and La protein, or their corresponding orthologs. This interaction occurs in a loop-dependent manner, which is crucial for its function and stability [83, 84]. Predominantly residing in the nucleus [85, 86], Y5 RNA can be transported to the cytoplasm under certain conditions. This cytoplasmic translocation can be initiated through a process of Ro60-dependent nuclear export, influenced by the lack of La protein binding at the RNA's 3′-end, or through specific trimming events of the RNA itself [87, 88]. The dynamic localization of Y5 RNA contrasts with other glycoRNA transcripts such as snRNAs and snoRNAs, which typically localize to the cytoplasm and nucleus, respectively. Flynn et al. have made significant findings regarding the association of glycoRNAs with the cellular membrane, rather than being free-floating in the soluble cytosol or confined within the nucleus. This suggests that the glycan components attached to these RNA transcripts may play a crucial role in determining their cellular localization. The presence of glycans might also drive specific functional changes in these RNAs, leading to alterations in their biological roles and interactions. This observation opens up new avenues for research into how glycosylation affects RNA function beyond mere localization, potentially impacting RNA stability, protein interactions, and the overall cellular response to environmental stimuli [11].

Research indicates that the glycan structures on glycoRNAs are structurally analogous to those found on proteins, suggesting a shared biosynthetic pathway and similar enzymatic processes. Therefore, examining the glycosyltransferases that facilitate protein glycosylation could provide valuable insights into the relationships between RNA molecules and glycoRNA-associated glycans. Specifically, glycoRNAs have been characterized by the presence of N-glycans, which are exclusively synthesized by N-glycosylation enzymes [30, 89, 90].

Experimental observations have demonstrated that the presence of glycoRNAs can be significantly reduced following treatment with endoglycosidases. These enzymes are highly selective for N-glycans and can effectively diminish glycoRNA labeling. Furthermore, the use of small molecule inhibitors that target N-glycan trimming enzymes also results in a dose-dependent decrease in glycoRNA labeling. In contrast, enzymes that are less specific to N-glycans, such as those that digest O-glycans or mucinases, tend to have only a partial or no effect on glycoRNA labeling. This specificity underscores the critical role of N-glycosylation enzymes in the biosynthesis and stability of glycoRNA complexes, which are crucial in both RNA and protein domains. Moreover, the genes encoding these glycosyltransferases are subject to epigenetic regulation, including DNA methylation and histone acetylation. Such epigenetic modifications lead to the production of specific N-glycans, which are vital for proper protein function within the proteome. Similarly, DNA methylation influences the subclasses of immunoglobulin G (IgG) by modulating the synthesis of IgG glycans [91]. This connection between glycosylation and epigenetic regulation extends beyond proteins to potentially influence glycoRNA biosynthesis. Given the structural homology between the glycans on glycoRNAs and proteins, it is plausible to hypothesize that DNA methylation and histone acetylation could similarly regulate the synthesis of glycoRNAs.

In a groundbreaking study, researchers utilized a sophisticated chemical tool known as peracetylated N-azidoacetylmannosamine (Ac4ManNAz), an azidosugar precursor of sialic acid, to specifically label living cells. This approach revealed that azide reactivity, indicative of the presence of this modified sugar, was primarily concentrated in highly purified RNA preparations rather than in other cellular lysates [11]. This finding suggested an intriguing specificity of glycan modifications in RNA molecules. The research team had previously developed and refined methods that employed metabolic labeling and bioorthogonal chemistry to explore glycan structures associated with proteins [33, 92, 93]. These techniques are founded on the principle that azido sugars—sugar molecules modified with an azide group—can be incorporated into the glycan components of cellular biomolecules. Once incorporated, these azide-modified sugars engage in bioorthogonal reactions, which are highly selective chemical reactions that occur under physiological conditions without interfering with native biochemical processes. Typically, these reactions involve a biotin probe that reacts with the azide group to form a stable covalent bond, enabling the subsequent detection and analysis of the labeled molecules in cells or even whole organisms. Building on this methodology, the team hypothesized that if azido sugars could be incorporated into protein-associated glycans, they might also integrate into RNA-associated glycans. The concentration of azide reactivity in RNA preparations, as opposed to other parts of the cell lysates, supported this hypothesis. It suggested a potential interaction between RNA molecules and cellular glycans that could be analogous to the known glycan modifications in proteins. This speculation opens up a new dimension in the study of RNA biology, hinting at the possibility that RNA, like proteins, may undergo glycosylation. This process could have profound implications for our understanding of RNA function, stability, and cellular localization. The use of bioorthogonal chemistry not only provides a powerful tool for investigating these potential glycosylation events in RNA but also paves the way for future studies to unravel the biological significance and mechanistic details of RNA–glycan interactions. They confirmed that RNA, but not other nucleotides such as DNA, extracted from labeled cells exhibited biotin reactivity, which was reversed by RNase treatment. This finding indicates that glycoRNAs, decorated with N-glycans, are highly fucosylated and sialylated. Since the study focused on labeling one glycan (sialic acid), the RNAs modified with sialoglycans might represent only one class of glycoRNAs. It remains to be determined whether all types of glycans, including those decorated with sialic acid and other glycol forms, can be conjugated to RNAs. Further research is needed to understand how RNA templates conjugate with carbohydrates and why RNA localizes to the cell membrane.

## 10.10 Glycan-RNA Vs. Glycosylation-Associated ncRNA

Prior to the discovery of glycan-RNAs, the understanding of the interplay between RNA and glycans was primarily focused on how ncRNAs influence protein glycosylation and the activity of glycosyltransferases. These enzymes play a crucial role in remodeling glycans, which in turn impacts various cellular activities. Protein glycosylation, recognized as a pivotal post-translational modification, has far-reaching effects on protein characteristics, including folding, localization, stability, and overall function. The modifications orchestrated by glycosylation are diverse, ranging from the addition of simple monosaccharides to nuclear transcription factors, to more complex, branched polysaccharide structures on cell-surface receptors, involving a wide array of sugar types being added to proteins. A specific instance illustrating the biological importance of these glycosyltransferases involves N-acetylgalactosamine transferase (GALNT), which catalyzes the initial step in mucin-type O-linked glycosylation. The expression and activity of this enzyme on the cell surface have been linked to various metabolic disorders and cancers [94].

A notable study by Li et al. [95] highlighted the intricate molecular mechanisms where the lncRNA SNHG7 interferes with miR-34a activity. SNHG7 acts as a ceRNA, effectively "sponging" miR-34a and preventing it from binding to and inhibiting GALNT. This interaction results in elevated levels of GALNT expression in cancerous tissues. The increased presence of this glycosyltransferase leads to enhanced cancer cell proliferation, invasion, and metastasis, driven by the aberrant patterns of O-glycosylation [96].

These findings reveal the intricate regulatory networks governing protein glycosylation and emphasize the crucial role of ncRNAs in influencing these pathways. By modulating the activity of glycosyltransferases, ncRNAs like SNHG7 can profoundly impact cellular behaviors and the progression of diseases, particularly cancer. This modulation is critical as specific fucosyltransferases (FUTs), which catalyze fucosylation processes, play pivotal roles in cancer's malignant characteristics. For instance, Xu et al. [97] demonstrated that exosome-derived lncRNA MALAT1 competes for miR-26a/26b binding sites, thereby upregulating FUT4 expression and enhancing fucosylation levels, which in turn promotes metastasis in colorectal cancer. Similarly, the lncRNA HOTAIR, known for its association with poor clinical outcomes, effectively "sponges" miR-326. This interaction leads to altered FUT levels and modifies the fucosylation of the E-selectin ligand, impacting cellular adhesion and metastasis [98]. Moreover, various microRNAs have been shown to regulate other members of the FUT family, thereby influencing the overall production of glycans [99, 100].

Sialylation plays a critical role in modulating cellular functions and signaling pathways. This modification is intricately regulated by specific enzymes known as sialyltransferases, with ST6GAL1 being a prominent example. The regulation and activity of ST6GAL1 are influenced by complex molecular interactions, such as those involving the ZFAS1/miR-150/ST6GAL1 axis, which significantly impacts cellular behavior including drug resistance mechanisms. ST6GAL1 is a sialyltransferase that catalyzes the addition of sialic acid to N-linked glycans on proteins. The activity and expression of ST6GAL1 are crucial for the sialylation processes that affect protein function and cell signaling. Through the ZFAS1/miR-150 axis, the upregulation of ST6GAL1 enhances the sialylation of specific cellular proteins, thereby modulating their activity and stability on the cell surface [101]. The interplay between miRNAs, lncRNAs, and sialyltransferases highlights a complex regulatory network that influences cancer progression and malignancy. Specifically, the interactions involving miR-214, the lncRNA HOTAIR, and the sialyltransferase ST6GAL1 underscore how RNA molecules can regulate enzyme expression and activity, ultimately affecting cell signaling pathways and cancer phenotypes. In the context of colorectal cancer, this regulatory axis has significant implications for disease progression and potential therapeutic targets [102]. This complex interplay also extends to the sialyltransferase family's role in altering glycosylation patterns, which, when interacted with by miRNAs, significantly affects breast cancer progression [99]. Table 10.2 summarizes the distinct yet interconnected roles that glycosylation-associated ncRNAs and glycoRNAs play in the broader context of cellular biology and their emerging significance in medical research.

**Table 10.2** Differences between glycosylation-associated ncRNAs vs. glycan-RNA

| Feature | Glycosylation-associated ncRNAs | Glycan-RNA |
|---|---|---|
| Definition | ncRNAs that are involved in the glycosylation process but are not directly glycosylated themselves. These ncRNAs regulate or influence the enzymes and pathways involved in glycosylation | RNAs that have glycans covalently attached to them, effectively making them a direct part of the glycosylation landscape |
| Role in glycosylation | Indirect involvement by regulating glycosylation enzymes and processes. These ncRNAs may affect the expression, stability, or activity of glycosyltransferases and other enzymes involved in the glycosylation pathway | Direct involvement as substrates of glycosylation. They participate actively in cellular processes by presenting glycosylated forms on the cell surface or within the cell |
| Cellular functions | Mainly regulatory; they can control the levels and activity of glycosylating enzymes, thereby influencing the glycosylation patterns and efficiencies of proteins and lipids | Diverse, ranging from cellular signaling to interactions with other cells and the environment, due to the presence of specific glycan structures |
| Examples of functions | Modulating the expression of genes involved in glycosylation, impacting glycan structures on proteins and lipids indirectly | Binding to specific receptors like Siglec receptors, participating in cell–cell recognition, and potentially modulating immune responses |
| Impact on research and medicine | Understanding these ncRNAs can lead to insights into the regulation of glycosylation, with implications for diseases where glycosylation is aberrant, such as cancer and autoimmune diseases | The study of glycoRNAs can expand our understanding of the glycome and its implications in health and disease, offering potential new targets for therapeutic intervention |
| Therapeutic potential | Potential targets for altering glycosylation patterns therapeutically, which could be beneficial in conditions where glycosylation is dysregulated | Potential for the development of biomarkers and glycan-based therapeutics, including novel approaches to target cellular interactions and pathways |

## 10.11 Future Prospects

This newly discovered glycoRNAs represent a distinct category from glycosylation-associated ncRNAs. While glycosylation-associated ncRNAs regulate the expression of glycan-related proteins and modify glycosylation patterns, thereby contributing to disease development, the sugars in these cases are not directly linked to ncRNAs but rather interact through related proteins. In contrast, glycoRNAs are ncRNAs directly linked with glycans, carrying genetic information and present on the cell surface, which sets them apart from glycosylation-associated ncRNAs. Traditionally, glycans were known to decorate proteins and lipids, enabling cells to generate diverse molecular forms from the same DNA blueprint. However, new research by Flynn et al. has revealed that RNA can also be glycosylated, with these modified nucleic acids found on cell surfaces. This groundbreaking discovery is the first of its kind in the RNA field. GlycoRNAs are located on the cell membrane's exterior, where they bind to Siglecs, a family of immune receptors linked to various diseases, such as systemic lupus erythematosus. Given their interaction with anti-dsRNA antibodies and Siglec receptors, glycoRNAs may become novel serum markers for rapid disease diagnosis. Moreover, RNA modified with glycans might be sensitive to immunotherapy drugs [67], serving as potential targets to enhance drug response and concentration in lesions. Therapeutic strategies like immunotherapy, radiotherapy, and chemotherapy could alter the genetic information or glycan structures of glycoRNAs on the cell surface, thus modifying the lesion microenvironment. GlycoRNAs hold promise as a new class of immune response signal receivers in drug therapy, extending beyond traditional protein receptors [103].

## 10.12 Conclusion

The discovery of glycoRNAs represents a paradigm shift in our understanding of molecular biology, particularly in the areas of glycosylation and RNA biology. This exciting development challenges the traditional view that glycosylation only pertains to proteins and lipids, highlighting RNA as a novel platform for glycosylation. Contrary to the typical intracellular localization of most RNAs, glycoRNAs are displayed on the cell surface. This unique positioning allows them to interact directly with cell-surface receptors, such as Siglecs (sialic-acid-binding immunoglobulin-type lectins), which are critical in mediating cell–cell interactions and signaling pathways. The potential for glycoRNAs to be further decorated with N-glycans adds another layer of complexity and suggests a significant regulatory capacity. This capability could influence various physiological and pathological processes, including immune responses and cancer cell interactions. The integration of glycoRNA studies into existing glycomics and broader omics workflows promises to expand our understanding of glycosylation beyond the conventional protein-centric view. This approach could unveil how glycoRNAs contribute to the overall glycan dynamics within the cell and impact cellular communication and disease pathology. Understanding the roles of glycoRNAs could lead to novel diagnostic markers for diseases where glycosylation patterns play a role, such as cancer, autoimmune diseases, and infectious diseases. The distinct glycosylation patterns of glycoRNAs could reflect specific disease states or stages. GlycoRNAs present new targets for therapeutic intervention. The development of glycoRNA mimetics or inhibitors could provide new modalities for treating diseases that are influenced by cellular glycosylation patterns. Moreover, targeting the interactions between glycoRNAs and Siglec receptors could modulate immune responses or inhibit pathological cell interactions in diseases like cancer. The potential to harness or mimic the biological functions of glycoRNAs opens up avenues for both natural therapeutics and the design of synthetic glycoRNA-based drugs. These therapies could specifically target glycan–receptor interactions, offering a high degree of specificity and reduced side effects compared to broader-acting drugs. The discovery of glycoRNAs significantly broadens the scope of glycoscience, merging it with RNA biology and offering profound new insights into cell biology and the molecular basis of diseases. As research progresses, the detailed understanding of glycoRNA structures, functions, and their interactions will undoubtedly influence future scientific directions and therapeutic innovations.

## References

1. Dunin-Horkawicz S, Czerwoniec A, Gajda MJ, Feder M, Grosjean H, Bujnicki JM. MODOMICS: a database of RNA modification pathways. Nucleic Acids Res. 2006;34(Database issue):D145–9.
2. Arango D, Sturgill D, Alhusaini N, Dillman AA, Sweet TJ, Hanson G, et al. Acetylation of cytidine in mRNA promotes translation efficiency. Cell. 2018;175(7):1872–86.e24.
3. Roundtree IA, Evans ME, Pan T, He C. Dynamic RNA modifications in gene expression regulation. Cell. 2017;169(7):1187–200.
4. Li X, Xiong X, Yi C. Epitranscriptome sequencing technologies: decoding RNA modifications. Nat Methods. 2016;14(1):23–31.
5. Hoernes TP, Clementi N, Faserl K, Glasner H, Breuker K, Lindner H, et al. Nucleotide modifications within bacterial messenger RNAs regulate their translation and are able to rewire the genetic code. Nucleic Acids Res. 2016;44(2):852–62.
6. Yildirim I, Kierzek E, Kierzek R, Schatz GC. Interplay of LNA and 2′-O-methyl RNA in the structure and thermodynamics of RNA hybrid systems: a molecular dynamics study using the revised AMBER force field and comparison with experimental results. J Phys Chem B. 2014;118(49):14177–87.
7. Charette M, Gray MW. Pseudouridine in RNA: what, where, how, and why. IUBMB Life. 2000;49(5):341–51.
8. Squires JE, Patel HR, Nousch M, Sibbritt T, Humphreys DT, Parker BJ, et al. Widespread occurrence of 5-methylcytosine in human coding and non-coding RNA. Nucleic Acids Res. 2012;40(11):5023–33.
9. Li X, Xiong X, Wang K, Wang L, Shu X, Ma S, et al. Transcriptome-wide mapping reveals reversible and dynamic N(1)-methyladenosine methylome. Nat Chem Biol. 2016;12(5):311–6.
10. Dominissini D, Nachtergaele S, Moshitch-Moshkovitz S, Peer E, Kol N, Ben-Haim MS, et al. The dynamic N(1)-methyladenosine methylome in eukaryotic messenger RNA. Nature. 2016;530(7591):441–6.
11. Spitale RC, Flynn RA, Zhang QC, Crisalli P, Lee B, Jung JW, et al. Erratum: Structural imprints in vivo decode RNA regulatory mechanisms. Nature. 2015;527(7577):264.
12. Boccaletto P, Machnicka MA, Purta E, Piatkowski P, Baginski B, Wirecki TK, et al. MODOMICS: a database of RNA modification pathways. 2017 update. Nucleic Acids Res. 2018;46(D1):D303–7.
13. Flynn RA, Pedram K, Malaker SA, Batista PJ, Smith BAH, Johnson AG, et al. Small RNAs are modified with N-glycans and displayed on the surface of living cells. Cell. 2021;184(12):3109–24.
14. Scheper AF, Schofield J, Bohara R, Ritter T, Pandit A. Understanding glycosylation: regulation through the metabolic flux of precursor pathways. Biotechnol Adv. 2023;67:108184.
15. Eichler J. Protein glycosylation. Curr Biol. 2019;29(7):R229–31.
16. An HJ, Froehlich JW, Lebrilla CB. Determination of glycosylation sites and site-specific heterogeneity in glycoproteins. Curr Opin Chem Biol. 2009;13(4):421–6.

17. Schjoldager KT, Narimatsu Y, Joshi HJ, Clausen H. Global view of human protein glycosylation pathways and functions. Nat Rev Mol Cell Biol. 2020;21(12):729–49.
18. Reily C, Stewart TJ, Renfrow MB, Novak J. Glycosylation in health and disease. Nat Rev Nephrol. 2019;15(6):346–66.
19. Stanley P. Golgi glycosylation. Cold Spring Harb Perspect Biol. 2011;3(4):a005199.
20. Wacker M, Linton D, Hitchen PG, Nita-Lazar M, Haslam SM, North SJ, et al. N-linked glycosylation in Campylobacter jejuni and its functional transfer into E. coli. Science. 2002;298(5599):1790–3.
21. Dove A. The bittersweet promise of glycobiology. Nat Biotechnol. 2001;19(10):913–7.
22. Rudd PM, Elliott T, Cresswell P, Wilson IA, Dwek RA. Glycosylation and the immune system. Science. 2001;291(5512):2370–6.
23. Merry CLR, Lindahl U, Couchman J, Esko JD. Proteoglycans and sulfated glycosaminoglycans. In: Varki A, Cummings RD, Esko JD, Stanley P, Hart GW, Aebi M, et al., editors. Essentials of glycobiology. 4th ed. Cold Spring Harbor: Cold Spring Harbor Laboratory Press; 2022. p. 217–32.
24. Lindahl U, Couchman J, Kimata K, Esko JD. Proteoglycans and sulfated glycosaminoglycans. In: Varki A, Cummings RD, Esko JD, Stanley P, Hart GW, Aebi M, et al., editors. Essentials of glycobiology. 3rd ed. Cold Spring Harbor: Cold Spring Harbor Laboratory Press; 2015. p. 207–21.
25. Esko JD, Kimata K, Lindahl U. Proteoglycans and sulfated glycosaminoglycans. In: Varki A, Cummings RD, Esko JD, Freeze HH, Stanley P, Bertozzi CR, et al., editors. Essentials of glycobiology. 2nd ed. Cold Spring Harbor: Cold Spring Harbor Laboratory Press; 2009.
26. Schnaar RL, Sandhoff R, Tiemeyer M, Kinoshita T. Glycosphingolipids. In: Varki A, Cummings RD, Esko JD, Stanley P, Hart GW, Aebi M, et al., editors. Essentials of glycobiology. 4th ed. Cold Spring Harbor: Cold Spring Harbor Laboratory Press; 2022. p. 129–40.
27. Fuster MM, Esko JD. The sweet and sour of cancer: glycans as novel therapeutic targets. Nat Rev Cancer. 2005;5(7):526–42.
28. Spiro RG. Protein glycosylation: nature, distribution, enzymatic formation, and disease implications of glycopeptide bonds. Glycobiology. 2002;12(4):43–56.
29. Varki A. Biological roles of oligosaccharides: all of the theories are correct. Glycobiology. 1993;3(2):97–130.
30. Anugraham M, Jacob F, Nixdorf S, Everest-Dass AV, Heinzelmann-Schwarz V, Packer NH. Specific glycosylation of membrane proteins in epithelial ovarian cancer cell lines: glycan structures reflect gene expression and DNA methylation status. Mol Cell Proteomics. 2014;13(9):2213–32.
31. van Leeuwen JE, Kearse KP. Deglucosylation of N-linked glycans is an important step in the dissociation of calreticulin-class I-TAP complexes. Proc Natl Acad Sci USA. 1996;93(24):13997–4001.
32. Zhu HJ, Liu D, Tran VP, Wu Z, Jiang K, Zhu H, et al. N-Linked glycosylation prevents deamidation of glycopeptide and glycoprotein. ACS Chem Biol. 2020;15(12):3197–205.
33. Hang HC, Yu C, Kato DL, Bertozzi CR. A metabolic labeling approach toward proteomic analysis of mucin-type O-linked glycosylation. Proc Natl Acad Sci USA. 2003;100(25):14846–51.
34. Van den Steen P, Rudd PM, Dwek RA, Opdenakker G. Concepts and principles of O-linked glycosylation. Crit Rev Biochem Mol Biol. 1998;33(3):151–208.
35. Minakata S, Manabe S, Inai Y, Ikezaki M, Nishitsuji K, Ito Y, et al. Protein C-mannosylation and C-mannosyl tryptophan in chemical biology and medicine. Molecules. 2021;26(17):5258.
36. Ihara Y, Inai Y, Ikezaki M, Matsui I-SL, Manabe S, Ito Y. C-Mannosylation: a modification on tryptophan in cellular proteins. In: Endo T, Seeberger PH, Hart GW, Wong C-H, Taniguchi N, editors. Glycoscience: biology and medicine. Tokyo: Springer; 2021. p. 1–8.
37. Kinoshita T. Glycosylphosphatidylinositol (GPI) anchors: biochemistry and cell biology: introduction to a thematic review series. J Lipid Res. 2016;57(1):4–5.
38. D'Angelo G, Capasso S, Sticco L, Russo D. Glycosphingolipids: synthesis and functions. FEBS J. 2013;280(24):6338–53.

39. Zhang T, de Waard AA, Wuhrer M, Spaapen RM. The role of glycosphingolipids in immune cell functions. Front Immunol. 2019;10:90.
40. Gagneux P, Varki A. Evolutionary considerations in relating oligosaccharide diversity to biological function. Glycobiology. 1999;9(8):747–55.
41. He M, Zhou X, Wang X. Glycosylation: mechanisms, biological functions and clinical implications. Signal Transduct Target Ther. 2024;9(1):194.
42. Arora K, Sherilraj PM, Abutwaibe KA, Dhruw B, Mudavath SL. Exploring glycans as vital biological macromolecules: a comprehensive review of advancements in biomedical frontiers. Int J Biol Macromol. 2024;268:131511.
43. Varki A, Cummings RD, Esko JD, Stanley P, Hart GW, Aebi M, et al., editors. Essentials of glycobiology. 4th ed. Cold Spring Harbor: Cold Spring Harbor Laboratory Press; 2022.
44. Varki A, Cummings RD, Esko JD, Stanley P, Hart GW, Aebi M, et al., editors. Essentials of glycobiology. 3rd ed. Cold Spring Harbor: Cold Spring Harbor Laboratory Press; 2015.
45. Varki A, Cummings RD, Esko JD, Freeze HH, Stanley P, Bertozzi CR, et al., editors. Essentials of glycobiology. 2nd ed. Cold Spring Harbor: Cold Spring Harbor Laboratory Press; 2009.
46. Brooks SA. Appropriate glycosylation of recombinant proteins for human use: implications of choice of expression system. Mol Biotechnol. 2004;28(3):241–55.
47. Shental-Bechor D, Levy Y. Effect of glycosylation on protein folding: a close look at thermodynamic stabilization. Proc Natl Acad Sci USA. 2008;105(24):8256–61.
48. Shental-Bechor D, Levy Y. Folding of glycoproteins: toward understanding the biophysics of the glycosylation code. Curr Opin Struct Biol. 2009;19(5):524–33.
49. McFadden PN, Clarke S. Conversion of isoaspartyl peptides to normal peptides: implications for the cellular repair of damaged proteins. Proc Natl Acad Sci USA. 1987;84(9):2595–9.
50. Tyler-Cross R, Schirch V. Effects of amino acid sequence, buffers, and ionic strength on the rate and mechanism of deamidation of asparagine residues in small peptides. J Biol Chem. 1991;266(33):22549–56.
51. Brodsky JL, Werner ED, Dubas ME, Goeckeler JL, Kruse KB, McCracken AA. The requirement for molecular chaperones during endoplasmic reticulum-associated protein degradation demonstrates that protein export and import are mechanistically distinct. J Biol Chem. 1999;274(6):3453–60.
52. Werner ED, Brodsky JL, McCracken AA. Proteasome-dependent endoplasmic reticulum-associated protein degradation: an unconventional route to a familiar fate. Proc Natl Acad Sci USA. 1996;93(24):13797–801.
53. Sanchez-Ruderisch H, Detjen KM, Welzel M, André S, Fischer C, Gabius HJ, et al. Galectin-1 sensitizes carcinoma cells to anoikis via the fibronectin receptor α5β1-integrin. Cell Death Differ. 2011;18(5):806–16.
54. Tanigaki K, Sacharidou A, Peng J, Chambliss KL, Yuhanna IS, Ghosh D, et al. Hyposialylated IgG activates endothelial IgG receptor FcγRIIB to promote obesity-induced insulin resistance. J Clin Invest. 2018;128(1):309–22.
55. Tangemann K, Bistrup A, Hemmerich S, Rosen SD. Sulfation of a high endothelial venule-expressed ligand for L-selectin. Effects on tethering and rolling of lymphocytes. J Exp Med. 1999;190(7):935–42.
56. Jajosky RP, Wu SC, Zheng L, Jajosky AN, Jajosky PG, Josephson CD, et al. ABO blood group antigens and differential glycan expression: perspective on the evolution of common human enzyme deficiencies. iScience. 2023;26(1):105798.
57. Varki A. Sialic acids in human health and disease. Trends Mol Med. 2008;14(8):351–60.
58. Li J, Hsu HC, Mountz JD, Allen JG. Unmasking fucosylation: from cell adhesion to immune system regulation and diseases. Cell Chem Biol. 2018;25(5):499–512.
59. Baum LG, Cobb BA. The direct and indirect effects of glycans on immune function. Glycobiology. 2017;27(7):619–24.
60. Linden SK, Sutton P, Karlsson NG, Korolik V, McGuckin MA. Mucins in the mucosal barrier to infection. Mucosal Immunol. 2008;1(3):183–97.

61. Vitiazeva V, Kattla JJ, Flowers SA, Lindén SK, Premaratne P, Weijdegård B, et al. The O-linked glycome and blood group antigens ABO on mucin-type glycoproteins in mucinous and serous epithelial ovarian tumors. PLoS One. 2015;10(6):e0130197.
62. Lowe JB, Marth JD. A genetic approach to mammalian glycan function. Annu Rev Biochem. 2003;72:643–91.
63. Sadasivan B, Lehner PJ, Ortmann B, Spies T, Cresswell P. Roles for calreticulin and a novel glycoprotein, tapasin, in the interaction of MHC class I molecules with TAP. Immunity. 1996;5(2):103–14.
64. Glithero A, Tormo J, Haurum JS, Arsequell G, Valencia G, Edwards J, et al. Crystal structures of two H-2Db/glycopeptide complexes suggest a molecular basis for CTL cross-reactivity. Immunity. 1999;10(1):63–74.
65. Abdel-Motal UM, Berg L, Rosén A, Bengtsson M, Thorpe CJ, Kihlberg J, et al. Immunization with glycosylated Kb-binding peptides generates carbohydrate-specific, unrestricted cytotoxic T cells. Eur J Immunol. 1996;26(3):544–51.
66. Hudrisier D, Riond J, Mazarguil H, Oldstone MB, Gairin JE. Genetically encoded and post-translationally modified forms of a major histocompatibility complex class I-restricted antigen bearing a glycosylation motif are independently processed and co-presented to cytotoxic T lymphocytes. J Biol Chem. 1999;274(51):36274–80.
67. Ma H, Zhou H, Song X, Shi S, Zhang J, Jia L. Modification of sialylation is associated with multidrug resistance in human acute myeloid leukemia. Oncogene. 2015;34(6):726–40.
68. Dagia NM, Gadhoum SZ, Knoblauch CA, Spencer JA, Zamiri P, Lin CP, et al. G-CSF induces E-selectin ligand expression on human myeloid cells. Nat Med. 2006;12(10):1185–90.
69. Delmotte P, Degroote S, Lafitte JJ, Lamblin G, Perini JM, Roussel P. Tumor necrosis factor alpha increases the expression of glycosyltransferases and sulfotransferases responsible for the biosynthesis of sialylated and/or sulfated Lewis x epitopes in the human bronchial mucosa. J Biol Chem. 2002;277(1):424–31.
70. Mondal N, Buffone A, Stolfa G, Antonopoulos A, Lau JT, Haslam SM, et al. ST3Gal-4 is the primary sialyltransferase regulating the synthesis of E-, P-, and L-selectin ligands on human myeloid leukocytes. Blood. 2015;125(4):687–96.
71. Barbier V, Erbani J, Fiveash C, Davies JM, Tay J, Tallack MR, et al. Endothelial E-selectin inhibition improves acute myeloid leukaemia therapy by disrupting vascular niche-mediated chemoresistance. Nat Commun. 2020;11(1):2042.
72. Pink M, Ratsch BA, Mardahl M, Durek P, Polansky JK, Karl M, et al. Imprinting of skin/inflammation homing in CD4+ T cells is controlled by DNA methylation within the fucosyltransferase 7 gene. J Immunol. 2016;197(8):3406–14.
73. Syrbe U, Jennrich S, Schottelius A, Richter A, Radbruch A, Hamann A. Differential regulation of P-selectin ligand expression in naive versus memory CD4+ T cells: evidence for epigenetic regulation of involved glycosyltransferase genes. Blood. 2004;104(10):3243–8.
74. Berditchevski F, Tolias KF, Wong K, Carpenter CL, Hemler ME. A novel link between integrins, transmembrane-4 superfamily proteins (CD63 and CD81), and phosphatidylinositol 4-kinase. J Biol Chem. 1997;272(5):2595–8.
75. Tominaga N, Hagiwara K, Kosaka N, Honma K, Nakagama H, Ochiya T. RPN2-mediated glycosylation of tetraspanin CD63 regulates breast cancer cell malignancy. Mol Cancer. 2014;13:134.
76. Contessa JN, Bhojani MS, Freeze HH, Ross BD, Rehemtulla A, Lawrence TS. Molecular imaging of N-linked glycosylation suggests glycan biosynthesis is a novel target for cancer therapy. Clin Cancer Res. 2010;16(12):3205–14.
77. Honma K, Iwao-Koizumi K, Takeshita F, Yamamoto Y, Yoshida T, Nishio K, et al. RPN2 gene confers docetaxel resistance in breast cancer. Nat Med. 2008;14(9):939–48.
78. Kramer R, Weber TK, Arceci R, Ramchurren N, Kastrinakis WV, Steele G, et al. Inhibition of N-linked glycosylation of P-glycoprotein by tunicamycin results in a reduced multidrug resistance phenotype. Br J Cancer. 1995;71(4):670–5.

79. Schinkel AH, Kemp S, Dollé M, Rudenko G, Wagenaar E. N-glycosylation and deletion mutants of the human MDR1 P-glycoprotein. J Biol Chem. 1993;268(10):7474–81.
80. Köhn M, Pazaitis N, Hüttelmaier S. Why YRNAs? about versatile RNAs and their functions. Biomol Ther. 2013;3(1):143–56.
81. Shi J, Zhou T, Chen Q. Exploring the expanding universe of small RNAs. Nat Cell Biol. 2022;24(4):415–23.
82. Boire G, Craft J. Human Ro ribonucleoprotein particles: characterization of native structure and stable association with the La polypeptide. J Clin Invest. 1990;85(4):1182–90.
83. O'Brien CA, Harley JB. A subset of hY RNAs is associated with erythrocyte Ro ribonucleoproteins. EMBO J. 1990;9(11):3683–9.
84. Wolin SL, Steitz JA. Genes for two small cytoplasmic Ro RNAs are adjacent and appear to be single-copy in the human genome. Cell. 1983;32(3):735–44.
85. Farris AD, Puvion-Dutilleul F, Puvion E, Harley JB, Lee LA. The ultrastructural localization of 60-kDa Ro protein and human cytoplasmic RNAs: association with novel electron-dense bodies. Proc Natl Acad Sci USA. 1997;94(7):3040–5.
86. Gendron M, Roberge D, Boire G. Heterogeneity of human Ro ribonucleoproteins (RNPS): nuclear retention of Ro RNPS containing the human hY5 RNA in human and mouse cells. Clin Exp Immunol. 2001;125(1):162–8.
87. Chen X, Quinn AM, Wolin SL. Ro ribonucleoproteins contribute to the resistance of Deinococcus radiodurans to ultraviolet irradiation. Genes Dev. 2000;14(7):777–82.
88. Sim S, Weinberg DE, Fuchs G, Choi K, Chung J, Wolin SL. The subcellular distribution of an RNA quality control protein, the Ro autoantigen, is regulated by noncoding Y RNA binding. Mol Biol Cell. 2009;20(5):1555–64.
89. Klasić M, Krištić J, Korać P, Horvat T, Markulin D, Vojta A, et al. DNA hypomethylation upregulates expression of the MGAT3 gene in HepG2 cells and leads to changes in N-glycosylation of secreted glycoproteins. Sci Rep. 2016;6:24363.
90. Horvat T, Mužinić A, Barišić D, Bosnar MH, Zoldoš V. Epigenetic modulation of the HeLa cell membrane N-glycome. Biochim Biophys Acta. 2012;1820(9):1412–9.
91. Wahl A, Kasela S, Carnero-Montoro E, van Iterson M, Štambuk J, Sharma S, et al. IgG glycosylation and DNA methylation are interconnected with smoking. Biochim Biophys Acta Gen Subj. 2018;1862(3):637–48.
92. Saxon E, Bertozzi CR. Cell surface engineering by a modified Staudinger reaction. Science. 2000;287(5460):2007–10.
93. Woo CM, Iavarone AT, Spiciarich DR, Palaniappan KK, Bertozzi CR. Isotope-targeted glycoproteomics (IsoTaG): a mass-independent platform for intact N- and O-glycopeptide discovery and analysis. Nat Methods. 2015;12(6):561–7.
94. Gaziel-Sovran A, Hernando E. miRNA-mediated GALNT modulation of invasion and immune suppression: a sweet deal for metastatic cells. Onco Targets Ther. 2012;1(5):746–8.
95. Li Y, Zeng C, Hu J, Pan Y, Shan Y, Liu B, et al. Long non-coding RNA-SNHG7 acts as a target of miR-34a to increase GALNT7 level and regulate PI3K/Akt/mTOR pathway in colorectal cancer progression. J Hematol Oncol. 2018;11(1):89.
96. Hollingsworth MA, Swanson BJ. Mucins in cancer: protection and control of the cell surface. Nat Rev Cancer. 2004;4(1):45–60.
97. Xu J, Xiao Y, Liu B, Pan S, Liu Q, Shan Y, et al. Exosomal MALAT1 sponges miR-26a/26b to promote the invasion and metastasis of colorectal cancer via FUT4 enhanced fucosylation and PI3K/Akt pathway. J Exp Clin Cancer Res. 2020;39(1):54.
98. Pan S, Liu Y, Liu Q, Xiao Y, Liu B, Ren X, et al. HOTAIR/miR-326/FUT6 axis facilitates colorectal cancer progression through regulating fucosylation of CD44 via PI3K/AKT/mTOR pathway. Biochim Biophys Acta, Mol Cell Res. 2019;1866(5):750–60.
99. Li Y, Sun Z, Liu B, Shan Y, Zhao L, Jia L. Tumor-suppressive miR-26a and miR-26b inhibit cell aggressiveness by regulating FUT4 in colorectal cancer. Cell Death Dis. 2017;8(6):e2892.

100. Liang L, Gao C, Li Y, Sun M, Xu J, Li H, et al. miR-125a-3p/FUT5-FUT6 axis mediates colorectal cancer cell proliferation, migration, invasion and pathological angiogenesis via PI3K-Akt pathway. Cell Death Dis. 2017;8(8):e2968.
101. Liu Q, Ma H, Sun X, Liu B, Xiao Y, Pan S, et al. Correction to: The regulatory ZFAS1/miR-150/ST6GAL1 crosstalk modulates sialylation of EGFR via PI3K/Akt pathway in T-cell acute lymphoblastic leukemia. J Exp Clin Cancer Res. 2019;38(1):357.
102. Liu B, Liu Q, Pan S, Huang Y, Qi Y, Li S, et al. The HOTAIR/miR-214/ST6GAL1 crosstalk modulates colorectal cancer procession through mediating sialylated c-Met via JAK2/STAT3 cascade. J Exp Clin Cancer Res. 2019;38(1):455.
103. Li X, Loh TJ, Lim JJ, Saw PE, Liao Y. Glycan-RNA: a new class of non-coding RNA. BIO Integration. 2022;3:124–31.

# Chapter 11
# Emerging ncRNAs: piRNAs, snoRNAs, Xist RNAs

**Abstract** The complex regulation of gene expression and cellular processes encompasses not only protein-coding transcripts but also a wide range of noncoding RNAs (ncRNAs). Recent research has highlighted the crucial roles of emerging ncRNA species, including PIWI-RNA, Xist RNA, small nuclear RNA (snRNA), small nucleolar RNA (snoRNA), and transfer RNA-derived small RNA (tsa-RNA). These ncRNAs are increasingly recognized as pivotal regulators of gene expression, influencing epigenetic modifications, transcriptional regulation, RNA processing, and the control of translation. Understanding the roles of these ncRNAs in various biological contexts has provided novel insights into fundamental cellular mechanisms and their dysregulation in disease states. Research has revealed that the misregulation of these ncRNAs is implicated in a broad spectrum of diseases, including cancer, neurodegenerative disorders, and metabolic syndromes. This deeper comprehension not only sheds light on the complex nature of cellular regulation but also highlights potential therapeutic targets for a variety of health conditions. Consequently, the elucidation of their molecular functions and regulatory networks presents promising avenues for the development of innovative therapeutic strategies. Harnessing the therapeutic potential of ncRNAs offers significant promise for precision medicine and the treatment of complex diseases. This chapter provides a comprehensive overview of the emerging landscape of the three emerging ncRNAs, including piRNAs, snoRNAs, and Xist RNAs, emphasizing their roles in health and disease and exploring the vast therapeutic opportunities they present. By delving into the mechanisms and applications of ncRNAs, we highlight their potential to revolutionize the treatment of diseases through targeted and precise molecular interventions.

## 11.1 Piwi-interacting RNA (piRNA)

Piwi-interacting RNA (piRNA), a small RNA species prevalent in the germline of various animal species, plays a crucial role in maintaining genomic integrity and regulating gene expression. In species such as fruit flies and mice, they are essential



P. E. Saw, E. Song, *RNA Therapeutics in Human Diseases*,
https://doi.org/10.1007/978-981-96-3041-7_11

for the preservation of genomic stability through the silencing of transposable elements [1]. piRNA has been identified predominantly in the germline, although in fruit flies they also occur in ovarian follicle cells. Research has also identified the presence of piRNA-like small RNA (pilRNA) in the heads of flies, with more limited occurrences observed in mouse tissues and human NK cells. These pilRNAs, like piRNAs, typically range from 24 to 32 nucleotides in length. Their discovery in these various tissues suggests that piRNA-like molecules may have broader functional roles beyond the germline, potentially contributing to regulatory processes in somatic cells and immune responses [2–10]. The generation of primary piRNA is believed to occur through the fragmentation of longer RNA sequences, enabling them to be associated with specific genomic areas known as piRNA clusters that exhibit pronounced strand specificity. A characteristic feature of primary piRNA is their significant bias toward 5′ uridine [2, 8, 11, 12]. The formation of secondary piRNAs occurs through an amplification process known as the ping-pong mechanism. This intricate process involves the base pairing of the first 10 nucleotides of primary piRNAs with complementary sequences on target transcripts. During this mechanism, primary piRNAs guide the cleavage of target RNAs, producing new piRNA precursors that are then processed into secondary piRNAs [2, 8, 9, 11]. This cycle of cleavage and amplification not only increases the number of piRNAs but also enhances the silencing of transposable elements and other genomic threats. The ping-pong mechanism is essential for maintaining genomic stability, particularly in the germline, by ensuring a robust and adaptive defense against potentially harmful genetic elements [2, 8–11].

In mammals, the male germline is characterized by a high concentration of piRNAs. Unlike in Drosophila, where piRNAs are primarily derived from transposable elements and repetitive sequences, piRNAs in adult mammalian testes predominantly originate from unannotated genomic regions that do not contain these elements. This difference in piRNA origin between species highlights the unique aspects of piRNA biogenesis and function in mammals. In mammalian testes, these unannotated regions give rise to piRNAs that play crucial roles in spermatogenesis, the process of sperm cell development. By arising from these specific genomic areas, mammalian piRNAs contribute to the regulation of gene expression and the maintenance of genomic integrity in the male germline [5, 6]. The potential functional significance of these piRNA, which do not originate from transposable elements, remains unclear. While the majority of studies suggest that piRNAs and Piwi family proteins are predominantly restricted to the germlines of mammals and flies, as well as to fly ovarian follicle cells, recent research has uncovered several dozen pilRNAs across various mouse tissues.

**Box 11.1 piRNA-Like Small RNAs (piLRNAs)**

In their research, Phey et al. conducted deep sequencing of small RNAs extracted from the sciatic nerve axoplasm and discovered a group of small noncoding RNAs (sncRNAs) that resemble mammalian piRNAs in size, referred to as piRNA-like sncRNAs (piLRNAs). There is compelling evidence suggesting that piRNAs expressed in neurons possess the ability to regulate genes near the cell bodies. However, there has been no prior indication of piLRNAs in the distal axons of neurons or any association with their inherent capacity to influence axonal growth in sensory neurons. The study by Phey et al. provided novel insights by showing that at least some of the piLRNAs identified within the sciatic nerve axoplasm exhibit typical piRNA characteristics, such as a preference for uracil at the 5′-end and a 2′-O-methylation at the 3′-end. Further exploration into the interactions of these piLRNAs revealed that they associate with the MIWI protein. Intriguingly, experiments involving the depletion of MIWI protein via siRNA demonstrated an increase in axon growth rates, a reduction in axon retraction post-injury, and enhanced axon regrowth following injury. This suggests that MIWI protein, in conjunction with piLRNAs, plays a significant role in modulating axonal dynamics, particularly in response to injury.

This research offers groundbreaking insights, marking the first evidence of axonal piRNA-like sncRNAs in adult sensory neurons. Adding to this, the team's previous findings indicated that 63 rat miRNAs, identified within the sciatic nerve, exhibited differential expression levels after nerve injury. This suggests a direct involvement in the regenerative response, likely through the regulation of axonal mRNA translation. Drawing parallels to axonal miRNAs, this study also found that the axonal levels of 18 piRNA-like sncRNAs were significantly altered following nerve injury. This alteration points to their unique regulatory roles in the axonal response to local extracellular signals, including those arising from injury. Together, these discoveries underline the complex regulatory landscape of RNA molecules in axonal injury response and regeneration, hinting at broader implications for therapeutic strategies aimed at enhancing nerve repair and functional recovery in sensory neurons [13].

Traditionally, piRNAs and their associated Piwi proteins have been studied extensively within the germline, where they play crucial roles in maintaining genomic integrity and regulating gene expression. However, groundbreaking research has revealed the presence of pilRNAs in a variety of somatic tissues in mice, challenging the notion that piRNAs are exclusive to the germline. These piRNA-like molecules have been detected in tissues such as the brain, liver, and other organs, indicating that they may have broader biological roles beyond what was initially understood [2, 4–6, 14]. Furthermore, recent findings have highlighted a pilRNA as to play an important a role in the transcriptional repression of immunoglobulin-like (IgG-like) receptors in human NK cells. This discovery is particularly significant as it expands our understanding of the functional roles of pilRNAs beyond their traditional association with germline cells. In the context of NK cells, which are crucial components of the immune system, this specific pilRNA contributes to the regulation of immune responses by modulating the expression of immunoglobulin-like receptors. These receptors are important for the recognition and targeting of infected or malignant cells by NK cells [15]. Additionally, numerous pilRNAs that exhibit characteristic piRNA sequence traits, such as the presence of a 2′-O-methyl group at the 3′-end, have been documented in the heads of Drosophila. This finding is significant because the 2′-O-methyl modification at the 3′-end is a hallmark of mature piRNAs, providing stability and resistance to exonucleases, which degrades RNA from the ends. The identification of pilRNAs with this modification in Drosophila heads suggests that these small RNAs may be functional and involved in regulatory processes beyond the germline and ovarian follicle cells where piRNAs are typically studied. The presence of these modified pilRNAs in the neural tissue of Drosophila implies potential roles in the nervous system, possibly in regulating gene expression, maintaining genomic integrity, or modulating neural development and function [16].

In a recent investigation, Zheng et al. discovered that small RNAs resembling piRNAs are prevalent across somatic tissues in flies, mice, and macaques [17]. Notably, the pilRNAs identified in the somatic tissues of all three species, with some exceptions in specific mouse cell types, closely resemble germline piRNAs rather than those found in the somatic cells of fly ovaries. This finding is significant as it suggests that pilRNAs might have conserved roles across different species and tissue types, extending beyond the traditional confines of germline cells. The resemblance of these pilRNAs to germline piRNAs, as opposed to the piRNAs in fly ovarian somatic cells, raises intriguing questions about their origin and function. Although the study did not explore the relationship between pilRNAs and Piwi proteins, which are typically associated with piRNA function in the germline, the widespread occurrence of pilRNAs in somatic tissues is noteworthy. This broad distribution implies that pilRNAs may play essential roles in various somatic processes, potentially including gene regulation, maintenance of genomic stability, and response to environmental stresses. For example, in the macaque epididymis, pilRNAs accounted for 15% of all mapped RNA sequences, and in the mouse pancreas, they made up about 3.5% of the total RNA reads. This percentage is notably higher than the 3% of total RNA reads that piRNAs comprise in mouse ovaries. In situ hybridization techniques have shown that pilRNA expression in somatic tissues is typically localized within specific cell types. This suggests that even tissues with relatively low overall levels of pilRNAs may exhibit high concentrations of these RNAs within particular cells. This similarity raises the intriguing possibility that pilRNAs in somatic tissues might perform functions analogous to those of germline piRNAs, such as regulating gene expression and maintaining genomic integrity.

## 11.2 Piwil 2 in Stem Cell and Cancer

The Piwi genes represent the first group of genes identified as crucial for stem cell self-renewal across a wide range of organisms. These genes are evolutionarily conserved and play pivotal roles in various biological processes, including stem cell self-renewal, gametogenesis, in many species ranging from Arabidopsis to humans. In mammals, the expression of piwi genes is predominantly observed in the testes, where they are crucial for spermatogenesis [18]. In humans, eight members of the Argonaute protein family have been identified, which are divided into two distinct subfamilies: the Piwi subfamily and the eIF2C/AGO subfamily. The Piwi subfamily includes Piwil1 (also known as Hiwi), Piwil2 (Hili), Piwil3, and Piwil4 (Hiwi2), predominantly expressed in the testes. The Piwi subfamily plays a critical role in the regulation of germline development and the maintenance of genomic integrity in sperm cells. Their expression in the testes suggests their involvement in spermatogenesis, where they likely contribute to the silencing of transposable elements and the regulation of gene expression necessary for the development of sperm cells. The other subfamily, eIF2C/AGO, includes proteins that are more broadly expressed in various tissues and are involved in RNAi and post-transcriptional gene silencing. These proteins play essential roles in the regulation of gene expression by guiding the RISC to target mRNAs for degradation or translational repression [19]. Piwil1 (hiwi) is specifically present in both normal and malignant male germ cells, displaying an expression pattern that varies with the stage of cell maturation. This pattern may be associated with the proliferation of germ cells [20]. Furthermore, Piwil1 (hiwi) has been identified in various primitive hematopoietic cells, suggesting it may influence or regulate the development of hematopoietic stem cells [21]. Collectively, data from multiple organisms underscore the vital role of piwi genes in stem-cell division.

Lee et al. elucidated the mechanistic pathway of Piwil2 in regulating apoptosis, demonstrating that Piwil2 upregulates the expression of Bcl-XL. Bcl-XL is a member of the bcl-2 gene family, which includes genes with roles in either promoting or inhibiting cell death [22–25]. The activation of STAT 3 in mice has been shown to be both necessary and sufficient for maintaining the undifferentiated state of embryonic stem (ES) cells [26, 27]. It has been observed that constitutively activated STAT proteins are present in a diverse array of human cancer cell lines and primary tumors, in contrast to normal cells and tissues [28]. The activation of STATs through tyrosine phosphorylation is typically a transient and tightly regulated event. However, in cancer cells, this activation becomes continuous due to the persistent activity of tyrosine kinases such as Src, EGFR, JAKs, and Bcr-Abl. These oncogenic tyrosine kinases can be activated by prolonged ligand/receptor interactions in autocrine or paracrine signaling, or by genetic mutations that result in constitutively active enzymes in tumors. Persistent activation of specific STATs, particularly STAT3 and STAT5, drive the upregulation of genes that encode apoptosis inhibitors and cell-cycle regulators, including Bcl-XL, Mcl-1, cyclins D1/D2, and c-Myc. This constant activation leads to an environment conducive to cancer development and progression, as the cells are continually pushed toward growth and survival, evading normal regulatory mechanisms that would typically keep proliferation in check and initiate cell death [28].

**Box 11.2 Cisplatin Induces Chromatin Modification in Piwil2**

The therapeutic effectiveness of cisplatin is primarily due to its ability to cause DNA damage by forming 1,2-intrastrand crosslinks, mainly between adjacent purine bases. These crosslinks, such as cis-Pt($NH_3$)2d(GpG) (Pt-GG) and cis-Pt($NH_3$)2d(ApG) (Pt-AG), account for 90% of the DNA damage induced by the drug. This DNA damage disrupts the replication and transcription processes, leading to cell death and contributing to cisplatin's anticancer activity [29]. The primary mechanism for removing cisplatin-induced crosslinks is nucleotide excision repair (NER). Increased NER capability is associated with heightened resistance to platinum-based chemotherapies, whereas reduced NER efficiency can increase the sensitivity of cancer cells to cisplatin. NER is notable for its adaptability in lesion recognition among all DNA repair mechanisms in living cells, being able to address a broad range of helix-distorting DNA lesions. In eukaryotic cells, the challenge of DNA repair is amplified by the structural complexity of chromatin, where DNA is tightly bound with histone and non-histone proteins, restricting accessibility to the damaged sites. To effectively repair DNA, cellular repair mechanisms must navigate these chromatin barriers, primarily influenced by histone modifications such as lysine acetylation, which regulate chromatin accessibility [30].

Wang et al. have identified Piwil2 expression in various human ovarian cancer cell lines and have elucidated its role in contributing to cisplatin resistance in these cells. Their research established that Piwil2 plays a crucial role in inhibiting apoptosis, thereby promoting cell survival, by activating the STAT3/Bcl-XL pathway. The discovery of Piwil2's involvement in the STAT3/Bcl-XL pathway highlights its potential impact on cancer cell resistance to chemotherapy, specifically to cisplatin. Cisplatin works by causing DNA damage that typically leads to apoptosis in cancer cells. However, the presence of Piwil2 interferes with this process, making the cancer cells more resistant to the drug [22]. Further, Wang et al. have uncovered an additional mechanism by which Piwil2 mediates cisplatin resistance. Specifically, they found that knocking down Piwil2 in human cancer cells and knocking out Mili in mouse embryonic fibroblasts (MEFs) impaired the removal of Pt-GG crosslinks after cisplatin treatment. This impairment suggests that Piwil2 plays a critical role in enhancing the repair of cisplatin-induced DNA intra-strand crosslinks, enabling cells to better withstand the DNA damage caused by platinum-based chemotherapy. Their findings indicate that Piwil2 supports the DNA repair machinery in effectively identifying and excising these Pt-GG crosslinks, a major form of DNA damage inflicted by cisplatin. By facilitating the repair of such lesions, Piwil2 helps cancer cells survive despite the presence of cisplatin, thus contributing to chemotherapy resistance. This ability to increase DNA repair in resistant cancer cells has been supported by various studies [31] and is recognized as a key mechanism for developing resistance to cisplatin.

## 11.3 Hiwi: Human Homolog of Piwi

Hiwi, a gene belonging to the Piwi family in humans, is situated on chromosome 12 at the band 12q24.33. This gene was first discovered by isolating it from a cDNA library derived from human testis tissue. As a member of the Piwi family, Hiwi is involved in the regulation of gene expression and the maintenance of genomic integrity, particularly in the germline [32]. Hiwi is characterized by a conserved structure that includes a centrally located PAZ motif and a Piwi motif situated at the C-terminal region. These structural features are crucial for its function in RNA binding and gene regulation. Recent research has shown that Hiwi is expressed in human CD34+ hematopoietic progenitor cells, which are early-stage cells capable of developing into various types of blood cells. However, Hiwi is not expressed in more differentiated cell types, indicating that its role is likely confined to the early stages of cell development and differentiation [33]. This selective expression pattern suggests that Hiwi may play a vital role in maintaining the stemness and proliferative capacity of progenitor cells, making it a key factor in the regulation of hematopoiesis [21]. Recent research has further implicated Hiwi in critical roles related to both stem cell proliferation and tumor development. Hiwi has been associated with various cancers, including pancreatic adenocarcinoma and gastric cancer. This suggests that Hiwi not only contributes to the maintenance and proliferation of stem cells but also plays a part in oncogenesis. Hiwi's expression in progenitor cells hints at its regulatory function, which is essential for maintaining a healthy stem cell population capable of supporting tissue regeneration and repair. However, the dysregulation of Hiwi expression can lead to aberrant cell proliferation and survival, which are hallmarks of cancer. In pancreatic adenocarcinoma and gastric cancer, overexpression of Hiwi has been observed, linking it to the promotion of tumor growth and progression. Hiwi may influence tumor development through its role in gene silencing and chromatin remodeling, impacting the expression of oncogenes and tumor suppressor genes [34, 35].

Previous studies have shown that elevated levels of Hiwi mRNA are commonly found in patients with soft-tissue sarcomas (STS). The overexpression of Hiwi in these patients has been identified as a marker associated with a high risk of tumor-related mortality. This suggests that Hiwi plays a significant role in the aggressiveness and progression of STS. The presence of high Hiwi mRNA levels could indicate more rapid tumor growth, increased resistance to apoptosis, and a greater potential for metastasis. As a result, patients with elevated Hiwi expression may have a poorer prognosis compared to those with lower levels of Hiwi [36]. Moreover, a significant upregulation of Hiwi has been observed in seminomas, setting them apart from non-seminomatous tumors, spermatocytic seminomas, and non-germ cell tumors of the adult testis. This distinct expression pattern suggests that Hiwi plays a specific role in the pathogenesis of germ cell malignancies. Seminomas are a type of testicular cancer that arises from germ cells, and the marked increase in Hiwi expression in these tumors indicates that Hiwi might be involved in the initiation and progression of these malignancies. In contrast, non-seminomatous tumors, spermatocytic seminomas, and non-germ cell tumors do not exhibit the same level of Hiwi upregulation, highlighting its unique association with seminomas [20]. More recent investigations have revealed that high levels of Hiwi expression are prevalent in human ESCC. This elevated expression of Hiwi has been significantly correlated with higher histological grades of the tumor, indicating a more aggressive disease state. Patients with tumors expressing high levels of Hiwi tend to experience poorer clinical outcomes compared to those with lower Hiwi expression [37].

In cancer research, scientists have discovered that Hiwi is highly expressed in the majority of malignant gliomas and glioblastoma cells. There is a positive association between its expression with tumor grade at both the protein and mRNA levels. This correlation suggests that higher Hiwi levels are associated with more aggressive and advanced tumor stages. Additionally, clinical outcomes vary significantly between patients with low and high Hiwi expression. Supporting this, recent work by Liu et al. demonstrated that targeting Hiwi using antisense adenovirus or RNA interference resulted in cell-cycle arrest in the G2/M phase in gastric cancer cells with high Hiwi expression. This finding indicates that Hiwi is not only a marker of tumor progression but also a potential therapeutic target. By inhibiting Hiwi, researchers were able to disrupt the cell cycle, preventing cancer cells from dividing and proliferating. The high expression of Hiwi in malignant gliomas and glioblastomas underscores its importance in the development and progression of these cancers [35].

## 11.4 Small Nucleolar RNAs (snoRNAs)

In eukaryotic organisms, the precise formation of the two primary types of modified nucleotides in rRNA—2′-O-methylated nucleotides and pseudouridines—is orchestrated by two major families of small nucleolar RNAs (snoRNAs): the box C/D snoRNAs and the H/ACA snoRNAs, respectively [38]. These snoRNAs facilitate the modifications by forming canonical guide RNA duplexes at specific modification sites on the rRNA. Box C/D snoRNAs are responsible for guiding the 2′-O-methylation of nucleotides. They typically do this by forming a guide RNA duplex with the target rRNA sequence, positioning the catalytic machinery precisely at the site of modification. Each box C/D snoRNA usually directs the methylation at one, or occasionally two, specific sites using single or paired antisense elements that match the target sequence. Like box C/D snoRNAs, they form guide RNA duplexes with the rRNA to ensure accurate modification. Each H/ACA snoRNA typically targets one or two specific uridine residues for conversion to pseudouridine, using single or paired antisense elements to recognize and bind the appropriate sequences [39, 40].

The role of snoRNAs extends beyond rRNA modifications, influencing various aspects of cellular metabolism and gene expression by affecting the maturation and function of other small RNAs as well. Their highly specific and regulated activities underscore the intricate control mechanisms in cellular RNA processing and modification. The functional scope of these guide RNA families has expanded beyond initial expectations, evidenced by the discovery of snoRNAs that target additional types of cellular RNAs such as snRNAs, tRNAs, and potentially mRNAs, along with the identification of numerous tissue-specific variants in mammals [41, 42]. The process of ribosome biogenesis in eukaryotes takes place in the nucleolus and involves a complex series of pre-rRNA processing steps. These steps are essential for removing extensive spacer regions from the primary rRNA transcript, resulting in the production of mature rRNAs for both the small (SSU) and large (LSU) ribosomal subunits in balanced quantities. Before the pre-rRNA is cleaved by endonucleases and exonucleases, it undergoes a series of nucleoside modifications across its SSU and LSU regions. The modifications are critical for the proper assembly and function of the ribosome, as they enhance the stability and performance of the rRNA [43–45].

The rRNA modifications, although not absolutely essential, are predominantly located within the most conserved and functionally critical regions of mature rRNAs, particularly in structural elements near the peptidyl transferase center. These areas are crucial for the ribosome's role in protein synthesis. While the positions of these modifications are largely conserved across diverse eukaryotic species, they are not perfectly identical, indicating some evolutionary variation [46]. The exact roles of these modifications remain somewhat elusive. However, it is believed that they play a significant part in fine-tuning the rRNA's folding and enhancing its interactions with ribosomal proteins. These modifications likely contribute to the precise assembly of ribosomal subunits and optimize the ribosome's functional capabilities. Specifically, the modifications such as 2′-O-ribose methylation and pseudouridylation are thought to stabilize the rRNA structure and promote correct folding patterns, which are essential for the ribosome's stability and function. By refining rRNA-protein interactions, these modifications may enhance the efficiency and accuracy of translation, ensuring that the ribosome can effectively synthesize proteins.

Directed by two major families of snoRNAs, these rRNA modifications occur at precise sites identified by the snoRNAs through the formation of a specific RNA duplex. The actual modification process, whether methylation or pseudouridylation, is carried out by a common protein enzyme associated with each snoRNA, namely a methylase or a pseudouridine synthase. Historically, it was anticipated that the combined number of snoRNA guides would near 200 in vertebrates, given that a single snoRNA typically guides one or two modifications. Recent discoveries, however, have revealed a significant underestimation in the complexity and diversity of these snoRNA guide families. This complexity extends beyond their genetic organization and synthesis to include variations in snoRNA structure and a broadened scope of RNA targets, encompassing roles beyond traditional ribosome biogenesis [41]. Recent discoveries have identified novel snoRNA guides that extend beyond the conventional targets of rRNAs. In vertebrates, snoRNAs have been found to target spliceosomal snRNAs. There is a growing number of "orphans" snoRNAs, which are snoRNAs without clearly identified RNA targets. Many of these orphan snoRNAs show tissue-specific expression patterns and are subject to genomic imprinting, indicating that they may play unique and specialized roles in different tissues, suggesting that they may be involved in regulating gene expression in a more nuanced and context-dependent manner [47, 48].

**Box 11.3 Brain-Specific snoRNAs and Genomic Imprinting**

Unlike the more commonly known snoRNAs that guide modifications of rRNAs or snRNAs, and distinct from most orphan snoRNAs without known targets, a significant subset of newly identified snoRNAs, particularly from the C/D family, exhibit a unique tissue-specific expression pattern [49]. These snoRNAs are predominantly expressed in the brain, highlighting their potential role in brain-specific functions. The brain-specific expression of these snoRNAs suggests that they may be involved in regulating processes that are critical for neural development, function, and maintenance. Additionally, some of these brain-specific snoRNAs are subject to genomic imprinting, a genetic phenomenon where only one allele of a gene is expressed while the other is silenced, depending on the parent of origin. This imprinting adds an extra layer of regulatory control, indicating that these snoRNAs might play roles in developmental processes that are sensitive to dosage and parental origin effects. The identification of these brain-specific and imprinted snoRNAs broadens our understanding of the diverse functional roles that snoRNAs can play. It also suggests that these snoRNAs could be critical for the proper functioning of the brain, potentially influencing neural plasticity, learning, memory, and other cognitive functions [50–55]. Intriguingly, the genes encoding these snoRNAs are regulated by genomic imprinting, a sophisticated epigenetic mechanism that permits gene expression exclusively from one parental allele—either paternal or maternal. Genomic imprinting is a process where certain genes are expressed in a parent-of-origin specific manner. This means that for each imprinted gene, only the allele from either the mother or the father is active, while the other allele is silenced. This selective expression plays a critical role in developmental processes and cellular functions. Imprinting is achieved through various epigenetic modifications, such as DNA methylation and histone modifications, which alter the chromatin structure around a gene without changing the genetic sequence itself. These modifications effectively "mark" the genes with a memory of their parental origin, which is crucial for normal development. The fact that snoRNA genes are subject to imprinting suggests they have roles that are sensitive to the dosage and timing of their expression, which could be vital for their functions in modifying rRNA, and possibly in regulating gene expression more broadly. Since snoRNAs contribute to the biochemical modification of other RNA molecules, primarily rRNAs and spliceosomal RNAs, their precise expression is likely crucial for maintaining the efficiency and accuracy of protein synthesis and other cellular processes [56].

(continued)

**Box 11.3** (continued)

An interesting departure from the usual pattern of brain-specific and imprinted snoRNAs is observed with the HBI-36/MBI-36 snoRNA. While this snoRNA is also uniquely expressed in the brain, it does not follow the common route of genomic imprinting. Instead, the expression of HBI-36/MBI-36 is restricted to a single chromosome because its host gene, the serotonin receptor 5-HT2c, resides on the X chromosome. This arrangement suggests that all brain-specific snoRNAs identified to date are transcribed from just one chromosome, adding to our understanding of their genetic organization. The exclusive presence of HBI-36/MBI-36 snoRNA in the brain suggests that it may have a specialized role in neural processes, which could be intimately linked to its host gene's involvement in neurotransmission and overall brain function. Since the 5-HT2c receptor is implicated in a variety of neurological functions and disorders, the associated snoRNA might influence these pathways as well. Additionally, the fact that HBI-36/MBI-36's expression is limited to the X chromosome introduces a complex layer of genetic regulation. In females, the X chromosome undergoes a process known as X inactivation, where one of the X chromosomes is largely silenced to balance gene dosage between males (XY) and females (XX). This mechanism means that the expression and potential impact of HBI-36/MBI-36 could vary significantly between males and females, affecting how this snoRNA influences brain function across sexes.

This finding highlights the intricate and diverse regulatory mechanisms governing snoRNA expression and function. The restriction of brain-specific snoRNAs to a single chromosome, whether through imprinting or chromosomal localization, underscores the specialized roles these RNAs may play in brain biology. It also opens up questions about how these snoRNAs contribute to the regulation of neural genes and their potential impact on brain function and behavior. Further research into the specific roles of HBI-36/MBI-36 and other brain-specific snoRNAs could provide valuable insights into the molecular underpinnings of brain function and the regulatory networks that sustain it.

Imprinted gene anomalies are also linked to behavioral disorders. Autism spectrum disorders, Turner syndrome, and Tourette syndrome have all been associated with improper imprinting. The precise mechanisms by which imprinted genes influence these disorders are still being explored, but it is clear that the delicate balance of gene expression controlled by imprinting is crucial for normal development and function. Furthermore, certain cancers, such as Wilms' tumor and a pediatric kidney cancer, have been associated with the disruption of imprinted genes. In these cases, the loss of normal imprinting control can lead to uncontrolled cell growth and tumor development. The functional haploidy of imprinted genes means that any genetic

(continued)

**Box 11.3** (continued)
anomaly affecting the active allele can have significant consequences. This sensitivity underscores the importance of proper genomic imprinting for maintaining normal cellular and developmental processes. Understanding these mechanisms not only provides insight into the etiology of various genetic disorders and cancers but also highlights potential targets for therapeutic intervention.

Interestingly, imprinted snoRNA genes have recently been discovered at two specific regions of the human genome: 15q11-q13 and 14q32. These regions exhibit a unique and notable genomic organization, suggesting that these snoRNA genes may play important roles in genomic imprinting and gene regulation. The 15q11-q13 region is well known for its involvement in several genetic disorders, including Prader-Willi and Angelman syndromes, both of which arise from disruptions in this imprinted region. Within this locus, imprinted snoRNA genes are organized in clusters which imply that their expression is tightly regulated and potentially coordinated to perform specific functions in cellular processes and development. Similarly, the 14q32 region is another critical imprinted locus associated with disorders such as Temple syndrome and Kagami-Ogata syndrome. The snoRNA genes in this region also display a clustered arrangement, indicating a complex regulatory network that governs their expression and function [50, 52, 54, 57].

## 11.5 Xist RNAs

X inactivation stands as a quintessential case of epigenetic regulation, where one of the two X chromosomes in female cells is randomly chosen to be transcriptionally silenced during the early phases of epiblast differentiation in development. This silenced state of the X chromosome is persistently maintained through all subsequent cell divisions and during the process of cellular differentiation. The mechanism initiating X inactivation is orchestrated by the noncoding RNA known as Xist (X-inactive specific transcript). Xist RNA blankets the designated X chromosome in a cis manner, directly affecting the chromosome it is produced from, leading to the suppression of gene expression and the establishment of heterochromatin, a more compact and transcriptionally inactive form of chromatin. This process plays a critical role in achieving dosage compensation—the equalization of gene expression from the X chromosome in both males and females. Since males possess only one X chromosome (XY), while females have two (XX), X inactivation is crucial to balance the expression levels of X-linked genes across the sexes, ensuring that females do not produce twice as many X-chromosome-derived proteins as males. This balancing act is vital for normal development and functioning, as discrepancies in gene dosage can lead to developmental abnormalities and various diseases. Understanding X inactivation not only illuminates a key area of genetic regulation but also enhances our insight into gender differences in disease presentation and response to treatment [58–61] Transcriptionally, the expression of Xist RNA is tightly controlled by various regulatory elements and factors that ensure it is activated only on the X chromosome destined for inactivation. These include promoter regions, enhancers, and silencers that modulate the transcriptional activity of the Xist gene. Additionally, chromatin modifications and DNA methylation patterns play significant roles in either promoting or repressing Xist transcription.

Post-transcriptionally, the stability, localization, and translation of Xist RNA are further regulated to ensure its proper function. RNA-binding proteins and other molecular machinery interact with Xist RNA to control its processing, stability, and localization to the X chromosome. These post-transcriptional modifications are crucial for the effective envelopment of the X chromosome by Xist RNA, which is indispensable for the commencement and propagation of gene silencing throughout the entire chromosome. These modifications alter the Xist RNA molecule itself, enhancing its ability to adhere to and cover the X chromosome. This coverage is a critical step in the process of X inactivation, as it sets the stage for the chromosome to undergo extensive changes in its chromatin structure. As Xist RNA spreads across the X chromosome, it recruits a variety of chromatin-modifying complexes. These complexes work to remodel the chromatin into a tightly packed state known as heterochromatin. This compacted state prevents the transcriptional machinery from accessing genes on the chromosome, effectively silencing their expression. The ability of Xist RNA to initiate this process depends heavily on specific post-transcriptional modifications that may influence its stability, localization, and interaction with other molecular components involved in gene silencing [62, 63]. Embryonic stem (ES) cells provide a valuable model for investigating the process of random X inactivation as they undergo this epigenetic phenomenon during the early stages of differentiation. These cells, which are pluripotent and capable of differentiating into various cell types, mimic the random selection of an X chromosome for silencing that occurs in early developmental stages in vivo. This makes ES cells an exemplary system for detailed studies of X inactivation.

Studying X inactivation in ES cells allows researchers to observe the initiation and progression of this complex regulatory mechanism in a controlled environment. They can analyze how one of the two X chromosomes is chosen for inactivation and how this choice is stably maintained as cells continue to divide and differentiate. This is crucial for understanding the intricate balance of gene expression that supports normal development [64]. In undifferentiated female ES cells, Xist RNA is transcribed at low levels from both X chromosomes. Upon differentiation, one Xist allele per diploid genome becomes activated, initiating X inactivation in cis, while the other Xist allele in females and the single Xist allele in males are deactivated. A distinctive feature of the Xist gene is its overlap with an antisense noncoding transcription unit known as Tsix, which acts to inhibit the accumulation of Xist RNA in cis. This process ensures that only one X chromosome in female cells is silenced, achieving dosage compensation. Tsix plays a crucial regulatory role by preventing Xist RNA accumulation on the chromosome it is transcribed from, thereby protecting that X chromosome from inactivation [65, 66]. Mutation studies in Tsix within both female [67–69] and male [70, 71] Studies in ES cells have shown that any disruption in Tsix transcription results in the systematic upregulation of Xist at the onset of differentiation, regardless of the cell's sex. This finding suggests that Tsix is essential for maintaining the randomness of Xist activation in female cells and for setting up the correct expression pattern of Xist in male cells to ensure proper X inactivation processes. Moreover, experiments using an inducible promoter to artificially sustain Tsix expression during the differentiation of female ES cells have demonstrated that maintaining Tsix activity prevents the mutated allele from activating Xist. This confirms Tsix's crucial role in regulating the timing of Xist upregulation during early differentiation [72].

Recent findings have unveiled the complex functions of Tsix in chromatin remodeling at the Xist/Tsix locus. Tsix triggers the dimethylation of histone H3 at lysine 4 (H3K4) throughout the locus while simultaneously preventing this modification from accumulating at the Xist promoter [62]. Additionally, male mouse embryonic fibroblasts (MEFs) derived from Tsix mutant lines exhibit abnormal chromatin structures at the Xist promoter. These atypical chromatin configurations are notably characterized by elevated levels of H3K4 dimethylation. This abnormal chromatin modification indicates that Tsix normally plays a crucial role in maintaining the correct chromatin state at the Xist promoter, ensuring that H3K4 dimethylation levels are kept in check. The elevated H3K4 dimethylation observed in the absence of functional Tsix suggests that Tsix is essential for repressing this specific histone modification, which is typically associated with active transcription [73]. Given the significant impact of Tsix loss on both chromatin modifications and Xist expression levels, and the crucial role of chromatin conformation in establishing specific gene expression patterns during differentiation, Navarro et al. proposed that Tsix-mediated regulation might create distinct epigenetic states at the Xist promoter on X chromosomes destined for inactivation and those remaining active. This, in turn, influences Xist expression pathways. Research using both wild-type and Tsix-deficient embryonic stem (ES) cells has demonstrated that Tsix inhibits the formation of euchromatin around a section of the Xist promoter encircled by CTCF binding sites. This inhibition prevents the transcriptional upregulation of Xist during differentiation. In contrast, truncation of Tsix leads to the establishment of a stable pseudo-euchromatic state at the 5′ region of Xist. This altered chromatin state effectively blocks the assembly of the transcription machinery at the Xist promoter, preventing the initiation of X inactivation. This contrasts sharply with findings from another recent investigation, which suggested that Tsix down-regulation prompts a heterochromatic state at Xist, subsequently leading to its transcriptional activation [63]. The research highlights Tsix's pivotal role in shaping the chromatin architecture of a specific 5′ region of the Xist gene bordered by CTCF binding sites, thus influencing the pattern of Xist expression. This underscores Tsix's crucial function in regulating Xist through epigenetic modifications, ensuring proper gene silencing and X chromosome inactivation. These insights emphasize the significant epigenetic impact of noncoding antisense transcription units like Tsix. It reveals that such elements play a more substantial role in genomic regulation than previously understood [74, 75].

In mammals, the principle of dosage compensation is vital for ensuring that the expression levels of X-linked genes are equivalent between males and females. This equilibrium is achieved through the transcriptional silencing of one of the two X chromosomes in female cells. The X chromosome that is silenced is known as the inactive X chromosome (Xi), and it undergoes a series of distinctive epigenetic modifications to maintain its inactivated state. These epigenetic changes are critical in transforming the active X chromosome into the Xi, effectively repressing its gene expression to balance with that of the single X chromosome in males. One of the primary modifications is the methylation of histone H3, a process that marks regions of the genome for gene repression. In particular, H3K27me3 serves as a key marker on the Xi. This modification is instrumental in establishing and maintaining a repressive chromatin state that is essential for the stable, long-term silencing of the X chromosome. In addition to H3K27me3, the Xi also exhibits hypoacetylation of histone H4. Acetylation of histones is generally associated with gene activation by making the chromatin structure more open and accessible for transcriptional machinery. Therefore, the hypoacetylation (reduction of acetylation) of histone H4 on the Xi contributes further to the condensed, transcriptionally inactive state of the chromosome. This reduction in acetylation levels is associated with tighter chromatin packing and gene silencing, as acetylation generally correlates with active transcription by loosening chromatin structure. The Xi is also enriched with the variant histone macroH2A, which further stabilizes the repressed state of the chromosome. MacroH2A is known to play a role in maintaining long-term gene silencing and contributing to the unique chromatin architecture of the Xi. Finally, DNA methylation is another crucial modification found on the Xi. Methylation of cytosine residues in CpG islands is associated with stable gene repression and is a key mechanism for maintaining the silenced state of the Xi through successive cell divisions [76, 77]. In mice, homozygous mutations in EED disrupt the maintenance of X chromosome silencing in extraembryonic tissues, although embryonic tissues remain unaffected [78]. Both EED and Ezh2 are notably present on the Xi in trophoblast stem (TS) cells [79], an extraembryonic cell type, indicating their involvement in regulating imprinted X inactivation through H3-K27 methylation.

Plath et al. have proposed that the initial coating of the X chromosome by Xist RNA is crucial for the temporary recruitment of the EED-Ezh2 complex. This complex then catalyzes H3K27me3. The presence of Xist RNA on the chromosome is both necessary and sufficient to attract the EED-Ezh2 complex and initiate this specific histone modification. This early recruitment marks the beginning of the silencing process, establishing the repressive chromatin state needed for dosage compensation. Interestingly, embryos with mutations in the *eed* gene can still initiate X inactivation across all cell types, indicating that the initial steps of Xist RNA coating and EED-Ezh2 complex recruitment are independent of *eed* function. However, these mutants fail to maintain X inactivation specifically in extraembryonic cells. This failure suggests that while the EED-Ezh2 complex is not essential for the initiation of X inactivation, it is crucial for the stable maintenance of the silenced state in certain cell types [78]. It remains uncertain whether embryonic cells in these *eed* mutant embryos exhibit defects in X inactivation.

## 11.6 Conclusion

The exploration and elucidation of emerging ncRNAs, including piRNAs, snoRNAs, and XistRNA, have exposed a new dimension of intricacy in gene regulation and cellular dynamics. These ncRNAs are integral to a variety of biological functions, such as chromatin remodeling, transcriptional control, mRNA processing, and protein synthesis. Grasping their roles illuminates key molecular pathways implicated in a range of health conditions, including cancers, neurological diseases, and metabolic disorders. Furthermore, ncRNAs present valuable avenues for therapeutic intervention and diagnostic applications. Their distinct expression profiles and functional importance in regulatory networks make them attractive candidates for the development of RNA-based treatment modalities, including RNA interference technologies, antisense oligonucleotides, and novel RNA-based vaccines. Leveraging the therapeutic capabilities of noncoding RNAs could significantly enhance personalized medicine approaches and offer new treatments for diseases that currently lack effective management options.

## References


1. Ozata DM, Gainetdinov I, Zoch A, O'Carroll D, Zamore PD. PIWI-interacting RNAs: small RNAs with big functions. Nat Rev Genet. 2019;20(2):89–108.
2. Brennecke J, Aravin AA, Stark A, Dus M, Kellis M, Sachidanandam R, et al. Discrete small RNA-generating loci as master regulators of transposon activity in Drosophila. Cell. 2007;128(6):1089–103.
3. Thomson T, Lin H. The biogenesis and function of PIWI proteins and piRNAs: progress and prospect. Annu Rev Cell Dev Biol. 2009;25:355–76.
4. Aravin A, Gaidatzis D, Pfeffer S, Lagos-Quintana M, Landgraf P, Iovino N, et al. A novel class of small RNAs bind to MILI protein in mouse testes. Nature. 2006;442(7099):203–7.
5. Girard A, Sachidanandam R, Hannon GJ, Carmell MA. A germline-specific class of small RNAs binds mammalian Piwi proteins. Nature. 2006;442(7099):199–202.
6. Grivna ST, Beyret E, Wang Z, Lin H. A novel class of small RNAs in mouse spermatogenic cells. Genes Dev. 2006;20(13):1709–14.
7. Vagin VV, Sigova A, Li C, Seitz H, Gvozdev V, Zamore PD. A distinct small RNA pathway silences selfish genetic elements in the germline. Science. 2006;313(5785):320–4.
8. Aravin AA, Sachidanandam R, Bourc'his D, Schaefer C, Pezic D, Toth KF, et al. A piRNA pathway primed by individual transposons is linked to de novo DNA methylation in mice. Mol Cell. 2008;31(6):785–99.
9. Gunawardane LS, Saito K, Nishida KM, Miyoshi K, Kawamura Y, Nagami T, et al. A slicer-mediated mechanism for repeat-associated siRNA 5′ end formation in Drosophila. Science. 2007;315(5818):1587–90.
10. Houwing S, Berezikov E, Ketting RF. Zili is required for germ cell differentiation and meiosis in zebrafish. EMBO J. 2008;27(20):2702–11.
11. Aravin AA, Sachidanandam R, Girard A, Fejes-Toth K, Hannon GJ. Developmentally regulated piRNA clusters implicate MILI in transposon control. Science. 2007;316(5825):744–7.
12. Lau NC, Robine N, Martin R, Chung WJ, Niki Y, Berezikov E, et al. Abundant primary piRNAs, endo-siRNAs, and microRNAs in a Drosophila ovary cell line. Genome Res. 2009;19(10):1776–85.

13. Phay M, Kim HH, Yoo S. Analysis of piRNA-like small non-coding RNAs present in axons of adult sensory neurons. Mol Neurobiol. 2018;55(1):483–94.
14. Ro S, Park C, Song R, Nguyen D, Jin J, Sanders KM, et al. Cloning and expression profiling of testis-expressed piRNA-like RNAs. RNA. 2007;13(10):1693–702.
15. Cichocki F, Lenvik T, Sharma N, Yun G, Anderson SK, Miller JS. Cutting edge: KIR antisense transcripts are processed into a 28-base PIWI-like RNA in human NK cells. J Immunol. 2010;185(4):2009–12.
16. Ghildiyal M, Seitz H, Horwich MD, Li C, Du T, Lee S, et al. Endogenous siRNAs derived from transposons and mRNAs in Drosophila somatic cells. Science. 2008;320(5879):1077–81.
17. Gou L-T, Zhu Q, Liu M-F. Small RNAs: an expanding world with therapeutic promises. Fundamental Res. 2023;3(5):676–82.
18. Kuramochi-Miyagawa S, Kimura T, Yomogida K, Kuroiwa A, Tadokoro Y, Fujita Y, et al. Two mouse piwi-related genes: miwi and mili. Mech Dev. 2001;108(1-2):121–33.
19. Sasaki T, Shiohama A, Minoshima S, Shimizu N. Identification of eight members of the Argonaute family in the human genome. Genomics. 2003;82(3):323–30.
20. Qiao D, Zeeman AM, Deng W, Looijenga LH, Lin H. Molecular characterization of hiwi, a human member of the piwi gene family whose overexpression is correlated to seminomas. Oncogene. 2002;21(25):3988–99.
21. Sharma AK, Nelson MC, Brandt JE, Wessman M, Mahmud N, Weller KP, et al. Human CD34(+) stem cells express the hiwi gene, a human homologue of the Drosophila gene piwi. Blood. 2001;97(2):426–34.
22. Lee JH, Schütte D, Wulf G, Füzesi L, Radzun HJ, Schweyer S, et al. Stem-cell protein Piwil2 is widely expressed in tumors and inhibits apoptosis through activation of Stat3/Bcl-XL pathway. Hum Mol Genet. 2006;15(2):201–11.
23. Zamzami N, Brenner C, Marzo I, Susin SA, Kroemer G. Subcellular and submitochondrial mode of action of Bcl-2-like oncoproteins. Oncogene. 1998;16(17):2265–82.
24. Reed JC. Bcl-2 family proteins. Oncogene. 1998;17(25):3225–36.
25. Ke H, Pei J, Ni Z, Xia H, Qi H, Woods T, et al. Putative tumor suppressor Lats2 induces apoptosis through downregulation of Bcl-2 and Bcl-x(L). Exp Cell Res. 2004;298(2):329–38.
26. Matsuda T, Nakamura T, Nakao K, Arai T, Katsuki M, Heike T, et al. STAT3 activation is sufficient to maintain an undifferentiated state of mouse embryonic stem cells. EMBO J. 1999;18(15):4261–9.
27. Niwa H, Burdon T, Chambers I, Smith A. Self-renewal of pluripotent embryonic stem cells is mediated via activation of STAT3. Genes Dev. 1998;12(13):2048–60.
28. Buettner R, Mora LB, Jove R. Activated STAT signaling in human tumors provides novel molecular targets for therapeutic intervention. Clin Cancer Res. 2002;8(4):945–54.
29. Jung Y, Lippard SJ. Direct cellular responses to platinum-induced DNA damage. Chem Rev. 2007;107(5):1387–407.
30. Zhu Q, Wani AA. Histone modifications: crucial elements for damage response and chromatin restoration. J Cell Physiol. 2010;223(2):283–8.
31. Saldivar JS, Wu X, Follen M, Gershenson D. Nucleotide excision repair pathway review I: implications in ovarian cancer and platinum sensitivity. Gynecol Oncol. 2007;107(1 Suppl 1):S56–71.
32. Cox DN, Chao A, Baker J, Chang L, Qiao D, Lin H. A novel class of evolutionarily conserved genes defined by piwi are essential for stem cell self-renewal. Genes Dev. 1998;12(23):3715–27.
33. Cerutti L, Mian N, Bateman A. Domains in gene silencing and cell differentiation proteins: the novel PAZ domain and redefinition of the Piwi domain. Trends Biochem Sci. 2000;25(10):481–2.
34. Grochola LF, Greither T, Taubert H, Möller P, Knippschild U, Udelnow A, et al. The stem cell-associated Hiwi gene in human adenocarcinoma of the pancreas: expression and risk of tumour-related death. Br J Cancer. 2008;99(7):1083–8.
35. Liu X, Sun Y, Guo J, Ma H, Li J, Dong B, et al. Expression of hiwi gene in human gastric cancer was associated with proliferation of cancer cells. Int J Cancer. 2006;118(8):1922–9.

36. Taubert H, Greither T, Kaushal D, Würl P, Bache M, Bartel F, et al. Expression of the stem cell self-renewal gene Hiwi and risk of tumour-related death in patients with soft-tissue sarcoma. Oncogene. 2007;26(7):1098–100.
37. He W, Wang Z, Wang Q, Fan Q, Shou C, Wang J, et al. Expression of HIWI in human esophageal squamous cell carcinoma is significantly associated with poorer prognosis. BMC Cancer. 2009;9:426.
38. Matera AG, Terns RM, Terns MP. Non-coding RNAs: lessons from the small nuclear and small nucleolar RNAs. Nat Rev Mol Cell Biol. 2007;8(3):209–20.
39. Brown JW, Echeverria M, Qu LH, Lowe TM, Bachellerie JP, Hüttenhofer A, et al. Plant snoRNA database. Nucleic Acids Res. 2003;31(1):432–5.
40. Lestrade L, Weber MJ. snoRNA-LBME-db, a comprehensive database of human H/ACA and C/D box snoRNAs. Nucleic Acids Res. 2006;34(Database issue):D158–62.
41. Kiss T. Small nucleolar RNAs: an abundant group of noncoding RNAs with diverse cellular functions. Cell. 2002;109(2):145–8.
42. Reichow SL, Hamma T, Ferré-D'Amaré AR, Varani G. The structure and function of small nucleolar ribonucleoproteins. Nucleic Acids Res. 2007;35(5):1452–64.
43. Maden BE. The numerous modified nucleotides in eukaryotic ribosomal RNA. Prog Nucleic Acid Res Mol Biol. 1990;39:241–303.
44. Bachellerie JP, Cavaillé J. Small nucleolar RNAs guide the ribose methylations of eukaryotic rRNAs. In: Grosjean H, Benne R, editors. Modification and editing of RNA: the alteration of RNA structure and function. Washington: ASM Press; 1998. p. 255–72.
45. Ofengand J, Fournier MJ. The pseudouridine residues of rRNA: number, location, biosynthesis and function. In: Grosjean H, Benne R, editors. Modification and editing of RNA: the alteration of RNA structure and function. Washington: ASM Press; 1998. p. 229–53.
46. Brimacombe R, Mitchell P, Osswald M, Stade K, Bochkariov D. Clustering of modified nucleotides at the functional center of bacterial ribosomal RNA. FASEB J. 1993;7(1):161–7.
47. Dennis PP, Omer A, Lowe T. A guided tour: small RNA function in Archaea. Mol Microbiol. 2001;40(3):509–19.
48. Uliel S, Liang XH, Unger R, Michaeli S. Small nucleolar RNAs that guide modification in trypanosomatids: repertoire, targets, genome organisation, and unique functions. Int J Parasitol. 2004;34(4):445–54.
49. Bachellerie JP, Cavaillé J, Hüttenhofer A. The expanding snoRNA world. Biochimie. 2002;84(8):775–90.
50. Cavaillé J, Buiting K, Kiefmann M, Lalande M, Brannan CI, Horsthemke B, et al. Identification of brain-specific and imprinted small nucleolar RNA genes exhibiting an unusual genomic organization. Proc Natl Acad Sci USA. 2000;97(26):14311–6.
51. Cavaillé J, Vitali P, Basyuk E, Hüttenhofer A, Bachellerie JP. A novel brain-specific box C/D small nucleolar RNA processed from tandemly repeated introns of a noncoding RNA gene in rats. J Biol Chem. 2001;276(28):26374–83.
52. de los Santos T, Schweizer J, Rees CA, Francke U. Small evolutionarily conserved RNA, resembling C/D box small nucleolar RNA, is transcribed from PWCR1, a novel imprinted gene in the Prader-Willi deletion region, which Is highly expressed in brain. Am J Hum Genet. 2000;67(5):1067–82.
53. Filipowicz W. Imprinted expression of small nucleolar RNAs in brain: time for RNomics. Proc Natl Acad Sci USA. 2000;97(26):14035–7.
54. Meguro M, Mitsuya K, Nomura N, Kohda M, Kashiwagi A, Nishigaki R, et al. Large-scale evaluation of imprinting status in the Prader-Willi syndrome region: an imprinted direct repeat cluster resembling small nucleolar RNA genes. Hum Mol Genet. 2001;10(4):383–94.
55. Runte M, Hüttenhofer A, Gross S, Kiefmann M, Horsthemke B, Buiting K. The IC-SNURF-SNRPN transcript serves as a host for multiple small nucleolar RNA species and as an antisense RNA for UBE3A. Hum Mol Genet. 2001;10(23):2687–700.
56. Reik W, Walter J. Genomic imprinting: parental influence on the genome. Nat Rev Genet. 2001;2(1):21–32.

57. Cavaillé J, Seitz H, Paulsen M, Ferguson-Smith AC, Bachellerie JP. Identification of tandemly-repeated C/D snoRNA genes at the imprinted human 14q32 domain reminiscent of those at the Prader-Willi/Angelman syndrome region. Hum Mol Genet. 2002;11(13):1527–38.
58. Heard E. Delving into the diversity of facultative heterochromatin: the epigenetics of the inactive X chromosome. Curr Opin Genet Dev. 2005;15(5):482–9.
59. Panning B, Dausman J, Jaenisch R. X chromosome inactivation is mediated by Xist RNA stabilization. Cell. 1997;90(5):907–16.
60. Sheardown SA, Duthie SM, Johnston CM, Newall AE, Formstone EJ, Arkell RM, et al. Stabilization of Xist RNA mediates initiation of X chromosome inactivation. Cell. 1997;91(1):99–107.
61. Ciaudo C, Bourdet A, Cohen-Tannoudji M, Dietz HC, Rougeulle C, Avner P. Nuclear mRNA degradation pathway(s) are implicated in Xist regulation and X chromosome inactivation. PLoS Genet. 2006;2(6):e94.
62. Navarro P, Pichard S, Ciaudo C, Avner P, Rougeulle C. Tsix transcription across the Xist gene alters chromatin conformation without affecting Xist transcription: implications for X-chromosome inactivation. Genes Dev. 2005;19(12):1474–84.
63. Sun BK, Deaton AM, Lee JT. A transient heterochromatic state in Xist preempts X inactivation choice without RNA stabilization. Mol Cell. 2006;21(5):617–28.
64. Chaumeil J, Okamoto I, Guggiari M, Heard E. Integrated kinetics of X chromosome inactivation in differentiating embryonic stem cells. Cytogenet Genome Res. 2002;99(1-4):75–84.
65. Ogawa Y, Lee JT. Antisense regulation in X inactivation and autosomal imprinting. Cytogenet Genome Res. 2002;99(1-4):59–65.
66. Rougeulle C, Avner P. The role of antisense transcription in the regulation of X-inactivation. Curr Top Dev Biol. 2004;63:61–89.
67. Debrand E, Chureau C, Arnaud D, Avner P, Heard E. Functional analysis of the DXPas34 locus, a 3′ regulator of Xist expression. Mol Cell Biol. 1999;19(12):8513–25.
68. Lee JT, Lu N. Targeted mutagenesis of Tsix leads to nonrandom X inactivation. Cell. 1999;99(1):47–57.
69. Luikenhuis S, Wutz A, Jaenisch R. Antisense transcription through the Xist locus mediates Tsix function in embryonic stem cells. Mol Cell Biol. 2001;21(24):8512–20.
70. Morey C, Navarro P, Debrand E, Avner P, Rougeulle C, Clerc P. The region 3′ to Xist mediates X chromosome counting and H3 Lys-4 dimethylation within the Xist gene. EMBO J. 2004;23(3):594–604.
71. Vigneau S, Augui S, Navarro P, Avner P, Clerc P. An essential role for the DXPas34 tandem repeat and Tsix transcription in the counting process of X chromosome inactivation. Proc Natl Acad Sci USA. 2006;103(19):7390–5.
72. Stavropoulos N, Lu N, Lee JT. A functional role for Tsix transcription in blocking Xist RNA accumulation but not in X-chromosome choice. Proc Natl Acad Sci USA. 2001;98(18):10232–7.
73. Sado T, Hoki Y, Sasaki H. Tsix silences Xist through modification of chromatin structure. Dev Cell. 2005;9(1):159–65.
74. Kiyosawa H, Yamanaka I, Osato N, Kondo S, Hayashizaki Y. Antisense transcripts with FANTOM2 clone set and their implications for gene regulation. Genome Res. 2003;13(6b):1324–34.
75. Numata K, Kanai A, Saito R, Kondo S, Adachi J, Wilming LG, et al. Identification of putative noncoding RNAs among the RIKEN mouse full-length cDNA collection. Genome Res. 2003;13(6b):1301–6.
76. Avner P, Heard E. X-chromosome inactivation: counting, choice and initiation. Nat Rev Genet. 2001;2(1):59–67.
77. Brockdorff N. X-chromosome inactivation: closing in on proteins that bind Xist RNA. Trends Genet. 2002;18(7):352–8.
78. Wang J, Mager J, Chen Y, Schneider E, Cross JC, Nagy A, et al. Imprinted X inactivation maintained by a mouse Polycomb group gene. Nat Genet. 2001;28(4):371–5.
79. Mak W, Baxter J, Silva J, Newall AE, Otte AP, Brockdorff N. Mitotically stable association of polycomb group proteins eed and enx1 with the inactive x chromosome in trophoblast stem cells. Curr Biol. 2002;12(12):1016–20.

# Chapter 12
# Extracellular RNAs: Biogenesis

**Abstract** The investigation of extracellular RNA (exRNA) has gained considerable momentum in recent years, underscoring its critical role in biomarker discovery and the development of innovative cancer therapies. exRNA comprises a diverse array of RNA subpopulations, each playing distinct biological roles. These RNA molecules are protected from enzymatic degradation through encapsulation within membranous vesicles, such as exosomes and microvesicles, or by forming complexes with lipids and proteins. exRNA is ubiquitously present in various biofluids, including blood, urine, saliva, and cerebrospinal fluid, making it an accessible and minimally invasive target for clinical diagnostics. This accessibility is particularly advantageous for cancer detection and monitoring, where non-invasive methods are highly sought after. The ability of exRNA to reflect the physiological and pathological state of cells and tissues positions it as a powerful tool for both diagnostic and prognostic purposes. Recent technological advancements have significantly broadened our understanding of exRNA biology. Key areas of progress include the elucidation of the mechanisms underlying exRNA biogenesis, the processes governing its compartmentalization within cells, and the sophisticated packaging systems that facilitate its incorporation into extracellular vesicles. These insights have paved the way for exploring exRNA not only as biomarkers but also as potential therapeutic agents, given their ability to mediate intercellular communication and influence various biological pathways. Despite these promising developments, several challenges remain in the field. The isolation and purification of exRNA present significant technical hurdles, as achieving high purity and yield is crucial for reliable downstream analyses. Variability in isolation methods can lead to inconsistencies in data, underscoring the need for standardized protocols to ensure reproducibility and accuracy in exRNA research. This chapter delves into the intricate processes involved in exRNA biogenesis, detailing the cellular machinery and pathways responsible for its production and release. We will explore the diverse sources of exRNA, the regulatory mechanisms that govern its expression and secretion, and the various types of extracellular vesicles that serve as carriers. The subsequent chapter will then transition to examining the functional roles of exRNA in human diseases, with a particular emphasis on its utility in diagnostic and therapeutic applications.

 
P. E. Saw, E. Song, *RNA Therapeutics in Human Diseases*,
https://doi.org/10.1007/978-981-96-3041-7_12

By providing a comprehensive overview of exRNA biology and its potential clinical implications, we aim to highlight the transformative impact of this emerging field on modern medicine.

## 12.1 Introduction

Extracellular RNA (exRNA) constitutes a pivotal avenue for conveying biological information, effectively reflecting both intra- and intercellular dynamics in real time. This category of RNAs, released in varied subpopulations across distinct cell lines, offers an innovative perspective into the intricate mechanisms of cellular communication. By traversing cellular boundaries, exRNAs participate in intercellular signaling, potentially influencing physiological processes and disease states. Despite their promising role as biomarkers and signaling molecules, the study of exRNAs is fraught with analytical challenges, primarily stemming from the diversity of isolation protocols. These protocols, designed to extract exRNAs from complex biological matrices, introduce variability that complicates their reliable and consistent analysis.

## 12.2 Biogenesis of exRNA and Their Role in Biology

exRNA predominantly evades degradation through integration into vesicles with membranes or creating chemical bond with lipids or proteins. Distinct EV subtypes, such as exosomes, microvesicles, and oncosomes, play varied roles in their biogenesis, each with unique budding processes and content compositions including mRNA, ncRNAs, DNA, proteins, and lipids. Recent technological advancements have facilitated the characterization of extracellular vesicle subtypes, revealing distinct subsets such as small exosomes (Exo-S), large exosomes (Exo-L), and exomeres. Notably, certain miRNAs and RNA-binding proteins (RBPs), pivotal to ncRNA sorting mechanisms, show a stronger association with exomeres. Identifying signature proteins capable of distinguishing EV subtypes remains a crucial challenge for leveraging the diagnostic and therapeutic potential of ncRNAs within different exosome subtypes (summarized in Table 12.1). These components facilitate intercellular communication, influencing a range of biological functions from developmental to pathological processes, including cancer metastasis (Fig. 12.1).

**Table 12.1** Differences between exosomes, microvesicles, and oncosomes in encapsulation of extracellular RNAs

| Feature | Exosomes | Microvesicles | Oncosomes | Small exosomes (Exo-S) | Large exosomes (Exo-L) | Exomeres |
|---|---|---|---|---|---|---|
| Size range | ~30–150 nm | ~100–1000 nm | ~500 nm to several micrometers | ~30–100 nm | ~150–500 nm | ~35–50 nm |
| Biogenesis | Endosomal origin; formed by inward budding of multivesicular bodies (MVBs) | Plasma membrane budding | Originates from cancer cells; may be larger | A subset of exosomes with smaller size | A subset of exosomes with larger size | Unique biogenesis involving direct budding from the plasma membrane |
| Cellular origin | Widely produced by various cell types | Derived from the outward budding of the plasma membrane | Specific to cancer cells, but can be found in other cells | Smaller exosomes, likely derived from the same process as regular exosomes | Larger exosomes, potentially derived from similar mechanisms as regular exosomes | Derived from a distinct, less well-characterized process |
| Molecular content | Protein, lipids, RNAs (mRNA, miRNA), lipids | Protein, lipids, RNAs (mRNA, miRNA) | Typically enriched with oncogenic factors, altered RNA and proteins | Similar to exosomes but smaller in size | Similar to exosomes but larger in size | Potentially enriched with different proteins and RNAs than other exosomes |
| Functional roles | Cell communication, immune modulation, biomarker potential | Cell signaling, tissue repair, and inflammation | Tumor progression, metastasis, and immune modulation | May have specialized roles in signaling due to their size | May have roles in larger-scale cell communication or tissue interactions | Potentially novel roles in cellular processes due to unique size and content |
| Isolation techniques | Ultracentrifugation, size exclusion chromatography, immunoaffinity capture | Ultracentrifugation, filtration, precipitation | Ultracentrifugation, density gradient centrifugation | Similar techniques as regular exosomes, with a focus on size fractionation | Similar techniques as regular exosomes, with adjustments for size | Specialized techniques needed for isolation, often involving specific density or size fractionation methods |

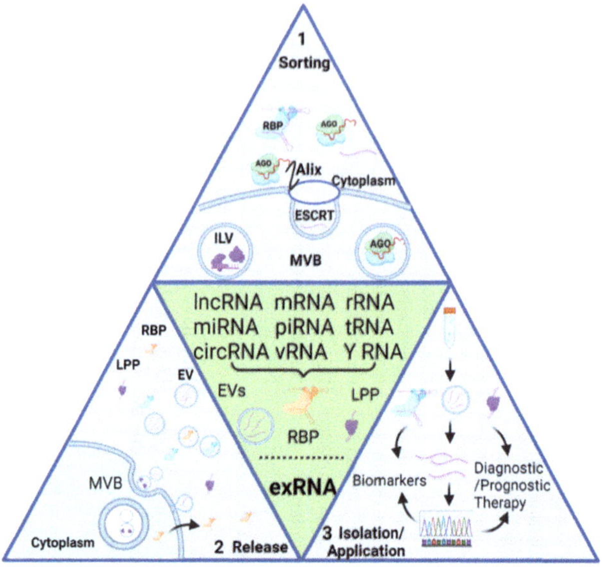


**Fig. 12.1** The process of sorting, releasing, and employing extracellular RNAs (exRNAs) involves complex mechanisms [1]. The central process involves various RNA types and fragments associating with extracellular vesicles (EVs), RNA-binding proteins (RBPs), and lipoproteins such as HDL. Specifically, the sorting of RNAs into EVs is mediated by RBPs recognizing specific motifs or structures, with RNA modifications like uridylation being particularly enriched in EVs [2]. Argonaute proteins, key partners in miRNA binding, and other components like hnRNPA2B1, are implicated in directing miRNAs into EVs. The release of exRNAs is facilitated through their combination with RBPs, lipoproteins, or EVs themselves [3]. For isolation and application, various techniques such as differential ultracentrifugation and immunocapture are employed to separate extracellular RBPs, lipoproteins, and vesicles, allowing for the purification of RNAs for sequencing or functional assays. These processes highlight the potential of exRNAs and their carriers as biomarkers or therapeutic agents

## 12.3 Coding RNAs: Extracellular mRNAs

mRNA represents a minor fraction of exRNA, making up roughly 2% of the total [1]. This scarcity of mRNA molecules presents a significant obstacle for sequencing efforts focused on extracellular mRNA. Such mRNA can be extracted from various biological fluids, including blood, urine, saliva, and cerebrospinal fluid, as well as from the supernatants of cell cultures in vitro. Most research on extracellular mRNA thus far has concentrated on mRNA associated with EVs. Yet, the composition of exRNA profiles may shift depending on the cell culture protocols or vesicle extraction methods utilized [1, 2]. The abundance of different exRNA subpopulations changes across fractions following ultracentrifugation [2], meaning the choice of EV extraction technique will undoubtedly influence the outcomes of exRNA sequencing. As much as 13% of RNA sequences from FBS can align with the human or mouse genomes, suggesting that RNA from FBS substantially contributes to inaccurate results in exRNA sequencing studies involving humans and mice [2]. Analysis of existing public exRNA sequencing data revealed that in exosomes, an estimated 2.6–17.2% of exRNA sequences were specific to bovines [2]. In the case of extracellular miRNA sequencing, the variety of unique miRNAs identified levels off once sequencing reaches around ten million reads [1]. The ideal sequencing depth for extracellular mRNA remains undetermined. The process of measuring mRNA levels necessitates the normalization against reliable reference controls, which have yet to be determined for any exRNA subpopulation, including mRNA [2]. Furthermore, for research into biomarkers for various diseases, there is a notable absence of a standard healthy control for comparison. The exploration into extracellular mRNA sequencing is thus fraught with significant hurdles, including low extraction yields, contamination from FBS-derived mRNA, and the absence of a unified approach for preparing mRNA-seq libraries.

## 12.4 Coding RNAs: Transfer RNAs

Transfer RNA (tRNA), comprising 78 to 90 nucleotides, plays a crucial role in protein synthesis by transporting amino acids to the ribosome. Here, the tRNA's anticodon pairs with the corresponding mRNA codon, facilitating the linking of amino acids into proteins [3]. Beyond its traditional role, recent studies highlight tRNA's involvement in various other cellular activities, including mRNA slicing, translation suppression, and inducing morphological alterations [4]. Notably, tRNAs are predominantly found in exosomes, surpassing the presence of other types of cellular RNAs [5–7]. Such a significant presence of tRNA in the EVs from diverse cell types, including breast cells, stem cells from bone marrow and adipose tissues, lung cells, and bodily fluids like semen, urine, and blood serum, has been documented [6–9]. Conversely, melanoma-derived EVs contain markedly less tRNA, with the tRNA content in microvesicles and apoptotic bodies these cells release mirroring that in the cytoplasm [10]. The selective presence of certain tRNAs in EVs underscores a specific pattern of tRNA release. For instance, analysis of EVs from mouse dendritic and T cells showed a sevenfold increase in tRNA-Lys-AAA sequences compared to those inside cells [5]. EVs are also rich in tRNA fragments (tRFs), although their precise roles remain largely unexplored. However, they are believed to participate in certain regulatory functions [11]. For example, the 5′ end fragment of tRNA-Gly can suppress protein synthesis, whereas smaller tRFs can non-specifically hinder translation [12, 13]. Similar enrichments of 5'-tRNA halves in EVs have been observed in various cell lines, including breast and lung cancer [6]. The mechanisms behind the enrichment of tRNAs and their halves in EVs, and their potential functional transfer to other cells, are areas of on-going investigation.

## 12.5 ncRNAs: Extracellular miRNAs

Extracellular miRNAs are short, ncRNAs crucial for post-transcriptional gene regulation. They influence a wide range of cellular mechanisms, serving both as regulators and stabilizers against transcriptional variability [14, 15]. It is estimated that miRNAs target around 60% of all mRNAs [16]. The biogenesis of miRNAs involves several steps, starting with the transcription of long precursor molecules, pri-miRNAs, which can be located near or far from their target mRNAs [17]. These pri-miRNAs, which form hairpin structures, are processed by Drosha to pre-miRNAs and then transported out of the nucleus. In the cytoplasm, Dicer cleaves these pre-miRNAs, producing mature miRNAs that integrate into the miRNA-induced silencing complex (miRISC) alongside Ago2 [18]. miRNAs regulate mRNA translation through competitive inhibition or destabilization, which may involve polyA tail shortening or direct cleavage by Ago2 for highly complementary targets [19–21]. The targeting is primarily determined by the miRNA's seed sequence, allowing a single miRNA to regulate multiple mRNAs [22]. Despite their cellular abundance, miRNAs are also found extracellularly, within vesicles or bound to lipoproteins, facilitating distant cell-to-cell communication and serving as potential biomarkers for various conditions [23–25].

## 12.6 Non-coding RNAs: Extracellular lncRNAs

Multiple studies have demonstrated that the transfer of lncRNAs via EVs plays a critical role in the development and progression of gastrointestinal cancers. Specifically, small EVs (sEVs) carrying the lncRNA ZFAS1 have been shown to boost the growth and mobility of gastric cancer cells [26]. Li and colleagues found that in pancreatic ductal adenocarcinoma (PDAC), highly metastatic cells distribute lncRNA-Sox2ot to other PDAC cells through sEVs. This lncRNA binds competitively to the miR-200 family, leading to an increase in Sox2 levels, which in turn facilitates the EMT and the enhancement of stem cell-like features, contributing to PDAC's aggressiveness and spread. Additionally, the presence of lncRNA-Sox2ot within sEVs in the bloodstream is associated with the TNM classification, presence of lymphatic or vascular invasion, and patient survival rates, suggesting its potential as a prognostic indicator for pancreatic cancer [27]. On the other hand, EV-carried lncRNAs also significantly affect the response of gastrointestinal cancers to chemotherapy via various mechanisms. For instance, sEVs from cisplatin-resistant gastric cancer cells can transfer the lncRNA HOTTIP to cisplatin-sensitive cells, where it targets miR-218 to activate HMGA1, thereby inducing cisplatin resistance [28]. The level of lncRNA HOTTIP in circulating sEVs is markedly higher in patients with cisplatin-resistant gastric cancer compared to those sensitive to cisplatin, highlighting its diagnostic and therapeutic potential. In colorectal cancer, sEVs containing lncRNA UCA1 from cetuximab-resistant cells can be assimilated by other cancer cells, increasing UCA1 expression and promoting resistance to cetuximab treatment in laboratory settings [29]. Furthermore, the progression and drug resistance of gastrointestinal cancers are influenced by lncRNAs transported by larger EVs. For example, hepatocellular carcinoma (HCC) cells can export lncRNA TUC339 via large EVs, which when absorbed by nearby cells, promote tumor growth, and reduce cell adherence to the ECM [30]. Takahashi and team observed that large EVs are also vehicles for the lincRNA-ROR in HCC, contributing to TGFβ-induced resistance to chemotherapy. Reducing lincRNA-ROR levels increases cell sensitivity to sorafenib, camptothecin, or doxorubicin, leading to enhanced apoptosis through the p53 pathway and a decrease in $CD133^{+}$ tumor-initiating cell numbers. This suggests that targeting lincRNA-ROR could enhance chemosensitivity in HCC [31].

## 12.7 Non-coding RNAs: Extracellular piRNAs

Piwi-interacting RNAs (piRNAs), the most extensive class of small non-coding RNA molecules within animal cells, are composed of potentially hundreds of thousands of unique species [32]. Measuring between 22 and 30 nucleotides in length, piRNAs play a pivotal role in safeguarding germ cells against transposons. They function by directing PIWI family Argonaute proteins to RNA sequences derived from these transposable elements, facilitating the cleavage and subsequent silencing of transposon RNA [33]. Unlike other small RNAs, piRNAs are produced from single-stranded precursors without the involvement of Dicer, utilizing two RNP complexes for their generation [34]. Within the cytoplasm, piRNAs and PIWI proteins congregate in specific structures known as germ granules, pi-bodies, and piP-bodies, which are analogous to processing bodies found in germ lines [35]. The occurrence of piRNAs outside cells has been less understood. A breakthrough came in 2016 when Freedman et al. discovered 144 small RNAs in the bloodstream that aligned with piRNAs, marking their notable presence in biological fluids [32]. This study also differentiated piRNAs from other extracellular RNAs through unique reverse transcriptase experiments, affirming the specificity of piRNA 3′-modifications via RT-qPCR analysis. The research showed that the expression levels of most piRNA species were comparable across B cells, neutrophils, peripheral mononuclear cells, platelets, and T cells, with significant upregulation observed in certain cell types compared to plasma. Notably, piRNA levels in exosomes did not significantly diverge from or were downregulated compared to plasma [32]. The mechanisms behind piRNA distribution into EVs and their potential functional transfer remain subjects of on-going inquiry.

## 12.8 ncRNAs: Extracellular YRNAs

YRNAs constitute a highly conserved class of small non-coding RNA, ranging from 84 to 113 nt in length in humans and surpassing the abundance of tRNA and U6 snRNA within cells [36]. Although their presence in extracellular vesicles, including microvesicles, exosomes, and RNPs, is reduced compared to cellular levels, YRNAs are more prevalent than most miRNAs. Despite their ubiquity across various biological contexts, only four YRNA species exist in humans: Y1, Y3, Y4, and Y5.

**Box 12.1 YRNAs: A Fascinating Class of Small ncRNAs**
YRNAs are involved in several fundamental biological processes, including DNA replication, RNA stability, and cellular stress responses [37, 38]. Their reduced presence in extracellular vesicles compared to cellular levels suggests that they might have distinct regulatory roles within the cell that are not entirely mirrored in the extracellular environment [39]. Interestingly, the four human YRNA species—Y1, Y3, Y4, and Y5—exhibit a high degree of conservation across different species, indicating their essential roles in cellular functions. In humans, there are four YRNA species: Y1, Y3, Y4, and Y5 [40]. Each of these YRNAs has unique characteristics and potential roles in cellular function, despite the limited understanding of their precise biological mechanisms. Most studied YRNA species are defined below [37–39].

**YRNA Y1**

YRNA Y1 is the most studied among the YRNA species. It is approximately 112 nucleotides long and plays a critical role in the initiation of DNA replication. Y1 interacts with various proteins, including those involved in the replication machinery, such as the origin recognition complex (ORC) and DNA polymerase alpha. The interaction between Y1 and these proteins is essential for the formation of the replication fork and the initiation of DNA synthesis.

**YRNA Y3**

YRNA Y3 is around 101 nucleotides in length and shares a similar role with Y1 in DNA replication. However, Y3 also appears to have unique functions that are distinct from Y1. It is involved in the regulation of RNA stability and may play a role in cellular stress responses. Y3 is known to form ribonucleoprotein complexes, which can influence the stability and translation of other RNAs within the cell. The exact mechanisms of how Y3 influences these processes are still under investigation.

**YRNA Y4**

YRNA Y4 is the least characterized of the four YRNAs, with a length of approximately 93 nucleotides. It is less abundant in cells compared to Y1 and Y3, and its specific functions are not well understood. Preliminary research

(continued)

**Box 12.1** (continued)
suggests that Y4 may be involved in the regulation of gene expression and cellular responses to environmental stress. Y4's interaction with proteins and other RNAs hints at a potential role in the broader regulatory networks within the cell, but more studies are needed to elucidate its precise functions.

**YRNA Y5**

YRNA Y5 is about 83 nucleotides long and is distinguished by its unique sequence and structure compared to the other YRNAs. Y5 has been implicated in the regulation of cellular differentiation and development. It may also play a role in the cellular response to DNA damage, helping to maintain genome stability. Y5's interactions with specific proteins involved in DNA repair pathways suggest that it could be crucial for protecting the cell from genomic instability.

***Common Characteristics and Research Challenges***

Despite their differences, all four YRNA species share some common features. They are highly conserved across different species, indicating their essential roles in cellular biology. YRNAs are also known to be associated with the Ro60 ribonucleoprotein, which is involved in autoimmune diseases like systemic lupus erythematosus. This association suggests that YRNAs might have broader implications in immune regulation and disease. Despite their conservation and prevalence, the precise mechanisms by which YRNAs operate and their specific targets within the cell remain largely unexplored. The current state of YRNA research is marked by a limited number of studies, underscoring the nascent stage of this field. The fewer than 100 published studies on YRNAs have primarily focused on their basic characterization, their biogenesis, and their involvement in fundamental cellular processes. However, recent advancements suggest that YRNAs might also have significant implications in disease contexts, particularly in cancer and autoimmune disorders, where their expression patterns and functional roles could provide insights into disease mechanisms and potential therapeutic targets.

As research progresses, it is anticipated that a deeper understanding of YRNAs will emerge, revealing their contributions to cellular homeostasis and their potential as biomarkers for various diseases [36]. YRNAs' lower stem domain is implicated in RNA quality control and the degradation of misfolded RNAs, engaging with the Ro60 chaperone and the PNPase exoribonuclease [41]. The upper stem may contribute to the initiation of chromosomal DNA replication [42]. YRNAs can be processed into shorter fragments, primarily featuring modifications at the 5′-end. Initially misidentified as miRNAs, these YRNA fragments have been recognized as distinct entities, with a Dicer-independent biogenesis and lacking association with Ago2 [43]. Contrary to miRNAs, YRNA fragments do not appear to silence gene expression similarly, as demonstrated in luciferase reporter assays [44]. Their biological roles remain largely undefined, although emerging evidence suggests involvement in cellular damage [45] and the processing of histone mRNA [46]. Interestingly, YRNA fragments are more prevalent in extracellular than cellular RNA, particularly within RNPs, where they constitute about 20% of non-rRNA small RNAs [47]. Future research into their creation, release, and functions promises to unveil new insights into RNA biology.

## 12.9 Other Emerging Extracellular ncRNAs

Among the newly identifiable exRNAs, there is notable expression of small nuclear RNAs (snRNAs), and small nucleolar RNAs (snoRNAs) [48–51]. snRNAs, in particular, have been localized within EVs. While their roles inside cells, primarily related to RNA splicing, methylation, and pseudouridylation, are well established [52], their functions in extracellular spaces remain largely unexplored. In 2011, Appaiah et al. discovered an upregulation of U6 small nuclear RNA in the serum of cancer patients [53]. Further analysis revealed a diverse array of snoRNAs within EV RNAs derived from human plasma, with C1orf213, LINC00324, and LOC388692 being notably abundant [48]. Despite scarce data on their expression and role in human tissues, recent studies utilizing the ion proton system in plasma samples from 40 individuals indicated that snoRNAs mainly facilitate chemical modifications of other RNAs, influencing alternative splicing and gene silencing [32].

For many years, the existence of RNA outside cells has been acknowledged, yet the functional roles of exRNAs remained a mystery until the discovery of functional transfers of mRNAs and miRNAs between mammalian cells [54–57]. Furthermore, exRNAs have emerged as valuable biomarkers for diseases, accessible through non-invasive liquid biopsies [58]. It is believed that exRNAs reside within membranous nanoparticles, such as EVs, which originate either from the cell's plasma membrane (e.g., ectosomes, microvesicles) or endosomal compartments (e.g., exosomes) [59]. Research to date has primarily examined the initiation and conclusion of the exRNA transfer process, focusing on how exRNAs are packaged and released by cells and their subsequent uptake by recipient cells, which is considered essential for the exRNAs' mechanisms of action. Recent studies on exRNA processing [60, 61], however, lead us to explore two previously under-considered aspects: first, beyond the selective inclusion into EVs, what alternative mechanisms account for the release of RNAs into the extracellular domain, and second, what occurs post-release and prior to cell uptake that influences exRNA dynamics?

Research indicated that the pathways through which RNAs are released are likely more varied and complex than previously understood, and that the processes occurring after release significantly affect exRNA levels. The notion that there is a universal model for these processes is increasingly contradicted by empirical data. While the selective incorporation of specific RNAs into EVs has been extensively documented, it is plausible that various intracellular RNAs might also be expelled into the extracellular space, either passively in response to concentration gradients or as a result of cell death, independently of vesicles. Additionally, extracellular modification of exRNAs represents a broad spectrum of activities, with certain exRNAs being more effectively shielded from degradation than others. Hence, an increase in the extracellular presence of a specific RNA type does not automatically indicate that it was selectively released, nor does the absence of selective packaging negate the possibility of the RNA serving a functional role.

## 12.10 Pathways for RNA Sorting in Extracellular Space

To elucidate the pathways through which RNAs are released into the extracellular environment, researchers have employed methods such as fluorescence microscopy to label and monitor RNAs. However, these cell biology-based strategies may not fully capture the diversity of RNAs found outside cells. A more comprehensive approach involves sequencing RNAs from both inside and outside cells and conducting differential expression analyses. This technique has led several research teams to discover that the profiles of extracellular microRNAs significantly differ from those within the parent cells, indicating a process of selective miRNA export [62–64]. Additionally, analyses focusing on motifs enriched in microRNAs either concentrated in or excluded from EVs have identified specific short sequences that predominate in one setting over the other [65]. The presence of these motifs strongly implies that RBPs play a crucial role in recognizing particular microRNAs and aiding their release, with a decade of research pinpointing several RBPs as key facilitators of this selective export [66, 67].

This hypothesis of selective export has been a prevailing theme in the study of extracellular miRNAs. Nonetheless, contradicting findings also exist. Notably, two separate studies with human MCF-7 cells cultured in serum-free medium showed a strong correlation between the levels of microRNAs inside and outside cells [6, 68]. Furthermore, an analysis targeting microRNAs in EV-depleted fetal bovine serum (FBS), which was used to culture cells, indicated that many microRNAs previously believed to be enriched in EVs might actually originate from the FBS [2, 69]. Given the high sequence similarity among miRNAs, distinguishing between those derived from FBS and human cells poses a significant challenge [70]. FBS is not the sole potential source of external microRNAs; recombinant protein supplements in serum-free media may also introduce microRNAs, such as miR-122-5p and miR-451a [71]. Consequently, without accounting for these sources of bias, differential abundance analyses between intracellular and extracellular samples may mistakenly identify these contaminant RNAs as being selectively released by cells.

## 12.11 The Extracellular Space as a Biogenetic Compartment

Reports have indicated that RNA processing occurs within EVs, suggesting a unique microenvironment within these vesicles that shields exRNAs from external RNases. RNase protection assays often serve as the preferred method to demonstrate the internal presence of exRNAs within EVs, safeguarded from extracellular RNase activity [72, 73]. Despite this, not all exRNAs are housed within EVs; those existing outside these vesicular confines, known as non-vesicular exRNAs, are significantly more susceptible to RNase degradation. Interestingly, a considerable portion of microRNAs found in cell culture media or human blood plasma exists in RNP complexes, notably with the AGO2 protein [68, 74], making non-vesicular RNAs a substantial, if not dominant, component of the exRNA landscape [6]. Despite their prevalence, non-vesicular exRNAs have historically received less attention compared to their vesicle-encased counterparts. However, recent studies have shed light on these RNAs, offering new insights into exRNA metabolism through the inhibition of extracellular RNases [75] or CRISPR/Cas9-mediated mutations in the RNase 1 gene [76]. These investigations revealed the presence of ribosomes, full-length tRNAs, and YRNAs in extracellular, non-vesicular samples, pointing to the role of RNases in fragmenting these RNAs into smaller pieces [75, 76]. The accumulation of certain ncRNA fragments, noted for their heightened extracellular stability, outlines the composition of the non-vesicular exRNAome under normal conditions [75]. Some fragments' durability, attributed to their ability to form dimers resistant to RNase, underscores the complex dynamics of non-vesicular ncRNA stability and biogenesis [77, 78].

## 12.12 The Dark Matter of the Extracellular RNAome

The term "dark matter of the transcriptome" refers to the vast portions of RNA transcripts that are produced by the genome but whose functions and significance remain largely unknown. This includes various types of ncRNAs that do not code for proteins but may have regulatory roles in gene expression and cellular processes. The "dark matter" concept highlights the complexity and the largely unexplored aspects of the transcriptome, much like how dark matter in cosmology refers to the unseen and poorly understood components of the universe. The exploration of the extracellular RNAome must expand beyond the notion of selective packaging as the sole mechanism influencing exRNA abundance. The limitations of current RNA sequencing technologies mean that our understanding of extracellular RNA compositions may not fully capture the diversity of exRNAs. Sequencing biases suggest we are only seeing a fraction of the "dark matter" within the RNA universe, overlooked due to incompatibilities with popular sequencing preparations [79]. Incorporating technical adjustments, such as demethylation or the use of thermostable reverse transcriptases, can alter the detected composition of extracellular samples [76, 80–83]. Furthermore, normalization challenges in exRNA studies highlight the need for a combination of advanced sequencing techniques and traditional methodologies, like Northern blot and RNA immunoprecipitation, to accurately profile the RNA and RNP contents of extracellular samples. The discovery of full ribosomes in non-vesicular fractions, only detectable with RNase inhibitors, points to a vast, uncharted territory of extracellular RNA entities awaiting discovery, underscoring the complex and dynamic nature of the extracellular RNAome.

**Box 12.2 Non-vesicular exRNAs: A Communication Wire from the Dead Cells?**

The inhibition of exRNA degradation has illuminated that profiles of non-vesicular exRNAs often resemble those found within intracellular lysates [75]. Though several mechanisms might account for this similarity [84–86], cell death emerges as a plausible contributor to the exRNA landscape. The prevalent oversight of cell death as a significant source of exRNAs likely stems from the rare detection of abundant species like full-length rRNAs and tRNAs in extracellular samples. Nonetheless, current evidence supports the initial presence of ribosomes and tRNAs in the extracellular space, with extracellular RNases playing a pivotal role in their swift degradation. This process disrupts the initial positive correlation between intracellular and non-vesicular exRNA profiles [75, 76]. The suggestion that extracellular ribosomes might exist in biofluids, such as blood serum [82], though not yet directly demonstrated, hints at cell death's potential role in shaping exRNA profiles beyond cell culture, possibly in vivo within solid tumors' hypoxic core or in cells attacked by lytic viruses. This situation could rationalize why immune cells might engage in exRNA surveillance, with pattern recognition receptors not only crucial for antiviral defences but also for identifying self-RNAs acting as damage-associated molecular patterns (DAMPs) [87]. Preliminary findings indicate that extracellular ribosomes might influence sterile inflammation when introduced to dendritic cells [75], suggesting a broader biological role for non-vesicular exRNAs.

Beyond serving as biomarkers, non-vesicular exRNAs might play various roles in biological processes. Their release could help balance RNA levels under normal conditions, alongside synthesis, degradation, excretion, and reuptake. The swift clearance of specific RNAs might trigger changes in cellular gene expression in response to stimuli [88]. Moreover, exRNA degradation could prevent the cytotoxic effects of uncontrolled RNase activity within the cytoplasm [89] and contribute to cellular energetics through the recycling of degradation products, similar to proteins [90]. Furthermore, non-vesicular exRNAs might facilitate intercellular communication, with extracellular RNases modulating this process by removing certain signals and generating others. For instance, RNase 1, which converts extracellular tRNAs into tRNA

(continued)

**Box 12.2** (continued)
halves [76] and is secreted by endothelial cells into the bloodstream [91], could shape the exRNAome and mitigate acute inflammation triggered by exRNAs acting as DAMPs. Considering other biofluids with potentially lower RNase activity than serum and plasma, it's worth noting the link between mutations in RNase A family members like Angiogenin (RNase 5) and neuro-inflammatory diseases [92]. While EVs are recognized for facilitating inter-cellular RNA transfer, specific tRNA halves can spontaneously enter cells from the extracellular space [93]. This raises questions about the potential roles of extracellular ribosomes and mRNAs [75]—could they convey messages from dying cells, serving as a final form of communication?

## 12.13 Conclusion

Exosomes, as nanoscale extracellular vesicles, play a crucial role in both intracellular and intercellular communication, safeguarding a variety of cargoes, including ncRNAs, from enzymatic degradation in the extracellular milieu. Their ability to be secreted through both ESCRT-dependent and ESCRT-independent pathways adds to their versatility in physiological and pathological contexts. NcRNAs within exosomes have emerged as key regulators in numerous processes, from pathogenic infections and inflammatory diseases to tumor dynamics, immunoregulation, and the modulation of treatment resistance, drawing significant attention in recent research. Despite the growing understanding of exosomal ncRNAs' involvement in these processes, the mechanisms governing their selective packaging and delivery to target cells remain elusive. This knowledge gap limits the clinical utility of exosomal ncRNAs as biomarkers for early disease diagnosis. Unraveling how cells specifically sort ncRNAs for exosomal inclusion could enable the identification of more representative exosomal RNAs, advancing early diagnostic capabilities. Exosomal ncRNAs have been shown to influence disease phenotypes through signaling pathways related to growth, invasion, metastasis, and angiogenesis. Animal studies confirm their significant impact, underscoring the importance of elucidating the early stages of disease-associated sorting, secretion, and dissemination of these ncRNAs for developing timely and effective interventions. While numerous methods exist for isolating exosomes, challenges persist due to the co-presence of extracellular ncRNAs with other vectors like lipoproteins and Argonaute proteins, complicating the precise study of exosomal ncRNAs. Efforts continue to refine exosome purification techniques to overcome these obstacles. Moreover, the recognition of exosomes as a diverse group of nanoparticles, each with unique extracellular vesicle properties, highlights the need for better characterization of exosome subtypes. Distinctions in the protein, lipid, and nucleic acid compositions of various exosome subtypes have been identified, emphasizing the heterogeneity of these secreted nanoparticles. With ongoing technological improvements for high-purity exosome separation, researchers are poised to delve deeper into the non-random distribution mechanisms of functional molecules within exosomes. Understanding the role and mechanistic underpinnings of exosomal ncRNAs holds promise for millions of patients, particularly those facing limited treatment options, potentially revolutionizing early disease detection and intervention strategies.

## References

1. Yuan T, Huang X, Woodcock M, Du M, Dittmar R, Wang Y, et al. Plasma extracellular RNA profiles in healthy and cancer patients. Sci Rep. 2016;6:19413.
2. Wei Z, Batagov AO, Carter DR, Krichevsky AM. Fetal bovine serum RNA interferes with the cell culture derived extracellular RNA. Sci Rep. 2016;6:31175.
3. Sharp SJ, Schaack J, Cooley L, Burke DJ, Söll D. Structure and transcription of eukaryotic tRNA genes. CRC Crit Rev Biochem. 1985;19(2):107–44.
4. Hurto RL. Unexpected functions of tRNA and tRNA processing enzymes. Adv Exp Med Biol. 2011;722:137–55.
5. Nolte-'t Hoen EN, Buermans HP, Waasdorp M, Stoorvogel W, Wauben MH, t Hoen PA. Deep sequencing of RNA from immune cell-derived vesicles uncovers the selective incorporation of small non-coding RNA biotypes with potential regulatory functions. Nucleic Acids Res. 2012;40(18):9272–85.
6. Tosar JP, Gámbaro F, Sanguinetti J, Bonilla B, Witwer KW, Cayota A. Assessment of small RNA sorting into different extracellular fractions revealed by high-throughput sequencing of breast cell lines. Nucleic Acids Res. 2015;43(11):5601–16.
7. Baglio SR, Rooijers K, Koppers-Lalic D, Verweij FJ, Pérez Lanzón M, Zini N, et al. Human bone marrow- and adipose-mesenchymal stem cells secrete exosomes enriched in distinctive miRNA and tRNA species. Stem Cell Res Ther. 2015;6(1):127.
8. Li M, Zeringer E, Barta T, Schageman J, Cheng A, Vlassov AV. Analysis of the RNA content of the exosomes derived from blood serum and urine and its potential as biomarkers. Philos Trans R Soc Lond Ser B Biol Sci. 2014;369:1652.
9. Vojtech L, Woo S, Hughes S, Levy C, Ballweber L, Sauteraud RP, et al. Exosomes in human semen carry a distinctive repertoire of small non-coding RNAs with potential regulatory functions. Nucleic Acids Res. 2014;42(11):7290–304.
10. Lunavat TR, Cheng L, Kim DK, Bhadury J, Jang SC, Lässer C, et al. Small RNA deep sequencing discriminates subsets of extracellular vesicles released by melanoma cells—evidence of unique microRNA cargos. RNA Biol. 2015;12(8):810–23.
11. Lee YS, Shibata Y, Malhotra A, Dutta A. A novel class of small RNAs: tRNA-derived RNA fragments (tRFs). Genes Dev. 2009;23(22):2639–49.
12. Ivanov P, Emara MM, Villen J, Gygi SP, Anderson P. Angiogenin-induced tRNA fragments inhibit translation initiation. Mol Cell. 2011;43(4):613–23.
13. Sobala A, Hutvagner G. Small RNAs derived from the 5′ end of tRNA can inhibit protein translation in human cells. RNA Biol. 2013;10(4):553–63.
14. Musilova K, Mraz M. MicroRNAs in B-cell lymphomas: how a complex biology gets more complex. Leukemia. 2015;29(5):1004–17.
15. Ebert MS, Sharp PA. Roles for microRNAs in conferring robustness to biological processes. Cell. 2012;149(3):515–24.
16. Friedman RC, Farh KK, Burge CB, Bartel DP. Most mammalian mRNAs are conserved targets of microRNAs. Genome Res. 2009;19(1):92–105.
17. Rodriguez A, Griffiths-Jones S, Ashurst JL, Bradley A. Identification of mammalian microRNA host genes and transcription units. Genome Res. 2004;14(10a):1902–10.

18. Pratt AJ, MacRae IJ. The RNA-induced silencing complex: a versatile gene-silencing machine. J Biol Chem. 2009;284(27):17897–901.
19. Filipowicz W, Bhattacharyya SN, Sonenberg N. Mechanisms of post-transcriptional regulation by microRNAs: are the answers in sight? Nat Rev Genet. 2008;9(2):102–14.
20. Eulalio A, Huntzinger E, Nishihara T, Rehwinkel J, Fauser M, Izaurralde E. Deadenylation is a widespread effect of miRNA regulation. RNA (New York, NY). 2009;15(1):21–32.
21. Hutvágner G, Zamore PD. A microRNA in a multiple-turnover RNAi enzyme complex. Science (New York, NY). 2002;297(5589):2056–60.
22. Nielsen CB, Shomron N, Sandberg R, Hornstein E, Kitzman J, Burge CB. Determinants of targeting by endogenous and exogenous microRNAs and siRNAs. RNA (New York, NY). 2007;13(11):1894–910.
23. Chen X, Liang H, Zhang J, Zen K, Zhang CY. Secreted microRNAs: a new form of intercellular communication. Trends Cell Biol. 2012;22(3):125–32.
24. Vickers KC, Remaley AT. Lipid-based carriers of microRNAs and intercellular communication. Curr Opin Lipidol. 2012;23(2):91–7.
25. Melman YF, Shah R, Danielson K, Xiao J, Simonson B, Barth A, et al. Circulating MicroRNA-30d is associated with response to cardiac resynchronization therapy in heart failure and regulates cardiomyocyte apoptosis: a translational pilot study. Circulation. 2015;131(25):2202–16.
26. Pan L, Liang W, Fu M, Huang ZH, Li X, Zhang W, et al. Exosomes-mediated transfer of long noncoding RNA ZFAS1 promotes gastric cancer progression. J Cancer Res Clin Oncol. 2017;143(6):991–1004.
27. Li Z, Jiang P, Li J, Peng M, Zhao X, Zhang X, et al. Tumor-derived exosomal lnc-Sox2ot promotes EMT and stemness by acting as a ceRNA in pancreatic ductal adenocarcinoma. Oncogene. 2018;37(28):3822–38.
28. Wang J, Lv B, Su Y, Wang X, Bu J, Yao L. Exosome-mediated transfer of lncRNA HOTTIP promotes cisplatin resistance in gastric cancer cells by regulating HMGA1/miR-218 Axis. Onco Targets Ther. 2019;12:11325–38.
29. Yang YN, Zhang R, Du JW, Yuan HH, Li YJ, Wei XL, et al. Predictive role of UCA1-containing exosomes in cetuximab-resistant colorectal cancer. Cancer Cell Int. 2018;18:164.
30. Kogure T, Yan IK, Lin WL, Patel T. Extracellular vesicle-mediated transfer of a novel long noncoding RNA TUC339: a mechanism of intercellular signaling in human hepatocellular cancer. Genes Cancer. 2013;4(7–8):261–72.
31. Takahashi K, Yan IK, Kogure T, Haga H, Patel T. Extracellular vesicle-mediated transfer of long non-coding RNA ROR modulates chemosensitivity in human hepatocellular cancer. FEBS Open Bio. 2014;4:458–67.
32. Ding L, Ren J, Zhang D, Li Y, Huang X, Hu Q, et al. A novel stromal lncRNA signature reprograms fibroblasts to promote the growth of oral squamous cell carcinoma via LncRNA-CAF/interleukin-33. Carcinogenesis. 2018;39(3):397–406.
33. Hu T, Hu J. Melanoma-derived exosomes induce reprogramming fibroblasts into cancer-associated fibroblasts via Gm26809 delivery. Cell Cycle (Georgetown, Tex). 2019;18(22):3085–94.
34. Cook KM, Figg WD. Angiogenesis inhibitors: current strategies and future prospects. CA Cancer J Clin. 2010;60(4):222–43.
35. Conigliaro A, Costa V, Lo Dico A, Saieva L, Buccheri S, Dieli F, et al. CD90+ liver cancer cells modulate endothelial cell phenotype through the release of exosomes containing H19 lncRNA. Mol Cancer. 2015;14:155.
36. Wu DM, Deng SH, Liu T, Han R, Zhang T, Xu Y. TGF-β-mediated exosomal lnc-MMP2-2 regulates migration and invasion of lung cancer cells to the vasculature by promoting MMP2 expression. Cancer Med. 2018;7(10):5118–29.
37. Guglas K, Kołodziejczak I, Kolenda T, Kopczyńska M, Teresiak A, Sobocińska J, et al. YRNAs and YRNA-derived fragments as new players in cancer research and their potential role in diagnostics. Int J Mol Sci. 2020;21:16.

38. Köhn M, Pazaitis N, Hüttelmaier S. Why YRNAs? About versatile RNAs and their functions. Biomol Ther. 2013;3(1):143–56.
39. Guglas K, Kolenda T, Stasiak M, Kopczyńska M, Teresiak A, Ibbs M, et al. YRNAs: new insights and potential novel approach in head and neck squamous cell carcinoma. Cells. 2020;9:5.
40. Valkov N, Das S. Y RNAs: biogenesis, function and implications for the cardiovascular system. In: Xiao J, editor. Non-coding RNAs in cardiovascular diseases. Singapore: Springer Singapore; 2020. p. 327–42.
41. Lang HL, Hu GW, Zhang B, Kuang W, Chen Y, Wu L, et al. Glioma cells enhance angiogenesis and inhibit endothelial cell apoptosis through the release of exosomes that contain long non-coding RNA CCAT2. Oncol Rep. 2017;38(2):785–98.
42. Lang HL, Hu GW, Chen Y, Liu Y, Tu W, Lu YM, et al. Glioma cells promote angiogenesis through the release of exosomes containing long non-coding RNA POU3F3. Eur Rev Med Pharmacol Sci. 2017;21(5):959–72.
43. Qiu JJ, Lin XJ, Tang XY, Zheng TT, Lin YY, Hua KQ. Exosomal metastasis-associated lung adenocarcinoma transcript 1 promotes angiogenesis and predicts poor prognosis in epithelial ovarian cancer. Int J Biol Sci. 2018;14(14):1960–73.
44. Wu Q, Wu X, Ying X, Zhu Q, Wang X, Jiang L, et al. Suppression of endothelial cell migration by tumor associated macrophage-derived exosomes is reversed by epithelial ovarian cancer exosomal lncRNA. Cancer Cell Int. 2017;17:62.
45. Chen C, Luo Y, He W, Zhao Y, Kong Y, Liu H, et al. Exosomal long noncoding RNA LNMAT2 promotes lymphatic metastasis in bladder cancer. J Clin Invest. 2020;130(1):404–21.
46. Eckert C, Klein N, Kornek M, Lukacs-Kornek V. The complex myeloid network of the liver with diverse functional capacity at steady state and in inflammation. Front Immunol. 2015;6:179.
47. Jenne CN, Kubes P. Immune surveillance by the liver. Nat Immunol. 2013;14(10):996–1006.
48. Fabbri M. Natural killer cell-derived vesicular miRNAs: a new anticancer approach? Cancer Res. 2020;80(1):17–22.
49. Fang P, Xiang L, Chen W, Li S, Huang S, Li J, et al. LncRNA GAS5 enhanced the killing effect of NK cell on liver cancer through regulating miR-544/RUNX3. Innate Immun. 2019;25(2):99–109.
50. Zheng Z, Chen M, Xing P, Yan X, Xie B. Increased expression of exosomal AGAP2-AS1 (AGAP2 antisense RNA 1) in breast cancer cells inhibits trastuzumab-induced cell cytotoxicity. Med Sci Monit. 2019;25:2211–20.
51. Xu CG, Yang MF, Ren YQ, Wu CH, Wang LQ. Exosomes mediated transfer of lncRNA UCA1 results in increased tamoxifen resistance in breast cancer cells. Eur Rev Med Pharmacol Sci. 2016;20(20):4362–8.
52. Xue Y, Tong L, LiuAnwei Liu F, Liu A, Zeng S, Xiong Q, et al. Tumor-infiltrating M2 macrophages driven by specific genomic alterations are associated with prognosis in bladder cancer. Oncol Rep. 2019;42(2):581–94.
53. Lu Y, Guo L, Ding G. PD1(+) tumor associated macrophages predict poor prognosis of locally advanced esophageal squamous cell carcinoma. Future Oncol (London, England). 2019;15(35):4019–30.
54. Park SA, Surh YJ. Modulation of tumor microenvironment by chemopreventive natural products. Ann N Y Acad Sci. 2017;1401(1):65–74.
55. Hanahan D, Coussens LM. Accessories to the crime: functions of cells recruited to the tumor microenvironment. Cancer Cell. 2012;21(3):309–22.
56. Shi Y, Du L, Lin L, Wang Y. Tumour-associated mesenchymal stem/stromal cells: emerging therapeutic targets. Nat Rev Drug Discov. 2017;16(1):35–52.
57. Kozlova N, Grossman JE, Iwanicki MP, Muranen T. The interplay of the extracellular matrix and stromal cells as a drug target in stroma-rich cancers. Trends Pharmacol Sci. 2020;41(3):183–98.

58. Ma P, Pan Y, Li W, Sun C, Liu J, Xu T, et al. Extracellular vesicles-mediated noncoding RNAs transfer in cancer. J Hematol Oncol. 2017;10(1):57.
59. Kagiya T. MicroRNAs and osteolytic bone metastasis: the roles of MicroRNAs in tumor-induced osteoclast differentiation. J Clin Med. 2015;4(9):1741–52.
60. Mohankumar S, Patel T. Extracellular vesicle long noncoding RNA as potential biomarkers of liver cancer. Brief Funct Genomics. 2016;15(3):249–56.
61. Kołat D, Hammouz R, Bednarek AK, Płuciennik E. Exosomes as carriers transporting long non-coding RNAs: molecular characteristics and their function in cancer (review). Mol Med Rep. 2019;20(2):851–62.
62. Villarroya-Beltri C, Baixauli F, Gutiérrez-Vázquez C, Sánchez-Madrid F, Mittelbrunn M. Sorting it out: regulation of exosome loading. Semin Cancer Biol. 2014;28:3–13.
63. Mittelbrunn M, Gutiérrez-Vázquez C, Villarroya-Beltri C, González S, Sánchez-Cabo F, González M, et al. Unidirectional transfer of microRNA-loaded exosomes from T cells to antigen-presenting cells. Nat Commun. 2011;2:282.
64. Cha DJ, Franklin JL, Dou Y, Liu Q, Higginbotham JN, Demory Beckler M, et al. KRAS-dependent sorting of miRNA to exosomes. elife. 2015;4:e07197.
65. Villarroya-Beltri C, Gutiérrez-Vázquez C, Sánchez-Cabo F, Pérez-Hernández D, Vázquez J, Martin-Cofreces N, et al. Sumoylated hnRNPA2B1 controls the sorting of miRNAs into exosomes through binding to specific motifs. Nat Commun. 2013;4:2980.
66. Mateescu B, Kowal EJ, van Balkom BW, Bartel S, Bhattacharyya SN, Buzás EI, et al. Obstacles and opportunities in the functional analysis of extracellular vesicle RNA—an ISEV position paper. J Extracell Vesicles. 2017;6(1):1286095.
67. Temoche-Diaz MM, Shurtleff MJ, Nottingham RM, Yao J, Fadadu RP, Lambowitz AM, et al. Distinct mechanisms of microRNA sorting into cancer cell-derived extracellular vesicle subtypes. elife. 2019:8.
68. Turchinovich A, Weiz L, Langheinz A, Burwinkel B. Characterization of extracellular circulating microRNA. Nucleic Acids Res. 2011;39(16):7223–33.
69. Tosar JP, Cayota A, Eitan E, Halushka MK, Witwer KW. Ribonucleic artefacts: are some extracellular RNA discoveries driven by cell culture medium components? J Extracell Vesicles. 2017;6(1):1272832.
70. Fromm B, Tosar JP, Lu Y, Halushka MK, Witwer KW. Human and cow have identical miR-21-5p and miR-30a-5p sequences, which are likely unsuited to study dietary uptake from cow Milk. J Nutr. 2018;148(9):1506–7.
71. Auber M, Fröhlich D, Drechsel O, Karaulanov E, Krämer-Albers EM. Serum-free media supplements carry miRNAs that co-purify with extracellular vesicles. J Extracell Vesicles. 2019;8(1):1656042.
72. Théry C, Witwer KW, Aikawa E, Alcaraz MJ, Anderson JD, Andriantsitohaina R, et al. Minimal information for studies of extracellular vesicles 2018 (MISEV2018): a position statement of the International Society for Extracellular Vesicles and update of the MISEV2014 guidelines. J Extracell Vesicles. 2018;7(1):1535750.
73. Driedonks TAP, Mol S, de Bruin S, Peters AL, Zhang X, Lindenbergh MFS, et al. Y-RNA subtype ratios in plasma extracellular vesicles are cell type- specific and are candidate biomarkers for inflammatory diseases. J Extracell Vesicles. 2020;9(1):1764213.
74. Arroyo JD, Chevillet JR, Kroh EM, Ruf IK, Pritchard CC, Gibson DF, et al. Argonaute2 complexes carry a population of circulating microRNAs independent of vesicles in human plasma. Proc Natl Acad Sci USA. 2011;108(12):5003–8.
75. Tosar JP, Segovia M, Castellano M, Gámbaro F, Akiyama Y, Fagúndez P, et al. Fragmentation of extracellular ribosomes and tRNAs shapes the extracellular RNAome. Nucleic Acids Res. 2020;48(22):12874–88.
76. Nechooshtan G, Yunusov D, Chang K, Gingeras TR. Processing by RNase 1 forms tRNA halves and distinct Y RNA fragments in the extracellular environment. Nucleic Acids Res. 2020;48(14):8035–49.

77. Gámbaro F, Li Calzi M, Fagúndez P, Costa B, Greif G, Mallick E, et al. Stable tRNA halves can be sorted into extracellular vesicles and delivered to recipient cells in a concentration-dependent manner. RNA Biol. 2020;17(8):1168–82.
78. Tosar JP, Gámbaro F, Darré L, Pantano S, Westhof E, Cayota A. Dimerization confers increased stability to nucleases in 5′ halves from glycine and glutamic acid tRNAs. Nucleic Acids Res. 2018;46(17):9081–93.
79. Giraldez MD, Spengler RM, Etheridge A, Godoy PM, Barczak AJ, Srinivasan S, et al. Comprehensive multi-center assessment of small RNA-seq methods for quantitative miRNA profiling. Nat Biotechnol. 2018;36(8):746–57.
80. Shurtleff MJ, Yao J, Qin Y, Nottingham RM, Temoche-Diaz MM, Schekman R, et al. Broad role for YBX1 in defining the small noncoding RNA composition of exosomes. Proc Natl Acad Sci USA. 2017;114(43):E8987–e95.
81. Qin Y, Yao J, Wu DC, Nottingham RM, Mohr S, Hunicke-Smith S, et al. High-throughput sequencing of human plasma RNA by using thermostable group II intron reverse transcriptases. RNA (New York, NY). 2016;22(1):111–28.
82. Akat KM, Lee YA, Hurley A, Morozov P, Max KE, Brown M, et al. Detection of circulating extracellular mRNAs by modified small-RNA-sequencing analysis. JCI Insight. 2019;5:9.
83. Chow FW, Koutsovoulos G, Ovando-Vázquez C, Neophytou K, Bermúdez-Barrientos JR, Laetsch DR, et al. Secretion of an Argonaute protein by a parasitic nematode and the evolution of its siRNA guides. Nucleic Acids Res. 2019;47(7):3594–606.
84. Lee KZ, Lestradet M, Socha C, Schirmeier S, Schmitz A, Spenlé C, et al. Enterocyte purge and rapid recovery is a resilience reaction of the gut epithelium to pore-forming toxin attack. Cell Host Microbe. 2016;20(6):716–30.
85. Lindenbergh MFS, Wubbolts R, Borg EGF, van't Veld EM, Boes M, Stoorvogel W. Dendritic cells release exosomes together with phagocytosed pathogen; potential implications for the role of exosomes in antigen presentation. J Extracell Vesicles. 2020;9(1):1798606.
86. Jeppesen DK, Fenix AM, Franklin JL, Higginbotham JN, Zhang Q, Zimmerman LJ, et al. Reassessment of exosome composition. Cell. 2019;177(2):428–45.e18
87. Gong T, Liu L, Jiang W, Zhou R. DAMP-sensing receptors in sterile inflammation and inflammatory diseases. Nat Rev Immunol. 2020;20(2):95–112.
88. Chiou NT, Kageyama R, Ansel KM. Selective export into extracellular vesicles and function of tRNA fragments during T cell activation. Cell Rep. 2018;25(12):3356–70.e4.
89. Thomas SP, Hoang TT, Ressler VT, Raines RT. Human angiogenin is a potent cytotoxin in the absence of ribonuclease inhibitor. RNA (New York, NY). 2018;24(8):1018–27.
90. Loo JM, Scherl A, Nguyen A, Man FY, Weinberg E, Zeng Z, et al. Extracellular metabolic energetics can promote cancer progression. Cell. 2015;160(3):393–406.
91. Koczera P, Martin L, Marx G, Schuerholz T. The ribonuclease a superfamily in humans: canonical RNases as the buttress of innate immunity. Int J Mol Sci. 2016;17:8.
92. van Es MA, Schelhaas HJ, van Vught PW, Ticozzi N, Andersen PM, Groen EJ, et al. Angiogenin variants in Parkinson disease and amyotrophic lateral sclerosis. Ann Neurol. 2011;70(6):964–73.
93. Ivanov P, O'Day E, Emara MM, Wagner G, Lieberman J, Anderson P. G-quadruplex structures contribute to the neuroprotective effects of angiogenin-induced tRNA fragments. Proc Natl Acad Sci USA. 2014;111(51):18201–6.

# Chapter 13
# Extracellular RNAs (exRNAs): Functions, Crosstalks, and Therapeutic Potential

**Abstract** We continue elucidating the biological role of exRNAs in this chapter. As mentioned in Chap. 12, exRNAs have emerged as groundbreaking circulating entities that serve as biomarkers across multiple disease states. Their unparalleled diversity and pivotal role in disease diagnosis underscore their importance as biomarkers. Extensive research has been dedicated to exploring exRNAs found in the bloodstream and various bodily fluids, aiming to evaluate their efficacy as indicators for health conditions and diseases. This chapter seeks to provide a comprehensive understanding of exRNAs, their biogenesis, and their function in human diseases, especially in cancer. Since most exRNAs are encapsulated within vesicles (i.e., exosomes, niosomes, EVs), we will also look at the significance of these vehicles in the function of exRNAs. Encapsulation protects exRNAs from degradation, facilitating their stable presence in circulation and enhancing their potential as reliable biomarkers. Exosomes and other EVs not only shield exRNAs but also aid in their targeted delivery to specific cells, thus playing a crucial role in intercellular communication and disease progression. In exploring the biogenesis of exRNAs, we will delve into the cellular mechanisms and pathways involved in their production, packaging, and release. Understanding these processes is essential for comprehending how exRNAs are generated and how their encapsulation within vesicles influences their stability and functionality. Additionally, we will discuss the role of exRNAs in various human diseases, with a particular focus on cancer. exRNAs have been implicated in tumorigenesis, metastasis, and the modulation of the tumor microenvironment. By acting as messengers between cancer cells and their surroundings, exRNAs can influence disease progression and response to treatment. We will examine specific examples where exRNAs serve as potential biomarkers for cancer diagnosis, prognosis, and therapeutic monitoring. Furthermore, we will address the prevailing challenges and obstacles encountered in the application of exRNAs as reliable disease markers. These challenges include technical issues related to the isolation and characterization of exRNAs, variability in detection methods, and the need for standardized protocols. We will incorporate insights from clinical and laboratory studies to highlight the current state of exRNA research and the efforts being made to overcome these hurdles. In summary, this chapter aims to provide a thorough understanding of the multifaceted roles of exRNAs, their

 
P. E. Saw, E. Song, *RNA Therapeutics in Human Diseases*,
https://doi.org/10.1007/978-981-96-3041-7_13

biogenesis, and their encapsulation within vesicles. By examining their significance in human diseases, particularly cancer, and addressing the challenges in their application as disease markers, we hope to shed light on the potential of exRNAs in advancing diagnostic and therapeutic strategies.

## 13.1 Introduction

Extracellular RNAs (exRNAs) represent a fascinating and diverse array of RNA molecules that extend beyond the confines of individual cells. These RNA species are present in virtually all bodily fluids, encompassing blood, saliva, urine, and cerebrospinal fluid, among others. This ubiquity underscores their potential importance in a wide range of biological processes and their utility in medical diagnostics [1]. The detection of exRNAs in these fluids is not a mere consequence of cellular turnover or random release; rather, it is the result of highly regulated processes [2]. Cells actively package and secrete exRNAs through various mechanisms, reflecting a sophisticated system of intercellular communication [3]. These exRNAs can convey information about the physiological state of the originating cells, making them invaluable indicators of both normal bodily functions and disease states.

One of the most striking features of exRNAs is their remarkable stability in the extracellular environment. Typically, RNA molecules are highly susceptible to degradation by ribonucleases (RNases), enzymes that are abundant in extracellular spaces. However, exRNAs are uniquely adapted to withstand these hostile conditions. This stability is largely attributed to their encapsulation within extracellular vesicles, such as exosomes and microvesicles, or their association with proteins and lipids. These protective measures shield exRNAs from enzymatic degradation, allowing them to persist in circulation and fulfill their roles as messengers. The stability and regulated presence of exRNAs in bodily fluids have significant implications for their use as biomarkers. Since they can reflect the physiological and pathological state of cells, exRNAs have emerged as promising candidates for non-invasive diagnostic and prognostic tools. For instance, specific exRNA profiles can indicate the presence of diseases such as cancer, cardiovascular disorders, and neurological conditions. Their detection in easily accessible fluids like blood and urine makes them particularly attractive for clinical applications, offering a less invasive alternative to traditional biopsy methods. Moreover, the biogenesis of exRNAs is an area of intense research interest. Understanding how exRNAs are produced, packaged, and released by cells can provide insights into their functional roles and mechanisms of action. This knowledge can further enhance their utility in clinical settings, enabling the development of targeted therapies that leverage exRNAs for disease treatment and management.

In addition to their potential as biomarkers, exRNAs play active roles in cellular communication. They can modulate gene expression in recipient cells, influencing various biological processes such as immune responses, cell proliferation, and apoptosis. This intercellular communication is crucial for maintaining homeostasis and orchestrating complex physiological responses. exRNAs are a diverse and stable group of RNA molecules that play critical roles in both normal physiology and disease. Their presence in bodily fluids and their resilience against degradation make them powerful tools for diagnostic and therapeutic applications. As research continues to uncover the intricacies of exRNA biogenesis and function, the potential for exRNAs to revolutionize medical practice becomes increasingly evident [4].

## 13.2 Part 1: Extracellular RNAs (exRNAs)

exRNAs are pivotal components of cellular communication networks, transcending the boundaries of their originating cells and tissues to mediate complex biological interactions. These RNA molecules are adept at conveying genetic and epigenetic information, thereby influencing cellular behavior and tissue function across great distances. This unique capability places exRNAs at the heart of numerous physiological and pathological processes. The transport mechanisms of exRNAs are particularly notable for their diversity and specificity. Vesicular carriers, such as exosomes and microvesicles, encapsulate exRNAs, providing a protective environment that shields them from enzymatic degradation while facilitating targeted delivery to recipient cells. These vesicles are produced by a variety of cell types and are known to carry a rich cargo of RNAs, proteins, lipids, and even DNA. Upon fusing with or being endocytosed by target cells, these vesicles release their contents, allowing exRNAs to modulate gene expression and influence cellular behavior directly. Non-vesicular transport represents another crucial pathway for exRNA mediation. In this mode, exRNAs are often bound to protein complexes. These RNA–protein complexes can include high-density lipoproteins (HDL) or specific RNA-binding proteins like Argonaute-2, which not only stabilize exRNAs but also facilitate their specific interaction with target cells. Through these complexes, exRNAs can be effectively sorted and directed to specific sites, enhancing the precision of intercellular signaling. We will detail these in the section below.

## 13.3 High-Density Lipoprotein (HDL)-Bound exRNAs

exRNAs that are associated with high-density lipoproteins (HDL) must be enclosed within exosomes to be effective. This encapsulation typically occurs through receptor-mediated endocytosis. The scavenger receptor class B type 1 (SR-B1) plays a crucial role in this process, facilitating the cellular uptake of miRNA complexes that are linked with HDL [5]. Recipient cells are equipped with a diverse array of membrane receptors that engage with the surface molecules on exosomes. These include tetraspanins, such as CD9, CD63, CD81, CD82, and CD151, which are integral in mediating various cellular interactions. Additionally, molecules involved in immune response, such as Class I and Class II major histocompatibility complex (MHC) proteins, play a role in antigen recognition. The presence of attachment proteins, including soluble NSF attachment protein receptors (SNAREs) and integrins, further facilitates the docking and fusion of exosomes with recipient cells [6–10]. When these receptors are activated, they can initiate a series of intracellular signaling pathways, integrate with the membrane of the target cell to release their contents internally, or aid the internalization of cargo into the recipient cell using various mechanisms. These mechanisms include clathrin- or caveolin-dependent endocytosis, which are processes involving the formation of small vesicles coated with specific proteins that facilitate the uptake of extracellular materials. Other methods such as phagocytosis and pinocytosis involve the engulfment of large particles or the ingestion of fluid, respectively. Micropinocytosis, a variant of pinocytosis, allows for the uptake of smaller volumes of extracellular fluid and its solutes. Lastly, lipid raft-mediated endocytosis involves lipid-rich microdomains within the cell membrane that help organize the entry of molecules into the cell. Each of these pathways plays a critical role in how cells receive and process external signals and materials, highlighting the complexity and precision of cellular communication and material transfer [11]. Other recently documented delivery routes include direct cell-to-cell transmission or through tunneling nanotubes [12].

## 13.4 exRNAs: Biological Significance in Cancer

### *13.4.1 Cancer-Derived exRNAs*

Studies have repeatedly shown that cancer cells emit significantly more exosomes than normal, healthy cells. This increased output of exosomes is recognized as a distinctive feature of cancer and is critically involved in the disease's advancement and underlying mechanisms. The excessive production of exosomes by cancer cells contributes to various aspects of cancer pathology, including the manipulation of the tumor microenvironment, promotion of metastasis, and evasion of the immune response. These exosomes carry cargos such as proteins, RNAs, and other molecules that can alter cell behavior, facilitate communication between cancerous and normal cells, and even prepare distant sites in the body for tumor metastasis [13–16]. A critical factor contributing to the elevated production of exosomes in cancer cells is the activation of specific cellular signaling pathways. For example, the p53-tumor suppression-activated pathway 6 (p53-TSA-6), also known as TP53-regulated inhibitor of apoptosis 1 (TRIAP1), plays a substantial role in this process. This pathway is involved in a variety of crucial cellular activities such as apoptosis (programmed cell death), cell survival, and the release of exosomes. In healthy cells, the p53 protein serves as a tumor suppressor, playing a key role in controlling the cell cycle and inducing apoptosis to prevent uncontrolled cell growth. However, in many cancerous cells, mutations in p53 or changes in related pathways may disrupt these regulatory mechanisms. Such alterations can lead to the aberrant secretion of exosomes, which in turn supports the processes of tumorigenesis and metastasis by enhancing the cancer cells' ability to communicate and manipulate their environment [14]. Furthermore, the exosome secretion is amplified by stress mechanisms activated under hypoxic conditions common within the TME [15], as well as the overexpression of heparanase, the only enzyme that degrades heparan sulfate, in numerous cancer cells [16]. However, it remains uncertain if tumor cells utilize varied strategies for exosome discharge based on the type of cancer or its level of aggression. Exosomes selected by cancer cells have the ability to influence adjacent cells through paracrine signaling or even affect cells in distant tissues via the circulatory system [17].

Cancer cells deploy exRNAs to modify the surrounding ECM into a pre-tumorigenic state [18]. These exRNAs could alter immune responses toward cancer cells, stimulate stromal fibroblasts to become tumorigenic, create a drug resistance environment, and trigger angiogenesis and metastasis. Cancer cells package hTERT mRNA into exosomes—a process that may be influenced by specific oncogenic signals or cellular stress responses. hTERT, the catalytic subunit of telomerase, is crucial for telomere maintenance and is commonly upregulated in cancer cells, contributing to their immortality. Once hTERT mRNA is loaded into exosomes, these vesicles are released from cancer cells into the surrounding microenvironment. Non-cancerous cells in the vicinity, such as fibroblasts, can then take up these exosomes through processes such as endocytosis. Upon delivery into recipient fibroblasts, the hTERT mRNA is translated into the telomerase enzyme. This expression of telomerase in fibroblasts, which typically have low or non-existent telomerase activity, can lead to the extension of their cellular lifespan and alterations in their properties to support tumor growth [19]. Similarly, glioblastoma cells utilize exosomes carrying exRNA that include mRNA that activates the EGFR, as a method of influencing the cells in their surrounding environment. By transferring these specific mRNA molecules to normal cells nearby, the cancer cells promote angiogenesis, which is the formation of new blood vessels. This process is crucial for providing the tumor with the necessary nutrients and oxygen for its growth. Additionally, these exosomes help suppress immune responses in the TME [20]. Lung cancer cells release exosomes that are enriched with miRNAs like miR-21 and miR-29a, which are then taken up by macrophages. miR-21 and miR 29a within these exosomes interact with TLR8, a pattern recognition receptor primarily found on myeloid cells like macrophages. TLR8 typically responds to viral RNA, suggesting that the engagement of miR-21 and miR-29a with TLR8 mimics a viral infection, triggering an immune response. The engagement of these miRNAs with TLR8 leads to the activation of the NF-κB pathway within macrophages to produce pro-inflammatory cytokines, including IL-6 and IL-8. The secretion of IL-6 and IL-8 not only helps in remodeling the TME but also supports the establishment of a pre-metastatic niche [21]. Similarly, extracellular miR-21 from CRC cells can initiate the TLR7-IL6 pathway, creating pro-inflammatory environment and encouraging the formation of pre-metastatic niches [22]. In addition, exRNAs from tumor cells are also implicated in creating niches resistant to chemotherapy. For example, gemcitabine in pancreatic cancer therapy triggers miR-155 expression, increasing exosome secretion, leading to increasing anti-apoptosis activity in neighboring cells, and conferring chemoresistance [23].

## 13.5 exRNAs Derived from Senescent Cells (SnCs)

Cellular senescence is a fundamental biological process where cells enter a state of permanent growth arrest without undergoing cell death. This process serves as a critical protective mechanism against uncontrolled cell proliferation and cancer development. Triggered by various stresses including oncogenic stimuli, DNA damage, or telomere shortening, senescence is characterized by specific molecular and cellular changes [24, 25]. One of the primary pathways activated during senescence, the p53 protein acts as a tumor suppressor by inducing the expression of p21, a cyclin-dependent kinase inhibitor. p21 then acts to inhibit cell cycle progression by deactivating cyclin-CDK complexes, leading to cell cycle arrest in the G1 phase. Another critical senescence pathway involves p16INK4a, which inhibits CDK4 and CDK6, leading to the activation of the retinoblastoma protein (pRB). Activated pRB represses E2F-target genes that are essential for S-phase entry, thus enforcing cell cycle arrest [26]. The absence of senescence mechanisms in mice and humans has been linked to a heightened susceptibility to cancer [27].

Cellular senescence is marked not only by a halt in cell division but also by a profound change in cellular activity, including alterations in the secretion profile of cells. This change is manifested as the senescence-associated secretory phenotype (SASP), which involves the release of a wide array of molecules, including cytokines, chemokines, growth factors, and proteases, as well as extracellular vesicles like exosomes. SASPs engage in communication with adjacent cells and tissues, thereby inducing localized and systemic inflammatory responses. SASPs can also spread the senescent phenotype to adjacent and neighboring cells, inducing damages [28, 29]. SASP can reinforce senescence and halt the proliferation of potentially malignant cells, acting as a tumor suppressor. However, the chronic secretion of SASP factors can create an environment conducive to tumor progression and metastasis by promoting inflammation, altering the immune response, and degrading the ECM. The accumulation of SnCs and their SASP components is linked to aging and age-related diseases. SASP factors can drive chronic inflammation, known as "inflammaging," a major factor leading to the development of many chronic diseases such as arthritis, diabetes, and neurodegeneration. SASP factors can help in wound healing and tissue regeneration by modulating the cellular environment. However, prolonged SASP activity can lead to fibrotic diseases by excessive deposition of extracellular matrix components.

### 13.5.1 Paradoxical Role of Senescent Cell-Derived Exosomes in Cancer Progression

While SASP has traditionally been viewed as a mechanism to reinforce cell cycle arrest and prevent tumorigenesis, emerging research suggests a more nuanced role, particularly with respect to the exosomes released by senescent cells. These exosomes can paradoxically promote cancer progression, highlighting the dual nature of senescence in tumor dynamics. Exosomes from SnCs contribute to an environment conducive to cancer progression through several mechanisms, including delivering inflammatory cytokines and growth factors, these exosomes can modify the extracellular matrix, promote angiogenesis, and recruit other cells that support tumorigenesis. Exosomes can carry molecules that modulate immune cell function, potentially suppressing the anti-tumor immune response and facilitating immune evasion by cancer cells [30]. Indeed, stress conditions including irradiation, oxidative stress, and chemotherapy could induce SnCs to elevate exosome secretion [30–32]. Paclitaxel, a widely used chemotherapeutic drug, targets and stabilizes microtubules, disrupting the normal function of the mitotic spindle and effectively blocking cell division. This disruption can trigger a DNA damage response (DDR) and activate p53 and other pathways, culminating in cellular senescence [33]. Recent studies have also found that SnC-exosomes have elevated level of EphA2, which could enhance cancer cell proliferation by stimulating the ERK pathway [34, 35]. Additionally, senescent human dermal fibroblasts have been observed to secrete miRNA-rich EVs, including miR-15b and miR-30a, both targeting pro-apoptotic proteins mRNAs [36]. Thus, through exRNAs, SnCs could also alter the physical properties of the tumor ECM, facilitating cancer cell movement and invasion. The role of exRNAs in cancer associated with cellular senescence is an emerging and compelling field of study, requiring further exploration and validation. Moreover, analyzing exRNA profiles from senescent-associated exosomes could provide insights into the mechanisms through which cancer develops resistance to chemotherapy.

**Box 13.1 exRNAs from Normal Cells Could Suppress Cancer Growth**
Exosomes play a complex role in the TME in influencing the response of cancer cells to therapies. A notable example of this is observed in breast cancer, where exosomes transferred from stromal cells to cancer cells can increase the resistance of these cancer cells to radiotherapy [37]. The role of exRNAs in intercellular communication extends beyond promoting disease; they can also convey anti-tumor properties. Notably, exRNAs derived from normal, non-malignant cells have the potential to suppress tumor growth when transferred to cancer cells. This capability highlights a promising avenue for developing RNA-based therapies that harness the natural tumor-suppressive properties of certain RNAs. Normal cells can release exosomes containing

(continued)

**Box 13.1** (continued)
exRNAs that have inherent tumor-suppressive functions. When cancer cells uptake these exosomes, the exRNAs can influence cancer cell behavior by modulating various cellular pathways. For example, PTEN is a critical tumor suppressor gene that negatively regulates the PI3K/Akt signaling pathway, which is often hyperactivated in many cancers. The transfer of PTEN mRNA through exosomes can have a therapeutic effect on cancer cells by restoring the expression of the PTEN gene. PTEN is a tumor suppressor gene that negatively regulates the PI3K/Akt signaling pathway, a critical pathway for cell survival and proliferation in many cancers. When PTEN mRNA is delivered via exosomes to cancer cells lacking this gene, it can lead to the inhibition of the PI3K/Akt pathway. miR-223 and miR-142 are notable for their regulatory roles. miR-223 has the capacity to target and suppress multiple oncogenic pathways, thereby inhibiting tumor growth and metastasis. Its regulatory actions help disrupt the cellular mechanisms that otherwise facilitate cancer progression. On the other hand, miR-142 plays a critical role in modulating the immune response against tumors. It suppresses the expression of specific oncogenic targets within cancer cells, contributing to a reduction in tumor viability and aggressiveness. By transferring miR-142 from non-malignant cells to cancer cells, it is possible to inhibit pathways that are crucial for cancer cell survival and proliferation [38, 39]. let-7 miRNAs target mRNAs of various oncogenic proteins for degradation or translational repression. By doing so, they inhibit pathways that are essential for cancer cell survival and proliferation, such as the Ras signaling pathway. In gastric cancer, as in many other cancers, the expression of let-7 is often downregulated, which correlates with increased tumor aggressiveness and poor prognosis. This downregulation can be due to genetic mutations, epigenetic alterations, or as recent findings suggest, through active depletion via exosomal export [40, 41]. The ability of cancer cells to autonomously deplete tumor-suppressive miRNAs via exosomes highlights the complexity of cancer biology and the innovative ways in which tumors maintain their growth and survival. By facilitating complex intercellular interactions, exRNAs contribute to the dynamic changes within the TME that support cancer cell survival and expansion. Delving into the mechanisms behind tumor cell exosome secretion and the journey of exRNAs to and within recipient cells could unveil new strategies for targeting cellular communication, ultimately aiming to halt cancer progression.

## 13.6 PART 2: exRNA in EVs

Recent investigations reveal that the composition of exRNAs varies by compartment. Exosomes were initially identified in reticulocytes [42, 43]. The genesis of exosomes begins with the inward budding of the plasma membrane, leading to the formation of an endosome. As this endosome matures, its membrane bulges inward and pinches off to form small vesicles within a structure called a multivesicular endosome (MVE). This compartmentalization process encapsulates nucleic acids, proteins, and lipids into the MVE. Subsequently, the MVE merges with the plasma membrane, releasing the enclosed vesicles—now termed exosomes—into the extracellular environment [44]. The process of exosome biogenesis is intricately structured and meticulously regulated. Factors such as Rab guanosine triphosphatases (GTPases), intracellular pH levels, and tumor suppressor proteins like TSAP6 (a p53-regulated protein involved in the tumor suppressor-activated pathway 6), along with enzymes such as heparanase, are known to influence exosome secretion.

Exosomes are small vesicles enclosed by a lipid bilayer membrane that encapsulates a portion of cytosol, the fluid part of a cell's interior. This membrane is characteristically enriched with tetraspanins—specifically proteins like CD9, CD63, and CD81. These tetraspanins are crucial for maintaining the structural integrity of the exosomes and enhancing their functional capabilities. They facilitate the sorting of cargo molecules into exosomes and play a pivotal role in the vesicles' ability to fuse with and enter recipient cells. Additionally, tetraspanins assist in the interactions between exosomes and target cells, which are vital for the effective transfer of signals and materials across cells. This composition highlights the complex architecture of exosomes and underscores their role as key players in cellular communication and molecular transport. Additionally, exosomes contain elements of the endosomal sorting complex required for transport (ESCRT), including ALIX, TSG101, along with proteins such as HSP70, HSP90, and tubulin. In exosomes derived from immune cells, antigen-presenting molecules like MHC are also prevalent. Beyond proteins, exosomes carry a range of nucleic acids, including genomic DNA (gDNA), mitochondrial DNA (mtDNA), various forms of RNA like mRNAs, miRNAs, lncRNAs, and circRNAs. These molecules are instrumental in facilitating cell-to-cell communication by transferring bioactive molecules from donor to recipient cells. The major components and cargoes of an exosome are highlighted in Table 13.1.

## 13.7 Role in the TME and Pre-Metastatic Niches (PMNs)

The recent discovery of the diverse roles and deregulation of extracellular vesicle RNAs (EV-RNAs) across various cancer types represent a significant advancement in cancer biology. Upon delivery, these RNAs can influence gene expression in recipient cells, modulating key processes like cell proliferation, migration, invasion, and even cell survival strategies. Upon delivery to target cells, EV-RNAs have the capacity to induce molecular and phenotypic alterations in these cells. The impact of EV-RNAs on cellular functions varies with the RNA type involved. EV-RNAs can support tumorigenic activities by enhancing oncogenic pathways, promoting angiogenesis, or conferring resistance to apoptosis and therapeutic agents. For instance, EV-derived miRNAs might downregulate tumor suppressor genes or upregulate oncogenes in recipient tumor cells. EV-RNAs contribute to the

**Table 13.1** Major components and cargoes of an exosome [44–47]

| Part of exosome | Molecule | Function |
|---|---|---|
| Bilayer/ ligand | CD9 [48] | CD9 is widely expressed on the surface of many cell types including leukocytes, platelets, endothelial cells, and epithelial cells. It is involved in several cellular processes such as cell adhesion, motility, and the fusion of cells and viruses. CD9 also plays a role in sperm-egg fusion during fertilization and is considered important in the process of muscle cell fusion and regeneration |
| | CD63 [49] | This tetraspanin is often used as a marker for exosomes due to its prevalent presence in the membranes of these vesicles. CD63 is implicated in the regulation of endosomal trafficking and lysosomal functions. It has roles in mediating signal transduction and modulating the immune response, particularly in antigen presentation. CD63 also influences melanogenesis and the metastatic potential of cancer cells, thereby affecting tumor progression |
| | CD81 [50] | CD81 is involved in a wide array of cellular functions, including cellular development, activation, proliferation, and differentiation. In the immune system, CD81 is essential for the proper formation of immune synapses between T cells and antigen-presenting cells. It also influences B cell activation and proliferation. Notably, CD81 is a receptor for the entry of certain pathogens, such as hepatitis C virus, into host cells |
| APCs/ immune cells | MHC [51] | The incorporation of MHC molecules on exosomes from immune cells plays a significant role in modulating the immune landscape. These exosomes can act as vehicles for antigen presentation, effectively extending the reach of antigen-presenting cells by delivering antigenic information directly to remote T cells |
| Endosomal sorting complex ESCRT | ALG-2-interacting protein X (ALIX) [52] | ALIX is involved in the endosomal sorting complex required for transport (ESCRT) machinery, which is critical for the budding and release of exosomes. ALIX binds to syndecan-syntenin complexes, facilitating the incorporation of specific cargos into exosomes. This protein is key in the mechanisms that determine the cargo selection and membrane abscission processes during exosome formation |
| | Tumor susceptibility gene 101 (TSG101) [53, 54] | TSG101 plays a vital role in vesicle formation and the sorting of ubiquitinated proteins into exosomes. It is crucial for the scission of the membrane neck during exosomal budding from the plasma membrane. TSG101's involvement in exosome formation and cargo sorting makes it a marker commonly used to identify exosomes |

(continued)

**Table 13.1** (continued)

| Part of exosome | Molecule | Function |
|---|---|---|
| | HSP70/90 [55, 56] | These chaperone proteins are involved in protein folding and protecting cells against stress. HSP70 and HSP90 are abundantly found in exosomes and play roles in stabilizing proteins during transport. Their presence on exosomes suggests a function in signaling, as they can activate immune responses when exosomes are taken up by recipient cells. These heat shock proteins also aid in the intercellular transfer of signaling molecules, enhancing cellular communication and potentially facilitating tumor progression |
| | Tubulin [57] | Tubulin is a structural protein that makes up microtubules, important for cell structure, chromosome separation in mitosis and meiosis, and intracellular transport. Its presence in exosomes can reflect its roles in cellular architecture and signaling pathways. Tubulin in exosomes might influence the stability and spatial orientation of proteins within the vesicle or participate in the delivery mechanisms of exosomal contents to recipient cells. |
| Cargo | Genomic DNA (gDNA), mitochondrial DNA (mtDNA), mRNAs, miRNAs, lncRNAs, circRNAs | These molecules are instrumental in facilitating cell-to-cell communication by transferring bioactive molecules from donor to recipient cells |

reprogramming of the stromal cells such as fibroblasts and endothelial cells, often converting them into supportive components of the tumor stroma, enhancing tumor growth and facilitating metastasis [58, 59].

**Pro-Tumoral EV-RNAs** EV-RNAs derived from cancer cells often carry oncogenic RNAs that directly contribute to the malignancy of other cancerous or precancerous cells. These EV-RNAs can enhance tumor proliferation, invasion, and migration by transferring genetic information that encourages these behaviors. Cancer-derived EV-RNAs can transform stromal cells into a tumor-promoting state. For example, fibroblasts can be converted into cancer-associated fibroblasts (CAFs) that secrete supportive growth factors, modify the ECM, and suppress immune surveillance, all of which nurture tumor growth and progression. EV-RNAs can confer drug resistance by transferring mRNAs or miRNAs that deactivate the pathways targeted by therapies or by enhancing the cancer cells' ability to evade apoptosis (Fig. 14.1).

**Anti-Tumoral EV-RNAs** Interestingly, EV-RNAs can also originate from non-cancerous and even healthy cells, carrying tumor-suppressive RNAs that inhibit cancer cell proliferation or induce cancer cell apoptosis. These EV-RNAs can coun-

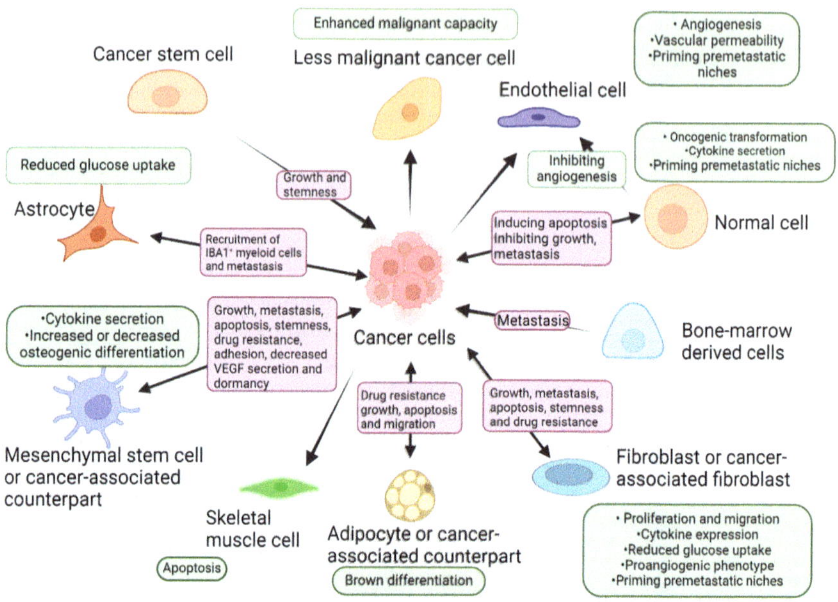


**Fig. 13.1** Communication within tumors and between tumors and their surrounding stroma, facilitated by extracellular vesicle RNAs (EV-RNAs), significantly influences the malignant traits of cancer cells. The onset, growth, and spread of cancer are largely due to complex and bidirectional interactions among diverse cellular entities. EV-RNAs from tumors can instigate oncogenic, pro-metastatic, pro-angiogenic, and differentiated states in stromal cells within the TME or at metastatic locations. Furthermore, these EV-RNAs from tumors can induce malignant behaviors in both normal and tumor-associated cells. Conversely, EV-RNAs from stromal or normal cells, modified by interactions with cancer, can also enhance the malignancy of cancer cells. Figure is adapted from [60] under Creative Commons CC-BY license

teract the oncogenic signals within the TME, providing a natural check against uncontrolled tumor growth. EV-RNAs can also play a role in activating immune responses against tumors. For instance, they can carry antigens that prime immune cells to recognize and destroy cancer cells, or they can carry immune-stimulatory miRNAs that enhance the efficacy of immune checkpoint therapies (Fig. 13.1).

## 13.8 Co-Operative Regulation of Malignancy by Tumor and Stromal EV-RNAs

In the dynamic environment of cancer progression, a competitive interaction exists between cancerous cells and the surrounding normal cells [61]. A natural defence mechanism involves the release of tumor-suppressive miRNAs by the abundant non-cancerous cells to counteract the malignant traits of nearby cancer cells [62]. A notable example of this phenomenon is EV-miR-143, which is derived from normal prostate epithelial cells and has been shown to have a tumor-suppressive effect on prostate cancer cells. EV-miR-143 is produced by normal prostate epithelial cells and packaged into extracellular vesicles. These vesicles are then released into the extracellular environment where they can be taken up by adjacent or distant prostate cancer cells. Upon uptake by prostate cancer cells, miR-143 delivered via EVs can influence cellular behavior. The genes targeted by miR-143 generally include those involved in cell proliferation and survival pathways [63]. EV-miR-145, derived from tumor-associated stromal cells, provides a compelling example of how stromal cell-derived miRNAs can actively contribute to the suppression of cancer cell growth, specifically in pancreatic ductal adenocarcinoma (PDAC) [64]. Recent research has illuminated how EVs containing miRNAs produced by liver stem cells can influence the behavior of HCC cells [65]. EV-miR-145 sourced from AMSCs has demonstrated promising effects in inhibiting proliferation and promoting apoptosis in prostate cancer cells [66]. EV-lncRNA-PTENP1, derived from normal cells, exemplifies the therapeutic potential of lncRNAs in cancer treatment, particularly in its ability to induce apoptosis and suppress tumorigenic characteristics in bladder cancer cells through interactions with miR-17 [67].

The roles of MSCs can vary significantly based on their source, the type of cancer, and the conditions within the TME, such as normoxia or hypoxia. A particular instance of this complexity is observed with EV-miR-410 derived from umbilical cord MSCs, which influences lung cancer progression by reducing apoptosis through its interaction with the PTEN signaling pathway [68]. Bone marrow-derived mesenchymal stem cells (BMSCs) play a pivotal role in the TME, particularly under hypoxic conditions—a common feature within solid tumors like lung cancer. Under hypoxic conditions, BMSCs are known to secrete EVs loaded with specific miRNAs, such as miR-193a-3p, miR-210-3p, and miR-5100, which have been shown to enhance EMT, and increase migration and invasion of lung cancer cells by modulating the STAT3 signaling pathway [69]. EV-miR-142-3p, which is secreted by BMSCs, promotes stem-like properties and drug resistance in colon cancer cells. miR-142-3p directly targets the mRNA of Numb, leading to its downregulation. Numb typically acts as a modulator of Notch signaling, a pathway crucial for cell differentiation and survival. By suppressing Numb, miR-142-3p disrupts normal cell differentiation pathways. The reduction in Numb levels correlates with an increase in stem-like properties of colon cancer cells. These cells exhibit enhanced self-renewal capabilities, which are characteristic of CSCs [70]. EV-miR-23b targets MARCKS, a protein important in cell motility and metastasis. By downregulating MARCKS, miR-23b can reduce the migration and invasive potential of BC cells at metastatic sites [71]. Table 13.2 summarized the potential therapeutic implications of EV-encapsulated RNAs in cancer therapies.

**Table 13.2** Potential therapeutic implications of modulating EV-associated RNAs in cancer treatment strategies

| Origin of EV | RNAs | Cancer type | Function | Ref |
|---|---|---|---|---|
| Stellate cells | miR-355-5p | HCC | Reducing the proliferation and invasiveness | [72] |
| Fibroblast | miR-195 | A rat model of cholangiocarcinoma | Tumor regression and improved survival rates | [73] |
| Bone marrow-derived mesenchymal stem cells (BMSCs) | miR-126-3p | Pancreatic cancer cells | Inhibiting the migration and invasion by targeting ADAM9 | [74] |
| γδ T cells | miR-138 | OSCC | Enhanced interferon-γ secretion, and cytotoxic activity of CD8+ T cells by suppressing PD-1 and CTLA-4 | [75] |
| Cancer cells | RISC-associated miRNAs | Various cancer | Oncogenic transformation of epithelial cells through RNA-processing enzyme, Dicer | [40] |
| Chronic myelogenous leukemia cells | EV-miR-146b-5p | CML | Transformation of mononuclear cells into leukemia-like cells and induce genomic instability | [76] |
| Arsenite-transformed hepatic epithelial cells | EV-miR-155 | HCC | Activation of IL-8 and IL-6/STAT3 signaling pathway | [77] |
| Arsenite-transformed cells | EV-circRNA-100 | HCC | Accelerates proliferation of normal hepatic cells by targeting miR-217 | [78] |

## 13.9 Reprogramming of Normal Cells in TME into Malignant Phenotype

Beyond the homeostatic interactions, the cancer microenvironment undergoes significant alterations to support tumor growth. The intercellular exchange of tumor-promoting RNAs via EVs leads to the reprogramming of normal cells, facilitating a symbiotic evolution with cancer cells and establishing a pro-tumorigenic dialogue. CAFs are a key player in the TME. While CAFs could originate from many sources, most are transformed from inactive fibroblasts into myofibroblasts. EVs carrying miR-105 from cancer cells illustrate a sophisticated mechanism of intercellular communication within the TME, particularly impacting the metabolic landscape. This specific miRNA plays a critical role in reprogramming CAFs to support tumor adaptation and survival under varying nutrient conditions. By targeting MXI1, a known inhibitor of the MYC oncogene, miR-105 facilitates a shift in the metabolic activities of CAFs, promoting a supportive environment for tumor growth [79]. EVs containing miR-211 from melanoma cells trigger the conversion of primary fibroblasts into CAFs by targeting IGF2R. This transformation likely aids in creating supportive niches that enhance tumor growth and invasion [80]. EVs carrying miR-9 from breast cancer (BC) cells illustrate a crucial mechanism by which cancer cells can manipulate their surrounding microenvironment to facilitate tumor progression and metastasis. By inducing a CAF-like phenotype in normal fibroblasts, EV-miR-9 plays a significant role in enhancing the invasive and migratory capabilities of breast cancer cells, particularly through the targeting of E-cadherin [81]. In another example, EV-lncRNA-CAF from OSCC triggers a CAF phenotype in normal fibroblasts by enhancing the secretion level of IL-33, promoting tumor proliferation [82]. CAFs and normal fibroblasts (NFs) differ significantly in their roles within the TME, especially in terms of the EVs and their miRNA contents they release. One of the key distinctions is observed in the secretion levels of EV-miR-320a [83]. Similarly, a reduction of EV-miR-3188 in CAFs augments the growth and survival of HNCC by acting on BCL2 [84]. Furthermore, levels of EV-miR-15a from BMSCs in multiple myeloma (MM) are reduced compared to those from healthy BMSCs, which contributes to the proliferation of MM cells [85].

## 13.10 EV-RNAs Mediating Cancer Therapy Resistance

EV-miR-21, sourced from cancer-associated adipocytes and CAFs, has been shown to confer resistance to paclitaxel in ovarian cancer cells by targeting apoptotic pathways. miR-21 in ovarian cancer cells targets APAF1, a critical component of the apoptosome, which plays a central role in the intrinsic pathway of apoptosis. By inhibiting APAF1, miR-21 disrupts the activation of caspase-9 and subsequently caspase-3, leading to a reduction in apoptosis. This mechanism is particularly detrimental when ovarian cancer cells are treated with paclitaxel [86]. EV-miR-196a, secreted by CAFs, affects the response of head and neck cancer cells resistant to the chemotherapy drug cisplatin. By targeting key tumor suppressor and cell cycle regulation genes, miR-196a enhances the proliferation of cancer cells and reduces their sensitivity to cisplatin. CAFs release EVs carrying miR-196a, which are then taken up by adjacent HNCC cells. The transfer of miR-196a via EVs ensures that it is delivered efficiently and is protected from enzymatic degradation in the extracellular environment. Also known as p27Kip1, CDKN1B is a crucial regulator of the cell cycle. miR-196a targets CDKN1B, leading to decreased expression of this inhibitor, which normally functions to halt cell cycle progression at the G1 phase. The suppression of CDKN1B facilitates cell cycle progression and proliferation. ING5 is involved in tumor suppression and apoptosis. By targeting ING5, miR-196a reduces the tumor-suppressive activities of this protein, thus decreasing apoptosis and enhancing cell survival [87]. A specific example of this dynamic involves EV-RN7SL1, an RNA component derived from activated fibroblasts, which, upon interaction with BC cells, can initiate a series of pro-tumorigenic processes through the activation of the RIG-I signaling pathway, a pattern recognition receptor (PRR) that typically recognizes viral RNA, leading to the activation of innate immune responses. In the context of cancer, the activation of RIG-I by RN7SL1 can paradoxically promote pro-inflammatory and pro-survival signaling pathways. The activation of RIG-I triggers the production of IFN-I and other inflammatory cytokines, which can foster a pro-tumorigenic environment within the TME [88]. lncRNA H19 has been found to play a critical role in promoting stemness and resistance to chemotherapy, specifically oxaliplatin, in colorectal cancer cells. lncRNA-H19 binds to miR-141 to regulate genes involved in maintaining cancer stem cell properties and drug sensitivity. By sponging miR-141, lncRNA-H19 prevents miR-141 from exerting its regulatory functions, leading to the upregulation of stemness-associated genes and pathways that contribute to chemoresistance, particularly against oxaliplatin [89]. EVs carrying miR-222 and miR-223 from BC cell-primed BMSCs play a critical role in modulating the behavior of breast cancer cells. These miRNAs contribute to the induction of quiescence and drug resistance, posing significant challenges to effective chemotherapy. miR-222 and miR-223 promote a state of quiescence in BC cells [90].

EV-miR-151a, derived from TMZ-resistant GBM cells, plays a crucial role in conferring chemoresistance to TMZ-sensitive cancer cells by targeting the DNA repair gene XRCC4. XRCC4 (X-ray repair cross-complementing protein 4) is a crucial component of the non-homologous end joining (NHEJ) pathway of DNA double-strand break repair. By ensuring effective DNA repair, XRCC4 helps maintain genomic stability, which is essential for cell survival under chemotherapy-induced DNA damage. Suppression by miR-151a: miR-151a targets and downregulates XRCC4 expression, impairing the DNA repair capacity of the cells. This suppression reduces the effectiveness of TMZ, which relies on inducing DNA damage to kill cancer cells [91]. EV-miR-222-3p, derived from gemcitabine-resistant lung cancer cells, exemplifies this process by promoting migration, invasion, and resistance to gemcitabine and anti-anoikis in gemcitabine-sensitive cancer cells. The downregulation of SOCS3 by miR-222-3p confers resistance to gemcitabine, making the cancer cells less susceptible to chemotherapy. miR-222-3p also promotes resistance to anoikis, allowing cancer cells to survive and disseminate through the bloodstream to distant sites [92]. Furthermore, cancer cells expel tumor-suppressive miRNAs through EVs to maintain and enhance their malignancy. For example, miR-23b, more abundant in EVs from metastatic than non-metastatic bladder cancer cells, reduces invasion and resistance to cell death, decreasing angiogenesis and lung colonization when its exocytosis is diminished [93]. miR-6126 is highly upregulated in ovarian cancer cell EVs compared to normal ovarian epithelial cells, which increases migration, invasion, and tumor growth [94]. Similarly, miR-940, elevated in EVs from ovarian cancer cells, deters proliferation, invasion, and migration, and promotes apoptosis by targeting SRC [95]. This intricate network of EV-mediated RNA transfer underscores potential therapeutic targets and mechanisms of resistance, offering insights into the development of novel cancer treatment strategies (Fig. 13.2).

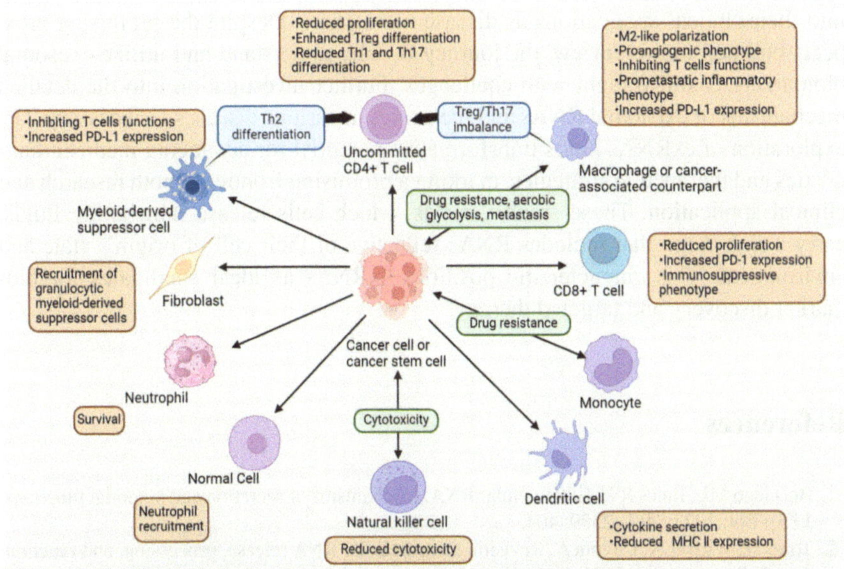


**Fig. 13.2** Illustrates the pivotal role of RNA molecules carried by EV-RNAs in the interactions between cancer cells and immune cells, as well as among immune cells themselves, which significantly influence the malignancy of cancer cells. RNAs from tumor-derived extracellular vesicles can interfere with the anti-tumor activities of various immune cells, thus promoting an immunosuppressive tumor microenvironment. This effect is achieved by diminishing the immune cells' ability to attack and destroy cancer cells effectively. For example, in the crosstak between cancer-cells and CAFs, these EV-RNAs contribute to creating an immunoinhibitory phenotype in CAFs and normal cells surrounding the tumor. CAFs, once altered by signals such as those from EV-RNAs, can support tumor growth and survival by secreting growth factors, modifying the extracellular matrix, and regulating immune responses. By influencing CAFs and normal cells to adopt characteristics that hinder the immune system's ability to combat the tumor, EV-RNAs enhance the tumor's ability to evade immune detection and destruction

## 13.11 Conclusion

Exosomal RNAs are important in cell-to-cell communication, holding significant promise for disease diagnosis and prognosis in the future. The exploration of exosomal biomarkers has already demonstrated their potential to transcend the limitations of traditional solid biopsies, notably by mitigating the invasive nature of surgical procedures. Despite this potential, the field of exosomal biomarker research is still in its nascent stages, with widespread clinical application facing substantial obstacles. While the convenience and non-invasive nature of exosomal biomarker analysis are clear advantages, challenges in biomarker selection persist, with no single biomarker yet achieving perfect diagnostic accuracy. Future research should prioritize the identification of biomarkers that exhibit significant changes in disease states to enhance diagnostic precision. For the broader implementation of exosomal biomarkers in clinical practice, advancements in exosome isolation techniques that are rapid, accurate, and efficient are crucial. Developing new methodologies to improve the isolation and purity of exosomes is essential for advancing research into their clinical applications as disease biomarkers. Despite the promising prospects outlined in this review, the journey to fully understand and utilize exosomal biomarkers is still fraught with challenges. Further investigation into the detailed mechanisms of exosomal RNAs and their roles in various diseases is necessary. The exploration of exRNAs holds transformative potential for advancing medical diagnostics and therapeutic strategies, marking a promising frontier in both research and clinical application. These small vesicles, which cells release into bodily fluids, carry a rich cargo that includes RNAs reflective of their cell of origin's state and environment. This characteristic positions exRNAs as ideal candidates for biomarker discovery and targeted therapy.

## References

1. Borniego ML, Innes RW. Extracellular RNA: mechanisms of secretion and potential functions. J Exp Bot. 2023;74(7):2389–404.
2. Tosar JP, Witwer K, Cayota A. Revisiting extracellular RNA release, processing, and function. Trends Biochem Sci. 2021;46(6):438–45.
3. Happel C, Ganguly A, Tagle DA. Extracellular RNAs as potential biomarkers for cancer. J Cancer Metastasis Treat. 2020:6.
4. Lässer C, Alikhani VS, Ekström K, Eldh M, Paredes PT, Bossios A, et al. Human saliva, plasma and breast milk exosomes contain RNA: uptake by macrophages. J Transl Med. 2011;9:9.
5. Vickers KC, Palmisano BT, Shoucri BM, Shamburek RD, Remaley AT. MicroRNAs are transported in plasma and delivered to recipient cells by high-density lipoproteins. Nat Cell Biol. 2011;13(4):423–33.
6. Baj-Krzyworzeka M, Szatanek R, Weglarczyk K, Baran J, Urbanowicz B, Brański P, et al. Tumour-derived microvesicles carry several surface determinants and mRNA of tumour cells and transfer some of these determinants to monocytes. Cancer Immunol Immunother. 2006;55(7):808–18.

7. Ekström K, Valadi H, Sjöstrand M, Malmhäll C, Bossios A, Eldh M, et al. Characterization of mRNA and microRNA in human mast cell-derived exosomes and their transfer to other mast cells and blood CD34 progenitor cells. J Extracell Vesicles. 2012:1.
8. Higginbotham JN, Demory Beckler M, Gephart JD, Franklin JL, Bogatcheva G, Kremers GJ, et al. Amphiregulin exosomes increase cancer cell invasion. Curr Biol. 2011;21(9):779–86.
9. Mittelbrunn M, Gutiérrez-Vázquez C, Villarroya-Beltri C, González S, Sánchez-Cabo F, González M, et al. Unidirectional transfer of microRNA-loaded exosomes from T cells to antigen-presenting cells. Nat Commun. 2011;2:282.
10. Montecalvo A, Larregina AT, Shufesky WJ, Stolz DB, Sullivan ML, Karlsson JM, et al. Mechanism of transfer of functional microRNAs between mouse dendritic cells via exosomes. Blood. 2012;119(3):756–66.
11. McKelvey KJ, Powell KL, Ashton AW, Morris JM, McCracken SA. Exosomes: mechanisms of uptake. J Circ Biomark. 2015;4:7.
12. Haimovich G, Ecker CM, Dunagin MC, Eggan E, Raj A, Gerst JE, et al. Intercellular mRNA trafficking via membrane nanotube-like extensions in mammalian cells. Proc Natl Acad Sci USA. 2017;114(46):E9873–E82.
13. Keustermans GC, Hoeks SB, Meerding JM, Prakken BJ, de Jager W. Cytokine assays: an assessment of the preparation and treatment of blood and tissue samples. Methods. 2013;61(1):10–7.
14. Lespagnol A, Duflaut D, Beekman C, Blanc L, Fiucci G, Marine JC, et al. Exosome secretion, including the DNA damage-induced p53-dependent secretory pathway, is severely compromised in TSAP6/Steap3-null mice. Cell Death Differ. 2008;15(11):1723–33.
15. Park JE, Tan HS, Datta A, Lai RC, Zhang H, Meng W, et al. Hypoxic tumor cell modulates its microenvironment to enhance angiogenic and metastatic potential by secretion of proteins and exosomes. Mol Cell Proteomics. 2010;9(6):1085–99.
16. Thompson CA, Purushothaman A, Ramani VC, Vlodavsky I, Sanderson RD. Heparanase regulates secretion, composition, and function of tumor cell-derived exosomes. J Biol Chem. 2013;288(14):10093–9.
17. Becker A, Thakur BK, Weiss JM, Kim HS, Peinado H, Lyden D. Extracellular vesicles in cancer: cell-to-cell mediators of metastasis. Cancer Cell. 2016;30(6):836–48.
18. Redzic JS, Balaj L, van der Vos KE, Breakefield XO. Extracellular RNA mediates and marks cancer progression. Semin Cancer Biol. 2014;28:14–23.
19. Gutkin A, Uziel O, Beery E, Nordenberg J, Pinchasi M, Goldvaser H, et al. Tumor cells derived exosomes contain hTERT mRNA and transform nonmalignant fibroblasts into telomerase positive cells. Oncotarget. 2016;7(37):59173–88.
20. Skog J, Würdinger T, van Rijn S, Meijer DH, Gainche L, Sena-Esteves M, et al. Glioblastoma microvesicles transport RNA and proteins that promote tumour growth and provide diagnostic biomarkers. Nat Cell Biol. 2008;10(12):1470–6.
21. Fang T, Lv H, Lv G, Li T, Wang C, Han Q, et al. Tumor-derived exosomal miR-1247-3p induces cancer-associated fibroblast activation to foster lung metastasis of liver cancer. Nat Commun. 2018;9(1):191.
22. Balacescu O, Sur D, Cainap C, Visan S, Cruceriu D, Manzat-Saplacan R, et al. The impact of miRNA in colorectal cancer progression and its liver metastases. Int J Mol Sci. 2018;19:12.
23. Mikamori M, Yamada D, Eguchi H, Hasegawa S, Kishimoto T, Tomimaru Y, et al. MicroRNA-155 controls exosome synthesis and promotes gemcitabine resistance in pancreatic ductal adenocarcinoma. Sci Rep. 2017;7:42339.
24. Campisi J. Senescent cells, tumor suppression, and organismal aging: good citizens, bad neighbors. Cell. 2005;120(4):513–22.
25. Hayflick L. The limited in vitro lifetime of human diploid cell strains. Exp Cell Res. 1965;37:614–36.
26. Campisi J. Cellular senescence, aging and cancer. ScientificWorldJournal. 2001;1:65.
27. Krtolica A, Parrinello S, Lockett S, Desprez PY, Campisi J. Senescent fibroblasts promote epithelial cell growth and tumorigenesis: a link between cancer and aging. Proc Natl Acad Sci USA. 2001;98(21):12072–7.

28. Jeon OH, David N, Campisi J, Elisseeff JH. Senescent cells and osteoarthritis: a painful connection. J Clin Invest. 2018;128(4):1229–37.
29. Jeon OH, Wilson DR, Clement CC, Rathod S, Cherry C, Powell B, et al. Senescence cell-associated extracellular vesicles serve as osteoarthritis disease and therapeutic markers. JCI. Insight. 2019;4:7.
30. Jakhar R, Crasta K. Exosomes as emerging pro-tumorigenic mediators of the senescence-associated secretory phenotype. Int J Mol Sci. 2019;20(10)
31. Atienzar-Aroca S, Flores-Bellver M, Serrano-Heras G, Martinez-Gil N, Barcia JM, Aparicio S, et al. Oxidative stress in retinal pigment epithelium cells increases exosome secretion and promotes angiogenesis in endothelial cells. J Cell Mol Med. 2016;20(8):1457–66.
32. Yu X, Harris SL, Levine AJ. The regulation of exosome secretion: a novel function of the p53 protein. Cancer Res. 2006;66(9):4795–801.
33. Kavanagh EL, Lindsay S, Halasz M, Gubbins LC, Weiner-Gorzel K, Guang MHZ, et al. Protein and chemotherapy profiling of extracellular vesicles harvested from therapeutic induced senescent triple negative breast cancer cells. Oncogenesis. 2017;6(10):e388.
34. Liang K, Liu F, Fan J, Sun D, Liu C, Lyon CJ, et al. Nanoplasmonic quantification of tumor-derived extracellular vesicles in plasma microsamples for diagnosis and treatment monitoring. Nat. Biomed Eng. 2017:1.
35. Takasugi M, Okada R, Takahashi A, Virya Chen D, Watanabe S, Hara E. Small extracellular vesicles secreted from senescent cells promote cancer cell proliferation through EphA2. Nat Commun. 2017;8:15729.
36. Terlecki-Zaniewicz L, Lämmermann I, Latreille J, Bobbili MR, Pils V, Schosserer M, et al. Small extracellular vesicles and their miRNA cargo are anti-apoptotic members of the senescence-associated secretory phenotype. Aging (Albany NY). 2018;10(5):1103–32.
37. Boelens MC, Wu TJ, Nabet BY, Xu B, Qiu Y, Yoon T, et al. Exosome transfer from stromal to breast cancer cells regulates therapy resistance pathways. Cell. 2014;159(3):499–513.
38. Aucher A, Rudnicka D, Davis DM. MicroRNAs transfer from human macrophages to hepato-carcinoma cells and inhibit proliferation. J Immunol. 2013;191(12):6250–60.
39. Putz U, Howitt J, Doan A, Goh CP, Low LH, Silke J, et al. The tumor suppressor PTEN is exported in exosomes and has phosphatase activity in recipient cells. Sci Signal. 2012;5(243):ra70.
40. Melo SA, Sugimoto H, O'Connell JT, Kato N, Villanueva A, Vidal A, et al. Cancer exosomes perform cell-independent microRNA biogenesis and promote tumorigenesis. Cancer Cell. 2014;26(5):707–21.
41. Ohshima K, Inoue K, Fujiwara A, Hatakeyama K, Kanto K, Watanabe Y, et al. Let-7 microRNA family is selectively secreted into the extracellular environment via exosomes in a metastatic gastric cancer cell line. PLoS One. 2010;5(10):e13247.
42. Pan BT, Johnstone RM. Fate of the transferrin receptor during maturation of sheep reticulocytes in vitro: selective externalization of the receptor. Cell. 1983;33(3):967–78.
43. Harding C, Heuser J, Stahl P. Receptor-mediated endocytosis of transferrin and recycling of the transferrin receptor in rat reticulocytes. J Cell Biol. 1983;97(2):329–39.
44. Théry C, Zitvogel L, Amigorena S. Exosomes: composition, biogenesis and function. Nat Rev Immunol. 2002;2(8):569–79.
45. Lee YJ, Shin KJ, Chae YC. Regulation of cargo selection in exosome biogenesis and its biomedical applications in cancer. Exp Mol Med. 2024;56(4):877–89.
46. Liu M, Wen Z, Zhang T, Zhang L, Liu X, Wang M. The role of exosomal molecular cargo in exosome biogenesis and disease diagnosis. Front Immunol. 2024;15:1417758.
47. Li J, Zhang Y, Dong P-Y, Yang G-M, Gurunathan S. A comprehensive review on the composition, biogenesis, purification, and multifunctional role of exosome as delivery vehicles for cancer therapy. Biomed Pharmacother. 2023;165:115087.
48. Qadir AS, Ceppi P, Brockway S, Law C, Mu L, Khodarev NN, et al. CD95/Fas increases stemness in cancer cells by inducing a STAT1-dependent type I interferon response. Cell Rep. 2017;18(10):2373–86.

49. Mathieu M, Névo N, Jouve M, Valenzuela JI, Maurin M, Verweij FJ, et al. Specificities of exosome versus small ectosome secretion revealed by live intracellular tracking of CD63 and CD9. Nat Commun. 2021;12(1):4389.
50. Ashraf Malik M, Ishtiyaq Ali Mirza J, Umar M, Manzoor S. CD81(+) exosomes play a pivotal role in the establishment of hepatitis C persistent infection and contribute toward the progression of hepatocellular carcinoma. Viral Immunol. 2019;32(10):453–62.
51. Lyu C, Sun H, Sun Z, Liu Y, Wang Q. Roles of exosomes in immunotherapy for solid cancers. Cell Death Dis. 2024;15(2):106.
52. Chatellard-Causse C, Blot B, Cristina N, Torch S, Missotten M, Sadoul R. Alix (ALG-2-interacting protein X), a protein involved in apoptosis, binds to endophilins and induces cytoplasmic vacuolization. J Biol Chem. 2002;277(32):29108–15.
53. Yang Y, Wang M, Zhang YY, Zhao SZ, Gu S. The endosomal sorting complex required for transport repairs the membrane to delay cell death. Front Oncol. 2022;12:1007446.
54. Gheytanchi E, Saeednejad Zanjani L, Ghods R, Abolhasani M, Shahin M, Vafaei S, et al. High expression of tumor susceptibility gene 101 (TSG101) is associated with more aggressive behavior in colorectal carcinoma. J Cancer Res Clin Oncol. 2021;147(6):1631–46.
55. Hu B, Liu G, Zhao K, Zhang G. Diversity of extracellular HSP70 in cancer: advancing from a molecular biomarker to a novel therapeutic target. Front Oncol. 2024;14:1388999.
56. Linder M, Pogge von Strandmann E. The role of extracellular HSP70 in the function of tumor-associated immune cells. Cancers. 2021;13:18.
57. Wang J, Yeung BZ, Cui M, Peer CJ, Lu Z, Figg WD, et al. Exosome is a mechanism of intercellular drug transfer: application of quantitative pharmacology. J Control Release. 2017;268:147–58.
58. Anastasiadou E, Jacob LS, Slack FJ. Non-coding RNA networks in cancer. Nat Rev Cancer. 2018;18(1):5–18.
59. Kim KM, Abdelmohsen K, Mustapic M, Kapogiannis D, Gorospe M. RNA in extracellular vesicles. Wiley Interdiscip Rev RNA. 2017;8:4.
60. Hu W, Liu C, Bi ZY, Zhou Q, Zhang H, Li LL, et al. Comprehensive landscape of extracellular vesicle-derived RNAs in cancer initiation, progression, metastasis and cancer immunology. Mol Cancer. 2020;19(1):102.
61. Kanada M, Bachmann MH, Contag CH. Signaling by extracellular vesicles advances cancer hallmarks. Trends Cancer. 2016:84–94.
62. Kosaka N, Yoshioka Y, Fujita Y, Ochiya T. Versatile roles of extracellular vesicles in cancer. J Clin Invest. 2016;126(4):1163–72.
63. Kosaka N, Iguchi H, Yoshioka Y, Hagiwara K, Takeshita F, Ochiya T. Competitive interactions of cancer cells and normal cells via secretory microRNAs. J Biol Chem. 2012;287(2):1397–405.
64. Han S, Gonzalo DH, Feely M, Rinaldi C, Belsare S, Zhai H, et al. Stroma-derived extracellular vesicles deliver tumor-suppressive miRNAs to pancreatic cancer cells. Oncotarget. 2018;9(5):5764–77.
65. Fonsato V, Collino F, Herrera MB, Cavallari C, Deregibus MC, Cisterna B, et al. Human liver stem cell-derived microvesicles inhibit hepatoma growth in SCID mice by delivering antitumor microRNAs. Stem Cells. 2012;30(9):1985–98.
66. Takahara K, Ii M, Inamoto T, Nakagawa T, Ibuki N, Yoshikawa Y, et al. microRNA-145 mediates the inhibitory effect of adipose tissue-derived stromal cells on prostate cancer. Stem Cells Dev. 2016;25(17):1290–8.
67. Zheng R, Du M, Wang X, Xu W, Liang J, Wang W, et al. Exosome-transmitted long non-coding RNA PTENP1 suppresses bladder cancer progression. Mol Cancer. 2018;17(1):143.
68. Dong L, Pu Y, Zhang L, Qi Q, Xu L, Li W, et al. Human umbilical cord mesenchymal stem cell-derived extracellular vesicles promote lung adenocarcinoma growth by transferring miR-410. Cell Death Dis. 2018;9(2):218.
69. Zhang X, Sai B, Wang F, Wang L, Wang Y, Zheng L, et al. Hypoxic BMSC-derived exosomal miRNAs promote metastasis of lung cancer cells via STAT3-induced EMT. Mol Cancer. 2019;18(1):40.

70. Li H, Li F. Exosomes from BM-MSCs increase the population of CSCs via transfer of miR-142-3p. Br J Cancer. 2018;119(6):744–55.
71. Ono M, Kosaka N, Tominaga N, Yoshioka Y, Takeshita F, Takahashi RU, et al. Exosomes from bone marrow mesenchymal stem cells contain a microRNA that promotes dormancy in metastatic breast cancer cells. Sci Signal. 2014;7(332):ra63.
72. Wang F, Li L, Piontek K, Sakaguchi M, Selaru FM. Exosome miR-335 as a novel therapeutic strategy in hepatocellular carcinoma. Hepatology. 2018;67(3):940–54.
73. Li L, Piontek K, Ishida M, Fausther M, Dranoff JA, Fu R, et al. Extracellular vesicles carry microRNA-195 to intrahepatic cholangiocarcinoma and improve survival in a rat model. Hepatology. 2017;65(2):501–14.
74. Wu DM, Wen X, Han XR, Wang S, Wang YJ, Shen M, et al. Bone marrow mesenchymal stem cell-derived Exosomal MicroRNA-126-3p inhibits pancreatic cancer development by targeting ADAM9. Mol Ther Nucleic Acids. 2019;16:229–45.
75. Li L, Lu S, Liang X, Cao B, Wang S, Jiang J, et al. γδTDEs: an efficient delivery system for miR-138 with anti-tumoral and immunostimulatory roles on oral squamous cell carcinoma. Mol Ther Nucleic Acids. 2019;14:101–13.
76. Zhang HM, Li Q, Zhu X, Liu W, Hu H, Liu T, et al. miR-146b-5p within BCR-ABL1-positive microvesicles promotes leukemic transformation of hematopoietic cells. Cancer Res. 2016;76(10):2901–11.
77. Chen C, Luo F, Liu X, Lu L, Xu H, Yang Q, et al. NF-kB-regulated exosomal miR-155 promotes the inflammation associated with arsenite carcinogenesis. Cancer Lett. 2017;388:21–33.
78. Dai X, Chen C, Yang Q, Xue J, Chen X, Sun B, et al. Exosomal circRNA_100284 from arsenite-transformed cells, via microRNA-217 regulation of EZH2, is involved in the malignant transformation of human hepatic cells by accelerating the cell cycle and promoting cell proliferation. Cell Death Dis. 2018;9(5):454.
79. Yan W, Wu X, Zhou W, Fong MY, Cao M, Liu J, et al. Cancer-cell-secreted exosomal miR-105 promotes tumour growth through the MYC-dependent metabolic reprogramming of stromal cells. Nat Cell Biol. 2018;20(5):597–609.
80. Dror S, Sander L, Schwartz H, Sheinboim D, Barzilai A, Dishon Y, et al. Melanoma miRNA trafficking controls tumour primary niche formation. Nat Cell Biol. 2016;18(9):1006–17.
81. Baroni S, Romero-Cordoba S, Plantamura I, Dugo M, D'Ippolito E, Cataldo A, et al. Exosome-mediated delivery of miR-9 induces cancer-associated fibroblast-like properties in human breast fibroblasts. Cell Death Dis. 2016;7(7):e2312.
82. Ding L, Ren J, Zhang D, Li Y, Huang X, Hu Q, et al. A novel stromal lncRNA signature reprograms fibroblasts to promote the growth of oral squamous cell carcinoma via LncRNA-CAF/interleukin-33. Carcinogenesis. 2018;39(3):397–406.
83. Zhang Z, Li X, Sun W, Yue S, Yang J, Li J, et al. Loss of exosomal miR-320a from cancer-associated fibroblasts contributes to HCC proliferation and metastasis. Cancer Lett. 2017;397:33–42.
84. Wang X, Qin X, Yan M, Shi J, Xu Q, Li Z, et al. Loss of exosomal miR-3188 in cancer-associated fibroblasts contributes to HNC progression. J Exp Clin Cancer Res. 2019;38(1):151.
85. Roccaro AM, Sacco A, Maiso P, Azab AK, Tai YT, Reagan M, et al. BM mesenchymal stromal cell-derived exosomes facilitate multiple myeloma progression. J Clin Invest. 2013;123(4):1542–55.
86. Au Yeung CL, Co NN, Tsuruga T, Yeung TL, Kwan SY, Leung CS, et al. Exosomal transfer of stroma-derived miR21 confers paclitaxel resistance in ovarian cancer cells through targeting APAF1. Nat Commun. 2016;7:11150.
87. Qin X, Guo H, Wang X, Zhu X, Yan M, Wang X, et al. Exosomal miR-196a derived from cancer-associated fibroblasts confers cisplatin resistance in head and neck cancer through targeting CDKN1B and ING5. Genome Biol. 2019;20(1):12.
88. Nabet BY, Qiu Y, Shabason JE, Wu TJ, Yoon T, Kim BC, et al. Exosome RNA Unshielding couples stromal activation to pattern recognition receptor signaling in cancer. Cell. 2017;170(2):352–66.e13

89. Ren J, Ding L, Zhang D, Shi G, Xu Q, Shen S, et al. Carcinoma-associated fibroblasts promote the stemness and chemoresistance of colorectal cancer by transferring exosomal lncRNA H19. Theranostics. 2018;8(14):3932–48.
90. Bliss SA, Sinha G, Sandiford OA, Williams LM, Engelberth DJ, Guiro K, et al. Mesenchymal stem cell-derived exosomes stimulate cycling quiescence and early breast cancer dormancy in bone marrow. Cancer Res. 2016;76(19):5832–44.
91. Zeng A, Wei Z, Yan W, Yin J, Huang X, Zhou X, et al. Exosomal transfer of miR-151a enhances chemosensitivity to temozolomide in drug-resistant glioblastoma. Cancer Lett. 2018;436:10–21.
92. Pardi N, Weissman D. Nucleoside modified mRNA vaccines for infectious diseases. Methods Mol Biol (Clifton, NJ). 2017;1499:109–21.
93. Ostenfeld MS, Jeppesen DK, Laurberg JR, Boysen AT, Bramsen JB, Primdal-Bengtson B, et al. Cellular disposal of miR23b by RAB27-dependent exosome release is linked to acquisition of metastatic properties. Cancer Res. 2014;74(20):5758–71.
94. Kanlikilicer P, Rashed MH, Bayraktar R, Mitra R, Ivan C, Aslan B, et al. Ubiquitous release of exosomal tumor suppressor miR-6126 from ovarian cancer cells. Cancer Res. 2016;76(24):7194–207.
95. Rashed MH, Kanlikilicer P, Rodriguez-Aguayo C, Pichler M, Bayraktar R, Bayraktar E, et al. Exosomal miR-940 maintains SRC-mediated oncogenic activity in cancer cells: a possible role for exosomal disposal of tumor suppressor miRNAs. Oncotarget. 2017;8(12):20145–64.

# Chapter 14
# Distinguishing Functional Roles of Protein-Coding Vs. Non-Coding RNAs: The Possibility of Bifunctional RNAs

**Abstract** The long-standing assumption in the field of biology, spanning nearly half a century, has been the dichotomous categorization of RNA into protein-coding or non-protein-coding entities. Traditionally, this distinction was relatively unambiguous: the easily identified protein-coding mRNAs make up a majority of transcripts while the rest, including spliceosomal RNAs, ribosomal RNAs, and transfer RNAs, belong to the ncRNAs category. However, with the latest advancements in genome-wide analyses, a plethora of non-coding transcripts have been discovered. Despite this, their significance and functional roles remain enigmatic. The revelation of this concealed transcriptome has posed a substantial challenge to our current understanding of genetic information regulation and expression. Consequently, the imperative to discriminate ncRNAs from mRNAs has grown not only in urgency but also complexity. Complicating matters, errors in annotations can occur both ways: certain transcripts presumed to encode proteins may not actually do so, while certain ncRNAs may, in fact, encode peptides. Furthermore, recent investigations have disclosed that certain RNAs can serve dual functions, operating intrinsically as functional ncRNAs, and encoding proteins. This phenomenon may be more widespread than previously thought. Moreover, it has been discovered that certain transcripts may serve as both RNAs and protein-encoding entities, which challenges the conventional classification of ncRNAs and mRNAs. As a result, we must not hastily disregard the functional potential of all RNA transcripts.

## 14.1 Introduction

The transcriptome—the comprehensive landscape of RNAs encrypted within the genome—surpasses prior conceptions in both scope and intricacy [1–3]. Studies have shown that a significant portion of the genome undergoes transcription, generating complex networks of overlapping transcripts (including both sense and antisense) that are frequently subjected to alternative splicing [1, 4–8]. Remarkably, however, conventional protein-coding mRNAs constitute only a mere fraction



P. E. Saw, E. Song, *RNA Therapeutics in Human Diseases*,
https://doi.org/10.1007/978-981-96-3041-7_14

[1, 9]. Consequently, much of this unanticipated transcription manifests as non-protein-coding entities that are colloquially denoted as "dark matter" [10, 11]. Therefore, it is unsurprising that substantial scientific attention has become directed towards the exploration of the non-coding transcriptome.

A prominent aspect of this investigation landscape has been the identification of numerous small RNAs, typically less than 200 nucleotides in length. Many of these have been categorized into different classes based on their functional attributes, length, biogenesis mechanisms, structural and sequence characteristics, and associations with protein partners. Some of these classes include endogenous siRNAs, piRNAs, and miRNAs [12]. Intriguingly, lncRNAs, defined as those exceeding 200 nt in length, appear to comprise a substantial proportion of the non-coding transcriptome in mammals. Recent studies, notably those involving tiling array analyses, have shown that most transcription events in the human genome produce lncRNAs [13]. Some lncRNAs can act as precursors to shorter RNA molecules, whereas others remain as relatively stable transcripts, either non-polyadenylated or polyadenylated [7, 14].

The role of lncRNAs in biology remains a subject of ongoing debate and scrutiny. Although there is increasing evidence of the diverse regulatory roles played by a growing number of lncRNAs [15, 16], most of them are still not well understood. Unlike smaller RNA molecules, lncRNAs do not have distinct structural or sequence features that make functional categorization or prediction easy. Moreover, the precise count of lncRNAs within the genome remains a matter of debate, making it challenging to ascertain their exact prevalence. Take the consortiums, H-Invitational and FANTOM, for example. These consortiums attempted to annotate the collections of full-length cDNA in humans and mice, respectively [6, 17]. Strikingly, they produced significantly different estimates for the prevalence of lncRNAs within the mammalian genome despite deploying similar methodologies in constructing cDNA library. In the case of mice, 33% of transcripts (34,030 out of 102,281) were designated as ncRNAs; whereas in humans, only 7% of transcripts (1377 out of 21,037) were identified as non-coding. These contrasting estimates highlight the challenge of differentiating between mRNAs and lncRNAs. The subsequent sections will explore the methods employed to differentiate between protein-coding and non-coding entities, providing a more comprehensive understanding of these complex challenges. Dinger et al. summarized an elegant review on the few main strategies to distinguish between coding and ncRNAs [18], which we will also elaborate below. Please note that none of the strategies that were put forward could be sufficient in being the sole determinant to distinguish these categories of RNAs.

## 14.2 Strategies to Distinguish Between Protein-Coding mRNAs and ncRNAs

### *14.2.1 Conservation of the Open Reading Frames (ORFs)*

In light of the limitations inherent in relying solely upon ORF size as a distinguishing criterion, an alternative strategy for discriminating ncRNAs from mRNAs involves the assessment of putative ORFs for homology with protein domains or proteins that are already known, which indirectly supports their functionality as mRNA. This is because a large proportion of putative human ORFs that are not conserved across species are likely to be random occurrences [19], which suggests they are not functional. Many researchers now use the absence of ORF conservation as evidence against the functionality of individual mRNAs. There are various resources and tools available to perform such analyses, conducting such analyses, including SUPERFAMILY [20], BLASTX [21], Pfam [22], and rsCDS [23]. Differentiating between mRNAs and ncRNAs on a transcriptome-wide scale can be achieved by using methods that are designed to identify the conversation of ORFs. Various techniques aimed at detecting ORF conservation can be utilized to distinguish ncRNAs from mRNAs across the entire transcriptome. Methods like CRITICA [24] and CSTminer [25, 26] use comparative approaches that rely on the tendency of protein-coding sequences to prefer changes that do not result in amino acid substitution (i.e., synonymous base changes) over those that do (i.e., non-synonymous changes). However, these approaches have limitations such as their dependence on the availability of genomes for comparison and the difficulty in detecting orthologous sequences due to the rapid evolution of many ncRNAs [27].

### *14.2.2 Lengths of the ORFs*

The length of ORF is one of the main strategies employed to differentiate between mRNAs and lncRNAs. To mitigate the possibility of erroneously categorizing long ncRNAs as mRNAs due to the occasional occurrence of short, putative ORFs within long non-coding sequences, researchers have traditionally applied minimum ORF length cutoffs. For example, as an effort to aid in the identification of potential mRNAs, the FANTOM consortium initially set 300 nucleotides (equivalent to 100 codons) as the cutoff point [28]. This threshold aligns with the observation that over 95% of proteins cataloged in the International Protein Index and Swiss-Prot are longer than 100 amino acids [29]. Subsequent studies have demonstrated that this particular threshold is highly effective in distinguishing between mRNAs and ncRNAs [30]. Nevertheless, relying solely on putative ORF length as a criterion presents certain challenges. Firstly, authentic lncRNAs may have putative ORFs of considerable lengths. For instance, well-established functional lncRNAs like KcnqOT1, H19, Mirg, Xist, and Gtl2 are unequivocally characterized as functional ncRNAs although all of them feature putative ORFs exceeding 100 codons [15]. Therefore, the application of a conventional ORF cutoff of 300 nucleotides could result in the misclassification of numerous lncRNAs as mRNAs, especially in the context of lncRNAs that are exceptionally long. Take murine Xist for instance, its length spans over 15 kilobases [31] and encompasses a putative ORF encoding 298 amino acids, resulting in an initial misclassification as a protein-coding gene upon its discovery [32].

Secondly, employing a 300-nucleotide ORF length threshold may inaccurately classify proteins smaller than 100 amino acids as ncRNAs. This potential for misclassification carries significant implications, given recent estimations suggesting the existence of nearly 3700 proteins in the mammalian proteome falling below this size threshold [29]. In order to mitigate such misclassifications, researchers have considered lowering the ORF length cutoffs, as exemplified by the H-Invitational consortium's implementation of a 60-nt threshold in their annotation [17]. Adopting a threshold so low may, however, lead to an underestimation of the number of identified ncRNAs, potentially explaining the relatively small numbers of ncRNAs annotated by the H-Invitational consortium.

Lastly, it is important to note that there is a possibility that certain atypical proteins may evade detection even when employing such low cutoffs. Take the tarsalless (*tal*) gene in Drosophila as an example. This gene, responsible for controlling tissue folding, encodes a transcript spanning approximately 1.5 kilobases [33]. Intriguingly, all putative ORFs within this transcript are tremendously short. Consequently, *tal* was originally categorized as being non-coding [34]. However, subsequent research uncovered that it indeed undergoes translation to yield multiple 11-amino-acid peptides, each fulfilling a functional role within the gene's framework [33]. These instances, such as the one presented by *tal,* underscore the dangers of disregarding even the smallest putative ORFs, raising the question of how many other ostensibly ncRNAs may, in fact, encode real albeit brief proteins.

### *14.2.3 Structural Approaches*

The methodologies previously delineated focused on the identification of mRNAs. Therefore, distinguishing lncRNAs usually occurs indirectly by identifying their lack of mRNA-like characteristics. On the contrary, some research sought to pinpoint ncRNAs by identifying RNA secondary structures that indicate potential functional properties. Noteworthy among these approaches are these methods are EvoFOLD [35], RNAz [36], and QRNA [37]. Nevertheless, it is important to acknowledge that ncRNA classification through these programs carries the risk of yielding significant false positives and false negatives. This is because conserved secondary structures are often found in mRNAs, especially in their 3'-UTRs. Additionally, functional ncRNAs can include secondary or tertiary structures featuring non-canonical base interactions [38]. Structural prediction programs typically do not consider these unconventional structural features, thus rendering the

**Table 14.1** Comparison of the three major methods for identifying RNA secondary structures

| Method | Description | Primary use | Unique point |
|---|---|---|---|
| EvoFOLD [35] | Utilizes evolutionary conservation of predicted RNA secondary structures across multiple species | Identifies RNA structural motifs that are evolutionarily conserved, suggesting functional relevance. Useful in identifying functional RNA elements across diverse genomes | EvoFOLD employs a phylogenetic hidden Markov model to predict conserved secondary structures. This method is particularly useful for discovering new ncRNA candidates that may have functional implications in various biological processes |
| RNAz [36] | Combines comparative sequence analysis with secondary structure prediction to identify conserved RNA structures | Predicts structurally conserved and thermodynamically stable RNA secondary structures in multiple sequence alignments, indicating potential functional ncRNAs | RNAz uses a statistical approach to evaluate whether the thermodynamic stability of a predicted RNA secondary structure is significantly higher than expected by chance. This approach is effective for finding RNA structures that are likely to be functional due to their stability and conservation across species |
| QRNA [37] | Compares nucleotide sequence alignments to predict conserved secondary structures and coding regions | Discriminates between coding and non-coding sequences based on the presence of conserved secondary structures, aiding in the identification of functional ncRNAs | QRNA utilizes a model that compares the likelihood of sequence conservation due to structural constraints versus coding constraints. This method is helpful in distinguishing ncRNAs from protein-coding sequences, especially in genomic regions where the distinction is ambiguous |

classification of transcripts as ncRNAs based solely on such predictions potentially flawed. Table 14.1 elaborate on the methods used for pinpointing ncRNAs based on RNA secondary structures.

### *14.2.4 Artifact Filtering*

It is crucial to accurately classify newly discovered transcripts as ncRNAs or mRNAs. However, this process hinges on the critical assumption that these transcripts are authentic and complete. Unfortunately, there are several factors that can cause problems and lead to the formation of truncated transcripts. These factors include contamination at the genomic level, internal priming of pre-mRNAs, and incomplete reverse transcription. As a result, some transcripts may be mistaken for ncRNAs even though they are not [39]. To mitigate these challenges, various strategies have been employed to eliminate possible experimental artifacts. For instance, the FANTOM3 consortium used a stringent methodology by applying strict criteria for support at both the 5′- and 3′-ends of transcripts. This rigorous approach reduced their initial catalog of 34,030 ncRNAs to a more reliable list of 3652 full-length ncRNAs [6]. Another approach entails excluding ncRNAs that map to the same genomic strand and locus as an mRNA, based on the assumption that these transcripts could be artificially truncated mRNAs [28]. It is important to note, however, that these filtering approaches may inadvertently exclude genuine ncRNAs. This can happen because current cloning and sequencing techniques have difficulty capturing lncRNAs that are overwhelming in length, such as Air and Xist (exceeding 5 kb) [40]. Additionally, a growing body of evidence indicates that many genomic loci contain both mRNAs and ncRNAs on the same strand, which increases the risk of misclassification [13]. Therefore, while these filtering strategies aim to enhance the accuracy of transcript classification, they must be applied judiciously to avoid the inadvertent exclusion of authentic ncRNAs.

### 14.2.5 Experimental Strategies

In addition to computational approaches, several experimental strategies have been employed in attempts to differentiate between mRNAs and ncRNAs. One such strategy involves in vitro translation assays, which have been conducted in isolated instances to ascertain if a putative ORF does indeed yield products in the form of proteins upon translation [33, 41, 42]. Positive outcomes from translation assays provide suggestive evidence that a given transcript functions as an mRNA. However, it is crucial to exercise caution in interpreting these results, as spurious ORFs have been known to produce translated products in vitro. Proteins resulting from such artifacts have even been found to cause the production of antibodies against them [43]. Alternatively, researchers have explored the association of transcripts with polysomes [31], which are indicative of actively translated mRNAs. Nevertheless, this method, too, presents challenges in definitively classifying transcripts. The presence or absence of polysome association may not unequivocally distinguish between mRNAs and ncRNAs, as the outcome can be influenced by various factors.

### 14.2.6 Combination Strategies

It is possible to combine many of the strategies discussed above to achieve better results. An example of this is CRITICA, which employs its comparative approach alongside statistical techniques and is considered the top-performing method among ten bioinformatics approaches utilized to differentiate between mRNAs and ncRNAs in the FANTOM cDNA collection [30]. Several other programs also utilize advanced statistical methodologies by incorporating various characteristic signatures of genes that code for proteins, such as splice donor/acceptor sites, ORF length, polyadenylation signals, and sequence homology. For instance, DIANA-EST combines artificial neural networks with statistical techniques to discriminate coding regions [44], while ESTScan utilizes a hidden Markov model approach [45]. CONC and CPC are two tools that have been introduced recently. These tools employ support vector machines (learning algorithms that are supervised) to effectively discriminate between ncRNAs and mRNAs [46, 47]. The presence of protein homologs, protein alignment information, protein secondary structure, peptide length, and amino acid composition are some of the factors that were taken into account. The accuracy of both CONC and CPC has been demonstrated through cross-validation against datasets for ncRNA and reference protein. Consequently, they represent the forefront of future methodologies for discriminating between these transcript classes.

**Box 14.1 Can Transcripts Truly Be Classified as Coding or ncRNAs—The Possibility of Bifunctional RNAs**

Recent reports have started to unveil a novel paradigm in the field of RNA, one that challenges the conventional categorization into mutually exclusive protein-coding and non-coding groups. These new findings describe a type of RNA that can either function as an independent RNA molecule, or readily undergo translation to produce protein. One such RNA is the human Steroid Receptor Activator (SRA). Initially classified as an ncRNA, SRA was recognized for its role in co-activating steroid hormone receptors at the transcript level [41]. Intriguingly, subsequent studies have demonstrated that SRA transcripts also encode a functional protein (SRAP) that works in opposition to SRA RNA in modulating steroid hormone receptors [48, 49]. This discovery is significant because it suggests that bifunctional transcripts may have the ability to negatively self-regulate, but further research is necessary to understand the mechanisms underlying this process and its regulation. Bifunctional RNAs have been found to serve multiple roles. For instance, the VegT RNA, originally identified for encoding a protein essential for forming primary germ layers in Xenopus, has also been found to have a structural role in the cytokeratin network of primordial germ cells [50]. Similarly, the Oskar RNA in Drosophila was first recognized for its role in being translated into proteins crucial for oocyte development [51, 52]. However, recent findings suggest that the 3'-UTR of the Oskar mRNA operates independently of the Oskar protein and plays an indispensable role in oogenesis [53]. Furthermore, some 3' UTRs can regulate cell proliferation and differentiation independently of their associated protein-coding sequences. Disruptions in these regulatory regions have been associated with certain cancers [54–59].

Bifunctional RNAs are currently rare, but given the abundance of conserved secondary structures in mRNAs, they may be more common than we think. Recent research indicates that up to 5% of mRNAs in yeast may serve

(continued)

**Box 14.1** (continued)
functional roles as RNA, and this percentage is expected to be significantly higher in more complex eukaryotes [60]. Additionally, mRNA can pair with antisense RNAs or pseudogenes to form duplex RNA structures, leading to the generation of endogenous siRNAs, thereby conferring additional functions to the mRNA [61–63]. Even synonymous sites within codons, which were previously deemed entirely redundant, can carry important information. "Silent" mutations can influence both splicing and co-translational folding, resulting in changes to protein function [64, 65]. These observations indicate that RNA may carry significantly more information on cis- and trans-acting elements than previously believed, both within and beyond the boundaries of protein-coding sequences.

## 14.3 lncRNAs Can Encode Peptides and Proteins

Initially, lncRNAs were classified based on the belief that they possessed little or no potential for coding proteins. The perception was that lncRNAs functioned primarily through their RNA sequences, influencing gene regulation and cellular processes without being translated into proteins. This view was shaped by the traditional understanding that the primary distinction between coding RNAs and non-coding RNAs lay in their ability to serve as templates for protein synthesis. However, recent advances in bioinformatics and ribosome profiling have significantly altered this perspective. Ribosome profiling, a technique that allows for the precise mapping of ribosome positions on mRNA during translation, has been instrumental in this shift. When applied to lncRNAs, these studies have frequently identified ribosomes associated with these supposedly non-coding transcripts. This association is not limited to a specific group of organisms but spans across various species, including both plants and animals, suggesting a fundamental, conserved feature of cellular biology. The presence of ribosomes on lncRNAs indicates that these RNA molecules may indeed possess ORFs capable of being translated into proteins or peptides. These ORFs are often small and were overlooked in initial analyses, perhaps due to their size or because they did not conform to conventional expectations for protein-coding genes, which typically include longer sequences. The discovery of these small ORFs challenges the traditional categorization of lncRNAs and suggests that the transcriptome may contain many more coding sequences than previously understood.

Bioinformatics has played a crucial role in uncovering the coding potential hidden within lncRNAs. Through sophisticated computational methods, researchers can predict ORFs within lncRNA sequences and assess their coding probability by analyzing codon usage, sequence conservation across species, and the presence of known protein domains. These analyses often reveal that certain segments within lncRNAs show hallmarks of evolutionary conservation and functional motifs typical of coding sequences. The peptides encoded by lncRNAs, though typically smaller than standard proteins, could have significant biological functions. These micropeptides might act in various cellular processes, including signaling pathways, cellular structure integrity, and regulation of larger proteins. The functions of these peptides are only beginning to be understood, but they could be crucial for cellular operations and organism development. This growing understanding that lncRNAs may have coding potential revises our comprehension of the genetic code and the complexity of gene regulation. It suggests that the transcriptome's functional landscape is much richer and more complex than previously thought. This revelation has broad implications for all areas of genetics and molecular biology, potentially affecting everything from developmental biology to the understanding and treatment of diseases [66–68].

**Box 14.2 The Steroid RNA Activator (SRA): Coding or Non-coding RNA?**

In recent years, the exploration of lncRNAs has unveiled their complex roles not only as regulators of gene expression but also as sources of functional micropeptides. A prime example of this dual functionality is found in the SRA, an lncRNA recognized for both its non-coding and coding capabilities. This duality illustrates the intricate nature of gene expression regulation and the versatility of genetic elements.

***Detailed Expansion on the SRA Gene***

The SRA gene was initially identified as a non-coding RNA with multiple RNA isoforms. It gained prominence due to its crucial roles in various biological processes [41, 69–71]. As a non-coding RNA, SRA notably acts as a regulator of gene expression influenced by steroid receptors. It also serves as a co-activator of nuclear receptors, which are pivotal in the transcriptional regulation mechanisms that control numerous cellular functions, including metabolism, reproduction, and development.

***Coding Functions and Micropeptide Production***

Beyond its non-coding roles, the SRA gene can encode a functional micropeptide known as Steroid Receptor RNA Activator Protein (SRAP). SRAP is notable for being conserved across different species, suggesting its evolutionary importance and functional significance. This micropeptide interacts specifically with a stem-loop structure within the SRA RNA itself. By binding to this structure, SRAP can negatively regulate the transcriptional activities of the SRA RNA, essentially modulating its own genetic regulator in a feedback loop mechanism. This interaction and its implications for gene regulation have been substantiated in various studies, underscoring the complex interplay between the SRA's coding and non-coding functions [69, 70, 72].

***Interchangeability of Coding and Non-coding Roles***

The ability of the SRA gene to switch between coding and non-coding roles is facilitated by alternative splicing events. These events involve the

(continued)

**Box 14.2** (continued)
inclusion or exclusion of certain introns and exons during the RNA processing stage. Alternative splicing allows the SRA gene to produce different RNA isoforms, some of which are translated into the SRAP peptide, while others function exclusively at the RNA level. This process is key to the gene's adaptability and functionality, allowing it to meet cellular demands under varying physiological conditions [48, 70, 73].

***Importance of Alternative Splicing***

The role of alternative splicing in generating functionally diverse RNA molecules from the SRA gene highlights the broader significance of this process in genetic regulation. By producing multiple RNA and protein products from a single gene, cells can effectively increase their functional repertoire without the need for additional genes. This efficiency is crucial in complex organisms, where cellular functions are highly diversified and tightly regulated [74].

Therefore, the SRA gene exemplifies the sophisticated regulatory mechanisms that can arise from the non-coding regions of the genome. Its ability to function both as a source of ncRNA and as a template for a coding micropeptide illustrates the dynamic nature of genomic expression and the potential hidden within the largely unexplored territories of lncRNAs. This duality also emphasizes the potential for discovering novel therapeutic targets and understanding disease mechanisms through the study of lncRNAs and their encoded peptides.

It is important to note that there are a limited number of known functional micropeptides. Table 14.2 below provided the expanded details provide a clearer view of how micropeptides, often overlooked due to their small size, play crucial roles in fundamental physiological processes across different species and tissue types. Their diverse functions underscore the complexity of genetic regulation and the potential for these peptides in therapeutic developments.

These findings suggest that certain sORFs that are located in non-coding regions of the genome may undergo more rigorous natural selection as compared to adjacent sequences. This raises intriguing questions about whether these sORFs have the potential to evolve into new genes or become part of novel genes during evolution. For instance, the *tal* gene in *Drosophila*, primarily associated with tarsal morphogenesis in fly legs, was initially thought to be a non-coding gene because none of its ORFs were more than 100 amino acids in length. However, detailed analysis has shown that the functionality of tal mainly relies on its ORF regions [82]. The *tal* gene contains five ORFs, four of which four contain a conserved 7-amino acid motif that is crucial for its function. The smallest peptide is only 11 amino acids long. Phylogenetic analysis has revealed that these *tal*-like peptides are present and consistent across metazoans, thereby establishing a new category of eukaryotic genes. This discovery broadens the potential range and function of lncRNA-encoded peptides hidden within the transcriptomes and genomes sequenced to date [33].

**Table 14.2** Specific roles and characteristics of various micropeptides

| Micropeptide | Length and origin | Expression and origin species | Interaction and function | Insights | References |
|---|---|---|---|---|---|
| MLN | 46 amino acids; Encoded by lncRNA LINC00948 in humans and 2310015B20Rik in mice | Specifically expressed in skeletal muscle | Directly interacts with SERCA, reducing its affinity for calcium ions, thus inhibiting calcium entry into the sarcoplasmic reticulum (SR), impacting muscle function and contraction | This peptide shows the complexity of interactions between non-coding RNAs and essential cellular mechanisms. By influencing SERCA's function, MLN modulates calcium ion dynamics in muscle cells, crucial for muscle contraction, and overall muscle physiology | [75–78] |
| Scl | Encoded by the non-coding gene pncr003:2 L | Regulates calcium trafficking in cardiac muscle of flies | A mutation in the Scl gene leads to an arrhythmic phenotype, indicating its critical role in cardiac rhythm regulation | The role of Scl in cardiac muscle highlights the importance of precise calcium regulation in heart function. The arrhythmic phenotype resulting from Scl mutations underscores the critical nature of its regulatory function and suggests potential therapeutic targets for arrhythmias | [79] |
| MOTS-C | Originates from mitochondrial 12S rRNA | Expressed in muscle cells' mitochondria | Regulates insulin sensitivity and metabolic homeostasis, highlighting a significant role in energy metabolism and cellular energy management | Beyond muscle contraction, MOTS-C illustrates the broader impact of micropeptides on cellular metabolism. Its role in regulating insulin sensitivity and metabolic processes connects mitochondrial function with systemic metabolic health, suggesting implications for diseases like diabetes and metabolic syndrome | [80, 81] |

## 14.4 CircRNAs Can Encode Peptides and Proteins

Bioinformatics analysis and high-throughput sequencing across various species have facilitated the identification of circRNAs [83–87]. However, their functions are not yet fully understood. circRNAs have been reported to exhibit diverse biological activities, including serving as transcriptional regulators in the nucleus [88], competing during pre-mRNA splicing [89], and acting as microRNA sponges [84, 90]. Chimeric circRNAs often originate from mRNA transcripts and may contain exonic regions of protein-coding genes as circRNAs, raising questions about their potential to encode proteins. Based on findings from numerous studies, circDNAs were found to be capable of coding both in vivo and in vitro, mainly though cap-independent translation mechanisms [83, 91–96]. To provide a structured view of the information regarding certain circRNAs that produce functional protein products, the following table summarizes key details about these circRNAs; we summarized some important circRNAs and their functional protein products in Table 14.3 below.

Although the idea that circRNAs can produce proteins was first proposed in studies of Archaea, where circularized introns generate site-specific endonucleases [101], there is currently limited experimental evidence supporting this theory. As a result, it is challenging to understand the functions of the proteins that circRNAs may translate to. Since most circRNAs arise from coding transcripts and contain full exons, it is possible that they and their translated products may provide novel mechanisms for regulating gene and protein expression [91]. Henceforth, further research is necessary to determine the potential functions associated with the coding capabilities of circRNA.

## 14.5 mRNAs Contain ncRNA Regulatory Functions

### 14.5.1 *The Role of mRNA in the Regulation of 3' UTR*

Recent research indicates that the 3'-UTRs of mRNAs are the primary location of non-coding regulatory functions. Although previously believed to be important only for mRNA self-stability and localization, the 3'-UTRs have emerged as key players in non-coding regulation. Unlike the highly conserved coding region, the 3'-UTR is more flexible and variable across species, with a length ranging from a few to hundreds of nucleotides, potentially correlating with biological complexity [102, 103]. Furthermore, alterations in sequence, apart from affecting base-pairing, are likely to induce corresponding structural changes, facilitating the transmission of information from RNA to protein [104]. It has been discovered that the 3'-UTR of mRNA plays a significant role in biological activities, even without being translated into proteins. This has become evident through an examination of modern cases. The 3' UTR can suppress the onset and progression of cancer cells, as exemplified by mRNAs like prohibitin mRNA, ribonucleotide reductase mRNA, and a-tropomyosin

**Table 14.3** circRNAs and their translated protein product

| circRNA | Origin and ORF details | Protein product and function | Additional insights and context | Implications | Reference |
|---|---|---|---|---|---|
| circ-ZNF609 | Contains an ORF that covers almost all the ORF regions of the host gene, ZNF609; differs slightly at the splice junction | The protein lacks the zinc-finger domain of its linear counterpart but influences myoblast proliferation | Heat shock can induce the translation of circ-ZNF609, suggesting a regulatory role in response to stress or stimuli. This adaptation may be crucial in muscle development and repair | The translation of circ-ZNF609 into a protein that affects myoblast proliferation is a fascinating example of how circRNAs can contribute to cell growth and differentiation. The induction of its translation by heat shock further emphasizes its potential role in cellular stress responses, which could be pivotal during muscle repair and regeneration processes | [97] |
| circ-FBXW7 | Encodes a novel 21 kDa protein isoform termed FBXW7-185aa, originating from a shorter ORF | This protein acts as a tumor suppressor by destabilizing c-Myc, thereby inhibiting the cell cycle and reducing tumor growth | The production of this protein demonstrates circRNAs' capacity to participate in critical cellular pathways such as cancer progression and cell cycle regulation | The ability of circ-FBXW7 to encode a protein that acts as a tumor suppressor illustrates the functional diversity of circRNAs beyond their previously assumed role as non-coding molecules. This protein's impact on c-Myc destabilization and tumor growth inhibition shows potential therapeutic avenues in oncology | [98, 99] |
| circ-SHPRH | Produces a protein using a rolling circular translation mechanism | The resultant protein, SHPRH-146aa, helps protect against genomic instability by ubiquitinating proliferating cell nuclear antigen (PCNA) in tumor cells | Indicates a protective role against cancer, enhancing our understanding of circRNAs in genomic maintenance and stability, particularly in oncogenic contexts | This circRNA's ability to produce a protein that stabilizes genomic integrity by interacting with PCNA underscores the potential protective roles circRNAs can play in preventing cancer progression. It offers a new layer of understanding of how circRNAs contribute to cellular defense mechanisms against DNA damage in cancer cells | [100] |
| circ-Mbl | Originates from the second exon of the splicing factor muscleblind (MBL/MBNL1) | Competes with pre-mRNA splicing and is involved in protein synthesis in the synaptosome, indicating roles in neural function and development | The presence of both circ-Mbl RNA and its related protein in the synaptosome, regulated by 4E-BP and FOXO, highlights the importance of circRNA translation in brain function and regulation | The involvement of circ-Mbl in both RNA and protein forms in the regulation of synaptic function through competition with pre-mRNA splicing and protein synthesis in the brain reveals the complex regulatory capabilities of circRNAs in neural pathways | [89, 91] |

mRNA [57, 58, 105]. Thus, it is clear that the 3'-UTRs of specific mRNAs can counteract tumorigenesis. It is probable that this occurs via RNA interactions with regulatory factors that participate in the post-transcriptional control of cellular growth. RNA-binding proteins can be recruited by the 3'-UTR, as seen in CD47 mRNA. CD47 mRNA has two variants with different 3'-UTRs: short (CD47-SU) and long (CD47-LU). Only the CD47-LU variant, characterized by its high AU content, can interact with the RNA-binding protein TIS11B to form a membraneless organelle. This organelle has a distinct biochemical and biophysical environment compared to the cytosol [106]. Additionally, the 3'-UTRs can function as competitive endogenous RNAs (ceRNAs), as seen in examples like CCR2 mRNA and Ube3a1 RNA, which operate similarly to lncRNAs [107, 108].

## 14.6 Non-coding Regulatory Functions of mRNA that Do Not Involve the 3' UTR Region

In addition to the 3'-UTR, the 5'-UTR and ORF can also participate in RNA-mediated regulatory functions although there have been relatively few recent reports on this phenomenon. It is interesting to note that two specific mRNAs, namely HIST1H1C mRNA and TP53 mRNA, have been identified for their contribution to ORF-mediated regulation. The TP53 protein is a well-known tumor suppressor pivotal in the onset and progression of tumors. Recently, a significant discovery was made regarding the TP53 mRNA. Specifically, a triple synonymous mutant (TriMp53) within codons caused a misshapen protein. This structural change led to the loss of internal ribosome entry site (IRES) activity of p47, an isoform of p53, and diminished the binding affinity of hnRNPC. However, it enhanced the binding to Mdm2, an E3 ubiquitin-protein ligase that helps to mediate p53/TP53 ubiquitination. Notably, this alteration also improved p53's ability to activate apoptosis. These findings collectively indicate that TP53 has multifaceted regulatory functions at both the RNA and protein levels, highlighting the intricate interconnection between the roles of RNA and protein molecules [109]. The mRNA of HIST1H1C plays a role in the regulation of telomere length. Beyond its protein-related product, a specific 15-nucleotide region within the ORF region (nt334–nt348) is crucial for the biological activity of HIST1H1C-mRNA. This region complements the P6b region's terminal stem-loop sequence through base-pairing, revealing the functional potential of non-traditional non-coding mRNA ORF regions [110].

There are only two notable examples of the 5'-UTR. One of them is the vascular endothelial growth factor (VEGF), which is a pivotal regulator of angiogenesis during embryonal development and cancer progression. The 5'-UTR of the VEGF mRNA is unusually long, spanning 1038 nucleotides, and it houses two IRES. These sites work together to regulate the expression of VEGF in a complex manner. Research has demonstrated that the presence of the 5'-UTR of VEGF mRNA in tumor cells can enhance the expression of genes that inhibit cell death and suppress the expression of genes that promote cell death. This indicates that the VEGF 5'

UTR plays a critical role in preventing cell death and may be a potential target for cancer treatment [111–113]. The c-MYC gene has several isoforms, including the P0 transcript which has an extended 5'-UTR of approximately 639 nucleotides compared to the two major isoforms (P1 and P2) of c-myc mRNA. Fascinatingly, expressing just the 5'-UTR of the c-MYC P0 transcript in HeLa cells significantly boosts the production of c-MYC2 (p64) and c-MYC1 (p67) proteins. This heightened protein expression is accompanied by increased sensitivity to apoptosis but reduced tumorigenicity, suggesting regulatory control over gene expression within the c-myc locus. These findings indicate the potential for the 5'-UTR to function in trans to exort regulatory control over gene expression [114].

**Box 14.3 Emergence of Bifunctional RNAs: A Myth or a Reality?**

Over the years, scientists have increasingly focused on the emergence of bifunctional RNAs and complex roles played by RNAs with multifaceted functions [66, 115–117]. Initially, these studies were limited to individual genes. However, with the advent of advanced techniques like ribosome profiling and mass spectrometry-based proteomics, researchers have been able to explore genomes and biological processes in a more comprehensive manner. These techniques have enabled scientists to uncover the intricate activities of various RNA molecules. The concept of bifunctional RNAs challenges traditional notions about genes. It also makes us wonder if RNA, which includes both coding and non-coding types, is an independent type of gene or a result of convergent evolution. Our objective in this review is to evaluate the present status of bifunctional RNAs, as documented in recent years, and to investigate the prevalence of bifunctional RNAs on a global level, leveraging large-scale datasets.

Recent insights from ribosome profiling have revealed that lncRNAs encode small peptides (micropeptide, see Table 14.2) that significantly expand the scope and diversity of the proteome. It has been predicted that a significant number of annotated lncRNAs in various eukaryotic organisms harbor small ORFs with translational potential [118–121]. Proteogenomic evidence confirms the expression of numerous small peptides originating from regions within lncRNA genes. These peptides exhibit differential expression across distinct cell types and during various developmental or disease stages. However, their precise functions are still not fully understood [122, 123]. Investigations have shown that mRNAs could also engage in cellular regulatory processes without coding for proteins [67]. Extensive RNA structure analyses have unveiled that mRNA secondary structures exert significant regulatory influence on their maturation and stability. This is true even for RNA silencing pathways that are conserved across eukaryotes. This implies that mRNAs share some of the functional properties observed in other RNA molecules [124–126].

(continued)

**Box 14.3** (continued)

Based on aforementioned findings, it can be suggested that the lines between coding and non-coding genes are becoming blurred. lncRNAs share many resemblances with traditional mRNAs, such as being transcribed by RNA polymerase II, possession of a 5′-cap and 3′-polyadenylated tail, and frequently being found in the cytoplasmic regions [127]. As a result, small sORFs found within lncRNAs are likely to be translated into peptides when they interact with ribosomes. These peptides tend to be shorter and show less conservation across species. This is consistent with the typical features of lncRNAs, which are usually marked by low expression levels, restricted phylogenetic conservation, and few introns [128–131]. According to studies, these peptides are regarded as an essential source of innovative proteins from an evolutionary standpoint [120]. Previous experimental findings have suggested that non-coding RNAs with low expression levels can play a role in the development of new protein-coding genes [132–134]. Given that peptides derived from lncRNAs have proven to be vital in numerous biological processes, more research is essential to explore their possible importance in gene evolution, expression, and regulation. Additionally, it has been shown that even random RNA sequences can transform into structurally complex and highly active RNA ligases. This demonstrates that apparent randomness may yield functional elements [135]. Thus, it is highly plausible that RNA molecules, in isolation, encompass a wealth of genetic information, including distinct structural features and ultra-conserved sequence elements. These elements have the potential to regulate the spatial aspects and timing of gene expression during cellular differentiation and development in a precise manner. In a very recent review article, Poliseno et al. elegantly summarized the coding and non-coding abilities of ncRNAs, which can also be seen in Fig. 14.1 below.

## 14.7 Conclusion

As the count of protein-coding genes continues to undergo downward revisions, the inventory of ncRNAs within the genome appears to expand incessantly. Nevertheless, a cloud of ambiguity still shrouds the accurate enumeration of ncRNAs in the genome due to inherent challenges in distinguishing them from mRNAs and experimental artifacts, particularly among the numerous long transcripts that are difficult to classify, even with the most advanced contemporary categorization methods. The complexity of the transcriptome is further compounded by the fact that it may not consist of distinct and separate entities. Instead, it comprises an intricate and overlapping cluster of multiple categories, which extend across significant genomic regions. This complex landscape forms an information continuum, where the boundaries between categories become increasingly blurred. This realization

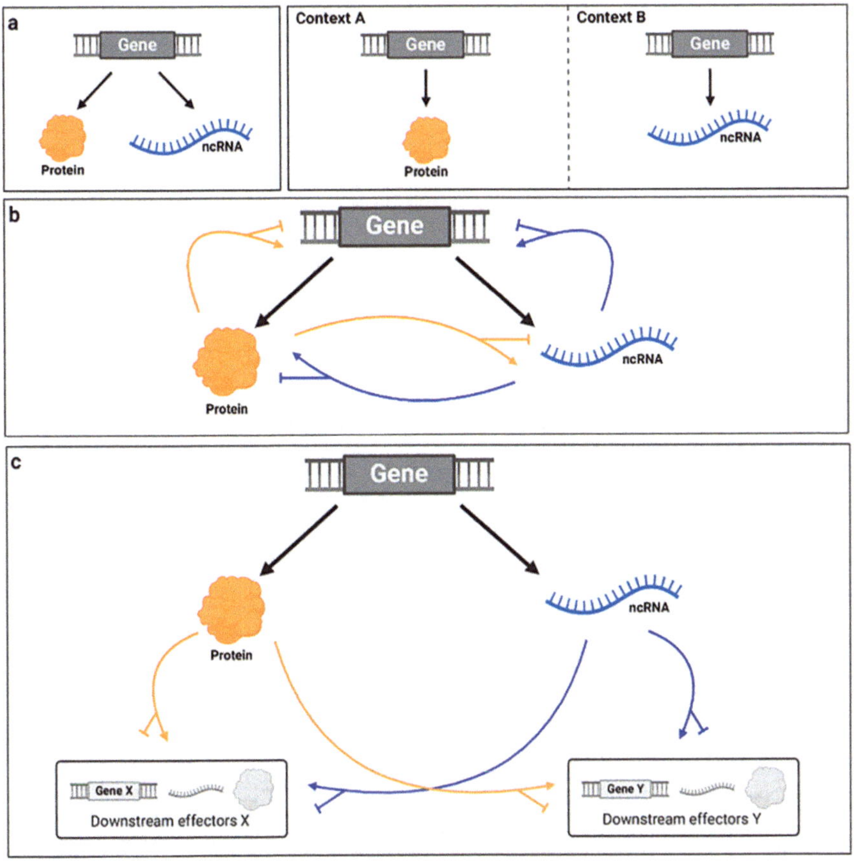


**Fig. 14.1** Schematic Representation of Gene Expression Regulation and ncRNA Function Across Different Contexts and Interactions. (**a**) Contextual Variation in Gene Expression. Context A: Depicts the expression of a gene leading to the production of a protein, demonstrating a direct transcriptional outcome. Context B: Illustrates the expression of the same gene leading to the production of an ncRNA, indicating an alternative transcriptional outcome based on differing cellular or environmental contexts. (**b**) Bidirectional regulation and feedback shows a gene simultaneously producing both protein and ncRNA. The ncRNA, in turn, regulates the expression of the gene either by enhancing or inhibiting transcription (as indicated by the orange and blue arrows). This bidirectional regulatory setup illustrates complex gene expression controls, possibly including feedback loops that modulate gene expression dynamics. (**c**) Crosstalk between gene products and downstream effects, displays a gene expressing both protein and ncRNA, where each of these products influences separate downstream genes or pathways. The protein activates downstream effectors X, associated with Gene X, while the ncRNA modulates downstream effectors Y linked to Gene Y. This figure summarizes the diverse roles of gene products in regulating multiple biological pathways, emphasizing the intricacy of cellular regulatory networks involving both coding and non-coding elements. Figure is reproduced from [136] under Creative Commons CC-BY 4.0 license

highlights the need for a deeper understanding of the transcriptome and its intricate functions. As we look to the future, it becomes imperative that we shed our historical biases, which are increasingly revealing themselves as outdated, and instead, adopt a perspective that acknowledges the RNA molecule as a versatile multitasker. It is becoming evident that RNA not only harbor multiple embedded roles but also give rise to diverse products. Embracing this multifaceted nature of RNA promises to unravel new layers of complexity in the biological information encoded within our genomes.

## References

1. Frith MC, Pheasant M, Mattick JS. The amazing complexity of the human transcriptome. Eur J Hum Genet. 2005;13(8):894–7.
2. Kapranov P, Willingham AT, Gingeras TR. Genome-wide transcription and the implications for genomic organization. Nat Rev Genet. 2007;8(6):413–23.
3. Mattick JS, Makunin IV. Non-coding RNA. Hum Mol Genet. 2006;15(Spec 1):R17–29.
4. Bertone P, Stolc V, Royce TE, Rozowsky JS, Urban AE, Zhu X, et al. Global identification of human transcribed sequences with genome tiling arrays. Science. 2004;306(5705):2242–6.
5. Birney E, Stamatoyannopoulos JA, Dutta A, Guigó R, Gingeras TR, Margulies EH, et al. Identification and analysis of functional elements in 1% of the human genome by the ENCODE pilot project. Nature. 2007;447(7146):799–816.
6. Carninci P, Kasukawa T, Katayama S, Gough J, Frith MC, Maeda N, et al. The transcriptional landscape of the mammalian genome. Science. 2005;309(5740):1559–63.
7. Cheng J, Kapranov P, Drenkow J, Dike S, Brubaker S, Patel S, et al. Transcriptional maps of 10 human chromosomes at 5-nucleotide resolution. Science. 2005;308(5725):1149–54.
8. Kapranov P, Cawley SE, Drenkow J, Bekiranov S, Strausberg RL, Fodor SP, et al. Large-scale transcriptional activity in chromosomes 21 and 22. Science. 2002;296(5569):916–9.
9. Finishing the euchromatic sequence of the human genome. Nature. 2004;431(7011):931–45.
10. Johnson JM, Edwards S, Shoemaker D, Schadt EE. Dark matter in the genome: evidence of widespread transcription detected by microarray tiling experiments. Trends Genet. 2005;21(2):93–102.
11. Riddihough. In the forests of RNA dark matter—introduction. Science. 2005;309(5740):1507.
12. Farazi TA, Juranek SA, Tuschl T. The growing catalog of small RNAs and their association with distinct Argonaute/Piwi family members. Development. 2008;135(7):1201–14.
13. Kapranov P, Cheng J, Dike S, Nix DA, Duttagupta R, Willingham AT, et al. RNA maps reveal new RNA classes and a possible function for pervasive transcription. Science. 2007;316(5830):1484–8.
14. Kiyosawa H, Mise N, Iwase S, Hayashizaki Y, Abe K. Disclosing hidden transcripts: mouse natural sense-antisense transcripts tend to be poly(A) negative and nuclear localized. Genome Res. 2005;15(4):463–74.
15. Prasanth KV, Spector DL. Eukaryotic regulatory RNAs: an answer to the 'genome complexity' conundrum. Genes Dev. 2007;21(1):11–42.
16. Amaral PP, Dinger ME, Mercer TR, Mattick JS. The eukaryotic genome as an RNA machine. Science. 2008;319(5871):1787–9.
17. Imanishi T, Itoh T, Suzuki Y, O'Donovan C, Fukuchi S, Koyanagi KO, et al. Integrative annotation of 21,037 human genes validated by full-length cDNA clones. PLoS Biol. 2004;2(6):e162.
18. Dinger ME, Pang KC, Mercer TR, Mattick JS. Differentiating protein-coding and noncoding RNA: challenges and ambiguities. PLoS Comput Biol. 2008;4(11):e1000176.

19. Clamp M, Fry B, Kamal M, Xie X, Cuff J, Lin MF, et al. Distinguishing protein-coding and noncoding genes in the human genome. Proc Natl Acad Sci USA. 2007;104(49):19428–33.
20. Gough J, Karplus K, Hughey R, Chothia C. Assignment of homology to genome sequences using a library of hidden Markov models that represent all proteins of known structure. J Mol Biol. 2001;313(4):903–19.
21. Gish W, States DJ. Identification of protein coding regions by database similarity search. Nat Genet. 1993;3(3):266–72.
22. Finn RD, Tate J, Mistry J, Coggill PC, Sammut SJ, Hotz HR, et al. The Pfam protein families database. Nucleic Acids Res. 2008;36(Database issue):D281–D8.
23. Furuno M, Kasukawa T, Saito R, Adachi J, Suzuki H, Baldarelli R, et al. CDS annotation in full-length cDNA sequence. Genome Res. 2003;13(6b):1478–87.
24. Badger JH, Olsen GJ. CRITICA: coding region identification tool invoking comparative analysis. Mol Biol Evol. 1999;16(4):512–24.
25. Mignone F, Grillo G, Liuni S, Pesole G. Computational identification of protein coding potential of conserved sequence tags through cross-species evolutionary analysis. Nucleic Acids Res. 2003;31(15):4639–45.
26. Castrignanò T, Canali A, Grillo G, Liuni S, Mignone F, Pesole G. CSTminer: a web tool for the identification of coding and noncoding conserved sequence tags through cross-species genome comparison. Nucleic Acids Res. 2004;32(Web Server issue):W624–W7.
27. Pang KC, Frith MC, Mattick JS. Rapid evolution of noncoding RNAs: lack of conservation does not mean lack of function. Trends Genet. 2006;22(1):1–5.
28. Okazaki Y, Furuno M, Kasukawa T, Adachi J, Bono H, Kondo S, et al. Analysis of the mouse transcriptome based on functional annotation of 60,770 full-length cDNAs. Nature. 2002;420(6915):563–73.
29. Frith MC, Forrest AR, Nourbakhsh E, Pang KC, Kai C, Kawai J, et al. The abundance of short proteins in the mammalian proteome. PLoS Genet. 2006;2(4):e52.
30. Frith MC, Bailey TL, Kasukawa T, Mignone F, Kummerfeld SK, Madera M, et al. Discrimination of non-protein-coding transcripts from protein-coding mRNA. RNA Biol. 2006;3(1):40–8.
31. Brockdorff N, Ashworth A, Kay GF, McCabe VM, Norris DP, Cooper PJ, et al. The product of the mouse Xist gene is a 15 kb inactive X-specific transcript containing no conserved ORF and located in the nucleus. Cell. 1992;71(3):515–26.
32. Borsani G, Tonlorenzi R, Simmler MC, Dandolo L, Arnaud D, Capra V, et al. Characterization of a murine gene expressed from the inactive X chromosome. Nature. 1991;351(6324):325–9.
33. Galindo MI, Pueyo JI, Fouix S, Bishop SA, Couso JP. Peptides encoded by short ORFs control development and define a new eukaryotic gene family. PLoS Biol. 2007;5(5):e106.
34. Tupy JL, Bailey AM, Dailey G, Evans-Holm M, Siebel CW, Misra S, et al. Identification of putative noncoding polyadenylated transcripts in Drosophila melanogaster. Proc Natl Acad Sci USA. 2005;102(15):5495–500.
35. Pedersen JS, Bejerano G, Siepel A, Rosenbloom K, Lindblad-Toh K, Lander ES, et al. Identification and classification of conserved RNA secondary structures in the human genome. PLoS Comput Biol. 2006;2(4):e33.
36. Washietl S, Hofacker IL, Stadler PF. Fast and reliable prediction of noncoding RNAs. Proc Natl Acad Sci USA. 2005;102(7):2454–9.
37. Rivas E, Eddy SR. Noncoding RNA gene detection using comparative sequence analysis. BMC Bioinformatics. 2001;2:8.
38. Leontis NB, Lescoute A, Westhof E. The building blocks and motifs of RNA architecture. Curr Opin Struct Biol. 2006;16(3):279–87.
39. Ravasi T, Suzuki H, Pang KC, Katayama S, Furuno M, Okunishi R, et al. Experimental validation of the regulated expression of large numbers of non-coding RNAs from the mouse genome. Genome Res. 2006;16(1):11–9.
40. Furuno M, Pang KC, Ninomiya N, Fukuda S, Frith MC, Bult C, et al. Clusters of internally primed transcripts reveal novel long noncoding RNAs. PLoS Genet. 2006;2(4):e37.

41. Lanz RB, McKenna NJ, Onate SA, Albrecht U, Wong J, Tsai SY, et al. A steroid receptor coactivator, SRA, functions as an RNA and is present in an SRC-1 complex. Cell. 1999;97(1):17–27.
42. Kohtz JD, Fishell G. Developmental regulation of EVF-1, a novel non-coding RNA transcribed upstream of the mouse Dlx6 gene. Gene Expr Patterns. 2004;4(4):407–12.
43. Glasgow E, Ryu SL, Yamashita M, Zhang BJ, Mutsuga N, Gainer H. APeg3, a novel paternally expressed gene 3 antisense RNA transcript specifically expressed in vasopressinergic magnocellular neurons in the rat supraoptic nucleus. Brain Res Mol Brain Res. 2005;137(1–2):143–51.
44. Hatzigeorgiou AG, Fiziev P, Reczko M. DIANA-EST: a statistical analysis. Bioinformatics. 2001;17(10):913–9.
45. Lottaz C, Iseli C, Jongeneel CV, Bucher P. Modeling sequencing errors by combining hidden Markov models. Bioinformatics. 2003;19(Suppl 2):ii103-12.
46. Liu J, Gough J, Rost B. Distinguishing protein-coding from non-coding RNAs through support vector machines. PLoS Genet. 2006;2(4):e29.
47. Kong L, Zhang Y, Ye ZQ, Liu XQ, Zhao SQ, Wei L, et al. CPC: assess the protein-coding potential of transcripts using sequence features and support vector machine. Nucleic Acids Res. 2007;35(Web Server issue):W345–W9.
48. Chooniedass-Kothari S, Hamedani MK, Troup S, Hubé F, Leygue E. The steroid receptor RNA activator protein is expressed in breast tumor tissues. Int J Cancer. 2006;118(4):1054–9.
49. Chooniedass-Kothari S, Emberley E, Hamedani MK, Troup S, Wang X, Czosnek A, et al. The steroid receptor RNA activator is the first functional RNA encoding a protein. FEBS Lett. 2004;566(1–3):43–7.
50. Kloc M, Wilk K, Vargas D, Shirato Y, Bilinski S, Etkin LD. Potential structural role of non-coding and coding RNAs in the organization of the cytoskeleton at the vegetal cortex of Xenopus oocytes. Development. 2005;132(15):3445–57.
51. Markussen FH, Michon AM, Breitwieser W, Ephrussi A. Translational control of oskar generates short OSK, the isoform that induces pole plasma assembly. Development. 1995;121(11):3723–32.
52. Ephrussi A, Lehmann R. Induction of germ cell formation by oskar. Nature. 1992;358(6385):387–92.
53. Jenny A, Hachet O, Závorszky P, Cyrklaff A, Weston MD, Johnston DS, et al. A translation-independent role of oskar RNA in early Drosophila oogenesis. Development. 2006;133(15):2827–33.
54. Amack JD, Paguio AP, Mahadevan MS. Cis and trans effects of the myotonic dystrophy (DM) mutation in a cell culture model. Hum Mol Genet. 1999;8(11):1975–84.
55. Jupe ER, Liu XT, Kiehlbauch JL, McClung JK, Dell'Orco RT. The 3′ untranslated region of prohibitin and cellular immortalization. Exp Cell Res. 1996;224(1):128–35.
56. Jupe ER, Liu XT, Kiehlbauch JL, McClung JK, Dell'Orco RT. Prohibitin in breast cancer cell lines: loss of antiproliferative activity is linked to 3′ untranslated region mutations. Cell Growth Differ. 1996;7(7):871–8.
57. Fan H, Villegas C, Huang A, Wright JA. Suppression of malignancy by the 3′ untranslated regions of ribonucleotide reductase R1 and R2 messenger RNAs. Cancer Res. 1996;56(19):4366–9.
58. Rastinejad F, Conboy MJ, Rando TA, Blau HM. Tumor suppression by RNA from the 3′ untranslated region of alpha-tropomyosin. Cell. 1993;75(6):1107–17.
59. Rastinejad F, Blau HM. Genetic complementation reveals a novel regulatory role for 3′ untranslated regions in growth and differentiation. Cell. 1993;72(6):903–17.
60. Warden CD, Kim SH, Yi SV. Predicted functional RNAs within coding regions constrain evolutionary rates of yeast proteins. PLoS One. 2008;3(2):e1559.
61. Watanabe T, Totoki Y, Toyoda A, Kaneda M, Kuramochi-Miyagawa S, Obata Y, et al. Endogenous siRNAs from naturally formed dsRNAs regulate transcripts in mouse oocytes. Nature. 2008;453(7194):539–43.

62. Tam OH, Aravin AA, Stein P, Girard A, Murchison EP, Cheloufi S, et al. Pseudogene-derived small interfering RNAs regulate gene expression in mouse oocytes. Nature. 2008;453(7194):534–8.
63. Ghildiyal M, Seitz H, Horwich MD, Li C, Du T, Lee S, et al. Endogenous siRNAs derived from transposons and mRNAs in Drosophila somatic cells. Science. 2008;320(5879):1077–81.
64. Komar AA. Silent SNPs: impact on gene function and phenotype. Pharmacogenomics. 2007;8(8):1075–80.
65. Kimchi-Sarfaty C, Oh JM, Kim IW, Sauna ZE, Calcagno AM, Ambudkar SV, et al. A "silent" polymorphism in the MDR1 gene changes substrate specificity. Science. 2007;315(5811):525–8.
66. Kageyama Y, Kondo T, Hashimoto Y. Coding vs non-coding: translatability of short ORFs found in putative non-coding transcripts. Biochimie. 2011;93(11):1981–6.
67. Nam JW, Choi SW, You BH. Incredible RNA: dual functions of coding and noncoding. Mol Cells. 2016;39(5):367–74.
68. Yeasmin F, Yada T, Akimitsu N. Micropeptides encoded in transcripts previously identified as Long noncoding RNAs: a new chapter in transcriptomics and proteomics. Front Genet. 2018;9:144.
69. Hube F, Guo J, Chooniedass-Kothari S, Cooper C, Hamedani MK, Dibrov AA, et al. Alternative splicing of the first intron of the steroid receptor RNA activator (SRA) participates in the generation of coding and noncoding RNA isoforms in breast cancer cell lines. DNA Cell Biol. 2006;25(7):418–28.
70. Hubé F, Velasco G, Rollin J, Furling D, Francastel C. Steroid receptor RNA activator protein binds to and counteracts SRA RNA-mediated activation of MyoD and muscle differentiation. Nucleic Acids Res. 2011;39(2):513–25.
71. Mattick JS. Challenging the dogma: the hidden layer of non-protein-coding RNAs in complex organisms. BioEssays. 2003;25(10):930–9.
72. Cooper C, Vincett D, Yan Y, Hamedani MK, Myal Y, Leygue E. Steroid receptor RNA activator bi-faceted genetic system: heads or tails? Biochimie. 2011;93(11):1973–80.
73. Emberley E, Huang GJ, Hamedani MK, Czosnek A, Ali D, Grolla A, et al. Identification of new human coding steroid receptor RNA activator isoforms. Biochem Biophys Res Commun. 2003;301(2):509–15.
74. Colley SM, Leedman PJ. Steroid receptor RNA activator—a nuclear receptor coregulator with multiple partners: insights and challenges. Biochimie. 2011;93(11):1966–72.
75. Anderson DM, Makarewich CA, Anderson KM, Shelton JM, Bezprozvannaya S, Bassel-Duby R, et al. Widespread control of calcium signaling by a family of SERCA-inhibiting micropeptides. Sci Signal. 2016;9(457):ra119.
76. Nelson BR, Makarewich CA, Anderson DM, Winders BR, Troupes CD, Wu F, et al. A peptide encoded by a transcript annotated as long noncoding RNA enhances SERCA activity in muscle. Science. 2016;351(6270):271–5.
77. Matsumoto A, Pasut A, Matsumoto M, Yamashita R, Fung J, Monteleone E, et al. mTORC1 and muscle regeneration are regulated by the LINC00961-encoded SPAR polypeptide. Nature. 2017;541(7636):228–32.
78. Anderson DM, Anderson KM, Chang CL, Makarewich CA, Nelson BR, McAnally JR, et al. A micropeptide encoded by a putative long noncoding RNA regulates muscle performance. Cell. 2015;160(4):595–606.
79. Magny EG, Pueyo JI, Pearl FM, Cespedes MA, Niven JE, Bishop SA, et al. Conserved regulation of cardiac calcium uptake by peptides encoded in small open reading frames. Science. 2013;341(6150):1116–20.
80. Slavoff SA, Heo J, Budnik BA, Hanakahi LA, Saghatelian A. A human short open reading frame (sORF)-encoded polypeptide that stimulates DNA end joining. J Biol Chem. 2014;289(16):10950–7.

81. Lee C, Zeng J, Drew BG, Sallam T, Martin-Montalvo A, Wan J, et al. The mitochondrial-derived peptide MOTS-c promotes metabolic homeostasis and reduces obesity and insulin resistance. Cell Metab. 2015;21(3):443–54.
82. Manak JR, Dike S, Sementchenko V, Kapranov P, Biemar F, Long J, et al. Biological function of unannotated transcription during the early development of Drosophila melanogaster. Nat Genet. 2006;38(10):1151–8.
83. Jeck WR, Sharpless NE. Detecting and characterizing circular RNAs. Nat Biotechnol. 2014;32(5):453–61.
84. Memczak S, Jens M, Elefsinioti A, Torti F, Krueger J, Rybak A, et al. Circular RNAs are a large class of animal RNAs with regulatory potency. Nature. 2013;495(7441):333–8.
85. Danan M, Schwartz S, Edelheit S, Sorek R. Transcriptome-wide discovery of circular RNAs in archaea. Nucleic Acids Res. 2012;40(7):3131–42.
86. Capel B, Swain A, Nicolis S, Hacker A, Walter M, Koopman P, et al. Circular transcripts of the testis-determining gene Sry in adult mouse testis. Cell. 1993;73(5):1019–30.
87. Sanger HL, Klotz G, Riesner D, Gross HJ, Kleinschmidt AK. Viroids are single-stranded covalently closed circular RNA molecules existing as highly base-paired rod-like structures. Proc Natl Acad Sci USA. 1976;73(11):3852–6.
88. Li Z, Huang C, Bao C, Chen L, Lin M, Wang X, et al. Exon-intron circular RNAs regulate transcription in the nucleus. Nat Struct Mol Biol. 2015;22(3):256–64.
89. Ashwal-Fluss R, Meyer M, Pamudurti NR, Ivanov A, Bartok O, Hanan M, et al. circRNA biogenesis competes with pre-mRNA splicing. Mol Cell. 2014;56(1):55–66.
90. Hansen TB, Jensen TI, Clausen BH, Bramsen JB, Finsen B, Damgaard CK, et al. Natural RNA circles function as efficient microRNA sponges. Nature. 2013;495(7441):384–8.
91. Pamudurti NR, Bartok O, Jens M, Ashwal-Fluss R, Stottmeister C, Ruhe L, et al. Translation of CircRNAs. Mol Cell. 2017;66(1):9–21.e7.
92. Wang Y, Wang Z. Efficient backsplicing produces translatable circular mRNAs. RNA. 2015;21(2):172–9.
93. Abe N, Matsumoto K, Nishihara M, Nakano Y, Shibata A, Maruyama H, et al. Rolling circle translation of circular RNA in living human cells. Sci Rep. 2015;5:16435.
94. Guo JU, Agarwal V, Guo H, Bartel DP. Expanded identification and characterization of mammalian circular RNAs. Genome Biol. 2014;15(7):409.
95. Li XF, Lytton J. A circularized sodium-calcium exchanger exon 2 transcript. J Biol Chem. 1999;274(12):8153–60.
96. Chen CY, Sarnow P. Initiation of protein synthesis by the eukaryotic translational apparatus on circular RNAs. Science. 1995;268(5209):415–7.
97. Legnini I, Di Timoteo G, Rossi F, Morlando M, Briganti F, Sthandier O, et al. Circ-ZNF609 is a circular RNA that can be translated and functions in myogenesis. Mol Cell. 2017;66(1):22–37.e9.
98. Yang Y, Gao X, Zhang M, Yan S, Sun C, Xiao F, et al. Novel role of FBXW7 circular RNA in repressing glioma tumorigenesis. J Natl Cancer Inst. 2018;110(3):304–15.
99. Chen H, Liu Y, Li P, Zhu D. RE: novel role of FBXW7 circular RNA in repressing glioma tumorigenesis. J Natl Cancer Inst. 2019;111(4):435.
100. Zhang M, Huang N, Yang X, Luo J, Yan S, Xiao F, et al. A novel protein encoded by the circular form of the SHPRH gene suppresses glioma tumorigenesis. Oncogene. 2018;37(13):1805–14.
101. Dalgaard JZ, Garrett RA, Belfort M. A site-specific endonuclease encoded by a typical archaeal intron. Proc Natl Acad Sci USA. 1993;90(12):5414–7.
102. Mayr C. Evolution and biological roles of alternative 3'UTRs. Trends Cell Biol. 2016;26(3):227–37.
103. Chen CY, Chen ST, Juan HF, Huang HC. Lengthening of 3'UTR increases with morphological complexity in animal evolution. Bioinformatics. 2012;28(24):3178–81.
104. Berkovits BD, Mayr C. Alternative 3' UTRs act as scaffolds to regulate membrane protein localization. Nature. 2015;522(7556):363–7.

105. Manjeshwar S, Branam DE, Lerner MR, Brackett DJ, Jupe ER. Tumor suppression by the prohibitin gene 3' untranslated region RNA in human breast cancer. Cancer Res. 2003;63(17):5251–6.
106. Ma W, Mayr C. A Membraneless organelle associated with the endoplasmic reticulum enables 3'UTR-mediated protein-protein interactions. Cell. 2018;175(6):1492–506.e19.
107. Hu J, Li X, Guo X, Guo Q, Xiang C, Zhang Z, et al. The CCR2 3'UTR functions as a competing endogenous RNA to inhibit breast cancer metastasis. J Cell Sci. 2017;130(19):3399–413.
108. Valluy J, Bicker S, Aksoy-Aksel A, Lackinger M, Sumer S, Fiore R, et al. A coding-independent function of an alternative Ube3a transcript during neuronal development. Nat Neurosci. 2015;18(5):666–73.
109. Candeias MM. The can and can't dos of p53 RNA. Biochimie. 2011;93(11):1962–5.
110. Ivanyi-Nagy R, Ahmed SM, Peter S, Ramani PD, Ong PF, Dreesen O, et al. The RNA interactome of human telomerase RNA reveals a coding-independent role for a histone mRNA in telomere homeostasis. elife. 2018:7.
111. Akiri G, Nahari D, Finkelstein Y, Le SY, Elroy-Stein O, Levi BZ. Regulation of vascular endothelial growth factor (VEGF) expression is mediated by internal initiation of translation and alternative initiation of transcription. Oncogene. 1998;17(2):227–36.
112. Huez I, Créancier L, Audigier S, Gensac MC, Prats AC, Prats H. Two independent internal ribosome entry sites are involved in translation initiation of vascular endothelial growth factor mRNA. Mol Cell Biol. 1998;18(11):6178–90.
113. Masuda K, Teshima-Kondo S, Mukaijo M, Yamagishi N, Nishikawa Y, Nishida K, et al. A novel tumor-promoting function residing in the 5′ non-coding region of vascular endothelial growth factor mRNA. PLoS Med. 2008;5(5):e94.
114. Blume SW, Miller DM, Guarcello V, Shrestha K, Meng Z, Snyder RC, et al. Inhibition of tumorigenicity by the 5′-untranslated RNA of the human c-myc P0 transcript. Exp Cell Res. 2003;288(1):131–42.
115. Dinger ME, Gascoigne DK, Mattick JS. The evolution of RNAs with multiple functions. Biochimie. 2011;93(11):2013–8.
116. Ulveling D, Francastel C, Hubé F. Identification of potentially new bifunctional RNA based on genome-wide data-mining of alternative splicing events. Biochimie. 2011;93(11):2024–7.
117. Hubé F, Francastel C. Coding and non-coding RNAs, the frontier has never been so blurred. Front Genet. 2018;9:140.
118. Olexiouk V, Crappé J, Verbruggen S, Verhegen K, Martens L, Menschaert G. sORFs.org: a repository of small ORFs identified by ribosome profiling. Nucleic Acids Res. 2016;44(D1):D324–9.
119. Mackowiak SD, Zauber H, Bielow C, Thiel D, Kutz K, Calviello L, et al. Extensive identification and analysis of conserved small ORFs in animals. Genome Biol. 2015;16:179.
120. Ruiz-Orera J, Messeguer X, Subirana JA, Alba MM. Long non-coding RNAs as a source of new peptides. elife. 2014;3:e03523.
121. Aspden JL, Eyre-Walker YC, Phillips RJ, Amin U, Mumtaz MA, Brocard M, et al. Extensive translation of small open Reading frames revealed by Poly-Ribo-Seq. elife. 2014;3:e03528.
122. Zhu Y, Orre LM, Johansson HJ, Huss M, Boekel J, Vesterlund M, et al. Discovery of coding regions in the human genome by integrated proteogenomics analysis workflow. Nat Commun. 2018;9(1):903.
123. Nesvizhskii AI. Proteogenomics: concepts, applications and computational strategies. Nat Methods. 2014;11(11):1114–25.
124. Katz L, Burge CB. Widespread selection for local RNA secondary structure in coding regions of bacterial genes. Genome Res. 2003;13(9):2042–51.
125. Li F, Zheng Q, Vandivier LE, Willmann MR, Chen Y, Gregory BD. Regulatory impact of RNA secondary structure across the Arabidopsis transcriptome. Plant Cell. 2012;24(11):4346–59.
126. Taggart AJ, DeSimone AM, Shih JS, Filloux ME, Fairbrother WG. Large-scale mapping of branchpoints in human pre-mRNA transcripts in vivo. Nat Struct Mol Biol. 2012;19(7):719–21.

127. van Heesch S, van Iterson M, Jacobi J, Boymans S, Essers PB, de Bruijn E, et al. Extensive localization of long noncoding RNAs to the cytosol and mono- and polyribosomal complexes. Genome Biol. 2014;15(1):R6.
128. Cabili MN, Trapnell C, Goff L, Koziol M, Tazon-Vega B, Regev A, et al. Integrative annotation of human large intergenic noncoding RNAs reveals global properties and specific subclasses. Genes Dev. 2011;25(18):1915–27.
129. Derrien T, Johnson R, Bussotti G, Tanzer A, Djebali S, Tilgner H, et al. The GENCODE v7 catalog of human long noncoding RNAs: analysis of their gene structure, evolution, and expression. Genome Res. 2012;22(9):1775–89.
130. Kutter C, Watt S, Stefflova K, Wilson MD, Goncalves A, Ponting CP, et al. Rapid turnover of long noncoding RNAs and the evolution of gene expression. PLoS Genet. 2012;8(7):e1002841.
131. Necsulea A, Soumillon M, Warnefors M, Liechti A, Daish T, Zeller U, et al. The evolution of lncRNA repertoires and expression patterns in tetrapods. Nature. 2014;505(7485):635–40.
132. Levine MT, Jones CD, Kern AD, Lindfors HA, Begun DJ. Novel genes derived from noncoding DNA in Drosophila melanogaster are frequently X-linked and exhibit testis-biased expression. Proc Natl Acad Sci USA. 2006;103(26):9935–9.
133. Cai J, Zhao R, Jiang H, Wang W. De novo origination of a new protein-coding gene in Saccharomyces cerevisiae. Genetics. 2008;179(1):487–96.
134. Reinhardt JA, Wanjiru BM, Brant AT, Saelao P, Begun DJ, Jones CD. De novo ORFs in Drosophila are important to organismal fitness and evolved rapidly from previously noncoding sequences. PLoS Genet. 2013;9(10):e1003860.
135. Ekland EH, Szostak JW, Bartel DP. Structurally complex and highly active RNA ligases derived from random RNA sequences. Science. 1995;269(5222):364–70.
136. Poliseno L, Lanza M, Pandolfi PP. Coding, or non-coding, that is the question. Cell Res. 2024;34(9):609–29.

# Chapter 15
# RNA Modifications in Human Diseases

**Abstract** The ever-expanding field of epitranscriptomics and RNA post-transcriptional modifications is delving into the significant role RNA modifications play in disease management. The therapeutic application of RNA modifications is emerging as a highly promising avenue for drug development, given the existence of over 170 distinct types of modifications across various RNA species. Critical biological processes implicated in numerous diseases have been found to be regulated by genes associated with these RNA modifications. Current research has identified inhibitors targeting genes that regulate RNA modifications, presenting a compelling proposition for RNA modification as a viable therapeutic target in disease treatment. This chapter aims to delve into the mechanisms by which molecular alterations target RNA modifications or their regulatory genes. It will also consolidate existing research on the impact of RNA modifications and regulatory genes within the sphere of drug development, highlighting the significant potential of targeting RNA modifications in the creation of innovative therapeutic strategies.

## 15.1 Introduction

In the vast and intricate world of RNA biology, modifications of RNA bases such as adenosine, guanosine, uridine, queuosine, along with the phosphate backbone and ribose ring, play pivotal roles across diverse organisms. These bases undergo a variety of chemical modifications, including the addition of aminoacyl, aminoalkyl, glycosyl groups, hydrocarbons, methyl groups, as well as sulfur and selenium elements [1]. The landscape of RNA modifications in eukaryotes is richly detailed, featuring modifications like 5-methylcytidine (m5C), 5-hydroxymethylcytidine (hm5C), N1-methyladenosine (m1A), N6-methyladenosine (m6A), N6,2'-O-dimethyladenosine, inosine, and pseudouridine, as outlined in seminal works [2–5]. These modifications are not merely epigenetic footnotes; they are fundamental to both the normal developmental processes [6] and the pathological progressions of diseases, including cancer and other nonmalignant conditions [7]. As the field progresses, the enzymes instrumental in RNA modification—categorized into

 
P. E. Saw, E. Song, *RNA Therapeutics in Human Diseases*,
https://doi.org/10.1007/978-981-96-3041-7_15

"writers" (methyltransferase enzymes), "readers" (binding proteins that recognize specific modifications), and "erasers" (demethylase enzymes)—have been increasingly identified, highlighting the dynamic nature of RNA modifications. Notably, modifications like m6A and m1A exemplify the known enzymatic interactions that underscore the reversible and regulated nature of these modifications. Despite advancements in identifying specific reader, writer, and eraser enzymes for some RNA modifications, a comprehensive understanding of the enzymatic landscape for many RNA modifications remains elusive, calling for further research. This chapter delves into the role of RNA modifications and their regulatory mechanisms as innovative targets in drug discovery, propelled by the burgeoning understanding of their significance in health and disease. The summary of various modifications and their implications in various diseases are summarized in Fig. 15.1 below.

## 15.2 Types of RNA Modification

The epitranscriptome encompasses the complex interplay of evolutionarily conserved elements that encode (writers), decode (readers), and remove (erasers) a myriad of chemical modifications on RNA bases [9] (see Fig. 15.2). To this point, researchers have cataloged over 100 distinct RNA modifications, with the

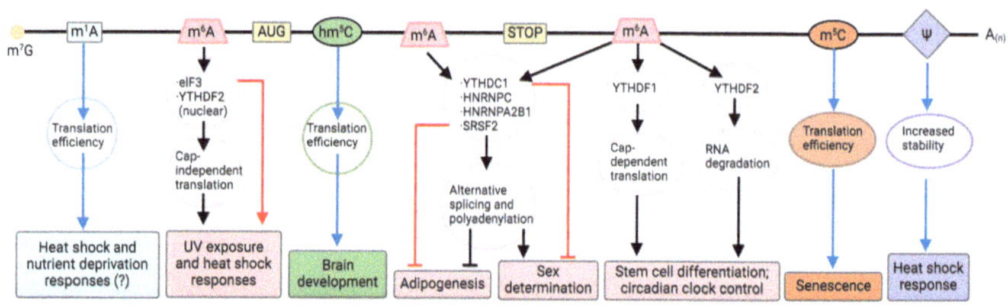


**Fig. 15.1** This figure illustrates the diverse biological functions regulated by different modifications on mRNA. m1A (N1-methyladenosine): Positioned on mRNA, m1A enhances translation efficiency, which is implicated in the cellular response to heat shock and nutrient deprivation. m6A (N6-methyladenosine): At AUG (start codon): m6A, facilitated by eIF3 and nuclear YTHDF2, promotes cap-independent translation, which is crucial for responses to UV exposure and heat shock. At stop codon: m6A influences alternative splicing and polyadenylation through proteins like YTHDC1, HNRNPC, HNRNPA2B1, and SRSF2. This modulation affects adipogenesis and sex determination. Additionally, m6A at the stop codon, recognized by YTHDF2, triggers RNA degradation. In other contexts, m6A regulates cap-dependent translation, impacting stem cell differentiation and circadian clock control. hm5C (5-hydroxymethylcytosine): Enhances translation efficiency, playing a significant role in brain development. m5C (5-methylcytosine): This modification improves translation efficiency and is linked to the cellular process of senescence. Ψ (Pseudouridine): Confers increased stability to mRNA, supporting the heat shock response. The figure demonstrates the intricate regulation of mRNA function through various modifications, each contributing to distinct cellular processes and stress responses. Figure is adapted from [8] under Springer Nature License 5901700114504

anticipation of discovering more [10]. Modifications such as N1-methyladenosine (m1A), N6-methyladenosine (m6A), 5-methylcytosine (m5C), and pseudouridine (Ψ) were initially found in abundant RNA molecules like tRNAs, snRNAs, and rRNAs. However, advancements in mass spectrometry and sequencing technologies have now enabled researchers to detect and characterize these modifications in RNAs that are less abundant, namely mRNAs (Fig. 15.2).

## 15.3 RNA N1-Methyladenosine Modification

The m1A modification, involving the addition of a methyl group to the first position of adenosine, was first identified in RNA studies dating back to 1961 [11], but it has only recently been scrutinized for its prevalence and distribution across various

a

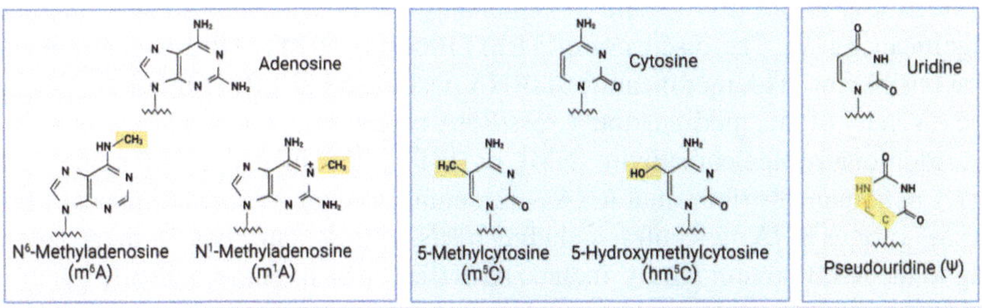


b

| Modification | Writer | Eraser | Reader |
|---|---|---|---|
| $N^6$-Methyladenosine | ·Methyltransferase-like 3 (METTL3) and METTL14<br>·Wilms tumour 1-associated protein (WTAP)<br>·RNA-binding motif protein 15 (RBM15) and RBM15B<br>·Protoin virilizer homologue (KIAA1429) | ·AlkB homologue 5 (ALKBH5)<br>·Fat mass and obesity-associated protein (FTO) | ·YTH domain family proteins 1-3 (YTHDF1-3)<br>·YTH domain-containing protein 1 (YTHDC1)<br>·Eukaryotic translation initiation factor 3 (eIF3) complex<br>·Heterogeneous nuclear ribonucleoproteins A2/B1(HNRNPA2B1) and HNRNPC<br>·Serine/arginine-rich splicing factor 2 (SRSF2) |
| $N^1$-Methyladenosine | Unkonwn | ALKBH3 | Unknown |
| 5-Methylcytosine | ·NOL1/NOP2/Sun domain family member 2 (NSUN2)<br>·DNA methyltransferase-like protein 2 (DNMT2) | Unkonwn | Unkonwn |
| 5-Hydroxymethylcytosine* | ·Ten-eleven translocation protein 1-3 (TET1-TET3) | Unkonwn | Unkonwn |
| Pseudouridine | ·Pseudouridine synthase 1-4 (PUS1-PUS4), PUS6, PUS7, PUS9<br>·Dyskeratosis congenita protein 1 (DKC1) | Unkonwn | Unkonwn |

*5-Hydroxymethylcytosine is formed by the hydroxylation of 5-methylcytosine.

**Fig. 15.2** RNA modifications and their writers, erasers, and readers. (**a**) Figures depicting both modified and unmodified RNA structures. (**b**) Detailed explanation of the writers, readers, and erasers during RNA modification. Figure is adapted from [8] under Springer Nature License 5901700114504

RNA molecules. Found internally within tRNA [12], rRNA [13, 14], mRNA [15], and mtRNA [16], m1A shares functional similarities with m6A, including the presence of specialized enzymes for methylation and demethylation processes. The existence of these eraser and writer enzymes for m1A categorizes it as a dynamic, reversible modification capable of post-transcriptionally modulating gene expression, akin to m6A [17]. In a pioneering transcriptome-wide analysis, Dominissini et al. [15] elucidated the distribution of m1A across eukaryotic cells, revealing its presence across the entire transcript spectrum, including the 3' UTR, CDS, and 5'-UTR. ALKBH5 was highlighted for its role in reversing m1A modifications in mRNA, underscoring the modification's dynamic nature [18]. Additionally, both TRMT61A and TRMT6 were identified as the enzymes responsible for adding methyl groups to form the m1A58 modification, further contributing to our understanding of RNA modification mechanisms and their biological implications [19]. This expanding knowledge of m1A and its role in RNA biology opens new avenues for research into RNA modifications' regulatory potential and their impact on gene expression and disease. The m1A RNA modification has been noted in mitochondrial tRNAs at positions 9 and 58, extending the understanding of its biological significance [19]. The discovery of TRMT10C and TRMT61B as writer enzymes responsible for m1A modification at tRNA position 58 [20, 21] further underscores the diversity of this modification across RNA types.

Additionally, nucleomethylin (NML or RRP8), identified as a writer enzyme for m1A in human 28s ribosomal RNAs [22], points to the widespread impact of m1A across various RNA molecules, including mRNA. Eraser enzymes for m1A, belonging to the AlkB protein family, include ALKBH1, which removes the methyl group from adenosine in tRNA, and its knockdown has been shown to increase m1A modification levels in tRNAi-Met [23]. Similarly, knockdown of ALKBH3 increases m1A levels in tRNA and decreases protein levels in cancer cells [24], illustrating the modification's role in cellular function and disease. The identification of nine potential m1A readers from the YTH domain family [25], including YTHDF3, which is suggested to bind directly to RNA modified by m1A, alongside YTHDF1-3 and YTHDC1 as direct binders to m1A-modified RNA [26], further elucidates the intricate network of m1A regulation and its impact on RNA biology. However, discrepancies in m1A detection across various studies, attributed to differences in methodological approaches, highlight the challenges in understanding the full extent of m1A modification. A recent study addressed these discrepancies, suggesting that contrary to earlier reports, m1A was not detected in certain transcript regions, pointing to potential issues with m1A mapping data related to mRNA caps [27]. This evolving landscape of m1A research underscores the complexity of RNA modifications and their critical role in gene expression regulation, cellular function, and the etiology of diseases.

### 15.3.1 RNA m1A: Biological Function and Implication in Disease

The investigation into m1A's biological role in both coding and ncRNAs represents a relatively nascent area of study. Despite extensive research across various organisms, insights into m1A functions within human RNA remain sparse. Currently, m1A modifications have been identified in mRNAs, tRNAs, and lncRNAs, with significant research focus on its role in tRNA. For instance, m1A58 is associated with the structural integrity of yeast tRNA [28], playing a crucial role in the stability of both the tRNA structure and the molecule as a whole [29]. Similarly, m1A58 is vital for the structure and stability of tRNA in human cells [23] and has been implicated in HIV replication, highlighting the biological significance of m1A modifications [30, 31]. Research indicates that the knockdown of ALKBH1, an m1A demethylase, increases the levels of tRNAi-Met (methionine transfer RNA), suggesting a functional role for m1A in tRNA metabolism. In mitochondrial messenger RNA (mt-mRNA), m1A's presence disrupts Watson–Crick base pairing, inhibiting the efficient translation of modified codons [16, 32]. Additionally, m1A modifications at the 5′ cap and in the 5'UTR of nuclear mRNA may enhance translation efficiency [16], indicating a regulatory role in gene expression. The link between m1A modifications and disease, particularly cancer, has begun to emerge, albeit with limited data. ALKBH3, known as prostate cancer antigen 1 (PCA-1), is overexpressed in various cancers [33, 34], and its upregulation in pancreatic cancer has been associated with cancer malignancy [35]. ALKBH3's role in altering RNA methylation levels suggests its involvement in tumor progression and the promotion of protein synthesis [24]. Furthermore, m1A regulatory genes have been linked to gastrointestinal cancers [36], underscoring the potential significance of m1A modifications in understanding and treating diseases. These findings highlight the need for continued exploration of m1A's biological functions and its implications for human health and disease.

## 15.4 RNA N6-Methyladenosine

m6A modification is an addition of a methyl group to the adenosine base at the sixth position. This modification is well-established across various RNA types, including rRNA, noncoding RNAs, and polyadenylated RNA. Transcriptome-wide m6A mapping studies conducted in 2012 highlighted m6A's significant role, revealing its presence in the mRNAs of over 7600 genes and more than 300 noncoding RNAs, distributed widely throughout the transcriptome. Contrary to being predominantly in the 5'-UTR, m6A modifications are notably enriched in the 3'-UTR regions, often located near stop codons and within the 3'-UTR, displaying a distinct distribution. Many transcripts exhibit one or multiple m6A peaks, with a minority showing four or more peaks, emphasizing the modification's ubiquity [3]. The m6A modification stands out as the first among RNA modifications to have its associated molecular machinery—comprising "erasers," "writers," and "reader" proteins—thoroughly identified [37]. Writers such as METTL3, METTL14, RBM15, VIRMA-KIAA1429, and WTAP facilitate the addition of methyl groups. Conversely, enzymes like ALKBH5 and FTO serve as "erasers," removing these methyl modifications, while "reader proteins—including eIF3, FMRP, HNRNPA2B1, HNRNPC, IGF2BP, LRPPRC, YTHDF1, YTHDF2, and YTHDF3—interpret and act upon these modifications, impacting RNA's function [37].

Beyond mRNAs, m6A modifications are prevalent across various ncRNA types. Initial findings of m6A in lncRNAs were reported by Meyer et al., with subsequent discoveries in transcripts like MEG3, MALAT1, GAS5, and XIST, highlighting m6A's wide-reaching effects across the transcriptome. Moreover, m6A modifications have been identified in over 1000 circRNAs in cancer cells, pointing to specific expression patterns and suggesting their important roles in immunity and protein translation initiation from circRNAs. Additionally, the presence of m6A has been associated with miRNAs, such as miR-17-5p, -21-5p, -200c-3p, and let-7a-5p, highlighting a potential link to increased cancer incidence. Notably, m6A residues have also been identified in the ribosomal RNAs, with specific positions in 18S and 28S rRNAs being methylated by METTL5 and ZCCHC4, respectively, underlining the modification's fundamental role in the broader RNA landscape. This extensive distribution and functional implication of m6A across various RNA molecules underscore its critical role in cellular processes and its potential as a target for therapeutic intervention (Fig. 15.3).

### *15.4.1 RNA N6-Methyladenosine: Biological Function and Implication in Disease*

Emerging research underscores the pivotal role of RNA modifications, particularly m6A modification, in a broad spectrum of biological functions and disease pathways. These modifications are integral to various fundamental processes, including embryonic stem cell differentiation, cellular development, meiotic progression, RNA-RNA interactions, RNA-protein binding dynamics, RNA stability regulation, temperature adaptation, and circadian rhythm management [37]. The modulation of these biological functions is largely influenced by the activity of m6A RNA modification especially in their writers, readers, and erasers [38]. Specifically, m6A modification has been linked to crucial mRNA processes such as stability [39, 40], translational efficiency [41, 42], splicing [43], and pri-miRNA processing [44]. The m6A reader YTHDF2, for instance, is known to facilitate mRNA degradation [40], while m6A erasers like FTO and ALKBH5 play roles in mRNA decay and stability [45, 46]. The writer WTAP is involved in mRNA splicing [47–49], highlighting the multifaceted influence of m6A on mRNA lifecycle. Furthermore, m6A readers such

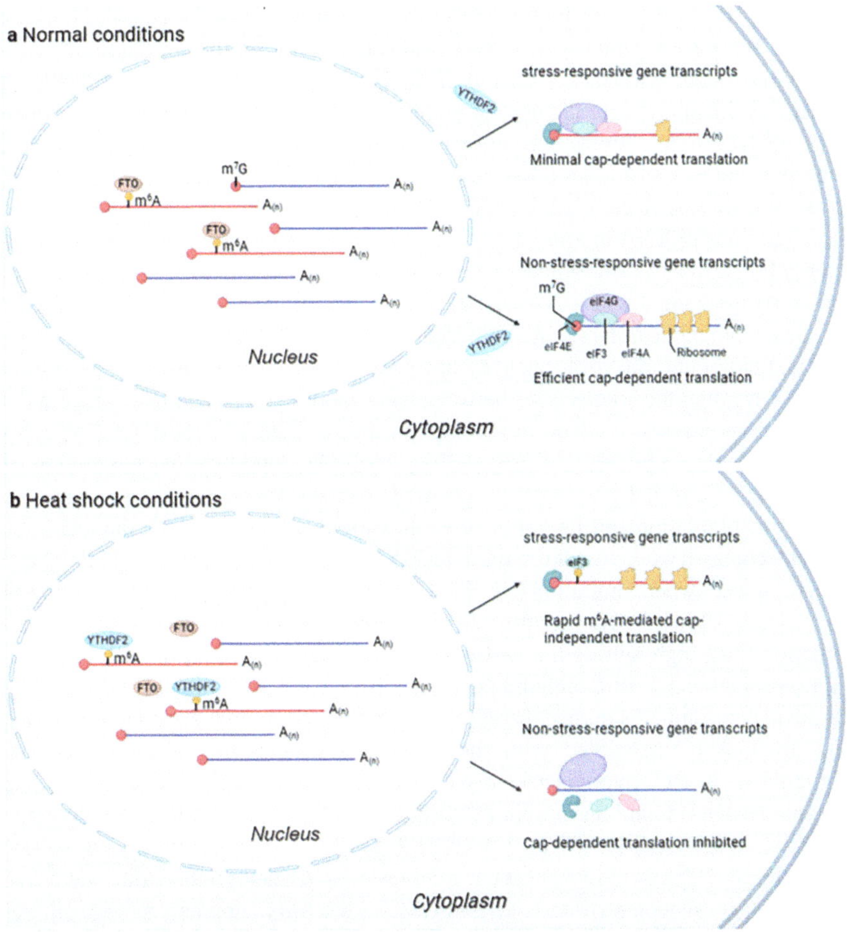


**Fig. 15.3** This figure illustrates the role of N6-methyladenosine (m6A) in regulating translation under normal and heat shock conditions. (**a**) Under normal conditions, m6A marks are deposited on mRNAs in the nucleus by methyltransferases, including METTL3/METTL14 (depicted as FTO for simplicity). Stress-responsive gene transcripts exhibit minimal cap-dependent translation, with YTHDF2 recognizing m6A modifications and leading to their degradation. Non-stress-responsive gene transcripts undergo efficient cap-dependent translation, facilitated by the canonical cap-binding complex (eIF4E, eIF4G, eIF4A) and assisted by YTHDF2 in stabilizing the process. (**b**) Upon heat shock, there is a rapid deposition of m6A on stress-responsive gene transcripts. Stress-responsive gene transcripts rapidly shift to m6A-mediated cap-independent translation, bypassing the need for the traditional cap-binding complex. This process is facilitated by eIF3, which directly interacts with m6A-modified transcripts to initiate translation quickly. Non-stress-responsive gene transcripts experience inhibited cap-dependent translation under heat shock conditions, leading to a preferential translation of stress-responsive proteins essential for coping with stress. This dynamic shift highlights the crucial role of m6A modification in regulating gene expression and protein synthesis in response to cellular stress, ensuring a swift and efficient response to environmental changes. Figure is adapted from [8] under Springer Nature License 5901700114504

as YTHDF1 and eIF3 are implicated in translation [41, 50–52], and YTHDF2 in mRNA decay [37], emphasizing the modification's role in the precision and efficiency of protein synthesis. Post-transcriptional modifications across various RNA types are critical for accurate protein production, reflecting the deep interconnectedness of RNA modifications with cellular machinery and disease development. There is ample evidence that links RNA modifications and their catalyzing enzymes with aggressive disease phenotypes and complex human disorders. Genetic defects causing alterations in RNA modifications can lead to severe outcomes affecting cellular energy homeostasis, development, apoptosis, and male fertility. Diseases directly or indirectly associated with RNA modifications, including those in tRNA, span cancer, neurological disorders, mitochondrial-linked diseases, and more [7].

The field of epitranscriptomics has unveiled significant insights into cancer biology by revealing the connections between RNA modifications and the disruption of enzymatic processes involved in reading, writing, and erasing these modifications with cancer development and progression. The m6A modification, possessing its unique set of enzymes (writers, erasers, and readers), showcases the hallmarks of epigenetic regulation and its implications in various cancers through the alteration in RNA chemistry (chemical modifications) [53]. Examples include melanoma and breast cancer, where mutations in the FTO eraser enzyme have been linked to increased risk [54]. Moreover, m6A modification plays a role in stem cell fate determination [39], highlighting the widespread implications of RNA modification disruptions in human health, including obesity [55–57] and neurological diseases [58]. Hypoxic environments induce the upregulation of hypoxia-inducible factors (HIFs) and are closely associated with the onset and progression of many cancers. Particularly, in BC, hypoxia triggers the upregulation of an m6A demethylating enzyme ALKBH5, and an inhibitor ZNF217, which inhibits the RNA methylation writer complex (RBM15–WTAP–METTL3–METTL14). Interestingly, ALKBH5 is also highly expressed in a subset of glioblastoma, termed glioblastoma stem-like cells (GSCs). Silencing of ALKBH5 in GSCs effectively suppressed the proliferation of patient-derived GSCs, illustrating its critical role in glioblastoma growth and progression [59]. Other alterations in m6A methylation, such as the silencing of METTL3 or METTL14, can also impact GSCs' self-renewal, growth, and tumor progression, while overexpression of METTL3 or inhibition of RNA demethylase FTO suppresses this phenomenon [60]. In lung cancer, METTL3 overexpression promotes survival, growth, invasion, and oncogene translation, with METTL3 depletion inhibiting cancer cell growth and increasing apoptosis [61]. Furthermore, RNA m6A modifying enzymes have been implicated in nonmalignant conditions like obesity and neurological diseases. This underscores the importance of m6A modification in both the development and progression of a wide range of diseases, emphasizing the potential for targeted therapies that modulate RNA modifications as a novel approach in cancer treatment and beyond.

## 15.5 RNA 5-Methylcytosine

m5C, known for its prominent role as an epigenetic marker in DNA, has been identified extensively across various classes of human RNA, including mRNAs, rRNAs, and tRNAs, due to recent advancements in antibody-based detection, bisulfite treatment, and deep sequencing analysis. As one of the most prevalent and recognized RNA modifications, m5C significantly contributes to a wide array of biological and pathological processes [62, 63]. In human tRNA, 28 m5C sites have been documented [64], with 27 of these sites subsequently confirmed through NGS techniques [4]. These modifications tend to cluster at the junction between the TC-stem and the variable region, notably at positions 48 and 49, and also within the anticodon loop at positions 34 and 38 [63]. An additional frequent site of m5C modification in higher eukaryotes is found at the tRNA acceptor stem position 72. Beyond tRNA, m5C modifications have been identified in other RNA classes, although they appear less frequently in eukaryotic rRNA. For instance, human 28S rRNA contains just two m5C nucleotides at positions 4413/4 and 3761, with no m5C detected in human 18S rRNA [65, 66]. Recently, using high-throughput RNA methylation using bisulfite sequencing in HeLa cells has verified the presence and extent of m5C modifications in mRNA, revealing a more widespread distribution than previously anticipated [4]. m5C was identified at 10,275 candidate sites across mRNA and noncoding RNAs, and at over 240 sites in tRNAs [4, 64]. In mouse embryonic cells, approximately 7500 m5C sites were detected in 1650 mRNAs, including 2075 sites within brain mRNAs, indicating significant variability in the methylome across different cell types [67]. Furthermore, around 3600 sites were identified in 2000 mRNAs in HeLa cells [68], underscoring the pervasive and critical nature of m5C modifications in regulating RNA function and gene expression.

### *15.5.1 RNA m5C: Biological Function and Implication in Disease*

The indispensable roles of rRNA and tRNA modifications in cellular adaptation to various stimuli are underpinned by their contributions to tRNA stability [69, 70], cell growth [71], and the cellular stress response [72–74]. Such modifications, particularly in tRNAs, influence translation's rate and accuracy by improving ribosome binding, adjusting frame-shifting rates, and minimizing reading errors [75]. m5C modifications, typically found around the anticodon loop and variable regions of tRNAs, have been demonstrated to stabilize tRNA's secondary structure, enhance its metabolic stability, affect aminoacylation, and assist in codon recognition, thereby impacting translation efficiency and fidelity [76]. Additionally, rRNA methylation is closely linked to translational fidelity and the precise recognition of tRNAs, further emphasizing the crucial role of these modifications in protein synthesis [76]. In NSun2-deficient mice, the absence of vault RNA methylation leads to improper vault processing into Ago-associated small RNAs, which leads to the abnormal expression of mRNAs [77]. Furthermore, DNMT2-mediated methylation has been shown to have a protective effect on tRNA substrates, which protects tRNAs from endonucleolytic cleavage, suggesting a protective mechanism conferred by specific methylation patterns [76]. The depletion of Dnmt2 and NSun2 enzymes results in tRNA hypomethylation, correlating with reduced tRNA levels and a significant decrease in the rates of protein translation [76]. These findings collectively underscore the profound impact of RNA modifications on cellular function, gene expression, and the overall proteome, marking them as crucial elements in understanding the complexity of cellular biology and the response to environmental challenges. Beyond methyltransferase enzymes, ALYREF (Aly/REF Export Factor) stands out as the sole reader identified for m5C modifications in RNA, with no eraser enzymes discovered for mammalian cells to date [68]. The methylation of cytosine across various RNA molecules is facilitated by a spectrum of enzymes, each localized to distinct subcellular regions including the nucleus, mitochondria, and cytoplasm [78]. This modification predominates in noncoding RNAs, encompassing a variety of mitochondrial tRNAs, tRNAs, and rRNAs.

The RNA methyltransferase enzyme DNMT2 is known for catalyzing the addition of a methyl group to tRNA at position C38 [79]. Among the NOL1/NOP2/Sun domain family, seven members are recognized as RNA methyltransferase genes. NOP2, a nucleolar protein, catalyzes methylation at position C4413 in 28S rRNA [80], while NSUN2 is responsible for methylation across various tRNA positions including C34, C48, C49, and C50 [4, 81–86]. Additionally, NSUN3 is tasked with methylating tRNA (Met) within mitochondria [87, 88], and NSUN4 methylates the 12S rRNA also in mitochondria [89]. NSUN5's deletion is linked to Williams–Beuren syndrome, affecting methylation at position C3782 of 28S rRNA [90]. NSUN6 has been demonstrated to catalyze methylation at position C72 in tRNA [91–94], and NSUN7 is involved in the methylation of enhancer RNAs (eRNAs) [95].

## 15.6 A-to-I Editing in RNA Modification

RNA editing is a prevalent modification process in both noncoding and coding RNAs, initially discovered in Xenopus laevis [96]. This modification process has since been observed in Alu elements and microRNAs, expanding our understanding of its ubiquity and functional implications. Similar to other RNA modifications, the A-to-I (adenosine to inosine) RNA editing is facilitated by the Adenosine Deaminase Acting on RNA (ADAR) family of enzymes. These enzymes alter RNA sequences, potentially leading to the production of proteins with amino acid sequences that differ from those encoded by the DNA template, thus introducing functional diversity within the proteome [97, 98]. This modification mechanism is akin to mutations in DNA, diversifying the proteins and transcripts necessary for normal cellular functions [99]. The two predominant types are adenosine to inosine (A-to-I) and cytidine to uridine (C-to-U), with ADAR enzymes catalyzing the A-to-I conversion in double-stranded RNAs. This conversion results in inosine being interpreted as guanine by the cellular machinery [100].

Currently, three ADAR family genes have been identified for A-to-I editing: ADAR1, ADAR2, and ADAR3. Only ADAR3 (also known as ADARB2) is considered to be catalytically inactive [97, 101] while ADAR1 [102] and ADAR2 [103] are enzymatically active in humans. ADAR1 and ADAR2 are ubiquitously expressed across various tissues in human, whereas ADAR3 expression is primarily restricted to the brain. Among these enzymes, ADAR2 exhibits significantly higher expression levels compared to others [101, 104, 105]. Structurally, ADAR genes share common features, including catalytic deaminase domains and an N-terminus [106]. RNA editing has been linked to gene silencing pathways in rats [107] and alternative splicing [108]. In humans, RNA editing can alter RNA structure folding and is associated with gene expression regulation [109], highlighting its critical role in genetic and epigenetic regulation mechanisms.

### *15.6.1 A-to-I RNA Editing: Biological Function and Implication in Disease*

The integration of deep sequencing technology with bioinformatics has revolutionized our understanding of RNA editing, revealing its widespread occurrence in both normal and pathological human tissues. Currently, a significant number of editing sites in human transcripts have been validated with strong evidence. Particularly, RNA editing events are notably prevalent in Alu sequences. These Alu sequences which are short interspersed nuclear elements comprising only about 10% of the human genome [110]. These millions of Alu sequences can form double-stranded RNA structures, making them prime targets for the active ADAR enzymes that would convert their adenosine to inosine [111]. RNA editing was found in many pre-mRNAs and identified 1445 human mRNAs with over 14,500 editing sites within introns, underscoring the significance of RNA editing in gene expression [112]. This is particularly evident in Alu elements, as it was discovered an astonishing 1.6 million editing sites in dsRNA, highlighting the extensive impact of RNA editing on the transcriptome [111]. RNA A-to-I editing also extends to primary transcripts of some miRNAs (namely, pri-miRNAs), impacting their maturation and function [113, 114]. Such editing has been reported in pri-miRNAs within the human brain, suggesting a common mechanism for gene regulation and silencing associated with miRNA activity [115]. The functional consequences of RNA editing was explored, which include the generation of new amino acid sequences, alterations in miRNA target specificity, changes in splicing patterns of pre-mRNA indicative of alternative splicing, and influences on RNA stability and degradation [116]. Together, these findings illustrate the critical roles of RNA editing in modulating gene expression and protein diversity, which have implications for understanding human biology and disease mechanisms.

## 15.7 Emerging RNA Modification Sites

### *15.7.1 2'-O-Methylation on RNA Ribose*

RNA ribose methylation refers to the specific 2'-O-methylation of the ribose 2'-OH moiety across all four standard nucleotides and various noncanonical nucleotides. This modification has been identified in over 40 positions within the human small subunit of 18S rRNAs (SSU-18S rRNA), highlighting its significance in the ribosomal machinery [117, 118]. Similar modifications have been observed in the human large subunit of 28S rRNA (LSU-28S rRNA), with recent high-throughput mapping techniques validating the existence of more than 60 2'-O-methylated sites. Additionally, human 5.8S rRNA hosts two such methylation positions [119], and human mitochondrial rRNA contains three identified sites [120, 121]. Beyond ribosomal RNAs, tRNAs and mRNAs also exhibit this modification, with five positions reported in tRNAs and potential sites within mRNA [122]. In human mitochondria, the methyltransferase enzymes responsible for these modifications have been identified as MRM1, MRM2, and MRM3 [120, 121, 123]. For tRNAs, enzymes such as TRMT13, TRMT44, TARBP1, and FTSJ1 have been noted for their regulatory roles in 2'-O-methylation [122]. This extensive catalog of 2'-O-methylated positions across various RNA molecules underscores their critical importance in the regulation of RNA function and stability, affecting processes ranging from protein synthesis to the fine-tuning of gene expression.

### *15.7.2 N7-Methylguanosine Modification at the 5′-Cap of Eukaryotic mRNA*

N7-Methylguanosine (m7G) plays a significant role in the transcription initiation phase, bearing a positive charge that impacts several aspects of mRNA metabolism [124, 125]. Besides its presence in mRNA, m7G modifications are also found in rRNA [126] and tRNAs [127], underscoring its ubiquity across different RNA types. This modification is instrumental in regulating transcription elongation, nuclear export, translation, polyadenylation [128], and stabilizing transcripts [129], highlighting its essential functions in RNA biology. The enzyme METTL1-WDR4 complex is known to catalyze the formation of m7G at position 46 of tRNAs, showcasing its critical role in tRNA biology. Additionally, the enzyme WBSCR22 is responsible for catalyzing m7G formation in 18S rRNA, with its involvement linked to various human disease states, including cancer and Williams–Beuren syndrome [130].

**Box 15.1 Defining RNA Modification in Human Diseases**

RNA modifications across various RNA types play pivotal roles in numerous biological processes and are implicated in a range of diseases. These modifications are catalyzed by a diversity of chemical groups, reflecting the complexity and dynamism of the RNA molecules involved. With the advancements in the field of epitranscriptomics, RNA modifications are increasingly recognized as promising targets for drug development, offering new avenues for therapeutic intervention [131]. Among these modifications, dynamic ones like m6A and m1A stand out due to their reversible nature, facilitated by specific enzymes responsible for adding and removing methyl groups. These modifications can significantly alter RNA function, affecting processes such as RNA stability, localization, splicing, and translation, thereby influencing cellular behavior and disease progression. The identification of enzymes, such as methyltransferases and demethylases, that mediate these modifications has opened up new possibilities for targeting these regulatory mechanisms with therapeutic agents. On the other hand, some RNA modifications are characterized by the presence of enzymes that specifically add methyl groups to nucleotides, highlighting the diversity of modification mechanisms and their potential as drug targets. Targeting the enzymes that regulate RNA modifications could lead to novel therapeutic strategies, aiming to modulate gene expression at the post-transcriptional level to combat various diseases. The potential of RNA modifications as therapeutic targets is supported by accumulating evidence from numerous studies. These investigations suggest that altering RNA modifications through specific compounds could offer a novel approach to treatment, particularly for conditions where current therapies are limited or ineffective. Given the crucial roles of RNA modifications in fundamental biological processes and their association with diseases, targeting these modifications presents a compelling opportunity for novel therapeutic interventions [14, 17, 59, 60, 128, 132]. As research in this area continues to evolve, it is expected that further insights will emerge, potentially leading to the development of innovative drugs that harness the power of epitranscriptomics to treat a wide array of conditions.

Certain RNA modifications, including m1A, m6A, m5C, and A-to-I editing, have well-established roles in a variety of biological functions [32, 58, 81, 112]. Nonetheless, the functions of many modifications in biological processes remain to be fully understood, indicating a rich area for further investigation. There is strong evidence linking specific RNA modifications, such as m6A, to diseases, including cancer [6, 7, 122]. However, the relationships between other RNA modifications and nonmalignant diseases warrant additional research to uncover broader therapeutic opportunities. The enzymes responsible for adding, removing, and recognizing RNA modifications (writers, erasers, and readers) represent precise molecular targets [133]. Modulating

(continued)

**Box 15.1** (continued)
the activity of these proteins offers a direct approach to alter RNA modification levels for therapeutic purposes. The ability to reversibly modify RNA suggests that drugs could be developed to dynamically influence these modifications in response to internal or external factors, potentially correcting dysregulated RNA modification patterns associated with disease. The nearly irreversible nature of some modifications, such as pseudouridine, presents unique challenges and opportunities in drug development, possibly allowing for more permanent therapeutic interventions. RNA modifications are found in a diverse array of RNA molecules, including circular RNAs, lncRNAs, mRNAs, rRNAs, and tRNAs. This broad presence expands the potential for drug development to target a wide range of diseases by modulating various aspects of RNA biology [7]. The occurrence of RNA modifications in distinct cellular locations, including the nucleus, cytoplasm, and mitochondria, allows for the development of targeted therapies [134]. This is particularly relevant for diseases with specific subcellular pathology, such as mitochondrial disorders [135]. These distinctive properties of RNA modifications highlight their potential as innovative targets for drug development. By further understanding the mechanisms, functions, and disease associations of RNA modifications, new therapeutic strategies may emerge, offering hope for treating a wide spectrum of conditions with precision and efficacy.

## 15.8 Techniques to Detect RNA Modification

The accumulating body of research suggests that genes regulating RNA modifications play a crucial role in cancer, and their association with other diseases is becoming increasingly apparent. A variety of compounds have been discovered that can modify the expression levels of m6A regulatory genes [136]. The Comparative Toxicogenomics Database (CTD) has compiled data from various studies on drugs capable of influencing the expression and/or DNA methylation of genes involved in RNA modifications. This summary includes research across a spectrum of methodologies, including cell culture, animal models, and human subjects [137].

### *15.8.1 Liquid Chromatography–Tandem Mass Spectrometry (LC-MS/MS)*

Recent advancements in liquid chromatography–tandem mass spectrometry (LC-MS/MS) have facilitated the accurate quantification of a wide array of RNA modifications under various conditions and across different cell types [138–141], proving to be effective in multiple species [139, 142, 143]. The strategy of combining specific enzymatic digestion of RNAs with an array of RNases, alongside LC–MS/MS techniques, has significantly aided in the comparative analysis of RNA modifications across different conditions and species [144]. Moreover, the utilization of isotopically labeled molecules ($^{13}C$,$^{15}N$,$^{18}O$) has allowed for detection down to the femtomole level, making it feasible to observe subtle differences in RNA modifications among RNA samples [144, 145]. Achieving meaningful insights using LC–MS/MS for RNA modifications necessitates the purification of specific RNA species free from contamination by other RNA types. Consequently, LC–MS/MS analysis has primarily been applied to tRNA and rRNA, owing to their rich modification landscape and relative ease of isolation [139, 140]. While there have been successful efforts to study modifications in other RNA classes, showing that such modifications are more widely distributed, a major limitation remains: the inability to determine the exact transcript and sequence location of these modifications using LC–MS/MS.

### 15.8.2 Next-Generation Sequencing (NGS)

The advent of NGS technologies has revolutionized the ability to map modified nucleosides across the entire transcriptome with precision [131, 146, 147]. These methods fall into three main categories: (a) Immunoprecipitation of fragmented RNAs using modification-specific antibodies followed by sequencing of the enriched RNA fragments (RIP-seq) to map modifications such as m1A [15, 148], m6A [2, 3], and hm5C [149]. (b) Application of chemical treatments to RNA before sequencing, leveraging the unique reactivity of modified bases, such as CMC [N-cyclohexyl-N99-(2-morpholinoethyl)-carbodiimidemetho-p-toluenesulfonate] for pseudouridine detection [150, 151](Chem-seq), or sodium bisulfite for m5C identification [4]. (c) Generation of distinct mismatch patterns in RNA sequencing data when reverse transcriptase synthesizes cDNA from RNA during library preparation, revealing the presence of modifications [148]. Current advancements in sequencing technologies and methodologies have led to the transcriptome-wide identification of six RNA modifications, which we have discussed above: m1A, m6A, m6Am, m5C, hm5C, pseudouridine (Ψ), and A-to-I editing, surpassing the capabilities of traditional RNA sequencing protocols [152, 153].

### 15.8.3 RIP-seq

The use of antibody-based methods, drawing inspiration from ChIP-seq techniques, for the transcriptome-wide identification of RNA modifications has been successfully demonstrated. These methods involve comparing the density of modified RNA that has been immunoprecipitated with that of an untreated control [2, 132]. While this approach is notably sensitive, its application is constrained by the limited range

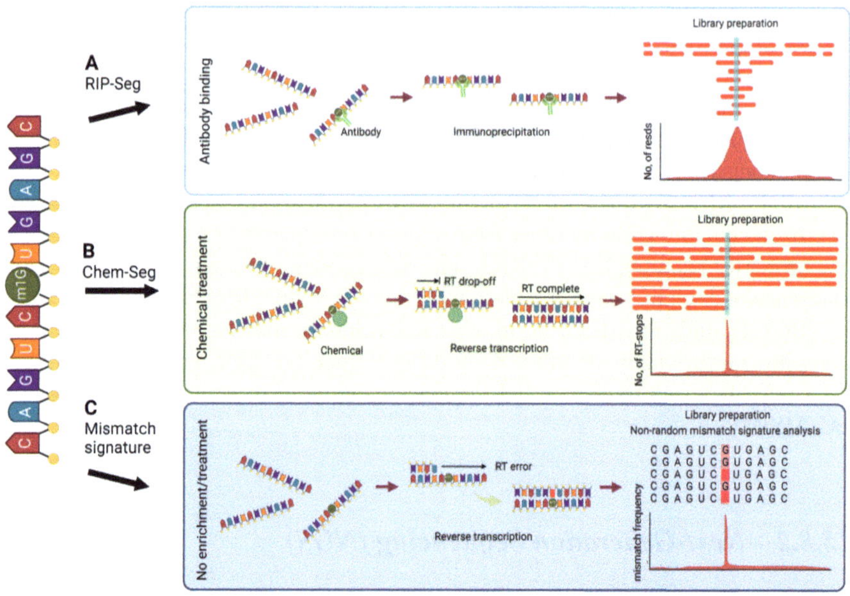


**Fig. 15.4** This figure outlines three primary methodologies for the genome-wide detection of RNA modifications, illustrating their workflows and outcomes: (A) RIP-Seq. Specific antibodies bind to modified RNA nucleotides, forming antibody-RNA complexes. The complexes are then precipitated, isolating the modified RNA fragments. These fragments undergo library preparation, followed by sequencing. The resulting read coverage indicates the presence and location of RNA modifications. Peaks in read distribution correspond to sites of modification. (B) Chem-Seq. RNA is treated with chemicals that selectively react with modified nucleotides, causing reverse transcriptase (RT) to either drop-off or complete the transcription process differently at modified sites. The treated RNA is reverse transcribed into cDNA, where RT stops or completes transcription can be detected. The cDNA is then prepared for sequencing. The number of RT stops (incomplete transcriptions) provides a signature indicating modification sites, with peaks representing the locations of chemical-induced stops. (C) Mismatch Signature Analysis. This method does not require prior enrichment or chemical treatment of RNA. During reverse transcription, modifications cause RT errors, resulting in mismatches in the cDNA sequence relative to the original RNA. Sequencing of the cDNA reveals non-random mismatch signatures. The frequency and location of these mismatches are analyzed to identify RNA modification sites, with peaks in mismatch frequency indicating the presence of modifications. These methodologies allow researchers to map RNA modifications across the transcriptome, providing insights into their roles in regulating gene expression and RNA function. Figure is adapted from [7] under Creative Commons CC-BY 4.0 license

of commercially available antibodies, which currently target only a few modifications such as m1A, m6A, m5C, and hm5C (Fig. 15.4(A)). Furthermore, one of the drawbacks of this method is the inability to achieve single-nucleotide resolution. However, this limitation can be addressed through minor adjustments to the existing protocol [154].

### 15.8.4 Chem-Seq

Chemical-based detection techniques rely on reagents that selectively interact with certain modified RNA nucleotides (Fig. 15.4(B)). Following reverse transcription, the presence of chemically modified nucleotides leads to an increase in RT termination at these sites, creating a concentration of reads ending at the modified positions. This accumulation enables the precise identification of chemically modified sites by observing the surge in reverse transcription termination sites (RTTS). An example of this method's effective application is Pseudo-seq, where CMC-modified pseudouridines block reverse transcription [150, 151]. Chemical-based approaches offer the advantage of mapping RNA modifications with single-nucleotide precision. However, several factors, including protein interactions [155], enhanced RNA secondary structures [156], and the presence of other RNA modifications [157, 158], can influence RT termination, potentially leading to false positives. Additionally, the tendency of RT enzymes to add non-template nucleotides [159] can further increase the likelihood of false positives.

Certain successful chemical detection methods do not exhibit RT drop-off; instead, they alter the base-pairing capability of the modified nucleotide. For instance, during RNA editing, glyoxal treatment protects guanosines from RNase T1 activity before reverse transcription, a protection not afforded to inosines, which are adenosines converted via deamination and mimic guanosine behavior [160]. Similarly, in bisulfite sequencing, uridine results from the sulfonation and subsequent deamination of unmethylated cytosine by sodium bisulfite, whereas methylated cytosine remains unchanged due to its resistance to this reaction [161]. During cDNA conversion, unmethylated cytosine is read as thymine. The potential for high false-positive rates in these methods arises from both the glyoxal protection of guanosines and the incomplete C-to-U conversion in bisulfite sequencing. To mitigate false positives, these methods require deep sequencing, often comparing wild-type RNAs to those from knockout or knockdown models of the relevant enzyme. 5-azacytidine-mediated RNA immunoprecipitation (Aza-IP) represents a hybrid approach, blending immunoprecipitation with chemical modification. This technique leverages the formation of covalent bonds between the cytidine analog 5-azacytidine and RNA methyltransferases, followed by immunoprecipitation to selectively enrich for m5C-modified RNA targets [85].

### 15.8.5 Mismatch Signature Analysis

Mismatch-signature-based analyses have enabled the creation of transcriptome-wide maps of RNA modifications with single-nucleotide precision [148]. This method detects unique non-random mismatch patterns at modified nucleotide positions during the enzymatic process of reverse transcription, which occur due to certain RNA modifications interfering with standard Watson–Crick (W–C) base pairing. A major limitation of this approach is the increase in mismatch rates when multiple RNA modifications disrupt W–C base pairing, complicating the accurate pinpointing of specific modifications. Efforts have been made to overcome this challenge through bioinformatics, specifically by developing algorithms trained on known mismatch patterns found in tRNA modifications to decipher these complex signatures and accurately identify the modifications involved [158]. However, since tRNA molecules are known for their extensive modification landscape, the mismatch signatures characteristic of tRNA modifications may not necessarily reflect those found in other RNA types or locations (Fig. 15.4(C)).

## 15.9 Future Approaches: Direct RNA Sequencing

Traditional genome-wide sequencing technologies, specifically those employing sequence-by-synthesis (SBS) methods, encounter a significant limitation in that they are generally unable to detect RNA and DNA modifications, with the exception of A-to-I editing, which manifests as an A-to-G mismatch. In the context of DNA, this limitation means that modifications such as m5C are read as C, subsequently being interpreted as G during sequencing. Similarly, in RNA sequencing, modifications are lost during the reverse transcription process, which converts RNA back to cDNA for sequencing. This not only eliminates the presence of edited bases and epigenetic markers but also has the potential to introduce significant errors [159]. Additionally, the efficiency of standard reverse transcription enzymes can be compromised by the length of RNA, leading to premature termination when encountering stable RNA structures [156]. Recently, the advent of third-generation sequencing (TGS) technologies has offered a promising avenue for mapping RNA modifications across the genome. Unlike their second-generation counterparts, TGS technologies are capable of producing much longer reads, sometimes exceeding 100 kb [162, 163]. Helicos Biosciences pioneered the use of a single-molecule SBS approach for single-molecule sequencing. Although it was an innovative attempt, it is unclear if this method could distinguish between modified and standard RNA bases, and it likely faced challenges similar to those of cDNA-based methods due to the reliance on polymerase for sequencing. Pacific Biosciences introduced a novel SBS technique capable of detecting RNA modifications through their single-molecule real-time (SMRT) sequencing platform. This technology demonstrated the ability to differentiate between m6A-modified and unmodified synthetic RNA sequences. However, further development was halted due to high costs and labor-intensive processes [131]. Oxford Nanopore Technologies (ONT) has developed another promising TGS method that directly measures changes in electrical current as DNA or RNA molecules traverse a nanopore formed by a bacterial transmembrane protein. Variations in current intensity allow for the identification of passing nucleotides, including those that are modified. ONT announced their initial results for direct RNA sequencing towards the end of 2016, initially focusing on DNA sequencing.

## 15.10 Conclusion

RNA modifications, which are explored in these studies have shown to be linked to various stages in the development of cancer. There is strong evidence backing the idea of modifying pathways associated with cancer by altering the involved RNA modifications or methylations installed de novo, hence leading to RNA modifications as potential targets in cancer therapy. In addition to that, RNA modifications were also found to be possible treatment targets of diseases apart from cancer from previous studies. For instance, certain RNA modifications were found to be concentrated in the brain may suggest the possibility of drug development targeting these specific RNA modifications in the treatment of neurological diseases. As mentioned, there exists various structures of RNA modifications in different types of RNA, such as coding or noncoding RNA, which opens a multitude of opportunity in drug development due its diversity. It can be said that there is a potential future for research in these approaches regarding RNA modifications. With the presently known properties of RNA modifications, drug development can utilize each of these properties in their research. The outcome from RNA modifications will depend on its availability as well as on the specific site or transcript of the modification level. Hence it is critical for RNA modification stoichiometry of each modification to be clarified. The site availability of the modification for each transcript and each type of RNA needed to be understood. However, with the current methodological approach, only the sites of RNA modification are known while the understanding of the dynamic changes of site-specific levels of modification in either noncoding RNA or individual transcript is still lacking. These hurdles in terms of technicality as well as financially have obstructed the further understanding of RNA modification functions, distribution, and implications in diseases. The knowledge of the level and site-specific modification is indeed crucial for the RNA modification based novel drug development. By further researching the RNA modification stoichiometry, specific sites of RNA modification can be determined to be targets of novel drug development.

## References

1. Boccaletto P, Machnicka MA, Purta E, Piatkowski P, Baginski B, Wirecki TK, et al. MODOMICS: a database of RNA modification pathways. 2017 update. Nucleic Acids Res. 2018;46(D1):D303–d7.
2. Dominissini D, Moshitch-Moshkovitz S, Schwartz S, Salmon-Divon M, Ungar L, Osenberg S, et al. Topology of the human and mouse m6A RNA methylomes revealed by m6A-seq. Nature. 2012;485(7397):201–6.
3. Meyer KD, Saletore Y, Zumbo P, Elemento O, Mason CE, Jaffrey SR. Comprehensive analysis of mRNA methylation reveals enrichment in 3' UTRs and near stop codons. Cell. 2012;149(7):1635–46.
4. Squires JE, Patel HR, Nousch M, Sibbritt T, Humphreys DT, Parker BJ, et al. Widespread occurrence of 5-methylcytosine in human coding and non-coding RNA. Nucleic Acids Res. 2012;40(11):5023–33.

5. Yoon KJ, Ringeling FR, Vissers C, Jacob F, Pokrass M, Jimenez-Cyrus D, et al. Temporal control of mammalian cortical neurogenesis by m(6)A methylation. Cell. 2017;171(4):877–89.e17.
6. Frye M, Harada BT, Behm M, He C. RNA modifications modulate gene expression during development. Science. 2018;361(6409):1346–9.
7. Jonkhout N, Tran J, Smith MA, Schonrock N, Mattick JS, Novoa EM. The RNA modification landscape in human disease. RNA. 2017;23(12):1754–69.
8. Lewis CJ, Pan T, Kalsotra A. RNA modifications and structures cooperate to guide RNA-protein interactions. Nat Rev Mol Cell Biol. 2017;18(3):202–10.
9. Li X, Xiong X, Yi C. Epitranscriptome sequencing technologies: decoding RNA modifications. Nat Methods. 2017;14(1):23–31.
10. Moshitch-Moshkovitz S, Dominissini D, Rechavi G. The epitranscriptome toolbox. Cell. 2022;185(5):764–76.
11. Dunn DB. The occurrence of 1-methyladenine in ribonucleic acid. Biochim Biophys Acta. 1961;46:198–200.
12. RajBhandary UL, Stuart A, Faulkner RD, Chang SH, Khorana HG. Nucleotide sequence studies on yeast phenylalanine sRNA. Cold Spring Harb Symp Quant Biol. 1966;31:425–34.
13. Peifer C, Sharma S, Watzinger P, Lamberth S, Kötter P, Entian KD. Yeast Rrp8p, a novel methyltransferase responsible for m1A 645 base modification of 25S rRNA. Nucleic Acids Res. 2013;41(2):1151–63.
14. Sharma S, Watzinger P, Kötter P, Entian KD. Identification of a novel methyltransferase, Bmt2, responsible for the N-1-methyl-adenosine base modification of 25S rRNA in Saccharomyces cerevisiae. Nucleic Acids Res. 2013;41(10):5428–43.
15. Dominissini D, Nachtergaele S, Moshitch-Moshkovitz S, Peer E, Kol N, Ben-Haim MS, et al. The dynamic N(1)-methyladenosine methylome in eukaryotic messenger RNA. Nature. 2016;530(7591):441–6.
16. Li X, Xiong X, Zhang M, Wang K, Chen Y, Zhou J, et al. Base-resolution mapping reveals distinct m(1)A methylome in nuclear- and mitochondrial-encoded transcripts. Mol Cell. 2017;68(5):993–1005.e9
17. Zhang C, Jia G. Reversible RNA modification N(1)-methyladenosine (m(1)A) in mRNA and tRNA. Genomics Proteomics Bioinformatics. 2018;16(3):155–61.
18. Li X, Xiong X, Wang K, Wang L, Shu X, Ma S, et al. Transcriptome-wide mapping reveals reversible and dynamic N(1)-methyladenosine methylome. Nat Chem Biol. 2016;12(5):311–6.
19. Ozanick S, Krecic A, Andersland J, Anderson JT. The bipartite structure of the tRNA m1A58 methyltransferase from S. cerevisiae is conserved in humans. RNA. 2005;11(8):1281–90.
20. Vilardo E, Nachbagauer C, Buzet A, Taschner A, Holzmann J, Rossmanith W. A subcomplex of human mitochondrial RNase P is a bifunctional methyltransferase—extensive moonlighting in mitochondrial tRNA biogenesis. Nucleic Acids Res. 2012;40(22):11583–93.
21. Chujo T, Suzuki T. Trmt61B is a methyltransferase responsible for 1-methyladenosine at position 58 of human mitochondrial tRNAs. RNA. 2012;18(12):2269–76.
22. Waku T, Nakajima Y, Yokoyama W, Nomura N, Kako K, Kobayashi A, et al. NML-mediated rRNA base methylation links ribosomal subunit formation to cell proliferation in a p53-dependent manner. J Cell Sci. 2016;129(12):2382–93.
23. Liu F, Clark W, Luo G, Wang X, Fu Y, Wei J, et al. ALKBH1-mediated tRNA demethylation regulates translation. Cell. 2016;167(3):816–28.e16.
24. Ueda Y, Ooshio I, Fusamae Y, Kitae K, Kawaguchi M, Jingushi K, et al. AlkB homolog 3-mediated tRNA demethylation promotes protein synthesis in cancer cells. Sci Rep. 2017;7:42271.
25. Zheng Q, Gan H, Yang F, Yao Y, Hao F, Hong L, et al. Cytoplasmic m(1)A reader YTHDF3 inhibits trophoblast invasion by downregulation of m(1)A-methylated IGF1R. Cell Discov. 2020;6:12.
26. Dai X, Wang T, Gonzalez G, Wang Y. Identification of YTH domain-containing proteins as the readers for N1-Methyladenosine in RNA. Anal Chem. 2018;90(11):6380–4.

27. Grozhik AV, Olarerin-George AO, Sindelar M, Li X, Gross SS, Jaffrey SR. Antibody cross-reactivity accounts for widespread appearance of m(1)A in 5'UTRs. Nat Commun. 2019;10(1):5126.
28. Basavappa R, Sigler PB. The 3 A crystal structure of yeast initiator tRNA: functional implications in initiator/elongator discrimination. EMBO J. 1991;10(10):3105–11.
29. Anderson JT, Droogmans L. Biosynthesis and function of 1-methyladenosine in transfer RNA. Top Curr Genet. 2005;12:121–39.
30. Ben-Artzi H, Shemesh J, Zeelon E, Amit B, Kleiman L, Gorecki M, et al. Molecular analysis of the second template switch during reverse transcription of the HIV RNA template. Biochemistry. 1996;35(32):10549–57.
31. Renda MJ, Rosenblatt JD, Klimatcheva E, Demeter LM, Bambara RA, Planelles V. Mutation of the methylated tRNA(Lys)(3) residue A58 disrupts reverse transcription and inhibits replication of human immunodeficiency virus type 1. J Virol. 2001;75(20):9671–8.
32. Safra M, Sas-Chen A, Nir R, Winkler R, Nachshon A, Bar-Yaacov D, et al. The m1A landscape on cytosolic and mitochondrial mRNA at single-base resolution. Nature. 2017;551(7679):251–5.
33. Konishi N, Nakamura M, Ishida E, Shimada K, Mitsui E, Yoshikawa R, et al. High expression of a new marker PCA-1 in human prostate carcinoma. Clin Cancer Res. 2005;11(14):5090–7.
34. Tasaki M, Shimada K, Kimura H, Tsujikawa K, Konishi N. ALKBH3, a human AlkB homologue, contributes to cell survival in human non-small-cell lung cancer. Br J Cancer. 2011;104(4):700–6.
35. Yamato I, Sho M, Shimada K, Hotta K, Ueda Y, Yasuda S, et al. PCA-1/ALKBH3 contributes to pancreatic cancer by supporting apoptotic resistance and angiogenesis. Cancer Res. 2012;72(18):4829–39.
36. Zhao Y, Zhao Q, Kaboli PJ, Shen J, Li M, Wu X, et al. m1A regulated genes modulate PI3K/AKT/mTOR and ErbB pathways in gastrointestinal cancer. Transl Oncol. 2019;12(10):1323–33.
37. Cao G, Li HB, Yin Z, Flavell RA. Recent advances in dynamic m6A RNA modification. Open Biol. 2016;6(4):160003.
38. Gilbert WV, Bell TA, Schaening C. Messenger RNA modifications: form, distribution, and function. Science. 2016;352(6292):1408–12.
39. Geula S, Moshitch-Moshkovitz S, Dominissini D, Mansour AA, Kol N, Salmon-Divon M, et al. Stem cells. m6A mRNA methylation facilitates resolution of naïve pluripotency toward differentiation. Science. 2015;347(6225):1002–6.
40. Wang X, Lu Z, Gomez A, Hon GC, Yue Y, Han D, et al. N6-methyladenosine-dependent regulation of messenger RNA stability. Nature. 2014;505(7481):117–20.
41. Meyer KD, Patil DP, Zhou J, Zinoviev A, Skabkin MA, Elemento O, et al. 5' UTR m(6)A promotes cap-independent translation. Cell. 2015;163(4):999–1010.
42. Zhou J, Wan J, Gao X, Zhang X, Jaffrey SR, Qian SB. Dynamic m(6)A mRNA methylation directs translational control of heat shock response. Nature. 2015;526(7574):591–4.
43. Zhao X, Yang Y, Sun BF, Shi Y, Yang X, Xiao W, et al. FTO-dependent demethylation of N6-methyladenosine regulates mRNA splicing and is required for adipogenesis. Cell Res. 2014;24(12):1403–19.
44. Alarcón CR, Lee H, Goodarzi H, Halberg N, Tavazoie SF. N6-methyladenosine marks primary microRNAs for processing. Nature. 2015;519(7544):482–5.
45. Li XC, Jin F, Wang BY, Yin XJ, Hong W, Tian FJ. The m6A demethylase ALKBH5 controls trophoblast invasion at the maternal-fetal interface by regulating the stability of CYR61 mRNA. Theranostics. 2019;9(13):3853–65.
46. Huang H, Weng H, Sun W, Qin X, Shi H, Wu H, et al. Recognition of RNA N(6)-methyladenosine by IGF2BP proteins enhances mRNA stability and translation. Nat Cell Biol. 2018;20(3):285–95.
47. Ping XL, Sun BF, Wang L, Xiao W, Yang X, Wang WJ, et al. Mammalian WTAP is a regulatory subunit of the RNA N6-methyladenosine methyltransferase. Cell Res. 2014;24(2):177–89.

48. Zheng G, Dahl JA, Niu Y, Fedorcsak P, Huang CM, Li CJ, et al. ALKBH5 is a mammalian RNA demethylase that impacts RNA metabolism and mouse fertility. Mol Cell. 2013;49(1):18–29.
49. Little NA, Hastie ND, Davies RC. Identification of WTAP, a novel Wilms' tumour 1-associating protein. Hum Mol Genet. 2000;9(15):2231–9.
50. Wang X, Zhao BS, Roundtree IA, Lu Z, Han D, Ma H, et al. N(6)-methyladenosine modulates messenger RNA translation efficiency. Cell. 2015;161(6):1388–99.
51. Zhuang M, Li X, Zhu J, Zhang J, Niu F, Liang F, et al. The m6A reader YTHDF1 regulates axon guidance through translational control of Robo3.1 expression. Nucleic Acids Res. 2019;47(9):4765–77.
52. Sheng H, Li Z, Su S, Sun W, Zhang X, Li L, et al. YTH domain family 2 promotes lung cancer cell growth by facilitating 6-phosphogluconate dehydrogenase mRNA translation. Carcinogenesis. 2020;41(5):541–50.
53. Lan Q, Liu PY, Haase J, Bell JL, Hüttelmaier S, Liu T. The critical role of RNA m(6)A methylation in cancer. Cancer Res. 2019;79(7):1285–92.
54. Iles MM, Law MH, Stacey SN, Han J, Fang S, Pfeiffer R, et al. A variant in FTO shows association with melanoma risk not due to BMI. Nat Genet. 2013;45(4):428–32. 32e1
55. Dina C, Meyre D, Gallina S, Durand E, Körner A, Jacobson P, et al. Variation in FTO contributes to childhood obesity and severe adult obesity. Nat Genet. 2007;39(6):724–6.
56. Frayling TM, Timpson NJ, Weedon MN, Zeggini E, Freathy RM, Lindgren CM, et al. A common variant in the FTO gene is associated with body mass index and predisposes to childhood and adult obesity. Science. 2007;316(5826):889–94.
57. Scuteri A, Sanna S, Chen WM, Uda M, Albai G, Strait J, et al. Genome-wide association scan shows genetic variants in the FTO gene are associated with obesity-related traits. PLoS Genet. 2007;3(7):e115.
58. Dh MG, Mcguinness D. m6a RNA methylation: the implications for health and disease. J Cancer Sci Clin Oncol. 2014;1(1):105.
59. Zhang S, Zhao BS, Zhou A, Lin K, Zheng S, Lu Z, et al. m(6)A demethylase ALKBH5 maintains Tumorigenicity of glioblastoma stem-like cells by sustaining FOXM1 expression and cell proliferation program. Cancer Cell. 2017;31(4):591–606.e6
60. Cui Q, Shi H, Ye P, Li L, Qu Q, Sun G, et al. m(6)A RNA methylation regulates the self-renewal and tumorigenesis of glioblastoma stem cells. Cell Rep. 2017;18(11):2622–34.
61. Lin S, Choe J, Du P, Triboulet R, Gregory RI. The m(6)a methyltransferase METTL3 promotes translation in human cancer cells. Mol Cell. 2016;62(3):335–45.
62. Legrand C, Tuorto F, Hartmann M, Liebers R, Jacob D, Helm M, et al. Statistically robust methylation calling for whole-transcriptome bisulfite sequencing reveals distinct methylation patterns for mouse RNAs. Genome Res. 2017;27(9):1589–96.
63. Motorin Y, Lyko F, Helm M. 5-methylcytosine in RNA: detection, enzymatic formation and biological functions. Nucleic Acids Res. 2010;38(5):1415–30.
64. Jühling F, Mörl M, Hartmann RK, Sprinzl M, Stadler PF, Pütz J. tRNAdb 2009: compilation of tRNA sequences and tRNA genes. Nucleic Acids Res. 2009;37(Database issue):D159–D62.
65. Maden BE. Identification of the locations of the methyl groups in 18 S ribosomal RNA from Xenopus laevis and man. J Mol Biol. 1986;189(4):681–99.
66. Maden BE. Locations of methyl groups in 28 S rRNA of Xenopus laevis and man. Clustering in the conserved core of molecule. J Mol Biol. 1988;201(2):289–314.
67. Amort T, Rieder D, Wille A, Khokhlova-Cubberley D, Riml C, Trixl L, et al. Distinct 5-methylcytosine profiles in poly(a) RNA from mouse embryonic stem cells and brain. Genome Biol. 2017;18(1):1.
68. Yang X, Yang Y, Sun BF, Chen YS, Xu JW, Lai WY, et al. 5-methylcytosine promotes mRNA export - NSUN2 as the methyltransferase and ALYREF as an m(5)C reader. Cell Res. 2017;27(5):606–25.
69. Motorin Y, Helm M. tRNA stabilization by modified nucleotides. Biochemistry. 2010;49(24):4934–44.

70. Alexandrov A, Chernyakov I, Gu W, Hiley SL, Hughes TR, Grayhack EJ, et al. Rapid tRNA decay can result from lack of nonessential modifications. Mol Cell. 2006;21(1):87–96.
71. Emilsson V, Näslund AK, Kurland CG. Thiolation of transfer RNA in Escherichia coli varies with growth rate. Nucleic Acids Res. 1992;20(17):4499–505.
72. Begley U, Dyavaiah M, Patil A, Rooney JP, DiRenzo D, Young CM, et al. Trm9-catalyzed tRNA modifications link translation to the DNA damage response. Mol Cell. 2007;28(5):860–70.
73. Netzer N, Goodenbour JM, David A, Dittmar KA, Jones RB, Schneider JR, et al. Innate immune and chemically triggered oxidative stress modifies translational fidelity. Nature. 2009;462(7272):522–6.
74. Thompson DM, Parker R. The RNase Rny1p cleaves tRNAs and promotes cell death during oxidative stress in Saccharomyces cerevisiae. J Cell Biol. 2009;185(1):43–50.
75. Chan CT, Pang YL, Deng W, Babu IR, Dyavaiah M, Begley TJ, et al. Reprogramming of tRNA modifications controls the oxidative stress response by codon-biased translation of proteins. Nat Commun. 2012;3:937.
76. Liebers R, Rassoulzadegan M, Lyko F. Epigenetic regulation by heritable RNA. PLoS Genet. 2014;10(4):e1004296.
77. Hussain S, Sajini AA, Blanco S, Dietmann S, Lombard P, Sugimoto Y, et al. NSun2-mediated cytosine-5 methylation of vault noncoding RNA determines its processing into regulatory small RNAs. Cell Rep. 2013;4(2):255–61.
78. Bohnsack KE, Höbartner C, Bohnsack MT. Eukaryotic 5-methylcytosine ($m^5C$) RNA methyltransferases: mechanisms, cellular functions, and links to disease. Genes (Basel). 2019;10:2.
79. Goll MG, Kirpekar F, Maggert KA, Yoder JA, Hsieh CL, Zhang X, et al. Methylation of tRNAAsp by the DNA methyltransferase homolog Dnmt2. Science. 2006;311(5759):395–8.
80. Bourgeois G, Ney M, Gaspar I, Aigueperse C, Schaefer M, Kellner S, et al. Eukaryotic rRNA modification by yeast 5-methylcytosine-methyltransferases and human proliferation-associated antigen p120. PLoS One. 2015;10(7):e0133321.
81. Brzezicha B, Schmidt M, Makalowska I, Jarmolowski A, Pienkowska J, Szweykowska-Kulinska Z. Identification of human tRNA:m5C methyltransferase catalysing intron-dependent m5C formation in the first position of the anticodon of the pre-tRNA Leu (CAA). Nucleic Acids Res. 2006;34(20):6034–43.
82. Blanco S, Kurowski A, Nichols J, Watt FM, Benitah SA, Frye M. The RNA-methyltransferase Misu (NSun2) poises epidermal stem cells to differentiate. PLoS Genet. 2011;7(12):e1002403.
83. Martinez FJ, Lee JH, Lee JE, Blanco S, Nickerson E, Gabriel S, et al. Whole exome sequencing identifies a splicing mutation in NSUN2 as a cause of a Dubowitz-like syndrome. J Med Genet. 2012;49(6):380–5.
84. Tuorto F, Liebers R, Musch T, Schaefer M, Hofmann S, Kellner S, et al. RNA cytosine methylation by Dnmt2 and NSun2 promotes tRNA stability and protein synthesis. Nat Struct Mol Biol. 2012;19(9):900–5.
85. Khoddami V, Cairns BR. Identification of direct targets and modified bases of RNA cytosine methyltransferases. Nat Biotechnol. 2013;31(5):458–64.
86. Motorin Y, Grosjean H. Multisite-specific tRNA:m5C-methyltransferase (Trm4) in yeast Saccharomyces cerevisiae: identification of the gene and substrate specificity of the enzyme. RNA. 1999;5(8):1105–18.
87. Nakano S, Suzuki T, Kawarada L, Iwata H, Asano K, Suzuki T. NSUN3 methylase initiates 5-formylcytidine biogenesis in human mitochondrial tRNA(met). Nat Chem Biol. 2016;12(7):546–51.
88. Van Haute L, Dietmann S, Kremer L, Hussain S, Pearce SF, Powell CA, et al. Deficient methylation and formylation of mt-tRNA(Met) wobble cytosine in a patient carrying mutations in NSUN3. Nat Commun. 2016;7:12039.
89. Metodiev MD, Spåhr H, Loguercio Polosa P, Meharg C, Becker C, Altmueller J, et al. NSUN4 is a dual function mitochondrial protein required for both methylation of 12S rRNA and coordination of mitoribosomal assembly. PLoS Genet. 2014;10(2):e1004110.

90. Janin M, Ortiz-Barahona V, de Moura MC, Martínez-Cardús A, Llinàs-Arias P, Soler M, et al. Epigenetic loss of RNA-methyltransferase NSUN5 in glioma targets ribosomes to drive a stress adaptive translational program. Acta Neuropathol. 2019;138(6):1053–74.
91. Krug AK, Kolde R, Gaspar JA, Rempel E, Balmer NV, Meganathan K, et al. Human embryonic stem cell-derived test systems for developmental neurotoxicity: a transcriptomics approach. Arch Toxicol. 2013;87(1):123–43.
92. Haag S, Warda AS, Kretschmer J, Günnigmann MA, Höbartner C, Bohnsack MT. NSUN6 is a human RNA methyltransferase that catalyzes formation of m5C72 in specific tRNAs. RNA. 2015;21(9):1532–43.
93. Long T, Li J, Li H, Zhou M, Zhou XL, Liu RJ, et al. Sequence-specific and shape-selective RNA recognition by the human RNA 5-methylcytosine methyltransferase NSun6. J Biol Chem. 2016;291(46):24293–303.
94. Waldmann T, Grinberg M, König A, Rempel E, Schildknecht S, Henry M, et al. Stem cell transcriptome responses and corresponding biomarkers that indicate the transition from adaptive responses to cytotoxicity. Chem Res Toxicol. 2017;30(4):905–22.
95. Aguilo F, Li S, Balasubramaniyan N, Sancho A, Benko S, Zhang F, et al. Deposition of 5-methylcytosine on enhancer RNAs enables the coactivator function of PGC-1α. Cell Rep. 2016;14(3):479–92.
96. Bass BL, Weintraub H. A developmentally regulated activity that unwinds RNA duplexes. Cell. 1987;48(4):607–13.
97. Uchida S, Jones SP. RNA editing: unexplored opportunities in the cardiovascular system. Circ Res. 2018;122(3):399–401.
98. Nishikura K. A-to-I editing of coding and non-coding RNAs by ADARs. Nat Rev Mol Cell Biol. 2016;17(2):83–96.
99. Picardi E, Manzari C, Mastropasqua F, Aiello I, D'Erchia AM, Pesole G. Profiling RNA editing in human tissues: towards the inosinome Atlas. Sci Rep. 2015;5:14941.
100. Bass BL. RNA editing by adenosine deaminases that act on RNA. Annu Rev Biochem. 2002;71:817–46.
101. Chen CX, Cho DS, Wang Q, Lai F, Carter KC, Nishikura K. A third member of the RNA-specific adenosine deaminase gene family, ADAR3, contains both single- and double-stranded RNA binding domains. RNA. 2000;6(5):755–67.
102. Kim U, Wang Y, Sanford T, Zeng Y, Nishikura K. Molecular cloning of cDNA for double-stranded RNA adenosine deaminase, a candidate enzyme for nuclear RNA editing. Proc Natl Acad Sci USA. 1994;91(24):11457–61.
103. Melcher T, Maas S, Herb A, Sprengel R, Higuchi M, Seeburg PH. RED2, a brain-specific member of the RNA-specific adenosine deaminase family. J Biol Chem. 1996;271(50):31795–8.
104. Melcher T, Maas S, Herb A, Sprengel R, Seeburg PH, Higuchi M. A mammalian RNA editing enzyme. Nature. 1996;379(6564):460–4.
105. O'Connell MA, Krause S, Higuchi M, Hsuan JJ, Totty NF, Jenny A, et al. Cloning of cDNAs encoding mammalian double-stranded RNA-specific adenosine deaminase. Mol Cell Biol. 1995;15(3):1389–97.
106. Mannion N, Arieti F, Gallo A, Keegan LP, O'Connell MA. New insights into the biological role of mammalian ADARs; the RNA editing proteins. Biomol Ther. 2015;5(4):2338–62.
107. Bass BL. How does RNA editing affect dsRNA-mediated gene silencing? Cold Spring Harb Symp Quant Biol. 2006;71:285–92.
108. Rueter SM, Dawson TR, Emeson RB. Regulation of alternative splicing by RNA editing. Nature. 1999;399(6731):75–80.
109. Wang IX, So E, Devlin JL, Zhao Y, Wu M, Cheung VG. ADAR regulates RNA editing, transcript stability, and gene expression. Cell Rep. 2013;5(3):849–60.
110. Cordaux R, Batzer MA. The impact of retrotransposons on human genome evolution. Nat Rev Genet. 2009;10(10):691–703.

111. Bazak L, Haviv A, Barak M, Jacob-Hirsch J, Deng P, Zhang R, et al. A-to-I RNA editing occurs at over a hundred million genomic sites, located in a majority of human genes. Genome Res. 2014;24(3):365–76.
112. Athanasiadis A, Rich A, Maas S. Widespread A-to-I RNA editing of Alu-containing mRNAs in the human transcriptome. PLoS Biol. 2004;2(12):e391.
113. Blow MJ, Grocock RJ, van Dongen S, Enright AJ, Dicks E, Futreal PA, et al. RNA editing of human microRNAs. Genome Biol. 2006;7(4):R27.
114. Luciano DJ, Mirsky H, Vendetti NJ, Maas S. RNA editing of a miRNA precursor. RNA. 2004;10(8):1174–7.
115. Kawahara Y, Megraw M, Kreider E, Iizasa H, Valente L, Hatzigeorgiou AG, et al. Frequency and fate of microRNA editing in human brain. Nucleic Acids Res. 2008;36(16):5270–80.
116. Wulff BE, Sakurai M, Nishikura K. Elucidating the inosinome: global approaches to adenosine-to-inosine RNA editing. Nat Rev Genet. 2011;12(2):81–5.
117. Incarnato D, Anselmi F, Morandi E, Neri F, Maldotti M, Rapelli S, et al. High-throughput single-base resolution mapping of RNA 2′-O-methylated residues. Nucleic Acids Res. 2017;45(3):1433–41.
118. Zhu Y, Pirnie SP, Carmichael GG. High-throughput and site-specific identification of 2'-O-methylation sites using ribose oxidation sequencing (RibOxi-seq). RNA. 2017;23(8):1303–14.
119. Nazar RN, Sitz TO, Busch H. Homologies in eukaryotic 5.8S ribosomal RNA. Biochem Biophys Res Commun. 1975;62(3):736–43.
120. Lee KW, Okot-Kotber C, LaComb JF, Bogenhagen DF. Mitochondrial ribosomal RNA (rRNA) methyltransferase family members are positioned to modify nascent rRNA in foci near the mitochondrial DNA nucleoid. J Biol Chem. 2013;288(43):31386–99.
121. Lee KW, Bogenhagen DF. Assignment of 2'-O-methyltransferases to modification sites on the mammalian mitochondrial large subunit 16 S ribosomal RNA (rRNA). J Biol Chem. 2014;289(36):24936–42.
122. Ayadi L, Galvanin A, Pichot F, Marchand V, Motorin Y. RNA ribose methylation (2'-O-methylation): occurrence, biosynthesis and biological functions. Biochim Biophys Acta Gene Regul Mech. 2019;1862(3):253–69.
123. Rorbach J, Boesch P, Gammage PA, Nicholls TJ, Pearce SF, Patel D, et al. MRM2 and MRM3 are involved in biogenesis of the large subunit of the mitochondrial ribosome. Mol Biol Cell. 2014;25(17):2542–55.
124. Cowling VH. Regulation of mRNA cap methylation. Biochem J. 2009;425(2):295–302.
125. Furuichi Y. Discovery of m(7)G-cap in eukaryotic mRNAs. Proc Jpn Acad Ser B Phys Biol Sci. 2015;91(8):394–409.
126. Sloan KE, Warda AS, Sharma S, Entian KD, Lafontaine DLJ, Bohnsack MT. Tuning the ribosome: the influence of rRNA modification on eukaryotic ribosome biogenesis and function. RNA Biol. 2017;14(9):1138–52.
127. Guy MP, Phizicky EM. Two-subunit enzymes involved in eukaryotic post-transcriptional tRNA modification. RNA Biol. 2014;11(12):1608–18.
128. Zhang LS, Liu C, Ma H, Dai Q, Sun HL, Luo G, et al. Transcriptome-wide mapping of internal N(7)-methylguanosine methylome in mammalian mRNA. Mol Cell. 2019;74(6):1304–16.e8.
129. Furuichi Y, LaFiandra A, Shatkin AJ. 5'-terminal structure and mRNA stability. Nature. 1977;266(5599):235–9.
130. Haag S, Kretschmer J, Bohnsack MT. WBSCR22/Merml is required for late nuclear pre-ribosomal RNA processing and mediates N7-methylation of G1639 in human 18S rRNA. RNA. 2015;21(2):180–7.
131. Saletore Y, Meyer K, Korlach J, Vilfan ID, Jaffrey S, Mason CE. The birth of the Epitranscriptome: deciphering the function of RNA modifications. Genome Biol. 2012;13(10):175.

132. Batista PJ, Molinie B, Wang J, Qu K, Zhang J, Li L, et al. m(6)A RNA modification controls cell fate transition in mammalian embryonic stem cells. Cell Stem Cell. 2014;15(6):707–19.
133. Esteve-Puig R, Bueno-Costa A, Esteller M. Writers, readers and erasers of RNA modifications in cancer. Cancer Lett. 2020;474:127–37.
134. Delaunay S, Helm M, Frye M. RNA modifications in physiology and disease: towards clinical applications. Nat Rev Genet. 2024;25(2):104–22.
135. Cui L, Ma R, Cai J, Guo C, Chen Z, Yao L, et al. RNA modifications: importance in immune cell biology and related diseases. Signal Transduct Target Ther. 2022;7(1):334.
136. Berdasco M, Esteller M. Towards a druggable epitranscriptome: compounds that target RNA modifications in cancer. Br J Pharmacol. 2022;179(12):2868–89.
137. Davis AP, Grondin CJ, Johnson RJ, Sciaky D, McMorran R, Wiegers J, et al. The comparative toxicogenomics database: update 2019. Nucleic Acids Res. 2019;47(D1):D948–d54.
138. Addepalli B, Limbach PA. Mass spectrometry-based quantification of pseudouridine in RNA. J Am Soc Mass Spectrom. 2011;22(8):1363–72.
139. Chan CT, Dyavaiah M, DeMott MS, Taghizadeh K, Dedon PC, Begley TJ. A quantitative systems approach reveals dynamic control of tRNA modifications during cellular stress. PLoS Genet. 2010;6(12):e1001247.
140. Su D, Chan CT, Gu C, Lim KS, Chionh YH, McBee ME, et al. Quantitative analysis of ribonucleoside modifications in tRNA by HPLC-coupled mass spectrometry. Nat Protoc. 2014;9(4):828–41.
141. Yan M, Wang Y, Hu Y, Feng Y, Dai C, Wu J, et al. A high-throughput quantitative approach reveals more small RNA modifications in mouse liver and their correlation with diabetes. Anal Chem. 2013;85(24):12173–81.
142. Patil A, Chan CT, Dyavaiah M, Rooney JP, Dedon PC, Begley TJ. Translational infidelity-induced protein stress results from a deficiency in Trm9-catalyzed tRNA modifications. RNA Biol. 2012;9(7):990–1001.
143. Begley U, Sosa MS, Avivar-Valderas A, Patil A, Endres L, Estrada Y, et al. A human tRNA methyltransferase 9-like protein prevents tumour growth by regulating LIN9 and HIF1-α. EMBO Mol Med. 2013;5(3):366–83.
144. Li S, Limbach PA. Method for comparative analysis of ribonucleic acids using isotope labeling and mass spectrometry. Anal Chem. 2012;84(20):8607–13.
145. Nikcevic I, Wyrzykiewicz TK, Limbach PA. Detecting low-level synthesis impurities in modified phosphorothioate oligonucleotides using liquid chromatography—high resolution mass spectrometry. Int J Mass Spectrom. 2011;304(2–3):98–104.
146. Novoa EM, Mason CE, Mattick JS. Charting the unknown epitranscriptome. Nat Rev Mol Cell Biol. 2017;18(6):339–40.
147. Helm M, Motorin Y. Detecting RNA modifications in the epitranscriptome: predict and validate. Nat Rev Genet. 2017;18(5):275–91.
148. Hauenschild R, Tserovski L, Schmid K, Thüring K, Winz ML, Sharma S, et al. The reverse transcription signature of N-1-methyladenosine in RNA-Seq is sequence dependent. Nucleic Acids Res. 2015;43(20):9950–64.
149. Delatte B, Wang F, Ngoc LV, Collignon E, Bonvin E, Deplus R, et al. RNA biochemistry. Transcriptome-wide distribution and function of RNA hydroxymethylcytosine. Science. 2016;351(6270):282–5.
150. Schwartz S, Bernstein DA, Mumbach MR, Jovanovic M, Herbst RH, León-Ricardo BX, et al. Transcriptome-wide mapping reveals widespread dynamic-regulated pseudouridylation of ncRNA and mRNA. Cell. 2014;159(1):148–62.
151. Carlile TM, Rojas-Duran MF, Zinshteyn B, Shin H, Bartoli KM, Gilbert WV. Pseudouridine profiling reveals regulated mRNA pseudouridylation in yeast and human cells. Nature. 2014;515(7525):143–6.
152. Shafik A, Schumann U, Evers M, Sibbritt T, Preiss T. The emerging epitranscriptomics of long noncoding RNAs. Biochim Biophys Acta. 2016;1859(1):59–70.

153. Ramaswami G, Zhang R, Piskol R, Keegan LP, Deng P, O'Connell MA, et al. Identifying RNA editing sites using RNA sequencing data alone. Nat Methods. 2013;10(2):128–32.
154. Linder B, Grozhik AV, Olarerin-George AO, Meydan C, Mason CE, Jaffrey SR. Single-nucleotide-resolution mapping of m6A and m6Am throughout the transcriptome. Nat Methods. 2015;12(8):767–72.
155. König J, Zarnack K, Rot G, Curk T, Kayikci M, Zupan B, et al. iCLIP reveals the function of hnRNP particles in splicing at individual nucleotide resolution. Nat Struct Mol Biol. 2010;17(7):909–15.
156. Aviran S, Pachter L. Rational experiment design for sequencing-based RNA structure mapping. RNA. 2014;20(12):1864–77.
157. Motorin Y, Muller S, Behm-Ansmant I, Branlant C. Identification of modified residues in RNAs by reverse transcription-based methods. Methods Enzymol. 2007;425:21–53.
158. Ryvkin P, Leung YY, Silverman IM, Childress M, Valladares O, Dragomir I, et al. HAMR: high-throughput annotation of modified ribonucleotides. RNA. 2013;19(12):1684–92.
159. Chen D, Patton JT. Reverse transcriptase adds nontemplated nucleotides to cDNAs during 5'-RACE and primer extension. BioTechniques. 2001;30(3):574–80. 82
160. Cattenoz PB, Taft RJ, Westhof E, Mattick JS. Transcriptome-wide identification of A > I RNA editing sites by inosine specific cleavage. RNA. 2013;19(2):257–70.
161. Schaefer M, Pollex T, Hanna K, Lyko F. RNA cytosine methylation analysis by bisulfite sequencing. Nucleic Acids Res. 2009;37(2):e12.
162. Rhoads A, Au KF. PacBio sequencing and its applications. Genomics Proteomics Bioinformatics. 2015;13(5):278–89.
163. Laver T, Harrison J, O'Neill PA, Moore K, Farbos A, Paszkiewicz K, et al. Assessing the performance of the Oxford Nanopore Technologies MinION. Biomol Detect Quantif. 2015;3:1–8.

# Chapter 16
# RNA Screening Design

**Abstract** In recent years, there has been notable progress in identifying compounds that specifically interact with and modulate RNA, paving the way for their potential clinical application. This section delves into the methodologies employed in the quest for small molecules targeting RNA, highlighting both successful strategies and challenges faced in this field. We examine the advantages and limitations associated with various screening approaches and discuss key factors influencing the outcomes of RNA-targeted screening projects. Effective screening methodologies have enabled the identification of RNA-binding small molecules capable of modulating RNA function. These methods include high-throughput screening (HTS), fragment-based screening, and structure-based design approaches. HTS, in particular, allows for the rapid screening of large compound libraries against RNA targets, facilitating the discovery of lead compounds with desired binding affinity and specificity. However, the efficacy of these screenings is heavily reliant on the quality and diversity of the compound libraries used. Currently, the availability of RNA-focused compound libraries remains limited, posing a significant hurdle in the search for potent RNA ligands. Moreover, understanding RNA structural dynamics is critical for designing effective RNA-targeting therapeutics. RNA molecules exhibit complex folding patterns and dynamics that influence their binding interactions with small molecules. Advances in biochemical and structural analyses, coupled with computational modeling, are essential for predicting and optimizing RNA–ligand interactions. Nonetheless, challenges persist in accurately predicting RNA structure and dynamics, which can impact the success of screening efforts. To overcome these challenges, integrating high-throughput screening techniques with enhanced biochemical, structural, and computational analyses of RNA targets is crucial. This integrative approach promises to accelerate the discovery and optimization of RNA-targeting therapeutics by providing comprehensive insights into RNA–ligand interactions and enhancing the predictability of compound efficacy. Ultimately, advancing our understanding of RNA biology and therapeutic development hinges on overcoming these obstacles and leveraging multidisciplinary approaches to unlock the full potential of RNA as a therapeutic target.

 
P. E. Saw, E. Song, *RNA Therapeutics in Human Diseases*,
https://doi.org/10.1007/978-981-96-3041-7_16

## 16.1 Introduction

RNA plays a crucial role in a multitude of vital physiological and pathophysiological processes within cells. It serves as a carrier of genetic information, catalyzes essential biological reactions like splicing and protein synthesis, regulates transcription and translation processes, and influences the metabolic functions of proteins [1]. Targeting RNA using small molecules represents a promising strategy in the quest to develop therapies for various human diseases. This approach leverages the ability of small molecules to interact selectively with RNA molecules, influencing their function and potentially correcting dysregulated biological processes at the molecular level [1]. Since the 1940s, RNA-binding molecules, notably aminoglycoside antibiotics that target rRNA, have been instrumental in medicine, identified through unguided phenotypic and genetic screenings [2]. Yet, the broad spectrum of RNA affinity displayed by aminoglycosides hampers their specificity [3], underscoring the necessity to broaden the range of compounds capable of selectively binding RNA.

The discovery of numerous new RNA categories, such as riboswitches and lncRNAs, many of which adopt specific functional structures, has greatly expanded the range of RNA molecules suitable for targeting in drug development. These diverse RNA types play crucial roles in regulating gene expression, cellular processes, and disease mechanisms. Riboswitches, for instance, are RNA elements that can change their structure upon binding specific small molecules, thereby controlling gene expression in response to cellular conditions. This unique property makes riboswitches attractive targets for developing small molecule therapeutics that can modulate gene expression pathways. lncRNAs are another class of RNA molecules that do not code for proteins but instead regulate various cellular processes through interactions with proteins, other RNAs, or DNA. They have been implicated in diverse biological functions, including chromatin remodeling, transcriptional regulation, and cellular signaling pathways. Targeting lncRNAs with small molecules holds potential for therapeutic interventions aimed at correcting dysregulated gene expression associated with diseases like cancer and neurological disorders [4–6]. Nevertheless, the path to creating RNA-targeted drugs is fraught with distinct obstacles, distinguishing it from the more advanced field of protein-targeted drug discovery [2]. Beyond the differences in structural and chemical attributes between protein and RNA ligands, discussed in depth elsewhere [2, 7], formulating and executing screening approaches tailored for RNA is a critical component of the success of discovering drugs that act on RNA targets (Fig. 16.1).

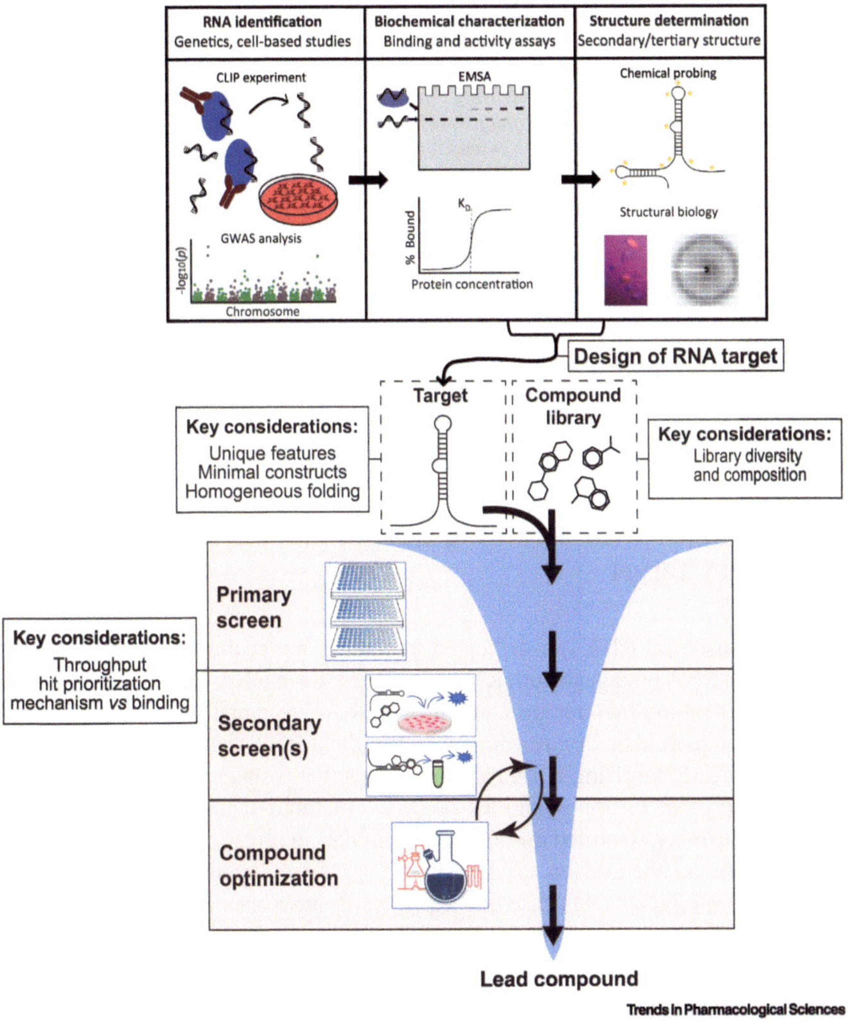


**Fig. 16.1** The systematic process involved in developing RNA screening methodologies. Beginning at the top, the identification of target RNAs is achieved through a combination of genetic techniques, such as genome-wide association studies and knockout models, and cell-based strategies like crosslinking and immunoprecipitation (CLIP) methods. These approaches enable

researchers to pinpoint RNAs crucial for specific biological processes or disease pathways. Once identified, RNA functions are elucidated using biochemical assays, such as electrophoretic mobility shift assays, to understand how RNAs interact with proteins or other molecules. Meanwhile, the structural features of RNAs are probed using chemical methods to map secondary structures and biophysical techniques to explore tertiary structures. This comprehensive analysis aids in constructing accurate RNA target models essential for subsequent screening steps. The initial high-throughput screening phase involves testing a large library of compounds against the RNA target model. Screening methods may be mechanism-focused, aiming to disrupt specific RNA functions, or affinity-based, seeking compounds that bind with high specificity and affinity to the RNA target. Hits from this screening are rigorously validated through orthogonal secondary assays to confirm their biological activity and specificity. Iterative refinements of hit compounds follow, aimed at optimizing their potency, selectivity, and pharmacokinetic properties. This iterative process leads to the development of lead compounds suitable for further preclinical and clinical evaluation as potential therapeutics. Throughout the RNA screening pipeline, specific considerations unique to RNA-targeted compounds are highlighted at each stage. These considerations include the challenges posed by RNA's dynamic structure, its potential for alternative conformations, and the need for compounds that can penetrate cellular membranes to reach intracellular RNA targets [8]. Adapted from [9]. Image is reproduced under Creative Commons CC-BY license

## 16.2 RNA-Focused Screening Design: Finding a Suitable RNA Target

Screens focused on RNA are dedicated to finding molecules that either directly interact with RNA or engage with RNP complexes. Such RNA-specific therapeutics could open new avenues for treating conditions where targeting proteins is challenging, such as genetic diseases linked to abnormal RNA transcripts or infectious diseases. Directly targeting RNA can pose difficulties, especially if the RNA transcripts are present in low quantities or lack structured segments. Consequently, RNPs, increasingly identified through advancements in proteomics and CLIP-based methods, present themselves as alternative targets for influencing the RNA life cycle and its various roles [10]. Historically, RNPs have been considered challenging targets due to their typically large and complex protein–RNA interaction sites, which are often unstructured. This obstacle is being overcome through the development of drugs that either competitively or allosterically inhibit RNA-binding domains or target RNPs for destruction [11]. Given that RNA-binding proteins may associate with multiple RNA types, targeting RNPs shows particular promise for systemic illnesses like cancer and autoimmune diseases [12].

In the design of RNA-targeting screens, it is crucial to distinguish between targeting microbial versus human RNA/RNP. Pathogens, for example, possess essential riboswitches and ribozymes that are well-structured, making them suitable for small molecule interaction, and are distinct from human sequences, thus reducing off-target effects [5, 13]. Conversely, targeting human RNAs (i.e., miRNAs, lncRNAs, or spliceosomal RNAs) necessitates a meticulous differentiation between disease-associated RNAs and similar RNAs that play roles in other cellular functions or are present in healthy cells. This distinction is crucial to avoid off-target effects and unintended interference with normal cellular processes. Recent advancements in biochemical, mechanistic, and cellular understanding of these ncRNAs have significantly contributed to the progress in their specific and selective targeting.

For example, precise targeting of disease-specific miRNAs involves designing molecules that can selectively bind to and modulate these miRNAs without affecting other miRNAs critical for normal cellular functions. Similarly, lncRNAs, which are involved in regulating gene expression at multiple levels, can exhibit highly specific roles in disease pathogenesis. Targeting lncRNAs requires an in-depth understanding of their structure, function, and interaction networks. Advances in RNA sequencing and bioinformatics have enabled researchers to identify lncRNAs with disease-specific expression patterns, paving the way for the development of targeted therapies. Spliceosomal RNAs are essential components of the splicing machinery, responsible for the accurate removal of introns from pre-mRNA transcripts. Aberrations in splicing can lead to various genetic disorders and cancers. Targeting spliceosomal RNAs or their interactions with other splicing factors requires high precision to correct splicing defects without disrupting normal splicing events. Enhanced biochemical techniques, such as high-resolution RNA structure probing and CLIP, provide detailed maps of RNA interactions and modifications. Mechanistic studies reveal how specific RNA molecules contribute to disease processes, guiding the design of selective inhibitors or modulators. Cellular insights, gained from advanced imaging techniques and scRNA-seq, offer a deeper understanding of RNA dynamics within different cellular contexts [14, 15].

## 16.3 Implementing a Suitable RNA Screening Method

Similar to screens designed for proteins, RNA/RNP screenings are conducted to observe the phenotypic or mechanistic impacts of chemical compounds in either a cellular or a cell-free setup (referred to as mechanistic screens) or to assess the direct interaction with the target on a molecular level (known as affinity screens). Recent advancements in screening platforms, the development of compound libraries, and the enhancement of validation assays have been specifically designed to address the unique biophysical and biochemical properties of RNA molecules. This specialized optimization, when combined with the discovery of novel and viable RNA/RNP targets, has significantly enhanced the effectiveness of screening methodologies. By tailoring screening platforms and validation assays to the specific characteristics of RNA molecules, researchers have been able to more accurately identify and target RNA interactions critical for disease processes. Advanced screening platforms have been developed to accommodate the unique structures and functions of RNA molecules. These platforms often include HTS techniques that are capable of rapidly testing large libraries of compounds for their ability to interact with RNA targets. Furthermore, the creation of compound libraries specifically designed for RNA targeting has been crucial. These libraries include small molecules, peptides, and other compounds that have been optimized for their ability to bind RNA structures with high affinity and specificity. Finally, validation assays have been refined to ensure that the interactions observed in screening are specific and biologically relevant. These assays often include biochemical methods such as electrophoretic mobility shift assays (EMSA) to confirm binding, as well as cellular assays to test the functional impact of these interactions in a biological context. This has also helped to identify new RNA/RNP targets and expanding their repertoire. Detailed mechanistic studies have provided insights into how these RNA/RNP targets contribute to disease. Understanding the functional roles of these molecules allows for the rational design of screening assays aimed at disrupting pathogenic RNA interactions while sparing normal cellular functions. Consequently, the ensuing segments will delve into specific instances of recent screening endeavors—both successful and unsuccessful—directed at RNA/RNPs.

## 16.4 Capturing Active Compounds through Mechanistic Screening

Gene-reporter assays that utilize luminescent markers, such as luciferase, or fluorescent markers are frequently employed for mechanistic screenings. These assays are invaluable tools in RNA research, as they allow for the monitoring of gene expression and RNA function in real-time, providing insights into the mechanisms of action of various compounds or genetic modifications. Gene-reporter assays enable the continuous observation of gene expression and RNA activity in living cells. The use of luminescent or fluorescent markers allows researchers to track these processes in real-time, providing dynamic insights into cellular responses. These assays offer quantitative data on gene expression levels. Luminescent markers like luciferase produce a light signal that can be measured with high sensitivity, while fluorescent markers can be quantified using fluorescence microscopy or flow cytometry. These reporters are highly sensitive, allowing for the detection of even low levels of gene expression. The specificity of these markers ensures that the signals observed are directly related to the target gene or RNA of interest [16]. Nonetheless, these reporter assays might reveal compounds acting through unwanted mechanisms, indicating the importance of applying such methodologies to disease-relevant systems with well-defined mechanisms that are amenable to mechanistic evaluation. Furthermore, mechanistic screenings can be carried out with isolated factors, which addresses concerns regarding nonspecific effects, though this approach may face challenges with in vivo reproducibility and a higher incidence of false positives.

Screens based on fluorescence using isolated RNA or RNPs are designed to detect structural shifts that trigger fluorescence resonance energy transfer (FRET). This approach is particularly relevant for studying catalytic RNAs and dynamic RNA/RNP entities, providing insights into their structural changes and interactions in real-time. FRET is a powerful technique used to study the dynamics of biomolecules. It involves the transfer of energy from a donor fluorophore to an acceptor fluorophore when they are in close proximity (typically 1–10 nm). This transfer results in a measurable change in fluorescence that can be used to infer molecular interactions and conformational changes. FRET-based screens are particularly effective for detecting structural shifts in RNA molecules or RNA–protein complexes. When the conformation of an RNA or RNP changes, the distance between the donor and acceptor fluorophores can also change, resulting in a detectable FRET signal. This enables the study of dynamic processes, such as folding, unfolding, and interaction with other molecules. As an example, FRET screens can be used to monitor the conformational changes that occur during the catalytic cycle of ribozymes. By attaching donor and acceptor fluorophores at strategic locations on the ribozyme, researchers can observe how the structure changes during catalysis and how these changes correlate with function. Furthermore, FRET is able to capture the dynamic interaction of RNA/RNP complex. Many RNA molecules and RNPs undergo significant conformational changes as part of their function. For example, splicing factors and ribonucleoprotein complexes involved in RNA processing are highly dynamic. FRET-based screens can provide real-time insights into these conformational changes, helping to elucidate the mechanisms underlying their function. For non-catalytic RNPs, alternative screening methods involve anchoring them to beads followed by assays based on radioactivity/fluorescence, detailed in Table 16.1.

Employed in the search for miRNA processing inhibitors or pathogenic RNP blockers [20, 21], these methodologies often make use of off-the-shelf screening kits, which allow for easy, sensitive, and automated evaluation of extensive compound libraries [18]. An alternative strategy involves affixing small molecules to beads [e.g., via small molecule microarrays (SMMs)], which are adaptable to both purified probes and cell lysates, utilizing minimal material, especially with robotic systems [22]. Although attaching a vast, varied collection of compounds to surfaces poses challenges and might impact compound effectiveness, advancements in attachment chemistry and SMM designs have mitigated these issues [22]. To circumvent the need for affixing targets or compounds, screenings can also utilize dried agarose plates, as demonstrated by AbsorbArray technology. This innovative approach simplifies the screening process and offers several advantages in the study of molecular interactions, particularly those involving RNA, proteins, and small molecules. AbsorbArray technology employs dried agarose plates to immobilize and screen various biological molecules without the need for complex attachment procedures. The agarose matrix naturally absorbs and holds the compounds, providing a stable environment for interactions to occur [23]. A drawback of fluorescence-based screenings is the high rate of false positives, often due to compounds interfering optically, hence the recommendation for their combined use. Even validated compounds might show reduced efficacy in vivo if they fail to alter target activity, possibly due to inaccurate target representation in assays or pharmacokinetic/pharmacodynamic factors. These challenges, while common to both RNA and protein targets, are especially pertinent to RNA, which may adopt different shapes in vivo and in vitro or have large inactive regions [4–6]. Nevertheless, improvements in compound design and the addition of functional groups influencing the RNA life cycle can address low efficacy. Moreover, RNA binders with limited therapeutic value can still be invaluable as research tools for probing RNA structure, function, and mechanisms. These RNA-binding molecules can help elucidate the complex dynamics and roles of RNA within cellular processes, even if they are not directly applicable as treatments. All these strategies are summarized in Fig. 16.2.

**Table 16.1** Alternatives for RNA screening methods

| | Overview | Application to RNA screening | Challenges | Ref |
|---|---|---|---|---|
| Scintillation proximity assay (SPA) | SPA relies on the principle that when a radiolabeled molecule comes into close proximity to a scintillant-embedded bead, the emitted beta particles (from the radiolabel) excite the scintillant, causing it to emit light. This light emission can be detected and measured, indicating the presence of a binding event or interaction | **Mechanism**<br>The target RNA or a binding partner (such as a protein or small molecule) is radiolabeled, typically with isotopes like tritium ($^{3}H$) or iodine-125 ($^{125}I$). The radiolabeled molecule is incubated with scintillant-coated beads and the molecule of interest. If an interaction occurs, the radiolabeled molecule comes into close proximity to the scintillant. The beta particles emitted by the radiolabel excite the scintillant, which then emits light. This light is measured using a scintillation counter, providing a quantitative readout of the interaction<br>**Applications in RNA Research**<br>SPA can be used to study interactions between RNA molecules and RBPs. The assay can be employed to screen for small molecules that bind to RNA targets, such as riboswitches or other regulatory RNAs. This is useful for drug discovery efforts aimed at targeting RNA. SPA can help investigate the folding and conformational changes of RNA molecules by monitoring interactions with other molecules or with itself in the case of RNA dimerization or higher-order structures<br>**Advantages**<br>SPA is a homogeneous assay, meaning it does not require separation of bound and free components. This simplifies the workflow and makes it suitable for high-throughput applications. The assay is highly sensitive, capable of detecting low levels of radiolabeled molecules, which is particularly useful for studying low-abundance RNAs or weak interactions. SPA can be applied to various types of interactions, including RNA–RNA, RNA–protein, and RNA–small molecule interactions | • The need for radiolabeling can pose safety and disposal challenges, as well as potential alterations to the molecule's natural behavior due to the radiolabel<br>• Care must be taken to minimize nonspecific binding and background signal, which can interfere with the detection of true interactions<br>• The assay conditions, such as buffer composition and incubation times, need to be carefully optimized to ensure accurate and reproducible results | [17] |

| | | | | |
|---|---|---|---|---|
| AlphaScreen | AlphaScreen utilizes donor and acceptor beads to detect interactions between biomolecules. When the donor and acceptor beads are brought into proximity (typically less than 200 nm) by a molecular interaction, a cascade of energy transfer events results in a detectable luminescent signal | **Mechanism**<br>Donor beads are conjugated with a molecule of interest (e.g., an RNA molecule or a protein), and acceptor beads are conjugated with a binding partner (e.g., a protein or small molecule). Upon excitation by a laser at 680 nm, donor beads produce singlet oxygen, which diffuses to nearby acceptor beads if they are within proximity due to an interaction. The acceptor beads then emit light at 520–620 nm, which can be detected and measured. The resulting luminescent signal is proportional to the degree of interaction between the molecules conjugated to the donor and acceptor beads<br>**Applications in RNA Research**<br>AlphaScreen can be used to study interactions between RNA molecules RBPs. The technology is well-suited for screening small molecules that bind to RNA targets, such as riboswitches or other regulatory RNAs. This application is valuable for drug discovery efforts targeting RNA. AlphaScreen can detect interactions between different RNA molecules, which is useful for studying RNA dimerization, tertiary structure formation, and RNA-based regulatory mechanisms<br>**Advantages**<br>AlphaScreen is a no-wash, homogeneous assay, which simplifies the workflow and is amenable to high-throughput screening. The assay is highly sensitive, capable of detecting low-abundance interactions and weak binding affinities. AlphaScreen can be adapted to study a wide range of interactions, including protein–protein, protein–DNA, and RNA–RNA interactions. The technology requires relatively small amounts of sample, making it cost-effective and suitable for precious or limited samples | • Assay conditions, including buffer composition, bead concentrations, and incubation times, need to be carefully optimized for each specific interaction being studied<br>• Nonspecific binding or high background signals can sometimes interfere with detection. Proper controls and optimization can help mitigate these issues<br>• The effectiveness of AlphaScreen relies on the close proximity of donor and acceptor beads. Therefore, the assay design must ensure that the interaction brings the beads into the required proximity | [18] |

(continued)

**Table 16.1** (continued)

| | Overview | Application to RNA screening | Challenges | Ref |
|---|---|---|---|---|
| Homogeneous time-resolved fluorescence (HTRF) | HTRF involves the use of donor and acceptor fluorophores that are linked to molecules of interest. When these molecules interact and bring the donor and acceptor into close proximity, energy transfer occurs, resulting in a measurable fluorescent signal. The time-resolved aspect of HTRF allows for the separation of the specific fluorescent signal from background fluorescence by measuring the emission at a specific time point after excitation | **Mechanism:**<br>Donor fluorophores, typically europium cryptate or terbium cryptate, are conjugated to one molecule (e.g., an RNA molecule). Acceptor fluorophores, such as XL665 or d2, are conjugated to the interacting partner (e.g., a protein). Upon excitation at a specific wavelength, the donor fluorophore emits energy that can be transferred to the nearby acceptor fluorophore if they are within close proximity (typically 1–10 nm). This energy transfer results in the emission of light from the acceptor fluorophore. The emission from the acceptor fluorophore is measured after a delay, typically a few hundred microseconds after excitation. This delay allows the short-lived background fluorescence to decay, leaving only the specific signal for detection<br>**Applications in RNA research**<br>HTRF is used to study interactions between RNA molecules and RBPs. This is crucial for understanding the regulation of RNA and identifying potential therapeutic targets. The technology can be applied to screen small molecules that bind to RNA targets, such as riboswitches or other regulatory RNAs. HTRF can be used to study conformational changes in RNA molecules. By labeling different parts of an RNA molecule with donor and acceptor fluorophores, researchers can monitor structural shifts in response to binding events or environmental changes<br>**Advantages**<br>HTRF provides high sensitivity and specificity due to the time-resolved measurement, which reduces background fluorescence and enhances the signal-to-noise ratio. HTRF is a homogeneous assay, meaning it does not require wash steps or separations. This simplifies the workflow and is suitable for high-throughput screening. HTRF assays typically have a wide dynamic range, allowing for the detection of both high and low affinity interactions | • Efficient and specific conjugation of fluorophores to the target molecules is essential for accurate measurements. Nonspecific labeling can lead to background signals and false positives<br>• Conditions such as buffer composition, incubation times, and fluorophore concentrations need to be optimized for each specific interaction being studied<br>• The success of HTRF relies on the close proximity of the donor and acceptor fluorophores. Therefore, the assay must be designed to ensure that the interacting molecules bring the fluorophores into the required proximity | [19], |

| | | | | |
|---|---|---|---|---|
| Catalytic enzyme-linked click chemistry assay (cat-ELCCA) | cat-ELCCA utilizes the bioorthogonal nature of click chemistry reactions to label and detect target molecules in a highly specific manner. This assay typically involves the use of an enzyme to catalyze a reaction that incorporates a clickable handle onto the target molecule, followed by a click reaction with a detection probe | **Mechanism**<br>An enzyme catalyzes the attachment of a clickable handle (such as an alkyne or azide group) onto the target molecule (e.g., RNA, protein, or small molecule). This step ensures that the target is specifically modified in a way that allows for subsequent detection. The modified target molecule undergoes a click reaction with a detection probe containing a complementary clickable group (such as an azide if the target has an alkyne or vice versa). The click reaction is highly specific, fast, and efficient, resulting in the covalent attachment of the detection probe to the target molecule. The detection probe typically contains a reporter group, such as a fluorophore or an enzyme, that allows for the quantification of the target molecule. The reporter group's signal can be measured using standard detection methods, such as fluorescence or colorimetric readouts<br>**Applications in RNA research**<br>cat-ELCCA can be used to study interactions between RNA molecules RBPs. By attaching clickable handles to RNAs and detecting their interaction with proteins, researchers can investigate binding affinities and interaction dynamics. The assay can be applied to screen for small molecules that bind to RNA targets. By incorporating clickable handles into RNA molecules and detecting their interaction with potential drug candidates, cat-ELCCA facilitates the discovery of RNA-targeting therapeutics. Cat-ELCCA can help study RNA structural changes by labeling specific RNA regions with clickable handles. Detecting these modifications can provide insights into RNA folding, conformational shifts, and structural dynamics in response to binding events or environmental changes<br>**Advantages**<br>Click chemistry reactions are highly specific and efficient, ensuring that the detection probe is accurately attached to the target molecule without significant background noise. Cat-ELCCA can be adapted to study a wide range of molecular interactions, including RNA–RNA, RNA–protein, and RNA–small molecule interactions. The use of catalytic enzymes and click chemistry enhances the sensitivity of the assay, allowing for the detection of low-abundance interactions and weak binding affinities. Similar to other enzyme-linked assays, cat-ELCCA can be performed in a homogeneous format, simplifying the workflow and making it suitable for high-throughput screening | • Choosing the appropriate enzyme to catalyze the attachment of clickable handles is crucial for the assay's success. The enzyme must be specific to the target molecule and efficient in catalyzing the modification<br>• Assay conditions, such as enzyme concentrations, reaction times, and click chemistry reagents, need to be carefully optimized for each specific application<br>• Minimizing nonspecific labeling and background signal is essential to ensure accurate and reliable results. Proper controls and optimization can help mitigate these issues | [20, 21] |

(continued)

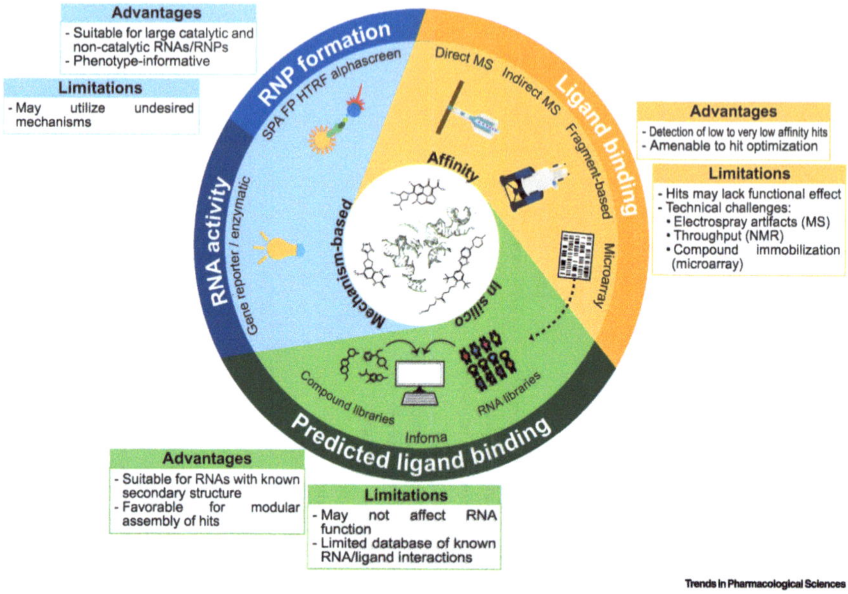


**Fig. 16.2** The diverse strategies for targeting RNA and RNPs through several screening methods, such as mechanism-based, affinity, and in silico approaches, each with distinct advantages and limitations as discussed in their sections. The integration of cell-based and in vitro assays in initial high-throughput screenings, followed by secondary orthogonal validations, along with computational analyses using virtual libraries of compounds and RNA motifs for experimental verification, is crucial for identifying, refining, and validating potent and specific RNA modulators. Mechanism-based screenings utilize various reporter and enzymatic activity assays to evaluate the impact of compounds on RNA function in both cell culture and biochemical contexts, alongside techniques to determine the formation or disruption of recombinant RNP complexes. Ligand-binding strategies focus on identifying high-affinity RNA ligands, employing methods such as mass spectrometry, NMR spectroscopy, or microarrays. More targeted approaches include fluorescence titration, secondary structure mapping, and fluorescence polarization with tagged compounds. The integration of affinity data from extensive microarray assays, using libraries of compounds and short RNAs, has led to the development of the virtual tool Inforna, which is used to design specific RNA ligands. Adapted from [9]. Image is reproduced under Creative Commons CC-BY license

## 16.5 Challenges of RNA-Targeted Screens

### *16.5.1 Selecting Screening Methods*

Achieving structurally uniform RNA in vitro, particularly with full-length transcripts, is more challenging than with proteins. Additionally, many noncoding RNAs lack mechanism-based assays for in vitro analysis, further complicating the study of their functions and interactions [14]. Thus, for initial high-throughput screening phases, cell-based methodologies might offer advantages over using recombinant entities due to the complexity of RNA structures, even though they might activate responses through mechanisms different from the intended target modulation. In vitro high-throughput screenings are valuable for identifying small molecule regulators of RNA–protein interactions. However, these agents might not always produce the anticipated phenotypic effects. To refine hits and validate leads, it is essential to use a combination of mechanistic and affinity tests in both cellular and in vitro settings.

### 16.5.2 Identifying Targetable Regions

RNA's propensity to adopt multiple stable shapes surpasses that of proteins, potentially complicating ligand-binding kinetics but also unveiling varied and selective binding sites. Often, RNA ligands function by stabilizing specific shapes rather than outright blocking an interaction site. Hence, techniques that trace RNA dynamics, such as chemical probing and NMR, are essential for the mechanistic elucidation and validation of interactions between small molecules and RNA on a molecular scale. These studies are typically conducted on simplified, yet functional RNA segments that encompass only the critical regions of the target RNA. Chemical probing techniques, such as SHAPE (Selective 2'-Hydroxyl Acylation analyzed by Primer Extension) and DMS (dimethyl ulfate) probing, provide insights into the secondary structure of RNA by modifying accessible nucleotides. These modifications can then be analyzed to infer the RNA's folding pattern and structural dynamics. Chemical probing helps identify specific binding sites where small molecules interact with RNA, aiding in the precise mapping of these interactions and understanding their impact on RNA structure. Conversely, NMR spectroscopy offers atomic-level resolution of RNA structures and their dynamics. It allows researchers to observe how small molecules affect RNA folding and stability in real-time. NMR can track conformational changes and interactions over time, providing a comprehensive view of how small molecules influence RNA behavior at a molecular level [24].

### 16.5.3 RNA-Focused Compound Libraries

Most existing compound libraries are designed with protein targets in mind, which results in RNA-oriented collections facing a limited chemical diversity and a propensity for containing non-selective, generic molecules. This discrepancy poses significant challenges for the discovery of specific and effective RNA-targeting compounds [25]. Unlike their protein-targeting counterparts, RNA ligands frequently feature elongated, nitrogen-rich structures and consist of heterocyclic rings and aromatics [26]. Efforts to expand RNA-specific compound libraries are currently underway, leveraging empirical research alongside advanced machine learning and statistical analysis techniques. Additionally, virtual libraries are being developed to further enhance the discovery of RNA-targeting compounds. Here, we highlight some strategies to expand RNA-specific compound libraries. (a) Empirical Research: Researchers are exploring natural products and their derivatives for novel RNA-binding compounds. Natural products often possess unique chemical scaffolds that can specifically interact with RNA structures. Existing compounds are being chemically modified to improve their specificity and affinity for RNA targets. This involves systematic alterations to identify structural features that enhance RNA binding. (b) Machine Learning and Statistical Analysis: Machine learning algorithms are used to analyze large datasets of RNA-binding compounds, identifying patterns and features that contribute to effective RNA interactions. These models can predict new compounds likely to bind RNA with high specificity. Statistical methods integrate data from various sources, including experimental screenings and structural databases, to refine the selection of potential RNA-targeting molecules. (c) Development of Virtual Libraries: Virtual libraries of compounds are generated and screened using computational techniques. These virtual screenings can rapidly assess the potential of thousands of compounds to interact with RNA targets, prioritizing candidates for further experimental validation. Virtual libraries are designed based on the structural analysis of RNA targets. By focusing on compounds that fit specific RNA secondary and tertiary structures, these libraries enhance the likelihood of identifying effective RNA modulators [25].

### *16.5.4 Commonalities with Protein Target Screening*

Notwithstanding the outlined distinctions, screening for RNA and protein ligands exhibits common ground. The thermodynamics governing the interaction with RNA and protein targets shares similarities; the binding of highly specific compounds to either target type is primarily enthalpy-driven, while the association of more general binders, such as aminoglycosides, tends to be entropy-led [27]. Furthermore, the principles steering the refinement of hits into lead compounds, including pharmacokinetic attributes and adherence to established guidelines like "Lipinski's rule of five," are analogous for both RNA and protein ligands.

**Box 16.1 Targeting RNA "Life Cycle" [9] and Not RNA itself**
Martin et al. highlighted the phrase of "targeting RNA life cycle" and not targeting RNA itself might prove to be the way forward [9]. Creating binders that specifically target RNA is intricate and may not directly result in the anticipated biological outcomes. As a result, strategies targeting various stages of the RNA life cycle—including biogenesis, maturation, localization,

(continued)

**Box 16.1** (continued)
functioning, and degradation—are being explored as complementary approaches to the development of direct RNA binders. These strategies offer additional avenues for therapeutic intervention by modulating RNA-related processes rather than directly binding to RNA molecules. These methods offer a way to influence cellular mechanisms associated with RNAs or proteins considered "undruggable" by targeting their respective mRNAs.

***Targeting RNA Biogenesis***

FDA-approved drugs have targeted components of the spliceosome. Compounds that interact with RNA G-quadruplexes (G4) or RNA/DNA hybrids (R-loops) have been identified through a range of screening methods, including mechanism-based, affinity-based, and computational strategies [28–30]. Moreover, drugs targeting RNA/RNP can have secondary effects on transcription. For example, inhibitors of the Rev-RRE interaction primarily affect the localization and processing of HIV-1 RNA but also impact transcription. Inhibitors of the Rev-RRE interaction are designed to disrupt the binding between the Rev. protein and the RRE in HIV-1 RNA. This interaction is crucial for the proper export of viral RNA from the nucleus to the cytoplasm, where it can be translated and packaged into new virus particles. By disrupting RNA localization and processing, these inhibitors can also affect the transcription of HIV-1. The altered localization may impact the feedback mechanisms that regulate viral RNA synthesis, leading to changes in transcriptional activity. This dual effect enhances the therapeutic potential of these inhibitors by targeting multiple stages of the viral life cycle. Additionally, the activity of CRISPR/Cas9 RNP can be modulated to regulate transcription, either by inhibition or enhancement. Catalytically inactive Cas9 (dCas9) can be fused with transcriptional repressors to inhibit gene expression. This approach can silence target genes by blocking the binding of transcription factors or recruiting repressive chromatin modifiers. Conversely, dCas9 can be fused with transcriptional activators to enhance gene expression. This method can upregulate target genes by recruiting transcriptional activators or modifying chromatin to a more open, transcriptionally active state [31, 32].

***Targeting RNA Maturation***

Many modifications in cellular RNAs have been linked to diseases, manipulating the RNA epigenome presents significant therapeutic opportunities. These modifications, collectively referred to as the "epitranscriptome," play crucial roles in regulating RNA stability, translation, localization, and splicing, making them attractive targets for therapeutic intervention [33, 34]. Current screening efforts are concentrating on two main strategies: targeting RNA-modifying enzymes to block natural RNA modifications and employing artificial RNPs such as CRISPR/Cas13a to introduce novel RNA

(continued)

**Box 16.1** (continued)
modifications. These approaches are pivotal in manipulating the RNA epigenome for therapeutic purposes. The dual approach of targeting RNA-modifying enzymes and utilizing artificial RNPs like CRISPR/Cas13a represents a powerful strategy for manipulating the RNA epigenome [35, 36].

***Targeting RNA Localization and Activity***

The organization of RNA into nuclear bodies and its export to the cytoplasm are crucial for maintaining chromatin stability and promoting efficient protein synthesis, respectively. Recent screening efforts have focused on identifying the machinery responsible for nuclear retention or export of RNA. These efforts have uncovered compounds that disrupt various RNA functional interactions, including those involving viral RNPs, riboswitches, and IRES motifs. Additionally, compounds that bind to G-quadruplex (G4) RNA structures can alter the folding of UTRs in mRNA, thereby affecting ribosome binding and translation [15].

***Targeting RNA Degradation***

Regulating RNA degradation is crucial for the precise control of mRNA levels. In this context, screenings have identified ribonuclease catalytic mimics designed to specifically degrade RNAs. For example, a mimic of ribonuclease A (RNase A) targeting RNA triplet repeats effectively degrades specific mRNAs. Precise control of mRNA degradation is essential for maintaining appropriate levels of gene expression. By regulating the stability and degradation of mRNAs, cells can quickly adapt to changes in their environment and modulate protein production accordingly [37]. Compounds derived from Inforna, when attached to synthetic RNase L mimics, have shown improved inhibition of miRNAs [38, 39]. Furthermore, the glycopeptide bleomycin A5, which catalytically degrades RNA, can enhance the effectiveness of RNA inhibitors when conjugated to them [40, 41]. RNase L mimics operate exclusively in the cytoplasm, where RNase L is found, whereas bleomycin A5 conjugates are active in both the nucleus and cytoplasm [42].

## 16.6 In Silico Screens Identify New Druggable RNA Motifs

Despite advancements in identifying compounds that target RNA/RNP through mechanistic or affinity screenings, the development of RNA/RNP-targeted drugs still encounters significant challenges. These include a limited number of high-resolution 3D RNA/RNP structures, which hinders the identification of potential druggable sites, and the intricate conformational dynamics of RNA, complicating the application of rational structure-based approaches during the optimization of compounds [43]. In this scenario, the emerging field of RNA computational biology offers valuable insights for guiding the design of drugs targeting RNA, particularly when combined with experimental methods [44, 45]. Molecular dynamics simulations, supported by data from cyroEM, NMR, and X-ray crystallography, enable the modeling of RNA's conformational flexibility. This approach enhances the probability of identifying potential binders through virtual screening efforts [43]. This approach has been recently applied to target the stem-loop II motif within the genomic RNA of SARS-CoV-2 [46]. Moreover, molecular dynamics and computational docking studies enhance the process of prioritizing and refining hit compounds by forecasting their potential binding sites and modes of interaction with the RNA target.

Additionally, computational analyses, validated through biochemical experiments, enable the assembly of virtual compound libraries. Notably, these libraries may include molecules that have been shown to act on natural RNAs/RNPs or those identified through "reverse" screening on synthetic RNA targets. For instance, the 2D combinatorial screening (2DCS) microarray platform enables the evaluation of interactions between immobilized compounds and various RNA sequences. The RNA molecules that bound to these compounds were then sequenced and cataloged in a database known as Inforna, which aligns the secondary structures of RNA targets of these compounds [47]. Inforna is particularly effective for RNAs with straightforward, well-defined secondary structures, such as miRNAs and RNA repeats [48, 49]. Consequently, Inforna serves as a bridge between virtual and experimental screening methods, offering initial leads for both novel compound development and the identification of new RNA targets amenable to drug development. Furthermore, Inforna-derived compounds targeting adjacent sites can sometimes be conjugated to produce more efficacious compounds. It is important to note that computational screenings heavily rely on the predicted structures of RNA, which may not always be accurate. Thus, integrating computational predictions with empirical structural determinations of RNA is crucial, especially for complex and dynamic RNA molecules.

## 16.7 In Silico Screens: Inforna-Derived Compounds Can be Assembled into High-Affinity Hits

In computational screenings, such as the case with Inforna, compounds can be synthesized into high-affinity agents with significant potential for therapeutic applications. An example of this involves the early stage of mature miRNA production where pri-miRNAs are cleaved by the Drosha enzyme. This process is critical, as mature miRNAs play a role in degrading specific mRNAs. Targeting the complexes formed between specific pri-miRNAs and the Drosha nuclease presents a promising strategy for treating certain cancers characterized by miRNA dysregulation [48]. Inforna's computational tools recently predicted that two benzimidazole-based compounds would bind to adjacent regions near the Drosha binding site on the cancer-related miRNA, miR-96 [48]. Experiments involving fluorescence-binding demonstrated that the combined application of these two compounds led to enhanced binding affinity. This increase in affinity was further supported by various assays, including orthogonal cell culture studies, in vitro Drosha processing experiments, and mouse models investigating cancer progression. These studies validated that the newly formed bivalent compound effectively inhibited the cleavage of miR-96 by Drosha, with a $K_d$ of 85 ± 1 nM and a $IC_{50}$ of approximately 50 nM.

**Box 16.2 Screening for RNA Interaction Disruptors**

Screening efforts have focused on identifying compounds that target the machinery involved in RNA nuclear retention or export. This includes proteins and complexes that mediate RNA transport between the nucleus and cytoplasm. These screenings have identified compounds that disrupt the functional interactions of RNA, leading to altered localization and processing [50, 51].

***Disrupting Viral RNPs*:** Compounds that disrupt viral RNA–protein interactions can inhibit viral replication [52]. For example, targeting the interactions between viral RNA and nucleocapsid proteins can prevent the assembly of viral RNPs, hindering the virus's ability to replicate. These compounds offer potential therapeutic strategies for treating viral infections by specifically targeting viral RNA interactions without affecting host cell RNA.

***Modulating Riboswitches and IRES Motifs***: Riboswitches are regulatory RNA elements that change conformation upon binding to specific metabolites, thereby controlling gene expression [53]. Compounds that bind to riboswitches can modulate their activity and influence gene expression. IRES motifs in viral and cellular mRNAs allow for translation initiation independent of the 5′ cap structure. Compounds that disrupt IRES function can inhibit translation initiation, providing a means to control protein synthesis.

***Targeting G-Quadruplex (G4) RNA Structures***: G4 structures are four-stranded RNA configurations formed by guanine-rich sequences [54]. These structures can influence RNA stability and translation by altering the folding of UTRs in mRNA. Compounds that bind to G4 RNA structures can affect the folding of mRNA, thereby influencing ribosome binding and translation efficiency. This modulation can regulate protein synthesis at the translational level.

## 16.8 Conclusion

Recently, high-throughput screening methods specifically designed for RNA targets have discovered new compounds that are now approved for clinical use. This development significantly broadens the scope of RNA-directed pharmaceuticals beyond traditional aminoglycoside antibiotics and expands the range of targetable RNAs beyond ribosomal units. Some strategies have focused on directly binding to RNAs, while others have aimed at intervening in the "life cycle" of functional RNAs, including their biogenesis, maturation, localization, or degradation processes. The driving forces behind these breakthroughs in RNA-targeted drug discovery have been technological advancements and the detailed molecular and functional elucidation of new RNAs, trends expected to continue progressing. Looking forward, computational innovations driven by machine learning and artificial intelligence are poised to make significant impacts. These advancements include the development of methodologies for the experimental determination of RNA/RNP 3D structures, bioinformatics tools for novel RNA structure predictions, algorithms dedicated to RNA modeling and dynamic simulations, and techniques for structure-based RNA-ligand docking, all of which raise pressing questions for the field.

Currently, no single screening methodology comprehensively addresses all aspects of RNA-targeted drug discovery. The choice of the most suitable assay depends on various factors, including the target RNA's action mechanism, size, and structure, as well as the available resources and infrastructure for conducting high-throughput screenings and subsequent validation tests. Successful examples from recent explorations include the identification of spliceosome modulators Risdiplam [55] and Branaplam, the group II intron inhibitor intronistat B, the ribosomal inhibitor 13, and the miR-96-targeting compound targaprimir-96 [56]. These successes result from targeting well-defined disease-related systems with appropriate phenotypic or mechanistic assays and focusing on RNA targets supported by substantial biochemical, biophysical, and structural insights. Therefore, the future of RNA drug development is likely to benefit from integrating novel high-throughput screening technologies and an expanded chemical landscape, alongside more sophisticated computational and experimental analyses of RNA targets. Additionally, the emergence of the next wave of RNA-directed compound libraries promises to further enhance this field. The insights gained from these endeavors are expected to guide future screening efforts and improve the efficacy of RNA-focused therapeutic strategies.

## References


1. Dammes N, Peer D. Paving the road for RNA therapeutics. Trends Pharmacol Sci. 2020;41(10):755–75.
2. Disney MD. Targeting RNA with small molecules to capture opportunities at the intersection of chemistry, biology, and medicine. J Am Chem Soc. 2019;141(17):6776–90.
3. Kelly ML, Chu CC, Shi H, Ganser LR, Bogerd HP, Huynh K, et al. Understanding the characteristics of nonspecific binding of drug-like compounds to canonical stem-loop RNAs and their implications for functional cellular assays. RNA. 2021;27(1):12–26.
4. Yamamoto H, Unbehaun A, Spahn CMT. Ribosomal chamber music: toward an understanding of IRES mechanisms. Trends Biochem Sci. 2017;42(8):655–68.
5. Breaker RR. Riboswitches and translation control. Cold Spring Harb Perspect Biol. 2018;10:11.
6. Uroda T, Anastasakou E, Rossi A, Teulon JM, Pellequer JL, Annibale P, et al. Conserved pseudoknots in lncRNA MEG3 are essential for stimulation of the p53 pathway. Mol Cell. 2019;75(5):982–95.e9
7. Warner KD, Hajdin CE, Weeks KM. Principles for targeting RNA with drug-like small molecules. Nat Rev Drug Discov. 2018;17(8):547–58.

8. Saw PE, Song E. Advancements in clinical RNA therapeutics: present developments and prospective outlooks. Cell Rep Med. 2024;5(5):101555.
9. Martin WJ, Grandi P, Marcia M. Screening strategies for identifying RNA- and ribonucleoprotein-targeted compounds. Trends Pharmacol Sci. 2021;42(9):758–71.
10. Hentze MW, Castello A, Schwarzl T, Preiss T. A brave new world of RNA-binding proteins. Nat Rev Mol Cell Biol. 2018;19(5):327–41.
11. Julio AR, Backus KM. New approaches to target RNA binding proteins. Curr Opin Chem Biol. 2021;62:13–23.
12. Wu P. Inhibition of RNA-binding proteins with small molecules. Nat Rev Chem. 2020;4(9):441–58.
13. Fedorova O, Jagdmann GE Jr, Adams RL, Yuan L, Van Zandt MC, Pyle AM. Small molecules that target group II introns are potent antifungal agents. Nat Chem Biol. 2018;14(12):1073–8.
14. Chillón I, Marcia M. The molecular structure of long non-coding RNAs: emerging patterns and functional implications. Crit Rev Biochem Mol Biol. 2020;55(6):662–90.
15. Donlic A, Hargrove AE. Targeting RNA in mammalian systems with small molecules. Wiley Interdiscip Rev RNA. 2018;9(4):e1477.
16. Green KM, Sheth UJ, Flores BN, Wright SE, Sutter AB, Kearse MG, et al. High-throughput screening yields several small-molecule inhibitors of repeat-associated non-AUG translation. J Biol Chem. 2019;294(49):18624–38.
17. Stohler P, Dawson RJP. Identifying conformation-selective heavy-chain-only antibodies against membrane proteins by a thermal-shift scintillation proximity assay. Methods Mol Biol. 2020;2127:185–90.
18. Yasgar A, Jadhav A, Simeonov A, Coussens NP. AlphaScreen-based assays: ultra-high-throughput screening for small-molecule inhibitors of challenging enzymes and protein-protein interactions. Methods Mol Biol. 2016;1439:77–98.
19. Katsuya K, Hori Y, Oikawa D, Yamamoto T, Umetani K, Urashima T, et al. High-throughput screening for linear ubiquitin chain assembly complex (LUBAC) selective inhibitors using homogenous time-resolved fluorescence (HTRF)-based assay system. SLAS Discov. 2018;23(10):1018–29.
20. Lorenz DA, Vander Roest S, Larsen MJ, Garner AL. Development and implementation of an HTS-compatible assay for the discovery of selective small-molecule ligands for pre-microRNAs. SLAS Discov. 2018;23(1):47–54.
21. Bell NM, L'Hernault A, Murat P, Richards JE, Lever AM, Balasubramanian S. Targeting RNA-protein interactions within the human immunodeficiency virus type 1 lifecycle. Biochemistry. 2013;52(51):9269–74.
22. Uttamchandani M, Yao SQ. The expanding world of small molecule microarrays. Methods Mol Biol. 2010;669:1–15.
23. Velagapudi SP, Costales MG, Vummidi BR, Nakai Y, Angelbello AJ, Tran T, et al. Approved anti-cancer drugs target oncogenic non-coding RNAs. Cell Chem Biol. 2018;25(9):1086–94.e7.
24. Chillón I, Marcia M, Legiewicz M, Liu F, Somarowthu S, Pyle AM. Native purification and analysis of long RNAs. Methods Enzymol. 2015;558:3–37.
25. Rizvi NF, Santa Maria JP Jr, Nahvi A, Klappenbach J, Klein DJ, Curran PJ, et al. Targeting RNA with small molecules: identification of selective, RNA-binding small molecules occupying drug-like chemical space. SLAS Discov. 2020;25(4):384–96.
26. Morgan BS, Forte JE, Culver RN, Zhang Y, Hargrove AE. Discovery of key physicochemical, structural, and spatial properties of RNA-targeted bioactive ligands. Angew Chem Int Ed Engl. 2017;56(43):13498–502.
27. Umuhire Juru A, Patwardhan NN, Hargrove AE. Understanding the contributions of conformational changes, thermodynamics, and kinetics of RNA-small molecule interactions. ACS Chem Biol. 2019;14(5):824–38.
28. Rocca R, Moraca F, Costa G, Nadai M, Scalabrin M, Talarico C, et al. Identification of G-quadruplex DNA/RNA binders: structure-based virtual screening and biophysical characterization. Biochim Biophys Acta Gen Subj. 2017;1861(5 Pt B):1329–40.

29. Lang X, Green MD, Wang W, Yu J, Choi JE, Jiang L, et al. Radiotherapy and immunotherapy promote tumoral lipid oxidation and ferroptosis via synergistic repression of SLC7A11. Cancer Discov. 2019;9(12):1673–85.
30. Miglietta G, Russo M, Capranico G. G-quadruplex-R-loop interactions and the mechanism of anticancer G-quadruplex binders. Nucleic Acids Res. 2020;48(21):11942–57.
31. Maji B, Gangopadhyay SA, Lee M, Shi M, Wu P, Heler R, et al. A high-throughput platform to identify small-molecule inhibitors of CRISPR-Cas9. Cell. 2019;177(4):1067–79.e19.
32. Qiu M, Li P. CRISPR/Cas-based diagnostics and gene therapy. J BIOI. 2021;2(3):121–9.
33. Yanas A, Liu KF. RNA modifications and the link to human disease. Methods Enzymol. 2019;626:133–46.
34. Huang H, Weng H, Deng X, Chen J. RNA modifications in cancer: functions, mechanisms, and therapeutic implications. Annu Rev Cancer Biol. 2020;4:221–40.
35. Abudayyeh OO, Gootenberg JS, Essletzbichler P, Han S, Joung J, Belanto JJ, et al. RNA targeting with CRISPR–Cas13. Nature. 2017;550(7675):280–4.
36. Cully M. Chemical inhibitors make their RNA epigenetic mark. Nat Rev Drug Discov. 2019;18(12):892–4.
37. Nguyen L, Luu LM, Peng S, Serrano JF, Chan HY, Zimmerman SC. Rationally designed small molecules that target both the DNA and RNA causing myotonic dystrophy type 1. J Am Chem Soc. 2015;137(44):14180–9.
38. Costales MG, Aikawa H, Li Y, Childs-Disney JL, Abegg D, Hoch DG, et al. Small-molecule targeted recruitment of a nuclease to cleave an oncogenic RNA in a mouse model of metastatic cancer. Proc Natl Acad Sci USA. 2020;117(5):2406–11.
39. Costales MG, Matsumoto Y, Velagapudi SP, Disney MD. Small molecule targeted recruitment of a nuclease to RNA. J Am Chem Soc. 2018;140(22):6741–4.
40. Angelbello AJ, Rzuczek SG, McKee KK, Chen JL, Olafson H, Cameron MD, et al. Precise small-molecule cleavage of an r(CUG) repeat expansion in a myotonic dystrophy mouse model. Proc Natl Acad Sci USA. 2019;116(16):7799–804.
41. Li Y, Disney MD. Precise small molecule degradation of a noncoding RNA identifies cellular binding sites and modulates an oncogenic phenotype. ACS Chem Biol. 2018;13(11):3065–71.
42. Liu X, Haniff HS, Childs-Disney JL, Shuster A, Aikawa H, Adibekian A, et al. Targeted degradation of the oncogenic MicroRNA 17-92 cluster by structure-targeting ligands. J Am Chem Soc. 2020;142(15):6970–82.
43. Ganser LR, Lee J, Rangadurai A, Merriman DK, Kelly ML, Kansal AD, et al. High-performance virtual screening by targeting a high-resolution RNA dynamic ensemble. Nat Struct Mol Biol. 2018;25(5):425–34.
44. Manigrasso J, Marcia M, De Vivo M. Computer-aided design of RNA-targeted small molecules: a growing need in drug discovery. Chem. 2021;7(11):2965–88.
45. De Vivo M, Masetti M, Bottegoni G, Cavalli A. Role of molecular dynamics and related methods in drug discovery. J Med Chem. 2016;59(9):4035–61.
46. Aldhumani AH, Hossain MI, Fairchild EA, Boesger H, Marino EC, Myers M, et al. RNA sequence and ligand binding alter conformational profile of SARS-CoV-2 stem loop II motif. Biochem Biophys Res Commun. 2021;545:75–80.
47. Disney MD, Winkelsas AM, Velagapudi SP, Southern M, Fallahi M, Childs-Disney JL. Inforna 2.0: a platform for the sequence-based design of small molecules targeting structured RNAs. ACS Chem Biol. 2016;11(6):1720–8.
48. Velagapudi SP, Cameron MD, Haga CL, Rosenberg LH, Lafitte M, Duckett DR, et al. Design of a small molecule against an oncogenic noncoding RNA. Proc Natl Acad Sci USA. 2016;113(21):5898–903.
49. Parkesh R, Childs-Disney JL, Nakamori M, Kumar A, Wang E, Wang T, et al. Design of a bioactive small molecule that targets the myotonic dystrophy type 1 RNA via an RNA motif-ligand database and chemical similarity searching. J Am Chem Soc. 2012;134(10):4731–42.
50. Zhu S, Rooney S, Michlewski G. RNA-targeted therapies and high-throughput screening methods. Int J Mol Sci. 2020;21:8.

51. Ursu A, Vézina-Dawod S, Disney MD. Methods to identify and optimize small molecules interacting with RNA (SMIRNAs). Drug Discov Today. 2019;24(10):2002–16.
52. Mahajan S, Choudhary S, Kumar P, Tomar S. Antiviral strategies targeting host factors and mechanisms obliging +ssRNA viral pathogens. Bioorg Med Chem. 2021;46:116356.
53. Mandal M, Breaker RR. Gene regulation by riboswitches. Nat Rev Mol Cell Biol. 2004;5(6):451–63.
54. Fay MM, Lyons SM, Ivanov P. RNA G-Quadruplexes in biology: principles and molecular mechanisms. J Mol Biol. 2017;429(14):2127–47.
55. Ratni H, Scalco RS, Stephan AH. Risdiplam, the first approved small molecule splicing modifier drug as a blueprint for future transformative medicines. ACS Med Chem Lett. 2021;12(6):874–7.
56. Singh RN, Ottesen EW, Singh NN. The first orally deliverable small molecule for the treatment of spinal muscular atrophy. Neurosci Insights. 2020;15:2633105520973985.

# Chapter 17
# Single-Cell RNA Sequencing (sc-RNAseq) in the Development of Biomedical Therapeutics

**Abstract** Single-cell technologies, especially single-cell RNA sequencing (scRNA-seq), along with the accompanying computational tools and an expanding repository of public data, are revolutionizing the process of drug discovery and development. These advancements are opening new avenues for target identification, thanks to enhanced understanding of diseases via cell subtyping. Additionally, the integration of scRNA-seq into highly multiplexed functional genomics screens is improving the processes of target credentialing and prioritization. scRNA-seq plays a crucial role in selecting appropriate preclinical disease models and offers fresh perspectives on the mechanisms of action of drugs. scRNA-seq contributes to decision-making processes by facilitating better biomarker identification for patient stratification, offering more accurate means to track drug responses and disease progression. This chapter sheds light on how scRNA-seq techniques are being implemented across various stages of drug discovery and development while also addressing the persistent challenges faced by the pharmaceutical industry in adopting these technologies.

## 17.1 Introduction

The drug discovery process is often marked by inefficiency, characterized by escalating costs, protracted development periods, and a high failure rate. These challenges stem, in part, from a constrained understanding of human biology, including the intricacies of disease mechanisms, the identification of viable therapeutic targets, and the variability in disease responses. The last decade has seen significant advancements in SC technologies, which have begun to address these limitations and offer complementary capabilities to traditional bulk analysis methods still in use. The development of SC technologies now extends to the analysis of DNA, epigenetic markers, proteins, and other cellular components, broadening the scope of possible insights into cellular function and disease mechanisms. ScRNA-seq has facilitated the discovery of molecular pathways, enabling the prediction of patient survival outcomes, therapeutic responses, potential resistance development, and

 
P. E. Saw, E. Song, *RNA Therapeutics in Human Diseases*,
https://doi.org/10.1007/978-981-96-3041-7_17

suitability for alternative treatments. In this chapter, we explore how SC technologies, with a focus on scRNA-seq methods, are being integrated into every stage of the drug development process, from identifying targets to making clinical decisions.

## 17.2 scRNA-seq: The Fundamentals

The scRNA-seq process encompasses three principal stages: (a) generating the library, (b) pre-processing, and (c) post-processing. Initially, the library generation phase involves isolating single cells or nuclei, capturing mRNA, and sequencing these genetic materials [1]. Following the acquisition of sequences, the workflow transitions to computational analyses. The pre-processing stage entails preliminary analysis for data counting and cleaning. During post-processing, the complexity of data is simplified through dimensionality reduction, enabling the identification of gene signatures and cell types and facilitating the creation of visual representations. Optional steps, such as data integration and batch correction, may enhance the analysis, supporting further inferential studies. These stages are often repeated in cycles to refine results.

## 17.3 Generation of RNA Library and Sequencing

This initial stage, sample preparation, is critical and typically necessitates the dissociation of tissue through mechanical or enzymatic means, tailored to the specific type of sample. This process inevitably results in the release of RNA into the surrounding suspension, potentially elevating background noise or interference if not adequately addressed during subsequent data processing. Utilizing fresh samples is crucial for obtaining high-quality scRNA-seq results, although single-nucleus RNA sequencing becomes the method of choice for samples that require freezing [2]. For RNA capture, samples are divided into individual reaction chambers where cell lysis and RNA capture occur. This is frequently achieved using technologies such as the 10x Chromium system, which employs a unique method of combining cells, barcoded primers within beads, lysis buffer, and reverse transcription enzymes with oil to create discrete microdroplet reaction chambers. In contrast, plate-based methods utilize microwells, and automated microfluidic devices leverage various microchamber designs. The essential requirement across these technologies is the isolation of individual cells within separate compartments to avoid cross-contamination. Following cell separation, RNA transcripts from each cell are labeled with a barcoded unique molecular identifier (UMI). UMIs are crucial for distinguishing genuine cell transcripts from unwanted PCR duplicates generated during sample processing. The next steps involve creating and amplifying a cDNA library through reverse transcription. The specific tagging and amplification approach may necessitate several rounds of amplification, and adapter sequences for flow cell attachment might be added. This cDNA is then processed in a manner similar to bulk RNA-seq protocols, which includes fragmenting the cDNA to achieve a uniform pool of molecules and appending index sequences. These indices are vital for tracking the origin of sequencing reads, enabling the simultaneous sequencing of multiple samples.

## 17.4 Pre-processing of Sequence Data

For plate-based technologies like SMARTseq2, the analysis of reads can proceed through conventional bulk genome or transcriptome alignment and quantification workflows. Conversely, droplet-based platforms necessitate specialized tools to accurately allocate UMI counts to cell barcodes due to their high degree of cell multiplexing. It is typical for RNA capture rates to range from 10% to 20% across methodologies, a factor that must be considered during data analysis. The Cell Ranger pipeline, developed by 10x Genomics, is a popular choice for processing data from 10x platforms. It employs the STAR algorithm for RNA-seq alignment, in addition to providing functionalities for cell counting and quality control reporting [1]. Following alignment, the primary objective is to compile counts for each gene across individual cells, forming a cell-by-gene matrix. In the case of droplet-based technologies, an initial step often involves filtering to separate genuine cells from empty droplets. Additional steps may include the filtering of ambient RNA and strategies to eliminate doublets, enhancing data cleanliness and reducing the volume of data to manage. Subsequent to these preparatory measures, the matrix undergoes normalization to address variations in RNA capture rates among cells. This normalization process is crucial for accurate downstream analysis. The identification and flagging of highly variable genes within the sample is a critical step that facilitates the focusing of further analyses on these genes, owing to their potential significance in understanding cellular heterogeneity and dynamics [3].

## 17.5 Post-Processing of Sequence Data

Following the creation and normalization of the cell-by-gene matrix, typical scRNA-seq analyses proceed with unsupervised clustering to categorize cells into groups based on similar expression patterns. This is accompanied by dimensionality reduction techniques, such as t-distributed stochastic neighbor embedding (t-SNE) or uniform manifold approximation and projection (UMAP), which facilitate the visualization of cellular clusters within two or three-dimensional spaces [4]. These visual representations help elucidate the complex relationships and diversity within cell populations. Marker genes, indicative of specific cellular identities or states, are identified through differential expression analysis among the clusters. Additionally, cell-type annotation is performed to label clusters with known cell identities. To further refine and contextualize the data, integrative analyses may be employed to correct for batch effects that could skew interpretation. Such corrections are essential when comparing datasets across different experiments or conditions. Trajectory mapping, another advanced analysis, traces the paths of cell differentiation and development, offering insights into real-time cellular dynamics or in response to external stimuli. Given the iterative nature of data analysis in scRNA-seq, these downstream processes often require repetition and adjustment to enhance the clarity and accuracy of the insights gained. This iterative refinement is a critical component of the workflow, ensuring that the analyses accurately reflect the underlying biological reality and contribute to a comprehensive understanding of the studied system (Fig. 17.1).

## 17.6 Applications of sc-RNAseq in Drug Discovery

SC technologies offer transformative potential across the entire spectrum of drug discovery and development. Through enhanced disease understanding, which includes the identification of altered cellular compositions and states, these technologies can facilitate the discovery of novel cellular and molecular targets. This advanced subtyping capability enables the pinpointing of specific disease subtypes, guiding the identification and credentialing of new targets. The integration of SC sequencing with high-throughput functional genomics screens, such as those combining CRISPR technology with SC sequencing (scCRISPR screening), can significantly increase the efficiency of target credentialing [5]. This approach not only enhances throughput but also enriches perturbation data with mechanistic insights, aiding in the prioritization of targets. SC sequencing can shed light on the cell-type-specific actions of compounds, their off-target effects, and the variability in drug responses, thereby informing the selection of drug candidates. In clinical development, SC technologies are instrumental in identifying biomarkers for patient stratification, understanding the mechanisms behind a drug's effect or resistance, and tracking the progression of drug responses and disease states. Additionally, these technologies hold promise for characterizing and refining engineered biologics and cell therapies, marking a significant step forward in personalized medicine and treatment modalities. The application of SC technologies, especially scRNA-seq, is highlighted in several published studies, demonstrating their critical role in various phases of drug discovery and development. These studies showcase the practical utility and impact of SC technologies in the pharmaceutical industry, from early target discovery to clinical trials and beyond [1].

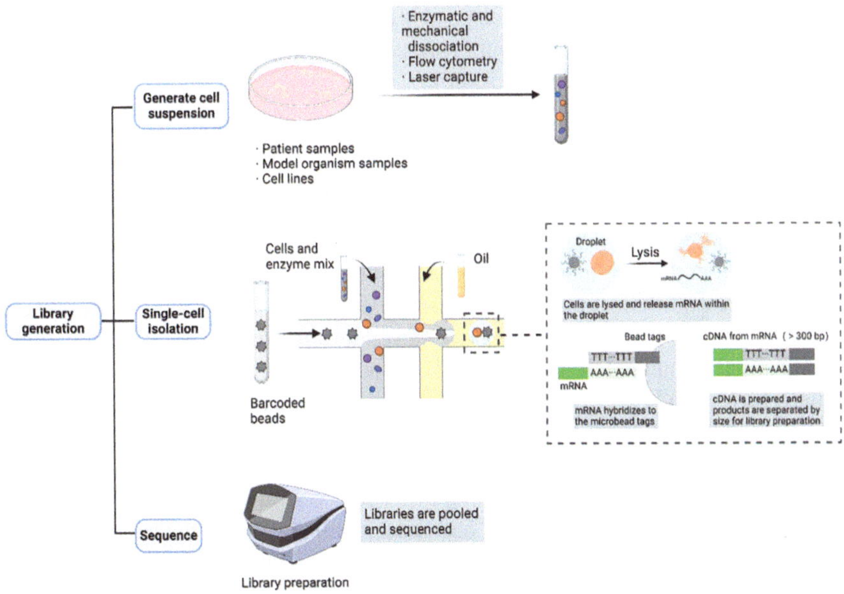


**Fig. 17.1** This figure illustrates the detailed workflow of RNA sequencing, highlighting the processes from generating cell suspensions to sequencing libraries. (i) Generation of Cell Suspension: Cells are obtained from patient samples, model organism samples, or cell lines. The cells are dissociated into a single-cell suspension using enzymatic and mechanical dissociation, flow cytometry, or laser capture. (ii) Single-Cell Isolation: The single-cell suspension is mixed with barcoded beads and enzymes. The cell-enzyme mixture is encapsulated into droplets using an oil phase. Each droplet contains a single cell and a barcoded bead. Cells within the droplets are lysed, releasing mRNA. (iii) Library Generation: Released mRNA hybridizes to barcoded bead tags within the droplet. mRNA is reverse transcribed to complementary DNA (cDNA). cDNA fragments longer than 300 base pairs are selected for library preparation. The cDNA is prepared and products are separated by size, pooled into libraries for sequencing. (iv) Sequencing: Prepared libraries are sequenced using high-throughput sequencing technologies. This workflow provides a comprehensive overview of the steps involved in RNA sequencing, from cell suspension generation to sequencing, enabling detailed analysis of gene expression at the single-cell level. Figure is drawn in Biorender.com

## 17.7 Emerging Single-Cell Technologies

***Technologies for CRISPR Screening*** The method of pooled CRISPR screening, known for its efficiency and scalability in the identification of drug targets, faces limitations due to its dependency on low-content readouts and its ability to only pinpoint genes associated with noticeable phenotypic changes. Addressing these constraints, advancements in single-cell CRISPR screening technologies like Perturb-seq [6–9] integrate pooled CRISPR screening with scRNA-seq or single-cell multi-omics analyses. Various computational tools such as MIMOSCA [6], scMAGeCK [10], MUSIC [11], and Mixscape [9], alongside a novel screening platform [12] have been developed. These resources facilitate the analysis of how specific genetic alterations influence gene activity, their interplay, or dependency on the cellular state and assist in identifying cell types most vulnerable to CRISPR-induced changes at the single-cell level.

***T Cell Receptor (TCR) and B Cell Receptor (BCR) Sequencing*** Single-cell TCR (scTCR-seq) and BCR sequencing (scBCR-seq) are pivotal in exploring the dynamics of T and B cell clones within tissues or circulating in peripheral blood by identifying clonotypes at a single-cell resolution. Clonotypes are cells of the adaptive immune system that originate from the same progenitor and share identical TCR or BCR sequences. Moreover, the reconstruction of TCR and BCR repertoires and the determination of clonality can also be achieved through scRNA-seq with the aid of computational methodologies [13–17]. Tools like scRepertoire [18] and CellaRepertorium [19] are employed to analyze clonotype dynamics. By integrating scTCR-seq or scBCR-seq with scRNA-seq, researchers can uncover the link between clonotype identity and the phenotypic or transcriptional states of T or B cells [20]. This comprehensive profiling of T and B cells using single-cell technologies has significantly contributed to our understanding of various diseases, including the intricacies of the cancer microenvironment and the identification of multiple sclerosis antigens.

***Single-Cell Epigenetics*** A myriad of single-cell technologies now allow for the detailed mapping of epigenetic features at an almost nucleotide-level precision. Techniques such as the single-cell sequencing assay for transposase-accessible chromatin (scATAC-seq) [21] unveil the open chromatin regions, while single-cell chromatin utility via transposase-activated tagmentation (scCUT&Tag) [22] and single-cell chromatin immunoprecipitation sequencing (scChIP-seq) [23] provide insights into chromatin histone modifications. Furthermore, single-cell bisulfite sequencing (scBS-seq) [24] elucidates DNA methylation patterns. Identifying activated promoters and enhancers specific to certain cell types or states aids in pinpointing the tissues, cell types, or biological conditions under which a gene is predominantly expressed, along with the transcriptional mechanisms driving such expression. These epigenetic mapping techniques also play a crucial role in linking non-coding genetic variants associated with diseases, as identified through genome-wide association studies (GWAS), to specific cell types.

***Single-Cell Proteomics*** The field of single-cell proteomics is emerging, with new methods enabling the exploration of proteome variations across individual cells [25]. Approaches in single-cell proteomics (sc-proteomics [26, 27]) vary from absolute quantification of select proteins to highly multiplexed protein analyses. A groundbreaking technique now allows for the detection of single proteins within cells, sensitive enough to discern single amino acid changes in peptides [28].

***Single-Cell Multi-Omics*** Technologies such as ECCITE-seq [29], scNMT-seq [30], and DOGMA-seq [31] facilitate the concurrent measurement of diverse biological markers (e.g., RNA expression, surface protein levels, clonotypes, DNA methylation, and chromatin accessibility) within the same single cells, marking significant advancements in the field.

***Advancements in Single-Cell Methods*** Innovations extend to the realms of single-cell miRNA [32] and lncRNA profiling [33], broadening the scope of RNA transcriptomics. Single-cell metabolomics (sc-metabolomics) approaches have been introduced for detailing the chemical compositions at the level of individual cells or organelles [34]. Additionally, scRibo-seq offers a glimpse into translation dynamics at a single-cell scale, achieving single codon precision when integrated with machine learning techniques [35]. Technologies such as scSPRITE [36] and Higashi [37] facilitate the investigation of the 3D genome organization within single cells.

***Spatially Resolved Omics*** Traditional single-cell technologies often overlook spatial information due to tissue dissociation processes. Recently developed spatially resolved omics techniques [38–40] aim to retain this crucial spatial context, with comprehensive reviews available on spatial transcriptomics and its computational methodologies [41–44]. Pre-processing of single-cell sequence data is essential for extracting meaningful insights, with a focus on refining methods in this rapidly evolving area of research.

***Single-Cell Multi-Omics Analysis*** The simultaneous analysis of single-cell multi-omics data stands as a powerful approach to intricately define cell types, states, and their links to disease progression and response to therapies [27]. Advanced analytical frameworks like the weighted nearest neighbor (WNN) method in Seurat [45], CiteFuse [28], MOFA+ [31], and totalVI [31–34, 45] have been developed to enhance the resolution of cell states and destinies through the integration of various single-cell data types. When dealing with multimodal datasets derived from different cells, computational approaches such as LIGER [33, 35] and canonical correlation analysis (CCA) in Seurat [34] facilitate the projection of these diverse datasets into a shared latent space. This methodological advance enables the joint analysis of variations across different sample groups and data modalities, enriching the understanding of complex biological phenomena.

***Annotation of Cell Types in scRNA-seq Data*** The annotation of cell types within scRNA-seq datasets often relies on the unsupervised clustering of cells followed by the identification of cluster-specific marker genes, a process that is notably time-consuming. To streamline cell-type annotation, several automated tools have been developed, such as Seurat label transfer [46], Garnett [47], scmap [48], SingleR [49], Cell-ID [50], and the more recent CellTypist [51]. With a dataset that is properly integrated, normalized, and annotated, a plethora of methods can be employed to extract insights.

***Trajectory Inference and Pseudo-Time Analysis*** Cells undergo dynamic processes like differentiation, treatment responses, and disease progression. A cell sample represents a snapshot of cells at different stages of these processes. Trajectory inference (TI) is utilized to map out these dynamic pathways. Key TI computational tools include Monocle [52], PAGA [53], Slingshot [54], STEMNET [55], and Scorpius [56], which generally necessitate a priori knowledge of the process topology and meticulous planning. Distinct from RNA velocity [57], which uses unspliced mRNA presence as an indicator of gene expression rate changes, methods such as Velocyto [57] and scVelo [58] build upon this concept. CellRank [59] merges TI with RNA velocity to predict cell fate decisions (Table 17.1).

**Box 17.1 Target Credentialing and Validation: Crucial Steps in scRNA-seq**

The process of target credentialing and validation involves building a robust case for a gene target by collating evidence from multiple fronts. This stage also critically assesses the translational validity of study models to bridge any discrepancies between model predictions and actual disease biology or therapeutic goals. scRNA-seq data play a crucial role in informing this process. Traditionally, target credentialing demands substantial resources, often leading to a compromise between the breadth of targets examined and the depth of analysis.

However, scCRISPR screening, especially when coupled with a genome-wide pooled screen, offers a solution to this dilemma by enabling the concurrent examination and profiling of numerous perturbations [6, 7, 12]. One notable application of this scCRISPR screening approach was the identification of key regulators of T cell stimulation and immunosuppression. This was achieved using a genome-wide pooled CRISPR screen, with subsequent Perturb-seq analysis on selected hits to delineate the affected gene programs, uncovering at least four promising anti-tumor targets [71]. This exemplifies how single-cell technologies, particularly in combination with CRISPR screening, can revolutionize the identification, credentialing, and validation of drug targets, streamlining the pathway from basic research to therapeutic development. The capabilities of CRISPR technology in target validation have significantly advanced with the integration of paired CRISPR activation (CRISPRa) and CRISPR interference (CRISPRi) screening alongside pooled scRNA-seq profiling.

**CRISPR Activation (CRISPRa)**

CRISPRa utilizes a dead or deactivated Cas9 (dCas9) protein fused with transcriptional activators. The system is designed to bind to specific promoter regions of target genes without cutting the DNA [8, 12]. The transcriptional activators then recruit the necessary cellular machinery to increase the transcription of the target gene. CRISPRa is primarily used to upregulate gene expression. This is particularly useful in research to determine the effects of increased levels of a gene's product on cellular function or phenotype. It can

(continued)

**Box 17.1** (continued)
also help in identifying the potential therapeutic effects of activating certain genes that might be underexpressed in diseases. Like other CRISPR systems, CRISPRa allows for highly specific targeting of genes due to the guide RNA that directs the dCas9-activator complex to precise genomic locations. It provides a controlled way to increase gene expression without permanently changing the genome, allowing reversible studies.

**CRISPR Interference (CRISPRi)**

CRISPRi also employs dCas9, but instead of a transcriptional activator, it is fused with a repressor domain or simply uses the dCas9 to physically block the transcriptional machinery access to the target gene's promoter [72]. This binding interferes with the transcription process, leading to reduced expression of the target gene. CRISPRi is used for gene knockdown experiments where reducing the expression of a gene is necessary to study its function or to model diseases associated with reduced gene function. It is especially useful in cases where complete knockout by CRISPR-Cas9 might result in lethality or complex phenotypes that are difficult to interpret. It offers a flexible approach to study gene function by partially reducing gene expression, which can be better than complete knockouts for essential genes [73]. Like CRISPRa, CRISPRi's effects are reversible, providing an advantage over permanent genetic modifications.

This expansion enhances the precision and depth with which potential drug targets can be investigated and validated. Furthermore, the application of Perturb-seq in vivo [74] opens new avenues for exploring gene functions across various cell types within their natural physiological environments, thus providing insights into the cellular heterogeneity of gene effects in disease contexts. The integration of these studies with comprehensive omics techniques offers a richer, more detailed view of the drug's mechanism of action (MoA) and potential toxicity, thus broadening the understanding of therapeutic effects and side effects [75–77].

## 17.8 Drug Screening and MoA Analysis

High-throughput screening (HTS) is a cornerstone in drug discovery, enabling the rapid evaluation of vast libraries of chemical compounds for their potential therapeutic effects. Traditionally, HTS has relied on either broad assessment of cell viability or proliferation or on more targeted measures such as the expression of specific markers. However, these approaches often fall short of capturing the full spectrum of cellular responses to drug treatment, largely because they treat cell populations as monolithic entities, disregarding the inherent cellular heterogeneity within samples. In contrast, SC transcriptomics revolutionizes how we can observe and interpret the effects of chemical compounds on cells. Unlike bulk RNA

**Table 17.1** Application of emerging scRNA-seq analysis in biology and medicine

| Application | Type of analysis | Methods and functions | Advantages | Ref |
|---|---|---|---|---|
| Analysis of biological pathways | Gene set enrichment analysis (GSVA) | GSEA works by ranking all genes in a dataset based on their differential expression between two phenotypes—Such as disease vs. healthy control—And then assessing whether the members of a predefined gene set (such as those linked to a specific metabolic pathway or a disease process) are randomly distributed throughout the ranked list or predominantly located at the top or bottom. GSEA works by ranking all genes in a dataset based on their differential expression between two phenotypes—Such as disease vs. healthy control—And then assessing whether the members of a predefined gene set (such as those linked to a specific metabolic pathway or a disease process) are randomly distributed throughout the ranked list or predominantly located at the top or bottom. This approach is non-parametric and does not require an arbitrary cutoff for differential expression, making it robust against noise and variability inherent in high-throughput data | The methodology of GSEA can be enhanced by integrating additional layers of biological data, such as protein–protein interaction networks or regulatory sequences, to provide more context to the enrichment analysis | [60] |
| | Single sample gene set enrichment analysis (ssGSEA) | Single sample gene set enrichment analysis (ssGSEA) is a variation of the traditional gene set enrichment analysis (GSEA) tailored to evaluate gene set enrichment at the level of individual samples. This adaptation is particularly useful in RNA-sequencing studies where the focus is on personalized medicine or when sample sizes are limited. Unlike traditional GSEA, which typically requires a comparison between groups of samples to identify differentially expressed gene sets, ssGSEA allows for the assessment of enrichment within each sample independently. This makes it an invaluable tool for constructing profiles of gene set activity for individual biological samples, facilitating more tailored analyses such as patient stratification in clinical trials or detailed cellular phenotype characterization | A key benefit of ssGSEA is its ability to capture the unique biological state of each sample, reflecting subtle variations in gene expression that might be diluted or overlooked in broader group analyses. This individualized approach is crucial in studies where heterogeneity within a population (such as tumors of the same type but from different patients) is high | [61] |

| | | | | |
|---|---|---|---|---|
| | Pagoda2 | Pagoda2 is an advanced computational framework designed specifically for the analysis of single-cell RNA-sequencing (scRNA-seq) data. Building on the foundations laid by its predecessor, this tool addresses several challenges associated with scRNA-seq data, including high levels of noise, sparse data points, and the intrinsic heterogeneity of cellular populations. The primary aim of Pagoda2 is to normalize and stabilize the variance in single-cell gene expression data, facilitating more accurate and reliable downstream analyses. It employs robust statistical methods to adjust for technical and biological noise, allowing researchers to capture true biological variation among cells more effectively. This is crucial for identifying distinct cellular subpopulations and understanding the complex molecular mechanisms underpinning cellular functions and interactions | One of the key features of Pagoda2 is its ability to perform a comprehensive analysis of cell heterogeneity. It does this by clustering cells based on their gene expression profiles, identifying characteristic patterns that signify different cell types or states within a sample. Moreover, Pagoda2 enhances the identification of regulatory networks and pathways. By analyzing over-dispersion and co-expression patterns among genes, it can suggest potential regulatory interactions and signaling pathways that might be active in certain cell types or conditions | [62] |

(continued)

**Table 17.1** (continued)

| Application | Type of analysis | Methods and functions | Advantages | Ref |
|---|---|---|---|---|
| Methods for cell-type deconvolution | Cell-type deconvolution techniques | The primary goal of cell-type deconvolution is to attribute the observed gene expression profiles to specific cell types within a heterogeneous sample without the need for physical separation of cells. This is particularly valuable in cases where sample acquisition is limited, such as in clinical settings where invasive procedures must be minimized. Deconvolution techniques enable researchers to infer the proportion and type of cells contributing to the overall gene expression, providing a clearer picture of the cellular dynamics within the sample. This approach utilizes reference gene expression profiles of known cell types to estimate the cell-type composition in a mixed sample. Tools like CIBERSORT or xCell use large reference panels of pure-cell RNA-seq data to match and quantify the proportions of each cell type in the bulk sample. These methods require accurate and comprehensive reference datasets to be effective | Cell-type deconvolution is extensively used in cancer research to understand the tumor microenvironment, including the interaction between tumor cells and immune cells. This can guide the development of targeted therapies and improve prognostic assessments. Understanding how cell-type compositions change during development can provide insights into developmental processes and congenital diseases. Identifying the cellular targets of drugs within complex tissues can enhance the effectiveness and reduce the side effects of therapeutic interventions | [63–68] |
| Mapping disease-associated variants | Genome-wide association studies (GWAS) | GWAS are typically conducted on samples from people with a particular disease and from healthy controls. Researchers use high-throughput genotyping technologies to examine hundreds of thousands to millions of genetic variants, such as SNPs, across the genome in a large group of individuals. The goal is to find SNPs that occur more frequently in people with the disease than in those without | Identifying genetic risk factors for diseases can help in understanding the biological pathways involved in disease progression, influencing drug target discovery and the development of precision medicine strategies | [69] |

| | | | | |
|---|---|---|---|---|
| | Single-cell disease relevance scoring (scDRS) | scDRS (single-cell disease relevance scoring) is a computational tool designed to connect human disease-associated genetic variants, identified through GWAS, with scRNA-seq data. This integration allows researchers to understand how genetic variants associated with diseases manifest at the level of individual cell types, potentially identifying which cell types are implicated in various diseases. The core functionality of scDRS is to map the results from GWAS to specific cell types within heterogeneous tissues. It achieves this by leveraging the fine-resolution data generated by scRNA-seq to identify the cell types most likely affected by genetic variants linked to diseases. This process helps in pinpointing cellular contexts in which disease-associated genes are active, providing insights into the cellular basis of genetic risk factors. | scDRS provides a more granular view of disease mechanisms at the cellular level, which is vital for complex diseases where multiple cell types and pathways might be involved. Understanding the cell-specific impacts of disease-associated genetic variants can lead to more precise therapeutic targets. Researchers can use scDRS to explore the cellular basis of complex traits and polygenic conditions, helping to decode the intricate biological networks that underpin these traits | [70] |

sequencing, which averages gene expression across all cells in a sample, SC transcriptomics delineates the responses of individual cell types, uncovering cell-type-specific changes in pathways, off-target effects, and dose–response relationships. This granularity allows researchers to disentangle complex biological responses and to pinpoint specific cellular behaviors that may be masked in bulk analyses. Recent advancements in HTS methodologies have embraced SC transcriptomics, merging the comprehensive screening capabilities of traditional HTS with the nuanced, high-resolution insights provided by SCRNA-seq. These novel HTS approaches enable the assessment of compounds across multiple doses and biological conditions simultaneously, making them exceptionally well-suited for elucidating drugs' mechanisms of action. By doing so, researchers can achieve a more comprehensive understanding of a compound's function, including its potential therapeutic value and unwanted side effects.

To overcome the cost barriers traditionally associated with SC transcriptomics and to enhance the throughput of chemical perturbation studies, researchers have developed multiplexing techniques. These innovations allow for the simultaneous profiling of hundreds of compounds, considering diverse doses, time points, and cell types, thereby enabling a detailed exploration of compound functions at an unprecedented scale. Techniques such as hashing, which involves pooling samples from different experimental conditions using unique molecular identifiers, further increase the efficiency and throughput of these analyses. One notable implementation of these advanced methodologies is MIX-seq, which leverages SNP-based demultiplexing to analyze the transcriptional impact of 13 drugs across up to 99 cell lines [78]. This exemplifies the power of integrating SC transcriptomics with HTS, offering a robust framework for discovering and evaluating new therapeutic agents with a level of detail and scale previously unattainable. This advanced approach, utilizing transient transfection with short oligo barcodes, demonstrates the expansive potential of multiplexing in drug discovery and cellular biology studies. By initially testing this method with cell samples from different species, and later applying it to various time exposures of a human chronic myelogenous leukemia cell line to imatinib, researchers have significantly enhanced our ability to understand drug perturbations at a granular level [79].

## 17.9 Understanding Diseases Through scRNA-seq

scRNA-seq illuminates the variations in cell-type composition and phenotypic shifts characteristic of disease states. Additionally, the comprehensive nature of scRNA-seq allows for the identification of rare cell types that may play pivotal roles in disease pathogenesis. Through detailed insights into disease mechanisms provided by single-cell technologies, new avenues for therapeutic intervention are being explored. While a comprehensive review exceeds the scope of this chapter, there are ample evidences of scRNA-seq utilization across various human diseases, underscoring the transformative potential of single-cell based approaches in advancing disease understanding and treatment.

## 17.10 Cancer

Single-cell molecular phenotyping has played a crucial role in unraveling the intricacies of cancer development. Utilization of single-cell technologies has facilitated the identification of cells implicated in the initiation of prostate cancer, the diverse cellular landscape of papillary renal cell carcinoma (pRCC), and the progression from Barrett's esophagus to esophageal adenocarcinoma [80–82]. Through scRNA-seq, a detailed mapping of cellular and transcriptional diversity within tumors has been achieved, shedding light on the heterogeneity of cancer cells. This approach has been coupled with immunophenotyping to delineate stromal-immune environments, or "ecotypes" [83]. This methodology has elucidated the complexity of various cancers, including diffuse large B cell lymphoma, breast cancer, esophageal squamous cell carcinoma, and papillary thyroid carcinoma, highlighting the relationship between tumor microenvironment and clinical outcomes [84–87].

Innovative techniques like Perturb-seq are pushing the boundaries further by linking genetic variants to phenotypic alterations across different diseases, including cancer. For instance, this technology has enabled high-throughput analysis of the effects of genetic variants in key oncogenes and tumor suppressor genes such as KRAS and TP53, offering insights into their roles in cancer pathology [88]. Beyond genetic diversity, cancer's transcriptional heterogeneity is often driven by epigenetic factors that allow malignant cells to adapt to their microenvironment and develop resistance to therapies. Multi-omics single-cell profiling has been instrumental in understanding these epigenetic underpinnings, particularly in glioma, where it has unveiled the epigenetic landscapes facilitating tumorigenesis [89, 90]. Longitudinal studies, especially those employing mouse models or patient-derived xenografts (PDXs), have been valuable in tracking tumor evolution and identifying therapeutic targets. For example, a longitudinal single-cell analysis of a myeloma mouse model pinpointed the GCN2 stress response pathway as a promising target [91]. Similarly, serial single-cell whole-genome sequencing of breast epithelium and TNBC PDXs has revealed the impact of TP53 mutations and chemotherapy on clonal dynamics [92], offering new perspectives on tumor adaptation and treatment resistance.

Single-cell (SC) research has significantly advanced our understanding of metastasis, revealing intricate details about its mechanisms. A pioneering application of a Cas9-based single-cell lineage tracer in a lung cancer xenograft mouse model has highlighted the heterogeneous nature of metastatic potential, which is rooted in pre-existing, inheritable differences in gene expression, and unveiled the previously unrecognized suppressive influence of KRT17 on metastasis [93]. This work underscores the precision of cancer progression tracking at a subclonal level, offering profound insights into the dynamics of cancer spread. Moreover, single-cell immune profiling of melanoma in sentinel lymph nodes (SLNs) has exposed immunological alterations that undermine anti-melanoma defenses, contributing to high rates of relapse [94]. The observed progressive immune impairment linked with micro-metastasis in early to intermediate-stage cutaneous melanoma lays the groundwork for innovative neoadjuvant therapies aimed at bolstering endogenous anti-tumor responses [95]. In acral melanoma, a distinctively suppressed immune milieu compared to cutaneous melanoma from non-acral skin has been documented, with the expression of several actionable immune checkpoints, pointing to novel therapeutic avenues that might have remained undiscovered without single-cell methodologies [96]. Further insights into metastasis have been gleaned from single-cell analyses of circulating tumor cells (CTCs). In hepatocellular carcinoma (HCC), scRNA-seq of CTCs dissected the spatial heterogeneity and immune evasion mechanisms, pinpointing chemokine CCL5 as a crucial facilitator of CTC immune escape and suggesting a potential anti-metastatic therapy [97]. Intriguingly, research has shown that breast cancer cell dissemination is predominantly nocturnal. Single-cell RNA sequencing of blood CTCs, which surge during sleep phases in both human patients and mouse models, identified a significant activation of mitotic genes during rest, enabling metastatic efficiency [98].

## 17.11 Neurodegenerative Diseases

Parkinson's disease is linked to the specific loss of dopaminergic neurons within the substantia nigra, though not all such neurons are affected. Single-cell genomic analysis of human dopamine-producing neurons has uncovered that among ten transcriptionally distinct dopaminergic neuron subpopulations in the substantia nigra, only one is particularly susceptible to degeneration in Parkinson's disease. This vulnerable group exhibits a transcriptional profile heavily enriched with genes linked to Parkinson's disease risk [99], offering a promising direction for developing targeted therapies. In Alzheimer's disease, an innovative study focused on somatic DNA alterations within individual neurons employed scWGS to compare neurons from the hippocampus and prefrontal cortex of Alzheimer's patients against those from control subjects. This comparison unearthed genomic changes related to nucleotide oxidation, shedding light on how such damage may progressively compromise neuronal function and contribute to the disease's pathogenic progression [100]. The involvement of immune cells in neurodegenerative disorders has been a focal point of recent investigations. scRNA-seq of brain tissue from healthy and Alzheimer's disease mouse models has pinpointed specific microglial states associated with the disease, suggesting that targeting these cell states could offer therapeutic advantages [101]. Furthermore, single-cell transcriptomic and TCR analyses have identified specific T cell subsets that become activated and proliferate in Parkinson's disease, indicating potential immune-targeted interventions [102]. Recent advancements in single-cell technologies have ushered in novel approaches for studying the brain. Patch-seq [103, 104], a technique merging scRNA-seq with electrophysiological recordings, and VINE-seq [105], based on single-nucleus RNA sequencing (snRNA-seq). These pioneering methods offer unprecedented insights into the cellular complexity of the brain and the intricate changes occurring in neurodegenerative diseases, paving the way for new diagnostic and therapeutic strategies.

## 17.12 Inflammatory and Autoimmune Diseases

In spondyloarthritis, scRNA-seq facilitated the characterization of a specific regulatory T cell population [106], while also uncovering the presence of cytotoxic T cells in the synovium of patients with psoriatic arthritis, with their clonal expansion confirmed through T cell receptor sequencing (TCR-seq) [107]. Moreover, the differentiation of peripheral blood mononuclear cell (PBMC) samples from patients with anticitrullinated peptide antibody-positive (ACPA+) and negative (ACPA−) rheumatoid arthritis at the single-cell level delineated immune correlates unique to each subtype of rheumatoid arthritis [108], while immune profiling of skin biopsies identified distinct T cell resident memory, innate lymphoid cell, and CD8+ T cell gene signatures across common dermatological inflammatory conditions [108, 109].

In the context of multiple sclerosis, single-cell analysis of PBMC samples from twins discordant for the disease highlighted an inflammatory shift in monocytes and identified a subset of naive helper T cells that are IL-2 hyper-responsive in the affected individuals [110]. Single-cell techniques have further elucidated the role of Epstein–Barr virus (EBV) as a potential etiological factor in multiple sclerosis. Specifically, single-cell B cell receptor sequencing (scBCR-seq) across cerebrospinal fluid and blood samples from patients indicated an expansion of B cell clones binding to antigens in both glia (GlialCAM) and EBV (EBNA1), supporting epidemiological evidence linking EBV to multiple sclerosis [111]. In rheumatoid arthritis, single-cell analyses modeling expression quantitative trait loci (eQTLs) in memory T cells have identified autoimmune variants predominantly enriched in cell-state-dependent eQTLs [112], pinpointing risk variants for the disease located near genes like ORMDL3 and CTLA4. This highlights the importance of considering the functional cell state, as eQTL identification becomes more complex in studies aggregating cells from different states.

## 17.13 Infectious Diseases

scRNA-seq of circulating immune cells combined with metabolic profiling in COVID-19 patients has exposed a complex relationship between immune cell phenotypes and metabolic changes. This includes the identification of metabolically dominant, yet rare, T cell subsets and the division of monocytes into two metabolically distinct groups linked with disease severity [113]. Mechanistic studies incorporating single-cell transcriptomics and proteomics have demonstrated how the production of the C3a complement fragment by SARS-CoV-2 leads to the emergence of a CD16-expressing T cell subset associated with severe COVID-19 [114]. Analysis of post-mortem lung tissue samples from COVID-19 patients has provided a molecular snapshot of hyperinflammation, exhaustion of alveolar epithelial cells, vascular modifications, and fibrosis, suggesting FOXO3A suppression as a key factor in the fibroblast-to-myofibroblast transition observed in COVID-19-associated pulmonary fibrosis [115]. Additionally, a comprehensive examination of multi-tissue single-cell datasets from deceased COVID-19 patients, integrating scRNA-seq and single-nucleus RNA sequencing (snRNA-seq), has highlighted defective alveolar type 2 cell differentiation, fibroblast expansion, and the emergence of TP63+ basal-like progenitor cells within the lungs [116].

In the field of infectious diseases beyond viral infections, notably in the study of bacterial infections and tuberculosis, the development of Parallel sequential fluorescence in situ hybridization (Par-seqFISH) has enabled the capture of gene expression profiles at the individual cell level within prokaryotic populations, maintaining the spatial organization of cells. This method has uncovered the presence of heterogeneous clonal evolution within *Pseudomonas aeruginosa* populations during growth, revealing that multicellular biofilms can harbor subpopulations engaging in different physiological activities yet coexist within the same structure [117]. Such insights are crucial for understanding the complexity of bacterial behavior, particularly in the context of infection and biofilm formation. In tuberculosis research, advanced single-cell analyses combined with in vivo observations of *Mycobacterium tuberculosis*-associated granulomas have shed light on the cellular and transcriptional dynamics underpinning a successful host immune response. Studies have differentiated between granulomas that fail to clear the infection, characterized by a type 2 immunity signature and wound-healing processes, and those that are effective in bacterial control, marked by pro-inflammatory type 1, type 17 immunity, and cytotoxic T cell dominance [118].

## 17.14 Biomarkers and Patient Stratification

In certain scenarios, it is possible to categorize patients into more specific groups based on their disease prognosis or the presence of markers that can forecast how well they will respond to a drug. These biomarkers, which can be either prognostic or predictive, are frequently utilized as selection criteria in clinical trials to pinpoint individuals who are at a higher risk of disease progression or who are more likely to benefit from a specific medication. While bulk transcriptomic profiling has traditionally been the method of choice for identifying cancer prognostic biomarkers—such as the four consensus molecular subtypes (CMS1–4) for colorectal cancer (CRC) established by an international consortium [119]—the practical utility of the CMS classification in clinical settings remains unproven [120]. The main limitation of bulk sequencing is its inability to detect key cellular populations within CRC tumors and their intricate environments, leaving significant gaps in our understanding of epithelial cell variability within the CMS categories. Recent advancements in scRNA-seq have led to the identification of more accurate prognostic biomarkers in CRC [121, 122]. This approach has enabled the distinction of two epithelial cell populations exhibiting unique intrinsic CMS signatures (dubbed iCMS2 and iCMS3). When these findings are combined with data on microsatellite instability and fibrosis, a novel classification system known as IMF emerges [122], offering a refined stratification into five subtypes characterized by distinct signaling pathways, mutation spectra, and gene expression patterns. While the IMF classification shows promise, its clinical efficacy remains to be established.

Immunotherapy with immune checkpoint inhibitors (ICIs) has shown potential for long-lasting responses in certain cancer patients. Nonetheless, the mechanisms underlying variable patient responses to ICIs are still being unraveled, and the hunt for reliable predictive biomarkers for ICI responsiveness continues. To date, various biomarkers, such as tumor mutation burden (TMB), have been identified [123, 124], but they do not account for all patient responses to ICI treatments. Recent studies utilizing single-cell sequencing have unveiled potential new biomarkers for predicting ICI treatment outcomes. For instance, an analysis revealed that patients who respond to checkpoint blockade therapy exhibit an enrichment of $TCF7^{+}CD8^{+}$ T cells [125], a finding consistent across different cancer types treated with ICIs. Beyond traditional $CD8^{+}$ T cell mechanisms, single-cell analyses have also spotlighted other immune cells, including $TREM2^{hi}$ macrophages, $\gamma\delta$T cells, and $CXCL9^{+}$ tumor-associated macrophages, among others, as significant contributors to the ICI response [126–131]. These cells and states, validated across various studies and tumor types [132], have shown promise in surpassing the predictive value of established markers like TMB, TIL levels, and PDL1 expression. A notable example includes its application in high-grade serous ovarian cancer (HGSOC), where scRNA-seq of samples before and after chemotherapy revealed pre-existing cancer cell populations associated with stress, which become more prevalent during treatment. These stress-related genetic expressions were linked to a worse prognosis in HGSOC patients [133]. Furthermore, scRNA-seq offers insights into the likelihood of disease relapse, such as in cases of MLL-rearranged acute lymphoblastic leukemia (ALL), by measuring the ratio of cells resistant or susceptible to therapy [134]. Impressively, this approach has demonstrated superior relapse prediction capabilities compared to existing risk evaluation methods [135].

Numerous studies employing scRNA-seq to analyze diseased tissues have identified biomarkers that could predict drug responsiveness or resistance [126, 136, 137]. Nevertheless, translating these discoveries into clinical practice remains challenging. While such studies are generally limited by the number of participants, conditions examined, and the diversity of samples, techniques like cell-type deconvolution are being used to supplement traditional bulk RNA-seq studies, which typically provide a broader scope of response and outcome data [138]. This synergistic approach between single-cell and bulk analyses holds promise for advancing our understanding of disease mechanisms and enhancing patient care.

## 17.15 Monitoring of Drug Response and Disease Progression

The implementation of SC sequencing techniques in monitoring disease progression and evaluating therapy responses is beginning to significantly impact clinical decisions, particularly within oncology. A key measure in this regard is minimal residual disease (MRD), which serves as an indicator of remaining cancer cells either during therapy or after its completion. This approach has become fundamental in assessing the effectiveness of treatments. In conditions like AML, where patients may present with a variety of subclones characterized by complex genetic abnormalities [139], traditional definitions of complete remission—based on less than 5% blasts in the bone marrow as determined by morphological analysis—fail to consider the presence or evolution of these subclonal molecular irregularities. Increasing evidence suggests that detecting MRD below the 5% threshold could serve as an important predictor of relapse risk, thereby offering critical insights for shaping subsequent treatment strategies [140]. SC mutational profiling surpasses traditional methods for minimal residual disease (MRD) detection by offering a deeper analysis that allows for the identification of subclonal populations at finer thresholds and sheds light on their evolutionary changes during therapy [141]. This refined method improves MRD detection accuracy and is pivotal in identifying resistant clones that may trigger relapse. Such advanced and sensitive techniques highlight the pivotal role of SC sequencing in refining oncological treatments and decision-making, propelling the field toward more individualized and effective therapeutic approaches. Moreover, to enhance our comprehension of therapeutic resistance in cancer, the PERSIST-SEQ consortium was established, aiming to develop a SC atlas of persister cells (https://persist-seq.org/). This initiative, along with others like HTAN46, seeks to systematically delineate persister cell states across different cancer stages and types, both in adults and children, thereby improving strategies to combat therapeutic resistance [142], emphasizing the importance of considering non-genetic factors in the development of drug resistance.

## 17.16 Current Challenges

The pharmaceutical industry faces several hurdles in fully leveraging the transformative potential of scRNA-seq technologies, necessitating significant alterations in both infrastructure and operational methodologies. The rapid accumulation of scRNA-seq data in the public domain far exceeds the internal capabilities of any single pharmaceutical entity, presenting a substantial challenge in effectively assimilating all pertinent scRNA-seq information. Additionally, due to the specific sample requirements and the high costs associated with generating scRNA-seq data, it is unlikely that this approach will swiftly supplant traditional bulk molecular profiling in the realms of early discovery or clinical sample analysis. Therefore, a seamless integration of scRNA-seq with bulk molecular profiling data is essential for advancing research and development efforts.

### 17.16.1 Study Design and Implementation

The standardization of study designs and the execution of SC experiments are currently in their developmental stages. Despite the unparalleled resolution SC technologies offer, enabling a deeper insight into cell states and the identification of rare cell populations, accurately and consistently classifying cell types across various studies remains a challenge. This is particularly true for rare cell populations where subtle differences are critical for classification. Establishing a universal analysis pipeline, alongside standardized methodologies and terminologies, is critical to overcome these obstacles [30, 143–146]. SC sequencing efforts are principally constrained by costs, alongside limitations related to sample processing and computational capabilities. For scRNA-seq, the necessity to dissociate and promptly process tissue samples post-collection to maintain RNA integrity poses logistical challenges [147, 148]. Additionally, the preparation of SC libraries can be problematic in clinical settings lacking personnel trained in specialized sample handling and equipment. Ensuring sample quality and consistency, especially in expansive, multi-site clinical trials, is a further challenge that needs addressing. Recent progress in single-nucleus sequencing technologies, which can analyze both cryopreserved and formalin-fixed paraffin-embedded (FFPE) samples, presents an exciting avenue for preserving biopsy specimens for future investigations [149–151]. This method enhances the utility of previously collected samples and expands the range of analyses that can be conducted. Nonetheless, selecting the appropriate technology is critical due to inherent limitations each method may have [152, 153], highlighting the importance of a thorough and knowledgeable strategy when incorporating scRNA-seq into the R&D processes of the pharmaceutical industry.

An online tool available at (https://satijalab.org/howmanycells/) provides assistance in estimating the requisite number of cells for analysis, based on initial assumptions about cell diversity and composition in the study's biological context. Recommendations for selecting suitable sequencing protocols or the extent of sequencing needed have been given [154]. Furthermore, the strategies for designing longitudinal SC studies have been outlined [155]. SC experimental design introduces distinctive advantages and challenges in comparison to conventional bulk transcriptomics. The abundance of SC samples in such studies enables the employment of machine learning techniques, which may not be suitable for bulk experiments typically limited by their scale. However, the applicability of SC study outcomes might be restricted due to the small number of biological samples used in generating SC data. In contrast to bulk RNA-seq, scRNA-seq incurs higher costs and involves more complex sample acquisition and processing protocols. Although bulk methods have adapted to accommodate suboptimal RNA, frozen, and even FFPE samples, SC methodologies are just beginning to extend to non-fresh tissue samples. New developments like cryopreservation [156] and snRNA-seq [151] are under significant refinement. Strategically combining bulk and scRNA-seq within the same study can strike a balance between complexity and cost. This approach allows for the computational separation of cell types in bulk samples by analyzing fewer SC samples alongside a greater number of bulk-sequenced samples. Additionally, the use of existing SC datasets can help alleviate financial limitations.

### 17.16.2 Data Accessibility

The present structure of public stem cell (SC) data often does not meet the FAIR principles of data management, notably in terms of data accessibility [157]. Efforts to catalogue data, such as the BROAD Single Cell Portal (https://singlecell.broadinstitute.org/single_cell) and the spreadsheet detailing dataset metadata [158], alongside global efforts to create reference databases for healthy samples, like the Human Cell Landscape (HCL) [159] and Tabula Sapiens [160] (https://tabula-sapiens-portal.ds.czbiohub.org/), serve as initial points for dataset discovery. Yet, these endeavors do not encompass everything, necessitating manual searches through scholarly databases (e.g., PubMed) and omics repositories (e.g., GEO). The lack of standardized metadata across these platforms requires adjusting search strategies for different databases to achieve thoroughness. Within organizations, access to certain data may be restricted to a specific group of analysts. The challenge of managing permissions for data use, whether indicated in the metadata or an external system, poses various obstacles regarding internal risk management, compliance, and the ability for scientists and analysts to utilize or expand upon existing analyses.

### 17.16.3 Data Interoperability and Reusability

Publicly available stem cell transcriptomics datasets from research publications exhibit significant inconsistency in data format and organization. There is no standardization for the digital representation of expression or count matrices and experimental metadata [161]. A prevalent issue is the incomplete sample metadata, which hampers the interoperability of these datasets. Furthermore, the variation in data processing methods, such as quality control, cell-type classification, and the absence of a universally accepted cell-type naming system (using either basic or superficial classifications with varying levels of granularity across studies), requires the reanalysis of datasets for new scientific inquiries. The pharmaceutical sector typically undertakes internal curation to integrate publicly available SC datasets into their collections, ensuring uniform processing, or they contract external companies for this task. The services offered by these external vendors vary widely in scope and sophistication, ranging from project-specific and occasional curation of a few datasets to comprehensive solutions featuring automated pipelines, SC data browsers, and platforms for exploratory research. The level of curation effort also differs significantly, with some vendors processing raw sequencing data and others utilizing previously published gene expression data and cell-type classifications. A major hurdle is the technical variability in SC data due to factors like differing laboratory practices and experimental conditions. To mitigate redundant efforts among businesses and educational organizations, a collaborative approach toward establishing and adopting universal standards would be beneficial for the community. The academic field has already demonstrated the advantages of compiling databases that consist of uniformly processed or integrated datasets. Portals managed by researchers enable swift verification of specific gene expressions but are limited in facilitating the broader reuse of datasets. Even sophisticated portals like Cellxgene [162] are designed for specific analytical purposes and cannot accommodate extensive research queries. These resources, often hosted on temporary web servers, serve best for immediate data access post-publication rather than long-term use.

## 17.17 Conclusion

The intricacies of complex diseases, which often elude effective treatment, typically originate from multicellular processes, making SC research pivotal for deciphering the disease mechanisms and identifying potent cellular or molecular therapeutic targets. Advances in SC methodologies, coupled with refined multiplexing techniques, have dramatically enhanced the depth and breadth of assay capabilities. SC analysis of widely utilized preclinical models also assists researchers in choosing the model that most accurately mirrors critical aspects of human disease pathology. Examining human samples at the cellular level can expedite the identification of novel biomarkers, facilitating patient stratification based on prognosis or predicted treatment response, thus pushing the boundaries of personalized medicine. Observing the progression of diseased tissues through a SC lens over the course of treatment offers clinicians a detailed and mechanistic insight into the therapeutic response.

## References


1. Van de Sande B, Lee JS, Mutasa-Gottgens E, Naughton B, Bacon W, Manning J, et al. Applications of single-cell RNA sequencing in drug discovery and development. Nat Rev Drug Discov. 2023;22(6):496–520.
2. Olsen TK, Baryawno N. Introduction to single-cell RNA sequencing. Curr Protoc Mol Biol. 2018;122(1):e57.
3. Jovic D, Liang X, Zeng H, Lin L, Xu F, Luo Y. Single-cell RNA sequencing technologies and applications: a brief overview. Clin Transl Med. 2022;12(3):e694.
4. Narayan A, Berger B, Cho H. Assessing single-cell transcriptomic variability through density-preserving data visualization. Nat Biotechnol. 2021;39(6):765–74.
5. Yang L, Chan AKN, Miyashita K, Delaney CD, Wang X, Li H, et al. High-resolution characterization of gene function using single-cell CRISPR tiling screen. Nat Commun. 2021;12(1):4063.

6. Dixit A, Parnas O, Li B, Chen J, Fulco CP, Jerby-Arnon L, et al. Perturb-Seq: dissecting molecular circuits with scalable single-cell RNA profiling of pooled genetic screens. Cell. 2016;167(7):1853–66.e17.
7. Datlinger P, Rendeiro AF, Schmidl C, Krausgruber T, Traxler P, Klughammer J, et al. Pooled CRISPR screening with single-cell transcriptome readout. Nat Methods. 2017;14(3):297–301.
8. Jaitin DA, Weiner A, Yofe I, Lara-Astiaso D, Keren-Shaul H, David E, et al. Dissecting immune circuits by linking CRISPR-pooled screens with single-cell RNA-Seq. Cell. 2016;167(7):1883–96.e15.
9. Papalexi E, Mimitou EP, Butler AW, Foster S, Bracken B, Mauck WM 3rd, et al. Characterizing the molecular regulation of inhibitory immune checkpoints with multimodal single-cell screens. Nat Genet. 2021;53(3):322–31.
10. Yang L, Zhu Y, Yu H, Cheng X, Chen S, Chu Y, et al. scMAGeCK links genotypes with multiple phenotypes in single-cell CRISPR screens. Genome Biol. 2020;21(1):19.
11. Duan B, Zhou C, Zhu C, Yu Y, Li G, Zhang S, et al. Model-based understanding of single-cell CRISPR screening. Nat Commun. 2019;10(1):2233.
12. Adamson B, Norman TM, Jost M, Cho MY, Nuñez JK, Chen Y, et al. A multiplexed single-cell CRISPR screening platform enables systematic dissection of the unfolded protein response. Cell. 2016;167(7):1867–82.e21.
13. Stubbington MJT, Lönnberg T, Proserpio V, Clare S, Speak AO, Dougan G, et al. T cell fate and clonality inference from single-cell transcriptomes. Nat Methods. 2016;13(4):329–32.
14. Lindeman I, Emerton G, Mamanova L, Snir O, Polanski K, Qiao SW, et al. BraCeR: B-cell-receptor reconstruction and clonality inference from single-cell RNA-seq. Nat Methods. 2018;15(8):563–5.
15. Song L, Cohen D, Ouyang Z, Cao Y, Hu X, Liu XS. TRUST4: immune repertoire reconstruction from bulk and single-cell RNA-seq data. Nat Methods. 2021;18(6):627–30.
16. Upadhyay AA, Kauffman RC, Wolabaugh AN, Cho A, Patel NB, Reiss SM, et al. BALDR: a computational pipeline for paired heavy and light chain immunoglobulin reconstruction in single-cell RNA-seq data. Genome Med. 2018;10(1):20.
17. Rizzetto S, Koppstein DNP, Samir J, Singh M, Reed JH, Cai CH, et al. B-cell receptor reconstruction from single-cell RNA-seq with VDJPuzzle. Bioinformatics. 2018;34(16):2846–7.
18. Borcherding N, Bormann NL, Kraus G. scRepertoire: an R-based toolkit for single-cell immune receptor analysis. F1000Res. 2020;9:47.
19. McDavid A, Gu Y, VonKaenel E. *CellaRepertorium: data structures, clustering and testing for single cell immune receptor repertoires (scRNAseq RepSeq/AIRR-seq)*. https://rdrr.io/bioc/CellaRepertorium. 2021.
20. Zhang Z, Xiong D, Wang X, Liu H, Wang T. Mapping the functional landscape of T cell receptor repertoires by single-T cell transcriptomics. Nat Methods. 2021;18(1):92–9.
21. Buenrostro JD, Wu B, Litzenburger UM, Ruff D, Gonzales ML, Snyder MP, et al. Single-cell chromatin accessibility reveals principles of regulatory variation. Nature. 2015;523(7561):486–90.
22. Wu SJ, Furlan SN, Mihalas AB, Kaya-Okur HS, Feroze AH, Emerson SN, et al. Single-cell CUT&tag analysis of chromatin modifications in differentiation and tumor progression. Nat Biotechnol. 2021;39(7):819–24.
23. Grosselin K, Durand A, Marsolier J, Poitou A, Marangoni E, Nemati F, et al. High-throughput single-cell ChIP-seq identifies heterogeneity of chromatin states in breast cancer. Nat Genet. 2019;51(6):1060–6.
24. Clark SJ, Smallwood SA, Lee HJ, Krueger F, Reik W, Kelsey G. Genome-wide base-resolution mapping of DNA methylation in single cells using single-cell bisulfite sequencing (scBS-seq). Nat Protoc. 2017;12(3):534–47.
25. Slavov N. Learning from natural variation across the proteomes of single cells. PLoS Biol. 2022;20(1):e3001512.
26. Vistain LF, Tay S. Single-Cell Proteomics. Trends Biochem Sci. 2021;46(8):661–72.
27. Perkel JM. Single-cell proteomics takes Centre stage. Nature. 2021;597(7877):580–2.

28. Brinkerhoff H, Kang ASW, Liu J, Aksimentiev A, Dekker C. Multiple rereads of single proteins at single-amino acid resolution using nanopores. Science. 2021;374(6574):1509–13.
29. Mimitou EP, Cheng A, Montalbano A, Hao S, Stoeckius M, Legut M, et al. Multiplexed detection of proteins, transcriptomes, clonotypes and CRISPR perturbations in single cells. Nat Methods. 2019;16(5):409–12.
30. Clark SJ, Argelaguet R, Kapourani CA, Stubbs TM, Lee HJ, Alda-Catalinas C, et al. scNMT-seq enables joint profiling of chromatin accessibility DNA methylation and transcription in single cells. Nat Commun. 2018;9(1):781.
31. Mimitou EP, Lareau CA, Chen KY, Zorzetto-Fernandes AL, Hao Y, Takeshima Y, et al. Scalable, multimodal profiling of chromatin accessibility, gene expression and protein levels in single cells. Nat Biotechnol. 2021;39(10):1246–58.
32. Hücker SM, Fehlmann T, Werno C, Weidele K, Lüke F, Schlenska-Lange A, et al. Single-cell microRNA sequencing method comparison and application to cell lines and circulating lung tumor cells. Nat Commun. 2021;12(1):4316.
33. Gawronski KAB, Kim J. Single cell transcriptomics of noncoding RNAs and their cell-specificity. Wiley Interdiscip Rev RNA. 2017;8(6)
34. Seydel C. Single-cell metabolomics hits its stride. Nat Methods. 2021;18(12):1452–6.
35. VanInsberghe M, van den Berg J, Andersson-Rolf A, Clevers H, van Oudenaarden A. Single-cell Ribo-seq reveals cell cycle-dependent translational pausing. Nature. 2021;597(7877):561–5.
36. Arrastia MV, Jachowicz JW, Ollikainen N, Curtis MS, Lai C, Quinodoz SA, et al. Single-cell measurement of higher-order 3D genome organization with scSPRITE. Nat Biotechnol. 2022;40(1):64–73.
37. Zhang R, Zhou T, Ma J. Multiscale and integrative single-cell hi-C analysis with Higashi. Nat Biotechnol. 2022;40(2):254–61.
38. Chen KH, Boettiger AN, Moffitt JR, Wang S, Zhuang X. RNA imaging. Spatially resolved, highly multiplexed RNA profiling in single cells. Science. 2015;348(6233):aaa6090.
39. Rodriques SG, Stickels RR, Goeva A, Martin CA, Murray E, Vanderburg CR, et al. Slide-seq: a scalable technology for measuring genome-wide expression at high spatial resolution. Science. 2019;363(6434):1463–7.
40. Vickovic S, Eraslan G, Salmén F, Klughammer J, Stenbeck L, Schapiro D, et al. High-definition spatial transcriptomics for in situ tissue profiling. Nat Methods. 2019;16(10):987–90.
41. Liu B, Li Y, Zhang L. Analysis and visualization of spatial transcriptomic data. Front Genet. 2021;12:785290.
42. Hu J, Schroeder A, Coleman K, Chen C, Auerbach BJ, Li M. Statistical and machine learning methods for spatially resolved transcriptomics with histology. Comput Struct Biotechnol J. 2021;19:3829–41.
43. Zeng Z, Li Y, Li Y, Luo Y. Statistical and machine learning methods for spatially resolved transcriptomics data analysis. Genome Biol. 2022;23(1):83.
44. Palla G, Fischer DS, Regev A, Theis FJ. Spatial components of molecular tissue biology. Nat Biotechnol. 2022;40(3):308–18.
45. Efremova M, Vento-Tormo M, Teichmann SA, Vento-Tormo R. CellPhoneDB: inferring cell-cell communication from combined expression of multi-subunit ligand-receptor complexes. Nat Protoc. 2020;15(4):1484–506.
46. Stuart T, Butler A, Hoffman P, Hafemeister C, Papalexi E, Mauck WM 3rd, et al. Comprehensive integration of single-cell data. Cell. 2019;177(7):1888–902.e21.
47. Pliner HA, Shendure J, Trapnell C. Supervised classification enables rapid annotation of cell atlases. Nat Methods. 2019;16(10):983–6.
48. Kiselev VY, Yiu A, Hemberg M. Scmap: projection of single-cell RNA-seq data across data sets. Nat Methods. 2018;15(5):359–62.
49. Aran D, Looney AP, Liu L, Wu E, Fong V, Hsu A, et al. Reference-based analysis of lung single-cell sequencing reveals a transitional profibrotic macrophage. Nat Immunol. 2019;20(2):163–72.

50. Cortal A, Martignetti L, Six E, Rausell A. Gene signature extraction and cell identity recognition at the single-cell level with cell-ID. Nat Biotechnol. 2021;39(9):1095–102.
51. Domínguez Conde C, Xu C, Jarvis LB, Rainbow DB, Wells SB, Gomes T, et al. Cross-tissue immune cell analysis reveals tissue-specific features in humans. Science. 2022;376(6594):eabl5197.
52. Qiu X, Mao Q, Tang Y, Wang L, Chawla R, Pliner HA, et al. Reversed graph embedding resolves complex single-cell trajectories. Nat Methods. 2017;14(10):979–82.
53. Wolf FA, Hamey FK, Plass M, Solana J, Dahlin JS, Göttgens B, et al. PAGA: graph abstraction reconciles clustering with trajectory inference through a topology preserving map of single cells. Genome Biol. 2019;20(1):59.
54. Street K, Risso D, Fletcher RB, Das D, Ngai J, Yosef N, et al. Slingshot: cell lineage and pseudotime inference for single-cell transcriptomics. BMC Genomics. 2018;19(1):477.
55. Velten L, Haas SF, Raffel S, Blaszkiewicz S, Islam S, Hennig BP, et al. Human haematopoietic stem cell lineage commitment is a continuous process. Nat Cell Biol. 2017;19(4):271–81.
56. Schlitzer A, Sivakamasundari V, Chen J, Sumatoh HR, Schreuder J, Lum J, et al. Identification of cDC1- and cDC2-committed DC progenitors reveals early lineage priming at the common DC progenitor stage in the bone marrow. Nat Immunol. 2015;16(7):718–28.
57. La Manno G, Soldatov R, Zeisel A, Braun E, Hochgerner H, Petukhov V, et al. RNA velocity of single cells. Nature. 2018;560(7719):494–8.
58. Bergen V, Lange M, Peidli S, Wolf FA, Theis FJ. Generalizing RNA velocity to transient cell states through dynamical modeling. Nat Biotechnol. 2020;38(12):1408–14.
59. Lange M, Bergen V, Klein M, Setty M, Reuter B, Bakhti M, et al. CellRank for directed single-cell fate mapping. Nat Methods. 2022;19(2):159–70.
60. Hänzelmann S, Castelo R, Guinney J. GSVA: gene set variation analysis for microarray and RNA-seq data. BMC Bioinform. 2013;14:7.
61. Subramanian A, Tamayo P, Mootha VK, Mukherjee S, Ebert BL, Gillette MA, et al. Gene set enrichment analysis: a knowledge-based approach for interpreting genome-wide expression profiles. Proc Natl Acad Sci USA. 2005;102(43):15545–50.
62. Fan J, Salathia N, Liu R, Kaeser GE, Yung YC, Herman JL, et al. Characterizing transcriptional heterogeneity through pathway and gene set overdispersion analysis. Nat Methods. 2016;13(3):241–4.
63. Wang X, Park J, Susztak K, Zhang NR, Li M. Bulk tissue cell type deconvolution with multi-subject single-cell expression reference. Nat Commun. 2019;10(1):380.
64. Erdmann-Pham DD, Fischer J, Hong J, Song YS. Likelihood-based deconvolution of bulk gene expression data using single-cell references. Genome Res. 2021;31(10):1794–806.
65. Wang J, Roeder K, Devlin B. Bayesian estimation of cell type-specific gene expression with prior derived from single-cell data. Genome Res. 2021;31(10):1807–18.
66. Sokolowski DJ, Faykoo-Martinez M, Erdman L, Hou H, Chan C, Zhu H, et al. Single-cell mapper (scMappR): using scRNA-seq to infer the cell-type specificities of differentially expressed genes. NAR Genom Bioinform. 2021;3(1):lqab011.
67. Newman AM, Steen CB, Liu CL, Gentles AJ, Chaudhuri AA, Scherer F, et al. Determining cell type abundance and expression from bulk tissues with digital cytometry. Nat Biotechnol. 2019;37(7):773–82.
68. Luca BA, Steen CB, Matusiak M, Azizi A, Varma S, Zhu C, et al. Atlas of clinically distinct cell states and ecosystems across human solid tumors. Cell. 2021;184(21):5482–96.e28.
69. Jagadeesh KA, Dey KK, Montoro DT, Mohan R, Gazal S, Engreitz JM, et al. Identifying disease-critical cell types and cellular processes by integrating single-cell RNA-sequencing and human genetics. Nat Genet. 2022;54(10):1479–92.
70. Zhang MJ, Hou K, Dey KK, Sakaue S, Jagadeesh KA, Weinand K, et al. Polygenic enrichment distinguishes disease associations of individual cells in single-cell RNA-seq data. Nat Genet. 2022;54(10):1572–80.

71. Shifrut E, Carnevale J, Tobin V, Roth TL, Woo JM, Bui CT, et al. Genome-wide CRISPR screens in primary human T cells reveal key regulators of immune function. Cell. 2018;175(7):1958–71.e15.
72. Ghavami S, Pandi A. CRISPR interference and its applications. Prog Mol Biol Transl Sci. 2021;180:123–40.
73. Larson MH, Gilbert LA, Wang X, Lim WA, Weissman JS, Qi LS. CRISPR interference (CRISPRi) for sequence-specific control of gene expression. Nat Protoc. 2013;8(11):2180–96.
74. Jin X, Simmons SK, Guo A, Shetty AS, Ko M, Nguyen L, et al. In vivo perturb-Seq reveals neuronal and glial abnormalities associated with autism risk genes. Science. 2020;370(6520)
75. Lazo JS, Sharlow ER, Cornelison R, Hart DJ, Llaneza DC, Mendelson AJ, et al. Credentialing and pharmacologically targeting PTP4A3 phosphatase as a molecular target for ovarian cancer. Biomol Ther. 2021;11(7)
76. Wang W, Han D, Cai Q, Shen T, Dong B, Lewis MT, et al. MAPK4 promotes triple negative breast cancer growth and reduces tumor sensitivity to PI3K blockade. Nat Commun. 2022;13(1):245.
77. Wang PX, Ji YX, Zhang XJ, Zhao LP, Yan ZZ, Zhang P, et al. Targeting CASP8 and FADD-like apoptosis regulator ameliorates nonalcoholic steatohepatitis in mice and nonhuman primates. Nat Med. 2017;23(4):439–49.
78. McFarland JM, Paolella BR, Warren A, Geiger-Schuller K, Shibue T, Rothberg M, et al. Multiplexed single-cell transcriptional response profiling to define cancer vulnerabilities and therapeutic mechanism of action. Nat Commun. 2020;11(1):4296.
79. Shin D, Lee W, Lee JH, Bang D. Multiplexed single-cell RNA-seq via transient barcoding for simultaneous expression profiling of various drug perturbations. Sci Adv. 2019;5(5):eaav2249.
80. Song H, Weinstein HNW, Allegakoen P, Wadsworth MH 2nd, Xie J, Yang H, et al. Single-cell analysis of human primary prostate cancer reveals the heterogeneity of tumor-associated epithelial cell states. Nat Commun. 2022;13(1):141.
81. Wang Q, Zhang Y, Zhang B, Fu Y, Zhao X, Zhang J, et al. Single-cell chromatin accessibility landscape in kidney identifies additional cell-of-origin in heterogenous papillary renal cell carcinoma. Nat Commun. 2022;13(1):31.
82. Nowicki-Osuch K, Zhuang L, Jammula S, Bleaney CW, Mahbubani KT, Devonshire G, et al. Molecular phenotyping reveals the identity of Barrett's esophagus and its malignant transition. Science. 2021;373(6556):760–7.
83. Chen X, Song E. The theory of tumor ecosystem. Cancer Commun. 2022;42(7):587–608.
84. Steen CB, Luca BA, Esfahani MS, Azizi A, Sworder BJ, Nabet BY, et al. The landscape of tumor cell states and ecosystems in diffuse large B cell lymphoma. Cancer Cell. 2021;39(10):1422–37.e10.
85. Wu SZ, Al-Eryani G, Roden DL, Junankar S, Harvey K, Andersson A, et al. A single-cell and spatially resolved atlas of human breast cancers. Nat Genet. 2021;53(9):1334–47.
86. Zhang X, Peng L, Luo Y, Zhang S, Pu Y, Chen Y, et al. Dissecting esophageal squamous-cell carcinoma ecosystem by single-cell transcriptomic analysis. Nat Commun. 2021;12(1):5291.
87. Pu W, Shi X, Yu P, Zhang M, Liu Z, Tan L, et al. Single-cell transcriptomic analysis of the tumor ecosystems underlying initiation and progression of papillary thyroid carcinoma. Nat Commun. 2021;12(1):6058.
88. Ursu O, Neal JT, Shea E, Thakore PI, Jerby-Arnon L, Nguyen L, et al. Massively parallel phenotyping of coding variants in cancer with perturb-seq. Nat Biotechnol. 2022;40(6):896–905.
89. Chaligne R, Gaiti F, Silverbush D, Schiffman JS, Weisman HR, Kluegel L, et al. Epigenetic encoding, heritability and plasticity of glioma transcriptional cell states. Nat Genet. 2021;53(10):1469–79.
90. Johnson KC, Anderson KJ, Courtois ET, Gujar AD, Barthel FP, Varn FS, et al. Single-cell multimodal glioma analyses identify epigenetic regulators of cellular plasticity and environmental stress response. Nat Genet. 2021;53(10):1456–68.

91. Croucher DC, Richards LM, Tsofack SP, Waller D, Li Z, Wei EN, et al. Longitudinal single-cell analysis of a myeloma mouse model identifies subclonal molecular programs associated with progression. Nat Commun. 2021;12(1):6322.
92. Salehi S, Kabeer F, Ceglia N, Andronescu M, Williams MJ, Campbell KR, et al. Clonal fitness inferred from time-series modelling of single-cell cancer genomes. Nature. 2021;595(7868):585–90.
93. Quinn JJ, Jones MG, Okimoto RA, Nanjo S, Chan MM, Yosef N, et al. Single-cell lineages reveal the rates, routes, and drivers of metastasis in cancer xenografts. Science. 2021;371(6532)
94. Yaddanapudi K, Stamp BF, Subrahmanyam PB, Smolenkov A, Waigel SJ, Gosain R, et al. Single-cell immune mapping of melanoma sentinel lymph nodes reveals an actionable Immunotolerant microenvironment. Clin Cancer Res. 2022;28(10):2069–81.
95. Lund AW. Standing watch: immune activation and failure in melanoma sentinel lymph nodes. Clin Cancer Res. 2022;28(10):1996–8.
96. Li J, Smalley I, Chen Z, Wu JY, Phadke MS, Teer JK, et al. Single-cell characterization of the cellular landscape of Acral melanoma identifies novel targets for immunotherapy. Clin Cancer Res. 2022;28(10):2131–46.
97. Sun YF, Wu L, Liu SP, Jiang MM, Hu B, Zhou KQ, et al. Dissecting spatial heterogeneity and the immune-evasion mechanism of CTCs by single-cell RNA-seq in hepatocellular carcinoma. Nat Commun. 2021;12(1):4091.
98. Diamantopoulou Z, Castro-Giner F, Schwab FD, Foerster C, Saini M, Budinjas S, et al. The metastatic spread of breast cancer accelerates during sleep. Nature. 2022;607(7917):156–62.
99. Kamath T, Abdulraouf A, Burris SJ, Langlieb J, Gazestani V, Nadaf NM, et al. Single-cell genomic profiling of human dopamine neurons identifies a population that selectively degenerates in Parkinson's disease. Nat Neurosci. 2022;25(5):588–95.
100. Miller MB, Huang AY, Kim J, Zhou Z, Kirkham SL, Maury EA, et al. Somatic genomic changes in single Alzheimer's disease neurons. Nature. 2022;604(7907):714–22.
101. Keren-Shaul H, Spinrad A, Weiner A, Matcovitch-Natan O, Dvir-Szternfeld R, Ulland TK, et al. A unique microglia type associated with restricting development of Alzheimer's disease. Cell. 2017;169(7):1276–90.e17.
102. Wang P, Yao L, Luo M, Zhou W, Jin X, Xu Z, et al. Single-cell transcriptome and TCR profiling reveal activated and expanded T cell populations in Parkinson's disease. Cell Discov. 2021;7(1):52.
103. Cadwell CR, Palasantza A, Jiang X, Berens P, Deng Q, Yilmaz M, et al. Electrophysiological, transcriptomic and morphologic profiling of single neurons using patch-seq. Nat Biotechnol. 2016;34(2):199–203.
104. Fuzik J, Zeisel A, Máté Z, Calvigioni D, Yanagawa Y, Szabó G, et al. Integration of electrophysiological recordings with single-cell RNA-seq data identifies neuronal subtypes. Nat Biotechnol. 2016;34(2):175–83.
105. Yang AC, Vest RT, Kern F, Lee DP, Agam M, Maat CA, et al. A human brain vascular atlas reveals diverse mediators of Alzheimer's risk. Nature. 2022;603(7903):885–92.
106. Simone D, Penkava F, Ridley A, Sansom S, Al-Mossawi MH, Bowness P. Single cell analysis of spondyloarthritis regulatory T cells identifies distinct synovial gene expression patterns and clonal fates. Commun Biol. 2021;4(1):1395.
107. Penkava F, Velasco-Herrera MDC, Young MD, Yager N, Nwosu LN, Pratt AG, et al. Single-cell sequencing reveals clonal expansions of pro-inflammatory synovial CD8 T cells expressing tissue-homing receptors in psoriatic arthritis. Nat Commun. 2020;11(1):4767.
108. Wu X, Liu Y, Jin S, Wang M, Jiao Y, Yang B, et al. Single-cell sequencing of immune cells from anticitrullinated peptide antibody positive and negative rheumatoid arthritis. Nat Commun. 2021;12(1):4977.
109. Liu Y, Wang H, Taylor M, Cook C, Martínez-Berdeja A, North JP, et al. Classification of human chronic inflammatory skin disease based on single-cell immune profiling. Sci Immunol. 2022;7(70):eabl9165.

110. Ingelfinger F, Gerdes LA, Kavaka V, Krishnarajah S, Friebel E, Galli E, et al. Twin study reveals non-heritable immune perturbations in multiple sclerosis. Nature. 2022;603(7899):152–8.
111. Lanz TV, Brewer RC, Ho PP, Moon JS, Jude KM, Fernandez D, et al. Clonally expanded B cells in multiple sclerosis bind EBV EBNA1 and GlialCAM. Nature. 2022;603(7900):321–7.
112. Nathan A, Asgari S, Ishigaki K, Valencia C, Amariuta T, Luo Y, et al. Single-cell eQTL models reveal dynamic T cell state dependence of disease loci. Nature. 2022;606(7912):120–8.
113. Lee JW, Su Y, Baloni P, Chen D, Pavlovitch-Bedzyk AJ, Yuan D, et al. Integrated analysis of plasma and single immune cells uncovers metabolic changes in individuals with COVID-19. Nat Biotechnol. 2022;40(1):110–20.
114. Georg P, Astaburuaga-García R, Bonaguro L, Brumhard S, Michalick L, Lippert LJ, et al. Complement activation induces excessive T cell cytotoxicity in severe COVID-19. Cell. 2022;185(3):493–512.e25.
115. Wang S, Yao X, Ma S, Ping Y, Fan Y, Sun S, et al. A single-cell transcriptomic landscape of the lungs of patients with COVID-19. Nat Cell Biol. 2021;23(12):1314–28.
116. Delorey TM, Ziegler CGK, Heimberg G, Normand R, Yang Y, Segerstolpe Å, et al. COVID-19 tissue atlases reveal SARS-CoV-2 pathology and cellular targets. Nature. 2021;595(7865):107–13.
117. Dar D, Dar N, Cai L, Newman DK. Spatial transcriptomics of planktonic and sessile bacterial populations at single-cell resolution. Science. 2021;373(6556)
118. Gideon HP, Hughes TK, Tzouanas CN, Wadsworth MH 2nd, Tu AA, Gierahn TM, et al. Multimodal profiling of lung granulomas in macaques reveals cellular correlates of tuberculosis control. Immunity. 2022;55(5):827–46.e10.
119. Guinney J, Dienstmann R, Wang X, de Reyniès A, Schlicker A, Soneson C, et al. The consensus molecular subtypes of colorectal cancer. Nat Med. 2015;21(11):1350–6.
120. Sarshekeh AM, Lam M, Zorrilla IR, Holliday EB, Das P, Kee BK, et al. Consensus molecular subtype (CMS) as a novel integral biomarker in colorectal cancer: a phase II trial of bintrafusp alfa in CMS4 metastatic CRC. J Clin Oncol. 2020;38(15_suppl):4084.
121. Khaliq AM, Erdogan C, Kurt Z, Turgut SS, Grunvald MW, Rand T, et al. Refining colorectal cancer classification and clinical stratification through a single-cell atlas. Genome Biol. 2022;23(1):113.
122. Joanito I, Wirapati P, Zhao N, Nawaz Z, Yeo G, Lee F, et al. Single-cell and bulk transcriptome sequencing identifies two epithelial tumor cell states and refines the consensus molecular classification of colorectal cancer. Nat Genet. 2022;54(7):963–75.
123. Litchfield K, Reading JL, Puttick C, Thakkar K, Abbosh C, Bentham R, et al. Meta-analysis of tumor- and T cell-intrinsic mechanisms of sensitization to checkpoint inhibition. Cell. 2021;184(3):596–614.e14.
124. Li H, van der Merwe PA, Sivakumar S. Biomarkers of response to PD-1 pathway blockade. Br J Cancer. 2022;126(12):1663–75.
125. Sade-Feldman M, Yizhak K, Bjorgaard SL, Ray JP, de Boer CG, Jenkins RW, et al. Defining T cell states associated with response to checkpoint immunotherapy in melanoma. Cell. 2018;175(4):998–1013.e20.
126. Leader AM, Grout JA, Maier BB, Nabet BY, Park MD, Tabachnikova A, et al. Single-cell analysis of human non-small cell lung cancer lesions refines tumor classification and patient stratification. Cancer Cell. 2021;39(12):1594–609.e12.
127. Xiong D, Wang Y, You M. A gene expression signature of TREM2(hi) macrophages and γδ T cells predicts immunotherapy response. Nat Commun. 2020;11(1):5084.
128. Kieffer Y, Hocine HR, Gentric G, Pelon F, Bernard C, Bourachot B, et al. Single-cell analysis reveals fibroblast clusters linked to immunotherapy resistance in cancer. Cancer Discov. 2020;10(9):1330–51.
129. Dominguez CX, Müller S, Keerthivasan S, Koeppen H, Hung J, Gierke S, et al. Single-cell RNA sequencing reveals stromal evolution into LRRC15(+) Myofibroblasts as a determinant of patient response to cancer immunotherapy. Cancer Discov. 2020;10(2):232–53.

130. Guo X, Zhang Y, Zheng L, Zheng C, Song J, Zhang Q, et al. Global characterization of T cells in non-small-cell lung cancer by single-cell sequencing. Nat Med. 2018;24(7):978–85.
131. Zheng C, Zheng L, Yoo JK, Guo H, Zhang Y, Guo X, et al. Landscape of infiltrating T cells in liver cancer revealed by single-cell sequencing. Cell. 2017;169(7):1342–56.e16.
132. Pittet MJ, Michielin O, Migliorini D. Clinical relevance of tumour-associated macrophages. Nat Rev Clin Oncol. 2022;19(6):402–21.
133. Zhang K, Erkan EP, Jamalzadeh S, Dai J, Andersson N, Kaipio K, et al. Longitudinal single-cell RNA-seq analysis reveals stress-promoted chemoresistance in metastatic ovarian cancer. Sci Adv. 2022;8(8):eabm1831.
134. Candelli T, Schneider P, Garrido Castro P, Jones LA, Bodewes E, Rockx-Brouwer D, et al. Identification and characterization of relapse-initiating cells in MLL-rearranged infant ALL by single-cell transcriptomics. Leukemia. 2022;36(1):58–67.
135. Pieters R, Schrappe M, De Lorenzo P, Hann I, De Rossi G, Felice M, et al. A treatment protocol for infants younger than 1 year with acute lymphoblastic leukaemia (Interfant-99): an observational study and a multicentre randomised trial. Lancet. 2007;370(9583):240–50.
136. Aissa AF, Islam A, Ariss MM, Go CC, Rader AE, Conrardy RD, et al. Single-cell transcriptional changes associated with drug tolerance and response to combination therapies in cancer. Nat Commun. 2021;12(1):1628.
137. Zhang Y, Narayanan SP, Mannan R, Raskind G, Wang X, Vats P, et al. Single-cell analyses of renal cell cancers reveal insights into tumor microenvironment, cell of origin, and therapy response. Proc Natl Acad Sci USA. 2021;118(24)
138. Jerby-Arnon L, Shah P, Cuoco MS, Rodman C, Su MJ, Melms JC, et al. A cancer cell program promotes T cell exclusion and resistance to checkpoint blockade. Cell. 2018;175(4):984–97.e24.
139. Tyner JW, Tognon CE, Bottomly D, Wilmot B, Kurtz SE, Savage SL, et al. Functional genomic landscape of acute myeloid leukaemia. Nature. 2018;562(7728):526–31.
140. Schuurhuis GJ, Heuser M, Freeman S, Béné MC, Buccisano F, Cloos J, et al. Minimal/measurable residual disease in AML: a consensus document from the European LeukemiaNet MRD working party. Blood. 2018;131(12):1275–91.
141. Ediriwickrema A, Aleshin A, Reiter JG, Corces MR, Köhnke T, Stafford M, et al. Single-cell mutational profiling enhances the clinical evaluation of AML MRD. Blood Adv. 2020;4(5):943–52.
142. Kim C, Gao R, Sei E, Brandt R, Hartman J, Hatschek T, et al. Chemoresistance evolution in triple-negative breast cancer delineated by single-cell sequencing. Cell. 2018;173(4):879–93.e13.
143. Stoeckius M, Hafemeister C, Stephenson W, Houck-Loomis B, Chattopadhyay PK, Swerdlow H, et al. Simultaneous epitope and transcriptome measurement in single cells. Nat Methods. 2017;14(9):865–8.
144. Peterson VM, Zhang KX, Kumar N, Wong J, Li L, Wilson DC, et al. Multiplexed quantification of proteins and transcripts in single cells. Nat Biotechnol. 2017;35(10):936–9.
145. Chen S, Lake BB, Zhang K. High-throughput sequencing of the transcriptome and chromatin accessibility in the same cell. Nat Biotechnol. 2019;37(12):1452–7.
146. Ma S, Zhang B, LaFave LM, Earl AS, Chiang Z, Hu Y, et al. Chromatin potential identified by shared single-cell profiling of RNA and chromatin. Cell. 2020;183(4):1103–16.e20.
147. Smillie CS, Biton M, Ordovas-Montanes J, Sullivan KM, Burgin G, Graham DB, et al. Intra- and inter-cellular rewiring of the human colon during ulcerative colitis. Cell. 2019;178(3):714–30.e22.
148. Ren X, Wen W, Fan X, Hou W, Su B, Cai P, et al. COVID-19 immune features revealed by a large-scale single-cell transcriptome atlas. Cell. 2021;184(23):5838.
149. Mathys H, Davila-Velderrain J, Peng Z, Gao F, Mohammadi S, Young JZ, et al. Single-cell transcriptomic analysis of Alzheimer's disease. Nature. 2019;570(7761):332–7.
150. Melms JC, Biermann J, Huang H, Wang Y, Nair A, Tagore S, et al. A molecular single-cell lung atlas of lethal COVID-19. Nature. 2021;595(7865):114–9.

151. Ding J, Adiconis X, Simmons SK, Kowalczyk MS, Hession CC, Marjanovic ND, et al. Systematic comparison of single-cell and single-nucleus RNA-sequencing methods. Nat Biotechnol. 2020;38(6):737–46.
152. Thrupp N, Sala Frigerio C, Wolfs L, Skene NG, Fattorelli N, Poovathingal S, et al. Single-nucleus RNA-Seq is not suitable for detection of microglial activation genes in humans. Cell Rep. 2020;32(13):108189.
153. Der E, Suryawanshi H, Morozov P, Kustagi M, Goilav B, Ranabothu S, et al. Tubular cell and keratinocyte single-cell transcriptomics applied to lupus nephritis reveal type I IFN and fibrosis relevant pathways. Nat Immunol. 2019;20(7):915–27.
154. Haque A, Engel J, Teichmann SA, Lönnberg T. A practical guide to single-cell RNA-sequencing for biomedical research and clinical applications. Genome Med. 2017;9(1):75.
155. Ding J, Sharon N, Bar-Joseph Z. Temporal modelling using single-cell transcriptomics. Nat Rev Genet. 2022;23(6):355–68.
156. Guillaumet-Adkins A, Rodríguez-Esteban G, Mereu E, Mendez-Lago M, Jaitin DA, Villanueva A, et al. Single-cell transcriptome conservation in cryopreserved cells and tissues. Genome Biol. 2017;18(1):45.
157. Wilkinson MD, Dumontier M, Aalbersberg IJ, Appleton G, Axton M, Baak A, et al. The FAIR guiding principles for scientific data management and stewardship. Sci Data. 2016;3:160018.
158. Svensson V, da Veiga BE, Pachter L. A curated database reveals trends in single-cell transcriptomics. Database (Oxford). 2020;2020
159. Han X, Zhou Z, Fei L, Sun H, Wang R, Chen Y, et al. Construction of a human cell landscape at single-cell level. Nature. 2020;581(7808):303–9.
160. Jones RC, Karkanias J, Krasnow MA, Pisco AO, Quake SR, Salzman J, et al. The tabula sapiens: a multiple-organ, single-cell transcriptomic atlas of humans. Science. 2022;376(6594):eabl4896.
161. Füllgrabe A, George N, Green M, Nejad P, Aronow B, Fexova SK, et al. Guidelines for reporting single-cell RNA-seq experiments. Nat Biotechnol. 2020;38(12):1384–6.
162. Megill C, Martin B, Weaver C, Bell S, Prins L, Badajoz S, et al. cellxgene: a performant, scalable exploration platform for high dimensional sparse matrices. bioRxiv. 2021; https://doi.org/10.1101/2021.04.05.438318.

# Chapter 18
# RNA–Ligand Molecular Docking

**Abstract** The rapid progression of computer technology, encompassing both software algorithms and hardware, has significantly sped up the process of virtual screening, markedly enhancing the identification of potent small molecules for drug development. The use of computational modeling to understand the interactions between RNAs and the interdependency of RNAs with other molecules have become a critical component in the search for RNA-targeted therapeutics. Presently, the primary approach to modeling RNA–ligand interactions relies on the docking-and-scoring technique. For docking and scoring to be precise, it must address four essential challenges: (1) the ligand's conformational flexibility, (2) the RNA's conformational flexibility, (3) effective exploration of binding sites and poses, and (4) precise evaluation of various binding configurations. Furthermore, the task of modeling ligand binding to RNA, a negatively charged macromolecule, entails additional complexities over protein-ligand docking, such as the influence of metal ions. Thermodynamic approaches, employing both physics-based and knowledge-based evaluation functions, have demonstrated promising outcomes in accurately predicting ligand binding poses and affinity. Kinetic modeling of ligand binding, introducing the concept of dissociation kinetics into the docking process, has been shown to enhance the prediction of in vivo drug effectiveness. The emergence of deep learning techniques has introduced novel predictive tools for RNA–small molecule interactions. This section provides a summary of the latest computational strategies for RNA–ligand docking, highlighting their strengths and limitations.

## 18.1 Introduction

The field of RNA research is rapidly evolving, with new RNA structures and functions being discovered and advanced methods for RNA structure determination being developed. This progress is paving the way for RNA-based therapies as innovative treatments for various diseases. RNA-based therapeutic strategies are essentially of two types: the first involves therapeutic RNAs like RNA aptamers, ASOs, siRNAs, and guide RNAs (gRNAs) that target RNA transcripts, DNA, or proteins to

 
P. E. Saw, E. Song, *RNA Therapeutics in Human Diseases*,
https://doi.org/10.1007/978-981-96-3041-7_18

modulate specific biochemical pathways. This method has seen significant interest and advancement in gene therapy. The second strategy focuses on using RNA molecules as targets for small molecule drugs, similar to the approach in protein-targeted drug discovery. Despite only about 1.5% of the human genome being protein-coding and only a fraction of these genes being associated with diseases, the potential for protein targets in drug discovery is limited by the necessity for structural and energetic compatibility for high-affinity drug binding. Alternatively, targeting mRNA sequences associated with genes that are hard to target at the protein level offers a broader spectrum for druggability [1]. Moreover, the critical role of ncRNAs in a wide range of diseases, from cancer to infectious diseases like COVID-19, suggests that focusing on the extensive array of ncRNAs presents significant new avenues for pharmaceutical research and drug development.

Indeed, antibiotics targeting rRNA have proven highly effective in combating bacterial infections by inhibiting bacterial protein synthesis [2]. Antibiotics such as macrolides, tetracyclines, and aminoglycosides target specific regions of bacterial rRNA, disrupting ribosome function and impeding protein synthesis. For instance, macrolides like erythromycin bind to the 50S subunit of the bacterial ribosome, blocking the exit tunnel through which newly synthesized proteins emerge. This impediment prevents elongation of the peptide chain and effectively halts bacterial protein synthesis. Similarly, tetracyclines bind to the 30S subunit of the ribosome, interfering with the binding of tRNA to the ribosomal A-site, thereby inhibiting the addition of amino acids to the growing peptide chain. Aminoglycosides, such as streptomycin, disrupt the accuracy of translation by causing misreading of mRNA, leading to the production of faulty proteins and ultimately inhibiting bacterial growth [3–6]. Riboswitches typically consist of two main structural domains: the aptamer domain and the expression platform. The aptamer domain is responsible for binding to the ligand, while the expression platform regulates gene expression based on the ligand-bound state of the aptamer domain. When the ligand is absent, the aptamer domain adopts a conformation that allows the expression platform to form a terminator hairpin or sequester the ribosome-binding site (RBS), effectively inhibiting translation initiation or promoting transcription termination. This results in repression of gene expression. Conversely, in the presence of the ligand, the ligand-bound aptamer domain undergoes conformational changes, altering the structure of the expression platform. This change may disrupt the terminator hairpin or expose the RBS, thereby promoting translation initiation or preventing transcription termination. As a result, gene expression is activated [7–21]. The process of screening small molecule compounds from a library against specific RNA structures, like the three-stemmed RNA pseudoknot found in the SARS-CoV RNA genome, illustrates a targeted approach in drug discovery, particularly relevant for viral infections. In this case, the RNA pseudoknot structure is involved in stimulating −1 programmed ribosomal frameshifting (−1 PRF), a mechanism used by many viruses, including SARS-CoV, to regulate the expression of their proteins. The −1 PRF process is critical for the proper synthesis of viral polyproteins that are necessary for the virus's life cycle. By shifting the reading frame during protein translation, the virus can produce different proteins from the same RNA sequence, enhancing its ability to control protein production based on its needs. When a compound from the screening library shows an inhibition of this frameshift with an $IC_{50}$ value of 210 μM, it indicates that the compound effectively interferes with this specific ribosomal activity at a medium efficacy. The $IC_{50}$ value represents the concentration of the compound required to inhibit 50% of the ribosomal frameshifting activity under the conditions of the assay. This specific measure of 210 μM suggests a potential therapeutic application where the compound could disrupt the virus's ability to properly synthesize essential proteins, thereby hindering its replication and propagation [22–25].

Modeling the interactions between RNA and small ligand molecules indeed presents unique challenges compared to the more commonly studied protein–ligand interactions. The complexities associated with RNA–ligand interactions derive from the inherent properties of RNA, leading to multiple factors that must be accounted for in computational models and experimental designs: (i) RNA's backbone is highly charged due to the presence of one electronic charge on each phosphate group. This negative charge not only influences the folding of the RNA molecule itself but also significantly affects its interaction with ligands. The charge density can lead to electrostatic repulsion, which might inhibit the approach of negatively charged or even neutral molecules unless there is a counterbalancing mechanism. (ii) The presence of metal ions, particularly divalent cations like $Mg^{2+}$, is crucial in stabilizing the RNA structure and facilitating ligand binding. These metal ions can shield the negative charges on the RNA backbone, allowing the RNA to fold into its active conformation and create a stable binding pocket for ligands. In modeling and experimental setups, the concentration and type of metal ions present can significantly affect the outcomes, requiring careful consideration and accurate representation in simulations. (iii) Importance of Water Molecules: Water molecules play a critical role in mediating interactions between RNA and ligands. They can form bridging hydrogen bonds that link ligands with non-adjacent bases or backbone phosphates of the RNA, stabilizing the complex. Additionally, water molecules can also contribute to the solvation dynamics around the RNA–ligand complex, influencing its overall stability and binding affinity. Accurately accounting for water effects, including its structural and dynamic roles, is a challenging aspect of RNA–ligand modeling [26–29]. RNA molecules are characterized by a high degree of flexibility, enabling them to adopt multiple stable conformations. This flexibility is not merely a structural attribute; it plays a critical role in the biological functions of RNA, including its interactions with small molecules. The ability of RNA to switch between different conformers upon ligand binding can be essential for its regulatory roles but poses significant challenges for modeling these interactions [1, 29, 30]. Third, unlike many protein binding sites, which are often well-defined pockets with specific hydrophobic and hydrophilic regions, RNA binding sites tend to be less deep and more exposed to the solvent. They are typically more polar and may involve numerous interactions with water molecules. This exposure means that the binding sites are generally more solvated, influencing how small molecules interact with RNA. Furthermore, the shallow nature of these sites means that the surface of interaction is more spread out and less defined, which adds to the complexity of designing molecules that can effectively target these areas [28, 31–43].

**Box 18.1 Specialized Approach for RNA Docking**

Given the unique challenges presented by RNA compared to proteins, developing computational tools specifically tailored for RNA–ligand interactions is essential [44]. The differences in chemical structure, binding site properties, and interaction dynamics necessitate a specialized approach.

1. **Chemical and Structural Differences**

RNA's backbone is composed of alternating sugar and phosphate groups, which imparts a significant negative charge, unlike the more neutral protein backbones. This fundamental difference affects how RNA interacts with ligands, particularly in terms of electrostatic interactions and the need for charge neutralization, often facilitated by metal ions like $Mg^{2+}$. Additionally, RNA's ability to form a variety of secondary structures such as hairpins, loops, and bulges, which are less common in proteins, also influences its interaction with small molecules.

2. **Development of Specialized Computational Tools**

***Docking Algorithms***

While traditional docking algorithms used for proteins focus heavily on hydrophobic and van der Waals interactions within defined pockets, RNA–ligand docking must account for electrostatic interactions and the highly dynamic, often shallow, and solvent-exposed binding sites. New algorithms need to be developed that can dynamically adjust to the flexible nature of RNA and accurately simulate the solvation effects and ionic conditions relevant to RNA structures.

***Molecular Dynamics (MD) Simulations***

Advanced MD simulations that incorporate explicit solvent and ion models are crucial for understanding how ligands interact with the flexible and conformationally diverse RNA molecules. These simulations must be capable of capturing the transient conformations of RNA and predicting how these conformations can be stabilized or altered by ligand binding.

(continued)

**Box 18.1** (continued)

***Quantum Mechanical/Molecular Mechanical (QM/MM) Methods***

These are needed to accurately model the electronic properties of RNA–ligand complexes, particularly in cases where covalent interactions or significant polarization effects occur upon ligand binding.

3. **Integration of Experimental Data**

Incorporating experimental data from techniques such as X-ray crystallography, NMR spectroscopy, and cryo-electron microscopy can significantly enhance the accuracy of computational models. Such data provide insights into the conformational dynamics of RNA and the structural nuances of RNA–ligand interactions that are critical for refining computational approaches.

4. **Machine Learning and AI**

Leveraging machine learning and artificial intelligence can help overcome some of the limitations of current modeling techniques by predicting interaction hotspots and likely conformational changes in RNA upon ligand binding. These technologies can learn from large datasets of successful and unsuccessful interactions to improve prediction accuracy over time.

Developing these tools and methodologies not only advances our understanding of RNA–ligand interactions but also aids in the design of novel therapeutics targeting RNA, a frontier that holds significant promise given the central role of RNA in many diseases [45–48].

## 18.2 Identifying Druggable Target RNAs

Determining the potential for drugging a specific RNA target hinges on three critical considerations. Initially, one must assess whether modulating the target RNA's function—either by inhibition or enhancement—will effectively manage the disease. The next question involves the target RNA's accessibility to small molecule binding within the cellular milieu, which can be influenced by various factors, including the target RNA's abundance and stability during disease-related cellular events. Finally, it is essential to determine if the target RNA has a binding site suitable for strong and precise small molecule interactions, because targeting RNA with high specificity helps to minimize unintended impacts on other sites. Achieving high specificity may involve targeting RNA sequences unique to disease cells, or pathogens like viruses and bacteria, with riboswitches serving as an example of targets more prevalent in bacteria than in eukaryotic cells. An alternative approach involves using computational techniques to identify RNA motifs capable of forming distinct, high-affinity pockets for small molecule attachment. Inforna, a template-based strategy, excels in selecting RNA targets based on their secondary structural motifs, such as hairpins and various loop and bulge structures. The latest Inforna 2.0

database includes data on 1936 RNA secondary structure motif-ligand complexes, allowing for the identification of RNA motifs and matching them with suitable ligand partners based on fitness scores. These scores gauge the RNA–ligand interaction's affinity and the specificity of the RNA motif against a range of small molecules, with higher scores indicating a greater likelihood of druggability. Utilizing lead compounds identified through high fitness scores, subsequent chemical similarity screening can unearth potent binding candidates. Inforna has demonstrated success in several research initiatives, including the discovery of small molecules targeting cancer-associated miRNA precursors and specific RNA structures linked to diseases such as Parkinson's.

Studies demonstrated the utility of information content [49, 50] in pinpointing RNA motifs amenable to drug targeting due to their potential for specificity binding potency [51]. Information content quantifies the necessary information (expressed in bits) to describe the complexity of an RNA motif's sequence and structure. Motifs exhibiting higher bits (30 bits) signify greater complexity and a higher probability of being distinct within the transcriptome [52]. In practical research, the determination of RNA structural information content is facilitated by chemical probing methods like selective 2′-hydroxyl acylation followed by primer extension (SHAPE) [53–56]. Targeting RNA motifs that exhibit significant complexity (noted by high information content) may direct us toward RNAs characterized by enhanced binding specificity and affinity, thus offering better targets for drug development. Illustratively, the binding efficacies of GTP [49, 57] and targaprimir-9693 have been linked closely to the information content within RNA motifs. Specifically, a 10-bit augmentation in information content is associated with a tenfold improvement in binding efficacy [52].

## 18.3 Identifying Binding Sites for Target RNA

### *18.3.1 Assessment of Binding Pockets*

Recent statistical studies reveal that a significant number of RNAs naturally adopt structures with pockets that are well-suited for specific small molecule interactions [58]. To explore the potential of RNAs for ligand attachment, RNA binding pockets were assessed using the PocketFinder tool [51] across 1552 structured RNAs and all proteins cataloged in the Protein Data Bank (PDB) [59]. This evaluation considered the pocket's volume, its exposure to solvent, and the proportion of the pocket that is hydrophobic. Findings indicate that despite RNA and protein ligand-bound pockets sharing broadly similar physical characteristics, pockets in RNAs tend to be less hydrophobic on average than those found in proteins [58]. Furthermore, when comparing unbound and ligand-bound RNA pockets, the latter are typically found to be larger, more enclosed, and possess a greater hydrophobic nature [58].

### 18.3.2 Geometry-Based Methods

In the quest to identify RNA binding sites that match the shape of potential ligands, DOCK 665 utilizes an approach where it selects binding pockets from an inverse model of the receptor's surface, defined by overlapping spheres within each cavity [60]. rDock employs a dual-sphere mapping strategy for pinpointing potential binding locations. This method begins by placing large spherical probes at every grid point within the designated docking area to eliminate sites that are too superficial or shallow. Next, smaller spherical probes are placed at the remaining open grid points to identify cavities that could serve as potential binding sites [40]. The broad and shallow minor grooves in RNA, capable of accommodating various ligand shapes, often emerge as candidate binding pockets. However, their ability to serve as non-specific binding sites also leads to the challenge of false-positive results in predictions.

### 18.3.3 Energy-Based Methods

Alternative approaches for identifying binding sites involve calculating the interaction energies between the receptor and virtual atoms, which act as probes moving across the receptor's surface. Tools like PocketFinder [51] within ICM [61] and AutoLigand [62] in AutoDock [63] employ this methodology. PocketFinder characterizes interactions using a Lennard-Jones (LJ) potential between the probe and receptor atoms, with the interaction data being translated into grid maps to locate potential binding sites. AutoLigand adopts a comparable strategy but incorporates an additional iterative process to pinpoint the most suitable binding locations on the grid maps. This method also takes into account the connectivity among adjacent potential pockets, enhancing the identification of optimal binding sites.

### 18.3.4 Machine Learning Approaches

Viewing RNA–ligand interactions as a network of atom contacts is indeed an effective approach for identifying and understanding functional sites for RNA–ligand binding. This perspective allows researchers to decipher complex interaction patterns and provides insights into the mechanisms underlying RNA's ability to bind with various small molecules. By analyzing the network of atom contacts, researchers can identify specific nodes (atoms) and edges (interactions) that are crucial for ligand binding. This network analysis helps in pinpointing which contacts are essential for stability and specificity of the RNA–ligand complex. Identifying these key contacts can aid in the design of molecules that can effectively target and bind to RNA. The atom contact network approach allows for the mapping of interaction hotspots on the RNA molecule. These hotspots are areas where key interactions occur, such as hydrogen bonding, van der Waals forces, and electrostatic interactions. By understanding where these hotspots are located and how they contribute to ligand binding, researchers can tailor ligands to enhance binding efficacy and specificity. RNA molecules are not static; they often exhibit significant conformational flexibility. An atom contact network can help model how these interactions change over time, providing a dynamic view of the binding process. Such dynamic modeling is crucial for understanding how conformational changes in RNA influence ligand binding and how ligands might stabilize certain RNA conformations over others. Data from atom contact networks can greatly enhance computational modeling techniques, such as docking and molecular dynamics simulations. By integrating detailed information about where and how ligands interact with RNA, computational models can become more accurate and predictive. This integration can improve the design of new therapeutics by allowing for the virtual screening of compounds based on how well they fit into the identified interaction network.

Tools like Rsite [64] and Rsite2 [65] utilize networks based on inter-nucleotide Euclidean (hamming) distances in either 3D or 2D structures to locate functional sites through the identification of maximally clustered nucleotides. However, the inability of Rsite and Rsite2 to differentiate between types of nucleotide connections often results in incorrect predictions. To overcome this, Rbind converts an RNA structure into a graphical representation, with nodes representing nucleotides and edges denoting noncovalent contacts between them [66, 67]. It identifies functional sites as areas where nucleotides exhibit maximum closeness. In evaluations involving 19 RNA–ligand complexes, Rbind's performance (average positive predictive value, PPV = 0.67) significantly surpassed that of Rsite (average PPV = 0.42) and Rsite2 (average PPV = 0.40), highlighting the value of distinguishing between connection types in predicting functional sites [67]. The use of machine learning tools like RNAsite, which incorporates a random forest algorithm, represents a sophisticated method to analyze and predict RNA–ligand binding sites. By leveraging a random forest approach, RNAsite assesses each nucleotide's potential to be part of a functional site based on its proximity to a target ligand—specifically, checking if any atom is within 4 Å of the ligand. This threshold highlights nucleotides that are likely to be directly involved in binding interactions [68]. There are four main features used, summarized below.

(i) Geometrical Characteristics of Surface Convexity/Concavity: Understanding the structural deformations in RNA that may accommodate or repel ligand binding. Convex areas might represent outward-facing surfaces suitable for interactions, while concave regions could potentially form binding pockets.
(ii) Topological Aspects (Similar to Those in Rbind): These include data on the connectivity and arrangement of atoms within the RNA structure, which are crucial for identifying how the nucleotide might engage with a ligand. This topological mapping can reveal interaction patterns and structural motifs conducive to binding.
(iii) Nucleotide-Specific Accessible Surface Areas: It measures the surface area of each nucleotide that is accessible to a solvent, which is indicative of how

exposed or buried the nucleotide is within the RNA structure. Higher accessibility might suggest a greater likelihood of interaction with ligands.

(iv) Evolutionary Conservation from Multiple Sequence Alignments: Nucleotides that are highly conserved across different species are often crucial for the structural or functional integrity of RNA.

Assessing conservation helps predict which regions of the RNA are essential for its function and likely to be involved in critical interactions [66, 67]. Tested on datasets of 19 (RB19) and 18 (TE18) RNAs, RNAsite [68] demonstrated superior performance over Rsite, Rsite2, and RBind with MCC values of 0.253 and 0.567 and AUC values of 0.776 and 0.877 for the TE18 and RB19 sets, respectively. These findings emphasize the importance of incorporating a range of independent features into predictive models. While these models are tailored for RNA–small molecule complexes, their accuracy could be further enhanced by integrating additional machine learning techniques [69–72]. Machine learning and deep learning methods will be discussed in detail in Chap. 30.

## 18.4 Efficient Sampling with Flexible Conformations

The detailed study of possible formations of RNA–ligand complexes poses considerable difficulties because both RNA molecules and small ligands naturally exhibit flexibility. Additionally, RNA targets often undergo structural changes when ligands attach, a phenomenon referred to as the induced-fit effect or the conformational selection mechanism [17, 73–75]. This dynamic interplay between RNA folding processes, including those that occur co-transcriptionally, and ligand binding complicates virtual screening efforts [17, 73–76]. One prevalent strategy to address this issue is ensemble docking [26, 77–79], which involves docking a ligand to a variety of RNA structural conformations. Another strategy involves dynamically sampling the RNA's conformational alterations during the docking procedure. Despite the development of numerous methods [33, 37, 40–43] aimed at accommodating docking flexibility, accurately predicting significant conformational shifts remains a formidable challenge, mainly due to the extensive computational resources required for large-scale virtual screening. In cases where RNA targets, like ribosomal RNAs and riboswitches, come with reliable binding site data [80, 81], local sampling complemented by a stringent energy evaluation function can yield precise predictions of the ligand's binding orientation [82, 83]. Given the scarcity of RNA–ligand bound structure examples and limited knowledge of binding sites for RNAs (as opposed to proteins), the significance of blind docking—wherein a ligand is docked across the entire receptor surface without pre-identified binding site information—is underscored. While several computational models originally designed for protein–ligand docking have been adapted for RNA targets, others have been specifically crafted for RNA. Table 18.1 is a detailed comparison that elaborates on these computational tools, highlighting their origins and particular effectiveness in RNA applications.

## 18.5 Sampling of Ligand Conformations

Docking methodologies to accommodate the flexibility of ligands during interaction with target sites can be broadly categorized into three main approaches [85–87]: (a) multi-conformer docking, (b) incremental construction, and (c) stochastic optimization. Each strategy offers a distinct balance between computational efficiency and the ability of the ligand to adjust its conformation to the specific geometric characteristics of the binding site.

### *18.5.1 Multi-conformer Docking*

The multi-conformer docking algorithm creates a set of different conformations for a ligand (small molecule) and carries out rigid docking of each conformer against the same target. This method can be notably efficient for a specific binding site when only a few conformers are used in the docking process. The success of this technique largely depends on including conformations close to the ligand's native structure within the ensemble of conformers. Presently, there are several tools available for generating ligand conformers, including OMEGA [88, 89], RDKit [90], and Open Babel [91]. These tools have demonstrated their capability to quickly generate accurate conformational ensembles of small molecules, as evidenced in benchmarking studies [91, 92]. A new conformer generator, integrating a molecular dynamics approach with a quantum mechanics refined force field for ligand–RNA interactions, has recently emerged, offering heightened accuracy for computational drug discovery [93, 94]. The process of docking involves creating a ligand conformer ensemble before conformational sampling. Thus, the ensemble can be tailored specifically for the small molecules under investigation. For instance, conformer ensembles can be designed to favor low-energy states or to encapsulate a wide range of conformational possibilities. The RNA–ligand DOCKing (RLDOCK) model [36, 37] employs the multi-conformer strategy for docking flexible ligands. In RLDOCK, the manner in which a ligand binds is determined by four variables: R (the reference point on the RNA), L (the ligand conformer), A (the ligand atom anchored at R), and O (the three-dimensional rotation of L around A at R). This effectively captures the dynamic nature of ligand–RNA interactions. The RLDOCK algorithm methodically produces a varied set of flexible ligand conformers and their potential binding poses for each RNA–ligand pair, following a sequence of meticulously organized steps.

Initially, the algorithm identifies a set of potential anchor sites R on the RNA. It uses two primary criteria to ensure viability: (a) the absence of steric clashes between ligand and RNA atoms and (b) the surrounding RNA structure at R must constitute a pocket-like geometry. Leveraging the viable anchor sites R determined earlier, RLDOCK exhaustively explores all possible combinations of (R, L, A, O) to map out all conceivable ligand binding sites and poses. These configurations are

**Table 18.1** Docking tools used in RNA docking

| Docking tool | Originally designed For | Adapted/ specific for RNA | Description and effectiveness |
|---|---|---|---|
| AutoDock Vina [41] | Protein-ligand docking | Adapted for RNA | A widely used tool that has been adapted to handle RNA targets. Known for its speed and accuracy in docking simulations, it applies to both protein and RNA molecules, making it versatile |
| GOLD [35] | Protein-ligand docking | Adapted for RNA | Originally designed for proteins, GOLD has been successfully used for RNA by considering the unique structural properties and flexibility of RNA molecules, providing reliable docking predictions |
| Glide [34] | Protein-ligand docking | Adapted for RNA | Glide is renowned for its high-throughput screening capabilities. Its application to RNA involves adjustments to accommodate RNA's structural dynamics, proving effective in RNA–ligand binding prediction |
| AutoDock [63] | Protein-ligand docking | Adapted for RNA | AutoDock includes algorithms that can be tuned to explore the conformational space of RNA, helping in identifying potential binding sites and ligand orientations suitable for RNA structures |
| ICM [61] | Protein-ligand docking | Adapted for RNA | Known for its accurate scoring functions, ICM has been adapted to consider RNA's unique characteristics, enhancing its utility in RNA-targeted drug design |
| DOCK [42, 84] | Protein-ligand docking | Adapted for RNA | DOCK has been utilized for RNA by modifying its scoring functions to better predict RNA–ligand interactions, benefiting from its robust algorithm that handles diverse molecular shapes |
| FITTED [32] | Protein-ligand docking | Adapted for RNA | FITTED has been adapted to RNA docking, providing detailed analyses of ligand binding modes and energetics tailored to RNA's specific needs, thereby increasing the accuracy of RNA–ligand interaction predictions |
| RiboDock [33] | N/A | Specifically for RNA | Designed specifically for RNA targets, RiboDock excels in handling the complex secondary structures of RNA, making it highly effective for RNA-focused studies |
| rDock [40] | N/A | Specifically for RNA | rDock is crafted for RNA applications, featuring specialized algorithms to navigate RNA's unique topology and flexibility, thus delivering high accuracy in RNA-specific docking scenarios |
| MORDOR [43] | N/A | Specifically for RNA | MORDOR focuses on RNA targets with a tailored approach that considers the diverse and dynamic nature of RNA structures, enhancing its predictive performance in RNA–ligand docking |
| RLDOCK [36, 37] | N/A | Specifically for RNA | RLDOCK stands out in RNA docking due to its specialized algorithms that optimize binding predictions based on RNA-specific features, proving particularly useful in RNA therapeutic development |

saved for further refinement. To improve efficiency, prior to applying a scoring function for ranking, RLDOCK filters out configurations with high Lennard-Jones (LJ) potential interactions between the RNA and ligand. The algorithm selects R sites characterized by low LJ potentials ULJ(R) (under a predefined threshold), indicating minimal LJ interaction for any combination of (L, A, O) with the specific R. For each chosen R site, it identifies preferred ligand conformers L, selected based on their low LJ potentials ULJ (R, L), which is the least LJ potential across all combinations of (A, O) for the given pair (R, L). Following this, for every chosen pair of R and L, preferred ligand atoms A are determined by the lowest LJ potentials ULJ (R, L, A), the minimum interaction energy across all possible orientations O for the trio (R, L, A). Subsequently, a preferred ensemble of (R, L, A, O) that includes all potential ligand orientations (O) is compiled and ranked using the scoring function. To expedite the calculation of LJ potentials, RLDOCK utilizes a grid-based approach for energy calculation, wherein each grid point records the LJ energy between RNA and a probe atom placed at the grid, enabling quick LJ energy estimations for any binding mode. Through this comprehensive procedure, RLDOCK is capable of producing millions of potential ligand configurations by thorough rotation and translation of each ligand conformer at each potential binding site. This exhaustive sampling of ligand conformers and binding modes sets RLDOCK apart from other models, highlighting its unique capacity for thorough exploration in ligand docking processes.

## 18.6 Incremental Construction

The approach of incremental construction for sampling ligand conformations uses geometric matching to anchor rigid fragments of the ligand, which then guides the expansion of the ligand structure according to the local features of the binding pocket. This method allows for the dynamic sampling of ligand conformers, with the binding site's environment directly influencing the ligand's structural development. However, a notable limitation of this technique is the potential for initial minor inaccuracies to be magnified as the ligand grows, particularly in the case of larger ligands. DOCK 665 utilizes this incremental approach for exploring ligand conformations and searching for optimal binding positions [24]. Departing from the original method that depended on a greedy algorithm for pruning sampled ligand structures through clustering and ranking at each stage, an improved algorithm has now been implemented. This revised algorithm omits the step of conformational clustering, thereby preserving the original, varied conformations of the flexible bonds within the ligands. This adjustment has yielded a 10% improvement in the accuracy of predicting binding poses [42], underscoring the significance of retaining a broad spectrum of ligand conformers during the docking process.

## 18.7 Stochastic Optimization

The stochastic optimization approach, particularly through the use of Monte Carlo (MC) algorithms, plays a critical role in identifying binding modes by dynamically optimizing the variables that govern molecular interactions, such as torsional angles, orientation, and position of small molecules [34, 41]. Integrating various stochastic methods often enhances both sampling and optimization outcomes. For instance, AutoDock Vina implements a hybrid strategy, using MC for broad optimization efforts and the Broyden–Fletcher–Goldfarb–Shanno (BFGS) algorithm for more focused, local optimization [41]. Other examples include ICM [61] and RiboDock [33], which pair MC with simulated annealing for effective optimization. Programs designed for protein docking like GOLD [35], AutoDock [63], and FITTED [32] engage GA to explore and identify ligand conformations. Tailoring these methods for RNA has yielded very encouraging results. Adjustments such as recalibrating the scoring function [95] or introducing a new solvation term [96] have allowed AutoDock to effectively manage RNA–ligand interactions. The FITTED docking algorithm, initially designed for protein–ligand interactions, can indeed be adapted to effectively handle RNA–ligand docking scenarios. Given the unique structural and chemical characteristics of RNA compared to proteins, several adjustments and enhancements are necessary to optimize FITTED for RNA targets [29].

Despite their utility, a limitation of MC and GA methods is the risk of optimization becoming ensnared in local minima, a significant concern given the complex energy landscapes of RNA–ligand complexes. This issue can be mitigated by performing multiple dockings with varied initial ligand placements and employing strategies like tabu search (neighboring algorithm search) [97, 98] and stochastic tunneling [99] to escape local minima more effectively. The rDock software, which is designed to handle docking simulations for both nucleic acids and proteins, addresses one of the major challenges in computational docking: the entrapment of ligands in local minima during the conformational exploration process [40]. To overcome this issue, rDock implements an innovative GA/MC (Genetic Algorithm/ Monte Carlo) hybrid method coupled with Simplex minimization. rDock utilizes a GA/MC hybrid method. This method conducts three cycles of GA, low-temperature MC, and Simplex minimization, each using a distinct scoring function. Each of these components plays a critical role in refining the docking process, ensuring a more thorough exploration of potential docking conformations and enhancing the accuracy and efficiency of the simulation.

### *18.7.1 Genetic Algorithm (GA)*

The GA is used initially to explore a wide conformational space. It simulates the process of natural evolution by using operations such as selection, crossover, and mutation to generate populations of molecule conformations that evolve over time.

In rDock, the GA helps to avoid premature convergence on local minima by generating diverse ligand conformations. This diversity ensures that the search space is extensively sampled, increasing the chances of finding global minima.

### *18.7.2 Low-Temperature Monte Carlo (MC)*

MC method at a low temperature is employed to fine-tune the conformations by making small, random changes to the conformational states produced by the GA. The low-temperature setting helps to reduce the acceptance rate of energetically unfavorable changes, promoting a more focused exploration around the promising areas identified by the GA, thereby refining the search toward the true minimum.

### *18.7.3 Simplex Minimization*

This numerical optimization technique is used to perform a local minimization of the energy, starting from the conformations refined by the GA and MC stages. Simplex minimization adjusts the variables of the system (such as bond angles and torsion angles of the ligand) to find the lowest energy conformation in the immediate neighborhood of the initial conformation provided by the previous steps.

### *18.7.4 Distinct Scoring Functions*

Each phase of the cycle—GA, MC, and Simplex—utilizes a different scoring function to evaluate the energy of the ligand conformations. This approach allows each method to focus on different aspects of the docking interaction. Using distinct scoring functions can provide a more balanced assessment of various physicochemical properties of the interaction, such as electrostatic interactions, hydrogen bonding, hydrophobic effects, and geometric fit.

## 18.8 Incorporation of RNA Flexibility

Addressing receptor flexibility is crucial in accurately predicting ligand binding modes, a principle well-documented within protein-ligand docking scenarios. Ignoring this aspect can lead to significant inaccuracies, as the receptor's binding conformation might differ from its unbound state [100]. This challenge is even more pronounced in RNA–ligand docking due to the inherently greater flexibility of RNA structures compared to proteins. To mitigate these issues, researchers have devised several effective strategies for integrating RNA conformational variability into docking algorithms. These methodologies fall into three primary categories: soft docking, ensemble docking, and fully flexible docking. Soft docking focuses on accommodating minor conformational changes by allowing some degree of flexibility or "softness" in the interactions between the RNA and the ligand. This method often involves creating a more forgiving interaction environment that can adjust to small shifts in the RNA structure, thereby enabling a closer approximation of the actual binding process without explicitly modeling large-scale conformational alterations. Ensemble docking takes into account a range of possible RNA conformations by docking the ligand against multiple RNA structures, either generated through experimental methods or computational simulations. This approach acknowledges the dynamic nature of RNA by considering various structural states the RNA might adopt, offering a broader perspective on potential binding sites and modes. Fully flexible docking represents the most comprehensive approach, where both the RNA and ligand are allowed to adjust their conformations dynamically during the docking process. This technique aims to simulate the entire interaction process as closely as possible, capturing both the RNA's conformational changes and the ligand's adjustments in real-time to predict the most accurate binding modes. Each of these approaches addresses the critical aspect of RNA flexibility to varying extents, aiming to improve the accuracy of RNA–ligand docking predictions by accounting for the dynamic nature of RNA structures.

## 18.9 Soft Docking

Soft docking simplifies the process of accommodating receptor flexibility by reducing the energy penalties associated with steric clashes, thereby permitting a certain degree of overlap between the RNA and ligand molecules. This approach, which makes the docking process mathematically more tractable, is supported by programs like Glide [34] and GOLD [35], offering users the capability to apply softer constraints for interactions. Moitessier et al. successfully utilized this method for docking aminoglycosides into the ribosomal A-site RNA [26]. By adopting a set of softened van der Waals potentials, they accounted for RNA flexibility, resulting in a notable improvement in docking accuracy. The appeal of soft docking lies in its ease of implementation, making it a viable option for quickly exploring potential binding interactions. However, its effectiveness is limited to minor conformational adjustments, as the method does not modify the RNA backbone. Consequently, soft docking may not be suitable for scenarios requiring the modeling of extensive conformational changes within the RNA structure.

## 18.10 Ensemble Docking

Ensemble docking entails the process of docking a ligand into a group of RNA conformations or a conformation averaged from an ensemble. This approach offers a dynamic view of possible binding interactions. This method has been instrumental in various RNA-targeting research efforts [26, 77, 78], and its utility extends to the realm of protein–small molecule docking as well [101, 102]. For instance, employing soft docking techniques on an ensemble-averaged RNA structure to model RNA aminoglycoside complexes resulted in predictions closely aligning with experimental data, showcasing an average RMSD (root mean square deviation) of 2.49 Å from the experimentally observed binding modes [26]. Additionally, screening a library of roughly 100,000 drug-like organic compounds against a conformational ensemble of experimentally determined TAR structures highlighted a group of TAR-binding compounds [78]. From a real-world standpoint, the number of RNA conformations considered in ensemble docking is typically constrained by computational resources, making the selection of receptor conformations a critical factor in the accuracy of docking predictions. Depending solely on the conformational basin with the lowest energy may not always yield the best results, as ligand binding can stabilize and selectively enhance conformations found in other energy basins [102–104]. Hence, it is important for ensemble docking approaches to also consider less populated, ligand-free RNA conformations to ensure a comprehensive exploration of potential binding interactions.

## 18.11 Molecular Dynamics (MD) Simulation for RNA–Ligand Refinement

Molecular dynamics (MD) simulations [105–112] are pivotal not just for refining RNA–ligand complex conformations and generating structures for both ligands and RNA but also for illuminating the dynamics and the folding or unfolding processes of potential metastable states within the docking interaction [24, 78, 113]. Nonetheless, the occurrence of ligand binding can span to seconds [73, 75, 114], rendering a comprehensive all-atom MD simulation of such processes unfeasible with current computational resources, particularly in the context of drug molecule virtual screening. Through the application of sophisticated sampling techniques, various computational strategies have successfully depicted RNA conformational alterations triggered by ligand attachment [115–119]. Both a non-equilibrium MD simulation [120] and an MD simulation employing umbrella sampling [121] have uncovered the competitive dynamics between kissing-loop formation and small molecule engagement. These findings underscore the distinct benefits of MD simulations in exploring the physical principles underlying RNA–ligand interactions.

## 18.12 Molecular Dynamics for Fully Flexible RNA

Molecular dynamics (MD) simulations, when paired with an appropriate force field, offer dependable sampling of RNA–ligand complex configurations. For instance, MOlecular Recognition with a Driven dynamics OptimizeR (MORDOR), which integrates an RMSD penalty with the standard potential energy framework, facilitates the simulation of ligand docking pathways, thus enabling the examination of conformations and identifying ligand-triggered structural alterations in RNA [43]. An illustrative use of MORDOR in virtual screening has successfully identified a specific group of molecules that bind to human telomerase RNAs (hTR) [122]. Nevertheless, the broader adoption of MORDOR is hampered by its substantial computational demands, with docking simulations extending to hours. In response, Supervised Molecular Dynamics (SuMD) has been introduced to expedite the conformation sampling of RNA–drug complexes [123]. SuMD enhances simulation speeds by utilizing a tabu-like algorithm to steer the docking as the ligand approaches the binding site, switching to traditional MD simulation closer to the site. This dual strategy efficiently simulates the binding activity within a feasible timeframe. Although SuMD's generated pathways may not precisely mirror the actual binding process, it effectively captures potential conformational shifts. SuMD's aptitude for RNA–ligand docking is validated through its successful prediction of binding configurations for various pharmacologically relevant RNAs, achieving a minimal RMSD of 0.34 Å in optimal scenarios [123]. Furthermore, a methodology employing elastic potential grids, originally designed for protein flexibility during docking [124], has been adapted for RNA targets [79]. This approach calculates a three-dimensional potential field grid of the initial RNA structure using DrugScoreRNA [125], followed by AutoDock [63] for docking with these pre-established elastic grids. This strategy significantly improves docking results by accommodating RNA flexibility, outperforming standard docking to static RNA structures. However, this method's drawback lies in its prerequisite knowledge of the RNA's potential deformation endpoints and its inability to account for rotational or 2D structural shifts. Therefore, while molecular dynamics-based techniques offer detailed insights, their extensive computational requirements limit their applicability for large-scale virtual screening. Rigid docking, though swift, falls short on precision. Soft and ensemble docking strike a balance by trading off some conformational sampling depth for computational efficiency. Currently, the development of a comprehensive method that accurately accommodates receptor flexibility remains an open challenge.

## 18.13 Conclusion

The burgeoning interest in RNA-targeted therapeutics has spurred a growing need for computational strategies to predict RNA–ligand interactions, making virtual screening a pivotal initial phase in the design of new drugs when only the target

RNA sequence is known. A variety of docking methodologies and scoring algorithms have been devised to expedite this process while simultaneously enriching our comprehension of the RNA–ligand binding dynamics. Both physics-driven and knowledge-informed techniques have demonstrated considerable promise in forecasting ligand binding orientations and affinities. Nevertheless, despite being relatively nascent in the context of RNA–ligand docking, machine learning approaches, bolstered by sophisticated algorithms, are starting to outperform conventional methods with their promising results. The journey of computer-assisted drug discovery has evolved significantly from its protein-centric origins, with novel RNA-focused therapeutic designs and computational approaches coming to the forefront. As the compendium of known structures, along with kinetic and thermodynamic data such as binding affinities, continues to expand, data-driven strategies, particularly those employing machine learning, are poised to play an increasingly crucial role. Additionally, the recognition of the significance of kinetic effects in RNA–ligand binding on drug efficacy suggests that models centered on kinetics, though currently underutilized for RNA–drug interactions, will likely see rapid development. These would involve blind tests and meticulously curated benchmark datasets, analogous to those established within the protein–ligand modeling sphere, to facilitate progress in this area.

## References


1. Hermann T. Small molecules targeting viral RNA. Wiley Interdiscip Rev RNA. 2016;7(6):726–43.
2. Wilson DN. Ribosome-targeting antibiotics and mechanisms of bacterial resistance. Nat Rev Microbiol. 2014;12(1):35–48.
3. Bulkley D, Innis CA, Blaha G, Steitz TA. Revisiting the structures of several antibiotics bound to the bacterial ribosome. Proc Natl Acad Sci USA. 2010;107(40):17158–63.
4. Deak D, Outterson K, Powers JH, Kesselheim AS. Progress in the fight against multidrug-resistant bacteria? A review of U.S. food and drug administration-approved antibiotics, 2010–2015. Ann Intern Med. 2016;165(5):363–72.
5. Lin J, Zhou D, Steitz TA, Polikanov YS, Gagnon MG. Ribosome-targeting antibiotics: modes of action, mechanisms of resistance, and implications for drug design. Annu Rev Biochem. 2018;87:451–78.
6. Poehlsgaard J, Douthwaite S. The bacterial ribosome as a target for antibiotics. Nat Rev Microbiol. 2005;3(11):870–81.
7. Roth A, Breaker RR. The structural and functional diversity of metabolite-binding riboswitches. Annu Rev Biochem. 2009;78:305–34.
8. Blount KF, Breaker RR. Riboswitches as antibacterial drug targets. Nat Biotechnol. 2006;24(12):1558–64.
9. Mandal M, Breaker RR. Gene regulation by riboswitches. Nat Rev Mol Cell Biol. 2004;5(6):451–63.
10. Montange RK, Batey RT. Riboswitches: emerging themes in RNA structure and function. Annu Rev Biophys. 2008;37:117–33.
11. Tucker BJ, Breaker RR. Riboswitches as versatile gene control elements. Curr Opin Struct Biol. 2005;15(3):342–8.

12. Berens C, Suess B. Riboswitch engineering—making the all-important second and third steps. Curr Opin Biotechnol. 2015;31:10–5.
13. Breaker RR. Riboswitches and the RNA world. Cold Spring Harb Perspect Biol. 2012;4(2):a003566.
14. Garst AD, Edwards AL, Batey RT. Riboswitches: structures and mechanisms. Cold Spring Harb Perspect Biol. 2011;3(6):a003533.
15. Peselis A, Serganov A. Themes and variations in riboswitch structure and function. Biochim Biophys Acta. 2014;1839(10):908–18.
16. Serganov A, Nudler E. A decade of riboswitches. Cell. 2013;152(1–2):17–24.
17. Bédard AV, Hien EDM, Lafontaine DA. Riboswitch regulation mechanisms: RNA, metabolites and regulatory proteins. Biochim Biophys Acta Gene Regul Mech. 2020;1863(3):194501.
18. Breaker RR. Riboswitches and translation control. Cold Spring Harb Perspect Biol. 2018;10(11):a032797.
19. Hallberg ZF, Su Y, Kitto RZ, Hammond MC. Engineering and in vivo applications of riboswitches. Annu Rev Biochem. 2017;86:515–39.
20. McCown PJ, Corbino KA, Stav S, Sherlock ME, Breaker RR. Riboswitch diversity and distribution. RNA. 2017;23(7):995–1011.
21. Pavlova N, Kaloudas D, Penchovsky R. Riboswitch distribution, structure, and function in bacteria. Gene. 2019;708:38–48.
22. Plant EP, Pérez-Alvarado GC, Jacobs JL, Mukhopadhyay B, Hennig M, Dinman JD. A three-stemmed mRNA pseudoknot in the SARS coronavirus frameshift signal. PLoS Biol. 2005;3(6):e172.
23. Su MC, Chang CT, Chu CH, Tsai CH, Chang KY. An atypical RNA pseudoknot stimulator and an upstream attenuation signal for −1 ribosomal frameshifting of SARS coronavirus. Nucleic Acids Res. 2005;33(13):4265–75.
24. Park SJ, Kim YG, Park HJ. Identification of RNA pseudoknot-binding ligand that inhibits the −1 ribosomal frameshifting of SARS-coronavirus by structure-based virtual screening. J Am Chem Soc. 2011;133(26):10094–100.
25. Ritchie DB, Soong J, Sikkema WK, Woodside MT. Anti-frameshifting ligand reduces the conformational plasticity of the SARS virus pseudoknot. J Am Chem Soc. 2014;136(6):2196–9.
26. Moitessier N, Westhof E, Hanessian S. Docking of aminoglycosides to hydrated and flexible RNA. J Med Chem. 2006;49(3):1023–33.
27. Li Y, Shen J, Sun X, Li W, Liu G, Tang Y. Accuracy assessment of protein-based docking programs against RNA targets. J Chem Inf Model. 2010;50(6):1134–46.
28. Chen L, Calin GA, Zhang S. Novel insights of structure-based modeling for RNA-targeted drug discovery. J Chem Inf Model. 2012;52(10):2741–53.
29. Luo J, Wei W, Waldispühl J, Moitessier N. Challenges and current status of computational methods for docking small molecules to nucleic acids. Eur J Med Chem. 2019;168:414–25.
30. Tessaro F, Scapozza L. How 'Protein-docking' translates into the new emerging field of docking small molecules to nucleic acids? Molecules. 2020;25(12):2749.
31. Huey R, Morris GM, Olson AJ, Goodsell DS. A semiempirical free energy force field with charge-based desolvation. J Comput Chem. 2007;28(6):1145–52.
32. Corbeil CR, Englebienne P, Moitessier N. Docking ligands into flexible and solvated macromolecules. 1. Development and validation of FITTED 1.0. J Chem Inf Model. 2007;47(2):435–49.
33. Morley SD, Afshar M. Validation of an empirical RNA-ligand scoring function for fast flexible docking using Ribodock. J Comput Aided Mol Des. 2004;18(3):189–208.
34. Friesner RA, Banks JL, Murphy RB, Halgren TA, Klicic JJ, Mainz DT, et al. Glide: a new approach for rapid, accurate docking and scoring. 1. Method and assessment of docking accuracy. J Med Chem. 2004;47(7):1739–49.
35. Jones G, Willett P, Glen RC, Leach AR, Taylor R. Development and validation of a genetic algorithm for flexible docking. J Mol Biol. 1997;267(3):727–48.

36. Jiang Y, Chen SJ. RLDOCK method for predicting RNA-small molecule binding modes. Methods. 2022;197:97–105.
37. Sun LZ, Jiang Y, Zhou Y, Chen SJ. RLDOCK: a new method for predicting RNA-ligand interactions. J Chem Theory Comput. 2020;16(11):7173–83.
38. Sugita M, Hamano M, Kasahara K, Kikuchi T, Hirata F. New protocol for predicting the ligand-binding site and mode based on the 3D-RISM/KH theory. J Chem Theory Comput. 2020;16(4):2864–76.
39. Bezerra KS, Fulco UL, Esmaile SC, Lima Neto JX, Machado LD, Freire VN, et al. Ribosomal RNA-aminoglycoside hygromycin B interaction energy calculation within a density functional theory framework. J Phys Chem B. 2019;123(30):6421–9.
40. Ruiz-Carmona S, Alvarez-Garcia D, Foloppe N, Garmendia-Doval AB, Juhos S, Schmidtke P, et al. rDock: a fast, versatile and open source program for docking ligands to proteins and nucleic acids. PLoS Comput Biol. 2014;10(4):e1003571.
41. Trott O, Olson AJ. AutoDock Vina: improving the speed and accuracy of docking with a new scoring function, efficient optimization, and multithreading. J Comput Chem. 2010;31(2):455–61.
42. Lang PT, Brozell SR, Mukherjee S, Pettersen EF, Meng EC, Thomas V, et al. DOCK 6: combining techniques to model RNA-small molecule complexes. RNA. 2009;15(6):1219–30.
43. Guilbert C, James TL. Docking to RNA via root-mean-square-deviation-driven energy minimization with flexible ligands and flexible targets. J Chem Inf Model. 2008;48(6):1257–68.
44. Manigrasso J, Marcia M, De Vivo M. Computer-aided design of RNA-targeted small molecules: a growing need in drug discovery. Chem. 2021;7(11):2965–88.
45. Sousa SF, Fernandes PA, Ramos MJ. Protein-ligand docking: current status and future challenges. Proteins. 2006;65(1):15–26.
46. Huang SY, Grinter SZ, Zou X. Scoring functions and their evaluation methods for protein-ligand docking: recent advances and future directions. Phys Chem Chem Phys. 2010;12(40):12899–908.
47. Colwell LJ. Statistical and machine learning approaches to predicting protein-ligand interactions. Curr Opin Struct Biol. 2018;49:123–8.
48. Li J, Fu A, Zhang L. An overview of scoring functions used for protein-ligand interactions in molecular docking. Interdiscip Sci. 2019;11(2):320–8.
49. Carothers JM, Oestreich SC, Davis JH, Szostak JW. Informational complexity and functional activity of RNA structures. J Am Chem Soc. 2004;126(16):5130–7.
50. Schneider TD, Stormo GD, Gold L, Ehrenfeucht A. Information content of binding sites on nucleotide sequences. J Mol Biol. 1986;188(3):415–31.
51. An J, Totrov M, Abagyan R. Pocketome via comprehensive identification and classification of ligand binding envelopes. Mol Cell Proteomics. 2005;4(6):752–61.
52. Warner KD, Hajdin CE, Weeks KM. Principles for targeting RNA with drug-like small molecules. Nat Rev Drug Discov. 2018;17(8):547–58.
53. Siegfried NA, Busan S, Rice GM, Nelson JA, Weeks KM. RNA motif discovery by SHAPE and mutational profiling (SHAPE-MaP). Nat Methods. 2014;11(9):959–65.
54. Loughrey D, Watters KE, Settle AH, Lucks JB. SHAPE-Seq 2.0: systematic optimization and extension of high-throughput chemical probing of RNA secondary structure with next generation sequencing. Nucleic Acids Res. 2014;42(21):e165.
55. Mortimer SA, Trapnell C, Aviran S, Pachter L, Lucks JB. SHAPE-Seq: high-throughput RNA structure analysis. Curr Protoc Chem Biol. 2012;4(4):275–97.
56. Merino EJ, Wilkinson KA, Coughlan JL, Weeks KM. RNA structure analysis at single nucleotide resolution by selective 2′-hydroxyl acylation and primer extension (SHAPE). J Am Chem Soc. 2005;127(12):4223–31.
57. Carothers JM, Davis JH, Chou JJ, Szostak JW. Solution structure of an informationally complex high-affinity RNA aptamer to GTP. RNA. 2006;12(4):567–79.
58. Hewitt WM, Calabrese DR, Schneekloth JS Jr. Evidence for ligandable sites in structured RNA throughout the protein data bank. Bioorg Med Chem. 2019;27(11):2253–60.

59. Berman HM, Westbrook J, Feng Z, Gilliland G, Bhat TN, Weissig H, et al. The protein data bank. Nucleic Acids Res. 2000;28(1):235–42.
60. Kuntz ID, Blaney JM, Oatley SJ, Langridge R, Ferrin TE. A geometric approach to macromolecule-ligand interactions. J Mol Biol. 1982;161(2):269–88.
61. Abagyan R, Totrov M, Kuznetsov D. ICM—A new method for protein modeling and design: applications to docking and structure prediction from the distorted native conformation. J Comput Chem. 1994;15(5):488.
62. Harris R, Olson AJ, Goodsell DS. Automated prediction of ligand-binding sites in proteins. Proteins. 2008;70(4):1506–17.
63. Morris GM, Goodsell DS, Halliday RS, Huey R, Hart WE, Belew RK, et al. Automated docking using a Lamarckian genetic algorithm and an empirical binding free energy function. J Comput Chem. 2015;19(14):1639–62.
64. Zeng P, Li J, Ma W, Cui Q. Rsite: a computational method to identify the functional sites of noncoding RNAs. Sci Rep. 2015;5:9179.
65. Zeng P, Cui Q. Rsite2: an efficient computational method to predict the functional sites of noncoding RNAs. Sci Rep. 2016;6:19016.
66. Wang H, Zhao Y. RBinds: a user-friendly server for RNA binding site prediction. Comput Struct Biotechnol J. 2020;18:3762–5.
67. Wang K, Jian Y, Wang H, Zeng C, Zhao Y. RBind: computational network method to predict RNA binding sites. Bioinformatics. 2018;34(18):3131–6.
68. Su H, Peng Z, Yang J. Recognition of small molecule-RNA binding sites using RNA sequence and structure. Bioinformatics. 2021;37(1):36–42.
69. Zhao J, Cao Y, Zhang L. Exploring the computational methods for protein-ligand binding site prediction. Comput Struct Biotechnol J. 2020;18:417–26.
70. Kozlovskii I, Popov P. Spatiotemporal identification of druggable binding sites using deep learning. Commun Biol. 2020;3(1):618.
71. Krivák R, Hoksza D. P2Rank: machine learning based tool for rapid and accurate prediction of ligand binding sites from protein structure. J Chem Inform. 2018;10(1):39.
72. Jiménez J, Doerr S, Martínez-Rosell G, Rose AS, De Fabritiis G. DeepSite: protein-binding site predictor using 3D-convolutional neural networks. Bioinformatics. 2017;33(19):3036–42.
73. Al-Hashimi HM, Walter NG. RNA dynamics: it is about time. Curr Opin Struct Biol. 2008;18(3):321–9.
74. Ganser LR, Kelly ML, Herschlag D, Al-Hashimi HM. The roles of structural dynamics in the cellular functions of RNAs. Nat Rev Mol Cell Biol. 2019;20(8):474–89.
75. Scull CE, Dandpat SS, Romero RA, Walter NG. Transcriptional riboswitches integrate timescales for bacterial gene expression control. Front Mol Biosci. 2020;7:607158.
76. Zhao P, Zhang W, Chen SJ. Cotranscriptional folding kinetics of ribonucleic acid secondary structures. J Chem Phys. 2011;135(24):245101.
77. Stelzer AC, Frank AT, Kratz JD, Swanson MD, Gonzalez-Hernandez MJ, Lee J, et al. Discovery of selective bioactive small molecules by targeting an RNA dynamic ensemble. Nat Chem Biol. 2011;7(8):553–9.
78. Ganser LR, Lee J, Rangadurai A, Merriman DK, Kelly ML, Kansal AD, et al. High-performance virtual screening by targeting a high-resolution RNA dynamic ensemble. Nat Struct Mol Biol. 2018;25(5):425–34.
79. Krüger DM, Bergs J, Kazemi S, Gohlke H. Target flexibility in RNA-ligand docking modeled by elastic potential grids. ACS Med Chem Lett. 2011;2(7):489–93.
80. Tok JB, Bi L. Aminoglycoside and its derivatives as ligands to target the ribosome. Curr Top Med Chem. 2003;3(9):1001–19.
81. Daldrop P, Reyes FE, Robinson DA, Hammond CM, Lilley DM, Batey RT, et al. Novel ligands for a purine riboswitch discovered by RNA-ligand docking. Chem Biol. 2011;18(3):324–35.
82. Göertz GP, Abbo SR, Fros JJ, Pijlman GP. Functional RNA during Zika virus infection. Virus Res. 2018;254:41–53.

83. Rangan R, Zheludev IN, Hagey RJ, Pham EA, Wayment-Steele HK, Glenn JS, et al. RNA genome conservation and secondary structure in SARS-CoV-2 and SARS-related viruses: a first look. RNA. 2020;26(8):937–59.
84. Crooke ST, Witztum JL, Bennett CF, Baker BF. RNA-targeted therapeutics. Cell Metab. 2018;27(4):714–39.
85. Rognan D. Docking methods for virtual screening: principles and recent advances. In: Virtual screening: principles, challenges, and practical guidelines. Weinheim: Wiley; 2011. p. 153–76.
86. Waszkowycz B, Clark DE, Gancia E. Outstanding challenges in protein–ligand docking and structure-based virtual screening. Wiley Interdiscip Rev Comput Mol Sci. 2011;1(2):229–59.
87. Wehler T, Brenk R. Structure-based discovery of small molecules binding to RNA. In: Garner A, editor. RNA therapeutics. Topics in medicinal chemistry, vol. 27. Cham: Springer; 2017.
88. Software OES. OpenEye scientific software. OpenEye Scientific Software; 2008.
89. Hawkins PC, Skillman AG, Warren GL, Ellingson BA, Stahl MT. Conformer generation with OMEGA: algorithm and validation using high quality structures from the protein databank and Cambridge structural database. J Chem Inf Model. 2010;50(4):572–84.
90. RDKit. Open-Source Cheminformatics Software. http://www.rdkit.org.
91. Yoshikawa N, Hutchison GR. Fast, efficient fragment-based coordinate generation for open babel. J Chem Inform. 2019;11(1):49.
92. Hawkins PCD. Conformation generation: the state of the art. J Chem Inf Model. 2017;57(8):1747–56.
93. Moreno D, Zivanovic S, Colizzi F, Hospital A, Aranda J, Soliva R, et al. DFFR: a new method for high-throughput recalibration of automatic force-fields for drugs. J Chem Theory Comput. 2020;16(10):6598–608.
94. Zivanovic S, Colizzi F, Moreno D, Hospital A, Soliva R, Orozco M. Exploring the conformational landscape of bioactive small molecules. J Chem Theory Comput. 2020;16(10):6575–85.
95. Barbault F, Zhang L, Zhang L, Fan BT. Parametrization of a specific free energy function for automated docking against RNA targets using neural networks. Chemom Intell Lab Syst. 2006;82(1–2):269–75.
96. Detering C, Varani G. Validation of automated docking programs for docking and database screening against RNA drug targets. J Med Chem. 2004;47(17):4188–201.
97. Kitchen DB, Decornez H, Furr JR, Bajorath J. Docking and scoring in virtual screening for drug discovery: methods and applications. Nat Rev Drug Discov. 2004;3(11):935–49.
98. Tongcheng CTL. A combination of numeric genetic algorithm and tabu search can be applied to molecular docking. Comput Biol Chem. 2004;28(4):303–12.
99. Flick J, Tristram F, Wenzel W. Modeling loop backbone flexibility in receptor-ligand docking simulations. J Comput Chem. 2012;33(31):2504–15.
100. Totrov M, Abagyan R. Flexible ligand docking to multiple receptor conformations: a practical alternative. Curr Opin Struct Biol. 2008;18(2):178–84.
101. Huang SY, Zou X. Ensemble docking of multiple protein structures: considering protein structural variations in molecular docking. Proteins. 2007;66(2):399–421.
102. Amaro RE, Baudry J, Chodera J, Demir Ö, McCammon JA, Miao Y, et al. Ensemble docking in drug discovery. Biophys J. 2018;114(10):2271–8.
103. Reining A, Nozinovic S, Schlepckow K, Buhr F, Fürtig B, Schwalbe H. Three-state mechanism couples ligand and temperature sensing in riboswitches. Nature. 2013;499(7458):355–9.
104. Zhao B, Hansen AL, Zhang Q. Characterizing slow chemical exchange in nucleic acids by carbon CEST and low spin-lock field R(1ρ) NMR spectroscopy. J Am Chem Soc. 2014;136(1):20–3.
105. Cornell WD, Cieplak P, Bayly CI, Gould IR, Merz KM, Ferguson DM, et al. A second generation force field for the simulation of proteins, nucleic acids, and organic molecules. J Am Chem Soc. 1995;117(19):5179–97.
106. Case DA, Cheatham TE 3rd, Darden T, Gohlke H, Luo R, Merz KM Jr, et al. The Amber biomolecular simulation programs. J Comput Chem. 2005;26(16):1668–88.

107. Brooks BR, Brooks CL 3rd, Mackerell AD Jr, Nilsson L, Petrella RJ, Roux B, et al. CHARMM: the biomolecular simulation program. J Comput Chem. 2009;30(10):1545–614.
108. Denning EJ, Priyakumar UD, Nilsson L, Mackerell AD Jr. Impact of 2′-hydroxyl sampling on the conformational properties of RNA: update of the CHARMM all-atom additive force field for RNA. J Comput Chem. 2011;32(9):1929–43.
109. Schmid N, Eichenberger AP, Choutko A, Riniker S, Winger M, Mark AE, et al. Definition and testing of the GROMOS force-field versions 54A7 and 54B7. Eur Biophys J. 2011;40(7):843–56.
110. Banáš P, Sklenovský P, Wedekind JE, Šponer J, Otyepka M. Molecular mechanism of preQ1 riboswitch action: a molecular dynamics study. J Phys Chem B. 2012;116(42):12721–34.
111. Šponer J, Bussi G, Krepl M, Banáš P, Bottaro S, Cunha RA, et al. RNA structural dynamics as captured by molecular simulations: a comprehensive overview. Chem Rev. 2018;118(8):4177–338.
112. Kührová P, Mlýnský V, Zgarbová M, Krepl M, Bussi G, Best RB, et al. Improving the performance of the Amber RNA force field by tuning the hydrogen-bonding interactions. J Chem Theory Comput. 2019;15(5):3288–305.
113. Paulsen RB, Seth PP, Swayze EE, Griffey RH, Skalicky JJ, Cheatham TE 3rd, et al. Inhibitor-induced structural change in the HCV IRES domain IIa RNA. Proc Natl Acad Sci USA. 2010;107(16):7263–8.
114. Haller A, Soulière MF, Micura R. The dynamic nature of RNA as key to understanding riboswitch mechanisms. Acc Chem Res. 2011;44(12):1339–48.
115. Sharma M, Bulusu G, Mitra A. MD simulations of ligand-bound and ligand-free aptamer: molecular level insights into the binding and switching mechanism of the add A-riboswitch. RNA. 2009;15(9):1673–92.
116. Villa A, Wöhnert J, Stock G. Molecular dynamics simulation study of the binding of purine bases to the aptamer domain of the guanine sensing riboswitch. Nucleic Acids Res. 2009;37(14):4774–86.
117. Priyakumar UD, MacKerell AD Jr. Role of the adenine ligand on the stabilization of the secondary and tertiary interactions in the adenine riboswitch. J Mol Biol. 2010;396(5):1422–38.
118. Gong Z, Zhao Y, Chen C, Xiao Y. Role of ligand binding in structural organization of add A-riboswitch aptamer: a molecular dynamics simulation. J Biomol Struct Dyn. 2011;29(2):403–16.
119. Di Palma F, Colizzi F, Bussi G. Ligand-induced stabilization of the aptamer terminal helix in the add adenine riboswitch. RNA. 2013;19(11):1517–24.
120. Nguyen PH, Derreumaux P, Stock G. Energy flow and long-range correlations in guanine-binding riboswitch: a nonequilibrium molecular dynamics study. J Phys Chem B. 2009;113(27):9340–7.
121. Allnér O, Nilsson L, Villa A. Loop-loop interaction in an adenine-sensing riboswitch: a molecular dynamics study. RNA. 2013;19(7):916–26.
122. Pinto IG, Guilbert C, Ulyanov NB, Stearns J, James TL. Discovery of ligands for a novel target, the human telomerase RNA, based on flexible-target virtual screening and NMR. J Med Chem. 2008;51(22):7205–15.
123. Bissaro M, Sturlese M, Moro S. Exploring the RNA-recognition mechanism using supervised molecular dynamics (SuMD) simulations: toward a rational design for ribonucleic-targeting molecules? Front Chem. 2020;8:107.
124. Kazemi S, Krüger DM, Sirockin F, Gohlke H. Elastic potential grids: accurate and efficient representation of intermolecular interactions for fully flexible docking. ChemMedChem. 2009;4(8):1264–8.
125. Pfeffer P, Gohlke H. DrugScoreRNA--knowledge-based scoring function to predict RNA-ligand interactions. J Chem Inf Model. 2007;47(5):1868–76.

# Chapter 19
# RNA Imaging Technologies

**Abstract** RNA processing is integral to the accurate translation of genetic blueprints into functional RNA molecules and regulatory proteins, playing a pivotal role in cellular function and organismal development. Mastery over this complex process is crucial for understanding the myriad biological functions and disease mechanisms. To effectively study the RNA lifecycle and its regulatory roles, researchers are in dire need of sophisticated tools capable of monitoring RNA molecules with high spatial and temporal precision. This chapter delves into the significant strides made in the field of RNA imaging technologies, which have revolutionized our ability to observe and understand RNA dynamics at unprecedented levels of detail. We extensively review the latest innovations that have pushed the boundaries of resolution, sensitivity, and specificity in RNA visualization, including fluorescent in situ hybridization techniques, CRISPR-based systems, and advanced microscopy modalities. Furthermore, we discuss the prevailing trends in RNA imaging, such as the integration of live-cell imaging and high-throughput genomic methods, which synergize to provide comprehensive insights into RNA behavior and function. The potential of these technologies to answer long-standing questions in RNA biology is immense, offering glimpses into previously unobservable aspects of RNA processing, transport, localization, and interaction dynamics within the cellular milieu. In this chapter, we not only assess the technical capabilities and advancements of these imaging tools but also explore their profound implications in unraveling the intricate dynamics of RNA within living cells. By highlighting key studies and seminal works, this chapter underscores the transformative impact of modern RNA imaging technologies on our understanding of cellular biology and opens the door to novel therapeutic avenues in the treatment of genetic disorders.

## 19.1 Introduction

The conversion of DNA to protein involves an intricate series of steps centered on RNA metabolism. This journey begins with transcription and continues through splicing, localization, translation, and ultimately degradation of RNA molecules.



P. E. Saw, E. Song, *RNA Therapeutics in Human Diseases*,
https://doi.org/10.1007/978-981-96-3041-7_19

Each of these phases is intricately choreographed and strictly controlled, both spatially and temporally. Traditional biochemical and genetic approaches have shed light on the roles and functions of proteins and non-coding RNAs (ncRNAs) in RNA processing, focusing primarily on the "what" and "how" of these processes. However, a comprehensive understanding of the "when" and "where"—the timing and localization of RNA events—necessitates the visualization of RNA within living cells. Over the past 40 years, substantial efforts have been directed toward developing and refining RNA imaging tools suitable for both fixed and live cell analysis. These tools have capitalized on advances in fluorescent microscopy, image processing technology, DNA chemistry, and next-generation sequencing. Such innovations have facilitated several key achievements in the field: single-molecule detection that allows for precise quantification and localization, super-resolution imaging that provides clarity beyond the diffraction limit of light, multiplexing capabilities that enable the simultaneous observation of multiple RNA species, and the ability to monitor RNA dynamics in real-time within live cells. This chapter delves into these technological advancements in RNA imaging and discusses their significant contributions to our ongoing exploration of RNA biology. By enhancing our ability to observe RNA molecules in their native cellular context, these tools not only enrich our understanding of RNA functions but also open new avenues for investigating the intricate regulatory networks that govern cellular life (Fig. 19.1).

## 19.2 Current Technologies in RNA Imaging

RNA imaging technologies have rapidly evolved, providing powerful tools for investigating RNA in both fixed and live cell contexts. In fixed cells, current methodologies deliver significant throughput, enabling comprehensive analysis. These techniques allow for detailed mapping of RNA localization and robust quantification of expression levels across the entire transcriptome, providing a snapshot of RNA presence and distribution within a cell at a specific point in time. For live cells, the scope of monitoring is generally limited to tracking the expression of a single gene per fluorescent label, due to current technological constraints. However, the enhancements in temporal resolution offered by live-cell RNA imaging technologies have substantially deepened our understanding of RNA processing dynamics. These advancements allow researchers to observe how RNA molecules behave and interact in real time, shedding light on the processes governing RNA stability, transportation, and localization during various cellular states and conditions. These live-cell techniques are particularly crucial for unraveling the dynamic aspects of RNA biology, such as transcriptional bursts or the rapid changes in RNA localization that occur in response to cellular signals or during developmental processes. By continuously advancing RNA imaging tools, we enhance our capacity to explore these fundamental processes in greater detail, paving the way for new discoveries in the field of molecular biology.

**Fig. 19.1** Advancements in RNA imaging technology. This timeline demonstrates the rapid evolution and increasing sophistication of RNA imaging technologies, enhancing our ability to study RNA dynamics and localization at high resolution and multiplexing capacity

### 19.2.1 Fluorescence in Situ Hybridization (FISH) for Fixed Cells

In 1982, Singer and Ward marked a significant milestone in RNA imaging by pioneering the use of fluorescence in situ hybridization (FISH) for detecting RNA. They developed a method where actin messenger RNA (mRNA) was specifically targeted. This technique involved using a rhodamine-tagged avidin, which bound to a DNA probe that had been modified with biotinylated 2′-deoxyuridine-5′-triphosphate (dUTP). This innovative approach allowed for the direct visualization of mRNA molecules within a cellular environment, highlighting the spatial distribution and localization of specific RNA transcripts. The integration of fluorescent tags with nucleic acid probes provided a powerful tool for molecular biology, enabling researchers to observe the dynamic processes of RNA within fixed cells under a microscope. The use of rhodamine as a fluorescent label, due to its bright and stable fluorescence properties, allowed for clear and precise imaging, thereby setting a foundation for numerous advancements in RNA detection techniques. This seminal work by Singer and Ward not only demonstrated the feasibility of using hybridization-based methods for RNA visualization but also opened new avenues for studying RNA function and regulation in situ. Their contributions continue to influence the development of more sophisticated RNA imaging methodologies, enhancing our ability to study cellular processes at the molecular level [1]. By 1998, the field of RNA imaging had taken another significant leap forward with the advent of single-molecule FISH (smFISH). This innovative technique expanded upon traditional FISH methods by introducing the use of complementary DNA (cDNA) oligonucleotide probes, each synthesized with five fluorophores. This enhancement significantly improved the sensitivity and specificity of RNA detection, allowing for the visualization and quantification of individual RNA molecules within cells. The incorporation of multiple fluorophores per probe amplified the fluorescent signal, thereby facilitating the direct observation of single RNA transcripts in their natural cellular environment. This level of resolution was pivotal for researchers aiming to understand complex gene expression patterns and the dynamics of RNA within individual cells. smFISH has since been instrumental in providing insights into the stochastic nature of gene expression, RNA distribution across different cell types, and the spatial organization of RNA within the cellular architecture [2]. By 2008, single-molecule FISH (smFISH) underwent a significant refinement, enhancing its capability to detect mRNAs at the single-molecule level with even greater precision and sensitivity. This advancement involved the strategic use of 48 DNA probes for each mRNA target, with each probe carrying a single fluorophore. This approach multiplied the fluorescent signals associated with each mRNA molecule, thereby dramatically improving the detectability and clarity of individual mRNA molecules within the cellular milieu. This enhancement allowed researchers to more accurately quantify and locate mRNA molecules, providing a detailed view of gene expression in various cellular contexts. The use of multiple probes per mRNA not only increased the robustness and signal-to-noise ratio of the detection but also enabled a more reliable differentiation between closely related mRNA sequences, thereby allowing for the nuanced study of gene expression variations and mRNA isoforms [3]. Rolling-circle amplification (RCA)-FISH represents a distinct and innovative approach in RNA imaging, diverging from the traditional multi-probe strategies typically used in single-molecule FISH. RCA-FISH begins with the hybridization of a specially designed padlock probe to a specific mRNA target. This probe is uniquely structured so that its ends are adjacent only when perfectly hybridized to the target mRNA, allowing them to be efficiently ligated into a circular form. Once the padlock probe is hybridized and ligated, the circularized probe serves as a template for the RCA process. During RCA, DNA polymerase is used to amplify the signal by rolling along the circular template, synthesizing a long, single-stranded DNA molecule containing many tandem repeats of the probe sequence. Each repeat can then be targeted by multiple fluorescently labeled oligonucleotides, thereby significantly amplifying the fluorescence signal associated with each mRNA molecule [4, 5].

**Box 19.1 Evolution and Advancement of Single-Molecule RNA Imaging Technologies** The evolution of single-molecule RNA imaging technologies has seen significant improvements, especially with advancements that build upon foundational techniques like smFISH and RCA-FISH.

**RNAscope**

RNAscope employs a unique and highly effective amplification strategy that involves a cascade of hybridization steps using tiled primary, secondary, and tertiary DNA oligonucleotide probes [6]. (i) Primary probes are designed to hybridize specifically to the target mRNA. Each primary probe is constructed to have a region that binds to the mRNA and a tail sequence that serves as a binding site for the secondary probes. This specificity is crucial for ensuring that the signal amplification is targeted only to the mRNA of interest. (ii) Following the binding of the primary probes, secondary probes are introduced. These probes are designed to hybridize to the tail sequences of the primary probes. Secondary probes carry multiple binding sites for the tertiary probes, effectively amplifying the potential signal. (iii) The final layer of amplification comes with the tertiary probes, which bind to the secondary probes. Tertiary probes are typically labeled with multiple fluorophores, dramatically increasing the fluorescence signal from each mRNA molecule. This layered approach allows RNAscope to produce a highly amplified signal from each mRNA molecule, significantly enhancing the visibility of single molecules even within densely packed cellular environments. The tiling of the probes ensures comprehensive coverage and robust signal amplification, which is particularly beneficial for detecting low-abundance transcripts or in tissues where background fluorescence is a challenge [6].

**Click-Amplifying FISH (clampFISH)**

clampFISH is a cutting-edge method in RNA imaging that utilizes a sophisticated iterative process to significantly enhance signal detection for mRNA molecules [7]. This technique employs a combination of bio-orthogonal click chemistry and FISH to achieve remarkable levels of signal amplification, making it an invaluable tool for researchers studying cellular gene expression at the single-molecule level. The process begins with the hybridization of padlock probes that are specifically designed to target and circularize around the mRNA of interest. These probes are uniquely structured to form a closed loop when perfectly aligned with their target sequence, facilitating the next step of ligation. The ends of the circularized padlock probes are covalently bonded using click chemistry—a highly efficient and specific chemical reaction that is bio-orthogonal, meaning it does not interfere with or perturb biological processes within the cell. This ligation step is crucial as it stabilizes the circularized probe, setting the stage for robust signal amplification. clampFISH amplifies the signal through an iterative process where additional oligonucleotides are hybridized and ligated to extend the chain, building a complex, branched structure. Each cycle of hybridization and ligation exponentially increases the number of available sites for fluorescent probe attachment. Following the construction of this amplified structure, fluorescently labeled FISH probes are hybridized to the branched DNA created by the previous steps. Because of the extensive branching, each mRNA target can bind multiple fluorescent probes, leading to substantial signal enhancement. The result of this technique is a more than 400-fold increase in signal amplification per RNA molecule compared to traditional FISH methods. This level of amplification allows for extremely precise and clear visualization of mRNA molecules in their cellular context, even when those mRNA molecules are present in very low numbers [8].

### *19.2.2 Multiplexed RNA Imaging in Fixed Cells*

Subcellular multiplexed RNA imaging methods generally fall into two categories: combinatorial FISH and in situ sequencing.

### *19.2.3 Combinatorial FISH*

Combinatorial FISH utilizes a "spectral barcode" for each distinct RNA target, where every bit of the barcode corresponds to a specific fluorochrome used in a particular imaging round. The number of unique transcripts that can be identified grows exponentially with each additional bit in a barcode. In 2002, the technique employed five pseudocolors and two imaging rounds to distinguish ten unique transcripts [9]. By 2014, the field of RNA imaging had seen significant advancements with the development of seqFISH. This method employed a strategic approach using four different fluorescent colors across two rounds of imaging, enabling the precise identification of 12 unique transcripts within budding yeast. The technique's ability to sequence through different rounds of hybridization and imaging allowed for the expansion of identifiable RNA species in a single cell, marking a significant step forward in multiplexing capabilities.

The innovation of seqFISH demonstrated how sequential imaging could enhance the depth and breadth of transcriptomic analysis in single cells, offering a glimpse into the complex regulatory networks at play within simple organisms like yeast [10]. Later enhancements in MERFISH and techniques have expanded the capability to detect up to 10,000 unique RNA targets. Following seqFISH, another groundbreaking advancement came with the introduction of multiplexed error-robust fluorescence in situ hybridization (MERFISH) [11]. This technique represented a monumental leap in RNA imaging capabilities, as it allowed for the detection of over 100 unique transcripts for the first time. MERFISH combines encoding schemes derived from information theory with robust error-detection algorithms to significantly amplify the number of distinct RNA molecules that can be imaged and identified within individual cells. MERFISH uses a combinatorial labeling strategy where each unique RNA molecule is identified by a specific combination of fluorescent labels across multiple rounds of hybridization. The technique incorporates error-robust encoding, which helps in distinguishing between signals from different transcripts and correcting potential errors in signal reading, enhancing both accuracy and reliability. Images are captured in sequential rounds, each adding a layer of data that contributes to the final decoding process to identify the RNA species present. However, these methods diverge in their strategies to overcome the issue of optical density. MERFISH employs expansion microscopy (ExM) [12] to alleviate crowding, whereas seqFISH+ adopts a sparse labeling strategy [13], choosing to identify only a subset of targets in each imaging round to avoid overcrowding [13, 14].

### *19.2.4 In Situ Sequencing (ISS)*

In situ sequencing (ISS) represents a sophisticated integration of rolling-circle amplification (RCA)-FISH with sequencing-by-ligation (SBL) methodologies. This innovative approach combines the strengths of robust signal amplification with precise sequence identification, enabling researchers to not only detect but also sequence mRNA molecules directly within their cellular context. ISS begins by employing RCA-FISH to amplify the target mRNA. In this step, padlock probes that are specific to the mRNA sequence of interest hybridize to their target, and upon successful hybridization, the probes are circularized and then amplified through rolling-circle amplification. This process results in a localized, amplified signal at the site of the mRNA molecule, enhancing detection capabilities. Following amplification, ISS uses sequencing-by-ligation, a method that sequences DNA by sequentially ligating fluorescently labeled oligonucleotides to a DNA template. In the context of ISS, this technique is applied to the amplified products from

RCA-FISH. As each oligonucleotide is ligated to the growing strand, it is identified by its fluorescent tag, allowing the sequence of the RNA to be determined directly on the slide. The combination of RCA and SBL enables the decoding of barcodes that are specific to each mRNA, thus identifying the exact sequence of the RNA molecule at its original location within the cell. This direct sequencing within tissues preserves spatial information, providing insights into where specific mRNAs are expressed [15]. A subsequent iteration, hybridization-based ISS (HybISS), introduced a revamped barcoding system and probe design enhancements for refined spatial resolution in RNA transcript detection [16]. BaristaSeq [17] employed a similar approach, utilizing Illumina's sequencing-by-synthesis (SBS) chemistry for its process. More recently, STARmap [18] enhanced ISS accuracy through the use of dual partially complementary probes for each target, a novel error-minimizing SBL method (SEDAL) for sequencing 5-nt barcodes, and hydrogel embedding to eliminate background autofluorescence, enabling the simultaneous measurement of 1020 genes within intact medial prefrontal cortex tissue with an error rate of approximately 1.8%. Fluorescent in situ RNA sequencing (FISSEQ) represents a revolutionary approach in the field of molecular biology, particularly in the study of RNA within its native cellular environment. This technique allows for the comprehensive and unbiased quantification and sequencing of all RNA types at the single-molecule level directly within fixed tissues and cells.

The process begins by crosslinking RNA within cells or tissues to preserve its spatial orientation and integrity. Reverse transcription is then performed directly in situ to convert these fixed RNA molecules into cDNA. The cDNA is then locally amplified within the cellular environment, typically using methods like rolling-circle amplification (RCA). This step generates dense clusters of cDNA, enhancing the signal strength from each original RNA molecule. Once amplified, the cDNA clusters are sequenced directly within the tissue using fluorescently labeled nucleotides. This sequencing typically involves cyclic methods where nucleotides are added one at a time, and their incorporation is recorded based on the fluorescence emitted. Advanced imaging and computational techniques are used to capture and decode the fluorescent signals, translating them into nucleotide sequences. This process allows researchers to map where each RNA molecule resides within the cells and what sequences they contain [19, 20]. Unlike previous methods, FISSEQ employs random hexamer primers for hybridization, with reverse-transcribed cDNA, then circularized using CircLigase II for RCA, and partially sequenced using SOLiD sequencing for genome alignment. Despite its groundbreaking approach to transcriptome-wide measurement, challenges such as optical crowding, the predominance of rRNA in reads, and low circularization efficiency have hindered its broader application. Expansion sequencing (ExSeq) sought to mitigate these issues by combining FISSEQ with expansion microscopy (ExM) and ex situ sequencing, enhancing detection efficiency and accuracy [21].

## 19.3 Understanding RNA Biology Through Advancement in Imaging Technologies

The advances in RNA imaging techniques have profoundly impacted our understanding of the RNA lifecycle, encompassing transcription, splicing, localization, translation, and degradation. Each stage of RNA's journey from the nucleus to its final degradation or recycling within the cell can now be visualized and studied in unprecedented detail, providing insights that were once beyond reach.

***Transcription*** Live-cell RNA imaging employing MS2 systems has facilitated the exploration of diverse transcriptional dynamics. Notably, it has been instrumental in characterizing transcriptional bursting phenomena [22, 23]. Utilizing fluorescently tagged RNA polymerase II in conjunction with MS2-labeled nascent mRNA has enabled the quantification of the elongation rate [24]. Through high-speed, time-lapse measurements, researchers have been able to determine the elongation rate and observe transcriptional bursting at multiple scales, governed by clusters of polymerases moving in close formation, known as "convoys" [25]. Fluorescence fluctuation microscopy with MS2 has provided insights into bursting at the level of individual genes [26]. Fluorescent dUTPs, when used in conjunction with fluorescence anisotropy imaging, have revealed that chromatin within transcriptionally active compartments adopts a more open conformation in living cells [27]. Advances in MERFISH and seqFISH+ technologies have enabled studies on the impact of chromosome three-dimensional (3D) organization on transcriptional activity. Modifications to seqFISH+ to target intronic regions of 10,421 genes have shown that nascent transcription sites tend to be positioned at the surfaces of chromosomes [28]. DNA-MERFISH has been developed to map chromatin at the genome-wide level, and in combination with MERFISH and immunofluorescence, it has been possible to concurrently image over 1000 gene loci, nascent transcripts from these loci, and nuclear structures like nuclear speckles and nucleoli [29]. Subcellular RNA imaging is poised to continue unraveling key aspects of transcription. The advancements in live-cell imaging of transcription sites and the imaging of chromosomal architecture with MERFISH and seqFISH+ are paving the way toward a deeper understanding of transcriptional bursting mechanisms. Furthermore, leveraging pooled genomic screens alongside RNA imaging [30] offers a promising approach to evaluating the roles of various transcriptional activators and repressors.

***Splicing*** The discovery that about 90% of pre-mRNAs undergo splicing either during or immediately after transcription has spurred researchers to delve deeper into the structural and kinetic relationships between these two fundamental processes. This exploration has been significantly enhanced by the advancements in RNA imaging technologies, particularly single-molecule fluorescence in situ hybridization (smFISH). Using smFISH, Tyagi and his team were able to make a critical observation regarding the impact of RNA secondary structure on the timing of splicing. They discovered that the presence of an intronic polypyrimidine tract within a strong secondary structure can disrupt the typical coupling of splicing with transcription. Specifically, such structural formations result in splicing being delayed until after the transcription has been fully completed [31]. Live-cell RNA imaging has provided invaluable insights into the dynamic processes of RNA transcription and processing within the cellular environment. A notable example of this is the use of the MS2 system in combination with fluorescence recovery after photobleaching (FRAP) techniques, as employed by Shav-Tal and collaborators. This approach has led to significant findings regarding the relationship between splicing events and the kinetics of RNA polymerase elongation [32]. By combining MS2 and PP7 systems to label introns and exons in different fluorescent colors, Larson and colleagues were able to simultaneously monitor transcription and splicing, revealing a coordinated mechanism through kinetic competition [33]. These investigations also highlighted the variability in splicing timing, ranging from 20 seconds to several minutes, and identified stochastic occurrences of alternative splicing at the single-cell level, demonstrating significant variability among cells [34]. Transcriptomic analyses have proposed alternative splicing as a pathway for RNA localization [35, 36]. The recent development of APEX-seq by Ting and associates represents a significant technological advancement in the field of RNA biology, particularly in enhancing the spatial resolution of transcriptome-wide isoform mapping. This innovative technique has extended the capabilities of RNA localization studies, allowing researchers to map RNA transcripts to nine distinct cellular organelles with unprecedented precision. APEX-seq leverages the APEX (ascorbate peroxidase) enzyme, which can be genetically targeted to specific cellular compartments. When activated, APEX catalyzes the biotinylation of nearby molecules, including RNA, which can then be selectively purified and sequenced. This method offers a unique combination of temporal resolution and spatial specificity, enabling the study of the dynamic distribution of RNA within the cellular landscape [37]. In highly asymmetric cells (i.e. neurons), where RNA localization along neurites correlates with specific functions, an enhanced spatial resolution for transcriptome mapping is crucial for a deeper understanding of the role of alternative splicing in subcellular localization.

***RNA Transport*** In mammalian cells, mRNA demonstrated velocities ranging from 0.1 to 1 μm/s [38]. Analysis of RNA motion in Cos cells identified four distinct mobility patterns with varying likelihoods: (1) stationary (33–40%), (2) directed movement (2–5%), (3) confined diffusion (40–45%), and (4) unrestricted diffusion (15–25%). Notably, mRNA exhibits the capacity to transition between these mobility states [39]. RNAs synthesized in the nucleus exit via the nuclear pore complex (NPC) for subsequent processing. The integration of the MS2 system and molecular beacons in RNA research has led to significant insights into the intracellular dynamics of RNP complexes, particularly concerning their movement within the nucleus and export through nuclear pore complexes (NPCs). These studies have refined our understanding of cellular RNA transport mechanisms, contrasting diffusion-based models with observed behaviors [40]. It was determined, through the use of MS2 coupled with super-registration techniques allowing 20-ms temporal and 26-nm spatial precision, that nucleocytoplasmic transport adheres to a three-phase model: docking (80 ms), transit (5–20 ms), and release (80 ms), with NPC transit not constituting the bottleneck [41]. While prior research posited that multiple mRNA species could be co-transported within a single RNP [42–44], smFISH and quantitative imaging have demonstrated that mRNA molecules primarily travel individually within neuronal dendrites [45]. This finding was corroborated by molecular beacon studies showing that over 70% of β-actin mRNA molecules independently traverse neuronal axons [46]. This research into RNA dynamics, particularly within neuronal axons, has illuminated the diversity of mechanisms employed for axonal transport of mRNA. The study identified several distinct transport mechanisms, each contributing to the intricate control of mRNA localization and function within neurons. These mechanisms include diffusion with entrapment, motor-driven directed movement, localized mRNA protection against degradation, and an innovative "sushi belt" model. Each of these mechanisms plays a critical role in ensuring the precise delivery and stabilization of mRNA necessary for neuronal function and plasticity.

Future research in the field of RNA biology is set to delve deeper into the intricate mechanisms of RNP assembly and transport, addressing critical gaps in our current understanding. These studies aim to unravel the complex biological principles that govern the assembly of RNPs, their interaction with cellular transport systems, and the roles of various molecular components in these processes [47]. The applicability of this transportation method across the transcriptome warrants further exploration. It has been proposed that mRNA molecules navigate dendrites and axons singularly [44, 46], a process studied in only a few RNA types. While transporting individual mRNAs might enhance specificity, it likely incurs higher energy costs compared to simultaneous multi-mRNA transport.

***RNA Localization*** The role of subcellular RNA localization in influencing cellular functions across various cell types has been thoroughly explored [48–52]. In 1986, Singer and Lawrence first identified specific localization patterns of actin, vimentin, and tubulin mRNAs in intact somatic cells through ISH [53]. This foundational work spurred further studies into the subcellular localization of individual mRNA species utilizing smFISH [54–57]. More recently, multiplexed error-robust FISH (MERFISH) has been applied to study the subcellular localization of transcripts from approximately 10,000 genes [11]. As part of this research, transcripts of well-characterized secretome genes were observed to predominantly localize to the ER [11]. Additionally, the researchers introduced a pseudotime analysis based on the enrichment of RNA in nuclear versus cytoplasmic compartments to deduce the cell-cycle stages of individual cells.

Localized translation is a crucial mechanism in highly polarized cells such as neurons, enabling these cells to rapidly respond to stimuli in a spatially specific manner. This capability is especially important in neuronal dendrites, where synaptic inputs require quick and localized responses for synaptic plasticity and other critical functions. In neurons, localized translation allows for the immediate synthesis of proteins at specific sites, such as near synapses, in response to signaling events. This process ensures that proteins crucial for synaptic modifications are quickly available, supporting changes in synaptic strength that underlie learning and memory [58]. Similarly, smFISH has shown that intestinal epithelial cells employ asymmetric subcellular RNA localization to modulate translational efficiency [59]. In axonal processes, the innovative use of fluorescent-UTP labeling alongside SunTag nascent protein labeling has provided groundbreaking insights into the dynamics of intracellular transport and local protein synthesis. This research has particularly highlighted the role of Rab7a endosomes, which are crucial for the transport and local translation of mRNA within axons. Rab7a endosomes are specialized organelles that play a key role in the late endosomal pathway. In neurons, these endosomes have been observed to carry not only mRNA but also ribosomes, thus serving as mobile sites for mRNA translation. This capacity is essential for maintaining and modifying neuronal function, especially in axons where the distance between the cell body and synapses can be extremely long [60]. This localized translation mechanism also appears to operate in non-polarized cells, implying a model of co-translational RNA targeting [61]. RNA mislocalization has been linked to various neurodegenerative conditions [62], with transcriptomic sequencing efforts identifying the implicated mislocalized mRNAs. The emergence of spatial transcriptomics and live-cell RNA imaging now provides enhanced tools to examine mRNA mislocalization with greater spatial and temporal accuracy, opening new avenues for understanding the complex dynamics of RNA within cells.

***Translation*** The TRICK (translating RNA imaging by coat protein knock-off) technique, developed by Chao and colleagues, represents a significant advancement in RNA biology, specifically in the study of mRNA translation dynamics. This method involves an innovative dual-labeling strategy that utilizes the PP7 bacteriophage coat protein to tag the coding sequence and the MS2 coat protein to tag the 3' UTR of mRNA molecules. This dual-tagging approach allows researchers to visualize and track the translation status of individual mRNA molecules within living cells with high precision. Chao and their team made a critical discovery using the TRICK technique: mRNAs are not translated within the nucleus. Instead, their findings demonstrated that there is a clear absence of translation within the nuclear compartment, as indicated by the retention of both PP7 and MS2 signals within this region, and that the translation of mRNAs commences swiftly after their export from the nucleus into the cytoplasm. This transition is visually captured by the loss of PP7 fluorescence in the cytoplasm, confirming the onset of translation within minutes following nuclear export [63]. The dual labeling of translating proteins and RNA using the SunTag and MS2 systems facilitated insights into translation within sub-dendritic compartments [64], the ribosome count per polysome [65], and the process by which ribosome occupancy unfolds mRNA [66]. The integration of SunTag and MS2 methodologies into RNA research has enabled significant advancements in our understanding of how cells regulate protein synthesis under stress conditions. Using the SunTag/MS2 system, researchers have observed critical behaviors of mRNA under stress conditions: (i) mRNAs can rapidly recommence translation upon recovery from cellular stress. This observation indicates that cells can swiftly re-initiate protein synthesis to restore normal function once the stress condition is mitigated, highlighting a dynamic regulatory mechanism that allows cells to adapt and recover from transient environmental challenges. (ii) Contrary to the previously held belief that stress granules generally serve as sites where translation is inhibited during stress, findings from SunTag/MS2 methodologies suggest that mRNAs within stress granules can indeed undergo active translation. This revelation challenges the traditional view of stress granules merely as storage sites for untranslated mRNAs and suggests a more nuanced role where these structures may also facilitate a selective translation under specific conditions [67, 68].

Live-cell RNA imaging and fixed-cell RNA imaging are complementary techniques, each offering unique insights into the dynamics of RNA translation. By utilizing these methods in tandem, researchers can gain a comprehensive understanding of how RNA functions within the cell across both spatial and temporal dimensions. Through the integration of single-molecule FISH (smFISH) with nascent protein detection via O-propargyl-puromycin staining, it was revealed that global mRNA localization within the intestinal epithelium exhibits polarization, consequently influencing translational efficiency polarization [59]. A similar strategy highlighted that mRNA localization necessitates active translation, supporting the concept of extensive co-translational RNA targeting [61]. These innovative imaging approaches are crucial for unraveling the intricate mechanisms of mRNA translation and localization, offering profound insights into cellular function and regulation.

### *19.3.1 RNA Degradation*

Using smFISH, Singer and team delved into the dynamics of cell-cycle-regulated RNA degradation in yeast, uncovering that promoter-driven activity has a direct impact on the cytoplasmic degradation timing and manner of an mRNA [69]. To achieve a higher temporal resolution in observing mRNA degradation within living cells, Chao and colleagues developed an innovative technique called 3'-RNA end accumulation during turnover (TREAT), which provides a novel approach for studying RNA stability and degradation dynamics. This method specifically allows researchers to examine the stability of RNA molecules and the mechanisms governing their turnover within the cellular environment. The TREAT method focuses on the accumulation of 3′ ends of RNA molecules during their degradation process. By selectively enriching and sequencing these 3′ ends, researchers can gain insights into how and when RNA molecules are degraded, as well as identify specific sequences or factors involved in this process [70]. Moreover, when processing bodies (P-bodies) were labeled in conjunction with TREAT, it was observed that the majority of TREAT-labeled mRNAs do not undergo degradation within P-bodies, challenging prior assumptions about the role of P-bodies as the primary sites of RNA degradation [71]. Additionally, mRNAs located in stress granules and P-bodies during stress exhibited degradation rates during recovery that were comparable to those of cytosolic mRNAs [67]. Beyond the degradation of standard transcripts, cells employ nonsense-mediated decay (NMD) to eliminate mRNAs containing premature termination codons. Using the MS2/SunTag system for imaging translating mRNA [72], it was shown that NMD efficiency is influenced by the presence of introns, and each translation cycle of the same RNA has an identical chance of triggering NMD [73]. mRNAs carried within RNP complexes are generally shielded from degradation, which ensures their accurate delivery to specific destinations. Future research, employing high spatial and temporal resolution imaging techniques, is poised to illuminate the complex relationship between RNA degradation and localization, offering deeper insights into cellular regulation and function.

Innovative imaging technologies allow scientists to visualize the initiation and progression of RNA transcription in real time. Techniques like live-cell imaging have made it possible to observe the dynamics of RNA polymerase as it synthesizes RNA, revealing the kinetics and regulation mechanisms underlying gene expression. RNA splicing is a critical process where introns are removed, and exons are joined to form mature mRNA. Advances in RNA imaging have enabled researchers to track splicing events as they occur within the nucleus, identifying the precise timing and location of these modifications. This visibility helps in understanding how alternative splicing contributes to protein diversity and how errors in splicing can lead to disease. Techniques such as smFISH and seqFISH have revolutionized our ability to map RNA molecules within cells, allowing scientists to see exactly where different RNAs accumulate. This information is crucial for understanding how RNAs influence cellular structure and function, including their roles in targeting protein synthesis to specific locations within the cell. Imaging methods that combine fluorescence microscopy with techniques like ribosome profiling provide a detailed view of translation. Researchers can now observe where, when, and how proteins are synthesized from RNA, gaining insights into the regulation of protein production and the impact of various stimuli or conditions on translation efficiency. Finally, advances in RNA imaging have also shed light on RNA degradation processes. Techniques that track RNA decay in real-time have elucidated the pathways and rates at which different RNAs are broken down, revealing the layers of regulation that determine RNA stability and lifespan in the cell.

## 19.4 Biomedical and Clinical Application

### *19.4.1 Imaging ncRNAs*

The development of intracellular single-molecule, high-resolution localization and counting (iSHiRLoC) represents a significant advancement in the field of cellular biology, particularly in the study of miRNAs and their roles in post-transcriptional gene regulation. This technique has been instrumental in visualizing and quantifying the behavior of fluorescently labeled miRNAs within living cells, such as HeLa and U2OS cells, enabling researchers to explore the dynamics of miRNA interaction with ribonucleoprotein (RNP) complexes in unprecedented detail [74–77]. iSHiRLoC combines single-molecule sensitivity with high-resolution imaging to track and count individual RNA molecules within cells. miRNAs are tagged with fluorescent markers, allowing them to be visualized using advanced microscopy techniques. Techniques such as total internal reflection fluorescence (TIRF) microscopy or super-resolution microscopy are used to achieve the spatial resolution necessary to distinguish individual miRNA molecules within the complex environment of the cell. By continuously imaging the fluorescently labeled miRNAs, iSHiRLoC allows researchers to track the movement and interactions of these molecules over time, providing insights into their kinetic behaviors and pathways [76]. Furthermore, iSHiRLoC studies, along with the tracking of fluorescently labeled P-bodies, indicated that miRNAs within P-bodies are predominantly non-functional [77]. An smFISH analysis of 61 lncRNAs demonstrated that nearly half are localized in the cytoplasm [78]. lncRNAs in the nucleus exhibit diverse distribution patterns that are critical for their functional roles in regulating gene expression, nuclear architecture, and cellular responses. lncRNAs like MALAT1 (metastasis associated lung adenocarcinoma transcript 1) and NEAT1 (nuclear enriched abundant transcript 1) are prime examples, each localized in distinct nuclear structures—speckles and paraspeckles, respectively. These structures are involved in various regulatory processes, including RNA splicing, nuclear organization, and stress responses [79]. smFISH studies also highlighted cell-to-cell expression variability among lncRNAs, similar to what is observed with mRNAs [78]. Moreover, smFISH investigations revealed that the imprinting lncRNAs Kcnqlot1 [80] and Air [81] localize specifically at their transcription sites on the same allele, suggesting these lncRNAs may exert gene silencing effects in cis. These findings underscore the complexity of ncRNA functions within cells and the advanced imaging techniques that continue to unravel this complexity.

### *19.4.2 Imaging Viral RNA*

The integration of RNA stem-loop systems and live-cell imaging techniques has significantly advanced our understanding of the dynamics of HIV-1, particularly the role and behavior of viral RNA within the host cells. One pivotal study utilizing these technologies revealed a critical aspect of HIV-1 biology: over 90% of HIV-1 particles contain RNA, highlighting the high efficiency of RNA encapsulation in the viral replication cycle [82]. The sophisticated technique of labeling individual RNA strands with different colors has significantly advanced our understanding of HIV-1 biology, particularly in terms of viral assembly and the packaging process of its genomic RNA. This method has provided clear visual evidence that the HIV-1 structural protein Gag encapsulates a dimeric RNA molecule, rather than two monomeric RNA molecules, during virion assembly [83]. The MS2/SunTag system indicated that approximately half of HIV-1 RNA undergoes active translation, and Gag selectively packages RNA that is not being translated [84]. RNA imaging techniques like FISH and RNAscope have been extended to the detection of various viruses, including SARS-CoV-2 [85–88], the single-stranded RNA virus responsible for the COVID-19 pandemic, which has caused over 6 million deaths globally.

The application of smFISH to study the dynamics of host mRNA during SARS-CoV-2 infection has provided significant insights into how the virus impacts crucial aspects of the host's immune response. By employing smFISH, researchers were able to visualize and quantitatively analyze the behavior and localization of mRNA molecules at a single-molecule level, revealing specific disruptions in the biogenesis pathways of IFN types I and II. These interferons are key components of the antiviral response, and their effective disruption by SARS-CoV-2 highlights a strategic viral mechanism to evade host immunity [89].

## 19.5 Challenges and Future Outlook

As the scope of mRNA imaging in fixed cells has transitioned from focusing on individual targets to encompassing the transcriptome, the speed of imaging and the complexity of image analysis have emerged as significant hurdles in the study of subcellular mRNA localization. Moreover, accurately distinguishing multiple mRNAs as diffraction-limited spots poses a challenge, hindering insights into potential co-processing of various mRNA species in identical locations. Advances in artificial intelligence (AI) are set to dramatically enhance the field of RNA research, particularly in areas such as cell segmentation, RNA localization assignment, and the detection and tracking of spots within cellular images. These AI-driven technologies promise to extend the limits of our current understanding of RNA localization and dynamics, offering more precise and efficient analysis methods [56, 90–92].

The evolution of live-cell RNA imaging techniques has been pivotal in advancing our understanding of RNA biology, especially in terms of spatiotemporal dynamics. However, as described in this chapter, current methodologies, while advanced, still exhibit limitations that restrict their broader application in understanding the complex interplay of RNA species within cells. Most live-cell RNA imaging studies to date have focused on a relatively small selection of mRNA species. This limitation is often due to the challenges of designing specific probes for a broader range of mRNAs or the technical limitations in simultaneously imaging multiple RNA types due to spectral overlap or signal interference. The tracking periods in many studies are generally short, primarily due to the rapid photobleaching of fluorescent probes, the toxicity of long-term light exposure to cells, or the dynamic nature of cellular processes that may move the RNA out of the focal plane. Achieving single-molecule resolution remains challenging. Current techniques may not distinguish individual RNA molecules, especially within densely packed cellular environments or when RNA clusters together. This resolution is crucial for understanding the exact behavior and processing of individual RNA molecules.

Phototoxicity remains a significant challenge in live-cell imaging, particularly when conducting long-term observations at the single-molecule level. This issue arises because the light required to excite fluorescent tags can also damage cellular components, alter cellular functions, and even lead to cell death. One straightforward approach to minimize phototoxicity is to reduce the intensity of the excitation light. Lower light levels can decrease the energy absorbed by the cells and, consequently, reduce damage. However, this often comes at the expense of signal intensity and can make it harder to detect and track individual molecules. The development and use of more photostable fluorescent probes that require less intense light for excitation can also help mitigate phototoxicity. Fluorescent proteins or dyes that are brighter and more photostable under reduced light conditions allow for longer observation periods without compromising cellular viability. Efforts to develop systems capable of mitigating this issue will be crucial for enabling comprehensive tracking of mRNA throughout its lifecycle, offering deeper insights into the intricate dynamics of RNA processing and localization within living cells. The convergence of RNA imaging and sequencing technologies represents a transformative leap forward in RNA biology, enabling researchers to explore RNA dynamics and function with unprecedented precision and scale. This fusion is particularly potent due to the combination of high-resolution imaging's detailed spatial insights and RNA sequencing's broad analytical scope [30, 93, 94].

## 19.6 Conclusion

Integrating advanced imaging techniques that can concurrently visualize RNA, DNA, and RBPs within their native cellular contexts offers a groundbreaking shift in how we study gene expression and RNA functions. This holistic view enables scientists to directly observe and understand the dynamic interplay between these crucial biomolecules during key cellular processes such as transcription, RNA splicing, and translation. Such insights are pivotal for deciphering the complex regulation mechanisms that underlie gene expression and the maintenance of cellular homeostasis. Furthermore, coupling this integrated imaging approach with high-throughput technologies like CRISPR screenings and systematic analyses of RBP-RNA interactions allows for an unprecedented scale of exploration into the genetic and molecular machinery of cells. This comprehensive approach not only enhances our understanding of RNA processing within the broader context of cellular function but also opens new avenues for therapeutic interventions targeting RNA and its associated proteins in various diseases.

## References

1. Singer RH, Ward DC. Actin gene expression visualized in chicken muscle tissue culture by using in situ hybridization with a biotinated nucleotide analog. Proc Natl Acad Sci USA. 1982;79(23):7331–5.
2. Femino AM, Fay FS, Fogarty K, Singer RH. Visualization of single RNA transcripts in situ. Science. 1998;280(5363):585–90.

3. Raj A, van den Bogaard P, Rifkin SA, van Oudenaarden A, Tyagi S. Imaging individual mRNA molecules using multiple singly labeled probes. Nat Methods. 2008;5(10):877–9.
4. Larsson C, Koch J, Nygren A, Janssen G, Raap AK, Landegren U, et al. In situ genotyping individual DNA molecules by target-primed rolling-circle amplification of padlock probes. Nat Methods. 2004;1(3):227–32.
5. Larsson C, Grundberg I, Söderberg O, Nilsson M. In situ detection and genotyping of individual mRNA molecules. Nat Methods. 2010;7(5):395–7.
6. Wang F, Flanagan J, Su N, Wang LC, Bui S, Nielson A, et al. RNAscope: a novel in situ RNA analysis platform for formalin-fixed, paraffin-embedded tissues. J Mol Diagn. 2012;14(1):22–9.
7. Dardani I, Emert BL, Goyal Y, Jiang CL, Kaur A, Lee J, et al. ClampFISH 2.0 enables rapid, scalable amplified RNA detection in situ. Nat Methods. 2022;19(11):1403–10.
8. Rouhanifard SH, Mellis IA, Dunagin M, Bayatpour S, Jiang CL, Dardani I, et al. ClampFISH detects individual nucleic acid molecules using click chemistry-based amplification. Nat Biotechnol. 2018;
9. Levsky JM, Shenoy SM, Pezo RC, Singer RH. Single-cell gene expression profiling. Science. 2002;297(5582):836–40.
10. Chen KH, Boettiger AN, Moffitt JR, Wang S, Zhuang X. RNA imaging. Spatially resolved, highly multiplexed RNA profiling in single cells. Science. 2015;348(6233):aaa6090.
11. Xia C, Fan J, Emanuel G, Hao J, Zhuang X. Spatial transcriptome profiling by MERFISH reveals subcellular RNA compartmentalization and cell cycle-dependent gene expression. Proc Natl Acad Sci USA. 2019;116(39):19490–9.
12. Wang G, Moffitt JR, Zhuang X. Multiplexed imaging of high-density libraries of RNAs with MERFISH and expansion microscopy. Sci Rep. 2018;8(1):4847.
13. Eng CL, Lawson M, Zhu Q, Dries R, Koulena N, Takei Y, et al. Transcriptome-scale super-resolved imaging in tissues by RNA seqFISH. Nature. 2019;568(7751):235–9.
14. Lubeck E, Coskun AF, Zhiyentayev T, Ahmad M, Cai L. Single-cell in situ RNA profiling by sequential hybridization. Nat Methods. 2014;11(4):360–1.
15. Ke R, Mignardi M, Pacureanu A, Svedlund J, Botling J, Wählby C, et al. In situ sequencing for RNA analysis in preserved tissue and cells. Nat Methods. 2013;10(9):857–60.
16. Gyllborg D, Langseth CM, Qian X, Choi E, Salas SM, Hilscher MM, et al. Hybridization-based in situ sequencing (HybISS) for spatially resolved transcriptomics in human and mouse brain tissue. Nucleic Acids Res. 2020;48(19):e112.
17. Chen X, Sun YC, Church GM, Lee JH, Zador AM. Efficient in situ barcode sequencing using padlock probe-based BaristaSeq. Nucleic Acids Res. 2018;46(4):e22.
18. Wang X, Allen WE, Wright MA, Sylwestrak EL, Samusik N, Vesuna S, et al. Three-dimensional intact-tissue sequencing of single-cell transcriptional states. Science. 2018;361(6400)
19. Lee JH, Daugharthy ER, Scheiman J, Kalhor R, Yang JL, Ferrante TC, et al. Highly multiplexed subcellular RNA sequencing in situ. Science. 2014;343(6177):1360–3.
20. Lee JH, Daugharthy ER, Scheiman J, Kalhor R, Ferrante TC, Terry R, et al. Fluorescent in situ sequencing (FISSEQ) of RNA for gene expression profiling in intact cells and tissues. Nat Protoc. 2015;10(3):442–58.
21. Alon S, Goodwin DR, Sinha A, Wassie AT, Chen F, Daugharthy ER, et al. Expansion sequencing: spatially precise in situ transcriptomics in intact biological systems. Science. 2021;371(6528)
22. Chubb JR, Trcek T, Shenoy SM, Singer RH. Transcriptional pulsing of a developmental gene. Curr Biol. 2006;16(10):1018–25.
23. Golding I, Paulsson J, Zawilski SM, Cox EC. Real-time kinetics of gene activity in individual bacteria. Cell. 2005;123(6):1025–36.
24. Darzacq X, Shav-Tal Y, de Turris V, Brody Y, Shenoy SM, Phair RD, et al. In vivo dynamics of RNA polymerase II transcription. Nat Struct Mol Biol. 2007;14(9):796–806.
25. Tantale K, Mueller F, Kozulic-Pirher A, Lesne A, Victor JM, Robert MC, et al. A single-molecule view of transcription reveals convoys of RNA polymerases and multi-scale bursting. Nat Commun. 2016;7:12248.

26. Larson DR, Zenklusen D, Wu B, Chao JA, Singer RH. Real-time observation of transcription initiation and elongation on an endogenous yeast gene. Science. 2011;332(6028):475–8.
27. Sinha DK, Banerjee B, Maharana S, Shivashankar GV. Probing the dynamic organization of transcription compartments and gene loci within the nucleus of living cells. Biophys J. 2008;95(11):5432–8.
28. Shah S, Takei Y, Zhou W, Lubeck E, Yun J, Eng CL, et al. Dynamics and spatial genomics of the nascent transcriptome by intron seqFISH. Cell. 2018;174(2):363–76.e16.
29. Hu W, Liu C, Bi ZY, Zhou Q, Zhang H, Li LL, et al. Comprehensive landscape of extracellular vesicle-derived RNAs in cancer initiation, progression, metastasis and cancer immunology. Mol Cancer. 2020;19(1):102.
30. Feldman D, Singh A, Schmid-Burgk JL, Carlson RJ, Mezger A, Garrity AJ, et al. Optical pooled screens in human cells. Cell. 2019;179(3):787–99.e17.
31. Vargas DY, Shah K, Batish M, Levandoski M, Sinha S, Marras SA, et al. Single-molecule imaging of transcriptionally coupled and uncoupled splicing. Cell. 2011;147(5):1054–65.
32. Brody Y, Neufeld N, Bieberstein N, Causse SZ, Böhnlein EM, Neugebauer KM, et al. The in vivo kinetics of RNA polymerase II elongation during co-transcriptional splicing. PLoS Biol. 2011;9(1):e1000573.
33. Coulon A, Ferguson ML, de Turris V, Palangat M, Chow CC, Larson DR. Kinetic competition during the transcription cycle results in stochastic RNA processing. elife. 2014;3:3.
34. Waks Z, Klein AM, Silver PA. Cell-to-cell variability of alternative RNA splicing. Mol Syst Biol. 2011;7:506.
35. Wu KE, Parker KR, Fazal FM, Chang HY, Zou J. RNA-GPS predicts high-resolution RNA subcellular localization and highlights the role of splicing. RNA. 2020;26(7):851–65.
36. Ciolli Mattioli C, Rom A, Franke V, Imami K, Arrey G, Terne M, et al. Alternative 3' UTRs direct localization of functionally diverse protein isoforms in neuronal compartments. Nucleic Acids Res. 2019;47(5):2560–73.
37. Fazal FM, Han S, Parker KR, Kaewsapsak P, Xu J, Boettiger AN, et al. Atlas of subcellular RNA localization revealed by APEX-Seq. Cell. 2019;178(2):473–90.e26.
38. Dynes JL, Steward O. Dynamics of bidirectional transport of arc mRNA in neuronal dendrites. J Comp Neurol. 2007;500(3):433–47.
39. Fusco D, Accornero N, Lavoie B, Shenoy SM, Blanchard JM, Singer RH, et al. Single mRNA molecules demonstrate probabilistic movement in living mammalian cells. Curr Biol. 2003;13(2):161–7.
40. Mor A, Suliman S, Ben-Yishay R, Yunger S, Brody Y, Shav-Tal Y. Dynamics of single mRNP nucleocytoplasmic transport and export through the nuclear pore in living cells. Nat Cell Biol. 2010;12(6):543–52.
41. Grünwald D, Singer RH. In vivo imaging of labelled endogenous β-actin mRNA during nucleocytoplasmic transport. Nature. 2010;467(7315):604–7.
42. Carson JH, Gao Y, Tatavarty V, Levin MK, Korza G, Francone VP, et al. Multiplexed RNA trafficking in oligodendrocytes and neurons. Biochim Biophys Acta. 2008;1779(8):453–8.
43. Gao Y, Tatavarty V, Korza G, Levin MK, Carson JH. Multiplexed dendritic targeting of alpha calcium calmodulin-dependent protein kinase II, neurogranin, and activity-regulated cytoskeleton-associated protein RNAs by the A2 pathway. Mol Biol Cell. 2008;19(5):2311–27.
44. Tübing F, Vendra G, Mikl M, Macchi P, Thomas S, Kiebler MA. Dendritically localized transcripts are sorted into distinct ribonucleoprotein particles that display fast directional motility along dendrites of hippocampal neurons. J Neurosci. 2010;30(11):4160–70.
45. Batish M, van den Bogaard P, Kramer FR, Tyagi S. Neuronal mRNAs travel singly into dendrites. Proc Natl Acad Sci USA. 2012;109(12):4645–50.
46. Turner-Bridger B, Jakobs M, Muresan L, Wong HH, Franze K, Harris WA, et al. Single-molecule analysis of endogenous β-actin mRNA trafficking reveals a mechanism for compartmentalized mRNA localization in axons. Proc Natl Acad Sci USA. 2018;115(41):E9697–e706.

47. Liao YC, Fernandopulle MS, Wang G, Choi H, Hao L, Drerup CM, et al. RNA granules hitchhike on lysosomes for long-distance transport, using Annexin A11 as a molecular tether. Cell. 2019;179(1):147–64.e20.
48. Medioni C, Mowry K, Besse F. Principles and roles of mRNA localization in animal development. Development. 2012;139(18):3263–76.
49. Kloc M, Zearfoss NR, Etkin LD. Mechanisms of subcellular mRNA localization. Cell. 2002;108(4):533–44.
50. Engel KL, Arora A, Goering R, Lo HG, Taliaferro JM. Mechanisms and consequences of subcellular RNA localization across diverse cell types. Traffic. 2020;21(6):404–18.
51. Fernandopulle MS, Lippincott-Schwartz J, Ward ME. RNA transport and local translation in neurodevelopmental and neurodegenerative disease. Nat Neurosci. 2021;24(5):622–32.
52. Thelen MP, Kye MJ. The role of RNA binding proteins for local mRNA Translation: implications in neurological disorders. Front Mol Biosci. 2019;6:161.
53. Lawrence JB, Singer RH. Intracellular localization of messenger RNAs for cytoskeletal proteins. Cell. 1986;45(3):407–15.
54. Stoeger T, Battich N, Herrmann MD, Yakimovich Y, Pelkmans L. Computer vision for image-based transcriptomics. Methods. 2015;85:44–53.
55. Samacoits A, Chouaib R, Safieddine A, Traboulsi AM, Ouyang W, Zimmer C, et al. A computational framework to study sub-cellular RNA localization. Nat Commun. 2018;9(1):4584.
56. Petukhov V, Xu RJ, Soldatov RA, Cadinu P, Khodosevich K, Moffitt JR, et al. Cell segmentation in imaging-based spatial transcriptomics. Nat Biotechnol. 2022;40(3):345–54.
57. Battich N, Stoeger T, Pelkmans L. Image-based transcriptomics in thousands of single human cells at single-molecule resolution. Nat Methods. 2013;10(11):1127–33.
58. Yoon YJ, Wu B, Buxbaum AR, Das S, Tsai A, English BP, et al. Glutamate-induced RNA localization and translation in neurons. Proc Natl Acad Sci USA. 2016;113(44):E6877–e86.
59. Moor AE, Golan M, Massasa EE, Lemze D, Weizman T, Shenhav R, et al. Global mRNA polarization regulates translation efficiency in the intestinal epithelium. Science. 2017;357(6357):1299–303.
60. Cioni JM, Lin JQ, Holtermann AV, Koppers M, Jakobs MAH, Azizi A, et al. Late endosomes act as mRNA Translation platforms and sustain mitochondria in axons. Cell. 2019;176(1–2):56–72.e15.
61. Chouaib R, Safieddine A, Pichon X, Imbert A, Kwon OS, Samacoits A, et al. A dual protein-mRNA localization screen reveals compartmentalized Translation and widespread co-translational RNA targeting. Dev Cell. 2020;54(6):773–91.e5.
62. Turner-Bridger B, Caterino C, Cioni JM. Molecular mechanisms behind mRNA localization in axons. Open Biol. 2020;10(9):200177.
63. Halstead JM, Lionnet T, Wilbertz JH, Wippich F, Ephrussi A, Singer RH, et al. Translation. An RNA biosensor for imaging the first round of translation from single cells to living animals. Science. 2015;347(6228):1367–671.
64. Wu B, Eliscovich C, Yoon YJ, Singer RH. Translation dynamics of single mRNAs in live cells and neurons. Science. 2016;352(6292):1430–5.
65. Morisaki T, Lyon K, DeLuca KF, DeLuca JG, English BP, Zhang Z, et al. Real-time quantification of single RNA translation dynamics in living cells. Science. 2016;352(6292):1425–9.
66. Adivarahan S, Livingston N, Nicholson B, Rahman S, Wu B, Rissland OS, et al. Spatial Organization of Single mRNPs at different stages of the gene expression pathway. Mol Cell. 2018;72(4):727–38.e5.
67. Wilbertz JH, Voigt F, Horvathova I, Roth G, Zhan Y, Chao JA. Single-molecule imaging of mRNA localization and regulation during the integrated stress response. Mol Cell. 2019;73(5):946–58.e7.
68. Mateju D, Eichenberger B, Voigt F, Eglinger J, Roth G, Chao JA. Single-molecule imaging reveals Translation of mRNAs localized to stress granules. Cell. 2020;183(7):1801–12.e13.
69. Trcek T, Larson DR, Moldón A, Query CC, Singer RH. Single-molecule mRNA decay measurements reveal promoter- regulated mRNA stability in yeast. Cell. 2011;147(7):1484–97.

70. Horvathova I, Voigt F, Kotrys AV, Zhan Y, Artus-Revel CG, Eglinger J, et al. The dynamics of mRNA turnover revealed by single-molecule imaging in single cells. Mol Cell. 2017;68(3):615–25.e9.
71. Sheth U, Parker R. Decapping and decay of messenger RNA occur in cytoplasmic processing bodies. Science. 2003;300(5620):805–8.
72. Tanenbaum ME, Gilbert LA, Qi LS, Weissman JS, Vale RD. A protein-tagging system for signal amplification in gene expression and fluorescence imaging. Cell. 2014;159(3):635–46.
73. Hoek TA, Khuperkar D, Lindeboom RGH, Sonneveld S, Verhagen BMP, Boersma S, et al. Single-molecule imaging uncovers rules governing nonsense-mediated mRNA decay. Mol Cell. 2019;75(2):324–39.e11.
74. Pitchiaya S, Androsavich JR, Walter NG. Intracellular single molecule microscopy reveals two kinetically distinct pathways for microRNA assembly. EMBO Rep. 2012;13(8):709–15.
75. Pitchiaya S, Krishnan V, Custer TC, Walter NG. Dissecting non-coding RNA mechanisms in cellulo by single-molecule high-resolution localization and counting. Methods. 2013;63(2):188–99.
76. Pitchiaya S, Heinicke LA, Park JI, Cameron EL, Walter NG. Resolving subcellular miRNA trafficking and turnover at single-molecule resolution. Cell Rep. 2017;19(3):630–42.
77. Pitchiaya S, Mourao MDA, Jalihal AP, Xiao L, Jiang X, Chinnaiyan AM, et al. Dynamic recruitment of single RNAs to processing bodies depends on RNA functionality. Mol Cell. 2019;74(3):521–33.e6.
78. Cabili MN, Dunagin MC, McClanahan PD, Biaesch A, Padovan-Merhar O, Regev A, et al. Localization and abundance analysis of human lncRNAs at single-cell and single-molecule resolution. Genome Biol. 2015;16(1):20.
79. Yang LZ, Wang Y, Li SQ, Yao RW, Luan PF, Wu H, et al. Dynamic imaging of RNA in living cells by CRISPR-Cas13 systems. Mol Cell. 2019;76(6):981–97.e7.
80. Terranova R, Yokobayashi S, Stadler MB, Otte AP, van Lohuizen M, Orkin SH, et al. Polycomb group proteins Ezh2 and Rnf2 direct genomic contraction and imprinted repression in early mouse embryos. Dev Cell. 2008;15(5):668–79.
81. Nagano T, Mitchell JA, Sanz LA, Pauler FM, Ferguson-Smith AC, Feil R, et al. The air noncoding RNA epigenetically silences transcription by targeting G9a to chromatin. Science. 2008;322(5908):1717–20.
82. Chen J, Nikolaitchik O, Singh J, Wright A, Bencsics CE, Coffin JM, et al. High efficiency of HIV-1 genomic RNA packaging and heterozygote formation revealed by single virion analysis. Proc Natl Acad Sci USA. 2009;106(32):13535–40.
83. Jouvenet N, Simon SM, Bieniasz PD. Imaging the interaction of HIV-1 genomes and gag during assembly of individual viral particles. Proc Natl Acad Sci USA. 2009;106(45):19114–9.
84. Chen J, Liu Y, Wu B, Nikolaitchik OA, Mohan PR, Chen J, et al. Visualizing the translation and packaging of HIV-1 full-length RNA. Proc Natl Acad Sci USA. 2020;117(11):6145–55.
85. Shaffer SM, Joshi RP, Chambers BS, Sterken D, Biaesch AG, Gabrieli DJ, et al. Multiplexed detection of viral infections using rapid in situ RNA analysis on a chip. Lab Chip. 2015;15(15):3170–82.
86. Liu F, Han K, Blair R, Kenst K, Qin Z, Upcin B, et al. SARS-CoV-2 infects endothelial cells in vivo and in vitro. Front Cell Infect Microbiol. 2021;11:701278.
87. Carossino M, Ip HS, Richt JA, Shultz K, Harper K, Loynachan AT, et al. Detection of SARS-CoV-2 by RNAscope(®) in situ hybridization and immunohistochemistry techniques. Arch Virol. 2020;165(10):2373–7.
88. Hepp C, Shiaelis N, Robb NC, Vaughan A, Matthews PC, Stoesser N, et al. Viral detection and identification in 20 min by rapid single-particle fluorescence in-situ hybridization of viral RNA. Sci Rep. 2021;11(1):19579.
89. Burke JM, St Clair LA, Perera R, Parker R. SARS-CoV-2 infection triggers widespread host mRNA decay leading to an mRNA export block. RNA. 2021;27(11):1318–29.
90. Caicedo JC, Cooper S, Heigwer F, Warchal S, Qiu P, Molnar C, et al. Data-analysis strategies for image-based cell profiling. Nat Methods. 2017;14(9):849–63.

91. Moen E, Bannon D, Kudo T, Graf W, Covert M, Van Valen D. Deep learning for cellular image analysis. Nat Methods. 2019;16(12):1233–46.
92. Ronneberger O, Fischer P, Brox T. U-net: convolutional networks for biomedical image segmentation. In: Medical image computing and computer-assisted intervention–MICCAI 2015: 18th international conference, Munich, Germany, October 5–9, 2015, proceedings, part III 18. Cham: Springer; 2015.
93. Wheeler EC, Vu AQ, Einstein JM, DiSalvo M, Ahmed N, Van Nostrand EL, et al. Pooled CRISPR screens with imaging on microraft arrays reveals stress granule-regulatory factors. Nat Methods. 2020;17(6):636–42.
94. Markmiller S, Soltanieh S, Server KL, Mak R, Jin W, Fang MY, et al. Context-dependent and disease-specific diversity in protein interactions within stress granules. Cell. 2018;172(3):590–604.e13.

# Chapter 20
# RNA-Based Imaging System

**Abstract** RNA transcripts serve as pivotal components of cellular function, engaging in dynamic interactions with various proteins and other RNA molecules to modulate a myriad of biological processes. A comprehensive grasp of RNA biosynthesis and processing, coupled with their functional roles within cells, is indispensable for deciphering the intricate mechanisms of cellular operation and the pathogenesis of diseases linked to RNA abnormalities. This chapter is dedicated to exploring advanced live RNA imaging techniques, providing an in-depth analysis of the principles that govern these methodologies, while critically assessing their strengths and identifying potential limitations. This chapter extends to a detailed examination of a diverse spectrum of applications that underscore the versatility of RNA imaging techniques across different biological organisms and cellular contexts. This approach not only illustrates the adaptability of RNA imaging tools but also enhances our understanding of cellular dynamics and regulatory mechanisms at the molecular level. Moreover, this chapter delves into the significant role of RNA imaging technologies in investigating the progression of diseases associated with mutant RNA transcripts, particularly those characterized by triplet repeat expansions. By delineating how these mutant RNAs contribute to disease pathology, RNA imaging emerges as a powerful tool, offering profound insights into the molecular foundations of genetic disorders. This exploration illuminates the potential of RNA imaging to transform our understanding of complex disease mechanisms, thereby advancing diagnostic approaches and refining therapeutic strategies. Through this comprehensive review, this chapter aims to highlight the transformative impact of RNA imaging in the field of molecular biology, demonstrating its critical role in advancing our understanding of health and disease.

## 20.1 Introduction

In cells, RNA molecules are generated through complex biogenesis pathways and, upon maturation, engage in regulating essential cellular functions. To thoroughly understand RNA's diverse roles, it is critical to study the entire life cycle of

 
P. E. Saw, E. Song, *RNA Therapeutics in Human Diseases*,
https://doi.org/10.1007/978-981-96-3041-7_20

transcripts, from their genesis to their degradation. Conventional molecular techniques in RNA biology enable the identification of RNA's length, sequence, and structure and facilitate the quantification of average RNA levels within populations of cells. Techniques such as pull-down assays offer insights into RNA–protein interactions but fall short of illustrating the dynamic interactions and spatial-temporal shifts experienced by ribonucleoprotein (RNP) complexes. Fluorescence in situ hybridization (FISH) enables the visualization of RNA distribution within cells, but its capabilities are confined to capturing static snapshots of RNA at specific moments in its life cycle. This limitation restricts the ability to observe dynamic changes and real-time interactions of RNA molecules. The key benefit of live RNA imaging technology lies in its capability to monitor RNA's movement and observe the temporal alterations in RNAs and RNPs. Table 20.1 is a detailed elaboration in a tabular format that highlights the various strategies employed in live RNA imaging, each with its key benefits and applications.

Live RNA imaging greatly enhances our understanding by offering a dynamic view that extends far beyond the static snapshots provided by traditional methods [9]. This dynamic perspective allows for real-time observation of RNA behaviors and interactions, significantly broadening the scope of accessible information [10]. Live RNA imaging systems facilitate the investigation of a wide range of cellular activities involving RNAs, encompassing the entire life cycle of eukaryotic mRNAs. This includes processes from Pol II transcription, pre-mRNA processing and splicing, to nuclear export, cytoplasmic transport, translation, and ultimate degradation. By enabling real-time tracking of a transcript's journey within the cell, live RNA imaging provides a comprehensive view in a single experiment, allowing for an in-depth exploration of each stage of RNA dynamics. This section delves into vector-based RNA imaging in living cells and how these systems are tailored for use in bacterial, fungal, plant, and animal cells. It discusses the design and key characteristics of these systems and critically evaluates their benefits and limitations. The application segment briefly mentions significant studies categorized by the cellular functions examined.

## 20.2 Hybridization Methods

In situ hybridization (ISH) is a technique employed to identify RNA sequences within cells. It involves the use of a DNA or RNA probe that binds to its target RNA through complementary base pairing, as outlined by Watson–Crick rules. The probe is tagged with identifiable markers, such as fluorophores, for visualization purposes (Fig. 20.1a). Fluorescence in situ hybridization (FISH) extends this capability, allowing for the differentiation of RNA molecules with minimal sequence differences, even by a single nucleotide [11]. Thanks to its high specificity, FISH enables the detection of individual RNA molecules in fixed cells, especially when paired with various amplification methods [11, 12]. Advancements have resulted in the development of improved in situ hybridization (ISH) derivatives that significantly

**Table 20.1** Current imaging technologies and their applications

| Imaging technology | Description | Key benefits | Applications |
|---|---|---|---|
| Direct RNA tagging [1, 2] | This method involves tagging RNA molecules directly with fluorophores or other labels to visualize them in live cells | Allows real-time tracking of RNA molecules, revealing insights into the dynamics of RNA transport and localization | Used to study RNA transport mechanisms, localization, and the spatial-temporal dynamics of RNA within the cellular environment |
| Molecular beacons [3–6] | Short oligonucleotide probes that form a hairpin structure; they fluoresce upon hybridization with target RNA | Provides high specificity and minimal background signal due to their quenching mechanism until binding occurs | Ideal for detecting specific RNA transcripts in vivo, including mRNA translation processes and gene expression regulation in response to environmental stimuli |
| MTRIPs (multiply labeled tetravalent RNA imaging probes) [7] | Probes designed with multiple fluorophores that increase signal intensity and stability | Enhances signal strength and stability, improving visualization of RNAs even at low concentrations | Useful in studying RNA interactions and dynamics in densely packed cellular environments, such as during cell division or in highly transcriptionally active cells |
| Pumilio-based systems [8] | Utilizes the Pumilio protein, which can be engineered to bind specific RNA sequences and is tagged with a fluorescent marker | Allows customizable targeting of RNA sequences through protein engineering, offering flexibility and specificity | Applied in tracking and quantifying the expression of specific RNAs, investigating RNA–protein interactions, and studying post-transcriptional regulation mechanisms |
| Vector-based systems | These systems use vectors to express RNA sequences fused to aptamers that bind to fluorescent proteins or dyes | Enables endogenous tagging of RNAs, preserving native RNA behavior and interactions | Suitable for long-term studies of RNA life cycle, monitoring RNA from synthesis to degradation and observing RNA behavior in development and disease contexts |

reduce background noise, such as molecular beacons, depicted in Fig. 20.1b. These innovative beacons consist of a DNA probe with a fluorophore on one end and a quencher on the other. When the target RNA is absent, the probe forms a hairpin structure, bringing the quencher and fluorophore close together and effectively quenching the fluorescence, thereby enhancing signal clarity. Binding to the target RNA causes the beacon to unfold, distancing the quencher from the fluorophore and restoring fluorescence [13]. Since their introduction in the 1990s, molecular beacons have evolved, including the diversification of quenchers to incorporate nanoparticles [14]. While microinjection of molecular beacons into live cells can

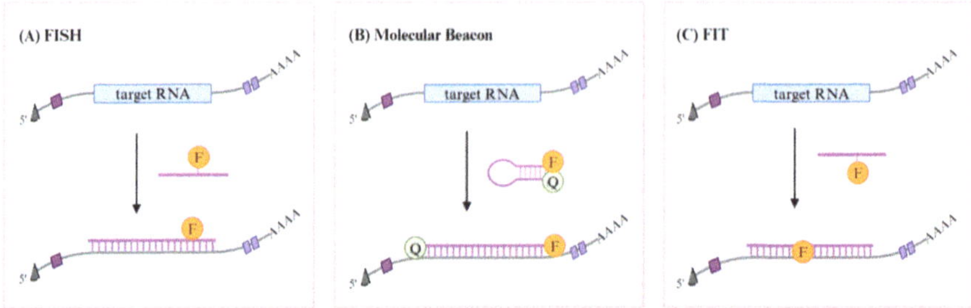


**Fig. 20.1** Various hybridization-based approaches for RNA visualization. (**a**) Traditional fluorescence in situ hybridization (FISH) uses an RNA probe tagged with a fluorophore that specifically pairs with the target RNA sequence. (**b**) Molecular beacon technology improves the signal-to-noise ratio compared to standard FISH probes by quenching the fluorescence when the probe is unbound to its target. (**c**) Forced intercalation (FIT) probes work by forcing a dye molecule to intercalate into the duplex formed between the probe and its target upon binding, significantly enhancing the fluorophore's fluorescence activation. Figure is drawn in Biorender.com

lead to their unintentional accumulation in the nucleus, the integration of a tRNA sequence has been found to effectively counteract this issue [15].

To enhance the specificity and intensity of the fluorescent signal upon target binding, an innovative method involving forced intercalation (FIT) probes has been developed (see Fig. 20.1c). These probes, made from peptide nucleic acid (PNA) or DNA strands, include a synthetic base mimic—often thiazole orange—which intercalates between the base pairs of Watson–Crick helices. This design ensures that fluorescence is activated exclusively upon accurate base pairing, enhancing specificity and reducing the likelihood of false-positive signals in RNA imaging [16, 17]. The dramatic activation effect—approximately a 30-fold increase—positions FIT probes as a significant advancement in the field of ISH. The scope of FIT probes has broadened to include variants that emit light in the blue and green spectra [18, 19], and their application has been extended to the visualization of mRNA in cellular contexts and within Drosophila embryos [16, 20]. Hybridization-based RNA detection is a potent method in fixed specimens and its utility extends to live cells and organisms when using probes like molecular beacons and FIT probes that emit strong activation signals. Yet, these probes, being synthetic nucleic acid modifications or analogs, face a significant challenge as they cannot autonomously cross cell membranes nor can they be synthesized within the cell. This limitation necessitates the development of innovative delivery mechanisms to fully exploit their potential in live-cell imaging. Additionally, successful hybridization necessitates target RNA regions devoid of secondary structures, and the hybridization conditions often require adaptation to suit the intracellular environment. Recent advancements have led to the development of probes and conditions for live-cell application, a technique referred to as fluorescence in vivo hybridization (FIVH) [21–23]. Probes modified with 2'-O-methyl groups display quicker hybridization rates, higher melting points, improved specificity and nuclease resistance, and the capability to bind to structured targets—qualities that are advantageous for FIVH [21]. These have been employed for the detection of various RNA forms, including snRNAs, rRNA, and poly(A) RNA [21]. Nonetheless, the introduction of these probes into cells necessitates temporary cell permeabilization, a critical step for implementing FIVH.

## 20.3 Particle-Associated Hybridization-Based Imaging Probes

Particle-associated hybridization-based imaging probes represent a sophisticated method in RNA visualization and molecular biology that leverages the advantages of nanotechnology. These probes consist of hybridization sequences attached to nanoparticles or other nanostructures, which serve multiple roles: enhancing the delivery and stability of the probes, increasing the specificity of the signal, and amplifying the detectable signal for better imaging and analysis. Typically, these particle-associated probes are composed of DNA or RNA sequences that are designed to hybridize specifically with their target RNA within the cell. The particles themselves may be made of gold, silver, quantum dots, or other materials that are biocompatible and enhance the optical properties of the attached probes. When these probes hybridize to their target RNA sequences, the particles facilitate a stronger signal through either enhanced fluorescence or other optical properties such as plasmon resonance. This results in a higher signal-to-noise ratio, making it easier to detect specific RNA molecules even at low concentrations. The use of nanoparticles can protect RNA probes from degradation by nucleases within the cell, extending their operational lifespan. Metallic nanoparticles, for instance, can enhance the fluorescence signals by localizing multiple fluorophores or enhancing the local electromagnetic field. Different particles can be used to simultaneously track multiple RNA targets within the same cell by using probes that fluoresce at different wavelengths [24, 25]. Despite their advantages, these probes face challenges such as potential cytotoxicity and the complexity of delivering them into living cells without disrupting normal cellular functions. Advances in bioconjugation chemistry and nanoparticle design are continually being developed to address these issues. Researchers are also exploring biodegradable particles and surface modifications that improve cellular uptake while minimizing potential adverse effects. Particle-associated hybridization-based imaging probes exemplify a dynamic area of research that combines molecular biology with materials science to push the boundaries of cellular imaging and diagnostics [26]. Innovative strides in RNA detection technology have led to the creation of sophisticated systems where gold nanoparticles are tethered to quantum dots via DNA strands that hybridize with specific targets. This design is particularly crafted for the sensitive and precise detection of microRNAs, which are crucial in regulating gene expression and are often implicated in various diseases, including cancer. The system employs gold nanoparticles and quantum dots connected by specially designed DNA strands. These strands serve as a bridge and contain sequences complementary to the target microRNA. Upon encountering the target microRNA, hybridization occurs, which brings the quantum dots and gold nanoparticles into close proximity. The proximity of quantum dots to gold nanoparticles significantly enhances the fluorescent signal through a process known as Förster resonance energy transfer (FRET) or by plasmonic enhancement. This increase in fluorescence makes it easier to detect even low levels of microRNAs within a sample [27]. The presence of the target microRNA induces the separation of quantum dots from the gold nanoparticles, eliminating quenching and thereby generating a detectable signal. This mechanism allows for the quantitative binding to target RNA in vitro, enabling the differentiation between cell lines that express a particular miRNA.

## 20.4 Covalent Modification of RNA in Cells

Covalent modification of RNA within cells is a critical area of study that touches on how RNA molecules are chemically altered post-transcriptionally, impacting their stability, localization, function, and participation in cellular processes. These modifications are enzymatically driven and can occur on various RNA species, including mRNA, tRNA, rRNA, and ncRNAs. The most common modification, especially m6A in mRNA, affects RNA stability and the efficiency of translation. In tRNA and rRNA, methylation typically enhances stability and function in translation. The conversion of uridine to pseudouridine alters the RNA structure, potentially affecting its function and interactions with other molecules. The conversion of uridine to pseudouridine alters the RNA structure, potentially affecting its function and interactions with other molecules, such as A-to-I (adenosine to inosine) editing mediated by ADAR enzymes, which can alter the coding potential and splicing of mRNAs. Furthermore, the incorporation of thio-modified bases in tRNAs might also affect the accuracy and efficiency of protein synthesis.

A method that preserves RNA functionality while enabling visualization involves attaching a small chemical group to the RNA, which can later be targeted for linking to a fluorophore using click chemistry techniques. Among these techniques, the copper(I)-catalyzed azide-alkyne cycloaddition (CuAAC) is particularly noteworthy for its rapidity and precision. In this chemical reaction, an azide reacts with an alkyne in the presence of a copper(I) catalyst. Although CuAAC is effective and exhibits high selectivity, the toxicity associated with copper(I) at millimolar concentrations limits its application to fixed cell samples. A notable advancement in this area was made by Jao and Salic, who developed a method to integrate ethynyl groups across the entire RNA pool within living cells. They achieved this by introducing 5-ethynyluridine (EU) into cells, which is then metabolically converted into its active triphosphate form. This technique allows for the selective labeling of RNA in a manner that is compatible with live-cell imaging, though with careful management of the copper(I) concentrations to minimize cytotoxic effects. This approach is a significant stride in RNA visualization, offering a powerful tool for studying RNA dynamics and function in a cellular context [28].

Several copper-free click reactions, more amenable for use in live-cell imaging, have been identified as bioorthogonal click reactions [29]. Sawant and colleagues created an azido-modified UTP analog suitable for the bioorthogonal strain-promoted azide-alkyne cycloaddition (SPAAC) [30]. This innovative technique allows for the execution of the click reaction directly within live cells. However, the necessary UTP analog, such as 5-ethynyluridine (EU), must be externally supplied to the cells because its uridine precursor cannot efficiently penetrate cell membranes nor is it effectively utilized by the cell's ribonucleoside salvage pathway, as depicted in Fig. 20.2a. This method of introducing modified nucleotides is crucial for enabling live-cell imaging but comes with its own set of challenges. One significant limitation of using modified nucleotides, whether for transcription processes or for adding poly(A) tails, is the difficulty in distinguishing between different RNA subtypes. Since the modification is uniformly introduced across all RNA molecules transcribed during the treatment, it does not discriminate among various classes of RNA, such as mRNA, tRNA, or rRNA. This lack of specificity can obscure important differences in RNA processing, localization, and function within the cell. To address these challenges, further refinement of labeling techniques or the development of more selective labeling compounds may be necessary to enhance our understanding of RNA dynamics and functions specific to different RNA types. An alternative strategy for specific subtype labeling involves the post-synthetic attachment of chemical groups suitable for click reactions, utilizing RNA-modifying enzymes. Different RNA subtypes can also be distinctly labeled by exploiting unique structures or modifications specific to an RNA type, such as specifically targeting the 5′-cap of mRNA using a tailored methyltransferase that acts exclusively on the mRNA cap (illustrated in Fig. 20.2b) [31–33]. This method is particularly well-suited for live-cell applications because it utilizes an S-adenosylmethionine (AdoMet) analog that can be synthesized intracellularly. This is achieved using methionine analogs that are cell-permeable and stable. These analogs are converted into the AdoMet analog by a specifically engineered version of the enzyme methionine adenosyltransferase (MAT), which is tailored to recognize and process these modified substrates. The use of a modified MAT enzyme to synthesize AdoMet from these analogs inside the cell ensures that the entire process is biocompatible and minimally disruptive to normal cellular functions. This capability is essential for studying methylation dynamics and other AdoMet-dependent processes in real time, providing insights into cellular metabolism and gene expression regulation under physiological conditions. Additionally, this approach can be expanded to include other types of chemical modifications, potentially opening new avenues for exploring cellular mechanisms and interactions at a molecular level in live cells [32]. Sequence-specific RNA modification using a propargyl group, followed by fluorescent labeling, represents a refined technique for studying RNA in vitro. This approach leverages the specificity of box C/D methyltransferase-guide RNA complexes, which are designed to recognize and bind to specific RNA sequences. When combined with a propargyl-adapted AdoMet analog, this complex facilitates the precise addition of a propargyl group to the target RNA sequence. The introduction of the propargyl group into RNA allows for subsequent attachment of a fluorophore through click chemistry, particularly via CuAAC, enabling the visual tracking of

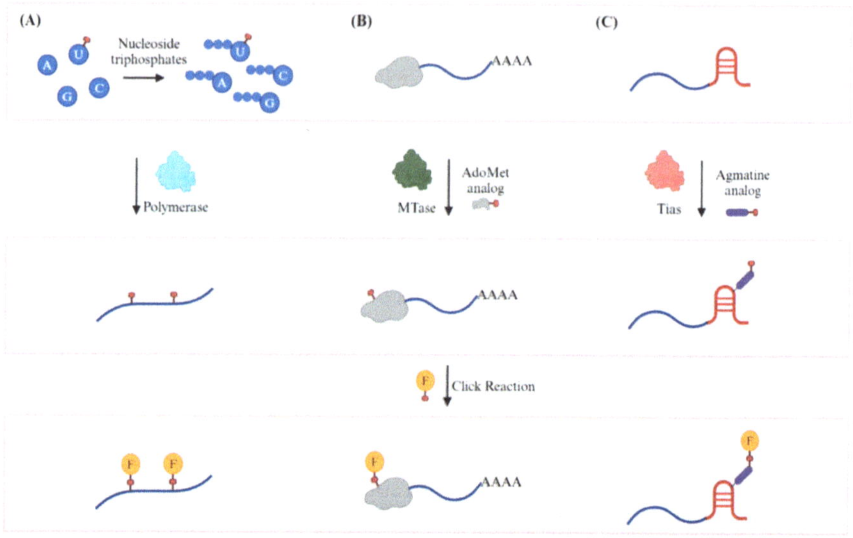


**Fig. 20.2** The methodology for introducing covalent modifications into RNA molecules and their subsequent visualization using click chemistry, highlighting three primary mechanisms: (**a**) Integration of Modified Nucleotides: This process involves the cellular RNA polymerases incorporating modified nucleotides into newly synthesized RNA. For instance, ethynyluridine, which is permeable to cell membranes, can be directly supplied to cells. Once inside, it is metabolized into its active triphosphate form, allowing for straightforward integration during RNA synthesis. In contrast, for modifications such as azido-uridine, which are not membrane permeable, cells require the introduction of the corresponding triphosphates via transfection methods. (**b**) Targeting Unique RNA Features: Specific features of RNA subtypes, such as the 5′ cap of mRNA, can be selectively targeted for modification. This is achieved using a specialized methyltransferase (MTase) variant engineered to modify the mRNA cap with a functional group that is compatible with click chemistry. This modification process is dependent on the availability of an appropriate S-adenosylmethionine (AdoMet) analog, which serves as the methyl donor in the reaction. (**c**) Targeted Addition to Specific Transcripts: To add a click-reactive group to a particular RNA transcript, a tRNA-like sequence can be attached to the RNA of interest. The enzyme tRNA^Ile2-agmatidine synthetase (Tias) is then used to attach the clickable group to this appended sequence, provided the appropriate agmatine analogs are available. This allows for precise labeling of specific RNA molecules within the cellular environment. Figure is drawn in Biorender.com

these modified RNA molecules. This method is particularly powerful for dissecting RNA functions and interactions within complex biological systems since it permits the direct visualization of RNA molecules in their native environments. This technology not only provides a high degree of specificity in RNA labeling but also opens up possibilities for multiplexing—where multiple RNAs can be simultaneously tagged with different fluorophores for comprehensive analysis. By facilitating detailed studies of RNA dynamics, localization, and interactions, this method enhances our understanding of RNA's role in cellular processes and holds potential for identifying novel therapeutic targets [34]. Li and colleagues have developed a system for RNA labeling that involves extending an RNA molecule with a tRNA-derived sequence and introducing an enzyme that specifically modifies this sequence ($tRNA^{Ile2}$-agmatidine synthetase, or Tias) into the cell (shown in Fig. 20.2c) [35].

The RNA-Tias system represents an innovative method for RNA labeling within cells, utilizing the capacity to incorporate agmatine analogs that are reactive in click chemistry. This system enables precise and efficient RNA labeling, leveraging the unique properties of these analogs to facilitate subsequent visualization and tracking of RNA molecules in a cellular context. Additionally, this system can be further tailored to target specific RNA sequences through the use of a tRNA-derived recognition motif. This motif can be selectively recognized and tagged by an engineered transglycosylase enzyme. This enzyme is specifically designed to attach large, visualizable groups to the RNA, enhancing the detectability and analysis of RNA molecules under cellular conditions. This capability is particularly valuable for studying RNA localization, turnover, and interaction with other biomolecules, providing crucial insights into RNA function and dynamics in living cells [36].

## 20.5 RNA-Binding Proteins

The use of RNA-binding proteins derived from bacteriophages for RNA labeling in cells has proven effective, with the MS2 coat protein system (MS2-MCP) being particularly noteworthy. This innovative system utilizes a version of the bacteriophage MS2 coat protein (MCP), an RNA-binding protein that specifically recognizes and binds to RNA sequences that form hairpin structures. The MS2 coat protein is genetically fused to green fluorescent protein (GFP). This fusion allows the MCP to retain its RNA-binding ability, while the GFP provides a fluorescent tag, enabling the visualization of RNA dynamics in real time within live cells. The target RNA is engineered to include multiple MS2-binding sites (MBS). These sites are sequences that form hairpin loops to which the MCP-GFP fusion protein can specifically bind. Once expressed in cells, the MCP-GFP fusion protein binds to the MBS sequences on the RNA of interest. Each RNA molecule can be tagged with multiple GFP molecules, significantly enhancing the fluorescent signal and allowing for easy detection and tracking of RNA movements and interactions within the cellular environment. This system is especially valuable for studying RNA localization, trafficking, and turnover in live cells. It provides a robust tool for observing the real-time behavior of RNA, offering insights into the complex regulation of gene expression at the RNA level [37]. Recently, the MS2 system has been utilized to visualize individual mRNA molecules in live mouse cells [38], showcasing the method's capability to track and resolve singular mRNA entities. However, the generation of transgenic models using this technique is notably labor-intensive. Another limitation includes the potential for the MS2-fusion tag and RNA extensions to disrupt the normal function or positioning of mRNA [39, 40]. Pumilio, a member of the PUF family of RNA-binding proteins, merits attention as well. Characterized by its modular structure, pumilio's binding specificity to RNA sequences can be modified through engineering [41–43]. An appealing feature of pumilio is its ability to directly bind to target RNA without the necessity for an RNA tag. Employing RNA-binding proteins for RNA visualization offers the advantage of targeting two distinct RNA sequences with different proteins, facilitating the observation of their interactions [44, 45]. However, the issue of high background fluorescence from unattached proteins can be mitigated by utilizing a split GFP system, which only emits light when dimerized [46, 47].

## 20.6 Reporter Protein Expression by Trans-splicing to Visualize RNA

The technique of modifying mRNA to include sequences that facilitate the expression of reporter proteins like luciferase has been instrumental in studying mRNA localization within specific tissues of living organisms. However, while effective for tissue-level analysis, this method does not provide subcellular resolution, limiting its usefulness for more detailed spatial investigations within cells. Expanding on this concept, Bhaumik and colleagues developed an innovative trans-splicing strategy that significantly enhances the scope of mRNA studies. This method involves the introduction of an exogenous RNA, which is processed into pre-mRNA, into the cells of living organisms via microinjection. The key feature of this approach is the strategic design of the pre-mRNA so that it undergoes trans-splicing with endogenous mRNA sequences. This splicing event results in the expression of luciferase directly within the cells, providing a luminous, quantifiable signal that indicates the presence and location of the mRNA [48]. This innovation enabled the detection of p53 at both organismal and cellular levels. Despite its potential for broad application [49], the method developed by the Gambhir laboratory has seen limited advancement since its initial presentation, with its primary drawback being the restriction to visualizing only exogenous RNA.

## 20.7 Aptamer-Based Transcript Imaging Systems

Aptamer-based systems for transcript imaging offer a powerful approach for visualizing specific RNA molecules within living cells [50–52]. These systems generally rely on the creation of a chimeric RNA transcript by incorporating an aptamer sequence into the exogenous transcript of interest. This integration enables the specific and dynamic visualization of single transcripts using fluorescence microscopy. The aptamer itself does not exhibit fluorescence; however, upon binding to a specific protein partner that is merged with an auto-fluorescent protein (forming a chimeric protein), fluorescence is activated under suitable excitation wavelengths. Alternatively, the chimeric transcript can be visualized using an organic dye that fluoresces upon binding to the aptamer. This approach is advantageous because it leverages the properties of organic dyes to achieve high sensitivity and specificity in RNA imaging. Several aptamer–dye systems have been developed to meet the stringent requirements necessary for effective RNA imaging in live cells. (as summarized in Table 20.2).

**Table 20.2** Some examples of aptamer-based in vitro RNA labeling methodologies

| System | Description | Dye characteristics | Applications | Advantages |
|---|---|---|---|---|
| II-mini3–4 system [86] | The II-mini3–4 system features a compact RNA aptamer that specifically binds to its target dye. The design of II-mini3–4 focuses on reducing the aptamer's size while maintaining high specificity and affinity for the dye | The dye used in the II-mini3–4 system fluoresces upon binding to the aptamer, allowing for real-time RNA visualization | This system is particularly useful for imaging small RNAs or RNA fragments within living cells. Its compact design minimizes interference with RNA function, making it ideal for detailed studies of RNA dynamics and localization | The small size of the II-mini3–4 aptamer reduces the likelihood of steric hindrance and functional disruption of the target RNA, enhancing the accuracy of RNA imaging |
| Dimethyl-indole red [87] | Aptamers specific to dimethyl-indole red (DIR) have been developed to bind this fluorescent dye with high affinity. These aptamers are engineered to ensure strong binding, resulting in significant fluorescence activation upon interaction with DIR | DIR is a red fluorescent dye that becomes highly fluorescent when bound to its aptamer. This dye is particularly useful for imaging in the red spectrum, which can provide better tissue penetration and reduced phototoxicity compared to shorter wavelength dyes | DIR-aptamer systems are suitable for imaging RNA in deep tissue samples or in applications requiring longer-wavelength fluorescence. This system is advantageous in reducing background fluorescence and photobleaching. | The use of DIR allows for imaging in the red spectral range, which is less damaging to cells and tissues and provides deeper penetration for in vivo studies |
| ASR7 [88] | The ASR7 system uses an aptamer specifically designed to bind to a novel dye, ASR7. The aptamer's structure ensures high specificity and strong binding, facilitating significant fluorescence activation upon interaction | ASR7 is a fluorescent dye that emits a bright signal upon binding to its corresponding aptamer. The fluorescence properties of ASR7 make it suitable for high-resolution imaging applications | This system can be used for high-resolution imaging of RNA molecules within living cells, providing detailed insights into RNA behavior, localization, and interactions | The ASR7 system's high specificity and brightness make it ideal for applications requiring detailed and precise RNA imaging, such as single-molecule studies |

(continued)

**Table 20.2** (continued)

| System | Description | Dye characteristics | Applications | Advantages |
|---|---|---|---|---|
| DCF-MPP [89] | The DCF-MPP system employs an aptamer that binds specifically to the DCF-MPP dye. The aptamer is designed to ensure that binding to DCF-MPP induces strong fluorescence activation | DCF-MPP is a fluorescent dye that becomes highly fluorescent upon binding to its aptamer. It is designed to provide a bright and stable signal for RNA imaging | DCF-MPP is particularly useful for visualizing RNA molecules in living cells, offering a robust tool for studying RNA transport, localization, and dynamics | The DCF-MPP system provides a strong and stable fluorescent signal, making it suitable for long-term imaging studies and applications requiring consistent fluorescence intensity |

### *20.7.1 Aptamer–Protein Systems*

Aptamer–protein systems exploit the naturally high-affinity interactions between specific RNA motifs and their corresponding binding proteins for RNA imaging and tracking within cells. These systems are often enhanced through engineering to improve the binding affinity and specificity of both the RNA motifs and the protein components [53]. The RNA motifs and proteins in these systems are frequently engineered to increase their binding affinity and reduce potential non-specific interactions. This enhancement is crucial for ensuring that the aptamer–protein interactions are both strong and specific. To minimize the size and potential steric hindrance of the protein–RNA complex, only the essential RNA-binding domain of the protein is typically retained. This reduction in size helps to maintain the functionality of the RNA and reduces any interference that might affect the RNA's natural behavior.

### *20.7.2 MS2 System*

One of the most well-known examples is the MS2 coat protein system. In this system, the MS2 coat protein, which specifically binds to MS2 RNA hairpins, is fused to a fluorescent protein. The RNA of interest is tagged with multiple MS2 binding sites, allowing the MS2 coat protein-fluorescent protein fusion to bind and illuminate the RNA [37], and subsequently utilized by Bloom and associates to explore mRNA localization in budding yeast [54]. Within the MS2 bacteriophage, genomic RNA interacts with its coat protein to form a viral capsid through the process of multimerization and RNA–protein binding. The MS2 leverages this natural interaction for RNA visualization and tracking within cells. This system consists of two main components: the RNA operator derived from the MS2 bacteriophage and a customized MS2 protein that binds this RNA operator as a dimer, providing enhanced RNA–protein affinity. The customized MS2 protein, often fused to a fluorescent marker such as GFP, specifically binds to the MS2 RNA operator sequences introduced into the RNA of interest. The RNA is typically engineered to contain multiple MS2 binding sites, allowing multiple MS2 protein.

The binding of multiple MS2 protein dimers to the RNA operator sequences facilitates multimerization, which is essential for forming a detectable fluorescent signal and for visualizing the RNA in live cells [55, 56]. The aptamer attached to a target transcript forms a 19-nucleotide RNA hairpin consisting of a 4-nucleotide loop and a 7 base pair stem, which includes a single adenine bulge. Notably, introducing a uridine (U) to cytidine (C) mutation within the loop's natural sequence significantly enhanced the protein binding affinity, increasing it by 50-fold. The aptamer's structure is crucial for its binding specificity and affinity. The 4-nucleotide loop and 7 base pair stem create a stable hairpin conformation, with a single adenine bulge contributing to the precise folding and interaction surface required for protein binding. The introduction of a U > C mutation in the loop sequence plays a pivotal role in boosting the binding affinity. This specific mutation alters the local RNA structure or interaction dynamics, facilitating a much stronger interaction with the binding protein [57]. To visualize individual transcript molecules, multiple hairpins are generally employed, owing to a single MS2 RNA hairpin's insufficiency for single-molecule imaging. The MS2 system's high specificity and sensitivity have established it as the most commonly employed mechanism for RNA labeling and imaging in live cells, as evidenced by over 100 studies utilizing MS2 across various biological systems to image a wide array of RNAs.

### *20.7.3 Lambda N22 System*

The lambda N22 system, developed by Daigle and Ellenberg, represents a significant advancement in the study of RNA within mammalian cells. By leveraging the high specificity and affinity of the lambda N22 protein for the box B RNA sequence, this system enables precise and real-time visualization of RNA dynamics. The lambda N22 protein is a truncated form of the lambda bacteriophage N protein, specifically retaining the RNA-binding domain. This domain is engineered to bind with high specificity to the box B RNA sequence, ensuring precise targeting. The target RNA is engineered to include multiple box B sequences, which are the specific binding sites for the lambda N22 protein. These sequences form RNA hairpins recognized by the lambda N22 protein, facilitating strong and specific interactions [58]. When expressed in mammalian cells, the lambda N22 protein, often fused to a fluorescent protein (such as GFP), binds specifically to the box B sequences inserted into the target RNA. This binding allows for the fluorescent tagging and visualization of the RNA in real time. By incorporating multiple box B sequences into the RNA, the binding of several lambda N22-GFP molecules is facilitated, amplifying the fluorescent signal and enabling the detection of even low-abundance RNA molecules [59]. A key advantage of this system is the compact size of the N protein fragment used, which minimizes potential impacts on the cellular dynamics of the RNA being studied [58]. The lambda N22–box B interaction is highly specific and stable, ensuring that the fluorescent signal accurately represents the target RNA's location and behavior. By using a specific RNA–protein interaction, the system reduces background fluorescence, providing a clear and precise imaging signal. The lambda N22 system can be adapted to study various RNAs by simply adding box B sequences to the RNA of interest, making it a versatile tool for RNA research [60].

### 20.7.4 BglG System

On another note, the BglG system, introduced by Hu et al., marked a significant advancement alongside the MS2 system in studying the heterogeneity of HIV-1 virions within human cells. This system utilizes the RNA-binding properties of the BglG protein to visualize and analyze specific RNA molecules [44]. BglG is an RNA-binding protein derived from the bacterial bgl operon. It recognizes and binds to specific RNA sequences, known as BglG-binding sites, which form stem-loop structures. The target RNA is engineered to include multiple BglG-binding sites, which are necessary for the BglG protein to specifically and efficiently bind to the RNA [61, 62]. The BglG protein, often fused to a fluorescent marker such as GFP, binds specifically to the engineered BglG-binding sites in the target RNA. This specific interaction allows for the fluorescent labeling of the RNA, enabling its visualization within the cell. By incorporating multiple BglG-binding sites into the RNA, several BglG-GFP molecules can bind simultaneously, enhancing the fluorescent signal and allowing for more precise detection of the RNA [63].

### 20.7.5 PP7 System

The PP7 system, unveiled in 2011 by Singer and his team, was developed to investigate the intricacies of transcription initiation and elongation processes in yeast. This system employs the PP7 bacteriophage coat protein, which specifically binds to RNA hairpin structures known as PP7-binding sites [64]. The PP7 coat protein is an RNA-binding protein derived from the PP7 bacteriophage. It is engineered to recognize and bind with high specificity to PP7 RNA hairpin sequences. The target RNA is engineered to include multiple PP7-binding sites. These sites form hairpin structures that the PP7 coat protein can bind to, allowing for the visualization of the RNA. The PP7 coat protein, often fused to a fluorescent marker such as GFP, binds specifically to the PP7-binding sites engineered into the RNA of interest. This binding allows for the fluorescent tagging of the RNA, making it visible under a fluorescence microscope. By incorporating PP7-binding sites into nascent RNA transcripts, researchers can monitor the transcription process in real time. The fluorescently tagged PP7 coat protein binds to the newly synthesized RNA, allowing for the visualization of transcription initiation and elongation [65]. In conjunction with the MS2 system, the PP7 setup facilitated bimolecular fluorescence complementation (BiFC) by allowing the reassembly of the EGFP protein from its separate fragments [47, 66].

### *20.7.6 U1Ap System*

The U1Ap system is another innovative tool designed for the visualization and study of RNA dynamics within living cells. It utilizes the specific interaction between the U1A protein, derived from the U1 small nuclear ribonucleoprotein (snRNP) complex, and its corresponding RNA-binding site. The U1A protein is an RNA-binding protein that specifically interacts with stem-loop structures in the RNA. It is a component of the U1 snRNP, which is involved in the splicing of pre-mRNA. The target RNA is engineered to include multiple U1A-binding sites, which are RNA sequences that form hairpin structures recognized by the U1A protein. Simultaneously introduced by Silver and Brodsky [67] and Takizawa and Vale [68], the U1Ap system has been applied in a limited number of studies, specifically four conducted in yeast cells [69–72]. The U1A protein, often fused to a fluorescent marker such as GFP, binds specifically to the U1A-binding sites engineered into the RNA. This binding enables the fluorescent tagging and visualization of the RNA within the cell. By incorporating multiple U1A-binding sites into the RNA, the binding of several U1A-GFP molecules is facilitated, significantly enhancing the fluorescent signal. This makes it easier to detect and study the RNA molecules in real time. For efficiency and minimal interference, only a fragment encompassing the first 94 amino acids—which includes the RNA recognition motif (RRM) domain responsible for RNA binding—was merged with a fluorescent marker [73, 74]. The U1Ap system allows researchers to visualize specific RNA molecules within living cells. This can be used to study the localization, dynamics, and interactions of RNA in various cellular contexts. Given U1A's role in the splicing machinery, the U1Ap system can be particularly useful for studying pre-mRNA splicing dynamics and the assembly of the spliceosome complex. The system can be employed to monitor the expression levels, stability, and transport of specific RNAs, providing insights into gene regulation mechanisms [67].

### *20.7.7 HTLV-1 Rex System*

The HTLV-1 Rex system, introduced by Broude et al., was employed alongside the lN22 system in a bimolecular fluorescence complementation (BiFC) experiment conducted in *E. coli* [75]. The HTLV-1 Rex system employs a notably small viral peptide, only 16 amino acids (aa) in length, making it the smallest peptide used in any RNA imaging system. This peptide binds specifically to a 36-nt hairpin derived from the HTLV-1 Rex responsive element (RxRE). The RxRE hairpin is characterized by a 4-nt loop and two bulges within the stem, creating a unique structure that facilitates high-affinity binding by the peptide. The 16-aa peptide's small size minimizes potential steric hindrance and interference with the RNA's natural functions, making it an ideal candidate for RNA imaging. The 36-nt RxRE hairpin includes specific structural features—a 4-nt loop and two bulges within the stem—that the viral peptide recognizes and binds to with high specificity [51]. In a physiological context, the Rex responsive element (RxRE) within viral RNA is targeted by the HTLV-1 Rex protein, facilitating the RNA's export from the nucleus. This interaction is critical for the life cycle of HTLV-1, as it ensures that the viral RNA is efficiently transported out of the nucleus to be translated and packaged into new virions. The HTLV-1 Rex protein binds specifically to the RxRE hairpin structure within the viral RNA. The RxRE is characterized by a 4-nucleotide loop and two bulges within the stem, providing a unique recognition site for the Rex protein. Once bound to the RxRE, the Rex protein promotes the export of the viral RNA from the nucleus to the cytoplasm. This process involves interactions with the cellular nuclear export machinery, ensuring that the viral RNA is efficiently transported through the nuclear pore complex [76].

### *20.7.8 TAT–TAR and REV–RRE Systems*

Cui and team have recently introduced two innovative systems derived from HIV sequences: the TAT–TAR and REV–RRE systems [77]. TAT–TAR System: The HIV-1 TAT (trans-activator of transcription) protein is a regulatory protein that binds specifically to the TAR (trans-activation response) element, a structured RNA sequence found at the 5′ end of all nascent viral transcripts. The TAR element is an RNA hairpin structure that includes a bulge and a loop, providing the necessary binding site for the TAT protein. The HIV-1 REV protein is another regulatory protein that binds specifically to the RRE (REV response element), a structured RNA sequence located within the env gene of the viral genome. The RRE is an RNA element that forms a complex secondary structure, which is recognized and bound by the REV protein. These systems leverage the natural binding affinities between the TAT and REV proteins and their respective TAR and RRE RNA sequences [78–80], to explore mRNA–protein dynamics. Both the TAT–TAR and REV–RRE systems offer high specificity due to the unique and strong interactions between the HIV proteins and their respective RNA elements. These systems enable real-time visualization of RNA molecules, allowing researchers to study dynamic processes such as transcription, RNA localization, and nuclear export. The systems can be adapted to various RNAs by incorporating the TAR or RRE elements, making them versatile tools for RNA research [77].

### 20.7.9 eIF4A System

The eIF4A system, introduced by Broude and colleagues in 2007, represents a significant advancement in the field of RNA imaging. This system was designed to image both mRNA and rRNA, leveraging the properties of the eIF4A peptide derived from mice. The eIF4A protein is known for its strong affinity toward RNA due to its two RNA-binding domains, making it particularly effective for binding to a relatively large 58 nucleotide (nt) aptamer [81]. The eIF4A system, introduced by Broude et al., offers significant advantages for fluorescence complementation studies, enabling the visualization of RNA dynamics within living cells. This system utilizes the eukaryotic initiation factor 4A (eIF4A) peptide from mice, which is distinguished by its two RNA-binding domains, and demonstrates strong affinity toward a relatively large 58 nt aptamer. The eIF4A peptide is part of the eukaryotic initiation factor 4A, essential for the initiation of translation. It is characterized by two RNA-binding domains, providing strong and specific binding to RNA sequences [82].

### 20.7.10 Aptamer–Dye Systems

In RNA imaging, aptamer–dye systems present an innovative alternative to protein–aptamer based methods. These systems rely on the interaction between RNA aptamers and specific dye molecules, allowing for the visualization of RNA molecules within living cells without the need for protein-based components [83]. Some examples for current aptamer–dye systems include: (i) Spinach Aptamer: Binds to the fluorophore DFHBI (3,5-difluoro-4-hydroxybenzylidene imidazolinone), which fluoresces green when bound to the aptamer. The Spinach aptamer allows for the visualization of tagged RNA in live cells. (ii) Mango Aptamer: Binds to a thiazole orange derivative, which emits bright fluorescence upon binding. The Mango aptamer provides a robust signal for RNA imaging. (iii) Broccoli Aptamer: An improved version of the Spinach aptamer, binding to DFHBI derivatives with enhanced brightness and stability. Unlike protein-based systems, aptamer–dye systems do not require the expression of exogenous proteins, reducing the potential for interference with cellular processes. The specific binding between the aptamer and dye ensures high sensitivity and specificity in RNA detection, providing clear and reliable imaging results. Multiple aptamer–dye pairs can be used simultaneously to study different RNA species, enabling multiplexed RNA imaging.

By substituting proteins with small organic dyes and integrating shorter sequences into the RNA under study, aptamer–dye systems mitigate concerns about the potential impact of protein or polypeptide binding on the properties of transcripts. This approach offers several advantages that enhance the utility and accuracy of RNA imaging. The promising capability of aptamer–dye systems for RNA visualization across various biological contexts has been underscored by numerous researchers [52, 84, 85]. Key advantages of this system include reducing steric hindrance, as small organic dyes cause less steric hindrance compared to larger protein or polypeptide molecules. This ensures that the natural conformation and function of the RNA are less likely to be disrupted. They have minimal interference with RNA function, as the use of short RNA sequences (aptamers) for dye binding minimizes the likelihood of altering the RNA's intrinsic properties, such as its stability, folding, and interactions with other cellular components. These systems also have improved cellular compatibility as small organic dyes are often more biocompatible than proteins, reducing the risk of eliciting an immune response or other cellular stress responses that could affect experimental outcomes. Although a variety of aptamer–dye pairs have been identified for in vitro RNA labeling, their application in live biological studies remains limited. Notable examples are given in Table 20.2.

Each of these aptamer–dye systems—II-mini3-4, dimethyl-indole red (DIR), ASR7, and DCF-MPP—offers unique advantages for RNA imaging, leveraging specific interactions between RNA aptamers and fluorescent dyes to enable precise and high-resolution visualization of RNA molecules within living cells. By minimizing the potential impact on RNA function and providing strong, specific fluorescence signals, these systems enhance our ability to study RNA dynamics, localization, and interactions in real time, contributing significantly to our understanding of RNA biology and gene regulation (Fig. 20.3).

Examples of Genetic Constructs Used for Imaging Experiments (O–R): O. DNA Construct for MS2 System: A genetic construct that includes the MS2 binding sites within the RNA of interest and the gene encoding the MS2-GFP fusion protein. Application: Imaging RNA with the MS2 system. P. DNA Constructs for BiFC with Two Systems: Genetic constructs that encode two complementary halves of a fluorescent protein fused to different RNA-binding proteins. Application: Visualization of RNA–protein interactions via BiFC. Q. System for Gene Locus, mRNA, and Protein Product Imaging: A comprehensive construct that allows for simultaneous imaging of gene loci, mRNA transcripts, and the resulting protein products. Application: Holistic studies of gene expression from DNA to RNA to protein. Figure is adapted from [90] under Creative Commons CC-BY 3.0 license.

## 20.8 Examples of Applications of Live RNA Imaging Systems

Live RNA imaging systems have revolutionized our ability to study RNA within the cellular context, providing critical insights into the dynamic processes of various RNA species across both prokaryotic and eukaryotic cells. These systems facilitate the visualization and analysis of mRNAs, ncRNAs, and RNA viruses, with a significant emphasis on investigating the mRNA life cycles within mammalian cells. Here are some specific applications:

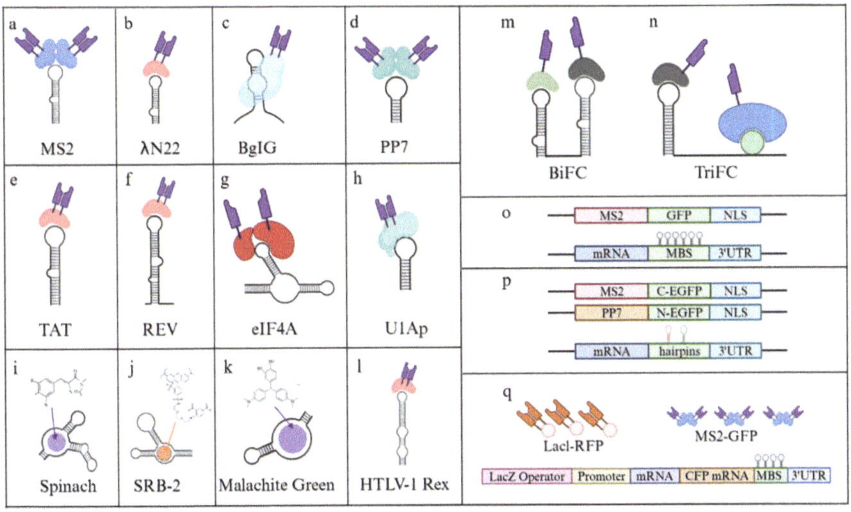


**Fig. 20.3** Vector-based systems for RNA live imaging. Schematic Structures of RNAs with Protein Partners or Fluorescent Dyes (**a**–**l**): (**a**) MS2 System: Utilizes the MS2 bacteriophage coat protein that binds to MS2 RNA hairpin structures. Application: Real-time visualization of RNA dynamics and localization. (**b**) λN22 System: Employs the λN22 protein, which binds specifically to the box B RNA hairpin sequence. Application: Studies on transcription initiation and elongation. (**c**) BglG System: Uses the BglG protein to bind specific RNA sequences derived from the bacterial bgl operon. Application: Imaging RNA within HIV-1 virions and analyzing RNA heterogeneity. (**d**) PP7 System: Involves the PP7 bacteriophage coat protein binding to PP7 RNA hairpins. Application: Investigating transcription processes in yeast. (**e**) U1Ap System: Utilizes the U1A protein from the U1 snRNP complex, binding to U1A-specific RNA sequences. Application: RNA imaging and splicing studies. (**f**) HTLV-1 Rex System: Uses the 16 amino acid HTLV-1 Rex peptide binding to the 36 nt RxRE RNA hairpin. Application: Studying RNA export from the nucleus. (**g**) TAT System: Involves the HIV-1 TAT protein binding to TAR RNA elements. Application: Visualization of transcription regulation. (**h**) REV System: Employs the HIV-1 REV protein binding to the RRE RNA element. Application: Imaging RNA nuclear export. (**i**) eIF4A System: Uses the eIF4A peptide, which binds to a 58 nt aptamer, attached to fluorescent protein halves. Application: Fluorescence complementation studies of RNA. (**j**) Spinach System: Involves the Spinach aptamer binding to the fluorophore DFHBI. Application: Real-time imaging of RNA with green fluorescence. (**k**) Malachite Green System: Uses the Malachite Green aptamer binding to the malachite green dye. Application: RNA visualization with specific dye binding. (**l**) SRB-2 System: Involves the SRB-2 aptamer that binds to a specific dye. Application: High-resolution RNA imaging. Additional Configurations (**m**–**n**): (**m**) BiFC with Two Systems: Bimolecular fluorescence complementation (BiFC) using two different protein systems to reconstitute a fluorescent protein. Application: Studying RNA–protein interactions. (**n**) TriFC (Tri-Molecular Fluorescence Complementation): Uses three fragments of a fluorescent protein for highly specific RNA–protein interaction studies. Application: Complex studies of RNA and protein interactions

### *20.8.1 Visualization of mRNA Life Cycles in Mammalian Cells*

Systems like the PP7 and MS2 allow for the real-time monitoring of mRNA synthesis, providing insights into the dynamics of transcription initiation and elongation processes. The U1Ap system can be used to study the splicing of pre-mRNA by visualizing the interactions between splicing factors and mRNA. The HTLV-1 Rex and REV systems facilitate the study of mRNA export from the nucleus to the cytoplasm, revealing mechanisms of RNA transport and regulation. Systems like Spinach and TAT-TAR enable the tracking of mRNA localization within the cytoplasm and its association with ribosomes for translation. Investigating the synthesis of transcripts initiates by assessing gene activity. This evaluation encompasses the examination of the onset of transcription and individual transcription occurrences [91]. It is possible to enumerate the quantity of transcripts produced anew over sequential intervals, enabling the observation of shifts in transcriptional activity when environmental conditions vary [92, 93]. Furthermore, the duration from the assembly of initiation factors to the commencement of transcription has been quantified [94]. Once the transcript undergoes nuclear processing fully, the mature mRNA is conveyed to the nuclear periphery for export out of the nucleus. Tracking the motion of RNA molecules revealed that their movement is stochastic rather than directed, being subject to energetic influences [95]. Various mechanisms are involved in the nuclear exit of transcripts [96], typically via nuclear pores, with the export process occurring more rapidly than the diffusion rates of mRNA within the nucleoplasm [97]. A reduction in ATP levels has been linked to the retention of mRNAs within cells, as well as disturbances in splicing, 3′ end processing [67], and the presence of triplet repeat expansion mutations [98, 99].

### *20.8.2 Studying ncRNAs*

The MS2 and BglG systems can be adapted to visualize snRNAs, which play crucial roles in RNA splicing and other regulatory processes. Aptamer–dye systems, such as Spinach and Broccoli, allow for the visualization of lncRNAs, providing insights into their roles in gene regulation and chromatin organization. The eIF4A system can be used to study miRNA processing and their interactions with target mRNAs, contributing to our understanding of post-transcriptional gene regulation. Utilizing live imaging technologies, researchers can ascertain the rate at which transcripts elongate and provide an intricate account of RNA polymerase II's functionality [64]. To differentiate between transcripts that are newly synthesized and those that have been disengaged from the transcription site, researchers employed the fluorescence recovery after photobleaching (FRAP) technique [100, 101]. The rate of transcript elongation was determined using RNA that was marked with two different labels, incorporating PP7 hairpins at the 5' UTR and MS2 hairpins at the 3' UTR. The interval needed for the elongation from PP7 to MS2 hairpins is reflected in the time taken to detect signals from these tags [102]. Subsequent nuclear processes post-transcription, such as splicing and the creation of 3' UTR, are pivotal as they may affect how long pre-mRNA remains at the transcription site. Unlike intron-less transcripts, those with introns are retained at the transcription site longer than the duration of polymerase action, suggesting that splicing restrains the pre-mRNA at the site [103]. In examining splicing dynamics, the duration of the signal from an aptamer sequence positioned within the intron was recorded [104].

### 20.8.3 Studying RNA–Protein Interaction

Marking both mRNAs and their corresponding protein outputs has provided insights into numerous aspects of translation. Research has established that during their journey through the cytoplasm, transcripts do not engage in translation. Investigating whether the positioning of transcripts is influenced by protein synthesis revealed that the localization of certain mRNAs is crucial for the proper placement of proteins [105], and reciprocally, that translational activity can dictate the positioning of transcripts through an SRP-dependent process [106]. Further studies have indicated that a variety of proteins and microRNAs play roles in controlling RNA stability [107] and affect the translation process of the transcripts under investigation [108–110].

### 20.8.4 Studying RNA Degradation Pathways

The degradation of RNA within a cell is a crucial process that ensures the proper regulation of gene expression and the removal of defective or unnecessary RNA molecules. Multiple pathways exist for RNA degradation, each tailored to specific RNA types and cellular conditions. Here, we focus on the degradation pathway of mRNA, which typically proceeds through a well-defined sequence of steps. The degradation of mRNA often begins with the shortening and eventual removal of the poly(A) tail at the 3′ end of the mRNA molecule. This process is catalyzed by deadenylase enzymes, such as the CCR4-NOT complex and the PAN2-PAN3 complex. The length of the poly(A) tail influences the stability of the mRNA; a longer poly(A) tail generally confers greater stability and translational efficiency, while deadenylation marks the mRNA for further degradation [111]. The study of mRNA degradation using live imaging techniques has significantly advanced our understanding of the dynamic processes involved in RNA decay. This approach often involves tracking the co-location of mRNA transcripts with cellular structures associated with degradation, such as Processing bodies (P bodies) [72, 112, 113]. The specific locations of transcript degradation have been identified by blocking nucleolytic decay [114], and the kinetics of RNA breakdown have been observed in scenarios where transcription is universally inhibited [115].

### 20.8.5 RNA Imaging for Neuronal Diseases

Genetic neurodegenerative conditions linked to simple repeat expansions, such as Huntington's disease, myotonic dystrophy, and certain types of spinocerebellar ataxia, are characterized by expansions of short nucleotide sequences within specific genes. These repeat expansions can lead to neurodegeneration through a variety of mechanisms that involve both the altered RNA transcripts and the mutant proteins they encode. Additionally, these diseases could result from a reduced level of the functional protein produced from the mutated gene. Mechanistically, expanded repeat sequences in RNA can lead to the formation of RNA foci, which are nuclear or cytoplasmic aggregates of RNA. These RNA foci can sequester RNA-binding proteins, disrupting normal cellular processes. Mutant RNA transcripts can interfere with normal splicing, transcription, and translation processes by binding to essential RNA-binding proteins and other regulatory factors. This sequestration can result in the misregulation of various cellular RNAs and proteins. The altered RNA may adopt abnormal secondary structures, such as hairpins or G-quadruplexes, which can interact aberrantly with cellular proteins, leading to toxicity. Neuronal cell dysfunction often emerges as an initial symptom, highlighting the importance of conducting pathogenesis research within appropriate experimental models. To date, a multitude of studies leveraging RNA imaging techniques in live neuronal cell lines have been undertaken.

The principal manifestation of RNA toxicity in neurological disorders resulting from trinucleotide repeat expansions is the formation of nuclear foci. These ribonuclear inclusions have been observed in patient cells and are a hallmark of several genetic neurodegenerative conditions. The presence of these inclusions is indicative of the pathological processes at play and contributes to the dysfunction and death of neurons. For example, (CAG in SCA3 and HD [116, 117], CTG in DM1 [118–120] and HDL2 [121], CGG in FXTAS [122, 123]), tetranucleotide repeats (CCTG in DM2 [124–126]), pentanucleotide repeats (TGGAA in SCA31 [127], ATTCT in SCA10 [128]), and hexanucleotide repeats (GGGGCC in ALS/FTD [129, 130]) are known to be mutations that lead to the production of mutant RNA transcripts containing expanded repeat sequences. The aggregates formed by transcripts with expanded repeats entrap various proteins. Through the use of FISH and immunofluorescence methods, researchers have been able to assess the dimensions and configurations of these inclusions and identify numerous proteins that are ensnared within them [131, 132]. FISH imaging has been instrumental in advancing our understanding of RNA toxicity in neurodegenerative diseases caused by trinucleotide repeat expansions. FISH allows for the precise localization and quantification of RNA foci within single nuclei and helps determine whether these foci contain entire mutant transcripts or merely fragments with expanded repeats [124, 133]. The relationship between mRNA expression levels and the development of foci [119, 134, 135], as well as the detrimental effects of cytoplasmic RNA foci, has been explored [136].

The process governing aggregation, whether it is entirely regulated or occurs by chance, remains unclear. Key questions surrounding nuclear foci encompass their dynamics and stability, compositional changes, factors that affect stability, and potential therapeutic agents aimed at RNA foci. The process of foci formation, including their formation dynamics, movement, stability, and protein interactions, can be examined through live RNA imaging techniques. In investigations into the mechanisms of neurodegenerative diseases linked to triplet repeat expansions, researchers have focused on two specific types of nuclear inclusions containing transcripts with expanded repeats. These studies have employed innovative RNA imaging systems like MS2 and Spinach2 to mark and visualize sections of the DMPK and FMR1 transcripts, which contain mutant CUG and CGG repeats, respectively [98, 99]. In myotonic dystrophy type 1 (DM1), the presence of expanded CUG repeats within the DMPK gene transcripts has significant cellular implications. Research focusing on the transition of these mutant transcripts into nuclear foci within the nucleoplasm has provided critical insights into their mobility and the resultant effects on cellular processes [99, 120]. Recent insights into the pathogenesis of myotonic dystrophy type 1 (DM1) underscore the pivotal role of the Muscleblind-like 1 (MBNL1) protein in the formation of nuclear foci associated with expanded CUG repeats in transcripts. These discoveries highlight a dynamic and active role for MBNL1 in the initiation of nuclear foci formation, rather than a passive entrapment within pre-existing aggregates. Supporting this, silencing the MBNL1 protein led to a reduction in the number of cellular foci. This interaction between the protein and transcripts not yet in aggregates aligns with observed changes in alternative splicing typical of DM1, even in cells without nuclear foci [137]. In experiments focusing on CGG repeat-containing transcripts, foci formation was noted shortly after transcription began [98], a rapid aggregation attributed to the employment of a potent promoter in the RNA imaging construct, corroborating FISH findings [121]. Utilizing RNA imaging alongside FRAP and fluorescence loss in photobleaching (FLIP) techniques, researchers were able to probe the formation dynamics of CUG foci [99], revealing their random and unpredictable nature. Spinach2 imaging revealed, for the first time, the impact of the cell cycle on foci formation [98]. During cell division, CGG-containing foci merged into a single large inclusion, which then was evenly distributed among progeny cells. Following division, these aggregates momentarily dispersed, with signals from the labeled transcripts spreading through the cytoplasm before quickly regrouping into foci.

The variability in structure and composition of nuclear foci presents a promising avenue for the development of therapeutic interventions aimed at nuclear foci, potentially facilitating the breakdown of pathological inclusions. Through the use of RNA imaging technologies, the impact of small molecular weight compounds on the formation and stability of foci has been monitored [98]. The compounds under investigation were found to either reduce the formation of new foci or cause existing ones to disperse. This capability of RNA imaging systems to track mutant triplet repeat transcripts during a crucial phase of their cellular life cycle that contributes to disease progression underscores the value of these technologies. A thorough comparison of the cellular pathways of normal and mutant transcripts is a crucial approach in neurogenetics, particularly for disorders characterized by repeat expansions like Huntington's disease, myotonic dystrophy (DM), and various forms of spinocerebellar ataxia. This comparative analysis is expected to uncover novel RNA-mediated pathological mechanisms and provide deeper insights into the cellular dysfunctions that drive these neurodegenerative disorders.

## 20.9 Conclusion

Understanding the complete life cycle of RNA transcripts—from their genesis and maturation, through their functional roles, to their eventual decay—represents a fundamental objective within cell biology. This aim, ambitious due to the diverse array of both coding and noncoding transcripts and their deep integration into a multitude of cellular functions, necessitates sustained and dedicated research efforts. As elucidated in this review, the advent of RNA live imaging technologies has propelled significant advancements in our comprehension of RNA dynamics. These technologies have unveiled intricate details about the synthesis, processing, movement, and complex interactions of RNA within live cells. Significantly, the field of RNA live imaging has evolved substantially, marked notably by the adoption of the MS2 system developed by Robert Singer and his team during the 1990s. This system, in particular, has enabled detailed visualization and analysis of RNA, enhancing our understanding of its cellular roles and the pathologies arising from its dysregulation. Looking ahead, the continued application of RNA live imaging promises to shed light on numerous RNA-related cellular phenomena. This ongoing exploration is poised to unravel further the complex roles of RNA in cellular operations and disease, potentially leading to novel therapeutic approaches and enhancing our understanding of fundamental biological processes.

## References

1. Lehmann R, Nüsslein-Volhard C. Abdominal segmentation, pole cell formation, and embryonic polarity require the localized activity of oskar, a maternal gene in drosophila. Cell. 1986;47(1):141–52.
2. Weil TT, Parton RM, Davis I. Making the message clear: visualizing mRNA localization. Trends Cell Biol. 2010;20(7):380–90.
3. Driever W, Nüsslein-Volhard C. The bicoid protein determines position in the drosophila embryo in a concentration-dependent manner. Cell. 1988;54(1):95–104.
4. Kellis M, Wold B, Snyder MP, Bernstein BE, Kundaje A, Marinov GK, et al. Defining functional DNA elements in the human genome. Proc Natl Acad Sci USA. 2014;111(17):6131–8.
5. Kindler S, Wang H, Richter D, Tiedge H. RNA transport and local control of translation. Annu Rev Cell Dev Biol. 2005;21:223–45.
6. Moore A, Medarova Z. Imaging of siRNA delivery and silencing. Methods Mol Biol. 2009;487:93–110.
7. Bassell GJ, Kelic S. Binding proteins for mRNA localization and local translation, and their dysfunction in genetic neurological disease. Curr Opin Neurobiol. 2004;14(5):574–81.
8. Jansson MD, Lund AH. MicroRNA and cancer. Mol Oncol. 2012;6(6):590–610.
9. Chen X, Zhang D, Su N, Bao B, Xie X, Zuo F, et al. Visualizing RNA dynamics in live cells with bright and stable fluorescent RNAs. Nat Biotechnol. 2019;37(11):1287–93.
10. Le P, Ahmed N, Yeo GW. Illuminating RNA biology through imaging. Nat Cell Biol. 2022;24(6):815–24.
11. Larsson C, Grundberg I, Söderberg O, Nilsson M. In situ detection and genotyping of individual mRNA molecules. Nat Methods. 2010;7(5):395–7.

57. Lowary PT, Uhlenbeck OC. An RNA mutation that increases the affinity of an RNA-protein interaction. Nucleic Acids Res. 1987;15(24):10483–93.
58. Daigle N, Ellenberg J. LambdaN-GFP: an RNA reporter system for live-cell imaging. Nat Methods. 2007;4(8):633–6.
59. Legault P, Li J, Mogridge J, Kay LE, Greenblatt J. NMR structure of the bacteriophage lambda N peptide/boxB RNA complex: recognition of a GNRA fold by an arginine-rich motif. Cell. 1998;93(2):289–99.
60. Schönberger J, Hammes UZ, Dresselhaus T. In vivo visualization of RNA in plants cells using the $\lambda N_{22}$ system and a GATEWAY-compatible vector series for candidate RNAs. Plant J. 2012;71(1):173–81.
61. Aymerich S, Steinmetz M. Specificity determinants and structural features in the RNA target of the bacterial antiterminator proteins of the BglG/SacY family. Proc Natl Acad Sci USA. 1992;89(21):10410–4.
62. Houman F, Diaz-Torres MR, Wright A. Transcriptional antitermination in the bgl operon of E. Coli is modulated by a specific RNA binding protein. Cell. 1990;62(6):1153–63.
63. Manival X, Yang Y, Strub MP, Kochoyan M, Steinmetz M, Aymerich S. From genetic to structural characterization of a new class of RNA-binding domain within the SacY/BglG family of antiterminator proteins. EMBO J. 1997;16(16):5019–29.
64. Larson DR, Zenklusen D, Wu B, Chao JA, Singer RH. Real-time observation of transcription initiation and elongation on an endogenous yeast gene. Science. 2011;332(6028):475–8.
65. Lim F, Peabody DS. RNA recognition site of PP7 coat protein. Nucleic Acids Res. 2002;30(19):4138–44.
66. Delebecque CJ, Lindner AB, Silver PA, Aldaye FA. Organization of intracellular reactions with rationally designed RNA assemblies. Science. 2011;333(6041):470–4.
67. Brodsky AS, Silver PA. Pre-mRNA processing factors are required for nuclear export. RNA. 2000;6(12):1737–49.
68. Takizawa PA, Vale RD. The myosin motor, Myo4p, binds Ash1 mRNA via the adapter protein, She3p. Proc Natl Acad Sci USA. 2000;97(10):5273–8.
69. Brengues M, Teixeira D, Parker R. Movement of eukaryotic mRNAs between polysomes and cytoplasmic processing bodies. Science. 2005;310(5747):486–9.
70. Chung S, Takizawa PA. Multiple Myo4 motors enhance ASH1 mRNA transport in Saccharomyces cerevisiae. J Cell Biol. 2010;189(4):755–67.
71. Kilchert C, Spang A. Cotranslational transport of ABP140 mRNA to the distal pole of S. Cerevisiae. EMBO J. 2011;30(17):3567–80.
72. Lavut A, Raveh D. Sequestration of highly expressed mRNAs in cytoplasmic granules, P-bodies, and stress granules enhances cell viability. PLoS Genet. 2012;8(2):e1002527.
73. Allain FH, Gubser CC, Howe PW, Nagai K, Neuhaus D, Varani G. Specificity of ribonucleoprotein interaction determined by RNA folding during complex formulation. Nature. 1996;380(6575):646–50.
74. Oubridge C, Ito N, Evans PR, Teo CH, Nagai K. Crystal structure at 1.92 a resolution of the RNA-binding domain of the U1A spliceosomal protein complexed with an RNA hairpin. Nature. 1994;372(6505):432–8.
75. Yiu HW, Demidov VV, Toran P, Cantor CR, Broude NE. RNA detection in live bacterial cells using fluorescent protein complementation triggered by interaction of two RNA aptamers with two RNA-binding peptides. Pharmaceuticals. 2011;4(3):494–508.
76. Nakano K, Watanabe T. HTLV-1 rex: the courier of viral messages making use of the host vehicle. Front Microbiol. 2012;3:330.
77. Yin J, Zhu D, Zhang Z, Wang W, Fan J, Men D, et al. Imaging of mRNA-protein interactions in live cells using novel mCherry trimolecular fluorescence complementation systems. PLoS One. 2013;8(11):e80851.
78. Cao H, Tamilarasu N, Rana TM. Orientation and affinity of HIV-1 tat fragments in tat-TAR complex determined by fluorescence resonance energy transfer. Bioconjug Chem. 2006;17(2):352–8.

79. Harada K, Martin SS, Frankel AD. Selection of RNA-binding peptides in vivo. Nature. 1996;380(6570):175–9.
80. Heaphy S, Dingwall C, Ernberg I, Gait MJ, Green SM, Karn J, et al. HIV-1 regulator of virion expression (rev) protein binds to an RNA stem-loop structure located within the rev response element region. Cell. 1990;60(4):685–93.
81. Oguro A, Ohtsu T, Svitkin YV, Sonenberg N, Nakamura Y. RNA aptamers to initiation factor 4A helicase hinder cap-dependent translation by blocking ATP hydrolysis. RNA. 2003;9(4):394–407.
82. Valencia-Burton M, Broude NE. Visualization of RNA using fluorescence complementation triggered by aptamer-protein interactions (RFAP) in live bacterial cells. Curr Protoc Cell Biol. 2007;Chapter 17:Unit 17.1.
83. Paige JS, Wu KY, Jaffrey SR. RNA mimics of green fluorescent protein. Science. 2011;333(6042):642–6.
84. Ehrhardt DW, Frommer WB. New technologies for 21st century plant science. Plant Cell. 2012;24(2):374–94.
85. Nevo-Dinur K, Govindarajan S, Amster-Choder O. Subcellular localization of RNA and proteins in prokaryotes. Trends Genet. 2012;28(7):314–22.
86. Sando S, Narita A, Aoyama Y. Light-up Hoechst-DNA aptamer pair: generation of an aptamer-selective fluorophore from a conventional DNA-staining dye. Chembiochem. 2007;8(15):1795–803.
87. Constantin TP, Silva GL, Robertson KL, Hamilton TP, Fague K, Waggoner AS, et al. Synthesis of new fluorogenic cyanine dyes and incorporation into RNA fluoromodules. Org Lett. 2008;10(8):1561–4.
88. Lee J, Lee KH, Jeon J, Dragulescu-Andrasi A, Xiao F, Rao J. Combining SELEX screening and rational design to develop light-up fluorophore-RNA aptamer pairs for RNA tagging. ACS Chem Biol. 2010;5(11):1065–74.
89. Sparano BA, Koide K. Fluorescent sensors for specific RNA: a general paradigm using chemistry and combinatorial biology. J Am Chem Soc. 2007;129(15):4785–94.
90. Urbanek MO, Galka-Marciniak P, Olejniczak M, Krzyzosiak WJ. RNA imaging in living cells - methods and applications. RNA Biol. 2014;11(8):1083–95.
91. Yunger S, Rosenfeld L, Garini Y, Shav-Tal Y. Quantifying the transcriptional output of single alleles in single living mammalian cells. Nat Protoc. 2013;8(2):393–408.
92. Barberis M, Beck C, Amoussouvi A, Schreiber G, Diener C, Herrmann A, et al. A low number of SIC1 mRNA molecules ensures a low noise level in cell cycle progression of budding yeast. Mol BioSyst. 2011;7(10):2804–12.
93. Yunger S, Rosenfeld L, Garini Y, Shav-Tal Y. Single-allele analysis of transcription kinetics in living mammalian cells. Nat Methods. 2010;7(8):631–3.
94. Dowen JM, Bilodeau S, Orlando DA, Hübner MR, Abraham BJ, Spector DL, et al. Multiple structural maintenance of chromosome complexes at transcriptional regulatory elements. Stem Cell Reports. 2013;1(5):371–8.
95. Shav-Tal Y, Darzacq X, Shenoy SM, Fusco D, Janicki SM, Spector DL, et al. Dynamics of single mRNPs in nuclei of living cells. Science. 2004;304(5678):1797–800.
96. Pitchiaya S, Heinicke LA, Custer TC, Walter NG. Single molecule fluorescence approaches shed light on intracellular RNAs. Chem Rev. 2014;114(6):3224–65.
97. Mor A, Suliman S, Ben-Yishay R, Yunger S, Brody Y, Shav-Tal Y. Dynamics of single mRNP nucleocytoplasmic transport and export through the nuclear pore in living cells. Nat Cell Biol. 2010;12(6):543–52.
98. Strack RL, Disney MD, Jaffrey SR. A superfolding Spinach2 reveals the dynamic nature of trinucleotide repeat-containing RNA. Nat Methods. 2013;10(12):1219–24.
99. Querido E, Gallardo F, Beaudoin M, Ménard C, Chartrand P. Stochastic and reversible aggregation of mRNA with expanded CUG-triplet repeats. J Cell Sci. 2011;124(Pt 10):1703–14.
100. Darzacq X, Shav-Tal Y, de Turris V, Brody Y, Shenoy SM, Phair RD, et al. In vivo dynamics of RNA polymerase II transcription. Nat Struct Mol Biol. 2007;14(9):796–806.

101. Maiuri P, Knezevich A, De Marco A, Mazza D, Kula A, McNally JG, et al. Fast transcription rates of RNA polymerase II in human cells. EMBO Rep. 2011;12(12):1280–5.
102. Hocine S, Raymond P, Zenklusen D, Chao JA, Singer RH. Single-molecule analysis of gene expression using two-color RNA labeling in live yeast. Nat Methods. 2013;10(2):119–21.
103. Brody Y, Neufeld N, Bieberstein N, Causse SZ, Böhnlein EM, Neugebauer KM, et al. The in vivo kinetics of RNA polymerase II elongation during co-transcriptional splicing. PLoS Biol. 2011;9(1):e1000573.
104. Martin RM, Rino J, Carvalho C, Kirchhausen T, Carmo-Fonseca M. Live-cell visualization of pre-mRNA splicing with single-molecule sensitivity. Cell Rep. 2013;4(6):1144–55.
105. Chang L, Shav-Tal Y, Trcek T, Singer RH, Goldman RD. Assembling an intermediate filament network by dynamic cotranslation. J Cell Biol. 2006;172(5):747–58.
106. Kraut-Cohen J, Afanasieva E, Haim-Vilmovsky L, Slobodin B, Yosef I, Bibi E, et al. Translation- and SRP-independent mRNA targeting to the endoplasmic reticulum in the yeast Saccharomyces cerevisiae. Mol Biol Cell. 2013;24(19):3069–84.
107. Yan D, Wu Z, Chisholm AD, Jin Y. The DLK-1 kinase promotes mRNA stability and local translation in C. Elegans synapses and axon regeneration. Cell. 2009;138(5):1005–18.
108. Lee EK, Kim HH, Kuwano Y, Abdelmohsen K, Srikantan S, Subaran SS, et al. hnRNP C promotes APP translation by competing with FMRP for APP mRNA recruitment to P bodies. Nat Struct Mol Biol. 2010;17(6):732–9.
109. Reich J, Snee MJ, Macdonald PM. miRNA-dependent translational repression in the drosophila ovary. PLoS One. 2009;4(3):e4669.
110. Tominaga K, Srikantan S, Lee EK, Subaran SS, Martindale JL, Abdelmohsen K, et al. Competitive regulation of nucleolin expression by HuR and miR-494. Mol Cell Biol. 2011;31(20):4219–31.
111. Houseley J, Tollervey D. The many pathways of RNA degradation. Cell. 2009;136(4):763–76.
112. Blanco FF, Sanduja S, Deane NG, Blackshear PJ, Dixon DA. Transforming growth factor β regulates P-body formation through induction of the mRNA decay factor tristetraprolin. Mol Cell Biol. 2014;34(2):180–95.
113. Jang LT, Buu LM, Lee FJ. Determinants of Rbp1p localization in specific cytoplasmic mRNA-processing foci. P-bodies J Biol Chem. 2006;281(39):29379–90.
114. Sheth U, Parker R. Decapping and decay of messenger RNA occur in cytoplasmic processing bodies. Science. 2003;300(5620):805–8.
115. Muramoto T, Cannon D, Gierlinski M, Corrigan A, Barton GJ, Chubb JR. Live imaging of nascent RNA dynamics reveals distinct types of transcriptional pulse regulation. Proc Natl Acad Sci USA. 2012;109(19):7350–5.
116. de Mezer M, Wojciechowska M, Napierala M, Sobczak K, Krzyzosiak WJ. Mutant CAG repeats of huntingtin transcript fold into hairpins, form nuclear foci and are targets for RNA interference. Nucleic Acids Res. 2011;39(9):3852–63.
117. Mykowska A, Sobczak K, Wojciechowska M, Kozlowski P, Krzyzosiak WJ. CAG repeats mimic CUG repeats in the misregulation of alternative splicing. Nucleic Acids Res. 2011;39(20):8938–51.
118. Davis BM, McCurrach ME, Taneja KL, Singer RH, Housman DE. Expansion of a CUG trinucleotide repeat in the 3′ untranslated region of myotonic dystrophy protein kinase transcripts results in nuclear retention of transcripts. Proc Natl Acad Sci USA. 1997;94(14):7388–93.
119. Mankodi A, Teng-Umnuay P, Krym M, Henderson D, Swanson M, Thornton CA. Ribonuclear inclusions in skeletal muscle in myotonic dystrophy types 1 and 2. Ann Neurol. 2003;54(6):760–8.
120. Taneja KL, McCurrach M, Schalling M, Housman D, Singer RH. Foci of trinucleotide repeat transcripts in nuclei of myotonic dystrophy cells and tissues. J Cell Biol. 1995;128(6):995–1002.
121. Wilburn B, Rudnicki DD, Zhao J, Weitz TM, Cheng Y, Gu X, et al. An antisense CAG repeat transcript at JPH3 locus mediates expanded polyglutamine protein toxicity in Huntington's disease-like 2 mice. Neuron. 2011;70(3):427–40.

122. Sellier C, Rau F, Liu Y, Tassone F, Hukema RK, Gattoni R, et al. Sam68 sequestration and partial loss of function are associated with splicing alterations in FXTAS patients. EMBO J. 2010;29(7):1248–61.
123. Tassone F, Iwahashi C, Hagerman PJ. FMR1 RNA within the intranuclear inclusions of fragile X-associated tremor/ataxia syndrome (FXTAS). RNA Biol. 2004;1(2):103–5.
124. Liquori CL, Ricker K, Moseley ML, Jacobsen JF, Kress W, Naylor SL, et al. Myotonic dystrophy type 2 caused by a CCTG expansion in intron 1 of ZNF9. Science. 2001;293(5531):864–7.
125. Margolis JM, Schoser BG, Moseley ML, Day JW, Ranum LP. DM2 intronic expansions: evidence for CCUG accumulation without flanking sequence or effects on ZNF9 mRNA processing or protein expression. Hum Mol Genet. 2006;15(11):1808–15.
126. Schoser BG, Ricker K, Schneider-Gold C, Hengstenberg C, Dürre J, Bültmann B, et al. Sudden cardiac death in myotonic dystrophy type 2. Neurology. 2004;63(12):2402–4.
127. Sato N, Amino T, Kobayashi K, Asakawa S, Ishiguro T, Tsunemi T, et al. Spinocerebellar ataxia type 31 is associated with "inserted" penta-nucleotide repeats containing (TGGAA)n. Am J Hum Genet. 2009;85(5):544–57.
128. White MC, Gao R, Xu W, Mandal SM, Lim JG, Hazra TK, et al. Inactivation of hnRNP K by expanded intronic AUUCU repeat induces apoptosis via translocation of PKCdelta to mitochondria in spinocerebellar ataxia 10. PLoS Genet. 2010;6(6):e1000984.
129. Mori K, Arzberger T, Grässer FA, Gijselinck I, May S, Rentzsch K, et al. Bidirectional transcripts of the expanded C9orf72 hexanucleotide repeat are translated into aggregating dipeptide repeat proteins. Acta Neuropathol. 2013;126(6):881–93.
130. Sareen D, O'Rourke JG, Meera P, Muhammad AK, Grant S, Simpkinson M, et al. Targeting RNA foci in iPSC-derived motor neurons from ALS patients with a C9ORF72 repeat expansion. Sci Transl Med. 2013;5(208):208ra149.
131. Fardaei M, Rogers MT, Thorpe HM, Larkin K, Hamshere MG, Harper PS, et al. Three proteins, MBNL, MBLL and MBXL, co-localize in vivo with nuclear foci of expanded-repeat transcripts in DM1 and DM2 cells. Hum Mol Genet. 2002;11(7):805–14.
132. Miller JW, Urbinati CR, Teng-Umnuay P, Stenberg MG, Byrne BJ, Thornton CA, et al. Recruitment of human muscleblind proteins to (CUG)(n) expansions associated with myotonic dystrophy. EMBO J. 2000;19(17):4439–48.
133. Cardani R, Mancinelli E, Giagnacovo M, Sansone V, Meola G. Ribonuclear inclusions as biomarker of myotonic dystrophy type 2, even in improperly frozen or defrozen skeletal muscle biopsies. Eur J Histochem. 2009;53(2):e13.
134. Furling D, Coiffier L, Mouly V, Barbet JP, St Guily JL, Taneja K, et al. Defective satellite cells in congenital myotonic dystrophy. Hum Mol Genet. 2001;10(19):2079–87.
135. Holt I, Jacquemin V, Fardaei M, Sewry CA, Butler-Browne GS, Furling D, et al. Muscleblind-like proteins: similarities and differences in normal and myotonic dystrophy muscle. Am J Pathol. 2009;174(1):216–27.
136. Dansithong W, Wolf CM, Sarkar P, Paul S, Chiang A, Holt I, et al. Cytoplasmic CUG RNA foci are insufficient to elicit key DM1 features. PLoS One. 2008;3(12):e3968.
137. Mahadevan MS, Yadava RS, Yu Q, Balijepalli S, Frenzel-McCardell CD, Bourne TD, et al. Reversible model of RNA toxicity and cardiac conduction defects in myotonic dystrophy. Nat Genet. 2006;38(9):1066–70.

# Chapter 21
# Targeting RNA with Small Molecules: Fundamental Principles

**Abstract** The field of RNA biology has recently made significant strides, revealing its vital involvement in various diseases, from infections caused by viruses and bacteria to cancer and neurological conditions. This insight has spurred interest across numerous research groups in crafting small molecules akin to drugs that can interact with RNA. Nevertheless, targeting RNA presents several unique obstacles. RNA's dynamic nature and its lack of chemical diversity pose significant challenges. Moreover, the process often leads to the identification of non-selective RNA-binding ligands during screening efforts. This section delves into crucial considerations and breakthroughs in the development of small molecules aimed at RNA, spanning the essential chemistry to notable examples of small molecules that have shown clinical success. We begin our exploration by delving into the various functional categories of RNA, as well as its structural complexities and dynamic properties. Next, we will discuss the fundamental principles of RNA recognition, the techniques used for screening small molecules, and the methodologies for determining RNA structure. Following this, we will tackle the unique challenges and innovative strategies involved in designing ligands that target RNA, considering the perspectives of both RNA and small molecules. Finally, we will highlight some of the most significant and impactful examples in this field to date.

## 21.1 Introduction

The use of antisense oligonucleotides (ASOs) and small interfering RNAs (siRNAs), which target specific sequences through complementary base pairing, has significantly advanced the identification and therapeutic targeting of RNA molecules, achieving success in several instances [1]. Despite these advances, small molecule drugs are often preferred for treating diseases due to their generally more favorable pharmacokinetic properties and the flexibility offered by medicinal chemistry for optimization. These small molecules have also proven to be effective in altering RNA structure, serving as a supplementary approach to ASOs, which recognize specific sequences [1]. The growing interest in targeting RNA has been

 
P. E. Saw, E. Song, *RNA Therapeutics in Human Diseases*,
https://doi.org/10.1007/978-981-96-3041-7_21

further invigorated by recent discoveries of a vast array of functional RNA molecules within human cells. A notable contributor to this field is the Encyclopedia of DNA Elements (ENCODE) project, which is an extension of the Human Genome Project. The primary goal of ENCODE is to identify and catalog all the functional elements within the human genome, including these newly discovered RNA molecules [2]. The ENCODE project revealed that only about 1.5% of the human genome is translated into proteins, while approximately 70–90% is transcribed into RNA. This suggests that a substantial portion of the genome produces ncRNAs rather than protein-coding mRNAs. Although various classes of ncRNAs, such as miRNAs, circRNAs, and lncRNAs, were already known, the ENCODE project significantly advanced our understanding by mapping numerous previously undiscovered lncRNAs, miRNAs, and other ncRNAs across different cell types [2].

## 21.2 Understanding RNA Structure and Properties

### *21.2.1 RNA Primary Structure*

The structural complexity of RNA is typically categorized into primary, secondary, and tertiary forms. At its core, the primary structure consists of a sequence of nucleotides, which are elaborated into three fundamental components [3]. A critical aspect of RNA structure is the glycosidic bond, which connects the ribose sugar to one of the four nitrogen-rich aromatic nucleobases: the purines adenine (A) and guanine (G) and the pyrimidines cytosine (C) and uracil (U). This linkage is illustrated in Fig. 21.1. The sugar-phosphate backbone of RNA, which carries a negative charge, is constructed from phosphodiester linkages and ribose units. The ribose units contain a 2′-hydroxyl group, resulting in unique sugar puckering patterns that distinguish RNA from DNA. The configuration of this backbone is significantly influenced by the surrounding solvent. Additionally, the folding process of RNA is challenged by the electrostatic repulsion between the negatively charged phosphate groups. However, this repulsion is mitigated by the binding of metal ions, such as $Mg^{2+}$ and monovalent $Na^{+}$ [3].

### *21.2.2 RNA Secondary Structure and Their Base-Pairing Patterns*

The secondary structure in RNA is characterized by localized folds resulting from base pairing, which involves interactions between hydrogen bond donor and acceptor groups on RNA bases. This folding predominantly occurs through the Watson–Crick–Franklin (WCF) hydrogen bonding arrangement, featuring canonical base pairs such as adenine–uracil (A–U) and guanine–cytosine (G–C). Similar to DNA, RNA base pairing enables the formation of a double helix or duplex structure. The unique orientation of the ribose sugar within the RNA backbone results in the

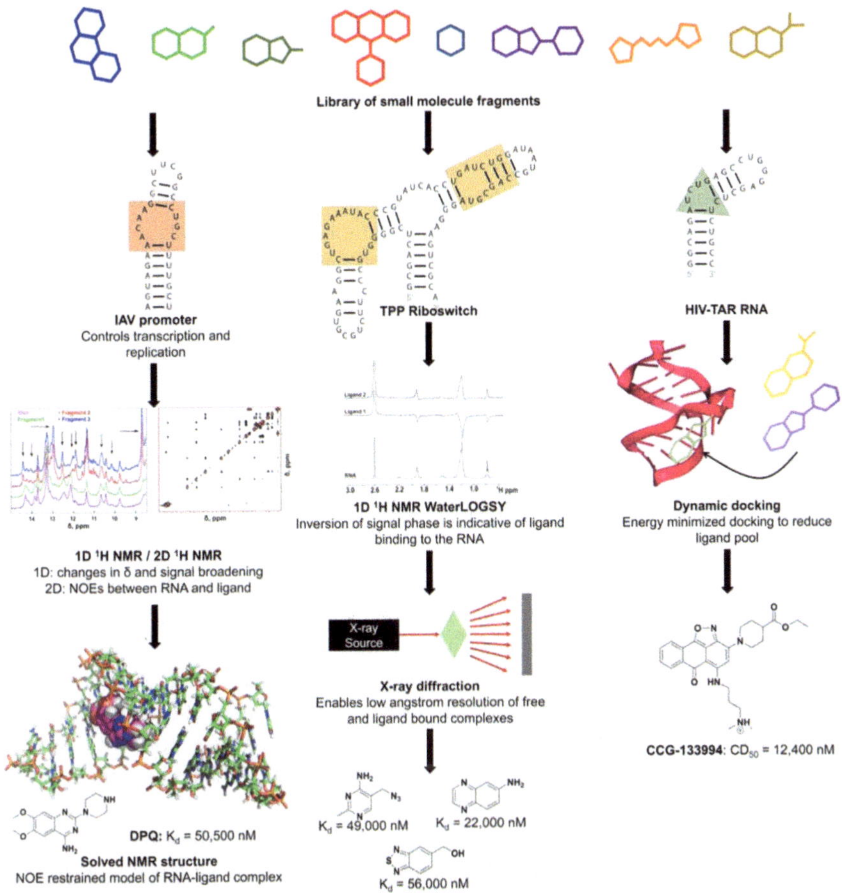


**Fig. 21.1** A comprehensive workflow for identifying and characterizing RNA-targeting small molecules. Starting with a library of small molecule fragments, the process involves targeting HIV-TAR RNA, the TPP riboswitch, and the IAV promoter. For HIV-TAR RNA, dynamic docking reduces the ligand pool, identifying CCG-133994 with a CD_50 of 12,400 nM. The TPP riboswitch is studied using 1D$^{1}$HNMR WaterLOGSY to detect ligand binding, followed by X-ray diffraction to obtain high-resolution structures, revealing ligands with $K_d$ values of 49,000 nM, 22,000 nM, and 56,000 nM. The IAV promoter is analyzed using 1D and 2D $^{1}$HNMR to observe binding-induced signal changes and NOEs, leading to the solved NMR structure of the RNA–ligand complex and the identification of DPQ with a $K_d$ of 50–500 nM. This workflow highlights the integration of dynamic docking, NMR spectroscopy, and X-ray diffraction to uncover RNA–small molecule interactions for therapeutic applications. Figure is reproduced from [22] with permission from Elsevier (License number: 5902940274150)

A-form duplex, which is distinct from the B-form structure of DNA. This A-form structure is stabilized by strong and cooperative stacking interactions between the planar and polarizable surfaces of nucleobases. Additionally, the presence of more hydrogen bonds in guanine–cytosine (G–C) base pairs makes them slightly more stable compared to adenine–uracil (A–U) pairs.

Beyond duplex regions, RNA's unique single-stranded secondary structures are formed by nucleotides that do not participate in hydrogen bonding. These structures include apical loops, internal loops, and bulges, which combine with base-paired regions to create RNA's complex topologies. These intricate structures can serve as binding sites for various ligands, including proteins and small molecules. The structural complexity of RNA is further enhanced by non-traditional base-pairing patterns. The stability of these non-canonical pairs can be influenced by several factors, such as the necessity for glycosidic bond rotation, which involves steric considerations [4, 5], base protonation, and metal binding. Non-standard pairings include uridines or protonated cytosines pairing with pyrimidines across the Watson–Crick–Franklin interface, known as the Hoogsteen pattern, which can form a base triple with a standard WCF pair. Uridines can also engage in "wobble" interactions with guanines and protonated adenosines on the WCF surface.

### *21.2.3 RNA Tertiary Structure*

RNA tertiary structures arise when elements of secondary structure interact spatially, bringing together regions that may be distant in the primary or secondary structures during the folding process. This folding process is driven by the desire to maximize base stacking interactions, similar to how protein folding is largely influenced by the hydrophobic effect. In RNA, these tertiary interactions play a crucial role in stabilizing complex three-dimensional architectures, essential for their biological functions and interactions with other molecules [5]. The architectural formation of RNA tertiary structures relies heavily on the coaxial stacking of helices (the alignment of two helices along their central axes), shaped significantly by the topology of the RNA junction, where the helices meet [5]. One of the initial groundbreaking discoveries was the structural revelation of yeast phenylalanine transfer RNA using X-ray crystallography, which exemplified this principle [6]. Beyond coaxial stacking, the compatibility of nucleotide sequences over long ranges facilitates base pairing that leads to the formation of structures like kissing loops and pseudoknots [4]. Additionally, a triple helix structure identified in MALAT1, resulting from consecutive base triples, exemplifies this, providing stability against degradation for the plentiful MALAT1 transcript [7]. RNA tertiary structure is also supported by backbone desolvation and the influence of metal ions through charge screening or direct interaction. For instance, guanine bases can assemble into G-quartets stabilized by potassium ions, and these units can stack together to form highly stable G-quadruplex structures. Such structures have been implicated in a variety of diseases due to their involvement in regulatory processes related to gene expression and telomere maintenance [8]. Together, these unique tertiary structures offer specific binding sites for interacting with small molecules, presenting opportunities to investigate the relationships between structure and function in RNA and explore their therapeutic potential.

### 21.2.4 RNA Quaternary Structure

The functional versatility of RNA arises from its ability to interact with a wide array of biomolecules, encompassing DNA, proteins, and other RNA molecules. These interactions play pivotal roles in various biological processes, such as gene regulation, RNA processing, and protein synthesis. Understanding the complexity of these interactions is essential for deciphering the intricate networks that govern cellular functions and for developing novel therapeutic strategies targeting RNA-related diseases [9]. These interactions give rise to what is known as RNA quaternary structures. Among these, the RNA–protein interactions stand out as a pivotal area of research, especially concerning their regulation by small molecules. For instance, numerous RBPs achieve binding through interactions such as nucleobase stacking with aromatic amino acids and electrostatic connections between the RNA backbone and cationic amino acids [3]. Furthermore, certain RBPs exhibit a preference for binding specific RNA tertiary structures. The PRC2 complex, for example, has been shown to bind various RNA molecules, often at G-quadruplex structures, suggesting a nuanced mechanism of RNA recognition by RBPs [10]. It is also noteworthy that RNAs can participate in complex higher-order structures with multiple RNA and protein components. Significant research, including studies by Yan et al. [11], has shed light on the intricate interplay between spliceosome components, such as RNA substrates, protein subunits, and snRNAs, which play a key role in splicing [11]. Another vital RNA–protein complex is the ribosome, crucial for protein synthesis. Research, exemplified by studies conducted by Ramakrishnan et al., has provided detailed insights into the interactions between RNA molecules and between RNA and proteins within the bacterial 70S ribosome. These investigations underscore the critical role of magnesium coordination and other electrostatic interactions in stabilizing the complex structure and function of the ribosome [12]. These complex structures not only inspire new RNA-targeted therapeutic strategies but have also led to the development of effective treatments.

## 21.3 RNA Dynamics

RNA structures at all levels are susceptible to alterations governed by diverse cellular signals, such as protein interactions and changes in the ionic milieu. Furthermore, the structural dynamics of RNA are influenced by posttranscriptional modifications—reversible changes to RNA nucleotides that can significantly affect their behavior [13, 14]. Recent advancements in structure determination techniques have revealed RNA's capability to adopt a range of conformations, shifting the perspective toward viewing these molecules as part of structural "ensembles" or collections of multiple structures, rather than being rigidly fixed in a single conformation. This concept is critical as different structures within these ensembles can lead to varied functional outcomes and offer novel binding sites for ligands, thus influencing RNA's structural equilibrium [15]. The dynamics of RNA can be significantly altered by posttranscriptional modifications and other regulatory mechanisms, which, in turn, can have profound effects on its function. For instance, N6-methyladenosine (m6A) has been shown to disrupt Watson–Crick–Franklin base pairing through steric hindrance while enhancing base stacking in single-stranded areas, thereby modifying the structure and dynamics of RNA [14]. For instance, the addition of m6A modification within a hairpin loop located at the 5′-end of the lncRNA MALAT1 was discovered to enhance its flexibility. This modification consequently strengthened its interaction with heterogeneous nuclear ribonucleoprotein C (hnRNP C). These findings highlight how RNA modifications can dynamically influence RNA–protein interactions, potentially modulating biological processes such as RNA localization, stability, and regulatory functions within cells. Understanding these molecular mechanisms could offer insights into therapeutic strategies targeting RNA modifications in various diseases [13]. Moreover, alterations in RNA secondary structure can impact protein interactions, exemplified by the HIV transactivation response (TAR) element RNA. This RNA segment demonstrates flexibility between ground state (GS) and excited state (ES) configurations, where only the GS conformation interacts with the Cyclin T1–Tat protein complex. The specific biological consequences of this structural flexibility are currently under investigation, but they underscore the potential for small molecules to modulate viral transcription and the production of viral particles [16].

The dynamic and adaptable structural complexity of RNA distinguishes it from other biomolecules. While RNA shares fundamental components with DNA, its potential for structural diversity surpasses that of DNA. Unlike proteins, which generally adopt a single, thermodynamically stable structure, RNA often exists as ensembles or populations of related structures. This inherent variability in RNA structure enables it to fulfill diverse biological roles, from acting as catalysts in enzymatic reactions to regulating gene expression through intricate folding mechanisms [17]. The RNA folding landscape is considered more "rugged" compared to that of proteins, characterized by the coexistence of several energetically favorable structures [17]. This conformational versatility of RNA, despite its limited chemical diversity relative to the 20 amino acids found in proteins, suggests that RNA's structural diversity may compensate for its simpler composition, offering distinct opportunities for regulation through small molecule ligands.

## 21.4 RNA–Small Molecule Recognition: Principles and Methods

Small molecules exhibit a unique capability to selectively identify and bind to specific pockets within RNA structures, thereby modifying the function of the transcript. Understanding the primary forces that facilitate this recognition has been advanced through structural and biophysical analyses of these binding sites. This discussion will explore the essential noncovalent interactions crucial for the recognition of small molecules by RNA. Additionally, it will outline both conventional and innovative techniques used to detect and characterize these binding events. These insights are pivotal for developing strategies that harness small molecules to target RNA effectively, offering potential therapeutic avenues for diseases where RNA dysregulation plays a significant role [1].

## 21.5 Recognition Principles and Noncovalent Interactions: The Fundamentals

The escalating efforts to characterize RNA–small molecule structures offer a chance to scrutinize these datasets for patterns in how small molecules recognize RNA. This analysis aims to hasten the formulation of general principles for RNA targeting and to delineate how small molecule recognition of RNA diverges from that of proteins, the more conventional targets for drug development. An illustrative study by the Hargrove group analyzed RNA–small molecule and protein–small molecule complexes in the Protein Data Bank (PDB), uncovering distinct interaction types that predominate in small molecule recognition of each biomolecule [18]. For RNA, stacking and hydrogen bonding emerge as primary mechanisms of recognition. Stacking is facilitated by the aromatic regions of small molecules, with ligands featuring large aromatic ring systems capable of intercalating between base pairs, thereby enhancing $\pi$–$\pi$ interactions [19]. Additionally, specific hydrogen bonds formed between small molecules and nucleotides play a crucial role in this interaction dynamic. Another illustrative case from the research involves paromomycin, which utilizes its hydroxyl and ether groups to establish hydrogen bonds with an adenine residue located within the HIV-1 dimerization initiation site. This site is crucial for regulating RNA function, showcasing the specificity and importance of hydrogen bonding in the interaction between small molecules and RNA. This example underscores how understanding these molecular interactions can inform the development of targeted therapies aimed at disrupting RNA-related processes in diseases like HIV/AIDS [20].

Unlike protein interactions, which often rely heavily on the hydrophobic effect, RNA interactions appear to be less influenced by hydrophobicity, though it remains an important factor [18]. This distinction was further supported by a study from Schneekloth and colleagues, who compared ligand-bound binding sites in RNA and proteins using data from the PDB. Their analysis revealed that while the size and solvent exposure (or "buriedness") of these binding sites were similar in both RNA and proteins, RNA binding sites exhibited significantly lower hydrophobicity. This finding aligns with RNA's inherently more polar nature, underscoring a unique avenue for specific targeting of RNA [21]. It is crucial to note, however, that the insights from such studies are based on RNA and proteins amenable to structural analysis and might not fully apply to more dynamic or disordered recognition areas within these biomolecules. Despite this limitation, the observed differences in interaction types and the hydrophobicity profile of the binding sites between RNA and proteins have led researchers to suggest that RNA-binding small molecules may possess privileged chemical features (Fig. 21.1).

## 21.6 Detection of RNA–Small Molecule Interactions: Methods and Considerations

Several in vitro techniques are utilized to identify interactions between RNA and small molecules, each offering distinct advantages and considerations for future research. These methods not only expand the array of detection capabilities but also integrate functional assessments, providing comprehensive insights into how small molecules interact with RNA structures. By employing a diverse range of techniques, researchers can uncover critical details about these interactions, paving the way for the development of novel therapeutic strategies that target RNA in various diseases.

### *21.6.1 Fluorescence-Based Methods*

The majority of current techniques for detecting small molecule binders to RNA leverage the sensitivity of fluorescence measurements. For example, researchers can quantify the relative binding affinities of small molecules by titrating them against RNA constructs labeled with fluorescent markers. This can be achieved using fluorescence indicator displacement (FID) assays, where the displacement of a fluorescent indicator molecule by the small molecule indicates binding affinity. Alternatively, the interaction can be monitored by observing changes in the fluorescence emission of ligands that inherently fluoresce when RNA is introduced [23–25] (Fig. 21.2a).

**Fluorescence Indicator Displacement (FID)** In FID assays, a fluorescent dye or probe molecule is initially bound to the RNA of interest. This binding typically causes the fluorescence of the dye to be quenched or altered due to proximity effects or energy transfer mechanisms. When a small molecule with higher affinity for the RNA binding site competes with the fluorescent indicator, it displaces the indicator molecule from the RNA. The displacement of the fluorescent indicator results in a change in its fluorescence properties (such as emission intensity or wavelength), which can be detected and quantified using fluorescence spectroscopy. FID assays provide quantitative data on the binding affinity of small molecules to RNA targets. The extent of displacement of the indicator molecule correlates with the binding

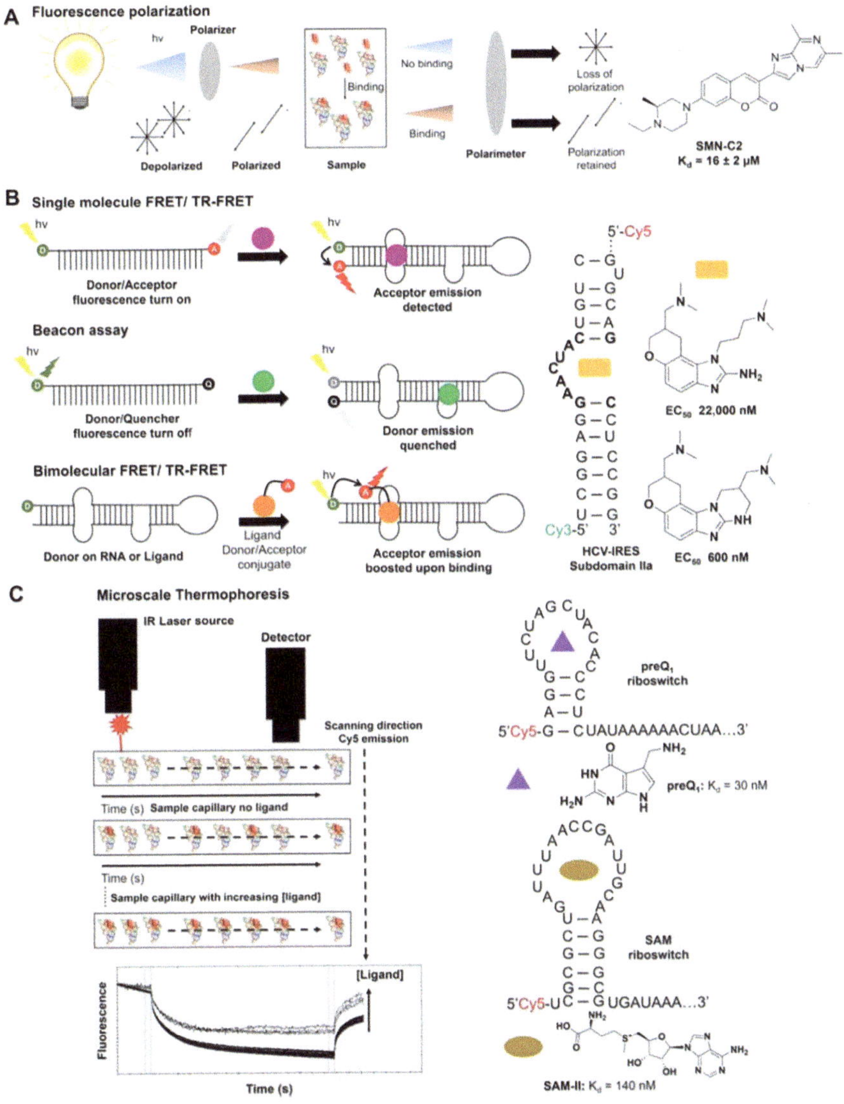


**Fig. 21.2** Various methods to detect small molecule binding to RNA molecules. (**a**) Fluorescence Polarization: A polarized light source is used to detect the binding of small molecules to RNA. When the small molecule binds to the RNA, the polarization of emitted light is retained, indicating binding. The binding of SMN-C2 to RNA is shown with a $K_d$ of 16 ± 2 μM. (**b**) Single Molecule FRET/TR-FRET and Beacon Assay: Fluorescence resonance energy transfer (FRET) and time-resolved FRET (TR-FRET) are used to detect RNA–small molecule interactions. The single molecule FRET/TR-FRET assay involves donor and acceptor fluorophores on RNA or ligands, with acceptor emission detected upon binding. The beacon assay uses a donor/quencher pair, where quenched emission indicates binding. In the bimolecular FRET/TR-FRET, ligand binding enhances acceptor emission. Examples include Cy5-labeled HCV-IRES subdomain IIa with ligands having $EC_{50}$ values of 22,000 nM and 600 nM. (**c**) Microscale Thermophoresis: This technique measures changes in fluorescence as RNA–small molecule complexes move through a temperature gradient. The preQ_1 riboswitch and SAM riboswitch are shown as examples, with $K_d$ values of 30 nM and 140 nM, respectively, for their small molecule ligands. Figure is reproduced from [22] with permission from Elsevier (License number: 5902940274150)

strength of the small molecule to the RNA. These assays are highly sensitive, capable of detecting even subtle changes in fluorescence upon displacement, which allows for precise determination of binding constants ($K_d$ values). FID assays can be adapted for various types of RNA targets and small molecules, making them versatile tools in drug discovery and RNA research. The high-throughput capability can be automated and scaled up for high-throughput screening of small molecule libraries, facilitating the identification of lead compounds for RNA-targeted therapies. However, there are some limitations to FID. Care must be taken to ensure that the fluorescent indicator and the small molecule do not interact directly, which could lead to false-positive results. While FID assays provide valuable binding affinity data, additional validation using orthogonal techniques may be necessary to confirm the specificity and biological relevance of the interactions observed (Fig. 21.2b) [23–25].

**Small Molecule Microarrays (SMMs)** SMMs consist of thousands to millions of spatially defined spots or features on a solid surface (such as glass slides or silicon wafers). Each spot contains a different small molecule immobilized in a controlled manner. Biomolecules of interest, such as RNA targets, are applied to the microarray. The array is then probed under controlled conditions to monitor interactions between the immobilized small molecules and the biomolecules. Various detection methods can be employed depending on the nature of the interaction being studied. For RNA interactions, fluorescence-based detection methods are often used, where fluorescently labeled RNA probes are used to detect binding to the small molecules on the array. The resulting data from SMM experiments typically include binding affinities, specificity, and structural preferences of small molecules toward RNA targets. Advanced data analysis techniques, including bioinformatics tools, are employed to interpret and derive meaningful insights from the vast amount of data generated. SMMs enable simultaneous screening of large libraries of small molecules against multiple RNA targets, accelerating the discovery of potential drug candidates or research probes. They can be used to study various types of interactions, including binding affinity, specificity, and selectivity of small molecules toward RNA structures or sequences. SMMs facilitate combinatorial chemistry approaches, where diverse chemical libraries can be rapidly screened to identify molecules with desired properties for RNA-targeted applications. They provide insights into the molecular recognition events between small molecules and RNA, contributing to a deeper understanding of RNA structure–function relationships and RNA-mediated biological processes. However, SMMs also have its own limitations. The immobilization of small molecules on the solid surface may alter their conformation or binding properties compared to their free form in solution, affecting the accuracy of binding affinity measurements. Results obtained from SMMs often require validation using orthogonal techniques to confirm the biological relevance and specificity of interactions observed on the microarray platform [23–25].

### 21.6.2 Isothermal Titration Calorimetry (ITC)

For detailed biophysical examination of RNA–small molecule interactions, including assessing the thermodynamics and kinetics of binding, isothermal titration calorimetry (ITC) is the method of choice [26]. ITC directly measures the heat change that occurs upon mixing a small molecule with an RNA solution, providing immediate insights into the binding constant ($K_d$). The resulting data allow for the calculation of all thermodynamic parameters related to the interaction. This information is crucial for understanding the enthalpic and entropic forces driving small molecule–RNA binding events, offering in-depth knowledge on how small molecules can modulate RNA structures and functions.

### 21.6.3 Surface Plasmon Resonance (SPR)

Surface plasmon resonance (SPR) is a sophisticated technique used for gaining kinetic insights into interactions, such as those between small molecules and RNA [26]. In SPR experiments, RNA is anchored onto a glass surface, and the binding of small molecules is detected through changes in the refractive index at the surface, effectively measuring the association ($k_{on}$) and dissociation ($k_{off}$) rates. This information is crucial for determining whether high-affinity interactions are primarily due to rapid binding or slow-release rates within small molecule–RNA systems. In an ideal scenario, binding assays would be paired with RNA-centric functional assays to thoroughly elucidate the nature of RNA–small molecule interactions and to link binding efficiencies with functional impacts—a growing area of interest in the field [1]. This approach becomes increasingly relevant with the rise of phenotypic screening efforts that pinpoint RNA as a target, as evidenced in recent research initiatives by major pharmaceutical companies like Novartis [27], Roche [28], and others (Fig. 21.2c).

## 21.7 Analyzing RNA–Small Molecule Interactions: Structural Approaches

Understanding interactions between RNA and small molecules goes beyond simple binding assays; structural insights play a crucial role. Unlike proteins, where there is an abundance of structural data—over 100,000 crystal and 10,000 NMR structures in the PDB—RNA structures are relatively underrepresented. Currently, there are just over 800 crystal and 500 NMR structures of RNA deposited in the PDB. This disparity is due to several challenges inherent in RNA structure determination, including its substantial negative charge and tendency for dynamic conformational changes. Methods for elucidating RNA structure can be broadly categorized into solid-state and solution-state approaches. Solid-state methods, such as X-ray crystallography and cyo-electron microscopy (cryo-EM), involve studying RNA molecules immobilized in a crystalline lattice or frozen in vitrified ice. These techniques provide high-resolution structural details but require RNA to adopt a stable, ordered conformation. In contrast, solution-state methods, such as NMR spectroscopy and small-angle X-ray scattering (SAXS), analyze RNA molecules in solution, reflecting their natural dynamic behavior. Solution-state approaches are advantageous for studying RNA flexibility and conformational changes but typically yield lower-resolution structures compared to solid-state methods.

### *21.7.1 Solid-State Methods*

**X-Ray Crystallography** In X-ray crystallography, the methodology revolves around exposing crystals containing the molecule of interest to X-rays. These X-rays diffract off the atoms within the crystal lattice, generating a diffraction pattern on a detector. From this diffraction pattern, an electron density map can be reconstructed, providing detailed spatial information about the arrangement of atoms within the biomolecule. As a cornerstone technique for the structural analysis of biomolecules, X-ray crystallography's notable early success was demonstrated in the determination of tRNA's structure [6]. X-ray crystallography can achieve atomic-level resolution, revealing detailed molecular interactions and structural features. Successful structure determination depends on obtaining high-quality crystals that diffract X-rays sufficiently, which can be challenging for some biomolecules like RNA due to their flexibility and negative charge. Despite challenges, X-ray crystallography remains a cornerstone technique for elucidating the three-dimensional structures of biomolecules. It provides crucial insights into biological functions, interactions with small molecules, and mechanisms of disease. Since then, it has provided invaluable insights into RNA structures in their unbound form, as well as their interactions with other RNA molecules and proteins, with the ribosome's structure being one of its most significant achievements [12]. The structure of the TERRA lncRNA complexed with an acridine analog is another example of the technique's application. Acridine analogs are small molecules known for their ability to interact with nucleic acids, including RNA. In this case, the acridine analog is designed to bind specifically to TERRA, potentially modulating its function or structure. To study the TERRA–acridine complex, researchers first crystallize the RNA–ligand complex under controlled conditions. The crystals must be carefully optimized to achieve high-resolution X-ray diffraction. The crystallized complex is exposed to X-rays, and the resulting diffraction pattern is captured by a detector. From this diffraction pattern, an electron density map is reconstructed using computational methods. This map reveals the precise arrangement of atoms within the RNA–ligand complex. With the electron density map as a guide, researchers can build a detailed three-dimensional model of the TERRA–acridine complex. This model illustrates how the acridine analog interacts with specific regions of the

TERRA RNA, such as through hydrogen bonding, stacking interactions, or other molecular contacts [29].

**Cryo-electron Microscopy (cryo-EM)** is a structural biology technique that allows researchers to visualize biological macromolecules, such as proteins and nucleic acids, at near-atomic resolution. It involves flash-freezing samples in vitrified ice, which preserves their natural structure and prevents artifacts that can occur with traditional sample preparation methods. It involves rapidly cooling a sample to cryogenic temperatures followed by electron microscopy to analyze the sample. This process leads to the formation of vitreous ice, preventing water molecules from arranging into a crystalline structure. Consequently, the biomolecule of interest is shielded from the damaging effects of freezing, all while being preserved in a hydrated state [29]. Cryo-EM has been instrumental in elucidating the structures of complex biomolecular assemblies, such as the ribosome [29] and the spliceosome [11], by allowing the observation of multiple conformational states within a single sample. This capability provides insights into RNA dynamics by facilitating the creation of structural ensembles, an approach not possible with X-ray crystallography [29]. Recent innovations in cryo-EM instrumentation and data processing algorithms have significantly improved the resolution capabilities of the technique. Cryo-EM can now resolve structural details down to the atomic level, providing insights into the precise arrangement of atoms within large macromolecular complexes. Although cryo-EM is a solid-state method and thus may still present limitations in accurately reflecting biologically relevant structures, the avoidance of crystal packing artifacts offers a potential advantage over crystallography in capturing structures that are more representative of natural conformations [30]. Future developments in cryo-EM may include advancements in sample preparation techniques, such as cryo-ET for studying dynamic RNA–ligand interactions in situ within cellular contexts.

### *21.7.2 Solution-State Methods*

**Nuclear Magnetic Resonance (NMR) spectroscopy** utilizes magnetic fields to probe molecules, where each atom within the molecule produces chemical shift values indicative of its chemical identity and surrounding environment. This technique, performed in solution, offers the advantage of capturing RNA conformations that closely resemble those found in biological contexts. However, it is important to consider that the choice of buffer conditions in NMR experiments can influence the accuracy and relevance of these conformations. Combining NMR with X-ray crystallography, cryo-EM, and computational modeling provides a comprehensive view of RNA structure and function, facilitating drug discovery and therapeutic development targeting RNA-related diseases. One challenge in NMR analysis arises from the overlapping chemical shifts due to the similar chemical properties of nucleobases, which generally restricts NMR studies to shorter RNA sequences, ideally those comprising 50 nucleotides or less, unless specific techniques like site-specific radiolabeling are applied [30]. A notable milestone in NMR spectroscopy was the elucidation of the structure of the HIV-1 5'-UTR, a segment of RNA consisting of 155 nucleotides. This achievement, accomplished by Telesnitsky, Summers, and their collaborators, stands out as one of the longest RNA complexes successfully analyzed using NMR, facilitated by innovative site-specific labeling techniques. Site-specific labeling techniques involved selectively introducing stable isotopes (such as ^13C and ^15N) into specific RNA residues. This approach enabled researchers to distinguish and characterize individual nucleotides within the large RNA molecule, overcoming traditional size limitations in NMR studies. The structural insights gained from the HIV-1 5'-UTR NMR study are valuable for understanding viral RNA biology, identifying potential drug targets, and designing RNA-targeted therapeutics. They contribute to broader efforts in combating HIV/AIDS and other RNA virus infections [31]. Beyond structural insights, NMR offers valuable data on ligand binding. Techniques like water–Ligand Observed via Gradient SpectroscopY (waterLOGSY) involve magnetization transfer from water to the ligand, mediated by intermolecular nuclear Overhauser effect (NOE) in both the presence and absence of an RNA target. This method capitalizes on the size disparity between the RNA–ligand and solely the ligand complex to deduce binding interactions [30].

**Chemical probing** is a technique that involves treating RNA with a reactive chemical, leading to the formation of adducts on nucleobases or sugar groups. The reactivity of these chemicals varies, depending on whether the nucleotides are base-paired or are specifically tailored to interact with specific RNA sites. For instance, SHAPE reagents (Selective 2'-Hydroxyl Acylation analyzed by Primer Extension) selectively modify the 2′-hydroxyl group of ribose sugars in nucleotides that exhibit conformational flexibility, often found in single-stranded regions. This method offers insights into RNA's structural dynamics, highlighting regions susceptible to structural changes or interactions with ligands. Conversely, dimethyl sulfate (DMS) targets adenine (A) and cytosine (C) nucleobases at the Watson–Crick–Franklin (WCF) interface. By methylating these bases, DMS provides valuable information about RNA's secondary structure, particularly its base-pairing patterns. This approach is crucial for identifying regions where nucleotides are involved in stable base pairs, revealing the spatial arrangement and stability of RNA double-stranded regions [32]. Following the probing step, the RNA is converted into complementary DNA (cDNA) through reverse transcription. Modifications lead to either truncation, where reverse transcriptase stops, or mutations, which can be identified by deep sequencing in the mutational profiling (MaP) method [32]. Schneekloth and colleagues utilized chemical probing to validate the binding site of a small molecule initially identified through the small molecule microarray (SMM) technique against the HIV-1 TAR RNA element. The study encompassed the entire 5'-UTR of HIV-1, including the TAR sequence, to ascertain that the small molecule's binding site remains stable within the broader RNA structure. Notable alterations in nucleotide reactivity were observed proximal to the TAR structure, providing conclusive evidence of the small molecule's specific interaction site. Understanding the precise binding site of small molecules on RNA, such as the HIV-1 TAR element, is critical for developing targeted therapies against RNA viruses. These insights aid in the design of compounds that disrupt viral replication or gene expression, potentially offering new treatment strategies for HIV/AIDS and related diseases [23]. Therefore, chemical probing emerges as a powerful method for validating small molecule binding sites and their specificity, particularly in the context of larger RNA molecules, offering a significant advantage in studying RNA–ligand interactions.

## 21.8 Current Challenges

Advancements in techniques for studying interactions between RNA and small molecules are paving the way for understanding selective recognition and potential clinical applications. However, despite these strides, many unanswered questions remain in this rapidly evolving field of research. New methods for analyzing RNA–small molecule interactions, such as chemical probing and small molecule microarrays, are enhancing our ability to investigate how molecules selectively recognize RNA targets. These advancements are crucial for developing RNA-targeted therapies and diagnostics. Insights gained from studying RNA–small molecule interactions hold promise for advancing clinical applications, potentially leading to the development of novel treatments for diseases where RNA plays a critical role, such as cancer, viral infections, and genetic disorders. Despite progress, several key questions remain unanswered. These include understanding the full range of structural motifs and non-canonical base pairs that RNA can adopt, deciphering the dynamic nature of RNA–small molecule complexes, and optimizing methods for precisely targeting RNA with small molecules. Examining RNA–small molecule interactions from both RNA and small molecule viewpoints will be essential. This dual approach allows for a comprehensive exploration of how RNA structures influence binding specificity and how small molecules can be designed to interact effectively with RNA targets.

## 21.9 RNA-Targeted Ligands: Current Status

Analysis of authoritative classic reviews on small molecules that target RNA shows that the search primarily yielded ligands that were highly basic (therefore, positively charged under physiological conditions) and flat, enabling them to insert between RNA bases and stack upon them [33–35]. These characteristics led to strong RNA binding but with minimal specificity. Such molecules often exhibited undesirable pharmacological traits, including poor cellular absorption and significant toxicity, which highlighted their non-ideal drug characteristics. Second, achieving both high drug-likeness and strong efficacy for molecules that aim at RNA secondary structures or repetitive sequences has proven challenging. Specifically, among molecules targeting secondary structures, a negative correlation exists between drug-likeness and RNA binding strength: the highest potencies are seen in the least drug-like compounds. Furthermore, in the limited instances where a comprehensive set of compounds within a discovery program has been analyzed, the resulting structure–activity relationship (SAR) tends to show minimal variance [23, 36–38], highlighting the difficulties in enhancing a lead molecule's potency. Hence, despite significant efforts and innovation, the majority of identified compounds do not possess the attributes of potent, bioavailable small molecule drugs nor target complex RNA structures akin to those targeted by high-specificity drugs. It is important to note our focus on RNA targeting with small molecules, particularly on identifying RNA with potential binding sites or pockets. In the realm of protein targeting, such compounds would analogously target protein–protein interactions [39] and be considered "beyond rule-of-five" molecules [40, 41]. Specifically targeting pathogenic trinucleotide repeats, such as the CUG and CAG repeats associated with myotonic dystrophy type 1 (DM1) and Huntington's disease, might pose a unique challenge. Although these repeats are structurally simple and lack apparent pockets suitable for ligands, the possibility remains that ligand binding might be enhanced by the repetitive sequence's overarching structure, potentially achieving necessary specificity and selectivity [42, 43]. While these approaches may hold promise, they typically entail a higher level of complexity and challenge.

**Box 21.1 "Ligandability": Drug-likeness and Lipinski's Rule of Five**

Small molecule drugs that demonstrate clinically effectiveness often exhibit physicochemical characteristics within a specific, narrow parameter range, aligning with what is termed "drug-likeness." This notion is critical as it acts as a shorthand for identifying molecules with favorable attributes for oral administration to humans, such as solubility, the ability to permeate cells and tissues, metabolic stability, and minimal toxicity. The "rule of five," introduced by Lipinski, serves as a prominent criterion for evaluating the drug-like properties of molecules. This guideline emerged from an analysis of established drugs and drug candidates. Lipinski proposed that molecules intended for oral bioavailability should adhere to specific thresholds: a molecular weight less than 500 Da, no more than five hydrogen bond donors, a $\log_P$ (octanol–water partition coefficient) below five, and a maximum of five hydrogen bond acceptors [44]. While the rule of five offers valuable guidelines, it is not an absolute rule and exceptions can be found among successful drugs. Modifications and refinements, such as the "rule of three" for fragment-based drug design, have been proposed to address specific contexts or compound classes. Adherence to the rule of five can expedite the drug development process by focusing resources on compounds with higher potential for oral bioavailability and therapeutic efficacy. It underscores the importance of optimizing molecular properties early in the drug discovery pipeline.

(continued)

**Box 21.1** (continued)

The concept of a "privileged chemical space" has been refined to include a singular measure, the quantitative estimate of drug-likeness (QED) score, which evaluates the overall drug-likeness of orally bioavailable molecules [45]. This metric permits certain deviations from Lipinski's criteria if the compound demonstrates superior overall quality. QED scores, which vary from zero (indicating entirely unfavorable properties) to one (signifying entirely favorable properties), offer a holistic assessment of a molecule's drug-likeness. This score effectively discriminates and aligns with the chemical intuition honed by medicinal chemists over years of practice [45, 46]. Displayed in the figure are bioactive molecules with both high and low QED scores. The median QED score among existing orally bioavailable human drugs is around 0.65, with 75% of such drugs scoring a QED of 0.5 or higher.

It is important to recognize that the concept of drug-likeness, including QED scoring, is predominantly based on small molecule drugs targeting proteins. However, certain successful drugs, including antibiotics and viral protease inhibitors, exhibit relatively low QED scores [45, 46], suggesting that a drug's potential utility is not solely determined by its drug-likeness. Additionally, there is growing evidence to suggest that RNA-targeting drugs may adhere to different principles than those targeting proteins, given their distinct structural [47] and kinetic [48] characteristics. Nevertheless, since drug-likeness significantly hinges on bioavailability and human physiological responses—which are constant across all drugs—it is proposed that RNA-targeting drugs will likely adhere to similar drug-likeness criteria as their protein-targeting counterparts, despite these differences.

## 21.10 Good RNA Targets Should Have High Information Content

A primary challenge in RNA targeting is balancing strong binding affinity with high selectivity amidst a cellular landscape populated by numerous RNA motifs with similar structures. Much of the research effort has been directed toward simple RNA configurations, including base-paired helices or duplexes with irregularities and bulges. In contrast, RNAs that are successfully targeted by drug-like molecules such as linezolid, ribocil, and riboflavin interact with intricate structures situated within cleft-like sites, mirroring those found in proteins. The complexity of RNA structures can be quantified using their information content, measured in bits [49–51], where one bit represents the information needed to choose between two alternatives. Advances in modeling RNA secondary structures with high precision are now possible, thanks to sequence covariation analysis (applicable when multiple alignable sequences exist) [52] and chemical probing methods like selective 2′-hydroxyl acylation analyzed by primer extension (SHAPE), which offer reliable estimates of an RNA's information content [53, 54]. The information content within an RNA target is estimated by aggregating the information from individual structural elements. This approach is particularly appealing for RNA because information content is relatively straightforward to calculate with a known secondary structure model and some biochemical data, offering an unbiased method to assess both current and future RNA targets. For example, a basic six base-pair RNA helix contains about 9 bits of information, whereas the binding sites for molecules like linezolid, ribocil, and branaplam are estimated to hold over 50 bits. The information content in a ligand-binding site encapsulates both direct interactions with the ligand and indirect contributions from surrounding nucleotides, indicating that nucleotides can influence the site's information content without directly contacting the ligand [55–57].

One of the clearest examples illustrating the link between RNA structural complexity and ligand binding comes from studies that explored RNA sequences identified through in vitro selection for their affinity to bind GTP [49]. These interactions predominantly involve the guanosine component of GTP [55] (molecular mass = 283 Da; QED score = 0.40). Notably, there is a significant positive correlation between the GTP binding affinity and both the complexity and information content of the RNA target. It has been experimentally determined that enhancing the target RNA's information content by 10 bits leads to an approximately tenfold increase in binding affinity [49]. This correlation also applies to another ligand, targaprimir-96 [58], where the line for targaprimir-96 is displaced from that of GTP because of its inherently higher affinity for the same level of information content, likely due to its larger size.

### 21.10.1 The "10 → 10 Relationship"

In general, RNA-targeted small molecule development involves designing molecules that can specifically bind to RNA molecules, influencing their function. This approach can be used for therapeutic purposes, such as targeting RNA structures involved in diseases. Small molecules that bind to RNA can potentially alter RNA stability, splicing, or translation, offering a way to modulate disease processes at the RNA level. The "10 → 10 relationship" in the context of RNA-targeted small molecule development refers to an observed correlation, where enhancing the information content of the target RNA by 10 bits results in an approximately tenfold increase in the binding affinity of a small molecule to that RNA. Therefore, it is crucial to provide a higher level of complexity or specificity of the RNA structure or sequence. Information content in a biological sense often relates to how structured or how many specific binding sites an RNA molecule might contain that a small molecule can recognize and bind to. Increasing the information content by 10 bits implies making the RNA structurally richer or more complex, which provides more specific and potentially stronger binding sites for small molecules. A tenfold increase in binding affinity suggests that the small molecule can bind more tightly to the RNA, which typically translates to higher efficacy in modulating the RNA's function or stability. This can be critical in therapeutic contexts, where higher affinity is often correlated with better drug performance. The "10 → 10 relationship" found in these studies lays a conceptual groundwork for understanding recent endeavors in RNA-targeted small molecule development [49]. For instance, ribocil and branaplam, which exhibit low-nanomolar binding affinities, interact with RNA and RNA–protein complexes characterized by high information content. Moreover, various projects aiming to discover ligands for targets with lower to moderate information content—such as a stem-loop (20 bits), the TAR hairpin (22 bits), and an asymmetric loop in the hepatitis C virus internal ribosome entry site (HCV-IRES) element (26 bits)—have identified molecules with binding affinities ranging from 1 to 100 μM. These affinities for lower-molecular-weight ligands align well with the "10 → 10 relationship." Significantly, all these target sites with lower to moderate affinity fall beneath the 30-bit threshold, which is approximately the level of specificity needed to uniquely identify an RNA target within the human transcriptome.

## 21.11 Provocative Guidelines

Molecules designed with the explicit intent to target RNA often compromise between drug-likeness and efficacy, a consequence attributed to their targeting of sites with limited information content and inferior pocket quality. Despite these challenges, the existing examples bolster the argument for RNA's inherent targetability. Prioritize RNA motifs with a complexity that ensures uniqueness, the ability to create high-quality pockets, and the potential to bind small molecule ligands with strong affinity. Effective RNA targets should exhibit at least 30 bits of information and possess a discernible pocket. Such motifs are expected to be prevalent in larger RNAs, underscoring the notion that the presence of substantial information content is crucial for identifying viable targetable pockets. Exercise caution in the preliminary designation of specific privileged chemical scaffolds for RNA targeting. While these may eventually be identified, a comprehensive understanding is necessary before promising scaffolds can be confidently determined. Focus should instead be placed on leveraging insights gained from the extensive history of protein-targeted drug development, as many of these principles are likely to be applicable to RNA-targeted therapeutics as well. Reconsider drawing direct parallels from the ribosome as a model for RNA targeting. Given their exceptionally high cellular concentration, ribosomal RNAs represent an atypical case. Ligands with micromolar affinity are typically suited for targets available at similar concentrations, which is why drugs targeting ribosomal RNAs can achieve specificity through comparatively weaker ribosome binding. Approach with caution the evaluation of studies involving compounds characterized by high basicity, capability for intercalation, strong stacking interactions, or significant hydrophobicity. Although such compounds may exhibit high affinity for RNA, their likelihood of discriminating effectively between intended and unintended RNA targets is low. Moreover, these compounds may pose significant challenges for optimization within a medicinal chemistry framework.

Aim for RNA–protein interactions or RNA elements that are stabilized through protein binding, similar to the approach used by branaplam and SMA-C5 [27, 59]. The strategy of targeting a specific RNA element along with a ubiquitous protein component of the cellular metabolic or regulatory systems could effectively yield both specificity (through the RNA-targeted component) and affinity (due to the complex interface with the protein). Leverage methodologies that facilitate the swift discovery of RNA motifs characterized by both high information content and the presence of high-quality pockets. The ability to quickly identify such RNA targets is a significant advantage RNA has over proteins, thanks to existing tools like quantitative chemical probing techniques [32, 54, 60]. These methods allow for the identification of RNA sites with rich information content [49, 51] and stable, well-defined structures [32, 61], assess how the cellular context affects a given RNA structure, and ensure that screening models accurately reflect the native RNA structure. Invest in the development and enhancement of tools, including sophisticated structural biology and modeling approaches tailored for RNAs with intricate structures [62–64]. Such advancements can help pinpoint complex RNA structures that are conducive to high-quality ligand binding, facilitating quick characterization of these targets and evaluation of how ligand interactions affect their functions. Establish consistent methods that link ligand binding with complex targets to definite therapeutic actions. There is emerging evidence suggesting that altering RNA structures near translation initiation sites can influence gene expression in ways that are beneficial therapeutically [65–67]. Additionally, there are indications that small molecules binding to mRNAs can affect translation efficiency [68, 69]. Both the disruption and stabilization of RNA–protein interactions show therapeutic potential. In parallel to the growing acceptance of covalent drugs targeting proteins, ligands capable of forming covalent bonds with or inducing cleavage in RNA targets [70, 71] also present a promising avenue.

Therefore, three fundamental elements are crucial for the success of RNA-targeted drug discovery that is effective, reproducible, and scalable. The foundational step involves choosing an RNA target with a clear linkage to disease, underscoring the need for a target that is not just biologically validated but also therapeutically significant. Second, to develop an optimized screening strategy, specific to various kinds of RNAs. This involves deploying a screening methodology capable of uncovering lead molecules that not only exhibit drug-like characteristics but also possess desirable pharmacological profiles. This is essential for transitioning from theoretical targets to practical therapeutic agents. Third, to identify RNA motifs with high information content, based on the 10 → 10 relationship. The key to high specificity and potency in RNA targeting lies in focusing on RNA motifs that are rich in information content. This ensures the identification of targets with high-quality pockets, enabling the effective and selective binding of small molecules.

## 21.12 Conclusion

This chapter advocates for a strategic shift in the development of RNA-targeted drugs, emphasizing the critical importance of focusing on the quality of RNA targets, which represents the third component in this approach. While identifying biologically validated RNA targets remains essential, there is a pressing need to intensify efforts aimed at assessing and enhancing the "ligandability" of these targets. The objective is not only to pinpoint RNA structures inherently suitable for drug-like interactions but also to ensure these structures can be effectively bound by small molecules with high affinity and selectivity. As our understanding deepens regarding what constitutes a high-quality RNA target, it becomes increasingly feasible that RNA could become as viable a target for drug discovery as proteins, if not more so. Given the widespread presence of targetable RNA motifs throughout the genome and the seemingly simpler principles governing RNA structure and folding compared to proteins, RNA targets offer potentially broader opportunities for therapeutic intervention than previously recognized. Successfully harnessing RNA-targeted therapeutic discovery has the potential to significantly expand our capacity to modulate pathways currently considered "undruggable," including the vast landscape of the non-coding transcriptome. This advancement promises to open new avenues for medical intervention, paving the way for innovative treatments that address previously unmet medical needs.

## References

1. Donlic A, Hargrove AE. Targeting RNA in mammalian systems with small molecules. Wiley Interdiscip Rev RNA. 2018;9(4):e1477.
2. ENCODE Project Consortium. An integrated encyclopedia of DNA elements in the human genome. Nature. 2012;489(7414):57–74.
3. Morgan BS, Hargrove AE. Chapter 7: Synthetic receptors for oligonucleotides and nucleic acids. In: Monographs in supramolecular chemistry, vol. 14. RSC Publishing; 2015. p. 253–325.
4. Blythe AJ, Fox AH, Bond CS. The ins and outs of lncRNA structure: how, why and what comes next? Biochim Biophys Acta. 2016;1859(1):46–58.
5. Butcher SE, Pyle AM. The molecular interactions that stabilize RNA tertiary structure: RNA motifs, patterns, and networks. Acc Chem Res. 2011;44(12):1302–11.
6. Sussman JL, Holbrook SR, Warrant RW, Church GM, Kim SH. Crystal structure of yeast phenylalanine transfer RNA. I. Crystallographic refinement. J Mol Biol. 1978;123(4):607–30.
7. Brown JA, Bulkley D, Wang J, Valenstein ML, Yario TA, Steitz TA, et al. Structural insights into the stabilization of MALAT1 noncoding RNA by a bipartite triple helix. Nat Struct Mol Biol. 2014;21(7):633–40.
8. Collie GW, Haider SM, Neidle S, Parkinson GN. A crystallographic and modelling study of a human telomeric RNA (TERRA) quadruplex. Nucleic Acids Res. 2010;38(16):5569–80.
9. Cech TR, Steitz JA. The noncoding RNA revolution-trashing old rules to forge new ones. Cell. 2014;157(1):77–94.
10. Wang X, Goodrich KJ, Gooding AR, Naeem H, Archer S, Paucek RD, et al. Targeting of Polycomb repressive complex 2 to RNA by short repeats of consecutive guanines. Mol Cell. 2017;65(6):1056–67.e5.

11. Zhang X, Yan C, Hang J, Finci LI, Lei J, Shi Y. An atomic structure of the human spliceosome. Cell. 2017;169(5):918–29.e14.
12. Selmer M, Dunham CM, Murphy FV 4th, Weixlbaumer A, Petry S, Kelley AC, et al. Structure of the 70S ribosome complexed with mRNA and tRNA. Science. 2006;313(5795):1935–42.
13. Liu N, Dai Q, Zheng G, He C, Parisien M, Pan T. N(6)-methyladenosine-dependent RNA structural switches regulate RNA-protein interactions. Nature. 2015;518(7540):560–4.
14. Roost C, Lynch SR, Batista PJ, Qu K, Chang HY, Kool ET. Structure and thermodynamics of N6-methyladenosine in RNA: a spring-loaded base modification. J Am Chem Soc. 2015;137(5):2107–15.
15. Ganser LR, Kelly ML, Herschlag D, Al-Hashimi HM. The roles of structural dynamics in the cellular functions of RNAs. Nat Rev Mol Cell Biol. 2019;20(8):474–89.
16. Dethoff EA, Petzold K, Chugh J, Casiano-Negroni A, Al-Hashimi HM. Visualizing transient low-populated structures of RNA. Nature. 2012;491(7426):724–8.
17. Chen SJ, Dill KA. RNA folding energy landscapes. Proc Natl Acad Sci USA. 2000;97(2):646–51.
18. Padroni G, Patwardhan NN, Schapira M, Hargrove AE. Systematic analysis of the interactions driving small molecule-RNA recognition. RSC Med Chem. 2020;11(7):802–13.
19. Collie GW, Sparapani S, Parkinson GN, Neidle S. Structural basis of telomeric RNA quadruplex--acridine ligand recognition. J Am Chem Soc. 2011;133(8):2721–8.
20. Vicens Q, Westhof E. Crystal structure of paromomycin docked into the eubacterial ribosomal decoding a site. Structure. 2001;9(8):647–58.
21. Hewitt WM, Calabrese DR, Schneekloth JS Jr. Evidence for ligandable sites in structured RNA throughout the protein data Bank. Bioorg Med Chem. 2019;27(11):2253–60.
22. Haniff HS, Knerr L, Chen JL, Disney MD, Lightfoot HL. Target-directed approaches for screening small molecules against RNA targets. SLAS Discov. 2020;25(8):869–94.
23. Sztuba-Solinska J, Shenoy SR, Gareiss P, Krumpe LR, Le Grice SF, O'Keefe BR, et al. Identification of biologically active, HIV TAR RNA-binding small molecules using small molecule microarrays. J Am Chem Soc. 2014;136(23):8402–10.
24. Donlic A, Morgan BS, Xu JL, Liu A, Roble C, Hargrove AE. Discovery of small molecule ligands for MALAT1 by tuning an RNA-binding scaffold. Angew Chem Int Ed. 2018;57:13242.
25. Wicks SL, Hargrove AE. Fluorescent indicator displacement assays to identify and characterize small molecule interactions with RNA. Methods. 2019;167:3–14.
26. Umuhire Juru A, Patwardhan NN, Hargrove AE. Understanding the contributions of conformational changes, thermodynamics, and kinetics of RNA-small molecule interactions. ACS Chem Biol. 2019;14(5):824–38.
27. Palacino J, Swalley SE, Song C, Cheung AK, Shu L, Zhang X, et al. SMN2 splice modulators enhance U1-pre-mRNA association and rescue SMA mice. Nat Chem Biol. 2015;11(7):511–7.
28. Campagne S, Boigner S, Rüdisser S, Moursy A, Gillioz L, Knörlein A, et al. Structural basis of a small molecule targeting RNA for a specific splicing correction. Nat Chem Biol. 2019;15(12):1191–8.
29. Brown A, Shao S. Ribosomes and cryo-EM: a duet. Curr Opin Struct Biol. 2018;52:1–7.
30. Thompson RD, Baisden JT, Zhang Q. NMR characterization of RNA small molecule interactions. Methods. 2019;167:66–77.
31. Keane SC, Heng X, Lu K, Kharytonchyk S, Ramakrishnan V, Carter G, et al. RNA structure. Structure of the HIV-1 RNA packaging signal. Science. 2015;348(6237):917–21.
32. Siegfried NA, Busan S, Rice GM, Nelson JA, Weeks KM. RNA motif discovery by SHAPE and mutational profiling (SHAPE-MaP). Nat Methods. 2014;11(9):959–65.
33. Aboul-ela F. Strategies for the design of RNA-binding small molecules. Future Med Chem. 2010;2(1):93–119.
34. Guan L, Disney MD. Recent advances in developing small molecules targeting RNA. ACS Chem Biol. 2012;7(1):73–86.
35. Thomas JR, Hergenrother PJ. Targeting RNA with small molecules. Chem Rev. 2008;108(4):1171–224.

36. Bottini A, De SK, Wu B, Tang C, Varani G, Pellecchia M. Targeting influenza a virus RNA promoter. Chem Biol Drug Des. 2015;86(4):663–73.
37. Joly JP, Mata G, Eldin P, Briant L, Fontaine-Vive F, Duca M, et al. Artificial nucleobase-amino acid conjugates: a new class of TAR RNA binding agents. Chemistry. 2014;20(7):2071–9.
38. Lee MK, Bottini A, Kim M, Bardaro MF Jr, Zhang Z, Pellecchia M, et al. A novel small-molecule binds to the influenza a virus RNA promoter and inhibits viral replication. Chem Commun (Camb). 2014;50(3):368–70.
39. Jin L, Wang W, Fang G. Targeting protein-protein interaction by small molecules. Annu Rev Pharmacol Toxicol. 2014;54:435–56.
40. Doak BC, Over B, Giordanetto F, Kihlberg J. Oral druggable space beyond the rule of 5: insights from drugs and clinical candidates. Chem Biol. 2014;21(9):1115–42.
41. Matsson P, Doak BC, Over B, Kihlberg J. Cell permeability beyond the rule of 5. Adv Drug Deliv Rev. 2016;101:42–61.
42. Childs-Disney JL, Disney MD. Approaches to validate and manipulate RNA targets with small molecules in cells. Annu Rev Pharmacol Toxicol. 2016;56:123–40.
43. Zimmerman SC. A journey in bioinspired supramolecular chemistry: from molecular tweezers to small molecules that target myotonic dystrophy. Beilstein J Org Chem. 2016;12:125–38.
44. Lipinski CA. Lead- and drug-like compounds: the rule-of-five revolution. Drug Discov Today Technol. 2004;1(4):337–41.
45. Bickerton GR, Paolini GV, Besnard J, Muresan S, Hopkins AL. Quantifying the chemical beauty of drugs. Nat Chem. 2012;4(2):90–8.
46. Ritchie TJ, Macdonald SJ. How drug-like are 'ugly' drugs: do drug-likeness metrics predict ADME behaviour in humans? Drug Discov Today. 2014;19(4):489–95.
47. Morgan BS, Forte JE, Culver RN, Zhang Y, Hargrove AE. Discovery of key physicochemical, structural, and spatial properties of RNA-targeted bioactive ligands. Angew Chem Int Ed Engl. 2017;56(43):13498–502.
48. Gleitsman KR, Sengupta RN, Herschlag D. Slow molecular recognition by RNA. RNA. 2017;23(12):1745–53.
49. Carothers JM, Oestreich SC, Davis JH, Szostak JW. Informational complexity and functional activity of RNA structures. J Am Chem Soc. 2004;126(16):5130–7.
50. Schneider TD, Stormo GD, Gold L, Ehrenfeucht A. Information content of binding sites on nucleotide sequences. J Mol Biol. 1986;188(3):415–31.
51. Witherell GW, Uhlenbeck OC. Specific RNA binding by Q beta coat protein. Biochemistry. 1989;28(1):71–6.
52. Shang L, Xu W, Ozer S, Gutell RR. Structural constraints identified with covariation analysis in ribosomal RNA. PLoS One. 2012;7(6):e39383.
53. Smola MJ, Rice GM, Busan S, Siegfried NA, Weeks KM. Selective 2′-hydroxyl acylation analyzed by primer extension and mutational profiling (SHAPE-MaP) for direct, versatile and accurate RNA structure analysis. Nat Protoc. 2015;10(11):1643–69.
54. Weeks KM, Mauger DM. Exploring RNA structural codes with SHAPE chemistry. Acc Chem Res. 2011;44(12):1280–91.
55. Carothers JM, Davis JH, Chou JJ, Szostak JW. Solution structure of an informationally complex high-affinity RNA aptamer to GTP. RNA. 2006;12(4):567–79.
56. Howe JA, Xiao L, Fischmann TO, Wang H, Tang H, Villafania A, et al. Atomic resolution mechanistic studies of ribocil: a highly selective unnatural ligand mimic of the E. Coli FMN riboswitch. RNA Biol. 2016;13(10):946–54.
57. Long KS, Vester B. Resistance to linezolid caused by modifications at its binding site on the ribosome. Antimicrob Agents Chemother. 2012;56(2):603–12.
58. Velagapudi SP, Cameron MD, Haga CL, Rosenberg LH, Lafitte M, Duckett DR, et al. Design of a small molecule against an oncogenic noncoding RNA. Proc Natl Acad Sci USA. 2016;113(21):5898–903.

59. Sivaramakrishnan M, McCarthy KD, Campagne S, Huber S, Meier S, Augustin A, et al. Binding to SMN2 pre-mRNA-protein complex elicits specificity for small molecule splicing modifiers. Nat Commun. 2017;8(1):1476.
60. Smola MJ, Calabrese JM, Weeks KM. Detection of RNA-protein interactions in living cells with SHAPE. Biochemistry. 2015;54(46):6867–75.
61. Mustoe AM, Busan S, Rice GM, Hajdin CE, Peterson BK, Ruda VM, et al. Pervasive regulatory functions of mRNA structure revealed by high-resolution SHAPE probing. Cell. 2018;173(1):181–95.e18.
62. Gilbert SD, Reyes FE, Edwards AL, Batey RT. Adaptive ligand binding by the purine riboswitch in the recognition of guanine and adenine analogs. Structure. 2009;17(6):857–68.
63. Miao Z, Westhof E. RNA structure: advances and assessment of 3D structure prediction. Annu Rev Biophys. 2017;46:483–503.
64. Warner KD, Homan P, Weeks KM, Smith AG, Abell C, Ferré-D'Amaré AR. Validating fragment-based drug discovery for biological RNAs: lead fragments bind and remodel the TPP riboswitch specifically. Chem Biol. 2014;21(5):591–5.
65. Corley M, Solem A, Phillips G, Lackey L, Ziehr B, Vincent HA, et al. An RNA structure-mediated, posttranscriptional model of human α-1-antitrypsin expression. Proc Natl Acad Sci USA. 2017;114(47):E10244–e53.
66. Kozak M. Regulation of translation via mRNA structure in prokaryotes and eukaryotes. Gene. 2005;361:13–37.
67. Liang XH, Shen W, Sun H, Migawa MT, Vickers TA, Crooke ST. Translation efficiency of mRNAs is increased by antisense oligonucleotides targeting upstream open reading frames. Nat Biotechnol. 2016;34(8):875–80.
68. Bandyopadhyay S, Cahill C, Balleidier A, Huang C, Lahiri DK, Huang X, et al. Novel 5′ untranslated region directed blockers of iron-regulatory protein-1 dependent amyloid precursor protein translation: implications for down syndrome and Alzheimer's disease. PLoS One. 2013;8(7):e65978.
69. Harvey I, Garneau P, Pelletier J. Inhibition of translation by RNA-small molecule interactions. RNA. 2002;8(4):452–63.
70. Nguyen L, Luu LM, Peng S, Serrano JF, Chan HY, Zimmerman SC. Rationally designed small molecules that target both the DNA and RNA causing myotonic dystrophy type 1. J Am Chem Soc. 2015;137(44):14180–9.
71. Rzuczek SG, Colgan LA, Nakai Y, Cameron MD, Furling D, Yasuda R, et al. Precise small-molecule recognition of a toxic CUG RNA repeat expansion. Nat Chem Biol. 2017;13(2):188–93.

# Chapter 22
# RNA Bioconjugation: Current Development in siRNA Therapeutics

**Abstract** The field of siRNA delivery has expanded rapidly over recent years, positioning siRNA as a powerful gene therapy tool for treating diverse diseases. Among the primary delivery approaches, siRNA bioconjugation stands out for its ability to enhance the pharmacological attributes of siRNA, while reducing stability-related toxicity and variability. This chapter presents an overview of the latest advancements in siRNA bioconjugation. It covers a range of conjugation approaches including those involving antibodies, peptides, aptamers, small molecules, lipidoids, cell-penetrating peptides, polymers, and nanoparticles. These bioconjugates, whether used independently or in combination with other compounds, have been shown to notably improve pharmacokinetics, extend biological half-life, and amplify targeting capabilities, all while ensuring effective gene silencing. These enhancements contribute to better therapeutic outcomes and reduce side effects. This chapter also highlights the critical role of conjugation techniques in maintaining the integrity, stability, and controlled release of siRNA bioconjugates and discusses the challenges associated with these methods.

## 22.1 Introduction

The emergence of siRNA as a transformative tool in gene therapy marks a significant stride in the fight against various diseases. With its ability to silence specific genes with precision, siRNA offers a promising avenue for addressing conditions that have long eluded traditional therapeutic approaches. Recent developments in siRNA delivery methods have focused on refining the efficacy and safety of these molecules through innovative bioconjugation strategies. By attaching siRNA to various molecules such as antibodies, peptides, aptamers, and nanoparticles, researchers have succeeded in enhancing the pharmacological profile of siRNA therapies. These conjugates not only improve the pharmacokinetics and biological half-life of siRNA but also augment its ability to target specific cells or tissues more effectively. Particularly, the focus on siRNA bioconjugates stems from their dominance in the field as the most prevalent form of RNA conjugates today [1]. Their

 
P. E. Saw, E. Song, *RNA Therapeutics in Human Diseases*,
https://doi.org/10.1007/978-981-96-3041-7_22

widespread use is attributed to their remarkable efficiency in bridging the gap between therapeutic potential and clinical application. Bioconjugation enhances siRNA's stability in the bloodstream, a critical factor often limiting the clinical utility of RNA-based therapies. By improving cellular uptake and targeting precision, these conjugates address key obstacles such as off-target effects and rapid degradation, which are common hurdles in the broader RNA therapeutic landscape [2].

In this chapter, we explore a spectrum of bioconjugation techniques that have shown potential in elevating the delivery and function of siRNA. From small molecules to complex polymers and lipidoids, each approach offers unique advantages and challenges. The integration of these bioconjugates into therapeutic regimes holds the promise of better patient outcomes by ensuring more controlled and precise gene silencing mechanisms. Furthermore, we will discuss the pivotal role of these conjugation strategies in maintaining the structural integrity and controlled release of siRNA, critical for maximizing therapeutic efficacy while minimizing adverse effects. Through an examination of the latest advancements and ongoing challenges, this chapter provides a comprehensive overview of the state of the art in siRNA bioconjugation and its implications for future therapeutic development.

## 22.2 Basic Understandings of siRNA

Unmodified siRNA consists of two strands of RNA that are susceptible to degradation by nucleases, enzymes that break down RNA molecules in biological environments. This susceptibility significantly limits the stability and half-life of unmodified siRNA in the bloodstream, making it less effective for therapeutic purposes. Additionally, unmodified siRNA can trigger immune responses and have off-target effects, leading to unintended gene silencing and potential toxicity (Fig. 22.1, left). To address these issues, siRNA molecules can be partially modified. These modifications often include the incorporation of chemically altered nucleotides that enhance stability and reduce immunogenicity. Partial modifications can improve the pharmacokinetic properties of siRNA, such as increasing resistance to nucleases and improving cellular uptake. However, partially modified siRNA might still face challenges related to off-target effects and incomplete protection from degradation (Fig. 22.1, middle). Fully modified siRNA incorporates extensive chemical

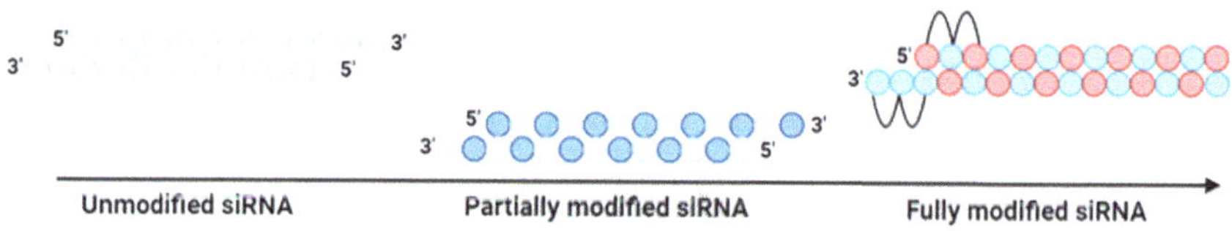


**Fig. 22.1** Basic structure of unmodified siRNA (left), partially modified (middle, presented by dark-blue circles), and fully modified siRNA (right, presented by red and dark-blue circles). (Figure is adapted from [1] under Creative Commons CC-BY license)

modifications throughout the RNA strands. These modifications can include the use of locked nucleic acids (LNAs), 2′-*O*-methyl modifications, and phosphorothioate linkages, among others (Fig. 22.1, right). Fully modified siRNA offers several advantages, such as significantly improve resistance to nuclease degradation, extending the siRNA's half-life in the bloodstream, extensive chemical modifications can help evade immune recognition, minimizing adverse immune responses, modifications can enhance the binding affinity to target RNA sequences, reducing off-target effects, and increasing the precision of gene silencing, with improved pharmacokinetic profiles, allowing for more effective delivery to target tissues.

**Box 22.1 Conjugation of RNA Via 2′-OH Acylation: Enhancing RNA Reactivity and Utility**

The 2′-OH group of ribonucleic acid (RNA) presents a unique site for chemical modification due to its accessibility and reactivity [3]. Acylation of the 2′-OH group is a novel approach that significantly enhances the versatility of RNA in various biochemical applications, including mapping structural interactions and creating labeled probes for research and diagnostic purposes [4].

**Overview of 2′-OH Acylation**

Acylation involves the introduction of an acyl group into a molecule, in this case, the RNA molecule. Specifically, the acylation of the 2′-OH group in RNA increases the molecule's overall reactivity, allowing for subsequent modifications and functionalization that are not possible with unmodified RNA. This chemical modification is particularly useful for attaching fluorescent labels or other molecular probes to RNA, which can then be used to track RNA dynamics within cells or to investigate RNA interactions with other biomolecules.

**Kinetics Studies and Mechanisms**

To refine and optimize 2′-OH acylation techniques, kinetics studies are essential. These studies assess the rate at which the acylation reaction occurs and identify the factors that influence this rate. Recent research in this area has revealed that both steric and inductive effects play significant roles in modulating the reactivity of the 2′-OH group during acylation.

***Steric Effects***

These refer to the influence of the physical size and spatial arrangement of atoms within the RNA molecule on the acylation reaction. Larger or more complex substituents near the 2′-OH group can hinder the approach of acylating agents, thus slowing the reaction. Conversely, modifications that reduce steric hindrance can enhance reactivity.

(continued)

**Box 22.1** (continued)

***Inductive Effects***

These are electronic effects transmitted through bonds affecting the electron density around the 2′-OH group. Electron-withdrawing groups in proximity to the 2′-OH group can increase its susceptibility to acylation by making it a better nucleophile. This effect can significantly affect the rate and efficiency of the acylation process.

**Applications and Implications**

The enhanced understanding of the kinetics and mechanisms of 2′-OH acylation offers new avenues for improving RNA labeling and mapping techniques. By selecting appropriate acylating agents and conditions that minimize steric hindrance and optimize inductive effects, researchers can achieve more efficient and selective labeling of RNA. This improvement is crucial for studies requiring high-resolution imaging of RNA within biological systems. Acylation of RNA can facilitate more detailed mapping of RNA structures and interactions. By attaching various chemical probes selectively to the 2′-OH group, scientists can use advanced analytical techniques, such as MS or NMR spectroscopy, to gain insights into RNA folding and function.

Further research into 2′-OH acylation will likely focus on developing more specific and efficient acylating agents that can be used under mild conditions to preserve RNA integrity. Additionally, exploring the acylation of RNA in the context of complex biological environments, such as within living cells, will be a significant step forward, potentially leading to new diagnostic and therapeutic tools. In conclusion, the conjugation of RNA via 2′-OH acylation represents a powerful method for enhancing RNA reactivity and expanding its utility in biological research and medical applications. Continued advancements in this area promise to yield even more sophisticated tools for studying and manipulating RNA [3].

Bioconjugation plays a critical role in enhancing the therapeutic potential of siRNA by addressing its inherent limitations such as instability and short half-life in the bloodstream. Conjugating siRNA with various linkers or biomolecules significantly improves its stability by protecting it from nuclease degradation and reducing renal clearance. Chemical modifications, such as PEGylation, not only shield siRNA from enzymatic breakdown but also increase its molecular weight, thereby prolonging its circulation time. Additionally, bioconjugation with cell-penetrating peptides (CPPs) enhances cellular uptake, ensuring more efficient delivery to target cells. Targeted delivery is further achieved by attaching siRNA to antibodies, aptamers, or specific ligands that recognize and bind to target cell surface receptors, thereby increasing local siRNA concentration at the desired site and minimizing off-target effects and systemic toxicity. Conjugation with polymers and encapsulation in biodegradable carriers like lipidoids or hydrogels provide controlled release mechanisms, maintaining therapeutic levels of siRNA over extended periods and reducing administration frequency. These modifications also help in evading the immune system, reducing immunogenicity and potential side effects. Overall, bioconjugation strategies are essential for improving the pharmacokinetics (PK), stability, and efficacy of siRNA, making it a more viable and effective tool for gene therapy applications. We will delve deeper into each section below.

## 22.3 Bioconjugates

Bioconjugation serves as a strategic approach for siRNA delivery to cells by crafting siRNA conjugates that interact with cellular components in three main ways: (1) conjugation with biomolecules such as folate [5], antibodies [6–8], aptamers [9, 10], certain peptides [11, 12], and carbohydrates [13] that specifically target cell membrane receptors; (2) linkage to molecules like cholesterol [14] and vitamins [15] that facilitate cell entry via natural transport mechanisms; and (3) attachment to entities that interact non-specifically with the cell membrane through positive electrostatic charges and hydrophobic characteristics [16, 17]. The configuration of the linker connecting siRNA with these biomolecules critically influences the siRNA's cellular accumulation and its biological efficacy. Importantly, linkers designed to cleave upon entering the cell—such as disulfide [18] and thioether bonds [17], pH-sensitive linkers like hydrazone [19] and carboxymethylmaleic anhydride [20], or light-sensitive linkages (β-[bis(4-methoxyphenyl)-phenylmethoxy]-2-nitrobenzeneethanol linker [21])—prevent the reduction in RNAi efficiency caused by RISC assembly inhibition. Furthermore, conjugates with stable linkers under experimental conditions are also extensively utilized [14], underlining the critical role of the conjugate's structure. The following discussion elaborates on these prevalent types of bioconjugates.

## 22.4 Lipophilic siRNA Derivatives

### *22.4.1 siRNA-Cholesterol Conjugate*

The concept of conjugating siRNAs with lipids and cholesterol stems from the need to enhance the delivery and efficacy of these therapeutic molecules in cells. siRNAs are powerful tools in gene therapy, used to silence specific genes that contribute to disease. However, their delivery to target cells in the human body presents significant challenges, primarily due to their rapid degradation by nucleases and poor cellular uptake due to their size and negative charge [22]. Cholesterol plays a crucial role in cellular functions, serving not only as a structural component of cell membranes but also in intracellular signaling and lipid metabolism. Its transport into cells is finely orchestrated through interactions with low-density lipoproteins (LDL) and high-density lipoproteins (HDL), which serve as its main carriers in the bloodstream. HDL plays a different role in cholesterol management, primarily involved in reverse cholesterol transport, which is the process of transporting cholesterol away from peripheral tissues back to the liver for excretion. HDL particles interact with cells via the scavenger receptor class B type I (SR-BI). SR-BI is integral in mediating the selective uptake of cholesterol esters from HDL particles into cells, unlike the endocytosis process seen with LDL. SR-BI does not mediate the endocytosis of the entire HDL particle; instead, it facilitates the transfer of cholesterol directly across the plasma membrane, thus enriching the cellular cholesterol pools [23, 24]. Research into the conjugation of siRNAs with cholesterol or other lipophilic derivatives represents a significant advancement in the field of targeted gene therapy. The efficiency of this binding and subsequent cellular activity is influenced by the degree of hydrophobicity of the lipophilic molecule involved [25]. However, the initial hypothesis that lipophilic siRNA derivatives enter cells predominantly through receptor-mediated endocytosis as part of these complexes was not supported by subsequent findings [25]. Instead, the transmembrane protein SIDT1 has been shown to facilitate the cellular entry of these lipophilic siRNA complexes. SIDT1 aids the penetration of double-stranded RNA (dsRNA) either by forming diffusion channels or by indirectly interacting with other proteins to enhance dsRNA uptake [25, 26]. SIDT1 is an intriguing protein involved in the cellular uptake of dsRNA, playing a significant role in RNAi mechanisms. SIDT1 contains a domain that can bind to steroid molecules. The interaction with steroids could potentially regulate the transport activity of SIDT1, possibly by altering its conformation or the efficiency with which it facilitates the uptake of dsRNA. This regulation may be crucial for maintaining cellular homeostasis and responding to environmental or intracellular signals [27]. Like SIDT1, SIDT2 facilitates the cellular uptake of dsRNA, playing a crucial role in RNAi, a key regulatory and defense mechanism against viral infections and in gene expression modulation. SIDT2 is also involved in transporting cholesterol within cells. This function is particularly noteworthy because it occurs even in the absence of lipids, suggesting a unique, lipid-independent mechanism for cholesterol handling by SIDT2 [28, 29]. Nevertheless, the precise function of SIDT1 in the uptake of lipophilic siRNA-lipoprotein complexes remains unclear.

One study provides valuable insights into the cellular uptake mechanisms of siRNA-cholesterol conjugates, particularly highlighting the roles of specific genes involved in endocytosis. Understanding these mechanisms is crucial for improving the delivery and efficacy of siRNA therapies, which are promising tools for gene silencing in various diseases. The study involved silencing key genes associated with endocytosis pathways, specifically DNM1L and CLTC, which are integral to clathrin-dependent endocytosis. Other genes involved in various endocytosis pathways, such as CAV1 (caveolin-mediated endocytosis), CDC42, and RAC1 (regulating actin dynamics), were also targeted. Silencing DNM1L and CLTC, which are critical for clathrin-dependent endocytosis, resulted in a significant reduction (~40%) in the uptake of the cholesterol conjugates. This suggests that clathrin-dependent pathways play a major role in the internalization of these conjugates. The implication is that silencing other genes like CAV1, CDC42, and RAC1 did not have as pronounced an effect as DNM1L and CLTC on the uptake of the conjugates. This observation can help prioritize which pathways might be most relevant for the optimized delivery of siRNA therapies [30]. The additional research into the clathrin-dependent intracellular transport of cholesterol-siRNA conjugates provides a deeper understanding of the dynamics and efficiency of siRNA delivery systems, crucial for optimizing therapeutic strategies. This research confirmed that clathrin-dependent pathways play a significant role in the intracellular transport of cholesterol-siRNA conjugates, contributing to 25% of the total internalization of these derivatives. This pathway involves the formation of clathrin-coated pits on the cellular membrane, which facilitate the entry of extracellular molecules into cells via vesicles [31]. Endocytosis not only involves the uptake of substances but also their sorting within endosomal compartments for either recycling back to the cell surface or degradation [32]. The sorting and recycling of the LDL receptor, for instance, typically occur within about 6 min [33]. In the initial 60 min following the exposure of HeLa cells to cholesterol-modified siRNA, only 5% of the conjugate was recycled, whereas 20% was directed toward the degradation pathway [11, 14, 25, 31, 34–40]. Major types of siRNA bioconjugation are summarized in Fig. 22.2 and the section below.

### 22.4.2 siRNA-Tocopherol Conjugate

Tocopherol (vitamin E) derivatives have been explored for siRNA delivery due to their antioxidant properties and involvement in cellular antioxidative pathways. However, tocopherol might not promote the same level of interaction with specific cellular receptors as cholesterol does, possibly resulting in lower cellular uptake efficiency [15, 25, 41–43]. The recent study's findings on the biodistribution of lipophilic siRNA conjugates, particularly highlighting the superior retention rate of cholesterol conjugates, underscore the efficiency of cholesterol as a delivery enhancer for siRNA therapies. The observed 62% retention rate for cholesterol conjugates suggests a more stable and prolonged presence within the body compared to other lipophilic conjugates. This higher retention rate can be attributed to cholesterol's natural integration into cellular and systemic lipid metabolism pathways [44]. Subsequent to subcutaneous administration, cholesterol conjugates were found to accumulate in virtually all major organs, including the liver, kidneys, adrenal glands, spleen, pancreas, heart, muscles, fat, thymus, lungs, the injection site, ovaries, and testes. Notably, the highest accumulation rates of cholesterol conjugates were observed in the liver, adrenal glands, spleen, and at the skin site of administration [44]. In comparison to other lipophilic conjugates, the accumulation of cholesterol derivatives in organs such as the liver, kidney, and spleen was either similar or lower. This finding suggests that specific modifications in RNA conjugates can influence their distribution and efficacy in different tissues. For example, a conjugate consisting of small interfering RNA (siRNA) linked to a saturated fatty acid (specifically docosanoic acid, also known as behenic acid, with a 22-carbon

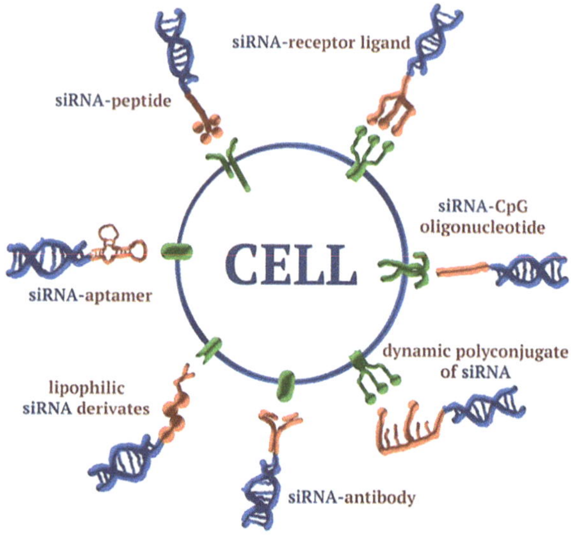


**Fig. 22.2** This diagram illustrates a variety of small interfering RNA (siRNA) formulations and their corresponding delivery strategies. Each siRNA complex is designed to enhance cellular uptake and improve therapeutic efficacy by engaging with specific cellular pathways or receptors. (1) siRNA-Receptor Ligand: Represents siRNAs conjugated to receptor-specific ligands that facilitate targeted delivery by binding to corresponding cell surface receptors. (2) siRNA-Peptide: Depicts siRNAs linked to peptides that facilitate penetration through cellular membranes or targeting specific cellular compartments. (3) siRNA-Aptamer: Shows siRNAs attached to aptamers, which are single-stranded DNA or RNA molecules capable of binding to specific molecular targets with high affinity, enhancing the specificity of siRNA delivery. (4) Lipophilic siRNA Derivatives: Illustrates siRNA molecules modified with lipophilic groups to improve membrane permeability and intracellular stability. (5) siRNA-Antibody: Depicts antibody-conjugated siRNAs targeting antigen-expressing cells, enabling precise delivery to cells expressing the target antigen. (6) Dynamic Polyconjugate of siRNA: Represents siRNAs incorporated into dynamic polyconjugates, which are synthetic carriers designed to release siRNA upon entering target cells. (7) siRNA-CpG Oligonucleotide: Shows siRNAs linked to CpG oligonucleotides, which can stimulate the immune response, potentially enhancing the therapeutic effects of the siRNA in immunomodulatory applications. (Figure is reproduced from [1] under Creative Commons CC-BY license)

chain, C22) showed markedly enhanced accumulation in muscle and adipose (fat) tissues. This conjugate's targeted delivery and uptake by these tissues were more efficient than those observed with the cholesterol-based conjugate. Upon subcutaneous administration at a dosage of 20 mg/kg, the fatty acid-conjugated siRNA achieved significant gene silencing. It inhibited the expression of the Huntington (Htt) and peptidylprolyl isomerase B (Ppib) genes by 20% and 30%, respectively, in muscle tissues. In adipose tissues, this conjugate was even more effective, leading to 50% and 30% inhibition of the Htt and Ppib genes, respectively. These results are indicative of the potential of fatty acid-siRNA conjugates to deliver therapeutic effects in specific tissues, highlighting their utility in treating conditions that require targeted gene silencing with minimal off-target effects. This selective distribution and action reinforce the importance of chemical composition and structure in the design of RNA-based therapeutics, where alterations in the conjugate's lipophilicity can significantly impact its biodistribution and efficacy [44]. This study highlighted that the degree of lipophilicity plays a crucial role in determining the biodistribution of these conjugates.

## 22.5 siRNA Conjugation with Low Lipophilic Lipids

siRNA technology harnesses the potential to selectively silence gene expression, making it a powerful tool for treating various genetic disorders and diseases. The effectiveness of siRNA is significantly enhanced by conjugation with lipids, which facilitates cellular uptake and delivery to target tissues. Conjugation with low lipophilic lipids like retinoic acid, lithocholic acid, and docosahexaenoic acid offers unique advantages due to their differing biological properties and interactions within the body [44].

### *22.5.1 siRNA-Retinoic Acid (RA) Conjugate*

Retinoic acid, a metabolite of vitamin A, is crucial for cellular growth and differentiation. Its role in gene expression is mediated through retinoic acid receptors in the nucleus, influencing cell differentiation, proliferation, and apoptosis [45]. When siRNA is conjugated with retinoic acid, it targets tissues where retinoic acid receptors are abundantly expressed, such as the skin and parts of the immune system. This targeting makes RA-siRNA conjugates particularly useful in dermatological conditions like psoriasis or acne, as well as in cancers where retinoic acid pathways are disrupted. The relatively low lipophilicity of RA compared to other lipids also means that these conjugates might have a better safety profile, reducing the risk of nonspecific lipid-associated side effects.

### *22.5.2 siRNA-Lithocholic Acid (LCA) Conjugate*

Lithocholic acid, a secondary bile acid, is produced in the liver and has a role in cholesterol metabolism and solubilization of fats [46]. siRNA conjugated with lithocholic acid can be particularly effective for targeting liver cells, leveraging LCA's natural liver tropism. This makes LCA-siRNA conjugates potential therapeutic agents for liver-specific diseases, such as hepatitis, liver fibrosis, or liver cancer. The hydrophobic yet somewhat less lipophilic nature of lithocholic acid compared to other bile acids might reduce systemic toxicity and enhance the delivery efficiency of siRNA to hepatic cells.

### 22.5.3 siRNA-Docosahexaenoic Acid (DHA)

DHA is an omega-3 fatty acid known for its benefits in brain health and cardiovascular systems [47]. Conjugating siRNA with DHA can harness its affinity for brain and retinal cells, thus opening pathways for treating neurological disorders and diseases of the eye. The specific targeting leveraged by DHA could facilitate the delivery of siRNA across the blood–brain barrier, a significant challenge in the treatment of neurodegenerative diseases like Alzheimer's or Parkinson's. Moreover, the anti-inflammatory properties of DHA may enhance the therapeutic profile of the siRNA conjugates, reducing potential inflammatory responses triggered by siRNA delivery systems. Recent research into conjugating siRNA with DHA has demonstrated a superior ability to suppress target gene expression compared to other lipophilic conjugates. This enhanced efficacy is likely due to DHA's inherent biological roles and its high affinity for brain tissue, which ensures more direct and effective delivery of siRNA into neuronal cells. When siRNA is conjugated with DHA, it leverages DHA's natural pathways and transport mechanisms across the blood–brain barrier, a challenging obstacle for most therapeutic agents. Experimental studies underscore the potential of DHA-siRNA conjugates. For instance, when these conjugates were injected into the brain striatum of mice at a dosage of approximately 1.25 mg/kg, there was a notable reduction in mRNA levels of the targeted gene. Specifically, in the striatum of the injected hemisphere, the mRNA levels of the Huntington gene—a gene implicated in Huntington's disease—were reduced by an impressive 73%. Furthermore, in the cortex, gene suppression was also significant, with a 52% reduction. These results not only highlight the direct impact of siRNA-DHA conjugates in the targeted region but also their influence on adjacent brain areas, suggesting effective diffusion and action of the conjugate within brain tissues [43].

The conjugation of siRNA with specific low lipophilic lipids like retinoic acid, lithocholic acid, and docosahexaenoic acid provides strategic advantages for targeted gene therapy. Each lipid's unique interaction with different cell types and tissues can be exploited to enhance the delivery and efficacy of siRNA, potentially transforming the landscape of therapeutic gene silencing [44]. This aligns with earlier observations that more lipophilic conjugates bind more efficiently to serum components and are less likely to be excreted by the kidneys [25, 48]. Despite their effectiveness, these lipophilic derivatives do not cross the BBB after subcutaneous or intravenous administration [44], leading researchers to explore direct brain injections to inhibit gene expression [35, 42, 43].

## 22.6 siRNA Conjugation with Peptide

Various proteins and peptides can intrude into cells leveraging endogenous transport mechanisms, facilitating the intracellular delivery of other molecules. The primary transport pathways for peptides involve attaching to surface proteins, glycoconjugates [49, 50], or anionic cell lipids, which then leads to uptake via endocytosis, direct membrane penetration [51–57], membrane disruption, or the formation of pores in the membrane [lytic peptides [58]]. Two principal methods are employed to obtain such peptides: phage display technology, or isolating segments of proteins that naturally perform similar roles. Herein, we summarize some strategies for delivering siRNAs via peptides, harnessing the unique and specific binding capacities of peptides to cell surface proteins. This specificity largely arises from the distinctive elements in their tertiary structures, which allow peptides to recognize and bind to particular cellular receptors or proteins with high affinity [11, 50, 59, 60].

### *22.6.1 siRNA-RGD and cRGD Peptides Conjugates*

The RGD peptide sequence and its cyclic derivative, cyclic RGD (cRGD), serve as prime examples of how peptide conjugation can enhance the targeted delivery of therapeutic agents like siRNA [61]. The cyclic form of RGD, derived from a phage display library optimized for binding to specific tumor cells, exemplifies a sophisticated approach in the field of targeted therapy. cRGD specifically targets $\alpha V\beta 3$ and $\alpha V\beta 5$ integrins, which are frequently overexpressed on the surface of tumor cells and vascular endothelial cells within the tumor microenvironment [62]. This overexpression provides a unique target for the delivery of therapeutic agents directly to tumors. The cyclic structure of cRGD enhances its stability and increases its binding affinity to these integrins, making it a highly effective targeting moiety for delivering drugs to tumor sites. The conjugation of siRNA with cRGD has shown significant promise in both in vitro and in vivo studies. In vitro, cRGD-siRNA conjugates have demonstrated the ability to efficiently penetrate tumor cells and silence target genes. The increased cellular uptake is facilitated by the specific interaction between cRGD and $\alpha V\beta 3/5$ integrins, leveraging the natural cellular processes of integrin-mediated endocytosis. In vivo studies further validate the effectiveness of this approach. For example, when cRGD-siRNA conjugates targeting the vascular endothelial growth factor receptor 2 (VEGFR2) were administered through intravenous injections to animals bearing A549 xenograft tumors, there was a notable 55% reduction in VEGFR2 mRNA levels. This significant decrease in a key angiogenesis mediator correlates directly with a reduction in tumor growth. The protocol involved six intravenous injections, highlighting the peptide's ability to consistently deliver siRNA and achieve sustained gene silencing with corresponding therapeutic effects [60, 63–65]. Efforts to enhance the efficacy by increasing the number of RGD peptides in the conjugate were made, though adding more peptides did not linearly enhance biological activity [66].

### 22.6.2 siRNA-CKSC Peptides Conjugate

CKSC peptides represent a novel approach in the field of targeted gene therapy, utilizing the unique properties of peptide sequences to enhance the delivery and efficacy of siRNA molecules. CSKC peptide is a specially designed sequence that mimics the insulin-like growth factor 1 (IGF-1). This mimicry allows CSKC to exploit the natural cellular pathways of IGF-1, specifically targeting cells that express the IGF-1 receptor, which is prevalent in various cancer cell types including breast cancer cells like MCF7 [67]. By binding to the IGF-1 receptor, the CSKC peptide can effectively introduce siRNA into cells through receptor-mediated endocytosis, a process by which cells absorb molecules by engulfing them in an energy-using process. This targeted approach is particularly beneficial as it ensures that the therapeutic siRNA is delivered directly to the intended cells, thereby maximizing the therapeutic impact while minimizing potential off-target effects [11].

### 22.6.3 siRNA-LEVDG Peptides Conjugate

The LEVDG peptide sequence is specifically designed to interact with cellular mechanisms involving caspase-4, an enzyme that plays a role in apoptotic pathways and is notably expressed in certain types of cancer cells, such as Jeg-3 choriocarcinoma cells. This peptide is unique because it is sensitive to cleavage by caspase-4, which is pivotal in its mechanism of action. When LEVDG is conjugated to siRNA, it serves a dual function: it targets and accumulates within cells expressing caspase-4, and it is subsequently activated by cleavage through caspase-4. In the context of Jeg-3 choriocarcinoma cells, the LEVDG peptide facilitates targeted delivery of antiSTAT3 siRNA. Once inside the cells, caspase-4 cleaves the LEVDG peptide, thereby enabling the siRNA to interfere with the RISC assembly process. This interference results in effective gene silencing, as evidenced by up to 70% suppression of STAT3 gene expression. STAT3 is a critical transcription factor involved in various cellular processes, including proliferation and survival, and its overexpression is linked to many cancers. What sets the LEVDG peptide apart is its high specificity for cells expressing caspase-4, which allows for selective targeting and minimizes effects on non-target cells. This was demonstrated in experiments where the same siRNA-LEVDG conjugates had no suppressive effect on HEK293 cells, which do not express caspase-4. This specificity is highly advantageous in therapeutic settings as it reduces the potential for off-target effects and increases the overall safety profile of the treatment [68].

### 22.6.4 siRNA-Lytic Peptides Conjugate

A lytic peptide, also known as an antimicrobial or cytolytic peptide, is a type of naturally occurring or synthetically engineered peptide that has the ability to disrupt the membrane integrity of target cells, leading to cell lysis and death. These peptides are part of the innate immune defense in many organisms, including humans, and are effective against a broad range of pathogens including bacteria, viruses, and fungi. Due to their broad-spectrum antimicrobial properties and mechanism of action that reduces the likelihood of resistance development, lytic peptides are considered promising candidates for therapeutic applications, especially as potential alternatives to traditional antibiotics in the fight against drug-resistant infections. One innovative strategy for delivering siRNA to target cells involves its conjugation with lytic peptides that facilitate siRNA release from the endosome [69]. This method leverages several mechanisms to enhance the efficiency of siRNA release from the endosome. (1) The formation of transmembrane pores by peptides due to their amphiphilic nature and ability to form complexes, such as melittin from bee venom, ricin, and other ribosome-inactivating proteins derived from oil [58, 70]. (2) The protonation of peptide groups leading to a decrease in endosomal pH, which subsequently increases osmotic pressure within the endosome and causes membrane rupture, exemplified by polyhistidine [71]. (3) Local destabilization of the membrane through the fusogenic properties of certain proteins, which then penetrate the endosome membrane, such as glycoprotein H from herpes virus, hemagglutinin-2 domain of influenza virus, and diphtheria toxin domain [72–75].

**Box 22.2 Cell-Penetrating Peptides (CPPs) in siRNA Delivery: Challenges and Future Outlook**

Cell-penetrating peptides (CPPs) are short peptides that facilitate the delivery of various molecular cargoes, including nucleic acids, proteins, and drugs, across cellular membranes. Their ability to translocate across biological membranes while carrying large molecules has made them highly valuable in the field of drug delivery and molecular biology. Among the most studied CPPs are penetratin, transportan, and the trans-activator of transcription (TAT) peptide. Each of these CPPs has unique properties and mechanisms of action, which enable them to effectively penetrate cellular barriers.

**Penetratin**, also known as Antennapedia homeodomain-derived peptide, is derived from the homeodomain of the Antennapedia protein in Drosophila. This peptide typically contains 16 amino acids and is rich in basic residues, which facilitate its interaction with the negatively charged cellular membranes [76]. Penetratin is believed to enter cells primarily through a direct translocation mechanism that does not necessarily require energy-dependent endocytosis. This ability to cross membranes without the involvement of cellular energy systems or receptors makes penetratin a powerful tool for delivering therapeutic agents directly into the cell cytoplasm and nucleus.

(continued)

**Box 22.2** (continued)

**Transportan** is a synthetic peptide originally derived from the fusion of the hormone galanin and mastoparan, a wasp venom peptide [49]. It is longer than most CPPs, usually around 27 amino acids, and has been shown to be highly efficient in translocating through cellular membranes. Transportan can carry a variety of different cargo molecules into a wide range of cell types. Unlike penetratin, transportan is believed to utilize both endocytic pathways and direct penetration, making it versatile for different types of cargo delivery, from small molecules to larger proteins.

**Trans-Activator of Transcription (TAT)** peptide is derived from the HIV-1 virus's Tat protein, which is involved in viral replication and transcription. It contains about 11 amino acids and is rich in basic residues, particularly arginine, which enhance its interaction with cell membranes and internalization [51]. Tat peptide is one of the most effective CPPs, known for its rapid and efficient uptake by cells. It is commonly used to deliver bioactive molecules, including peptides, proteins, nucleic acids, and even nanoparticles, into cells. Tat enters cells predominantly via energy-dependent endocytosis, leveraging the cell's natural uptake mechanisms to transport large and complex cargoes [55, 77, 78]. Multiple studies have demonstrated that CPP-siRNA conjugates are biologically active when administered to cells [11, 78]. Another promising application of CPPs in siRNA delivery involves the skin penetrating and cell entering (SPACE) peptide, identified through phage display technology for its ability to traverse the epidermis [79]. When siRNA was conjugated with the SPACE peptide and applied topically, it successfully penetrated both the epidermis and dermis. A single topical application of approximately 12 mg/kg of the siRNA-SPACE peptide conjugates reduced IL10 and GAPDH expression in the epidermis by 28% and 47%, respectively. Enhancing this conjugate with lipoplexes containing the SPACE peptide led to more significant suppression of GAPDH expression.

Despite these positive in vitro results, the in vivo application of CPP-siRNA conjugates remains limited due to their potential toxicity and ability to provoke immune responses [34, 80, 81]. One notable challenge in employing CPPs as delivery vehicles for siRNA is their tendency to form complexes with siRNA due to their positive charge, which subsequently hampers the interaction of siRNA with the RISC components. To mitigate this issue, strategies such as neutralizing or shielding the negative charge of siRNA have been developed, including chemical modifications like the *tert*-butyl-*S*-acyl-2--thioethyl phosphotriester (tBu-SATE) [17, 82]. Implementing tBu-SATE modifications in siRNA conjugates allows for the incorporation of multiple cationic peptides without interfering with the RNA interference mechanism. Research indicates that the biological efficacy of such conjugates is influenced by the quantity of CPP molecules, specifically the Tat peptide. For instance, conjugates containing four TAT peptides were found to be more effective in suppressing gene expression in cells than those with only two or three TAT peptides [17]. Despite encouraging in vitro results, the in vivo applications of these conjugates have yet to be explored [83].

## 22.7 siRNA and Receptor Ligand Conjugates

Key determinants of the effectiveness of targeted delivery to specific cells include the binding efficiency of the ligand to its receptor and the receptor's expression level on the cell membrane. Typically characterized by high specificity, the interaction between ligands and receptors significantly enhances targeted delivery when the ligands are covalently bound to therapeutic molecules [84].

### *22.7.1 siRNA-Asialoglycoprotein Receptor (ASGPR) Conjugate*

ASGPR is a C-type lectin, which means it is a calcium-dependent carbohydrate-binding protein. It is composed of several subunits that collectively function to bind glycoproteins containing the exposed galactose or *N*-acetylgalactosamine (GalNAc) residues once the terminal sialic acids are removed. In hepatocytes, ASGPR is responsible for mediating the endocytosis of these glycoproteins, which are then degraded in lysosomes. The receptor's expression level is notably high on hepatocytes, with estimates of 0.5–1 million molecules per cell, illustrating its significant presence and potential as a target for liver-specific therapeutic strategies. One of the most successful applications of targeting ASGPR for drug delivery involves the use of *N*-acetylgalactosamine (GalNAc). GalNAc is a derivative of galactose and specifically binds to ASGPR with a very high affinity, characterized by a dissociation constant ($K_d$) of approximately 2.5 nM. This high-affinity interaction ensures that therapeutic compounds such as siRNA conjugated with GalNAc are effectively recognized and internalized by hepatocytes. The conjugation of siRNA with GalNAc exploits this natural cellular pathway, providing a direct and efficient route for delivering RNA-based therapeutics into liver cells [85]. Conjugating siRNA with *N*-acetylgalactosamine significantly enhances siRNA delivery, reducing expression of the target gene (PCSK9) by 70% after a single subcutaneous injection of approximately 6 mg/kg [86]. Remarkably, even a 50% reduction in ASGPR expression does not diminish the biological activity of the siRNA and *N*-acetylgalactosamine conjugate; only a reduction of 95% impairs its effectiveness [87]. Owing to these factors and its cost-effective synthesis, siRNA conjugated with *N*-acetylgalactosamine stands out as a leading candidate for clinical development, particularly for treating liver diseases.

### *22.7.2 siRNA-Folic Acid Conjugate*

Another promising approach involves the use of folic acid for the targeted accumulation of siRNA in tumor cells [5]. The folic acid receptor is a high-affinity glycoprotein that binds to folate and its derivatives. It operates predominantly through a specific cellular uptake mechanism known as receptor-mediated endocytosis [88]. This process involves the receptor binding folate at the cell surface, followed by the internalization of the folate-receptor complex into the cell. The binding affinity of FAR for folic acid is exceptionally high, with a dissociation constant ($K_d$) of about $10^{10}$ M, indicating a very tight and specific interaction [88, 89]. This selective receptor expression suggests that folate-conjugated siRNAs are preferentially internalized by tumor cells through receptor-mediated endocytosis, leading to higher penetration efficiency compared to normal cells [90]. Indeed, in vivo studies have shown that fluorescently labeled siRNA conjugated with folate accumulates more efficiently in mouse tumors than does unmodified siRNA [5]. However, the practical use of siRNA and folate conjugates remains largely experimental due to the complexities involved in their synthesis.

## 22.8 siRNA and Aptamer Conjugates

Aptamers, often referred to as "chemical antibodies," are synthetic oligonucleotides (either DNA or RNA) that are engineered through an iterative process known as SELEX. These molecules typically weigh between 6 and 30 kDa and are favored for their ability to fold into intricate tertiary structures, which enable highly specific binding to target molecules. This specificity is akin to that of antibodies, but with the added benefits of chemical synthesis such as reproducibility, stability, and ease of modification. Aptamers can be designed to bind with high affinity to a variety of targets, ranging from small molecules to proteins and even whole cells. Their binding ability is driven by their unique three-dimensional structures, which precisely fit their target molecules. This high degree of specificity and affinity is what makes aptamers excellent candidates for therapeutic applications, particularly in targeting and regulating disease-related biomolecules [91, 92].

### *22.8.1 siRNA-A10 Aptamer Conjugate*

The A10 aptamer provides a remarkable example of the potential of aptamer technology in targeted cancer therapy. Specifically designed to bind to the prostate-specific membrane antigen (PSMA), which is overexpressed in prostate cancer cells, the A10 aptamer serves as a precise delivery vehicle for therapeutic agents. PSMA is a cell surface glycoprotein that is not only abundant in prostate cancer cells but also associated with tumor growth and progression [93]. The A10 aptamer binds specifically to PSMA, enabling targeted therapeutic delivery directly to prostate cancer cells while sparing normal tissues. This specificity is crucial for reducing systemic side effects and increasing the efficacy of the treatment. In a significant study using a xenograft LNCaP tumor-bearing mouse model—a system that closely mimics human prostate cancer—the A10 aptamer-siRNA conjugate demonstrated potent therapeutic efficacy. The treatment involved ten intratumoral injections of the conjugate, which resulted in high levels of gene silencing: Up to 85% silencing efficiency was observed in PLK1 silencing. PKL1 is involved in cell cycle regulation and its downregulation can lead to inhibited tumor growth. Achieving up to 90% silencing efficiency, targeting the Bcl2 gene is particularly impactful as it is involved in cell survival and apoptosis resistance. Silencing Bcl2 leads to increased apoptosis within tumors [10].

### 22.8.2 siRNA-A10-3 Aptamer Conjugate

The A10-3 aptamer, a derivative or variant of the A10 aptamer, has been tailored to enhance binding specificity and efficiency to the PSMA. This aptamer has been used to develop a conjugate with short hairpin RNA (shRNA), targeting the PRKDC gene—a key player in DNA damage response and repair. PRKDC (protein kinase, DNA-activated, catalytic subunit) is crucial for the DNA damage repair process, particularly in response to double-strand breaks. Silencing PRKDC hampers the cancer cell's ability to repair DNA damage effectively, rendering it more susceptible to treatments that induce such damage, like ionizing radiation. The shRNA-A10-3 aptamer conjugate was tested in a xenograft model using LNCaP cells. While the conjugate alone inhibited PRKDC expression by 65%, a significant finding itself, notable tumor regression was observed only when this genetic intervention was combined with ionizing radiation. This combination therapy leverages the reduced DNA repair capacity induced by PRKDC silencing, thereby enhancing the cytotoxic effects of radiation on tumor cells [94].

### 22.8.3 siRNA-A-1 Aptamer Conjugate

The A-1 aptamer targets gp120, a glycoprotein essential for HIV-1's ability to infect host cells. Gp120 facilitates the virus's entry into cells by binding to the CD4 receptor and co-receptors on the surface of T cells, making it a prime target for therapeutic intervention [95]. The conjugation of siRNA to the A-1 aptamer allows for the simultaneous targeting and knockdown of essential viral genes, thereby disrupting the viral life cycle. In a study conducted using humanized $Rag2^{-/-}$ $\gamma c^{-/-}$ (RAG-hu) mice, which are engineered to mimic human immune system responses, the efficacy of this therapeutic strategy was tested. The mice, infected with HIV-1, received weekly intravenous administrations of the A-1 aptamer-siRNA conjugate at a dose of 0.38 mg/kg. This treatment resulted in a significant, 105-fold reduction in viral RNA concentration in the plasma, showcasing the potential of the conjugate to dramatically lower the viral load in an infected organism [96].

### 22.8.4 *siRNA-Gint4.T/GL21.T Aptamers Conjugates*

Recent research has shown that certain aptamers, like Gint4.T and GL21.T, which target the beta-type platelet-derived growth factor receptor (PDGFR-β), can cross the BBB. This capability is significant because the BBB often presents a formidable obstacle to the delivery of therapeutic agents to the brain [97]. Following several intravenous injections, this siRNA conjugate was able to accumulate in xenograft glioblastomas, suppress the expression STAT3 gene by 60%, and significantly slow tumor growth [98]. These findings underscore the potential of aptamer-siRNA conjugates as a versatile platform for targeted gene therapy, capable of addressing a variety of disease states through specific molecular interactions.

## 22.9 siRNA and Antibody Conjugates

Antibody-siRNA Conjugates (ARCs) represent a novel class of therapeutic agents that harness the specificity of antibodies to deliver siRNAs directly to cells expressing particular receptor antigens. In HEK293 cells, which are human embryonic kidney cells commonly used in biological research, the ARC targeting the insulin receptor was able to reduce the expression of a target gene by 90% at a concentration of 115 nM. This high level of gene silencing illustrates the potent efficacy of ARCs when the target receptor is abundantly expressed on the cell surface [8]. Conversely, an ARC targeting the Lewis-Y protein only achieved a 60% inhibition at 300 nM when cells were treated with chloroquine, a compound known to inhibit endosome maturation [99]. ARCs, both covalent and non-covalent, are innovative tools for targeted gene silencing, capable of delivering therapeutic siRNA specifically to cells expressing certain receptors. Recently, interest has grown in non-covalent complexes of siRNA with antibodies, which leverage natural or engineered interactions for conjugate formation [6, 100–103]. However, a comprehensive comparison examining the effectiveness of ARCs with different target antigens revealed that receptor expression levels and the type of intracellular receptor transport affect ARC efficacy, although no direct correlation was found between the mode of receptor-antibody complex entry and biological activity [104]. The study did not evaluate antibody binding efficiency to receptors or the conjugates' endosomal escape, which are likely critical factors affecting ARC activity.

Further insights into the cellular entry mechanisms of ARCs were gained in studies involving the TENB2 antibody, known for its high biological activity. It was demonstrated that disrupting genes associated with clathrin-dependent receptor-mediated endocytosis decreased the gene silencing effectiveness of the ARC. Conversely, inhibiting genes linked to intracellular transport pathways, such as RAB5C and HPS4, enhanced ARC silencing activity [104]. RAB5C typically directs endosomal sorting toward the recycling pathway, whereas HPS4 regulates RNAi activity by promoting the merging of lysosomes with multivesicular bodies, potentially affecting the availability of RNAi-related proteins [105]. Recent insights suggest that suppressing the expression of HPS4 (Hermansky-Pudlak syndrome 4) can increase siRNA efficiency. HPS4 is involved in the biogenesis of lysosome-related organelles complex-3 (BLOC-3), which is important in trafficking molecules to lysosomes. The suppression of HPS4 might reduce siRNA degradation in endosomes or lysosomes, thus enhancing the delivery and activity of the siRNA within the ARC. In the xenograft model utilizing PC3-TENB2-high tumor-bearing nude mice, the ARC targeting the TENB2 antibody achieved a gene silencing efficacy of 33% after 3 intravenous injections of 24 mg/kg [104].

## 22.10 siRNA and CpG Oligonucleotide Conjugates

The strategic conjugation of siRNA with CpG-containing oligodeoxyribonucleotides (ODNs) represents a novel approach in the realm of gene therapy, particularly for enhancing the efficacy of siRNA in immune cells and certain cancer types that express the TLR9. This method leverages the innate immune system's mechanisms to achieve targeted and efficient gene silencing. TLR9 is an intracellular receptor predominantly expressed in B-lymphocytes, dendritic cells, macrophages, and some cancer cells. It is known for its role in recognizing unmethylated CpG motifs commonly found in bacterial and viral DNA, which activates immune responses [106–109]. However, the use of CpG oligonucleotides is limited by their propensity to activate cytokines and interleukins, restricting their application mainly to local injections due to rapid degradation in serum. Further refining the therapeutic approach, researchers have introduced chemical modifications to siRNA conjugates to enhance nuclease resistance, though this may alter interactions with Dicer [106]. Additionally, to target the gene expression of STAT3 specifically, CpG oligonucleotides were conjugated with a DNA duplex from the promoter region of the STAT3 gene. This innovative strategy allows the duplex, once inside the nucleus of target cells, to bind specific transcription factors and inhibit transcription [110]. Such a conjugate has shown therapeutic potential in a mouse model of acute myeloid leukemia following several intravenous injections (5 mg/kg) [111].

## 22.11 Dynamic Polyconjugates

Rozema and colleagues advanced the field of gene silencing with their development of dynamic polyconjugate (DPC) technology. This innovative approach is characterized by several critical features: a membrane-active polymer that can effectively engage with cellular membranes, a mechanism to reversibly mask the polymer's activity until it encounters the acidic conditions within endosomes, and the capacity to specifically target the polymer and its siRNA cargo to hepatocytes.

This targeting is achieved via a straightforward, low-pressure intravenous injection. Utilizing this sophisticated delivery system, the research demonstrated successful knockdown of two endogenous liver genes, apolipoprotein B (apoB) and peroxisome proliferator-activated receptor alpha (PPAR-α), in mice. This technology represents a significant step forward in the targeted delivery of siRNA, offering potential for enhanced therapeutic applications [20]. Arrowhead Pharmaceuticals has advanced siRNA delivery technologies with the development of dynamic polyconjugates, which incorporate dual cleavable bonds to promote the endosomal escape of siRNA [20].

**Box 22.3 Immune Modulating siRNA?**

A key component of the innate immune system is its ability to identify pathogen-associated molecular patterns through extra- and intracellular pattern-recognition receptors. An important receptor in this context is the endosomal TLR7, which is activated by single-stranded RNA (ssRNA). This activation can be modulated by specific sequence structures and posttranscriptional modifications in the RNA. Additionally, small molecule TLR7 agonists (smTLRa) are utilized as adjuvants to enhance the effectiveness of vaccines, with some evidence suggesting that covalent conjugation between the adjuvant and the vaccine components can yield synergistic effects [112].

In this research, they explored a novel approach by chemically integrating three therapeutic functionalities into a single RNA bioconjugate. This includes TLR7 stimulation via both ssRNA and smTLRa, along with the RNA's inherent therapeutic role, whether for vaccination or gene silencing purposes [113]. mRNA and siRNA bioconjugates were synthesized with covalently attached smTLRa and evaluated their efficacy in activating TLR7. However, surprisingly, the results revealed that these bioconjugates led to decreased TLR7 stimulation, rather than the anticipated synergistic increase. This result indicates that while the concept of integrating multiple therapeutic modalities into a single RNA bioconjugate is promising, the actual execution in the case of TLR7 activation presents challenges. The bioconjugation of small molecules to RNA, in particular, can inhibit rather than enhance TLR7 activation, highlighting the complexity of designing multifunctional therapeutic agents in the realm of innate immunity. Thus, these findings underscore the importance of considering how any modification, such as labeling or conjugation, might impact the immunostimulatory capabilities of RNA-based therapeutics. This knowledge is vital for the development of RNA bioconjugates that are not only effective in their therapeutic action but also refined in their interaction with the immune system. Further exploration into the nuances of how such modifications affect RNA functionality will be critical as these compounds advance through clinical development.

## 22.12 Conclusion

In conclusion, the field of siRNA delivery has underscored the potential of siRNA as a formidable gene therapy tool for various diseases. This chapter has detailed the significant advancements in siRNA bioconjugation, which markedly enhance the pharmacological profiles of siRNA while mitigating stability-related toxicity and variability. Through various conjugation approaches, including those involving antibodies, peptides, aptamers, small molecules, lipidoids, cell-penetrating peptides, polymers, and nanoparticles, siRNA bioconjugates have demonstrated improved pharmacokinetics, prolonged biological half-life, and enhanced targeting capabilities. These benefits collectively contribute to more effective gene silencing, superior therapeutic outcomes, and reduced side effects. This chapter has also highlighted the pivotal role of bioconjugation techniques in preserving the integrity, stability, and controlled release of siRNA, while acknowledging the challenges inherent in these methods. Overall, the advancements in siRNA bioconjugation represent a significant leap forward in the development of efficient and reliable siRNA-based therapeutics.

## References

1. Chernikov IV, Vlassov VV, Chernolovskaya EL. Current development of siRNA bioconjugates: from research to the clinic. Front Pharmacol. 2019;10:444.
2. Tai W. Current aspects of siRNA bioconjugate for in vitro and in vivo delivery. Molecules. 2019;24(12):2211.
3. Jash B, Kool ET. Conjugation of RNA via 2′-OH acylation: mechanisms determining nucleotide reactivity. Chem Commun (Camb). 2022;58(22):3693–6.
4. Velema WA, Kool ET. The chemistry and applications of RNA 2′-OH acylation. Nat Rev Chem. 2020;4(1):22–37.
5. Thomas M, Kularatne SA, Qi L, Kleindl P, Leamon CP, Hansen MJ, et al. Ligand-targeted delivery of small interfering RNAs to malignant cells and tissues. Ann N Y Acad Sci. 2009;1175:32–9.
6. Song E, Zhu P, Lee SK, Chowdhury D, Kussman S, Dykxhoorn DM, et al. Antibody mediated in vivo delivery of small interfering RNAs via cell-surface receptors. Nat Biotechnol. 2005;23(6):709–17.
7. Dassie JP, Liu XY, Thomas GS, Whitaker RM, Thiel KW, Stockdale KR, et al. Systemic administration of optimized aptamer-siRNA chimeras promotes regression of PSMA-expressing tumors. Nat Biotechnol. 2009;27(9):839–49.
8. Xia CF, Boado RJ, Pardridge WM. Antibody-mediated targeting of siRNA via the human insulin receptor using avidin-biotin technology. Mol Pharm. 2009;6(3):747–51.
9. Aronin N. Target selectivity in mRNA silencing. Gene Ther. 2006;13(6):509–16.
10. McNamara JO 2nd, Andrechek ER, Wang Y, Viles KD, Rempel RE, Gilboa E, et al. Cell type-specific delivery of siRNAs with aptamer-siRNA chimeras. Nat Biotechnol. 2006;24(8):1005–15.
11. Cesarone G, Edupuganti OP, Chen CP, Wickstrom E. Insulin receptor substrate 1 knockdown in human MCF7 ER+ breast cancer cells by nuclease-resistant IRS1 siRNA conjugated to a disulfide-bridged D-peptide analogue of insulin-like growth factor 1. Bioconjug Chem. 2007;18(6):1831–40.

12. Lau S, Graham B, Cao N, Boyd BJ, Pouton CW, White PJ. Enhanced extravasation, stability and in vivo cardiac gene silencing via in situ siRNA-albumin conjugation. Mol Pharm. 2012;9(1):71–80.
13. Nair JK, Willoughby JL, Chan A, Charisse K, Alam MR, Wang Q, et al. Multivalent N-acetylgalactosamine-conjugated siRNA localizes in hepatocytes and elicits robust RNAi-mediated gene silencing. J Am Chem Soc. 2014;136(49):16958–61.
14. Lorenz C, Hadwiger P, John M, Vornlocher HP, Unverzagt C. Steroid and lipid conjugates of siRNAs to enhance cellular uptake and gene silencing in liver cells. Bioorg Med Chem Lett. 2004;14(19):4975–7.
15. Nishina K, Unno T, Uno Y, Kubodera T, Kanouchi T, Mizusawa H, et al. Efficient in vivo delivery of siRNA to the liver by conjugation of alpha-tocopherol. Mol Ther. 2008;16(4):734–40.
16. Kwiatkowska A, Sobczak M, Mikolajczyk B, Janczak S, Olejniczak AB, Sochacki M, et al. siRNAs modified with boron cluster and their physicochemical and biological characterization. Bioconjug Chem. 2013;24(6):1017–26.
17. Meade BR, Gogoi K, Hamil AS, Palm-Apergi C, van den Berg A, Hagopian JC, et al. Efficient delivery of RNAi prodrugs containing reversible charge-neutralizing phosphotriester backbone modifications. Nat Biotechnol. 2014;32(12):1256–61.
18. Turner JJ, Ivanova GD, Verbeure B, Williams D, Arzumanov AA, Abes S, et al. Cell-penetrating peptide conjugates of peptide nucleic acids (PNA) as inhibitors of HIV-1 tat-dependent trans-activation in cells. Nucleic Acids Res. 2005;33(21):6837–49.
19. Dovydenko I, Tarassov I, Venyaminova A, Entelis N. Method of carrier-free delivery of therapeutic RNA importable into human mitochondria: lipophilic conjugates with cleavable bonds. Biomaterials. 2016;76:408–17.
20. Rozema DB, Lewis DL, Wakefield DH, Wong SC, Klein JJ, Roesch PL, et al. Dynamic PolyConjugates for targeted in vivo delivery of siRNA to hepatocytes. Proc Natl Acad Sci U S A. 2007;104(32):12982–7.
21. Yang J, Chen C, Tang X. Cholesterol-modified caged siRNAs for photoregulating exogenous and endogenous gene expression. Bioconjug Chem. 2018;29(4):1010–5.
22. Letsinger RL, Zhang GR, Sun DK, Ikeuchi T, Sarin PS. Cholesteryl-conjugated oligonucleotides: synthesis, properties, and activity as inhibitors of replication of human immunodeficiency virus in cell culture. Proc Natl Acad Sci U S A. 1989;86(17):6553–6.
23. Goldstein JL, Brown MS, Anderson RG, Russell DW, Schneider WJ. Receptor-mediated endocytosis: concepts emerging from the LDL receptor system. Annu Rev Cell Biol. 1985;1:1–39.
24. Yvan-Charvet L, Pagler TA, Wang N, Senokuchi T, Brundert M, Li H, et al. SR-BI inhibits ABCG1-stimulated net cholesterol efflux from cells to plasma HDL. J Lipid Res. 2008;49(1):107–14.
25. Wolfrum C, Shi S, Jayaprakash KN, Jayaraman M, Wang G, Pandey RK, et al. Mechanisms and optimization of in vivo delivery of lipophilic siRNAs. Nat Biotechnol. 2007;25(10):1149–57.
26. Feinberg EH, Hunter CP. Transport of dsRNA into cells by the transmembrane protein SID-1. Science. 2003;301(5639):1545–7.
27. Méndez-Acevedo KM, Valdes VJ, Asanov A, Vaca L. A novel family of mammalian transmembrane proteins involved in cholesterol transport. Sci Rep. 2017;7(1):7450.
28. Nguyen TA, Smith BRC, Tate MD, Belz GT, Barrios MH, Elgass KD, et al. SIDT2 transports extracellular dsRNA into the cytoplasm for innate immune recognition. Immunity. 2017;47(3):498–509.e6.
29. Takahashi M, Contu VR, Kabuta C, Hase K, Fujiwara Y, Wada K, et al. SIDT2 mediates gymnosis, the uptake of naked single-stranded oligonucleotides into living cells. RNA Biol. 2017;14(11):1534–43.
30. Gilleron J, Paramasivam P, Zeigerer A, Querbes W, Marsico G, Andree C, et al. Identification of siRNA delivery enhancers by a chemical library screen. Nucleic Acids Res. 2015;43(16):7984–8001.

31. Ly S, Navaroli DM, Didiot MC, Cardia J, Pandarinathan L, Alterman JF, et al. Visualization of self-delivering hydrophobically modified siRNA cellular internalization. Nucleic Acids Res. 2017;45(1):15–25.
32. Lakadamyali M, Rust MJ, Zhuang X. Ligands for clathrin-mediated endocytosis are differentially sorted into distinct populations of early endosomes. Cell. 2006;124(5):997–1009.
33. Brown MS, Goldstein JL. Receptor-mediated control of cholesterol metabolism. Science. 1976;191(4223):150–4.
34. Moschos SA, Jones SW, Perry MM, Williams AE, Erjefalt JS, Turner JJ, et al. Lung delivery studies using siRNA conjugated to TAT(48-60) and penetratin reveal peptide induced reduction in gene expression and induction of innate immunity. Bioconjug Chem. 2007;18(5):1450–9.
35. Alterman JF, Hall LM, Coles AH, Hassler MR, Didiot MC, Chase K, et al. Hydrophobically modified siRNAs silence huntingtin mRNA in primary neurons and mouse brain. Mol Ther Nucleic Acids. 2015;4(12):e266.
36. Chernikov IV, Gladkikh DV, Meschaninova MI, Karelina UA, Ven'yaminova AG, Zenkova MA, et al. Fluorophore labeling affects the cellular accumulation and gene silencing activity of cholesterol-modified siRNAs in vitro. Nucleic Acid Ther. 2019;29(1):33–43.
37. Soutschek J, Akinc A, Bramlage B, Charisse K, Constien R, Donoghue M, et al. Therapeutic silencing of an endogenous gene by systemic administration of modified siRNAs. Nature. 2004;432(7014):173–8.
38. Byrne M, Tzekov R, Wang Y, Rodgers A, Cardia J, Ford G, et al. Novel hydrophobically modified asymmetric RNAi compounds (sd-rxRNA) demonstrate robust efficacy in the eye. J Ocul Pharmacol Ther. 2013;29(10):855–64.
39. Khan T, Weber H, DiMuzio J, Matter A, Dogdas B, Shah T, et al. Silencing myostatin using cholesterol-conjugated siRNAs induces muscle growth. Mol Ther Nucleic Acids. 2016;5(8):e342.
40. Haraszti RA, Roux L, Coles AH, Turanov AA, Alterman JF, Echeverria D, et al. 5′-Vinylphosphonate improves tissue accumulation and efficacy of conjugated siRNAs in vivo. Nucleic Acids Res. 2017;45(13):7581–92.
41. Prakash TP, Lima WF, Murray HM, Li W, Kinberger GA, Chappell AE, et al. Identification of metabolically stable 5′-phosphate analogs that support single-stranded siRNA activity. Nucleic Acids Res. 2015;43(6):2993–3011.
42. Nikan M, Osborn MF, Coles AH, Biscans A, Godinho B, Haraszti RA, et al. Synthesis and evaluation of parenchymal retention and efficacy of a metabolically stable O-phosphocholine-N-docosahexaenoyl-l-serine siRNA conjugate in mouse brain. Bioconjug Chem. 2017;28(6):1758–66.
43. Nikan M, Osborn MF, Coles AH, Godinho BM, Hall LM, Haraszti RA, et al. Docosahexaenoic acid conjugation enhances distribution and safety of siRNA upon local administration in mouse brain. Mol Ther Nucleic Acids. 2016;5(8):e344.
44. Biscans A, Coles A, Haraszti R, Echeverria D, Hassler M, Osborn M, et al. Diverse lipid conjugates for functional extra-hepatic siRNA delivery in vivo. Nucleic Acids Res. 2019;47(3):1082–96.
45. Lavudi K, Nuguri SM, Olverson Z, Dhanabalan AK, Patnaik S, Kokkanti RR. Targeting the retinoic acid signaling pathway as a modern precision therapy against cancers. Front Cell Dev Biol. 2023;11:1254612.
46. Sheng W, Ji G, Zhang L. The effect of lithocholic acid on the gut-liver axis. Front Pharmacol. 2022;13:910493.
47. Li J, Pora BLR, Dong K, Hasjim J. Health benefits of docosahexaenoic acid and its bioavailability: a review. Food Sci Nutr. 2021;9(9):5229–43.
48. Osborn MF, Coles AH, Biscans A, Haraszti RA, Roux L, Davis S, et al. Hydrophobicity drives the systemic distribution of lipid-conjugated siRNAs via lipid transport pathways. Nucleic Acids Res. 2019;47(3):1070–81.
49. Pooga M, Hällbrink M, Zorko M, Langel U. Cell penetration by transportan. FASEB J. 1998;12(1):67–77.

50. Alberici L, Roth L, Sugahara KN, Agemy L, Kotamraju VR, Teesalu T, et al. De novo design of a tumor-penetrating peptide. Cancer Res. 2013;73(2):804–12.
51. Vivès E, Brodin P, Lebleu B. A truncated HIV-1 Tat protein basic domain rapidly translocates through the plasma membrane and accumulates in the cell nucleus. J Biol Chem. 1997;272(25):16010–7.
52. Thorén PE, Persson D, Karlsson M, Nordén B. The antennapedia peptide penetratin translocates across lipid bilayers—the first direct observation. FEBS Lett. 2000;482(3):265–8.
53. Console S, Marty C, García-Echeverría C, Schwendener R, Ballmer-Hofer K. Antennapedia and HIV transactivator of transcription (TAT) "protein transduction domains" promote endocytosis of high molecular weight cargo upon binding to cell surface glycosaminoglycans. J Biol Chem. 2003;278(37):35109–14.
54. Heitz F, Morris MC, Divita G. Twenty years of cell-penetrating peptides: from molecular mechanisms to therapeutics. Br J Pharmacol. 2009;157(2):195–206.
55. van den Berg A, Dowdy SF. Protein transduction domain delivery of therapeutic macromolecules. Curr Opin Biotechnol. 2011;22(6):888–93.
56. Lee SH, Castagner B, Leroux JC. Is there a future for cell-penetrating peptides in oligonucleotide delivery? Eur J Pharm Biopharm. 2013;85(1):5–11.
57. Gagat M, Zielińska W, Grzanka A. Cell-penetrating peptides and their utility in genome function modifications (review). Int J Mol Med. 2017;40(6):1615–23.
58. Meyer M, Dohmen C, Philipp A, Kiener D, Maiwald G, Scheu C, et al. Synthesis and biological evaluation of a bioresponsive and endosomolytic siRNA-polymer conjugate. Mol Pharm. 2009;6(3):752–62.
59. Alam MR, Ming X, Fisher M, Lackey JG, Rajeev KG, Manoharan M, et al. Multivalent cyclic RGD conjugates for targeted delivery of small interfering RNA. Bioconjug Chem. 2011;22(8):1673–81.
60. Liu X, Wang W, Samarsky D, Liu L, Xu Q, Zhang W, et al. Tumor-targeted in vivo gene silencing via systemic delivery of cRGD-conjugated siRNA. Nucleic Acids Res. 2014;42(18):11805–17.
61. Thirumalai A, Girigoswami K, Pallavi P, Harini K, Gowtham P, Girigoswami A. Cancer therapy with iRGD as a tumor-penetrating peptide. Bull Cancer. 2023;110(12):1288–300.
62. Kesharwani P, Chandra J, Karim S, Gupta G, Karwasra R, Sharma A. αvβ3 integrin targeting RGD peptide-based nanoparticles as an effective strategy for selective drug delivery to tumor microenvironment. J Drug Deliv Sci Technol. 2024;96:105663.
63. Sugahara KN, Teesalu T, Karmali PP, Kotamraju VR, Agemy L, Girard OM, et al. Tissue-penetrating delivery of compounds and nanoparticles into tumors. Cancer Cell. 2009;16(6):510–20.
64. Dubey PK, Mishra V, Jain S, Mahor S, Vyas SP. Liposomes modified with cyclic RGD peptide for tumor targeting. J Drug Target. 2004;12(5):257–64.
65. Weis SM, Cheresh DA. αV integrins in angiogenesis and cancer. Cold Spring Harb Perspect Med. 2011;1(1):a006478.
66. Yin L, Li X, Wang R, Zeng Y, Zeng Z, Xie T. Recent research progress of RGD peptide–modified nanodrug delivery systems in tumor therapy. Int J Pept Res Ther. 2023;29(4):53.
67. Lowman HB, Chen YM, Skelton NJ, Mortensen DL, Tomlinson EE, Sadick MD, et al. Molecular mimics of insulin-like growth factor 1 (IGF-1) for inhibiting IGF-1: IGF-binding protein interactions. Biochemistry. 1998;37(25):8870–8.
68. Koehn S, Schaefer HW, Ludwig M, Haag N, Schubert US, Seyfarth L, et al. Cell-specific RNA interference by peptide-inhibited-peptidase-activated siRNAs. J RNAi Gene Silencing. 2010;6(2):422–30.
69. Varkouhi AK, Scholte M, Storm G, Haisma HJ. Endosomal escape pathways for delivery of biologicals. J Control Release. 2011;151(3):220–8.
70. Sun J, Pohl EE, Krylova OO, Krause E, Agapov II, Tonevitsky AG, et al. Membrane destabilization by ricin. Eur Biophys J. 2004;33(7):572–9.
71. Chen G, Wang Y, Xie R, Gong S. Tumor-targeted pH/redox dual-sensitive unimolecular nanoparticles for efficient siRNA delivery. J Control Release. 2017;259:105–14.

72. Tu Y, Kim JS. A fusogenic segment of glycoprotein H from herpes simplex virus enhances transfection efficiency of cationic liposomes. J Gene Med. 2008;10(6):646–54.
73. Wadia JS, Stan RV, Dowdy SF. Transducible TAT-HA fusogenic peptide enhances escape of TAT-fusion proteins after lipid raft macropinocytosis. Nat Med. 2004;10(3):310–5.
74. Lee YJ, Johnson G, Peltier GC, Pellois JP. A HA2-fusion tag limits the endosomal release of its protein cargo despite causing endosomal lysis. Biochim Biophys Acta. 2011;1810(8):752–8.
75. Barati S, Chegini F, Hurtado P, Rush RA. Hybrid tetanus toxin C fragment-diphtheria toxin translocation domain allows specific gene transfer into PC12 cells. Exp Neurol. 2002;177(1):75–87.
76. Balayssac S, Burlina F, Convert O, Bolbach G, Chassaing G, Lequin O. Comparison of penetratin and other homeodomain-derived cell-penetrating peptides: interaction in a membrane-mimicking environment and cellular uptake efficiency. Biochemistry. 2006;45(5):1408–20.
77. Chiu YL, Ali A, Chu CY, Cao H, Rana TM. Visualizing a correlation between siRNA localization, cellular uptake, and RNAi in living cells. Chem Biol. 2004;11(8):1165–75.
78. Muratovska A, Eccles MR. Conjugate for efficient delivery of short interfering RNA (siRNA) into mammalian cells. FEBS Lett. 2004;558(1–3):63–8.
79. Hsu T, Mitragotri S. Delivery of siRNA and other macromolecules into skin and cells using a peptide enhancer. Proc Natl Acad Sci U S A. 2011;108(38):15816–21.
80. Boeckle S, Wagner E, Ogris M. C-versus N-terminally linked melittin-polyethylenimine conjugates: the site of linkage strongly influences activity of DNA polyplexes. J Gene Med. 2005;7(10):1335–47.
81. El-Andaloussi S, Järver P, Johansson HJ, Langel U. Cargo-dependent cytotoxicity and delivery efficacy of cell-penetrating peptides: a comparative study. Biochem J. 2007;407(2):285–92.
82. Hamil AS, Dowdy SF. Synthesis and conjugation of small interfering ribonucleic neutral SiRNNs. Methods Mol Biol. 2016;1364:1–9.
83. Kolosenko I, Edsbäcker E, Björklund AC, Hamil AS, Goroshchuk O, Grandér D, et al. RNAi prodrugs targeting Plk1 induce specific gene silencing in primary cells from pediatric T-acute lymphoblastic leukemia patients. J Control Release. 2017;261:199–206.
84. Nikam RR, Gore KR. Journey of siRNA: clinical developments and targeted delivery. Nucleic Acid Ther. 2018;28(4):209–24.
85. Spiess M. The asialoglycoprotein receptor: a model for endocytic transport receptors. Biochemistry. 1990;29(43):10009–18.
86. Ray KK, Landmesser U, Leiter LA, Kallend D, Dufour R, Karakas M, et al. Inclisiran in patients at high cardiovascular risk with elevated LDL cholesterol. N Engl J Med. 2017;376(15):1430–40.
87. Willoughby JLS, Chan A, Sehgal A, Butler JS, Nair JK, Racie T, et al. Evaluation of GalNAc-siRNA conjugate activity in pre-clinical animal models with reduced asialoglycoprotein receptor expression. Mol Ther. 2018;26(1):105–14.
88. Xia W, Low PS. Folate-targeted therapies for cancer. J Med Chem. 2010;53(19):6811–24.
89. Parker N, Turk MJ, Westrick E, Lewis JD, Low PS, Leamon CP. Folate receptor expression in carcinomas and normal tissues determined by a quantitative radioligand binding assay. Anal Biochem. 2005;338(2):284–93.
90. Low PS, Henne WA, Doorneweerd DD. Discovery and development of folic-acid-based receptor targeting for imaging and therapy of cancer and inflammatory diseases. Acc Chem Res. 2008;41(1):120–9.
91. Zhou J, Rossi J. Aptamers as targeted therapeutics: current potential and challenges. Nat Rev Drug Discov. 2017;16(6):440.
92. Catuogno S, Esposito CL, Condorelli G, de Franciscis V. Nucleic acids delivering nucleic acids. Adv Drug Deliv Rev. 2018;134:79–93.
93. Unterrainer LM, Calais J, Bander NH. Prostate-specific membrane antigen: gateway to management of advanced prostate cancer. Annu Rev Med. 2024;75:49–66.

94. Ni X, Zhang Y, Ribas J, Chowdhury WH, Castanares M, Zhang Z, et al. Prostate-targeted radiosensitization via aptamer-shRNA chimeras in human tumor xenografts. J Clin Invest. 2011;121(6):2383–90.
95. Yoon V, Fridkis-Hareli M, Munisamy S, Lee J, Anastasiades D, Stevceva L. The GP120 molecule of HIV-1 and its interaction with T cells. Curr Med Chem. 2010;17(8):741–9.
96. Neff CP, Zhou J, Remling L, Kuruvilla J, Zhang J, Li H, et al. An aptamer-siRNA chimera suppresses HIV-1 viral loads and protects from helper CD4(+) T cell decline in humanized mice. Sci Transl Med. 2011;3(66):66ra6.
97. Esposito CL, Nuzzo S, Kumar SA, Rienzo A, Lawrence CL, Pallini R, et al. A combined microRNA-based targeted therapeutic approach to eradicate glioblastoma stem-like cells. J Control Release. 2016;238:43–57.
98. Esposito CL, Nuzzo S, Catuogno S, Romano S, de Nigris F, de Franciscis V. STAT3 gene silencing by aptamer-siRNA chimera as selective therapeutic for glioblastoma. Mol Ther Nucleic Acids. 2018;10:398–411.
99. Ma Y, Kowolik CM, Swiderski PM, Kortylewski M, Yu H, Horne DA, et al. Humanized Lewis-Y specific antibody based delivery of STAT3 siRNA. ACS Chem Biol. 2011;6(9):962–70.
100. Xia CF, Zhang Y, Zhang Y, Boado RJ, Pardridge WM. Intravenous siRNA of brain cancer with receptor targeting and avidin-biotin technology. Pharm Res. 2007;24(12):2309–16.
101. Bäumer S, Bäumer N, Appel N, Terheyden L, Fremerey J, Schelhaas S, et al. Antibody-mediated delivery of anti-KRAS-siRNA in vivo overcomes therapy resistance in colon cancer. Clin Cancer Res. 2015;21(6):1383–94.
102. Sugo T, Terada M, Oikawa T, Miyata K, Nishimura S, Kenjo E, et al. Development of antibody-siRNA conjugate targeted to cardiac and skeletal muscles. J Control Release. 2016;237:1–13.
103. Ibtehaj N, Huda R. High-dose BAFF receptor specific mAb-siRNA conjugate generates Fas-expressing B cells in lymph nodes and high-affinity serum autoantibody in a myasthenia mouse model. Clin Immunol. 2017;176:122–30.
104. Cuellar TL, Barnes D, Nelson C, Tanguay J, Yu SF, Wen X, et al. Systematic evaluation of antibody-mediated siRNA delivery using an industrial platform of THIOMAB-siRNA conjugates. Nucleic Acids Res. 2015;43(2):1189–203.
105. Lee YS, Pressman S, Andress AP, Kim K, White JL, Cassidy JJ, et al. Silencing by small RNAs is linked to endosomal trafficking. Nat Cell Biol. 2009;11(9):1150–6.
106. Nechaev S, Gao C, Moreira D, Swiderski P, Jozwiak A, Kowolik CM, et al. Intracellular processing of immunostimulatory CpG-siRNA: Toll-like receptor 9 facilitates siRNA dicing and endosomal escape. J Control Release. 2013;170(3):307–15.
107. Kortylewski M, Swiderski P, Herrmann A, Wang L, Kowolik C, Kujawski M, et al. In vivo delivery of siRNA to immune cells by conjugation to a TLR9 agonist enhances antitumor immune responses. Nat Biotechnol. 2009;27(10):925–32.
108. Zhang Q, Hossain DM, Nechaev S, Kozlowska A, Zhang W, Liu Y, et al. TLR9-mediated siRNA delivery for targeting of normal and malignant human hematopoietic cells in vivo. Blood. 2013;121(8):1304–15.
109. Hossain DM, Dos Santos C, Zhang Q, Kozlowska A, Liu H, Gao C, et al. Leukemia cell-targeted STAT3 silencing and TLR9 triggering generate systemic antitumor immunity. Blood. 2014;123(1):15–25.
110. Sen M, Thomas SM, Kim S, Yeh JI, Ferris RL, Johnson JT, et al. First-in-human trial of a STAT3 decoy oligonucleotide in head and neck tumors: implications for cancer therapy. Cancer Discov. 2012;2(8):694–705.
111. Zhang Q, Hossain DM, Duttagupta P, Moreira D, Zhao X, Won H, et al. Serum-resistant CpG-STAT3 decoy for targeting survival and immune checkpoint signaling in acute myeloid leukemia. Blood. 2016;127(13):1687–700.
112. Hellmuth I, Freund I, Schlöder J, Seidu-Larry S, Thüring K, Slama K, et al. Bioconjugation of small molecules to RNA impedes its recognition by Toll-like receptor 7. Front Immunol. 2017;8:312.
113. Martin WJ, Grandi P, Marcia M. Screening strategies for identifying RNA- and ribonucleoprotein-targeted compounds. Trends Pharmacol Sci. 2021;42(9):758–71.

# Chapter 23
# RNA Nanotechnology: The Fundamentals

**Abstract** Over the past decade, RNA nanotechnology has witnessed significant advancements, with the synthesis of diverse RNA nanoparticles programmable in shape, size, and stoichiometry for broad nanobiotechnology applications. This progress is driven by breakthroughs in nucleotide chemical modifications that reduce RNA degradation, enhanced understanding of RNA's high thermodynamic stability suitable for in vivo applications, and improved control over its immunogenic properties. Innovations in defining RNA's complex 4D structure and dynamic intermolecular interactions, alongside reduced synthesis costs and demonstrated therapeutic safety and specificity, particularly in cancer treatment, further underscore its potential. This chapter will detail these advancements, illustrating RNA nanoparticles as programmable smart complexes and exploring future research directions in creating complex structures for applications ranging from tissue engineering to computer logic components, marking RNA nanotechnology as a transformative field in both biotechnology and medicine.

## 23.1 Introduction

Over the last decade, the discipline of RNA nanotechnology has witnessed remarkable progress [1–6], with the ability to fold into unique and intricate 3D shapes that are crucial for its diverse functions. This inherent structural versatility is one of the primary reasons why RNA is such an effective material for creating nanoparticles used in drug delivery systems. RNA's ability to self-assemble from the bottom up allows for precise control over the nanoparticle design, ensuring consistent size, structure, and functional properties. This attribute is essential for the high reproducibility required in therapeutic applications [2, 7–11], engaging in base stacking [12, 13], and establishing both canonical Watson–Crick (A–T, G–C) and noncanonical (G–U wobble, sheared G–A pair, G–A imino pair, A–U reverse Hoogsteen) base pairs [14, 15], which contribute to its diverse natural structural motifs [16, 17]. These attributes are leveraged in RNA nanotechnology to develop nanoparticles for nanomedicine and bio-nanotechnology. Moreover, RNA nanoparticles exhibit

 
P. E. Saw, E. Song, *RNA Therapeutics in Human Diseases*,
https://doi.org/10.1007/978-981-96-3041-7_23

well-defined size and structure, which are critical in navigating the complex in vivo environment of the human body. These characteristics facilitate enhanced cellular uptake, improved biodistribution, and reduced immunogenicity, making RNA an advantageous material for medical applications. Additionally, the structural attributes of RNA allow for the encapsulation and precise delivery of therapeutic agents, including small molecules, proteins, and other nucleic acids, directly to targeted cells. The therapeutic potential of RNA is significant, not only because of its role in gene regulation and expression but also due to its ability to interact with cellular machinery in ways that can be harnessed for disease treatment. RNA nanoparticles can be engineered to interfere with viral replication, modulate immune responses, or correct genetic mutations, which provides a versatile platform for treating a wide array of diseases, from infectious diseases to genetic disorders and cancers.

RNA designers can efficiently navigate through an array of RNA secondary and tertiary structural motifs, including bulges, stems, hairpins, loops, and junctions. These elements serve as fundamental components for creating nanoparticles with varied sizes and shapes through the manipulation of motif angles and sequence lengths [18–22]. Additionally, modifications in nucleotide types can modify nanoparticle characteristics, such as enhancing chemical stability through base or backbone alterations, or by adding fluorescent dyes to facilitate tracking. The physicochemical properties of RNA nanoparticles can be precisely adjusted to custom-fit them for specific applications both in vitro and in vivo. This chapter will present recent developments in RNA nanotechnology, highlighting novel technologies used to engineer stable RNA nanoparticles with diverse structures and functions, which is pioneered by Peixuan Guo and his team [1, 21, 23]. Furthermore, we will explore the burgeoning role of RNA in the fields of medicine and nanotechnology.

## 23.2 Definition of RNA Nanotechnology

RNA nanotechnology is a specialized field that focuses on designing, constructing, and manipulating RNA structures at the nanoscale level. This discipline harnesses RNA's unique molecular properties, such as its ability to form stable, intricate three-dimensional shapes through both canonical and noncanonical base pairing, as well as through base stacking and extensive networks of tertiary interactions. The primary building blocks of RNA nanoparticles include a core scaffold that provides structural integrity, targeting ligands that direct the nanoparticles to specific cells or tissues, regulatory elements that control the biological activity and timing of RNA function, and therapeutic modules that carry out desired therapeutic effects.

These RNA-based nanoparticles emulate the fundamental base-pairing mechanisms observed in DNA, offering a reliable framework for self-assembly. However, RNA's versatility surpasses that of DNA by exhibiting a structural adaptability and functional diversity comparable to proteins, which is significantly enhanced by its ability to form noncanonical base pairs and complex tertiary structures. This not only enriches the structural complexity of RNA nanoparticles but also improves their thermodynamic stability, making them exceptionally suitable for various biomedical applications, including targeted drug delivery, gene regulation, and vaccine development. Diverging from traditional RNA biology, which primarily investigates intra-RNA interactions and the structure–function correlations within 2D and 3D RNA formations, RNA nanotechnology is geared toward leveraging RNA's properties to craft architectures aimed at nanomedical applications. It emphasizes inter-RNA interactions and the quaternary structures formed by RNA motifs. Nonetheless, the two fields of RNA research are complementary rather than isolated. RNA nanotechnology frequently harnesses insights from conventional RNA biology; for instance, the use of functional RNAs like ribozymes [24], riboswitches [25], and miRNAs [26], which were initially identified through traditional studies. Furthermore, RNA nanotechnology benefits significantly from the structural motifs identified by RNA structural biologists, applying these discoveries to advance nanoparticle designs.

## 23.3 Properties of RNA Structure and Chemistry Favorable for Nanoparticle Construction

Nanobiotechnology encompasses the manipulation of biological materials at the nanoscale, including nucleic acid nanotechnology, which emerged over three decades ago [27]. Initially, the field leveraged DNA self-assembly, utilizing canonical base-pairing mechanisms (A–T, G–C) to create nanoparticles [28, 29]. The development of DNA nanostructures has marked a significant advancement in the field of nanotechnology, particularly in the creation of drug delivery systems and complex structural designs. Following the initial breakthroughs in DNA nanotechnology, researchers successfully engineered various DNA-based nanostructures such as nanocapsules and other nanocarriers specifically designed for drug delivery. These structures are tailored to encapsulate therapeutic agents effectively, protecting them from degradation while ensuring targeted delivery to specific tissues or cells, thereby enhancing the efficacy and safety of treatments [30–32]. One of the most groundbreaking methods in this arena is DNA origami, which has proven to be a robust technique for constructing intricate nanostructures. DNA origami involves the folding of a long single-stranded DNA, typically from a viral genome, into a predefined structure using a multitude of short staple strands. These staple strands are designed to bind at specific locations on the long strand, guiding it to fold into the desired shape [33], as summarized below.

### *23.3.1 Tetrahedrons*

These are one of the simplest 3D forms, consisting of four triangular faces. DNA tetrahedrons can encapsulate molecules, potentially serving as carriers for drugs or other small molecules in therapeutic applications [34].

### 23.3.2 Nanorobots

Constructed from DNA, these nanorobots can be programmed to open and release their contents only in response to specific molecular signals. This capability makes them highly suitable for targeted drug delivery, where the drug is released only in the vicinity of target cells, such as cancerous cells [35].

### 23.3.3 Helix Bundles

These structures are formed by multiple helices packed together, which can be used to create rigid frameworks for larger nanostructures or to organize functional components at nanoscale proximity, useful in nanoelectronics and photonics.

## 23.4 Structures Based on Tensegrity Principles

Tensegrity structures are composed of components in tension and compression and are known for their mechanical stability. In DNA nanotechnology, these principles are applied to create structures that maintain their integrity under physical stress, making them ideal for constructing robust frameworks in nanomechanical systems [36]. While DNA nanostructures effectively utilize base pairing for structural configurations, RNA surpasses DNA in several crucial aspects, including thermostability and structural and functional versatility. RNA and DNA are both nucleic acids composed of nucleotide subunits connected by phosphodiester bonds, featuring a sugar (ribose in RNA and deoxyribose in DNA), a phosphate group, and a nitrogenous base. In RNA, the bases include cytosine (C), guanine (G), adenine (A), and uracil (U), replacing thymine (T) found in DNA.

RNA's secondary structure is primarily formed through hydrogen bonds between complementary nucleotides (G–C and A–U), akin to the Watson–Crick base pairing seen in DNA. However, RNA's structural complexity is further enriched by base stacking interactions, which significantly contribute to the overall stability of nucleic acid structures. The presence of the 2′-OH group on the ribose sugar is a distinctive feature that not only differentiates RNA from DNA but also imparts unique chemical properties. This hydroxyl group increases RNA's susceptibility to hydrolysis, thus influencing its reactivity and the dynamics of its interactions with other biomolecules. Moreover, the sugar in RNA naturally adopts a C3′-endo conformation, which favors the formation of the A-type helix, characterized by 11 base pairs per turn. This helical structure is more compact and thermodynamically stable than DNA's B-type helix, which has about 10.5 base pairs per turn. The A-type helix in RNA enhances its thermostability, making

RNA structures particularly robust at higher temperatures. This property is pivotal for RNA's function in various cellular environments and under the stress conditions where DNA might denature. Originally identified as the intermediary between genomic DNA and protein synthesis, transferring genetic codes to the cell's translational machinery, RNA's role expanded dramatically with the discovery of its catalytic capabilities through ribozymes [37]. This groundbreaking revelation—previously thought exclusive to proteins—transformed our understanding of RNA's functions. Beyond its structured complexity, RNA also mediates specific interactions through single-stranded, unstructured regions, thereby broadening its functional landscape [38]. Decades of research have illuminated RNA's multifaceted roles in biological systems, spanning catalysis, protein synthesis, and gene regulation.

As we mentioned in previous chapters, RNA-based therapies hold great promise for treating a variety of diseases by targeting specific genetic pathways. However, their systemic delivery is challenging due to several inherent disadvantages that necessitate the use of nanoparticle platforms for effective delivery. First, RNA molecules are inherently unstable in the bloodstream, where they are susceptible to rapid degradation by nucleases. This degradation significantly reduces their half-life and effectiveness. Second, RNA molecules have negative charges due to their phosphate backbone, which leads to poor cellular uptake because of the repulsive interaction with the negatively charged cell membranes. Additionally, the immune system recognizes RNA molecules as foreign, potentially triggering immune responses that can lead to clearance before reaching their target. Nanoparticle platforms protect RNA from degradation, enhance cellular uptake by facilitating endocytosis, and can be engineered to evade the immune system, thereby improving the bioavailability and therapeutic efficacy of RNA-based treatments. Therefore, protection of being a "delivery system" would greatly enhance the stability and translational efficacy of RNA therapeutics. The idea of "cargo-to-carrier" has been reviewed in detailed in [39]. The specifics of RNA-based platform are summarized in Table 23.1.

## 23.5 Chemical Modification of RNA

The intrinsic instability of RNA previously limited its utility as a material for constructing therapeutics. Although the size and structure of RNA nanoparticles confer some resistance to nucleases, this protection is inadequate for in vivo applications. To bolster the stability of RNA-based therapeutics, researchers have devised a series of strategies that modify the RNA molecule at multiple levels, addressing its inherent vulnerabilities. These chemical modifications are essential for enhancing the stability and efficacy of RNA molecules when deployed in therapeutic applications, particularly in clinical settings where RNA's natural instability could limit its usefulness.

**Table 23.1** The purpose of using nanoparticle system for RNA therapies

| | |
|---|---|
| Negative charge of RNA nanoparticles repels cell entry | Numerous nanoparticle systems utilize passive targeting mechanisms, notably the EPR effect [40, 41]. These systems often employ positively charged nanoparticles that interact with the negatively charged cell membranes to facilitate cellular entry through fusion and electrostatic interactions. However, the positive charge on these nanoparticles frequently results in toxicity and unintended off-target effects [41, 42]. In contrast, RNA nanoparticles are characterized by their negatively charged phosphate backbones, making them highly anionic and less prone to nonspecific interactions. This polyanionic nature significantly enhances hydration and reduces the likelihood of protein corona formation, which can interfere with targeted delivery |
| RNA nanoparticles harbor multiple functionalities | RNA nanoparticles are crafted using functional RNA elements such as RNA aptamers, ribozymes, riboswitches, siRNA, and various ncRNAs. These components are effortlessly attached to the core sequences of the RNA scaffold, showcasing the inherent modularity of RNA nanoparticles. This modularity facilitates the self-assembly of these elements into multifunctional structures. Moreover, each helical branch of the RNA motif can be adorned with different functional units. For example, the pRNA-3WJ scaffold [43] is designed to accommodate up to three distinct functional modules. The inherent thermodynamic stability of the 3WJ scaffold ensures proper folding and functional integrity of the incorporated RNA modules [44]. Furthermore, chemical ligands such as fluorescent dyes, chemotherapeutic agents, or biotin can be straightforwardly attached to one of the RNA strands using established chemical conjugation methods [45]. This allows for the integration of multiple functionalities—targeting, therapeutic, and tracking—within a single nanoparticle. Alternatively, for enhanced or synergistic effects, multiple units with the same function, like identical siRNAs, or different siRNAs targeting various genes, can be incorporated into the same nanoparticle. In the context of medical applications, meticulous characterization of the nanoparticle's payload is crucial. RNA nanoparticles are assembled from the bottom up, allowing for precise control over every aspect of their design—from the selection of size, shape, and oligomers to the choice of functional units and tracking molecules |
| RNA nanoparticles possess favorable PK and PD profiles | The PK/PD profiles of nanoparticles are crucial attributes, second only to their therapeutic efficacy. The controlled synthesis of RNA nanoparticles results in uniform assemblies with consistent size and shape distributions. This consistency ensures reproducible PK/PD characteristics, facilitating systematic investigations into how size and shape affect PK/PD profiles and enabling some predictability in their behavior. Such uniformity in RNA nanoparticle assembly is advantageous for streamlining FDA approval as they advance toward clinical application. A critical consideration for in vivo applications is the thermodynamic stability of RNA nanoparticles. Typically, these nanoparticles require metal ions in millimolar concentrations for assembly due to intermolecular interactions. However, this limitation has been overcome with developments such as the pRNA-3WJ, which can assemble without metal ions, withstands denaturation by urea, and remains stable even at very low concentrations [43]. RNA nanoparticles are designed with enhanced in vivo longevity; those modified with nucleotides like 2′-F typically exhibit an extended half-life of about 5–12 h [46], compared to just 0.25–0.75 h for unmodified siRNA [47]. After systemic administration in models such as orthotopic and subcutaneous xenografts, or in mice with metastatic tumors, pRNA-3WJ nanoparticles demonstrate precise targeting of cancer cells with minimal or no accumulation in healthy organs and tissues [43, 46, 48–53] |

(continued)

**Table 23.1** (continued)

| | |
|---|---|
| RNA nanoparticle size and shape tune immunogenic potency for use in immunotherapy | Immune reactions serve as the body's defense mechanisms. Cancer cells can rapidly mutate, often developing resistance to protein therapies after multiple treatments. While traditional chemotherapy targets a single pathway of tumor proliferation, leading to the development of chemoresistant tumors, combination therapies and repeated treatments can help overcome this resistance. RNA nanoparticles, benefiting from their extended circulation time, do not induce antibody production as protein therapies do, making them viable for repeated treatments of chronic diseases. However, RNA nanoparticles can be engineered to be highly immunogenic by incorporating CpG DNA, a recognized FDA-approved immunological adjuvant [54–56]. The immunostimulatory effects of these particles vary depending on the shape of the RNA structures and the amount of CpG per structure, demonstrating that RNA polygons with different shapes coupled with CpG can significantly enhance cancer immunotherapy [57]. Chemical modifications significantly mitigate the immunostimulatory properties of RNA, reducing activation from various receptors. For instance, siRNA constructs modified with 2′-F or 2′-*O*-me exhibit minimal cytokine or interferon induction compared to their unmodified counterparts [58]. Remarkably, modifications such as 2′-F- or 2′-*O*-me-modified U nucleotides effectively prevent immune-related off-target effects, including those dependent on and independent of TLR pathways [59, 60]. Additionally, modifying the siRNA's terminal ends with LNA blocks interferon-α immunostimulatory activity while maintaining robust gene silencing effects [61]. Similarly, modifications like pseudouracil or 2-thiouracil eliminate RIG-1-mediated immune responses, particularly from RNA with a 5′-triphosphate group [59]. These chemical alterations serve dual purposes: They prevent undesired immune stimulation and enhance therapeutic efficacy, making them crucial for advancing cancer treatment and managing chronic viral infections |

**Box 23.1 pRNA-3WJ Scaffold**

The pioneer of the pRNA-3WJ (phi29 RNA three-way junction) scaffold is Peixuan Guo's research team. pRNA-3WJ is a critical component in the field of RNA nanotechnology, derived from the bacteriophage phi29 DNA packaging motor [43]. This RNA structure is naturally involved in the packaging of DNA into the phage's capsid and has been harnessed in nanotechnology due to its unique structural and functional properties. The pRNA-3WJ is characterized by its three-way junction structure, where three RNA strands come together to form a highly stable and specific geometric configuration. This structure is inherently stable, maintaining its integrity even under varying environmental conditions, such as changes in temperature and pH. The robustness of the 3WJ structure is largely attributed to the noncanonical interactions and tertiary contacts that stabilize the junction. One of the remarkable features of the pRNA-3WJ scaffold is its modularity. Each arm of the junction can be engineered to include additional functional elements without disrupting the core structure. This modularity allows for the attachment of various functional groups, such as siRNA, miRNA, ribozymes, aptamers, or other molecules that can target specific cells, deliver drugs, or modulate biological

(continued)

**Box 23.1** (continued)
pathways. The pRNA-3WJ scaffold has been utilized in several major key applications, including targeted drug delivery, gene regulation (the scaffold can be designed to carry siRNA or miRNA sequences that can specifically knock down the expression of target genes), imaging and diagnostics where modified pRNA-3WJ structures can be conjugated with fluorescent markers or other imaging agents. They can be used to track the distribution and localization of RNA nanoparticles in vivo, providing insights into their biodistribution and pharmacokinetics. They can also be used as a platform for vaccine development, where the intrinsic immunogenic properties of RNA make the pRNA-3WJ scaffold a potential candidate for vaccine development.

The use of pRNA-3WJ in nanotechnology offers several advantages, including increased biocompatibility and biodegradability. As a naturally occurring molecule, RNA is biocompatible and degradable within the body, reducing the risk of toxicity. The ability to attach a wide variety of functional molecules to the scaffold allows for multifunctional nanoparticles capable of performing several tasks simultaneously. The structural stability of the pRNA-3WJ ensures that the scaffold remains intact in the biological environment, providing a reliable framework for therapeutic applications. Overall, the pRNA-3WJ scaffold is a powerful tool in RNA nanotechnology, offering a versatile and stable platform for a wide range of biomedical applications. Its ability to be easily modified and functionalized makes it an invaluable asset in the development of targeted therapies, diagnostics, and potentially new vaccines.

**Sugar Modifications** Among the most common strategies for enhancing RNA stability is the modification of the sugar moiety within nucleotides, which significantly improves both thermal [62] and enzymatic resilience without compromising the formation of the A-form RNA helix, essential for RNAi mechanisms. While the 2′-OH group is not crucial for the RNA-induced silencing complex, modifications at this position allow siRNA to maintain its gene silencing function [63]. Modifications such as 2′-fluorine (2′-F), 2′-*O*-methyl (2′-*O*-Me), and 2′-amine (2′-$NH_2$) are relatively small and do not significantly alter the duplex's ability to fold properly [64, 65]. In contrast, bulkier substitutions like 2′-*O*-methoxyethyl (2′-*O*-MOE) are typically used only as termini modifiers because their size can disrupt RNA folding when used internally. Locked nucleic acid (LNA) involves a structural constraint of the ribose sugar, locking it into the A-form helix (3′-endo) configuration through a methylene bridge linking the 2′-oxygen and 4′-carbon [66]. Although extensive LNA modification can alter RNA folding, only a minimal number of modifications are generally required to achieve the desired stability enhancement.

**Backbone Modifications** Modifying the linkage between bases enhances RNA's resistance to nuclease activity. A common approach involves phosphorothioate modifications, where a sulfur atom replaces a nonbridging oxygen in the phosphodiester backbone [67]. This alteration not only boosts the thermal stability of RNA but also enhances the efficacy of siRNAs when placed strategically within the siRNA sequence. Phosphorothioate bonds have also facilitated the creation of a library of DNA aptamers, known as X-aptamers, which are employed for cell-specific targeting [68].

**Boranophosphate Substitution** Another modification, the boranophosphate substitution, replaces an oxygen atom in the phosphate backbone with borane, leading to improved siRNA performance and markedly greater serum stability compared to unmodified RNA [69]. A more drastic modification involves using peptide nucleic acids (PNAs), where the traditional negatively charged phosphate backbone is replaced by a neutrally charged amide linkage, enhancing the binding affinity to complementary RNA or DNA strands. PNAs are particularly resistant to degradation as their unique structure is not recognized by either nucleases or proteases.

**Synthetic Nucleic Acid Mimics** Bifacial peptide nucleic acids (bPNA) represent a cutting-edge development in the field of nucleic acid analogues, designed to enhance interactions with both DNA and RNA strands. These novel molecules are constructed with an α-peptide backbone, which differs significantly from the natural sugar-phosphate backbone of DNA and RNA, providing several biochemical advantages, including increased stability and resistance to enzymatic degradation. The core structure of bPNAs includes triaminotriazine (melamine) bases, which are engineered to interact specifically with the Watson–Crick faces of T or U. This unique capability allows bPNAs to form a stable triplex structure with T/U-rich DNA or RNA sequences. Triplex formation involves the binding of bPNA to the major groove of a pre-existing nucleic acid duplex, establishing hydrogen bonds with the Hoogsteen faces of the purine bases while the existing Watson–Crick base pairing remains intact. This interaction facilitates a stable, yet dynamic complex that can significantly influence the structure and function of the target nucleic acids [70–72]. This triplex formation is versatile, compatible with various backbones including peptides [73, 74], peptoids [75], and polyacrylates [70]. The application of bPoNAs in cellular systems, such as HEK293 and MCF7 cells, demonstrates their potential in both passive and active targeted drug delivery. Passive uptake relies on the natural cellular processes of endocytosis, where cells nonspecifically engulf nanoparticles. Conversely, targeted ligand-mediated entry involves the modification of bPoNAs with specific ligands that can bind to cell surface receptors, facilitating receptor-mediated endocytosis. A notable example of the efficacy of bPoNAs in targeted gene silencing involves their use in HeLa-Luc cells, which express both firefly and renilla luciferase. By combining bPoNA with cholesterol-modified siRNA targeting firefly luciferase, researchers achieved targeted delivery into the cells and a significant suppression of luciferase expression, demonstrating up to 40% reduction. This suppression indicates effective siRNA delivery and gene knockdown, validating the potential of bPoNAs in gene therapy applications [70–72].

**Base Modifications** Base modifications in RNA serve as crucial tools in enhancing the chemical stability, therapeutic efficacy, and functional versatility of RNA molecules, particularly in therapeutic applications. These modifications not only improve RNA's resistance to enzymatic degradation but can also influence its interaction with proteins and other cellular components, thereby altering its biological activities.

1. **Substitution of Uracil with Halogenated Analogs**
   5-Bromouracil and 5-Iodouracil: These are halogenated analogs of uracil that can enhance the chemical stability of RNA and are used to study RNA dynamics and structure. They are particularly useful in probing the mechanisms of RNA-mediated biological processes because they can be incorporated into RNA sequences in place of uracil without drastically altering the RNA's overall structure.
2. **Substitution of Adenine**
   Diaminopurine: Replacing adenine with diaminopurine increases hydrogen bonding potential, thereby enhancing the stability of RNA duplexes. This can be particularly beneficial in therapeutic RNA structures, such as siRNA, where increased duplex stability can enhance the efficacy of gene silencing.
3. **Thiouracil Modifications**
   2-Thiouracil and 4-Thiouracil: The incorporation of sulfur atoms in place of oxygen in the uracil base can significantly increase the photostability and nuclease resistance of RNA. These modifications are valuable in various RNA-based therapies and in fluorescence studies due to their enhanced photophysical properties.
4. **C-linked Pseudouracil**
   C-linked Pseudouracil: This modification involves altering the bonding structure in the uracil base, potentially leading to changes in the RNA's interaction with proteins and other molecules. Such modifications can be utilized to fine-tune the structural and functional properties of RNA for specific therapeutic applications.
5. **Methylation of Pyrimidines**
   5-Methylation of Pyrimidines: Methylation typically improves the stability of RNA molecules by making them less prone to enzymatic degradation. It can also affect how RNA interacts with other biomolecules, impacting its role in gene expression and regulation.

6. **Nonaromatic Base Modification**
   Dihydrouracil: This nonaromatic base can confer unique structural properties to RNA, potentially enhancing its flexibility and facilitating interactions that are different from those of standard RNA. This can be utilized to modulate the activity of RNA molecules in a controlled manner.

While these modifications generally enhance the chemical stability of RNA, their impact on RNA's biochemical functionality varies greatly depending on where the modifications are located within the RNA sequence. For example, modifications near active sites or binding regions can alter RNA's interaction with other molecules, influencing its role in biological processes [64].

## 23.6 Structural and Functional Implications

The overarching aim of chemically modifying RNA is to enhance its resistance to nucleases, thereby prolonging its circulation time in vivo and improving its therapeutic effectiveness. Over the last decade, additional structural and functional benefits of these chemical modifications have emerged, as detailed below.

**Modulation of Thermal Stability** One key advantage of RNA nanoparticles as a delivery system is the control over in vivo circulation time. For instance, replacing an RNA strand with DNA in the construction of RNA nanoparticles reduces thermal stability, whereas substituting unmodified RNA with 2′-F increases it [62]. This ability to fine-tune the circulation time of the delivery vector is particularly valuable in various therapeutic contexts. In some applications, it is advantageous for RNA nanoparticles to exhibit enhanced thermal stability. Extensive research has established a hierarchy of thermodynamic stability as follows: LNA/LNA > LNA/2′-F-RNA > 2′-F-RNA/2′-F-RNA > 2′-F-RNA/RNA > RNA/RNA > RNA/DNA > DNA/DNA [1, 62, 76–78]. By employing different thermodynamic properties within a single RNA nanoparticle, designers can create sophisticated delivery and drug release mechanisms. A practical application of this principle was demonstrated in a recent study where an 8 nucleotide (nt) LNA-modified RNA fragment, complementary to the "seed region" of a miRNA [50, 79], was integrated into RNA nanoparticles. Once delivered into the cytosol, this 8 nt LNA fragment effectively competed with and bound to the corresponding 8 nt seed region of the target miRNA, inhibiting its function [50]. This example underscores how RNA nanoparticles can be engineered to achieve specific biological outcomes by leveraging the unique properties conferred by chemical modifications.

**Box 23.2 Locked Nucleic Acid (LNA) in RNA Nanotechnology**

Locked nucleic acid (LNA) modifications represent a powerful strategy for enhancing the thermal stability and binding affinity of RNA molecules, making them especially useful in therapeutic applications such as gene silencing and regulation. The integration of LNA-modified RNA fragments into RNA nanoparticles, as demonstrated in the cited study, illustrates a practical application of this technology in modulating the function of miRNAs within cells. LNA is a modified RNA nucleotide with a methylene bridge connecting the 2′-oxygen and the 4′-carbon of the ribose ring. These bridging locks the ribose in the 3′-endo conformation, which is characteristic of the A-form geometry of RNA. This conformational restriction enhances the hybridization properties of the LNA molecule, increasing its thermal stability and affinity for complementary RNA sequences. These properties make LNA a potent tool for targeting specific RNA molecules in biological systems.

In the specific study referenced, an 8-nucleotide LNA-modified RNA fragment was designed to be complementary to the seed region of a target miRNA. The seed region of a miRNA (typically nucleotides 2–7 from the 5′ end) is crucial for its binding to target mRNAs, guiding the RISC to specific mRNAs to inhibit their expression. By designing an LNA-modified RNA that is complementary to this seed region, researchers can effectively create a competitor that binds more tightly and stably to the seed region than the natural targets of the miRNA. Once this LNA-modified fragment is incorporated into an RNA nanoparticle and delivered into the cytosol, the LNA fragment competes with the natural mRNA targets for binding to the miRNA seed region. Due to its enhanced thermal stability and binding affinity, the LNA fragment can effectively outcompete the natural mRNA targets. By binding to the miRNA, the LNA fragment prevents the miRNA from associating with its natural targets, thereby inhibiting its function. This disruption of miRNA activity can lead to the upregulation of genes that are typically downregulated by the target miRNA, offering a mechanism to control gene expression therapeutically. This approach has significant therapeutic implications. It allows for the precise modulation of gene expression by targeting specific miRNAs involved in disease processes. Diseases where miRNA activity is dysregulated, such as certain cancers, viral infections, and genetic disorders, could potentially be treated by selectively inhibiting problematic miRNAs using LNA-modified RNA fragments delivered via RNA nanoparticles. The use of LNA-modified RNA fragments within RNA nanoparticles exemplifies a sophisticated application of nucleic acid chemistry and nanotechnology to control and modulate cellular functions at the molecular level.

**Modulation of Serum Stability** The use of RNA in therapeutic applications, especially as a scaffold for nanoparticles, faces a major challenge due to its inherent instability in biological environments like blood serum. RNA molecules are naturally susceptible to degradation by nucleases present in the serum, which can rapidly degrade unmodified RNA, typically within minutes after introduction into the bloodstream. This rapid degradation limits the therapeutic potential of RNA-based treatments because it reduces the effective concentration of the RNA and shortens its functional lifespan in the body. To address this issue, significant efforts have been made to chemically modify RNA to enhance its stability without compromising its functionality. One of the most effective modifications is the introduction of a 2′-fluoro (2′-F) group on the ribose sugar of the RNA nucleotide. This modification has been shown to greatly increase the resistance of RNA to nuclease degradation. The 2′-F modification involves replacing the hydroxyl group at the 2′-position of the ribose sugar with a fluorine atom. The presence of the fluorine atom introduces steric hindrance, which can inhibit the action of nucleases that would typically cleave the phosphodiester backbone of the RNA. Fluorine atoms contribute to the hydrophobic character of the modified nucleotide, which can further deter water-soluble nucleases and reduce hydrolytic degradation. Crucially, the 2′-F modification does not significantly alter the ability of the RNA to form proper Watson–Crick base pairs with complementary RNA or DNA strands, preserving the fundamental properties necessary for the RNA's biological function [65]. Furthermore, alterations to the backbone connectivity of RNA, such as replacing traditional phosphodiester bonds with phosphorothioate, boranophosphate, or amide linkages (as in peptide nucleic acids or PNAs), can markedly increase the circulation time of RNA in a degradative serum environment [64]. This capacity to adjust stability in serum through strategic modifications is demonstrated in the construction of RNA nanosquares. By varying the ratios of different oligonucleotides, such as substituting DNA and 2′-F modified oligonucleotides, researchers can either destabilize or stabilize the RNA nanosquares [22]. This approach allows for precise tuning of nanoparticle stability to meet specific therapeutic needs, enhancing their efficacy and viability in biomedical applications.

**Increase in Potency** The cellular environment and the nature of the target molecule play significant roles in the effectiveness of RNA therapeutics. Different cell types may exhibit varying levels of accessibility to RNA molecules due to differences in membrane composition, endocytic activity, and the presence of specific receptors. Similarly, the target mRNA's location within the cell, its abundance, and its turnover rate can affect how effectively an RNA therapeutic can modulate gene expression. The sequence of the RNA therapeutic must be meticulously designed to ensure specific and efficient binding to the target mRNA, minimizing off-target effects. Chemical modifications are employed to enhance the pharmacological attributes of RNA molecules, addressing challenges like nuclease degradation, immune recognition, and off-target effects. In a specific study, an siRNA construct that incorporated both 2′-O-Me and 2′-F modifications demonstrated a potency that was 500 times greater than its unmodified counterpart. This dramatic increase in efficacy can be attributed to the synergistic effect of these modifications, which enhance the siRNA's stability and binding affinity while reducing its immunogenicity and susceptibility to degradation [80]. This substantial increase in potency highlights the significant impact that strategic chemical modifications can have on the therapeutic effectiveness of RNA. This approach involves the use of a low-affinity dihydrouracil base at the 3′-end of the sense strand paired with a high-affinity 2′-thiouracil base at the 3′-end of the antisense strand [81]. This suggests that even subtle changes in the chemical structure of RNA therapeutics can drastically alter their biological interactions and enhance their overall therapeutic impact. These findings emphasize the importance of carefully considered chemical modifications in the development of effective RNA-based treatments.

## 23.7 Creation of RNA Nanoparticles: The Details

The development of RNA nanoparticles involves an intricate, multi-step process beginning with the initial concept phase. At this stage, the specific attributes of the nanoparticles are defined, considering their structural configuration and the intended functional applications. Next, computational methods are utilized to predict the folding patterns of the individual components and to evaluate how interactions among the RNA molecules influence the ultimate quaternary structure. Following this design phase, the monomeric units are synthesized either through in vitro transcription using RNA polymerase or by chemical synthesis methods. These units are then assembled into the desired quaternary structures using either templated or spontaneous assembly techniques, adhering to precise stoichiometric proportions. After assembly, the RNA nanoparticles are subject to thorough characterization to ensure their correct folding and to confirm that they meet the initially specified structural and functional criteria. Techniques such as atomic force microscopy (AFM), electron microscopy (EM), gel electrophoresis, and chromatography are employed for this purpose. Upon successful validation, these RNA nanoparticles are applied in diverse fields including therapeutic interventions, diagnostics, and cellular function regulation. In this special issue of the journal *Methods*, esteemed experts provide an in-depth examination of each of these six critical stages in the realm of RNA nanotechnology, sharing their insights and recent advancements.

### *23.7.1 Designing the Structure and Function of RNA Nanoparticles*

The construction of RNA nanoparticles commences with the strategic design of building blocks that are engineered to self-assemble into specific quaternary structures. These building blocks are meticulously designed to include various functional components, such as chemical moieties, therapeutic modules, and other functional elements necessary for their intended applications. This design process is critical as it determines the ultimate functionality and targeting capabilities of the RNA nanoparticles. By precisely controlling the composition and configuration of these building blocks, researchers can ensure that the final assembled structures possess the desired properties and functionalities needed for specific biomedical and technological applications [1]. This initial stage involves developing a comprehensive design strategy that defines both the structural configuration and functional goals of the nanoparticles. Key insights into this process are provided in reviews by several leading researchers:

(a) Shu et al. explore the innovative use of pRNA derived from the bacteriophage phi29 DNA packaging motor for constructing chimeric pRNA constructs. These multivalent constructs are meticulously engineered to facilitate the targeted delivery of various therapeutic agents, including siRNA and ribozymes. By harnessing the unique assembly properties of pRNA, the team is able to design nanoparticles that can carry multiple therapeutic molecules simultaneously. This approach not only enhances the efficiency of delivery but also allows for the co-delivery of different types of therapeutic agents, potentially increasing therapeutic efficacy and specificity. The specificity and versatility of these pRNA-based delivery systems hold great promise for precision medicine, offering targeted therapeutic strategies that can be customized to individual patient needs and conditions [82].
(b) Dua et al. delve into the SELEX process, an advanced technique for isolating aptamers that are highly specific to disease-associated membrane proteins on cell surfaces. This methodology involves iterative rounds of selection, where nucleic acid sequences are enriched that bind with high affinity to target molecules. The end result is a collection of aptamers, short single-stranded RNA or DNA molecules, which can specifically bind to and inhibit the function of cellular targets. This approach is particularly useful in identifying potential therapeutic agents that can interact with proteins typically found in pathological states, thereby providing avenues for targeted therapy [83].
(c) Zhou et al. discuss their research on using the phi29 pRNA system for delivering a combination therapy against HIV. Their work involves a dual inhibitory anti-gp120 aptamer–siRNA chimera, where both components are engineered to suppress HIV activity effectively. This innovative delivery system utilizes the pRNA of the phi29 bacteriophage to transport the therapeutic agents directly into the infected cells, where the aptamer and siRNA work synergistically to inhibit HIV by blocking gp120, a glycoprotein essential for the virus's ability to infect host cells. This study exemplifies a strategic approach to viral inhibition that could be adapted for other viral infections [84].
(d) Ishikawa et al. review the significance of natural and synthetic GNRA/receptor interacting modules, which are integral to the structure and function of catalytic RNAs and riboswitches. These modules are pivotal in facilitating long-range tertiary interactions that are essential for the proper folding and functionality of RNA molecules. The review highlights how these interactions contribute to the complex dynamics of RNA structures, influencing their regulatory and catalytic activities. Understanding these modules enhances our knowledge of RNA machinery and opens up potential for designing RNA-based therapeutics and regulatory systems [85].

### 23.7.2 Computational Modeling to Predict RNA Secondary Structure and Folding

RNA's natural propensity to fold into complex structures through self-assembly is enabled by both standard and non-standard base pairings, along with intricate tertiary interactions like pseudoknots, kissing-loops, and stem stacking. This complex folding process, driven by a variety of interaction types, poses significant challenges in accurately predicting RNA structures. According to the RNA 2D prediction program developed by Zuker, experimental validations indicate that these predictions achieve about 70% accuracy for 2D structures. However, the accuracy drops further when it comes to predicting more complex 3D and 4D structures. This variability underscores the intricate nature of RNA architecture and the difficulties encountered in modeling its dynamic conformational changes [1].

Kasprzak and colleagues explore a range of computational methods used for designing RNA nanostructures, exemplified by the tectosquare—a nanostructure whose design and function have been experimentally validated. Critical to these designs are RNA junctions, which include elements like internal and multibranch loops, as well as kissing-loops. These junctions are sourced from specialized resources such as the RNAJunction database, serving as essential structural connectors that facilitate the assembly of larger RNA nanoconstructs with precise geometric attributes. To effectively design and optimize these complex RNA structures, advanced software tools are employed. RNA2D3D and NanoTiler are notable examples of such tools, enabling the transition from two-dimensional to three-dimensional conceptualizations of RNA structures. These tools help researchers visualize and iteratively refine the RNA constructs, ensuring that the designed models are both structurally viable and functionally capable. Additionally, molecular dynamics (MD) simulations play a pivotal role in this process, providing detailed insights into the dynamic behavior of RNA nanostructures under various physiological conditions. By simulating interactions at the atomic level, MD helps predict how nanoconstructs will behave in real-world biological environments, enhancing the predictability and reliability of RNA-based designs for practical applications. This integrated approach, combining RNA structural elements with sophisticated modeling and simulation tools, is critical for advancing the field of RNA nanotechnology [86]. These methodologies are also applicable for the design and evaluation of modular nanostructures incorporating a variety of natural RNA motifs, enhancing the repertoire of RNA structural elements available for nanoengineering. Additionally, interactions such as loop/loop contacts, tertiary architectural contacts, and specialized motifs like three-way or four-way junctions are crucial for the design of RNA nanoparticles. A wealth of information housed in well-established RNA structure databases provides a valuable resource for extracting known RNA motifs and structural units, which can then be repurposed for the development of innovative RNA nanoparticles tailored to specific functional requirements.

### 23.7.3 Synthesis and Modification of RNA Building Blocks for Nanoparticle Construction

The fabrication of functional RNA nanoparticles entails multiple sophisticated techniques, including the synthesis, conjugation, and sometimes crosslinking of various modules. Additionally, labeling each subunit and chemically modifying nucleotides are critical steps. These modifications are indispensable for adding site-specific labels or reactive groups that facilitate further labeling or conjugation with other biomolecules. These complex processes can be conducted through synthetic or enzymatic methods, tailored to meet the requirements for specificity and functionality in the RNA nanoparticles.

Paredes and colleagues provide a comprehensive review of the methodologies employed in RNA labeling, conjugation, and ligation. These techniques are essential for several key aspects of RNA nanoparticle research and application. They enable the visualization of RNA constructs, which is crucial for monitoring their interaction within biological systems. Moreover, these methods aid in the structural analysis of RNA, allowing researchers to understand and optimize the stability and function of RNA nanoparticles. Additionally, they assist in localizing RNA constructs within biological contexts and determining their biodistribution, which is vital for assessing their therapeutic potential and safety [45].

Among the array of techniques for labeling RNA, enzymatic methods stand out for their precision and versatility, particularly when employing enzymes like T7 RNA polymerase. This enzyme is instrumental for labeling RNA at the 5′- or 3′-ends with modified nucleotides either in vitro or in vivo. Laing and colleagues have refined this method to specifically optimize the conjugation of ligands, such as folate, to the 5′-end of AMP. This scalable strategy involves several key steps. Initially, the AMP is modified to include a folate group, after which it is purified with high yield using boronate affinity gel chromatography—a technique that exploits the affinity of boronate for cis-diols commonly found in nucleotide sugars. The purified folate-AMP conjugate is then incorporated at the 5′-end of RNA during in vitro transcription, a process facilitated by T7 RNA polymerase [87]. Additionally, Ponchon and Dardel outline contemporary strategies for in vivo RNA expression aimed at producing recombinant, stable RNA [88]. These methodologies contribute significantly to the development of RNA nanoparticles by providing robust and versatile ways to produce and modify RNA building blocks tailored to specific needs in nanomedicine and biotechnology.

### 23.7.4 RNA Nanoparticles: Assembly

The construction of RNA nanoparticles through self-assembly, whether templated or non-templated, relies on utilizing well-defined and predictable building blocks [1]. Templated assembly refers to the organization of RNA molecules under specific external influences, such as force, structural guidance, or spatial constraints that dictate the assembly process. In contrast, non-templated assembly allows the spontaneous formation of larger structures from individual components without any external scaffolding or influence. In this special issue, the non-templated self-assembly method is extensively reviewed by Shu et al. [82], Dua et al. [83], Zhou et al. [84], and Ponchon and Dardel [88]. Notably, both Shu et al. [82] and Zhou et al. [84] discuss the use of pRNA from the bacteriophage phi29 DNA packaging motor, which naturally forms a hexameric RNA ring that functions within the viral assembly mechanism. This pRNA has been ingeniously reengineered to create various oligomeric forms such as dimers, trimers, tetramers, hexamers, and arrays through specific interactions like hand-in-hand or foot-to-foot binding between interlocking loops [89]. These reconfigured pRNA structures serve as versatile platforms for delivering multiple therapeutic agents and constructing RNA arrays, demonstrating the utility of RNA nanoparticles as polyvalent vehicles in nanomedicine. Ponchon and Dardel [88] provide insights into recent advancements in in vivo RNA expression strategies that involve embedding the RNA of interest within the highly conserved frameworks of tRNA and 5S rRNA. This method effectively camouflages the therapeutic RNA, enhancing its stability and expression in biological environments. Such innovative approaches in RNA nanoparticle assembly are critical for the development of effective therapeutic delivery systems and for advancing the field of RNA nanotechnology.

### 23.7.5 RNA Nanoparticles: Characterization

Accurate characterization plays a pivotal role in confirming the structural integrity and functional properties of RNA nanoparticles after their assembly. Among the various techniques available, AFM stands out as a particularly effective method for this purpose. AFM offers the unique advantage of being able to provide high-resolution images of the surface topology of biomolecules, including the structure and dynamics of nano-assemblies and arrays composed of DNA or RNA [90]. A critical aspect of using AFM is the immobilization of the sample on a substrate for imaging. This step is essential because it directly affects the clarity and quality of the images obtained. For effective imaging using techniques like atomic force microscopy (AFM), the substrate on which nucleic acids are deposited plays a crucial role. The choice and preparation of these substrates are essential to ensure that the nucleic acids, such as RNA nanoparticles, are not only stably anchored but also retain their native conformations during imaging. This often requires significant modifications to the surface of the substrates. The ability to visualize these complex nanostructures provides invaluable insights into their geometric configuration and aids in assessing their potential for applications in fields such as targeted drug delivery and diagnostic imaging. The detailed structural data gathered through AFM can also guide improvements in nanoparticle design and optimization, ensuring that the nanoparticles function as intended in biological systems.

### 23.7.6 *RNA Nanoparticles: Biomedical Applications*

The utility of RNA nanoparticles in medical applications is well-demonstrated by the work of Zhou et al. [84], Shu et al. [82], and Dua et al. [83], who have explored the potential of these nanoparticles in diagnosing and treating a variety of diseases. Their research has shown promising results in the development of RNA-based therapeutic agents for several cancers, including prostate, lung, ovarian, brain, breast, and pancreatic cancers, as well as for viral infections such as hepatitis and HIV. The novel roles of RNA nanoparticles in nanomedicine are multifaceted. RNA nanoparticles can be designed to specifically bind to certain cell types, making them excellent tools for targeted diagnostic applications. Leveraging receptor-mediated endocytosis, RNA nanoparticles can deliver therapeutic molecules directly into specific types of cells, enhancing the efficacy and reducing side effects of treatments. Through mechanisms such as gene silencing and regulation, RNA nanoparticles can control gene expression within cells, offering a way to manipulate biological pathways and cellular behavior in disease treatment. RNA nanoparticles can be engineered to cross nuclear membranes, facilitating the delivery of therapeutic agents directly to cell nuclei where they can alter genetic material. One of the most promising capabilities of RNA nanoparticles is their potential to traverse the blood–brain barrier, a critical challenge in treating neurological disorders and brain cancers.

## 23.8 Most Recent Approaches in RNA Nanoparticle Construction

### 23.8.1 *Precise Size Control of RNA Nanoparticles*

The phi29 pRNA-3WJ (three-way junction) scaffold has recently emerged as a groundbreaking tool for the precise control of RNA nanoparticle sizes, demonstrating remarkable versatility and robustness in the construction of complex RNA polygons. This method capitalizes on the inherent 60° inner angle of the 3WJ scaffold to create various geometric shapes through a multi-strand assembly process. Initially, the natural configuration of the pRNA-3WJ, with its 60° angle, is ideally suited for forming equilateral triangles. This feature allows for the straightforward assembly of triangle-shaped RNA nanoparticles, a process that involves linking three 3WJ junctions to form the triangle's three corners. To expand the geometric possibilities, adjustments can be made to the inner angle of the 3WJ scaffold. By introducing an additional external strand and lengthening the inner strand, the inner angle can be manipulated to reach 90°, which is perfect for constructing square geometries. This adjustment is crucial for creating four-sided polygons, allowing for the formation of squares without compromising the structural integrity or thermal stability of the RNA scaffold. Further versatility is demonstrated by modifying the 3WJ to form pentagon geometries. This involves additional angle modifications to accommodate the five sides of a pentagon, showcasing the scaffold's adaptability in forming multiple-sided structures. One of the notable strengths of this approach is that these modifications do not affect the thermal stability of the resulting RNA polygons, maintaining their functional integrity under varying conditions. RNA squares, for example, have been designed and successfully assembled with precise edge lengths of 5, 10, and 20 nm. The uniformity and accuracy of these dimensions were confirmed using advanced characterization techniques such as atomic force microscopy (AFM) and dynamic light scattering, which demonstrated their consistent size distributions [20–22].

### 23.8.2 Construction of Unique 3D RNA Architectures and Containers

**RNA Tetrahedrons** The development of 3D RNA nanoparticles, particularly RNA tetrahedrons, represents a significant advancement in RNA nanotechnology, extending from previous 2D architectural designs. RNA tetrahedrons represent a fascinating class of RNA nanostructures that are defined by their geometric precision and robust architecture. Consisting of four triangular faces and six edges, these tetrahedral structures exhibit exceptional mechanical rigidity and structural stability, attributes that are highly prized in the realms of nanomedicine and nanomaterials. The mechanical rigidity of RNA tetrahedrons comes from their closed 3D structure, which inherently resists deformation under stress, making them more durable in physiological environments compared to linear or less structured RNA forms. This rigidity ensures that the tetrahedrons maintain their shape and functionality when navigating the dynamic and often harsh conditions of the human body. One of the noteworthy applications of RNA tetrahedrons is their use in targeted cancer therapy. These structures have been successfully employed to target triple-negative breast cancer cells by integrating an EGFR aptamer into the tetrahedron structure. This design enables specific binding to cancer cells, potentially reducing the side effects associated with nonspecific distribution of therapeutic agents [91].

**RNA Nanoprisms** Further expanding the utility of 3D nanostructures, RNA nanoprisms have been designed to encapsulate small molecule drugs, enhancing their stability and efficacy. For example, two sizes of nanoprisms have been constructed to protect and stabilize RNA aptamers that bind to fluorogenic malachite green molecules. In this configuration, the binding of malachite green to its aptamer results in a significant increase in fluorescence, which is further enhanced by the nanoprism. The smaller of the two nanoprisms effectively prevents the degradation of the aptamer by blocking the access of degrading enzymes through steric hindrance, nearly doubling the fluorescent half-life of the malachite green dye [18]. This encapsulation strategy not only protects the integrity of the small molecule drugs but also minimizes the leakage of potentially toxic cancer drugs to healthy cells.

This advancement is particularly crucial in the development of safer and more effective therapeutic delivery systems, where the selective targeting and controlled release of drugs can significantly improve treatment outcomes for cancer patients. The ability of RNA nanoparticles to be engineered into precise 3D structures, coupled with their biocompatibility and degradability, positions them as powerful tools in the advancement of nanomedicine, offering innovative solutions for drug delivery and disease treatment.

**Dendrimer-Like RNA Architectures** The phi29 pRNA-3WJ motif has been utilized to create highly branched, dendrimer-like RNA architectures. The construction process culminates in the formation of generation-4 (G4) RNA dendrimers. These advanced structures are notable for their uniform size of approximately 65 nm and their globular three-dimensional shape, which together ensure consistent performance in biological applications. A key feature of these dendrimers is their exact stoichiometry, with each dendrimer featuring 32 terminal units. This precise architecture is crucial for their functionality, particularly in therapeutic contexts. Furthermore, these RNA dendrimers can be densely loaded with various imaging agents, significantly enhancing the resolution of targeted cells or tumors when visualized via modern imaging techniques such as MRI or SPECT/PET. This attribute makes G4 RNA dendrimers particularly valuable in the field of medical imaging, providing detailed and precise visualization that is critical for both diagnostic and therapeutic applications [92].

## 23.9 Conclusion

To summarize the attributes of RNA molecules, RNA can fold into a diverse array of intricate three-dimensional shapes and complex structures, which are critical in various biological functions. This structural versatility enables the design of RNA nanoparticles that can mimic natural biomolecules or create entirely new molecular structures with tailored functionalities. RNA strands can hybridize or anneal at relatively low free-energy costs compared to other nucleic acids like DNA. This feature facilitates easier and more efficient formation of stable duplexes or complexes, essential for the reliable assembly of functional nanoparticles. RNA sequences can be synthesized and modified with high precision using current biochemical technologies. Modifications can include the addition of fluorophores, ligands, or other functional groups that enhance the therapeutic or diagnostic capabilities of RNA nanoparticles. The ability to predict and design the folding patterns of RNA allows for precise control over the final structure of the nanoparticles. This control is crucial for ensuring that they perform their intended functions, whether it is targeting specific cells, delivering drugs, or regulating gene expression. RNA can self-assemble into predefined structures under appropriate conditions, a property that simplifies the manufacturing process and can be harnessed to produce large quantities of uniform nanoparticles. This self-assembly is driven by the natural propensity of RNA to form base pairs and various secondary structures such as loops, bulges, and junctions. These attributes collectively make RNA an excellent candidate for developing sophisticated nano-tools. The flexibility to design and construct RNA nanoparticles with specific shapes and functions opens up vast possibilities in medicine, such as targeted drug delivery systems, therapeutic gene regulation, and advanced diagnostics. As such, RNA nanoparticles stand at the forefront of innovative approaches to tackle complex challenges in health care and biotechnology.

## References

1. Guo P. The emerging field of RNA nanotechnology. Nat Nanotechnol. 2010;5(12):833–42.
2. Guo P, Zhang C, Chen C, Garver K, Trottier M. Inter-RNA interaction of phage phi29 pRNA to form a hexameric complex for viral DNA transportation. Mol Cell. 1998;2(1):149–55.
3. Guo S, Tschammer N, Mohammed S, Guo P. Specific delivery of therapeutic RNAs to cancer cells via the dimerization mechanism of phi29 motor pRNA. Hum Gene Ther. 2005;16(9):1097–109.
4. Delebecque CJ, Lindner AB, Silver PA, Aldaye FA. Organization of intracellular reactions with rationally designed RNA assemblies. Science. 2011;333(6041):470–4.
5. Lee JB, Hong J, Bonner DK, Poon Z, Hammond PT. Self-assembled RNA interference microsponges for efficient siRNA delivery. Nat Mater. 2012;11(4):316–22.
6. Zuo H, Wu S, Li M, Li Y, Jiang W, Mao C. A case study of the likes and dislikes of DNA and RNA in self-assembly. Angew Chem Int Ed Engl. 2015;54(50):15118–21.
7. Laurenti E, Barde I, Verp S, Offner S, Wilson A, Quenneville S, et al. Inducible gene and shRNA expression in resident hematopoietic stem cells in vivo. Stem Cells. 2010;28(8):1390–8.
8. Hoeprich S, Zhou Q, Guo S, Shu D, Qi G, Wang Y, et al. Bacterial virus phi29 pRNA as a hammerhead ribozyme escort to destroy hepatitis B virus. Gene Ther. 2003;10(15):1258–67.
9. Chang KY, Tinoco I Jr. Characterization of a "kissing" hairpin complex derived from the human immunodeficiency virus genome. Proc Natl Acad Sci U S A. 1994;91(18):8705–9.
10. Bindewald E, Hayes R, Yingling YG, Kasprzak W, Shapiro BA. RNAJunction: a database of RNA junctions and kissing loops for three-dimensional structural analysis and nanodesign. Nucleic Acids Res. 2008;36(Database issue):D392–7.
11. Wagner C, Ehresmann C, Ehresmann B, Brunel C. Mechanism of dimerization of bicoid mRNA: initiation and stabilization. J Biol Chem. 2004;279(6):4560–9.
12. Sugimoto N, Nakano S, Katoh M, Matsumura A, Nakamuta H, Ohmichi T, et al. Thermodynamic parameters to predict stability of RNA/DNA hybrid duplexes. Biochemistry. 1995;34(35):11211–6.
13. Searle MS, Williams DH. On the stability of nucleic acid structures in solution: enthalpy-entropy compensations, internal rotations and reversibility. Nucleic Acids Res. 1993;21(9):2051–6.
14. Lemieux S, Major F. RNA canonical and non-canonical base pairing types: a recognition method and complete repertoire. Nucleic Acids Res. 2002;30(19):4250–63.
15. Leontis NB, Westhof E. Geometric nomenclature and classification of RNA base pairs. RNA. 2001;7(4):499–512.
16. Leontis NB, Lescoute A, Westhof E. The building blocks and motifs of RNA architecture. Curr Opin Struct Biol. 2006;16(3):279–87.
17. Leontis NB, Westhof E. Analysis of RNA motifs. Curr Opin Struct Biol. 2003;13(3):300–8.
18. Khisamutdinov EF, Jasinski DL, Li H, Zhang K, Chiu W, Guo P. Fabrication of RNA 3D nanoprisms for loading and protection of small RNAs and model drugs. Adv Mater. 2016;28(45):10079–87.

19. Khisamutdinov EF, Bui MN, Jasinski D, Zhao Z, Cui Z, Guo P. Simple method for constructing RNA triangle, square, pentagon by tuning interior RNA 3WJ angle from 60° to 90° or 108°. Methods Mol Biol. 2015;1316:181–93.
20. Khisamutdinov EF, Jasinski DL, Guo P. RNA as a boiling-resistant anionic polymer material to build robust structures with defined shape and stoichiometry. ACS Nano. 2014;8(5):4771–81.
21. Khisamutdinov EF, Li H, Jasinski DL, Chen J, Fu J, Guo P. Enhancing immunomodulation on innate immunity by shape transition among RNA triangle, square and pentagon nanovehicles. Nucleic Acids Res. 2014;42(15):9996–10004.
22. Jasinski DL, Khisamutdinov EF, Lyubchenko YL, Guo P. Physicochemically tunable polyfunctionalized RNA square architecture with fluorogenic and ribozymatic properties. ACS Nano. 2014;8(8):7620–9.
23. Guo S, Li H, Ma M, Fu J, Dong Y, Guo P. Size, shape, and sequence-dependent immunogenicity of RNA nanoparticles. Mol Ther Nucleic Acids. 2017;9:399–408.
24. Scott WG. Ribozymes. Curr Opin Struct Biol. 2007;17(3):280–6.
25. Breaker RR. Riboswitches and the RNA world. Cold Spring Harb Perspect Biol. 2012;4(2):a003566.
26. Ha M, Kim VN. Regulation of microRNA biogenesis. Nat Rev Mol Cell Biol. 2014;15(8):509–24.
27. Seeman NC. Nucleic acid junctions and lattices. J Theor Biol. 1982;99(2):237–47.
28. Jones MR, Seeman NC, Mirkin CA. Nanomaterials. Programmable materials and the nature of the DNA bond. Science. 2015;347(6224):1260901.
29. Seeman NC. Nanomaterials based on DNA. Annu Rev Biochem. 2010;79:65–87.
30. Andersen ES, Dong M, Nielsen MM, Jahn K, Subramani R, Mamdouh W, et al. Self-assembly of a nanoscale DNA box with a controllable lid. Nature. 2009;459(7243):73–6.
31. Goodman RP, Heilemann M, Doose S, Erben CM, Kapanidis AN, Turberfield AJ. Reconfigurable, braced, three-dimensional DNA nanostructures. Nat Nanotechnol. 2008;3(2):93–6.
32. Yang YR, Liu Y, Yan H. DNA nanostructures as programmable biomolecular scaffolds. Bioconjug Chem. 2015;26(8):1381–95.
33. Rothemund PW. Folding DNA to create nanoscale shapes and patterns. Nature. 2006;440(7082):297–302.
34. Ke Y, Sharma J, Liu M, Jahn K, Liu Y, Yan H. Scaffolded DNA origami of a DNA tetrahedron molecular container. Nano Lett. 2009;9(6):2445–7.
35. Douglas SM, Bachelet I, Church GM. A logic-gated nanorobot for targeted transport of molecular payloads. Science. 2012;335(6070):831–4.
36. Liedl T, Högberg B, Tytell J, Ingber DE, Shih WM. Self-assembly of three-dimensional prestressed tensegrity structures from DNA. Nat Nanotechnol. 2010;5(7):520–4.
37. Zaug AJ, Grabowski PJ, Cech TR. Autocatalytic cyclization of an excised intervening sequence RNA is a cleavage-ligation reaction. Nature. 1983;301(5901):578–83.
38. Jády BE, Kiss T. A small nucleolar guide RNA functions both in 2′-O-ribose methylation and pseudouridylation of the U5 spliceosomal RNA. EMBO J. 2001;20(3):541–51.
39. Saw PE, Lee S, Jon S. Naturally occurring bioactive compound-derived nanoparticles for biomedical applications. Adv Ther. 2019;2(5):1800146.
40. Matsumura Y, Maeda H. A new concept for macromolecular therapeutics in cancer chemotherapy: mechanism of tumoritropic accumulation of proteins and the antitumor agent smancs. Cancer Res. 1986;46(12 Pt 1):6387–92.
41. Prabhakar U, Maeda H, Jain RK, Sevick-Muraca EM, Zamboni W, Farokhzad OC, et al. Challenges and key considerations of the enhanced permeability and retention effect for nanomedicine drug delivery in oncology. Cancer Res. 2013;73(8):2412–7.
42. Lammers T, Kiessling F, Hennink WE, Storm G. Drug targeting to tumors: principles, pitfalls and (pre-) clinical progress. J Control Release. 2012;161(2):175–87.

43. Shu D, Shu Y, Haque F, Abdelmawla S, Guo P. Thermodynamically stable RNA three-way junction for constructing multifunctional nanoparticles for delivery of therapeutics. Nat Nanotechnol. 2011;6(10):658–67.
44. Shu D, Khisamutdinov EF, Zhang L, Guo P. Programmable folding of fusion RNA in vivo and in vitro driven by pRNA 3WJ motif of phi29 DNA packaging motor. Nucleic Acids Res. 2014;42(2):e10.
45. Paredes E, Evans M, Das SR. RNA labeling, conjugation and ligation. Methods. 2011;54(2):251–9.
46. Abdelmawla S, Guo S, Zhang L, Pulukuri SM, Patankar P, Conley P, et al. Pharmacological characterization of chemically synthesized monomeric phi29 pRNA nanoparticles for systemic delivery. Mol Ther. 2011;19(7):1312–22.
47. Behlke MA. Progress towards in vivo use of siRNAs. Mol Ther. 2006;13(4):644–70.
48. Haque F, Shu D, Shu Y, Shlyakhtenko LS, Rychahou PG, Evers BM, et al. Ultrastable synergistic tetravalent RNA nanoparticles for targeting to cancers. Nano Today. 2012;7(4):245–57.
49. Shu Y, Haque F, Shu D, Li W, Zhu Z, Kotb M, et al. Fabrication of 14 different RNA nanoparticles for specific tumor targeting without accumulation in normal organs. RNA. 2013;19(6):767–77.
50. Shu D, Li H, Shu Y, Xiong G, Carson WE 3rd, Haque F, et al. Systemic delivery of anti-miRNA for suppression of triple negative breast cancer utilizing RNA nanotechnology. ACS Nano. 2015;9(10):9731–40.
51. Cui D, Zhang C, Liu B, Shu Y, Du T, Shu D, et al. Regression of gastric cancer by systemic injection of RNA nanoparticles carrying both ligand and siRNA. Sci Rep. 2015;5:10726.
52. Rychahou P, Haque F, Shu Y, Zaytseva Y, Weiss HL, Lee EY, et al. Delivery of RNA nanoparticles into colorectal cancer metastases following systemic administration. ACS Nano. 2015;9(2):1108–16.
53. Lee TJ, Haque F, Shu D, Yoo JY, Li H, Yokel RA, et al. RNA nanoparticle as a vector for targeted siRNA delivery into glioblastoma mouse model. Oncotarget. 2015;6(17):14766–76.
54. Krieg AM. CpG motifs in bacterial DNA and their immune effects. Annu Rev Immunol. 2002;20:709–60.
55. Krieg AM. Therapeutic potential of Toll-like receptor 9 activation. Nat Rev Drug Discov. 2006;5(6):471–84.
56. Vollmer J, Krieg AM. Immunotherapeutic applications of CpG oligodeoxynucleotide TLR9 agonists. Adv Drug Deliv Rev. 2009;61(3):195–204.
57. Vanneman M, Dranoff G. Combining immunotherapy and targeted therapies in cancer treatment. Nat Rev Cancer. 2012;12(4):237–51.
58. Morrissey DV, Lockridge JA, Shaw L, Blanchard K, Jensen K, Breen W, et al. Potent and persistent in vivo anti-HBV activity of chemically modified siRNAs. Nat Biotechnol. 2005;23(8):1002–7.
59. Hornung V, Ellegast J, Kim S, Brzózka K, Jung A, Kato H, et al. 5′-Triphosphate RNA is the ligand for RIG-I. Science. 2006;314(5801):994–7.
60. Cekaite L, Furset G, Hovig E, Sioud M. Gene expression analysis in blood cells in response to unmodified and 2′-modified siRNAs reveals TLR-dependent and independent effects. J Mol Biol. 2007;365(1):90–108.
61. Hornung V, Guenthner-Biller M, Bourquin C, Ablasser A, Schlee M, Uematsu S, et al. Sequence-specific potent induction of IFN-alpha by short interfering RNA in plasmacytoid dendritic cells through TLR7. Nat Med. 2005;11(3):263–70.
62. Binzel DW, Khisamutdinov EF, Guo P. Entropy-driven one-step formation of Phi29 pRNA 3WJ from three RNA fragments. Biochemistry. 2014;53(14):2221–31.
63. Behlke MA. Chemical modification of siRNAs for in vivo use. Oligonucleotides. 2008;18(4):305–19.
64. Watts JK, Deleavey GF, Damha MJ. Chemically modified siRNA: tools and applications. Drug Discov Today. 2008;13(19–20):842–55.

65. Liu J, Guo S, Cinier M, Shlyakhtenko LS, Shu Y, Chen C, et al. Fabrication of stable and RNase-resistant RNA nanoparticles active in gearing the nanomotors for viral DNA packaging. ACS Nano. 2011;5(1):237–46.
66. Elmén J, Thonberg H, Ljungberg K, Frieden M, Westergaard M, Xu Y, et al. Locked nucleic acid (LNA) mediated improvements in siRNA stability and functionality. Nucleic Acids Res. 2005;33(1):439–47.
67. Eckstein F. Phosphorothioates, essential components of therapeutic oligonucleotides. Nucleic Acid Ther. 2014;24(6):374–87.
68. Somasunderam A, Thiviyanathan V, Tanaka T, Li X, Neerathilingam M, Lokesh GL, et al. Combinatorial selection of DNA thioaptamers targeted to the HA binding domain of human CD44. Biochemistry. 2010;49(42):9106–12.
69. Hall AH, Wan J, Shaughnessy EE, Ramsay Shaw B, Alexander KA. RNA interference using boranophosphate siRNAs: structure-activity relationships. Nucleic Acids Res. 2004;32(20):5991–6000.
70. Zhou Z, Xia X, Bong D. Synthetic polymer hybridization with DNA and RNA directs nanoparticle loading, silencing delivery, and aptamer function. J Am Chem Soc. 2015;137(28):8920–3.
71. Xia X, Piao X, Bong D. Bifacial peptide nucleic acid as an allosteric switch for aptamer and ribozyme function. J Am Chem Soc. 2014;136(20):7265–8.
72. Piao X, Xia X, Mao J, Bong D. Peptide ligation and RNA cleavage via an abiotic template interface. J Am Chem Soc. 2015;137(11):3751–4.
73. Zeng Y, Pratumyot Y, Piao X, Bong D. Discrete assembly of synthetic peptide-DNA triplex structures from polyvalent melamine-thymine bifacial recognition. J Am Chem Soc. 2012;134(2):832–5.
74. Piao X, Xia X, Bong D. Bifacial peptide nucleic acid directs cooperative folding and assembly of binary, ternary, and quaternary DNA complexes. Biochemistry. 2013;52(37):6313–23.
75. Mao J, Bong D. Synthesis of DNA-binding peptoids. Synlett. 2015;26:1581–5.
76. Bondensgaard K, Petersen M, Singh SK, Rajwanshi VK, Kumar R, Wengel J, et al. Structural studies of LNA:RNA duplexes by NMR: conformations and implications for RNase H activity. Chemistry. 2000;6(15):2687–95.
77. Kaur H, Arora A, Wengel J, Maiti S. Thermodynamic, counterion, and hydration effects for the incorporation of locked nucleic acid nucleotides into DNA duplexes. Biochemistry. 2006;45(23):7347–55.
78. Kurreck J, Wyszko E, Gillen C, Erdmann VA. Design of antisense oligonucleotides stabilized by locked nucleic acids. Nucleic Acids Res. 2002;30(9):1911–8.
79. Obad S, dos Santos CO, Petri A, Heidenblad M, Broom O, Ruse C, et al. Silencing of microRNA families by seed-targeting tiny LNAs. Nat Genet. 2011;43(4):371–8.
80. Allerson CR, Sioufi N, Jarres R, Prakash TP, Naik N, Berdeja A, et al. Fully 2′-modified oligonucleotide duplexes with improved in vitro potency and stability compared to unmodified small interfering RNA. J Med Chem. 2005;48(4):901–4.
81. Sipa K, Sochacka E, Kazmierczak-Baranska J, Maszewska M, Janicka M, Nowak G, et al. Effect of base modifications on structure, thermodynamic stability, and gene silencing activity of short interfering RNA. RNA. 2007;13(8):1301–16.
82. Shu Y, Cinier M, Shu D, Guo P. Assembly of multifunctional phi29 pRNA nanoparticles for specific delivery of siRNA and other therapeutics to targeted cells. Methods. 2011;54(2):204–14.
83. Dua P, Kim S, Lee DK. Nucleic acid aptamers targeting cell-surface proteins. Methods. 2011;54(2):215–25.
84. Zhou J, Shu Y, Guo P, Smith DD, Rossi JJ. Dual functional RNA nanoparticles containing phi29 motor pRNA and anti-gp120 aptamer for cell-type specific delivery and HIV-1 inhibition. Methods. 2011;54(2):284–94.
85. Ishikawa J, Fujita Y, Maeda Y, Furuta H, Ikawa Y. GNRA/receptor interacting modules: versatile modular units for natural and artificial RNA architectures. Methods. 2011;54(2):226–38.

86. Kasprzak W, Bindewald E, Kim TJ, Jaeger L, Shapiro BA. Use of RNA structure flexibility data in nanostructure modeling. Methods. 2011;54(2):239–50.
87. Laing BM, Guo P, Bergstrom DE. Optimized method for the synthesis and purification of adenosine—folic acid conjugates for use as transcription initiators in the preparation of modified RNA. Methods. 2011;54(2):260–6.
88. Ponchon L, Dardel F. Large scale expression and purification of recombinant RNA in Escherichia coli. Methods. 2011;54(2):267–73.
89. Shu D, Moll WD, Deng Z, Mao C, Guo P. Bottom-up assembly of RNA arrays and superstructures as potential parts in nanotechnology. Nano Lett. 2004;4(9):1717–23.
90. Lyubchenko YL, Shlyakhtenko LS, Ando T. Imaging of nucleic acids with atomic force microscopy. Methods. 2011;54(2):274–83.
91. Li H, Zhang K, Pi F, Guo S, Shlyakhtenko L, Chiu W, et al. Controllable self-assembly of RNA tetrahedrons with precise shape and size for cancer targeting. Adv Mater. 2016;28(34):7501–7.
92. Sharma A, Haque F, Pi F, Shlyakhtenko LS, Evers BM, Guo P. Controllable self-assembly of RNA dendrimers. Nanomedicine. 2016;12(3):835–44.

# Chapter 24
# RNA Nanotechnology: Biomedical Application

**Abstract** Recent developments in nanotechnology have highlighted the innovative use of RNA molecules in techniques such as replication, molding, and embossing, enhancing the diversity and resolution achievable in nanomaterial fabrication. The inherent adaptability of RNA allows for the creation of complex, programmable nanostructures through 3D fabrication, reversible self-assembly, and self-repair capabilities, transforming its application in integrated biological nanofabrication. With a potential library of 100-nucleotide RNA sequences yielding up to $1.6 \times 10^{60}$ distinct structures, RNA surpasses the complexity and diversity of many existing nanoscale systems. These advancements have also propelled RNA nanoparticles into emerging applications such as pathogen detection, where their ability to specifically recognize viral or bacterial sequences is critical; drug and gene delivery, leveraging their structural versatility for targeted therapeutic delivery; and direct therapeutic applications, particularly in cancer therapy to selectively target and modulate malignant cells. This chapter builds on previous discussions of RNA structure and function, integrating recent breakthroughs that establish RNA nanoparticles as a cornerstone of future nanotechnological and medical solutions.

## 24.1 Introduction

Recent advancements in nanotechnology have led to the creation of various nanoparticle platforms such as liposomes [1–3], polymers [4, 5], viral nanoparticle [6, 7], dendrimers [8–11], and inorganic nanoparticles [9, 12]. These platforms, all sub-micron in size, are promising for enhancing therapeutic approaches [13, 14]. A major challenge in the clinical application of therapeutic nanoparticles, including those made from RNA, is their tendency to accumulate non-specifically in essential healthy organs such as the liver, lungs, kidneys, and spleen. This nonspecific accumulation significantly diminishes the efficiency of targeted drug delivery, as a lower proportion of nanoparticles reach the intended disease-affected cells. Additionally, this can greatly increase the risk of toxicity and side effects for patients, which are critical concerns in



P. E. Saw, E. Song, *RNA Therapeutics in Human Diseases*,
https://doi.org/10.1007/978-981-96-3041-7_24

treatment safety and effectiveness. Variability in size, shape, and surface properties of nanoparticles can lead to inconsistent behavior within the body. This heterogeneity can affect how nanoparticles are distributed, absorbed, metabolized, and excreted, all of which can contribute to undesirable accumulation in healthy organs. Nanoparticles can aggregate, forming larger clumps, or dissociate into smaller components. Both phenomena can alter the intended function of the nanoparticles, potentially leading to reduced efficacy or increased toxicity. Aggregated particles might be more readily captured by the body's reticuloendothelial system (RES), particularly in organs like the liver and spleen, while dissociated particles may become more reactive and toxic. Effective nanoparticle therapies require optimal pharmacokinetics and pharmacodynamics. Inadequate PK/PD profiles can lead to rapid clearance from the bloodstream, inadequate tissue penetration, or excessive accumulation in non-target tissues [15]. Therapeutic nanoparticles must often navigate complex biological barriers, such as the blood–brain barrier or the dense extracellular matrix of tumors. Failure to effectively cross these barriers can prevent the nanoparticles from reaching their target cells within the tumor microenvironment, limiting their therapeutic potential and increasing the likelihood of off-target effects [16–18]. Moreover, several significant barriers hinder the broader clinical adoption of nanoparticle-based therapies, which include the high cost of production, unstable thermodynamic and chemical properties, and the lack of controlled-release mechanisms. These challenges play a crucial role in limiting the practicality and effectiveness of these advanced therapeutic technologies [19]. The intricate compositions and varied functionalities of nanocarriers, which often combine elements like inorganic/organic scaffolds, RNAi mechanisms, protein-targeting antibodies, and chemical drugs, present substantial challenges in gaining regulatory approvals. The multifaceted nature of these nanocarriers increases the complexity of their evaluation due to the broad range of interactions and behaviors they exhibit in biological systems [16, 20]. Nonetheless, recent research indicates that RNA nanoparticles can potentially address many of these challenges, facilitating significant advancements in nanomedicine for applications in cancer therapy [21–29], viral infections [30–32], and eye diseases [33]. Figure 24.1 depicts current delivery methods for RNA (example is focused on mRNA).

## 24.2 Application of RNA Biochemical Properties in RNA Nanotechnology

### *24.2.1 Applying RNA Aptamers for Targeted Therapeutic Delivery or as Potent Inhibitors*

The strategic delivery of nanoparticles plays a pivotal role in minimizing off-target toxicity and preventing undesired buildup in healthy tissues—challenges typically associated with passive delivery systems that depend solely on the EPR effects. This

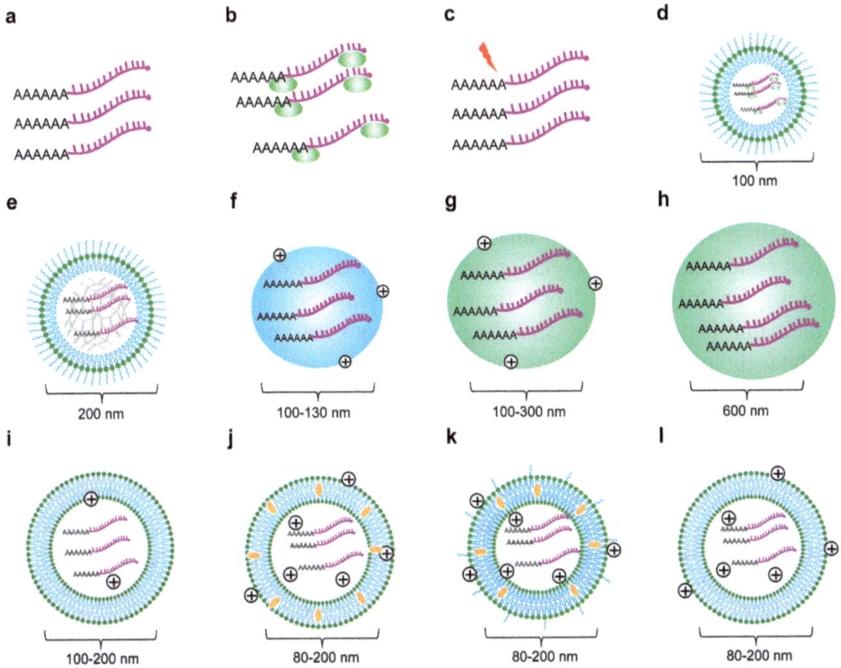


**Fig. 24.1** RNA delivery methods, illustrating various RNA delivery methods used to enhance the stability and cellular uptake of RNA molecules. (**a**) Naked mRNA: RNA molecules are delivered without any carrier, relying on direct cellular uptake. (**b**) Protamine: RNA is complexed with protamine, a positively charged protein that enhances RNA stability and cellular delivery. (**c**) Electroporation: Electrical pulses create temporary pores in the cell membrane, allowing RNA molecules to enter cells. (**d**) Protamine Liposome: RNA–protamine complexes are encapsulated within liposomes, providing an additional protective lipid bilayer. (**e**) Modified Dendrimer Nanoparticle: RNA is encapsulated within dendrimer nanoparticles, offering a highly branched polymeric structure for delivery. (**f**) Cationic Nanoemulsion: RNA is incorporated into cationic nanoemulsions, which are stabilized lipid droplets with positive charges. (**g**) Cationic Polymer: RNA is complexed with cationic polymers, forming nanoparticles through electrostatic interactions. (**h**) Polysaccharide Particle: RNA is encapsulated within polysaccharide-based particles, providing a biodegradable carrier. (**i**) Cationic Polymer Liposome: RNA is delivered via liposomes composed of cationic polymers, enhancing cellular interaction and uptake. (**j**) Cationic Lipid, Cholesterol Nanoparticle: RNA is encapsulated within nanoparticles made from cationic lipids and cholesterol, enhancing stability and delivery efficiency. (**k**) Cationic Lipid, Cholesterol, and PEG Nanoparticle: RNA is delivered using nanoparticles that include cationic lipids, cholesterol, and polyethylene glycol (PEG) for improved circulation time and reduced immunogenicity. (**l**) Cationic Lipid Nanoparticle: RNA is encapsulated within nanoparticles composed solely of cationic lipids, facilitating efficient delivery to cells. Each method offers unique advantages in terms of RNA protection, stability, and delivery efficiency, contributing to the development of effective RNA-based therapeutics. (Figure is adapted from [34] with permission from Springer Nature (License number: 5902231235843))

**Box 24.1 Anti-PEG Antibodies Due to PEGylation in Nanoparticle System: A Clinical Dilemma**

The practice of PEGylation, which involves the modification of pharmaceutical substances with polyethylene glycol (PEG), a flexible and hydrophilic polymer, has become a widely adopted strategy in drug delivery systems. This technique enhances the pharmacological properties of drugs, particularly protein-based therapies, by attaching PEG molecules to them through covalent bonds. Initially, PEGylation was developed to extend the circulation time of protein drugs in the bloodstream, improving their solubility and reducing their immunogenicity by shielding them from the immune system [35–37]. The technique of PEGylation, initially applied to protein drugs, was later discovered to be equally beneficial for enhancing the properties of lipid nanoparticles (LNPs). By attaching PEG to the surface of LNPs, a significant improvement in their surface characteristics was observed. This modification primarily contributes to providing steric hindrance, a physical barrier that prevents other molecules from coming into close proximity to the nanoparticle surface [38–41]. PEGylated lipids also prevent LNP aggregation, enhancing their overall stability. The circulatory half-life of LNPs is influenced by various PEG characteristics such as chain length and density on the LNP surface, and stable LNPs can be crafted by fine-tuning these factors [42]. This enhancement in circulation time, due to the steric stabilization provided by PEGylation, also increases the accumulation of LNPs in cancerous tissues via the enhanced permeation and retention (EPR) effect, boosting their therapeutic efficacy [43]. Consequently, PEGylation has emerged as a pivotal technique in the creation of advanced drug delivery systems, markedly improving their circulation duration, stability, and pharmacokinetics [44, 45].

While the integration of PEG conjugates has proven effective in prolonging circulation kinetics and enhancing pharmacokinetics, unexpected immune responses have also been observed with PEG-conjugate-bearing nanocarriers. Notably, a phenomenon known as accelerated blood clearance (ABC) occurs, where PEGylated nanocarriers are rapidly cleared from the bloodstream upon repeated administration. This response is initiated by the production of anti-PEG antibodies following the first injection, leading to faster clearance in subsequent injections [46–49]. In addition to the benefits of PEGylation, there are notable safety concerns associated with this technology, particularly the risk of hypersensitivity reactions. One such reaction is known as complement activation-related pseudoallergy (CARPA), which poses significant challenges to the safety and effectiveness of PEGylated nanocarriers. CARPA is an acute immune response that is triggered by the activation of the body's complement system—a part of the immune system that enhances (complements) the ability of antibodies and phagocytic cells to clear pathogens and

(continued)

**Box 24.1** (continued)
damaged cells [50–52]. These immunogenic and adverse reactions to PEGylated nanocarriers raise concerns regarding the clinical application of PEGylated therapies. Increased recognition of the potential clinical implications of anti-PEG antibodies has prompted the FDA to mandate the assessment of anti-PEG antibody responses in new drug applications involving PEG [53]. Exploration into viable alternatives to PEG is ongoing, although results have been mixed. To identify a suitable substitute, collaborative efforts between polymer chemists and drug delivery experts are essential to design, synthesize, and evaluate effective alternatives to PEG [54].

It is important to recognize that humans are frequently exposed to PEG through everyday items, including cosmetics, household products, and common medications like laxatives, as well as food packaging. This routine exposure may partially explain why PEG-specific antibodies are found in the blood of otherwise healthy individuals, suggesting a form of "immunization." Developing strategies to mitigate the immunogenicity of PEGylated nanocarriers while preserving their therapeutic efficacy is essential for the progression of this advantageous drug delivery technology. Achieving this requires a comprehensive understanding of the relationship between the structural characteristics of PEG-lipids and the immune-related adverse effects observed with PEGylated LNPs. Additionally, it is crucial to comprehend their biophysical interactions and physiological behaviors. Insights into these aspects are vital for informing the development of pharmaceutical formulations that are designed to minimize adverse immune reactions while enhancing the safety and effectiveness of PEGylated drugs. Such strategically designed formulations would not only improve therapeutic outcomes but also address the significant challenge of immunogenicity, which currently hinders the broader clinical adoption of these drug delivery systems. This approach promises to refine the integration of PEGylated nanocarriers in clinical settings, ensuring they deliver their full potential without compromising patient safety.

issue is significant because while EPR-based systems take advantage of abnormal vascular architectures in diseased tissues to enhance drug accumulation, they lack the specificity required to completely avoid healthy cells, often leading to systemic side effects. In the world of targeted therapies, a notable advancement involves the utilization of RNA aptamers. These are single-stranded RNA sequences designed to fold into unique three-dimensional structures, enabling them to selectively and robustly bind to specific extracellular domains of cell surface receptors. The binding affinity and specificity of RNA aptamers are derived from a combination of molecular interactions: electrostatic interactions, hydrogen bonding, van der Waals forces, and base stacking. These interactions allow aptamers to recognize and attach to their targets with high precision, akin to a lock and key mechanism, making them highly effective as targeted therapeutic agents [55].

#### 24.2.1.1 Aptamers

RNA aptamers represent a versatile tool in nanotechnology for drug delivery, allowing for the precise targeting of diseased cells without affecting surrounding healthy cells. This specificity not only enhances the therapeutic efficacy of the drug delivery system but also significantly reduces the risk of side effects associated with broader, less discriminative methods. The development and implementation of RNA aptamers in nanoparticle design are therefore seen as a major leap forward in the creation of highly efficient and safer therapeutic modalities in the field of RNA nanotechnology. Aptamers provide several compelling advantages over traditional protein antibodies, which have long been the cornerstone of targeted therapeutic applications. These advantages are significant in the context of developing more efficient and cost-effective treatments. First, aptamers have lower production costs compared to protein antibodies. This economic benefit arises because aptamers can be synthesized chemically, avoiding the complex and resource-intensive biological production systems required for antibodies. Furthermore, the selection and optimization process for aptamers is notably faster. Using systematic evolution of ligands by exponential enrichment (SELEX), aptamers can be developed and optimized within weeks, compared to the several months often required to generate and refine antibodies [56, 57].

The synthesis of aptamers is not only rapid but also allows for high reproducibility and consistency in batch quality. Unlike biological systems that can have batch-to-batch variability, the chemical synthesis of aptamers ensures that each batch is uniform, reducing the risk of discrepancies in therapeutic applications. Another critical advantage of aptamers is their minimal immunogenicity. Since aptamers are nucleic acid-based, they typically induce a lower immune response in the human body compared to protein-based antibodies. This characteristic is particularly beneficial in therapeutic contexts, where minimizing the immune response can be crucial to the effectiveness and safety of the treatment. Additionally, aptamers demonstrate efficient tissue penetration. Their relatively small size compared to antibodies allows them to more easily traverse biological membranes and reach target cells and tissues. This feature is particularly valuable in delivering therapeutic agents to dense or poorly vascularized tissues. Lastly, aptamers exhibit enhanced stability. They can withstand a range of environmental conditions that would typically denature proteins, such as changes in temperature and pH. This stability facilitates easier storage and handling, and the potential for modifications to enhance their stability further, such as chemical modifications to their backbone, ensures that aptamers maintain their functional integrity under physiological conditions [58, 59].

Currently, numerous RNA aptamers are available for targeting specific cell surface receptors, facilitating the subsequent internalization of RNA nanoparticles in various cancers, including glioblastoma (e.g., EGFRvIII) [60], breast cancer (e.g., EGFR, HER2, HER3) [61–63], prostate cancer (e.g., PSMA) [64], ovarian cancer (e.g., E-selectin) [65], colon cancer (e.g., EpCAM) [66], and lymphoma (e.g., CD19) [67], as well as in cells infected by viruses such as HIV (e.g., CD4) [68]. By integrating cell internalizing RNA aptamers into the structure of RNA nanoparticles, targeted delivery of therapeutic molecules like siRNA, miRNA, and chemotherapeutic agents into the cytosol is achieved, enhancing the specificity and efficiency of treatments. Aptamer-based therapeutics can also function through the direct interaction of an aptamer with a therapeutic target, which influences downstream signaling. Notable applications include targeting receptor tyrosine kinases (e.g., anti-EGFR inhibitors), cell adhesion molecules (e.g., E-selectin), immune modulators (e.g., anti-CTLA-4 inhibitors), and factors involved in cell growth (e.g., α-fetoprotein) for cancer treatment [55, 69, 70]. Additionally, aptamers are being developed to address neurological conditions such as multiple sclerosis (targeting IL-17), Alzheimer's disease (targeting β-secretase1), Parkinson's disease (targeting AMPA), and stroke (targeting Factor IXa) [55, 69, 70].

#### 24.2.1.2 Ribozymes

Ribozyme-based therapeutic strategies are increasingly recognized as powerful tools in the treatment of diseases like cancer and viral infections. Ribozymes are unique RNA motifs that possess enzymatic properties similar to those of proteins, enabling them to catalyze specific biochemical reactions within the cell. This capability makes them particularly suited for gene therapy applications, where precise intervention at the molecular level is required. In recent studies, ribozymes have demonstrated significant potential in addressing complex diseases such as breast cancer and hepatocellular carcinoma in animal models. Their mode of action typically involves the selective cleavage of target RNA molecules, which can effectively disrupt the expression of disease-associated genes. For instance, by targeting RNA transcripts that encode for critical oncogenic proteins, ribozymes can inhibit tumor growth and progression [71]. One of the standout features of ribozyme therapy is its specificity. While RNAi technologies also target RNA molecules for degradation and have been used extensively for gene silencing, ribozymes offer a distinct advantage due to their lower propensity for off-target effects. RNAi technologies can sometimes inadvertently affect the expression of other genes, which are not intended targets, leading to undesirable side effects. Ribozymes, by contrast, can be engineered to have high specificity for their RNA targets, thereby minimizing unintended interactions within the cellular environment [72].

One of the primary hurdles in the practical implementation of ribozyme therapy lies in the development of effective nonviral vectors for the targeted in vivo delivery of ribozymes to specific cells. This challenge stems from the need to transport therapeutic ribozymes into cells in a manner that is both safe and efficient, without relying on viral vectors which may pose risks of immunogenicity and insertional mutagenesis. To overcome this issue, innovative approaches have been explored. One such strategy involves integrating ribozyme sequences into RNA scaffolds, which serve as specialized delivery platforms. These platforms are designed to enhance the stability and targeting capability of ribozymes, ensuring that they are delivered efficiently and accurately to their intended cellular targets. This method not only protects the ribozymes from degradation in the biological environment but also facilitates their entry into specific cells. A compelling example of this approach is the development of pRNA nanoparticles engineered to carry a hepatitis B virus (HBV) specific hammerhead ribozyme. The hammerhead ribozyme is a well-known motif characterized by its ability to cleave RNA sequences at designated sites, making it particularly useful for therapeutic applications. In the case of HBV, the ribozyme is designed to target and cleave the poly(A) signal on HBV mRNA. This poly(A) signal is crucial for the stability and translation of HBV transcripts, and its disruption by the ribozyme leads to a decrease in viral mRNA levels and, consequently, a reduction in viral protein synthesis and replication. The use of pRNA nanoparticles as a vector for this ribozyme capitalizes on the unique structural properties of pRNA, which include its robustness and flexibility in forming complex structures. These nanoparticles can be customized to incorporate not just the ribozyme but also other targeting ligands or therapeutic molecules, providing a versatile and potent platform for addressing viral infections like HBV [22, 30].

### *24.2.2 Applying Riboswitches for Modulating Gene Expression and as Antimicrobial Targets*

Riboswitches offer a sophisticated and promising therapeutic approach to treating various diseases by directly modulating gene expression. These naturally occurring RNA elements are found in a wide array of organisms, from bacteria and archaea to plants, fungi, and algae, underscoring their evolutionary significance and versatility in biological regulation. The structure of riboswitches typically consists of two main domains: the sensor domain and the expression platform. The sensor domain functions as a natural aptamer, adept at detecting specific molecular ligands. This sensitivity to specific ligands allows riboswitches to act as molecular switches, changing conformation upon ligand binding. The second domain, the expression platform, is capable of adopting two distinct structural conformations depending on whether or not the ligand is bound to the sensor domain. This dual conformation capacity is pivotal, as it enables the riboswitch to toggle gene expression on or off without the need for additional protein cofactors. The mechanism of action for riboswitches is both direct and efficient, making them particularly attractive for therapeutic applications. In the absence of ligands, a riboswitch might maintain a conformation that either promotes or inhibits the formation of mRNA structures necessary for translation initiation or other aspects of gene expression. Upon ligand binding, the riboswitch undergoes a conformational shift that alters these critical RNA structures, thus modulating the expression of the genes under its control. This unique property of riboswitches to directly control gene expression makes them highly desirable targets for developing novel antimicrobial agents. Many pathogenic bacteria rely on specific metabolic pathways that are unique to microbial organisms and regulated by riboswitches. By designing molecules that mimic or block the natural ligands of these riboswitches, it is possible to create targeted therapies that disrupt these pathways, leading to the inhibition of bacterial growth or virulence without affecting eukaryotic cells [73].

#### 24.2.2.1 Metabolite-Binding Riboswitches as Antimicrobial Targets

Metabolite-binding riboswitches present a particularly promising avenue for developing novel antimicrobial agents, especially in the fight against multi-drug resistant pathogens. These riboswitches are capable of directly regulating essential metabolic pathways by binding to specific metabolites, making them critical control points in bacterial viability and virulence. One exemplary riboswitch is the TPP (thiamine pyrophosphate) riboswitch, which plays a pivotal role in the biosynthesis and transport of thiamine (vitamin B1) in bacteria. Thiamine is vital for bacterial metabolism, particularly in the function of enzymes that manage DNA synthesis and repair. By regulating this riboswitch, it might be possible to disrupt thiamine homeostasis, thereby inhibiting bacterial growth. Similarly, the lysine riboswitch controls both the biosynthesis and transport of lysine, an essential amino acid critical for protein synthesis in many bacterial species. Targeting this riboswitch could stifle lysine production or its movement within the cell, leading to protein synthesis disruption and, ultimately, bacterial growth inhibition. Another significant example is the FMN riboswitch, which controls the synthesis of riboflavin (vitamin B2). Riboflavin is another crucial molecule that plays a central role in numerous cellular processes, including energy production and electron transport. Disrupting the FMN riboswitch could cripple these fundamental processes, rendering bacterial cells nonviable [74–76].

### *24.2.3 Applying Thermodynamic Properties for Controlled Release of Functional Agents*

The application of thermodynamic properties in RNA nanotechnology provides a sophisticated means of controlling the release or activation of functional agents, such as therapeutic drugs or diagnostic molecules. By understanding and manipulating the stability of nucleic acid structures, researchers can devise RNA-based nanoparticles that release their cargo under specific conditions, enhancing the efficacy and safety of treatments. Nucleic acids, including RNA and DNA, exhibit unique thermodynamic characteristics that can be fine-tuned through nucleotide composition and sequence. These properties influence how tightly the nucleic acid strands bind together, which in turn affects the stability of the entire nanoparticle structure. By selecting specific nucleotides or modifying nucleic acid backbones, scientists can design nanoparticles that are stable under certain physiological conditions but disassemble to release their functional payloads when triggered by specific environmental changes. One innovative example of this approach is the development of pRNA-based nanosquares. These nanosquares are engineered from RNA molecules that form a square-shaped structure, serving as a stable platform for attaching therapeutic agents. The core strands of these nanosquares, which hold the structure together, can be made from RNA, DNA, or chemically modified nucleic acids like 2′-F RNA (2′-fluoro RNA) [77]. Incorporating different nucleic acids in the design of RNA nanoparticles provides a strategic approach to modulate their stability and, consequently, the release dynamics of encapsulated or attached drugs.

The choice between using natural nucleic acids like DNA or modified versions such as 2′-F RNA (2′-fluoro RNA) can significantly affect both the physical properties of the nanoparticles and their interaction with the biological environment [78–81]. Additionally, functionalities within RNA nanoparticles can be distributed across different helices, which are activated through toe-hold interactions. To track the functionality reassembly and activation, each segment can be labeled with a pair of Förster resonance energy transfer (FRET) molecules. Upon the merging of these functionalities, there is a significant increase in the FRET signal, providing a visual cue of the process. This method not only ensures targeted drug delivery but also enhances the monitoring of therapeutic activity in real-time, significantly improving the safety and efficacy of RNA-based therapies.

### *24.2.4 Applying RNAi Therapy: siRNA, miRNA, and LNA*

The discovery of the RNAi pathway in 1998 marked a transformative moment in molecular biology and medicine, providing a powerful tool for gene regulation that could potentially be harnessed for therapeutic purposes across a wide array of diseases, including cancer. RNAi operates through mechanisms that involve short double-stranded RNA molecules, which are typically 21–23 nucleotides in length, and play a crucial role in regulating gene expression [82, 83]. MiRNAs, naturally occurring in cells, can regulate multiple genes, while siRNAs are primarily synthetic, specifically engineered to target particular mRNA sequences, although some naturally occurring siRNAs have been identified [84]. RNAi technology harnesses the ability to target virtually any mRNA sequence with siRNAs and miRNAs, making it possible to treat diseases previously considered untreatable with traditional drugs.

## 24.3 pRNA-3WJ Nanoparticles for the Delivery of siRNA In Vivo

### *24.3.1 Therapeutic Potential of pRNA-3WJ Nanoparticles in Cancer Treatment*

The pRNA-3WJ (three-way junction) nanoparticle system represents a significant advancement in RNA nanotechnology, particularly for the delivery of siRNA in vivo. This system leverages the unique structural properties of the phi29 bacteriophage packaging RNA (pRNA), which naturally forms a stable three-way junction. These nanoparticles can be engineered to carry various therapeutic agents, including siRNA, making them highly versatile and effective for targeted therapy applications. The pRNA-3WJ structure is ideally suited for medical applications due to its biocompatibility, modular design, and intrinsic ability to form nanoparticles. These features allow for the precise assembly of complex structures capable of carrying multiple therapeutic agents simultaneously. In the case of cancer treatment, pRNA-3WJ nanoparticles can be specifically designed to include not only siRNA but also targeting ligands that enhance delivery to cancer cells. This has been detailed in the previous chapter (please refer to Chap. 25).

A notable example of the application of pRNA-3WJ nanoparticles is in targeted gastric cancer therapy. In one study, these nanoparticles were engineered to include folate as a targeting ligand. Folate receptors are commonly overexpressed on the surface of cancer cells, including many types of gastric cancer cells, making folate a useful ligand for targeting these cells specifically. By attaching folate to the pRNA-3WJ nanoparticles, researchers could direct the nanoparticles more efficiently to cancer cells while sparing healthy cells, thereby increasing therapeutic efficacy and reducing side effects. Furthermore, the same nanoparticles were equipped with siRNA targeting BRCAA1 (breast cancer associated antigen 1), a gene implicated in the progression and metastasis of gastric cancer. The siRNA serves to silence the expression of this gene, potentially inhibiting the growth and spread of cancer cells. The delivery of BRCAA1 siRNA via pRNA-3WJ nanoparticles ensures that the siRNA reaches the target cells in a protected form, increasing the likelihood of successful gene silencing [26]. The application of pRNA-3WJ nanoparticles in cancer therapy, particularly in gastric cancer, demonstrates the potential of advanced RNA nanotechnology in achieving targeted therapeutic effects. In an illustrative study, these nanoparticles were systemically administered to mice bearing subcutaneous gastric tumor xenografts.

Another investigation utilized pRNA-3WJ nanoparticles carrying both folate and luciferase siRNA to target intracranial glioblastoma xenografts in mice [28]. Post-systemic injection, these nanoparticles successfully silenced luciferase gene expression in the brain tumors, demonstrating the nanoparticles' capability to cross biological barriers and deliver siRNA effectively. A different approach involves RNAi microsponges, which are crafted from tandem repeats of luciferase siRNA synthesized via RCT. These are designed to be stable hairpin RNAs that are only converted into active siRNA by Dicer following cellular uptake, ensuring the specificity and efficacy of the gene silencing. This system was successfully tested in mice with subcutaneous ovarian cancer xenografts, where luciferase expression was effectively silenced. Expanding on this technology, further advancements have enabled these systems to carry multiple siRNAs to simultaneously target several genes [85], enhancing the therapeutic impact against complex diseases. Moreover, the application of RNA nanoparticles has also extended to viral therapy. For example, RNA nanoparticles composed of a CD4 aptamer coupled with siRNAs targeting HIV coreceptors CCR5, gag, and vif have shown promise in preventing the sexual transmission of HIV in humanized mice [86]. This demonstrates the potential of RNA nanoparticles not only in treating existing diseases but also in preventing new infections, thereby broadening the scope of RNA-based therapeutic strategies.

Although siRNA therapy has shown promise for gene knockdown, achieving effective and consistent results has sometimes been challenging. To overcome issues with efficacy, strategies involving multivalent siRNA delivery have been explored. One strategy to enhance siRNA efficacy is to increase the number of siRNA copies delivered to the target. For instance, augmenting the copy number of luciferase siRNA on the pRNA X-way motif led to a synergistic enhancement in gene knockdown efficiency [23]. This approach leverages the increased local concentration of siRNA within the cell to boost the likelihood of mRNA target interactions and subsequent gene silencing. The deployment of RNA nanotechnology in combating viral infections such as HIV-1 showcases its potential to revolutionize antiviral therapies. An innovative approach involving RNA nanorings is particularly noteworthy. These structures are engineered to deliver a comprehensive attack on the virus by carrying multiple siRNAs each designed to target different regions of the HIV-1 genome. RNA nanorings are a type of structured nanoparticle formed from RNA that can be engineered to hold multiple therapeutic agents. In this instance, the nanorings were designed to carry six distinct siRNAs, each targeting a unique sequence within the HIV-1 viral genome. The strategic selection of target regions across the genome is crucial for minimizing the potential for viral escape through mutations, a common issue in HIV therapy due to the high mutation rate of the virus. The incorporation of multiple siRNAs into a single delivery system ensures simultaneous downregulation of several critical genes involved in the viral life cycle. This multi-target approach is designed to hinder various stages of HIV-1 replication, from entry and reverse transcription to integration and assembly. By attacking the virus at multiple points in its life cycle, the likelihood of viral resistance decreases significantly [87]. By simultaneously attacking multiple genomic targets, this strategy aims to prevent the virus from escaping inhibition through mutation.

The pRNA-3WJ nanoparticles are engineered for high specificity and efficiency in cancer therapy. They incorporate EGFR-targeting RNA aptamers that allow the nanoparticles to bind specifically to cancer cells overexpressing EGFR, a common trait in many cancers, including TNBC. This targeting strategy ensures that the nanoparticles are preferentially internalized by the tumor cells, minimizing effects on normal tissues. Once internalized, the nanoparticles deliver an 8-mer anti-miR-21 locked nucleic acid (LNA) agent. The LNA design enhances the stability and binding affinity of the anti-miR-21, allowing it to effectively bind to and inhibit the seed region of miR-21. miR-21 is a well-known oncogenic miRNA associated with the suppression of tumor suppressor pathways, promotion of proliferation, and inhibition of apoptosis in various cancers. By targeting miR-21, the anti-miR-21 LNA agent counteracts these oncogenic effects, thereby facilitating the reactivation of tumor suppressor pathways and inducing apoptosis in cancer cells [25, 88]. Similarly, pRNA-3WJ nanoparticles featuring a PSMA aptamer and anti-miR-21 have been employed to target and inhibit growth in subcutaneous prostate cancer xenografts [29]. Beyond cancer, this nanoparticle design has also been adapted to silence viral genes effectively by targeting the 3′-UTR of the coxsackievirus genome [89, 90], showcasing the versatility of RNA nanoparticles in both oncologic and antiviral therapies. These strategies illustrate the ongoing advancements in RNA nanoparticle technology, demonstrating its potential to overcome the limitations of siRNA therapy and expand the possibilities for treating a wide array of diseases effectively.

## 24.4 RNA Nanoparticles for Immunotherapy

RNA nanoparticles, given their adaptability and precision, present a compelling framework for immunotherapy applications. One of the most innovative uses of this technology involves the integration of immune-stimulating motifs, such as CpG sequences, to create powerful immuno-adjuvants. This approach leverages the intrinsic immune-modulatory properties of CpG DNA, known to activate immune system pathways that are critical in fighting infections and cancer. CpG DNA sequences refer to short DNA stretches rich in cytosine and guanine bases separated by a phosphate, which are recognized by the TLR9 found predominantly in dendritic cells and B cells. The engagement of CpG sequences with TLR9 triggers a cascade of immune responses, notably the activation and maturation of these immune cells, leading to enhanced antigen presentation and the production of various cytokines. In the context of RNA nanoparticles, these CpG sequences can be functionally integrated into the design of RNA polygons, creating a hybrid nucleic acid construct that combines the structural advantages of RNA nanoparticles with the potent immune-stimulatory effects of CpG DNA. These RNA polygons can be tailored to various sizes and shapes, depending on the therapeutic need, and can be modified to include other functional elements, such as targeting ligands or therapeutic agents. The practical application of these CpG-functionalized RNA polygons has been demonstrated in preclinical models, such as in mice, where they have shown significant potential to boost the immune response. When administered, these nanoparticles have triggered a robust immune reaction, marked by a 100-fold increase in the induction of pro-inflammatory cytokines such as TNFα and IL-6. These cytokines play critical roles in mediating and regulating immune responses, enhancing the body's ability to fight off cancer cells and pathogens [91].

The integration of innovative structural and sequence modifications in RNA therapeutics is proving to be a powerful approach to enhance immunostimulatory effects. These advancements are particularly significant in the context of siRNA technologies, where subtle alterations can profoundly influence the immune system's response. A notable example of such an innovation involves the introduction of a miRNA-like non-pairing uridine bulge in the passenger strand of an siRNA, which has been shown to boost its immunostimulatory activity on human immune cells. The structure of an siRNA typically consists of a sense strand and an antisense strand, both of which are crucial for the RNA interference pathway. Introducing a non-pairing uridine bulge into the passenger strand of an siRNA involves strategically placing one or more uridines that do not participate in base pairing with the complementary strand. This bulge creates a localized distortion in the siRNA structure, which can affect how the siRNA interacts with cellular machinery and immune receptors [92]. Such modifications not only enhance the therapeutic potential of RNA nanoparticles but also broaden their applicability in fields like antiviral therapies and cancer immunotherapy.

In the field of immunotherapy, aptamers have been developed to target proteins that play crucial roles in cancer immunotherapy, particularly those that help tumor cells evade immune detection [93]. These aptamers offer several advantages over traditional monoclonal antibody therapy, including the potential to bypass dose-limiting autoimmune responses. Aptamers can be engineered to target a wide array of molecules, including cytokines (e.g., IFN-γ, TGF-β), co-stimulatory molecules (e.g., CTLA-4, CD28), adhesion molecules (e.g., VCAM-1, Selectin-P/L), and molecules involved in tumor immunosuppression. They can also target activation and inhibitory receptors on various immune cells such as T cells, NK cells, macrophages, and dendritic cells. Bispecific aptamers are engineered to connect two distinct cell surface antigens, thereby facilitating the convergence of T cells and tumor cells, which promotes the elimination of the latter. RNA-based aptamers have emerged as powerful tools in precision medicine, particularly in the field of cancer immunotherapy. An innovative illustration of their potential is the design of a dual-targeting aptamer that binds concurrently to the CD16α receptor on NK cells and the c-Met receptor on tumor cells. This dual engagement strategy leverages the natural immune mechanisms of NK cells to direct a potent antitumor response through antibody-dependent cellular cytotoxicity (ADCC). Challenges in creating bispecific aptamers, such as issues with refolding, instability, diminished binding specificity when merging two aptamers, and the need for size adjustment for simultaneous target engagement, can be addressed using RNA nanotechnology [22, 94].

## 24.5 Other Promising Approaches

**Molecular Beacons** Significant efforts are being directed toward the development of image-guided vectors capable of tracking RNA movements, detecting native gene activity, and silencing genes within living organisms. Among the prevalent techniques for imaging RNA expression in vitro is the use of molecular beacons that operate via competitive hybridization [95–103]. These beacons are designed as stem-loop (hairpin) structures in oligonucleotides, equipped with a fluorophore at one end and a quencher at the other. In their resting state, the proximity of the fluorophore to the quencher prevents fluorescence. However, when these beacons hybridize with a target RNA, the structure alters, distancing the fluorophore from the quencher and thus restoring fluorescence [104]. Numerous studies have demonstrated the potential of these beacons in tracking mRNA presence and RNA transport within cells [105–107], yet the development of robust in vivo delivery systems is necessary to extend this technology's application across wider research fields or into standardized clinical practices. RNA nanotechnology holds promise in enhancing the thermodynamic and chemical stability of molecular beacon-based RNA nanoparticles, thereby optimizing their hybridization efficiency with mRNA targets. For further insights into molecular beacon design and their applications, readers should consult the cited literature [108, 109].

**In Vitro** *Diagnostics* The early detection of diseases is crucial, as it allows for prompt medical intervention and often leads to improved survival outcomes. Detecting minimal concentrations of disease markers, such as proteins or small molecules, is key to early diagnosis. Aptamer-based biosensors, or aptasensors, represent a significant advancement over traditional antibodies and enzymes, particularly with the enhancements provided by nanotechnology [110]. These sensors can achieve remarkable sensitivity, detecting picomolar levels of an analyte through the use of amplification techniques such as real-time PCR and rolling circle amplification following aptamer binding. Integration with microarray technology enables the quantification of thousands of different targets simultaneously [111].

**RNA Computing and Biomemory** As computational demands escalate and transistor miniaturization continues, alternative computing methodologies are essential to maintain alignment with Moore's law, which forecasts a doubling of transistor density every 2 years. Nucleic acid-based computation, utilizing the atomic-scale dimensions of DNA and RNA, presents a promising avenue for sustaining this rate of advancement. The inception of DNA computing dates back to 1989 with the design of the first DNA logic gate [112], and a significant milestone was reached in 1994 with a DNA-based system solving a complex seven-city Hamiltonian problem [113]. This type of molecular computing operates through the hybridization of input DNA strands to complementary sequences immobilized on a chip, utilizing the inherent stability and precise control over DNA hybridization. Given RNA's rich variety of structural motifs, which differs significantly from DNA, it offers potential benefits for advancing nucleic acid computing by exploiting its structural complexity. Additionally, RNA's ability to function as an insulator in biomemory applications enhances its utility. An example of this is the employment of the pRNA-3WJ motif in RNA–quantum dot conjugates, serving both insulating and mediating roles in biomemory technologies [114]. The tunable size of RNA nanoparticles enables precise control over the distances between conductive surfaces in circuits, ensuring consistent electrical current across a chip. This adaptability and functional diversity make RNA particularly suitable for molecular-scale computing applications.

**RNA Nanotechnology and CRISPR** The CRISPR-Cas9 system, known for its ability to precisely edit, insert, or delete genes within a genome, has garnered significant interest due to its potential to target specific genetic sequences [115]. By utilizing tailored guide RNAs, this technology can be directed to modify genes, potentially rendering them harmless, or to introduce new genetic elements that confer specific traits [116, 117]. For instance, it has been employed to inhibit the hepatitis B virus in cells with chronic infections [118, 119]. Despite its promising capabilities, similar to the challenges faced with in vivo delivery of naked siRNA, unmodified guide RNAs also encounter stability issues when delivered into cells [120].

## 24.6 Current Challenges and Outlook

RNA nanotechnology is increasingly recognized for its potential in drug delivery, particularly in the treatment of diseases like cancer. However, significant advancements are necessary to advance this promising technology from the laboratory to clinical trials and ultimately to market availability. The challenges in this domain and potential solutions include the following:

**Limited Small Molecule Payload**: RNA nanoparticles currently face limitations in their capacity to load small molecule drugs. Attaching single drugs to the terminal ends of RNA strands provides precise drug release mechanisms. Although extending labeling across the entire chain can enhance drug payload, this approach might lead to structural issues, such as misfolding due to steric hindrance, and could also interfere with drug release. Computational strategies could be instrumental in pinpointing suitable sites for drug attachment that do not affect the nanoparticle's folding, ensuring that drugs are oriented out of the RNA structure's plane. Intercalation offers a way to boost drug loading but requires careful management to prevent premature drug release, which could cause nonspecific side effects. For instance, when employing pRNA-3WJ nanoparticles as delivery vectors for solid tumors, it is critical that the drug's release half-life exceeds 4 h, aligning with the time frame in which these nanoparticles predominantly localize at the tumor site, typically within 1–4 h after administration. Furthermore, RNA microsponges show promise for delivering high payloads of siRNAs [121], and their potential in broader preclinical applications could be enhanced by minimizing particle size to prevent unwanted organ accumulation. This balancing act between efficacy and safety highlights the nuanced approach needed in the development of RNA-based therapeutics.

***New Generation of Carriers***: Recent advancements include the development of up to fourth-generation RNA dendrimers [122]. In the quest to enhance targeted drug delivery and therapeutic efficacy, advanced nanocarriers such as dendrimers and 3D RNA polygons have shown significant promise. These nanostructures are engineered to carry multiple drugs and targeting agents, providing a multifaceted approach to disease treatment. 3D RNA polygons are sophisticated RNA nanostructures designed to form polygonal shapes, such as tetrahedrons, cubes, or cages. These structures are assembled from RNA strands that fold and hybridize into stable, predefined geometries. 3D RNA polygons can function as RNA cages or containers, capable of encapsulating substantial quantities of small molecule drugs. This encapsulation leverages the natural stability and folding properties of RNA to create secure and efficient drug carriers. The modularity of RNA polygons allows for the incorporation of multiple functional modules, including therapeutic RNAs (like siRNA or miRNA), targeting ligands, and imaging agents. This versatility enables thc design of highly specific and multifunctional drug delivery systems. RNA-based nanostructures are inherently biocompatible and biodegradable, minimizing the risk of long-term toxicity and ensuring safe elimination from the body after they have fulfilled their therapeutic role [123, 124]. These RNA structures maintain their intrinsic properties in both organic and aqueous solvents and offer biocompatibility and biodegradability, making them favorable over synthetic nanocarriers.

***Endosome Escape to Fulfill the Promise of RNAi***: Like many nanoparticle systems, RNA nanoparticles typically enter cells via receptor-mediated endocytosis, leading to a series of intracellular trafficking challenges. Initially, these nanoparticles are directed to early endosomal vesicles. From there, they are sorted and transported to late endosomes and lysosomes, often becoming confined there and failing to reach their target sites within the cell. A promising strategy to overcome this barrier involves the use of short eight nucleotide anti-miRNA locked nucleic acid (LNA) fragments, which have demonstrated successful endosomal escape and significant cancer regression when included in RNA nanoparticle delivery systems [25], though the efficiency of siRNA's endosomal escape remains less understood, as indicated by the generally low cancer regression rates following siRNA delivery through receptor-mediated endocytosis [26, 125]. Currently, our understanding of the various cellular pathways for endocytosis and the subsequent intracellular processing and endosomal escape of RNA nanoparticles is quite limited. However, there are established methods available that enhance the escape of nanoparticles from endosomes. These include the use of chemical functional groups such as acid-cleavable linkers—acetal, hydrazone, and maleic amides—or other acid-protonating groups like amino esters and sulfonamide [126]. These pH-sensitive materials are highly prone to protonation under the acidic conditions typical of endosomes, which leads to an ion influx that causes osmotic swelling and ultimately the rupture of the endosomal membrane, a process known as the "proton-sponge" effect. Moreover, a variety of chemical conjugation techniques exist for the precise incorporation of these endosome-disrupting chemicals onto RNA nanoparticles. These strategies offer a pathway to improve the delivery efficiency of RNA-based therapeutics by facilitating their release from endosomal compartments into the cytoplasm, where they can effectively perform their intended functions.

***Large-Scale Production and Purification of RNA Nanoparticles***: The advancement of RNA nanoparticle technology toward clinical applications faces several key challenges, particularly in the areas of large-scale production, purification, and cost-efficiency. RNA nanoparticles are typically constructed from multiple short strands, each less than 80 nucleotides in length, which is well-suited for chemical synthesis. Over time, the cost of synthesizing RNA oligonucleotides has decreased significantly, thanks to advancements in chemical synthesis techniques that utilize various 2′-protecting groups such as ether, acetal, orthoester, ester, *O*-acetalester, and pivaloyloxymethyl (PivOM) [127, 128]. These improvements have enabled commercial vendors to produce RNA in large quantities, up to tens of grams per synthesis cycle, while meeting good manufacturing practice (GMP) standards. Despite these advances, the large-scale purification of RNA nanoparticles remains a considerable obstacle. Traditional methods like high-performance liquid chromatography (HPLC) and gel electrophoresis often yield suboptimal results due to their limitations in separating fully assembled nanoparticles from their smaller constituent strands. Given the distinct size difference between the assembled nanoparticles and their individual components, preparative ultracentrifugation has been introduced as a viable method for the purification of these structures. Recent applications of this technique have shown promise [129], indicating that it could be an effective strategy for achieving high yields of properly assembled RNA nanoparticles. Addressing these production and purification challenges will be crucial for scaling up RNA nanotechnology to meet clinical demand, thereby enabling its potential as a transformative tool in therapeutic applications.

***Rational Incorporation of Chemically Modified Nucleotides***: Chemical modifications of RNA, while providing numerous benefits such as enhanced serum and thermodynamic stability and reduced immunogenicity, can sometimes adversely affect the structural integrity of RNA motifs. A notable example is the malachite green (MG) fluorogenic RNA aptamer, which becomes dysfunctional after 2′-F modification due to misfolding issues [130]. This underscores that rational chemical modification is not a one-size-fits-all solution; it depends heavily on the specific RNA sequence and its structural context. Effective patterns of chemical modifications might be suitable for one RNA system but entirely inappropriate for another. The overarching challenge lies in identifying universal chemical modification strategies that are effective across various RNA sequences. This requires a deep understanding of how modifications impact RNA folding and function in different contexts. For RNA nanoparticle designs aimed at clinical applications, there remains a critical need for general strategies to improve the integration of chemical modifications without compromising the natural function and structure of the RNA. Addressing these issues is essential for the successful clinical translation of RNA-based therapeutics, making it possible to harness their full potential while mitigating risks related to structural alterations.

***Computational Guide to RNA Nanoparticle Assembly***: The accurate folding of RNA constructs is crucial for ensuring their functional integrity, particularly in the context of multifunctional RNA nanoparticles. RNA molecules are inherently dynamic, capable of transitioning between multiple three-dimensional structures that often only slightly differ in their Gibbs free energy ($\Delta G$). This variability can influence the overall stability and functionality of larger RNA assemblies, which are subject to more complex thermodynamic and kinetic folding constraints. To address these challenges, several online tools have been developed to assist researchers in predicting RNA folding and designing RNA structures. These include Mfold [131], RNA Designer [132], Sfold [133], NUPACK [134], Nanofolder [135], and Hyperfold [81]. Each of these tools offers unique capabilities, from basic secondary structure prediction to more complex analyses that consider both thermodynamic stability and kinetic folding pathways. Despite these advancements, accurately predicting RNA structure and its assembly into nanoparticles remains a formidable challenge. The field of RNA nanotechnology urgently requires more sophisticated prediction software that can effectively model inter-RNA interactions and analyze three-dimensional (3D) and four-dimensional (4D) structures within nanoparticle designs. While there has been some progress in moving RNA computational models from traditional intramolecular interactions to more complex intermolecular interactions, the field is still relatively nascent [136–138]. Continued development in this area is essential for advancing RNA nanotechnology. Improved computational tools that can handle the intricacies of RNA structure and interaction will facilitate the design of more effective and reliable RNA-based nanoparticles, potentially revolutionizing their use in medical and biotechnological applications.

## 24.7 Conclusion

Since its inception in 1998, RNA nanotechnology has achieved significant strides, particularly over the past 5 years. The field has seen the development of sophisticated nanostructures that boast precisely organized functional modules within three-dimensional spaces, alongside programmable self-assembly into four-dimensional structures. Such advancements are pushing the boundaries of structure-based design, enhancing our ability to craft increasingly complex RNA nanoparticles. These developments hold the promise of 1 day replicating the functionalities of naturally occurring RNA nanomachines, such as ribosomes. Despite these advances, the use of RNA nanoparticles in clinical settings is not as advanced as other nanoparticle systems like liposomes and polymers, which have been in clinical trials for several years. For RNA nanoparticles to move into clinical applications, efforts must extend beyond merely constructing these nanoparticles. It is essential to leverage the unique advantages of RNA nanotechnology in practical applications. The potential of RNA nanotechnology has not gone unnoticed in the business world. The establishment of several startup companies dedicated to transitioning RNA nanotechnology from laboratory research to clinical practice is a testament to the growing interest from investors. These companies aim to harness the unique properties of RNA nanoparticles to create innovative therapeutic solutions, accelerating the path from bench to bedside. This shift toward commercial development underscores the need for continued innovation and application-focused research within the field.

## References

1. Yingchoncharoen P, Kalinowski DS, Richardson DR. Lipid-based drug delivery systems in cancer therapy: what is available and what is yet to come. Pharmacol Rev. 2016;68(3):701–87.
2. Niu Z, Conejos-Sánchez I, Griffin BT, O'Driscoll CM, Alonso MJ. Lipid-based nanocarriers for oral peptide delivery. Adv Drug Deliv Rev. 2016;106(Pt B):337–54.
3. Liu D, He C, Wang AZ, Lin W. Application of liposomal technologies for delivery of platinum analogs in oncology. Int J Nanomedicine. 2013;8:3309–19.
4. Pack DW, Hoffman AS, Pun S, Stayton PS. Design and development of polymers for gene delivery. Nat Rev Drug Discov. 2005;4(7):581–93.

5. Duncan R. Polymer therapeutics as nanomedicines: new perspectives. Curr Opin Biotechnol. 2011;22(4):492–501.
6. Lee KL, Twyman RM, Fiering S, Steinmetz NF. Virus-based nanoparticles as platform technologies for modern vaccines. Wiley Interdiscip Rev Nanomed Nanobiotechnol. 2016;8(4):554–78.
7. Smith MT, Hawes AK, Bundy BC. Reengineering viruses and virus-like particles through chemical functionalization strategies. Curr Opin Biotechnol. 2013;24(4):620–6.
8. Hong CA, Eltoukhy AA, Lee H, Langer R, Anderson DG, Nam YS. Dendrimeric siRNA for efficient gene silencing. Angew Chem Int Ed Engl. 2015;54(23):6740–4.
9. Sun W, Mignani S, Shen M, Shi X. Dendrimer-based magnetic iron oxide nanoparticles: their synthesis and biomedical applications. Drug Discov Today. 2016;21(12):1873–85.
10. Chaplot SP, Rupenthal ID. Dendrimers for gene delivery—a potential approach for ocular therapy? J Pharm Pharmacol. 2014;66(4):542–56.
11. Hsu HJ, Bugno J, Lee SR, Hong S. Dendrimer-based nanocarriers: a versatile platform for drug delivery. Wiley Interdiscip Rev Nanomed Nanobiotechnol. 2017;9(1):e1409.
12. Chen S, Hao X, Liang X, Zhang Q, Zhang C, Zhou G, et al. Inorganic nanomaterials as carriers for drug delivery. J Biomed Nanotechnol. 2016;12(1):1–27.
13. Lammers T, Kiessling F, Hennink WE, Storm G. Drug targeting to tumors: principles, pitfalls and (pre-) clinical progress. J Control Release. 2012;161(2):175–87.
14. van der Meel R, Vehmeijer LJ, Kok RJ, Storm G, van Gaal EV. Ligand-targeted particulate nanomedicines undergoing clinical evaluation: current status. Adv Drug Deliv Rev. 2013;65(10):1284–98.
15. Saw PE, Jon S. Understanding of the entry mechanism of nanoparticles into tumors determines the future direction of nanomedicine development. BIO Integr. 2021;1(4):193–5.
16. Grodzinski P, Torchilin V. Advanced drug delivery reviews: cancer nanotechnology. Amsterdam: Elsevier; 2014.
17. Grodzinski P, Farrell D. Future opportunities in cancer nanotechnology—NCI strategic workshop report. Cancer Res. 2014;74(5):1307–10.
18. Zamboni WC, Torchilin V, Patri AK, Hrkach J, Stern S, Lee R, et al. Best practices in cancer nanotechnology: perspective from NCI nanotechnology alliance. Clin Cancer Res. 2012;18(12):3229–41.
19. Peer D, Karp JM, Hong S, Farokhzad OC, Margalit R, Langer R. Nanocarriers as an emerging platform for cancer therapy. Nat Nanotechnol. 2007;2(12):751–60.
20. Tandon P, Farahani K. NCI image-guided drug delivery summit. Cancer Res. 2011;71(2):314–7.
21. Abdelmawla S, Guo S, Zhang L, Pulukuri SM, Patankar P, Conley P, et al. Pharmacological characterization of chemically synthesized monomeric phi29 pRNA nanoparticles for systemic delivery. Mol Ther. 2011;19(7):1312–22.
22. Shu D, Shu Y, Haque F, Abdelmawla S, Guo P. Thermodynamically stable RNA three-way junction for constructing multifunctional nanoparticles for delivery of therapeutics. Nat Nanotechnol. 2011;6(10):658–67.
23. Haque F, Shu D, Shu Y, Shlyakhtenko LS, Rychahou PG, Evers BM, et al. Ultrastable synergistic tetravalent RNA nanoparticles for targeting to cancers. Nano Today. 2012;7(4):245–57.
24. Shu Y, Haque F, Shu D, Li W, Zhu Z, Kotb M, et al. Fabrication of 14 different RNA nanoparticles for specific tumor targeting without accumulation in normal organs. RNA. 2013;19(6):767–77.
25. Shu D, Li H, Shu Y, Xiong G, Carson WE 3rd, Haque F, et al. Systemic delivery of anti-miRNA for suppression of triple negative breast cancer utilizing RNA nanotechnology. ACS Nano. 2015;9(10):9731–40.
26. Cui D, Zhang C, Liu B, Shu Y, Du T, Shu D, et al. Regression of gastric cancer by systemic injection of RNA nanoparticles carrying both ligand and siRNA. Sci Rep. 2015;5:10726.
27. Rychahou P, Haque F, Shu Y, Zaytseva Y, Weiss HL, Lee EY, et al. Delivery of RNA nanoparticles into colorectal cancer metastases following systemic administration. ACS Nano. 2015;9(2):1108–16.

28. Lee TJ, Haque F, Shu D, Yoo JY, Li H, Yokel RA, et al. RNA nanoparticle as a vector for targeted siRNA delivery into glioblastoma mouse model. Oncotarget. 2015;6(17):14766–76.
29. Binzel DW, Shu Y, Li H, Sun M, Zhang Q, Shu D, et al. Specific delivery of MiRNA for high efficient inhibition of prostate cancer by RNA nanotechnology. Mol Ther. 2016;24(7):1267–77.
30. Hoeprich S, Zhou Q, Guo S, Shu D, Qi G, Wang Y, et al. Bacterial virus phi29 pRNA as a hammerhead ribozyme escort to destroy hepatitis B virus. Gene Ther. 2003;10(15):1258–67.
31. Zhou J, Shu Y, Guo P, Smith DD, Rossi JJ. Dual functional RNA nanoparticles containing phi29 motor pRNA and anti-gp120 aptamer for cell-type specific delivery and HIV-1 inhibition. Methods. 2011;54(2):284–94.
32. Ma X, Wang D, Wu Y, Ho RJ, Jia L, Guo P, et al. AIDS treatment with novel anti-HIV compounds improved by nanotechnology. AAPS J. 2010;12(3):272–8.
33. Feng L, Li SK, Liu H, Liu CY, LaSance K, Haque F, et al. Ocular delivery of pRNA nanoparticles: distribution and clearance after subconjunctival injection. Pharm Res. 2014;31(4):1046–58.
34. Pardi N, Hogan MJ, Porter FW, Weissman D. mRNA vaccines—a new era in vaccinology. Nat Rev Drug Discov. 2018;17(4):261–79.
35. Davis FF. The origin of pegnology. Adv Drug Deliv Rev. 2002;54(4):457–8.
36. Davis FF, Van Es T, Palczuk NC. Non-immunogenic polypeptides. Google Patents; 1979.
37. Gregoriadis G. Derivatization of proteins for prolonged circulation and enhanced storage stability. Patent WO2001087922; 2001.
38. Klibanov AL, Maruyama K, Torchilin VP, Huang L. Amphipathic polyethyleneglycols effectively prolong the circulation time of liposomes. FEBS Lett. 1990;268(1):235–7.
39. Zalipsky S. Chemistry of polyethylene glycol conjugates with biologically active molecules. Adv Drug Deliv Rev. 1995;16(2–3):157–82.
40. Senior J, Delgado C, Fisher D, Tilcock C, Gregoriadis G. Influence of surface hydrophilicity of liposomes on their interaction with plasma protein and clearance from the circulation: studies with poly(ethylene glycol)-coated vesicles. Biochim Biophys Acta. 1991;1062(1):77–82.
41. Blume G, Cevc G. Liposomes for the sustained drug release in vivo. Biochim Biophys Acta. 1990;1029(1):91–7.
42. Hassan S, Prakash G, Ozturk A, Saghazadeh S, Sohail MF, Seo J, et al. Evolution and clinical translation of drug delivery nanomaterials. Nano Today. 2017;15:91–106.
43. Andresen TL, Jensen SS, Jørgensen K. Advanced strategies in liposomal cancer therapy: problems and prospects of active and tumor specific drug release. Prog Lipid Res. 2005;44(1):68–97.
44. D'Souza AA, Shegokar R. Polyethylene glycol (PEG): a versatile polymer for pharmaceutical applications. Expert Opin Drug Deliv. 2016;13(9):1257–75.
45. Torchilin VP. Polymer-coated long-circulating microparticulate pharmaceuticals. J Microencapsul. 1998;15(1):1–19.
46. Dams ET, Laverman P, Oyen WJ, Storm G, Scherphof GL, van Der Meer JW, et al. Accelerated blood clearance and altered biodistribution of repeated injections of sterically stabilized liposomes. J Pharmacol Exp Ther. 2000;292(3):1071–9.
47. Abu Lila AS, Kiwada H, Ishida T. The accelerated blood clearance (ABC) phenomenon: clinical challenge and approaches to manage. J Control Release. 2013;172(1):38–47.
48. Laverman P, Boerman OC, Storm G. Radiolabeling of liposomes for scintigraphic imaging. Methods Enzymol. 2003;373:234–48.
49. Wang X, Ishida T, Kiwada H. Anti-PEG IgM elicited by injection of liposomes is involved in the enhanced blood clearance of a subsequent dose of PEGylated liposomes. J Control Release. 2007;119(2):236–44.
50. Dézsi L, Fülöp T, Mészáros T, Szénási G, Urbanics R, Vázsonyi C, et al. Features of complement activation-related pseudoallergy to liposomes with different surface charge and PEGylation: comparison of the porcine and rat responses. J Control Release. 2014;195:2–10.
51. Kozma GT, Mészáros T, Vashegyi I, Fülöp T, Örfi E, Dézsi L, et al. Pseudo-anaphylaxis to polyethylene glycol (PEG)-coated liposomes: roles of anti-PEG IgM and complement activation in a porcine model of human infusion reactions. ACS Nano. 2019;13(8):9315–24.

52. Szebeni J. Complement activation-related pseudoallergy: a new class of drug-induced acute immune toxicity. Toxicology. 2005;216(2–3):106–21.
53. Verhoef JJ, Carpenter JF, Anchordoquy TJ, Schellekens H. Potential induction of anti-PEG antibodies and complement activation toward PEGylated therapeutics. Drug Discov Today. 2014;19(12):1945–52.
54. Hoang Thi TT, Pilkington EH, Nguyen DH, Lee JS, Park KD, Truong NP. The importance of poly(ethylene glycol) alternatives for overcoming PEG immunogenicity in drug delivery and bioconjugation. Polymers. 2020;12(2):298.
55. Sundaram P, Kurniawan H, Byrne ME, Wower J. Therapeutic RNA aptamers in clinical trials. Eur J Pharm Sci. 2013;48(1–2):259–71.
56. Ellington AD, Szostak JW. In vitro selection of RNA molecules that bind specific ligands. Nature. 1990;346(6287):818–22.
57. Tuerk C, Gold L. Systematic evolution of ligands by exponential enrichment: RNA ligands to bacteriophage T4 DNA polymerase. Science (New York, NY). 1990;249(4968):505–10.
58. Guo P. The emerging field of RNA nanotechnology. Nat Nanotechnol. 2010;5(12):833–42.
59. Guo P, Haque F, Hallahan B, Reif R, Li H. Uniqueness, advantages, challenges, solutions, and perspectives in therapeutics applying RNA nanotechnology. Nucleic Acid Ther. 2012;22(4):226–45.
60. Liu Y, Kuan CT, Mi J, Zhang X, Clary BM, Bigner DD, et al. Aptamers selected against the unglycosylated EGFRvIII ectodomain and delivered intracellularly reduce membrane-bound EGFRvIII and induce apoptosis. Biol Chem. 2009;390(2):137–44.
61. Esposito CL, Passaro D, Longobardo I, Condorelli G, Marotta P, Affuso A, et al. A neutralizing RNA aptamer against EGFR causes selective apoptotic cell death. PLoS One. 2011;6(9):e24071.
62. Thiel KW, Hernandez LI, Dassie JP, Thiel WH, Liu X, Stockdale KR, et al. Delivery of chemo-sensitizing siRNAs to HER2+-breast cancer cells using RNA aptamers. Nucleic Acids Res. 2012;40(13):6319–37.
63. Chen CH, Chernis GA, Hoang VQ, Landgraf R. Inhibition of heregulin signaling by an aptamer that preferentially binds to the oligomeric form of human epidermal growth factor receptor-3. Proc Natl Acad Sci U S A. 2003;100(16):9226–31.
64. Rockey WM, Hernandez FJ, Huang SY, Cao S, Howell CA, Thomas GS, et al. Rational truncation of an RNA aptamer to prostate-specific membrane antigen using computational structural modeling. Nucleic Acid Ther. 2011;21(5):299–314.
65. Mann AP, Somasunderam A, Nieves-Alicea R, Li X, Hu A, Sood AK, et al. Identification of thioaptamer ligand against E-selectin: potential application for inflamed vasculature targeting. PLoS One. 2010;5(9):e13050.
66. Shigdar S, Lin J, Yu Y, Pastuovic M, Wei M, Duan W. RNA aptamer against a cancer stem cell marker epithelial cell adhesion molecule. Cancer Sci. 2011;102(5):991–8.
67. Mayer G, Ahmed MS, Dolf A, Endl E, Knolle PA, Famulok M. Fluorescence-activated cell sorting for aptamer SELEX with cell mixtures. Nat Protoc. 2010;5(12):1993–2004.
68. Davis KA, Lin Y, Abrams B, Jayasena SD. Staining of cell surface human CD4 with 2′-F-pyrimidine-containing RNA aptamers for flow cytometry. Nucleic Acids Res. 1998;26(17):3915–24.
69. Kang KN, Lee YS. RNA aptamers: a review of recent trends and applications. Adv Biochem Eng Biotechnol. 2013;131:153–69.
70. Zhou J, Bobbin ML, Burnett JC, Rossi JJ. Current progress of RNA aptamer-based therapeutics. Front Genet. 2012;3:234.
71. Mulhbacher J, St-Pierre P, Lafontaine DA. Therapeutic applications of ribozymes and riboswitches. Curr Opin Pharmacol. 2010;10(5):551–6.
72. Fanning G, Amado R, Symonds G. Gene therapy for HIV/AIDS: the potential for a new therapeutic regimen. J Gene Med. 2003;5(8):645–53.
73. Serganov A. The long and the short of riboswitches. Curr Opin Struct Biol. 2009;19(3):251–9.

74. Blount KF, Breaker RR. Riboswitches as antibacterial drug targets. Nat Biotechnol. 2006;24(12):1558–64.
75. Suess B, Weigand JE. Engineered riboswitches: overview, problems and trends. RNA Biol. 2008;5(1):24–9.
76. Berens C, Groher F, Suess B. RNA aptamers as genetic control devices: the potential of riboswitches as synthetic elements for regulating gene expression. Biotechnol J. 2015;10(2):246–57.
77. Jasinski DL, Khisamutdinov EF, Lyubchenko YL, Guo P. Physicochemically tunable polyfunctionalized RNA square architecture with fluorogenic and ribozymatic properties. ACS Nano. 2014;8(8):7620–9.
78. Afonin KA, Bindewald E, Kireeva M, Shapiro BA. Computational and experimental studies of reassociating RNA/DNA hybrids containing split functionalities. Methods Enzymol. 2015;553:313–34.
79. Afonin KA, Viard M, Kagiampakis I, Case CL, Dobrovolskaia MA, Hofmann J, et al. Triggering of RNA interference with RNA-RNA, RNA-DNA, and DNA-RNA nanoparticles. ACS Nano. 2015;9(1):251–9.
80. Afonin KA, Viard M, Martins AN, Lockett SJ, Maciag AE, Freed EO, et al. Activation of different split functionalities on re-association of RNA-DNA hybrids. Nat Nanotechnol. 2013;8(4):296–304.
81. Bindewald E, Afonin KA, Viard M, Zakrevsky P, Kim T, Shapiro BA. Multistrand structure prediction of nucleic acid assemblies and design of RNA switches. Nano Lett. 2016;16(3):1726–35.
82. Fire A, Xu S, Montgomery MK, Kostas SA, Driver SE, Mello CC. Potent and specific genetic interference by double-stranded RNA in Caenorhabditis elegans. Nature. 1998;391(6669):806–11.
83. Aagaard L, Rossi JJ. RNAi therapeutics: principles, prospects and challenges. Adv Drug Deliv Rev. 2007;59(2–3):75–86.
84. Ni JZ, Chen E, Gu SG. Complex coding of endogenous siRNA, transcriptional silencing and H3K9 methylation on native targets of germline nuclear RNAi in C. elegans. BMC Genomics. 2014;15(1):1157.
85. Roh YH, Deng JZ, Dreaden EC, Park JH, Yun DS, Shopsowitz KE, et al. A multi-RNAi microsponge platform for simultaneous controlled delivery of multiple small interfering RNAs. Angew Chem Int Ed Engl. 2016;55(10):3347–51.
86. Wheeler LA, Vrbanac V, Trifonova R, Brehm MA, Gilboa-Geffen A, Tanno S, et al. Durable knockdown and protection from HIV transmission in humanized mice treated with gel-formulated CD4 aptamer-siRNA chimeras. Mol Ther. 2013;21(7):1378–89.
87. Afonin KA, Viard M, Koyfman AY, Martins AN, Kasprzak WK, Panigaj M, et al. Multifunctional RNA nanoparticles. Nano Lett. 2014;14(10):5662–71.
88. Obad S, dos Santos CO, Petri A, Heidenblad M, Broom O, Ruse C, et al. Silencing of microRNA families by seed-targeting tiny LNAs. Nat Genet. 2011;43(4):371–8.
89. Ye X, Hemida M, Zhang HM, Hanson P, Ye Q, Yang D. Current advances in Phi29 pRNA biology and its application in drug delivery. Wiley Interdiscip Rev RNA. 2012;3(4):469–81.
90. Ye X, Liu Z, Hemida MG, Yang D. Targeted delivery of mutant tolerant anti-coxsackievirus artificial microRNAs using folate conjugated bacteriophage Phi29 pRNA. PLoS One. 2011;6(6):e21215.
91. Khisamutdinov EF, Li H, Jasinski DL, Chen J, Fu J, Guo P. Enhancing immunomodulation on innate immunity by shape transition among RNA triangle, square and pentagon nanovehicles. Nucleic Acids Res. 2014;42(15):9996–10004.
92. Gantier MP, Tong S, Behlke MA, Irving AT, Lappas M, Nilsson UW, et al. Rational design of immunostimulatory siRNAs. Mol Ther. 2010;18(4):785–95.
93. Khedri M, Rafatpanah H, Abnous K, Ramezani P, Ramezani M. Cancer immunotherapy via nucleic acid aptamers. Int Immunopharmacol. 2015;29(2):926–36.

94. Boltz A, Piater B, Toleikis L, Guenther R, Kolmar H, Hock B. Bi-specific aptamers mediating tumor cell lysis. J Biol Chem. 2011;286(24):21896–905.
95. Chen AK, Behlke MA, Tsourkas A. Avoiding false-positive signals with nuclease-vulnerable molecular beacons in single living cells. Nucleic Acids Res. 2007;35(16):e105.
96. Chen AK, Behlke MA, Tsourkas A. Efficient cytosolic delivery of molecular beacon conjugates and flow cytometric analysis of target RNA. Nucleic Acids Res. 2008;36(12):e69.
97. Mhlanga MM, Vargas DY, Fung CW, Kramer FR, Tyagi S. tRNA-linked molecular beacons for imaging mRNAs in the cytoplasm of living cells. Nucleic Acids Res. 2005;33(6):1902–12.
98. Rhee WJ, Santangelo PJ, Jo H, Bao G. Target accessibility and signal specificity in live-cell detection of BMP-4 mRNA using molecular beacons. Nucleic Acids Res. 2008;36(5):e30.
99. Kim JK, Choi KJ, Lee M, Jo MH, Kim S. Molecular imaging of a cancer-targeting theragnostics probe using a nucleolin aptamer- and microRNA-221 molecular beacon-conjugated nanoparticle. Biomaterials. 2012;33(1):207–17.
100. Peng XH, Cao ZH, Xia JT, Carlson GW, Lewis MM, Wood WC, et al. Real-time detection of gene expression in cancer cells using molecular beacon imaging: new strategies for cancer research. Cancer Res. 2005;65(5):1909–17.
101. Bryson JM, Fichter KM, Chu WJ, Lee JH, Li J, Madsen LA, et al. Polymer beacons for luminescence and magnetic resonance imaging of DNA delivery. Proc Natl Acad Sci U S A. 2009;106(40):16913–8.
102. Qiu L, Wu C, You M, Han D, Chen T, Zhu G, et al. A targeted, self-delivered, and photocontrolled molecular beacon for mRNA detection in living cells. J Am Chem Soc. 2013;135(35):12952–5.
103. Liu TW, Akens MK, Chen J, Wise-Milestone L, Wilson BC, Zheng G. Imaging of specific activation of photodynamic molecular beacons in breast cancer vertebral metastases. Bioconjug Chem. 2011;22(6):1021–30.
104. Mhlanga MM, Tyagi S. Using tRNA-linked molecular beacons to image cytoplasmic mRNAs in live cells. Nat Protoc. 2006;1(3):1392–8.
105. Amrute-Nayak M, Bullock SL. Single-molecule assays reveal that RNA localization signals regulate dynein-dynactin copy number on individual transcript cargoes. Nat Cell Biol. 2012;14(4):416–23.
106. Tadakuma H, Ishihama Y, Shibuya T, Tani T, Funatsu T. Imaging of single mRNA molecules moving within a living cell nucleus. Biochem Biophys Res Commun. 2006;344(3):772–9.
107. Vargas DY, Raj A, Marras SA, Kramer FR, Tyagi S. Mechanism of mRNA transport in the nucleus. Proc Natl Acad Sci U S A. 2005;102(47):17008–13.
108. Zheng J, Yang R, Shi M, Wu C, Fang X, Li Y, et al. Rationally designed molecular beacons for bioanalytical and biomedical applications. Chem Soc Rev. 2015;44(10):3036–55.
109. Han SX, Jia X, Ma JL, Zhu Q. Molecular beacons: a novel optical diagnostic tool. Arch Immunol Ther Exp. 2013;61(2):139–48.
110. Kim YS, Raston NH, Gu MB. Aptamer-based nanobiosensors. Biosens Bioelectron. 2016;76:2–19.
111. Kraemer S, Vaught JD, Bock C, Gold L, Katilius E, Keeney TR, et al. From SOMAmer-based biomarker discovery to diagnostic and clinical applications: a SOMAmer-based, streamlined multiplex proteomic assay. PLoS One. 2011;6(10):e26332.
112. Okamoto A, Tanaka K, Saito I. DNA logic gates. J Am Chem Soc. 2004;126(30):9458–63.
113. Adleman LM. Molecular computation of solutions to combinatorial problems. Science (New York, NY). 1994;266(5187):1021–4.
114. Lee T, Yagati AK, Pi F, Sharma A, Choi JW, Guo P. Construction of RNA-quantum dot chimera for nanoscale resistive biomemory application. ACS Nano. 2015;9(7):6675–82.
115. Zalatan JG, Lee ME, Almeida R, Gilbert LA, Whitehead EH, La Russa M, et al. Engineering complex synthetic transcriptional programs with CRISPR RNA scaffolds. Cell. 2015;160(1–2):339–50.
116. Yin H, Song CQ, Dorkin JR, Zhu LJ, Li Y, Wu Q, et al. Therapeutic genome editing by combined viral and non-viral delivery of CRISPR system components in vivo. Nat Biotechnol. 2016;34(3):328–33.

117. Jiang W, Marraffini LA. CRISPR-Cas: new tools for genetic manipulations from bacterial immunity systems. Ann Rev Microbiol. 2015;69:209–28.
118. Ramanan V, Shlomai A, Cox DB, Schwartz RE, Michailidis E, Bhatta A, et al. CRISPR/Cas9 cleavage of viral DNA efficiently suppresses hepatitis B virus. Sci Rep. 2015;5:10833.
119. Kennedy EM, Bassit LC, Mueller H, Kornepati AVR, Bogerd HP, Nie T, et al. Suppression of hepatitis B virus DNA accumulation in chronically infected cells using a bacterial CRISPR/Cas RNA-guided DNA endonuclease. Virology. 2015;476:196–205.
120. Hendel A, Bak RO, Clark JT, Kennedy AB, Ryan DE, Roy S, et al. Chemically modified guide RNAs enhance CRISPR-Cas genome editing in human primary cells. Nat Biotechnol. 2015;33(9):985–9.
121. Lee JB, Hong J, Bonner DK, Poon Z, Hammond PT. Self-assembled RNA interference microsponges for efficient siRNA delivery. Nat Mater. 2012;11(4):316–22.
122. Sharma A, Haque F, Pi F, Shlyakhtenko LS, Evers BM, Guo P. Controllable self-assembly of RNA dendrimers. Nanomedicine. 2016;12(3):835–44.
123. Li H, Zhang K, Pi F, Guo S, Shlyakhtenko L, Chiu W, et al. Controllable self-assembly of RNA tetrahedrons with precise shape and size for cancer targeting. Adv Mater. 2016;28(34):7501–7.
124. Khisamutdinov EF, Jasinski DL, Li H, Zhang K, Chiu W, Guo P. Fabrication of RNA 3D nanoprisms for loading and protection of small RNAs and model drugs. Adv Mater. 2016;28(45):10079–87.
125. Guo S, Tschammer N, Mohammed S, Guo P. Specific delivery of therapeutic RNAs to cancer cells via the dimerization mechanism of phi29 motor pRNA. Hum Gene Ther. 2005;16(9):1097–109.
126. Kwon YJ. Before and after endosomal escape: roles of stimuli-converting siRNA/polymer interactions in determining gene silencing efficiency. Acc Chem Res. 2012;45(7):1077–88.
127. Lavergne T, Bertrand JR, Vasseur JJ, Debart F. A base-labile group for 2′-OH protection of ribonucleosides: a major challenge for RNA synthesis. Chemistry. 2008;14(30):9135–8.
128. Beaucage SL, Reese CB. Recent advances in the chemical synthesis of RNA. Curr Protoc Nucleic Acid Chem. 2009;Chapter 2:Unit 2.16 1–31.
129. Jasinski DL, Schwartz CT, Haque F, Guo P. Large scale purification of RNA nanoparticles by preparative ultracentrifugation. Methods Mol Biol. 2015;1297:67–82.
130. Flinders J, DeFina SC, Brackett DM, Baugh C, Wilson C, Dieckmann T. Recognition of planar and nonplanar ligands in the malachite green-RNA aptamer complex. Chembiochem. 2004;5(1):62–72.
131. Zuker M. Mfold web server for nucleic acid folding and hybridization prediction. Nucleic Acids Res. 2003;31(13):3406–15.
132. Andronescu M, Fejes AP, Hutter F, Hoos HH, Condon A. A new algorithm for RNA secondary structure design. J Mol Biol. 2004;336(3):607–24.
133. Ding Y, Chan CY, Lawrence CE. Sfold web server for statistical folding and rational design of nucleic acids. Nucleic Acids Res. 2004;32(Web Server issue):W135–41.
134. Zadeh JN, Steenberg CD, Bois JS, Wolfe BR, Pierce MB, Khan AR, et al. NUPACK: analysis and design of nucleic acid systems. J Comput Chem. 2011;32(1):170–3.
135. Bindewald E, Afonin K, Jaeger L, Shapiro BA. Multistrand RNA secondary structure prediction and nanostructure design including pseudoknots. ACS Nano. 2011;5(12):9542–51.
136. Bindewald E, Grunewald C, Boyle B, O'Connor M, Shapiro BA. Computational strategies for the automated design of RNA nanoscale structures from building blocks using NanoTiler. J Mol Graph Model. 2008;27(3):299–308.
137. Yingling YG, Shapiro BA. Computational design of an RNA hexagonal nanoring and an RNA nanotube. Nano Lett. 2007;7(8):2328–34.
138. Afonin KA, Bindewald E, Yaghoubian AJ, Voss N, Jacovetty E, Shapiro BA, et al. In vitro assembly of cubic RNA-based scaffolds designed in silico. Nat Nanotechnol. 2010;5(9):676–82.

# Chapter 25
# Current Development of RNA Delivery Systems

**Abstract** Precise and targeted delivery remains a significant challenge in the field of RNA therapeutics, essential for their successful clinical application. RNA molecules are inherently unstable in biological environments due to their susceptibility to degradation by RNases. To circumvent this challenge, RNA therapeutics require effective delivery systems that can protect and deliver the therapeutic RNA to its intended destination within cells. The development of innovative RNA delivery vehicles is therefore crucial. These vehicles serve multiple roles: shielding RNA from extracellular RNases, facilitating its cellular uptake, and promoting its release into the cytoplasm where it can undergo translation and produce therapeutic proteins. Various delivery strategies have been explored, including lipid nanoparticles, polymer-based carriers, viral vectors, and peptide-based complexes. Each approach offers distinct advantages in terms of stability, efficacy, and specificity in targeting different tissues or cell types. Recent advancements in RNA delivery technologies have focused on enhancing delivery efficiency, reducing immunogenicity, and improving tissue-specific targeting. Nanoparticle-based systems, for instance, can encapsulate RNA molecules, protecting them during transit through the bloodstream and enhancing their uptake by target cells through receptor-mediated endocytosis. Similarly, viral vectors engineered for RNA delivery can exploit natural mechanisms of cell entry, although concerns about safety and immune responses persist. Moreover, the ability to precisely engineer delivery systems allows for the customization of RNA therapeutics tailored to specific diseases or patient populations. This personalized approach holds promise for treating a wide range of disorders, from genetic diseases to cancers, by delivering therapeutic RNA directly to affected tissues while minimizing off-target effects. The ongoing advancements in RNA delivery systems are paving the way for the clinical translation of RNA therapeutics. By addressing the challenges of stability, delivery efficiency, and specificity, these innovations promise to unlock the full potential of RNA-based therapies in improving patient outcomes and advancing personalized medicine.



P. E. Saw, E. Song, *RNA Therapeutics in Human Diseases*,
https://doi.org/10.1007/978-981-96-3041-7_25

## 25.1 Introduction

RNA therapeutics represent a promising frontier in modern medicine, offering new avenues for treating diseases at the genetic level. Unlike traditional small molecule drugs, RNA-based therapies harness the body's own cellular machinery to target specific genes or proteins involved in disease processes. This approach holds immense potential for treating a wide range of disorders, including genetic diseases, infectious diseases, and cancers, by correcting aberrant gene expression or protein production. However, the journey from theory to clinical application faces several formidable challenges, primarily centered around the unique properties of RNA molecules. Currently, mRNAs, siRNAs, and ASOs are among the key types of RNA used in therapeutic applications. These molecules are highly susceptible to degradation by RNases present in biological fluids, limiting their stability and efficacy when administered systemically [1]. Additionally, their large size and negative charge hinder their cellular uptake and penetration across cellular membranes. To overcome these barriers, researchers have developed sophisticated RNA delivery systems designed to protect RNA molecules during transit, facilitate their uptake by target cells, and ensure their efficient release into the cytoplasm where they can exert their therapeutic effects [2]. Current RNA delivery systems encompass a variety of platforms, each tailored to address specific challenges and optimize therapeutic outcomes.

In drug delivery, various smart biomaterials and nanoplatforms have been designed within these few decades [3–5]. Taking lessons and insights from these, researchers used similar tactics to develop RNA carriers. Lipid nanoparticles (LNPs) represent one of the most advanced and widely studied RNA delivery systems. LNPs can encapsulate RNA molecules within their lipid bilayers, shielding them from RNase degradation in the bloodstream. These nanoparticles are designed to exploit natural processes of cellular uptake, such as receptor-mediated endocytosis, enabling efficient delivery of RNA into the cytoplasm where the RNA can be translated into therapeutic proteins or regulate gene expression. Polymer-based delivery systems offer another versatile approach. Polymers can electrostatically interact with RNA molecules to form nanoparticles or complexes that protect RNA from degradation and facilitate cellular uptake. These systems can be finely tuned to control the release kinetics of RNA within target cells, enhancing therapeutic efficacy while minimizing off-target effects. Viral vectors, such as adeno-associated viruses (AAVs) and lentiviruses, represent a third category of RNA delivery systems [6]. These vectors leverage the natural ability of viruses to infect cells and deliver genetic material. Engineered viral vectors can be modified to carry therapeutic RNA payloads, delivering them directly into target cells' nuclei for sustained gene expression or silencing. While viral vectors offer efficient delivery capabilities, concerns about immunogenicity and potential side effects necessitate careful design and testing in clinical settings. Peptide/protein-based delivery systems have also emerged as promising alternatives [7]. These systems utilize peptides or peptide conjugates to facilitate RNA binding, cellular uptake, and endosomal escape, ensuring efficient delivery of RNA payloads to the cytoplasm. Peptide-based approaches offer flexibility in design and can be engineered for enhanced specificity to target tissues or cell types, reducing systemic toxicity and improving therapeutic outcomes. Despite significant progress, challenges persist in optimizing RNA delivery systems for clinical use. Issues such as immunogenicity, off-target effects, and scalability of production remain critical considerations [8–10]. Moreover, the development of personalized RNA therapies tailored to individual genetic profiles represents a future frontier, requiring precise delivery systems capable of navigating the complexities of personalized medicine.

## 25.2 Lipid-Based Nanoparticles

Liposomes form by dispersing phospholipids, which have polar head groups and non-polar tails, in water. This unique structure gives liposomes versatility in drug delivery applications [9]. They can undergo surface modifications, allowing for enhanced delivery of various therapeutic agents. A prominent instance is Doxil, a liposomal formulation of doxorubicin employed in cancer therapy. Doxil's design reduces the drug's cardiac toxicity, showcasing how nanoparticles like liposomes can alter drug distribution in the body, thereby improving safety profiles and therapeutic outcomes [11].

**Box 25.1 Doxil: The First FDA-Approved Nanocarrier Encapsulating Drug for Cancer Therapy**

Doxil, originally developed by Johnson & Johnson, marked a significant milestone as the first FDA-approved liposomal drug in the United States. It received approval in 1995 for the treatment of AIDS-related Kaposi's sarcoma, followed by approval for ovarian cancer in 1999 [11]. The development of Doxil stemmed from the desire to mitigate the severe side effects associated with conventional doxorubicin, a potent chemotherapy agent used in cancer treatment but notorious for its cardiotoxicity. The formulation of Doxil encapsulates doxorubicin within a liposomal membrane, composed of phospholipids similar to those found in cell membranes. This encapsulation serves several crucial purposes: it alters the pharmacokinetics of doxorubicin, prolonging its circulation time in the bloodstream while reducing its exposure to normal tissues. By selectively accumulating in tumor tissues with leaky vasculature (a phenomenon known as the enhanced permeability and retention effect), Doxil enhances the drug's concentration at the tumor site while sparing healthy tissues from damage. Over the years, Doxil has undergone various clinical trials and refinements to optimize its efficacy and safety profile. It has shown efficacy not only in Kaposi's sarcoma and ovarian cancer but also in other solid tumors, such as breast cancer and multiple myeloma. Despite its

(continued)

**Box 25.1** (continued)
success, Doxil faced challenges, including supply shortages due to manufacturing issues and the emergence of generic versions following patent expiration. Today, liposomal formulations continue to be a promising approach in drug delivery, with Doxil serving as a pioneering example. Ongoing research explores new applications for liposomal technologies, including the development of targeted therapies and combination treatments. The legacy of Doxil underscores the transformative potential of liposomal drug delivery systems in improving cancer treatment outcomes and reducing the toxicity associated with conventional chemotherapy.

Liposomes are particularly advantageous because they mimic the composition of cell membranes, allowing them to effectively encapsulate mRNA when combined with cationic lipids [12]. The addition of cholesterol-modified lipids to these complexes enhances stability and improves transfection efficacy. For instance, Hattori et al. utilized cationic cholesterol-modified liposomes for successful siRNA delivery to mouse liver [13]. Liposomes were employed to deliver intravenous mRNA vaccines targeting DCs with tumor antigens such as NY-ESO-1, MAGE-A3, tyrosinase, and TPTE in advanced malignant melanoma patients, achieving antigen expression in spleen cells [14]. Despite these advantages, liposome delivery systems have notable limitations: (1) their instability can lead to fusion or leakage of RNA, reducing delivery efficiency; (2) they capture less RNA; (3) they pose risks if oxidized; and (4) their size can vary significantly, leading to increased variability between batches [9, 10, 15]. Several challenges linked to liposome stability and RNA protection have been mitigated through advancements in their surface engineering. Lipid nanoparticles (LNPs), composed of cationic lipids, cholesterol, polyethylene glycol (PEG), and possessing a hydrophilic core, not only safeguard anionic RNA from degradation but also extend its bloodstream presence.

## 25.3 Polymer Nanomaterials

Polymer nanomaterials are man-made substances composed of various building blocks that form complex structures. They often include synthetic polymers such as PLGA (poly(lactic-*co*-glycolic acid)), PLA (polylactic acid), chitosan, gelatin, polycaprolactone, and polyalkyl-cyanoacrylates [16, 17]. In more detail, these polymer nanomaterials are engineered at the nanoscale to achieve specific biomedical objectives. They can be tailored to degrade at controlled rates, ensuring sustained release of drugs or therapeutic molecules over extended periods. This capability is crucial for maintaining therapeutic levels of drugs in the body, thereby enhancing treatment efficacy and reducing side effects. Moreover, their versatility allows for the incorporation of various therapeutic agents, making them valuable in targeted drug delivery systems. Furthermore, the ability of these materials to protect sensitive substances from degradation in biological environments enhances their utility in medicine. Their biocompatibility ensures minimal toxicity and immunogenicity, crucial for applications ranging from cancer treatment to vaccine delivery. This combination of properties underscores the significance of polymer nanomaterials in advancing medical therapies through precise and effective delivery systems [18, 19]. Polymeric nanomaterials owe their adaptability to the capability of finely tuning polymer properties through minor chemical modifications of their basic units. This refinement allows for precise control over how therapeutic agents are released. Moreover, these polymers can be specifically designed to create injectable nanoparticles suitable for diverse administration routes, including intravenous, intramuscular, subdermal, or intraperitoneal delivery. This versatility ensures sustained release of drugs or therapeutic molecules over extended periods, ranging from weeks to months. In practice, these modifications in polymer chemistry enable customization of degradation rates and release kinetics, aligning with therapeutic needs. For instance, slow-release formulations can maintain steady drug levels in the bloodstream or target tissues, optimizing treatment outcomes while minimizing side effects. This capability is crucial in oncology, infectious diseases, and chronic conditions where controlled drug delivery is paramount. Furthermore, the development of injectable nanoparticle systems underscores their utility in clinical settings. By tailoring polymer characteristics, such as biocompatibility and biodegradability, researchers can advance the field of nanomedicine, offering innovative solutions for precise and sustained drug delivery across diverse medical applications [20].

However, the application of polymers in delivery systems is not without challenges, notably their potential to induce inflammation due to non-degradation or byproduct formation. For instance, PLGA degrades into lactic acid and glycolic acid, which can lower pH locally and trigger inflammation [21]. Yet, these issues can be overcome, as evidenced by the development of lipid-polymer hybrid nanoparticles by Zhang et al. [22]. These hybrids combine a polymer core with a lipid outer layer, mirroring the cellular membrane and allowing for enhanced control over drug release and improved targeting capabilities. This design has shown promise in preclinical studies for RNA therapeutics delivery, particularly in xenograft tumor models [23].

## 25.4 Silica Nanoparticles

Mesoporous silica nanoparticles (MSNPs) include an amorphous silicon dioxide matrix characterized by highly ordered and tunable mesoporous porosity [24]. Key attributes of MSNPs are their extensive surface areas, voluminous pore capacities, and the ease with which they can be chemically modified, thanks to well-established silanol chemistry. The surfaces of these nanoparticles can be tailored with positively charged groups to bind and transport negatively charged RNA molecules.

Furthermore, the uniformity of the pores allows for precise control over particle properties, enhancing their capacity to encapsulate nucleic acids and ensure efficient delivery. One challenge with MSNPs is their tendency to aggregate, which can pose risks such as thrombosis or tissue injury when injected. However, modifying the nanoparticle surfaces with PEG has proven effective in reducing aggregation and minimizing potential tissue damage. MSNPs are biocompatible [25, 26].

Innovative applications of MSNP are exemplified in the work by Meka et al., who engineered MSNPs with large pores and a bicontinuous cubic mesostructure specifically for delivering siRNA to human colon cancer cells (HCT116). Their research demonstrated the nanoparticles' impressive drug delivery capabilities, emphasizing their efficacy in suppressing tumor proteins. The design of MSNPs with large pores and a bicontinuous cubic mesostructure is particularly advantageous for drug delivery. These characteristics facilitate high loading capacity and controlled release of therapeutic agents like siRNA, which are crucial for targeted cancer therapy. By efficiently delivering siRNA to HCT116 cells, the MSNPs enabled significant suppression of tumor proteins, highlighting their potential as a promising tool in oncology research [27]. In another innovative study, Lee et al. developed hybrid mesoporous silica nanoparticles (MSNPs) featuring a dual-pore architecture to address complex cancer therapies. Their design involved integrating positively charged structures with large pores for loading anionic siRNA drugs targeting B cell lymphoma 2 (Bcl-2) and negatively charged structures with smaller pores for encapsulating the chemotherapeutic agent doxorubicin. This dual-pore strategy enabled the simultaneous delivery of genetic and chemotherapeutic treatments within a single nanoparticle platform. By incorporating siRNA against Bcl-2, known for its role in cancer cell survival, and doxorubicin, a potent chemotherapy drug, the hybrid MSNPs demonstrated versatility in enhancing therapeutic outcomes for cancer treatment [28].

## 25.5 Inorganic Nanostructures

Some inorganic nanoparticles in preclinical and clinical studies including gold nanoparticles (GNPs), quantum dots (QDots), nanographene oxide (GO), silver nanoparticles (AuNPs), and carbon nanotubes (CNTs) are prominent nanostructures synthesized for their capability to encapsulate RNA, providing protection against enzymatic degradation and enabling targeted delivery to precise disease sites [29, 30]. Many researchers are also developing novel inorganic nanoparticles that could be more biocompatible toward clinical translation [31, 32]. One notable application is demonstrated by Zheng et al., who utilized gold nanoparticles to deliver therapeutic nucleic acids, specifically siRNA targeting EGFR (epidermal growth factor receptor), via topical administration on both mouse and human skin. The study by Zheng et al. exemplifies a practical approach for achieving direct and localized gene silencing using gold nanoparticles. By encapsulating siRNA within these nanoparticles, they effectively shielded the nucleic acids from degradation while facilitating their targeted delivery to skin cells expressing EGFR. This targeted approach is critical in dermatological treatments where precise gene regulation can influence therapeutic outcomes significantly [33]. Li et al. demonstrated the efficacy of quantum dot-siRNA complexes in downregulating the HPV18 E6 gene in HeLa cells. This study underscores the potential of quantum dots in precise gene targeting applications. Quantum dots are semiconductor nanoparticles known for their unique optical and electronic properties, making them suitable for a variety of biomedical applications, including gene delivery. In the study by Li et al., quantum dots were utilized to complex with siRNA specifically targeting the HPV18 E6 gene, a known oncogene involved in cervical cancer progression. By effectively delivering and releasing the siRNA payload to HeLa cells, the quantum dot-siRNA complexes successfully suppressed the expression of HPV18 E6, highlighting their potential for precise and targeted gene silencing [34].

Further expanding the scope of these nanostructures, Yang et al. modified nanographene oxide with gadolinium (Gd-NGO) for dual purposes: delivering the anticancer drug epirubicin alongside Let-7g miRNA, which acts as a tumor suppressor by reducing expression levels of the Ras oncogene family. This strategy not only treated glioblastoma in mice but also enabled imaging of the treatment process, showcasing the multifunctional capabilities of nanographene oxide [35]. Additionally, Kam et al. explored the utility of functionalized carbon nanotubes for gene silencing, where they successfully delivered siRNA targeting lamin A/C. This intervention effectively suppressed the expression of this structural protein in HeLa cells, illustrating the carbon nanotubes' efficacy in gene delivery applications [36]. Each of these examples underscores the versatility and potential of nanostructures in advanced therapeutic and diagnostic strategies.

## 25.6 Biomimetic Nanoparticles

Advanced surface modifications in nanotechnology have enabled the creation of biomimetic nanoparticles, which mimic biological structures by incorporating functional membrane proteins on their exteriors. Molinaro et al. have pioneered this approach by developing "leukosomes," where leukocyte membrane proteins are integrated into the surfaces of liposomes. This innovative design enhances the nanoparticles' ability to target inflammatory sites effectively. Leukosomes derive their name from their composition, resembling leukocyte membranes. They leverage the presence of specific membrane proteins, such as LFA-1 and CD-45, which facilitate adhesion to activated endothelial cells found at inflammatory sites. This targeted interaction enhances the nanoparticles' specificity for delivering therapeutic payloads precisely where they are needed most [37]. This biomimetic approach also demonstrates natural anti-inflammatory properties through interactions with endothelial macrophages. In a mouse model of sepsis induced by lipopolysaccharide (LPS), the application of these macrophage-derived biomimetic nanoparticles led to a reduction in the expression of pro-inflammatory genes, including IL-6,

IL-1β, and TNF-α. Concurrently, there was an enhancement in the expression of anti-inflammatory genes such as IL-10 and TGF-β. The interaction of biomimetic nanoparticles with endothelial macrophages mimics the natural behavior of leukocytes in modulating immune responses. By integrating membrane proteins like LFA-1 and CD-45, these nanoparticles effectively adhere to activated endothelial cells, facilitating targeted delivery to sites of inflammation. This targeted approach not only improves the efficacy of drug delivery but also enhances the nanoparticles' ability to regulate the inflammatory environment by modulating gene expression [38]. The anti-inflammatory capabilities of these leukosomes have also been validated in various other conditions, such as inflammatory bowel disease (IBD) [39], atherosclerosis [40, 41], and cancer [41].

## 25.7 Dynamic Polyplexes

The natural evolution of lipid-free viruses, which utilize bare protein capsids for nucleic acid transfer, underscores the feasibility and effectiveness of lipid-free approaches in nucleic acid delivery. These viruses have evolved to efficiently protect and deliver genetic material, demonstrating a biological precedent for lipid-free nucleic acid transfer strategies. Inspired by this evolutionary success, researchers have developed synthetic systems such as cationic polymer-based complexes known as "polyplexes" and "polymer micelleplexes." These systems mimic the protective and delivery capabilities of viruses by forming stable complexes with nucleic acids. Cationic polymers interact electrostatically with negatively charged nucleic acids, forming complexes that protect the cargo from enzymatic degradation and facilitate cellular uptake. Polyplexes and polymer micelleplexes offer several advantages over lipid-based systems, including enhanced stability, tunable properties, and reduced immunogenicity [42–47]. These systems are adept at spontaneously encapsulating polyanionic nucleic acids into a compact, protected format.

When designing polymeric carriers for nucleic acid delivery, ensuring timely disassembly of nanoparticles during intracellular delivery is crucial, especially given the differences from natural lipid-based systems. Similar to viruses, which release their genetic material only upon reaching the appropriate intracellular site for replication, polymer-based carriers must maintain stability until they achieve targeted delivery. The stability of polymeric carriers relies heavily on the design of polyelectrolyte complexes formed through purely electrostatic interactions. Polymeric carriers are engineered to protect nucleic acids from degradation during circulation and facilitate their uptake into cells. Once inside cells, the carriers must efficiently release their cargo to exert therapeutic effects. This controlled release mechanism ensures that the nucleic acids are delivered intact to their intracellular targets, minimizing premature degradation and maximizing therapeutic efficacy [48].

For small nucleic acid cargoes like siRNA, incorporating hydrophobic domains into polymeric carriers can significantly enhance stability. This approach helps to shield the siRNA molecules from degradation and stabilize the polyplexes during circulation and cellular uptake. Conversely, larger nucleic acid molecules such as mRNA require different considerations. Compacting mRNA into smaller nanosized structures is crucial, especially in the initial stages of delivery, to facilitate efficient cellular uptake and protect the fragile mRNA strands. Moreover, the specific intracellular targeting and functional requirements of different nucleic acid cargoes profoundly influence the choice of polymeric carriers. While carriers initially designed for DNA delivery may be effective for some applications, modifications are often necessary to optimize delivery efficiency for mRNA or siRNA. These modifications can include adjusting the polymer structure, incorporating targeting ligands for specific cell types, or fine-tuning the release kinetics to match the cargo's therapeutic requirements [49, 50].

Polymeric carriers demonstrate enhanced efficacy through their bioresponsive and dynamic behaviors, which are crucial for overcoming two major cellular barriers encountered during nucleic acid delivery. (a) Polymeric carriers can be designed with pH-sensitive or redox-sensitive properties that enable them to respond to the acidic environment within endosomes. This responsiveness triggers the carriers to destabilize and facilitate escape from the endosomal compartment into the cytosol. This step is essential because endosomes typically degrade nucleic acids, preventing their therapeutic action. (b) Once inside the cytosol, polymeric carriers must efficiently disassemble to release their nucleic acid cargo at the site of action. This disassembly process can be engineered into the carrier design through stimuli-responsive mechanisms. For instance, changes in pH, reduction in redox potential, or enzymatic cleavage can trigger the breakdown of polymeric complexes, ensuring timely release of the nucleic acids. While effective, this process lacks the precise control and responsiveness that polymeric carriers can offer. Polymeric carriers can be finely tuned to respond to specific cues within the cellular environment, thereby enhancing their ability to deliver nucleic acids intact and functional to their intracellular targets [51, 52].

The initial generations of cationizable polymers, such as polyethylenimine (PEI) or polyamidoamine dendrimers, have shown significantly higher transfection efficiency compared to permanently cationic counterparts. This superiority is primarily attributed to their enhanced capability to escape endosomes, a crucial step in delivering nucleic acids into the cytosol. These polymers are characterized by a high density of amine groups that are partially protonated under physiological pH conditions. During the acidification process within endosomes, these groups undergo further protonation. This increased cationization strengthens the interaction between the polymer and the negatively charged endosomal membrane. The mechanism by which these polymers disrupt endosomal membranes is triggered by their charge density and the pH-dependent protonation of amine groups. This process allows them to efficiently escape the degradative environment of endosomes and protect nucleic acids from enzymatic degradation. In contrast, permanently cationic polymers lack the pH-dependent protonation capability and often exhibit lower transfection efficiency due to their inability to effectively disrupt endosomal membranes. Thus, the dynamic nature of cationizable polymers like PEI or polyamidoamine dendrimers, which can adjust their charge density in response to endosomal pH changes, enhances their utility in nucleic acid delivery systems [53–58]. Another crucial yet often overlooked aspect of these polymers is their protonation-induced increase in aqueous solubility, which plays a vital role in their functionality. For instance, Zuber and colleagues employed tyrosine-modified PEI (PEI-Y) to stabilize siRNA polyplexes hydrophobically. The effectiveness of PEI-Y hinged on its ability to dissolve self-aggregates in response to the acidic pH within endosomes [54].

Polyplex delivery systems face numerous challenges that require dynamic and adaptive solutions. One challenge, referred to as the "polyplex dilemma," involves maintaining a delicate balance where polyplexes must be stable enough to survive the extracellular environment yet disassemble efficiently upon reaching the intracellular target to release their nucleic acid cargo. Itaka et al. highlighted differences in the behavior of linear (LPEI) and branched PEI (BPEI) polyplexes; although both types facilitate endosomal escape, LPEI was more effective at disassembling and releasing plasmid DNA (pDNA) inside cells, which led to higher gene expression rates compared to BPEI polyplexes [59]. Additionally, unlike BPEI, LPEI could also transfect non-dividing cells [60]. Furthermore, the suitability of different PEI structures varies depending on the type of nucleic acid being delivered. LPEI is a preferable carrier for pDNA but less effective for siRNA or mRNA. In contrast, modifications such as succinylated branched PEI show greater efficacy for delivering siRNA and mRNA [61, 62]. Additionally, small chemical modifications to polymeric carriers can significantly enhance their delivery capabilities. Dohmen et al. reported that introducing a methylene group into every second aminoethylene unit of PEI-like oligoamines markedly improved mRNA transfer. This modification achieved a more favorable balance of cationic properties, enhancing the overall effectiveness of mRNA delivery [63]. These examples underscore the importance of fine-tuning the chemical structure and configuration of polyplexes to optimize their function for specific types of nucleic acid cargo.

## 25.8 Dynamic Polyplexes: Case Studies

### *25.8.1 Virus-Inspired Polymer for Endosomal Release (VIPER)*

Pun and colleagues innovatively developed a bioresponsive polymer known as VIPER [64, 65]. The copolymer comprises several distinct blocks: (a) Cationic poly(2-(dimethylamino)ethyl methacrylate) (PDMAEMA): This segment binds to and compacts nucleic acids, facilitating their encapsulation and protection. (b) Poly(oligo(ethylene glycol) monomethyl ether methacrylate) (pOEGMA): This block provides hydration and surface shielding to the nanoparticles, enhancing biocompatibility and reducing non-specific interactions. Reversibly hydrophobic domain with poly(2-diisopropylaminoethyl methacrylate) (pDIPAMA) and poly(pyridyl disulfide ethyl methacrylate) (pDSEMA): pDIPAMA undergoes a pH-dependent transition from hydrophobic to hydrophilic states, aiding in endosomal escape. The pDSEMA is utilized for the conjugation of melittin, a lytic peptide from bee venom, which facilitates the disruption of the endosomal lipid membrane, thereby enhancing the release of the polyplex into the cytoplasm. (c) Upon internalization and acidification within endosomes, pDIPAMA undergoes a sharp phase transition that triggers the exposure of pDSEMA-linked melittin peptides. This leads to disruption of the endosomal membrane and release of the nucleic acids into the cytoplasm. VIPER formulations have demonstrated efficacy in gene transfer to KB tumors through intratumoral application of plasmid DNA and gene silencing in the lungs via intratracheal administration of siRNA [64, 65]. However, the use of melittin raises concerns due to its potent cytotoxic and pore-forming capabilities [66]. To address this, studies on VIPER analogs that incorporate pH-specific fusion peptides instead of melittin have been conducted. These studies revealed that the pH-triggered exposure of peptides, akin to the "spring-loaded" mechanism seen in influenza hemagglutinin, helps to mitigate the cytotoxicity by ensuring that the disruptive activity of melittin is limited to the acidic environment within endosomes [67]. This innovative use of pDIPAMA's endosomal responsiveness has been further explored by other researchers, such as in the development of PD-L1 siRNA micelleplexes aimed at cancer immunotherapy [68], showcasing the broad potential applications of such pH-responsive polymers in medical science (Fig. 25.1).

### *25.8.2 mRNA Transfer by Charge-Altering Releasable Transporters (CARTs)*

Waymouth, Wender, and their colleagues innovatively created a series of amphiphilic diblock copolymers specifically designed for mRNA transfer, known as CARTs (cationic aliphatic random terpolymers) [69–71]. These CARTs consist of a polycarbonate block that features esters derived from lipophilic alcohols and cationic hydroxyethyl glycine repeat units, as illustrated in Fig. 25.3 of their publications. These structural features enable the CARTs to form complexes with mRNA through both electrostatic and hydrophobic interactions. A critical feature of these charge-altering releasable transporters (CARTs) is their capability to undergo intracellular polyplex disassembly, which enables the efficient release of mRNA into the cytoplasm. This process is facilitated by a charge-neutralizing intramolecular rearrangement mechanism. Upon cellular uptake and acidification within endosomes, the intramolecular rearrangement is triggered. This results in the degradation of the cationic polymer into smaller, neutral fragments. This process destabilizes the polyplex structure, facilitating the release of mRNA into the cytoplasm where it can

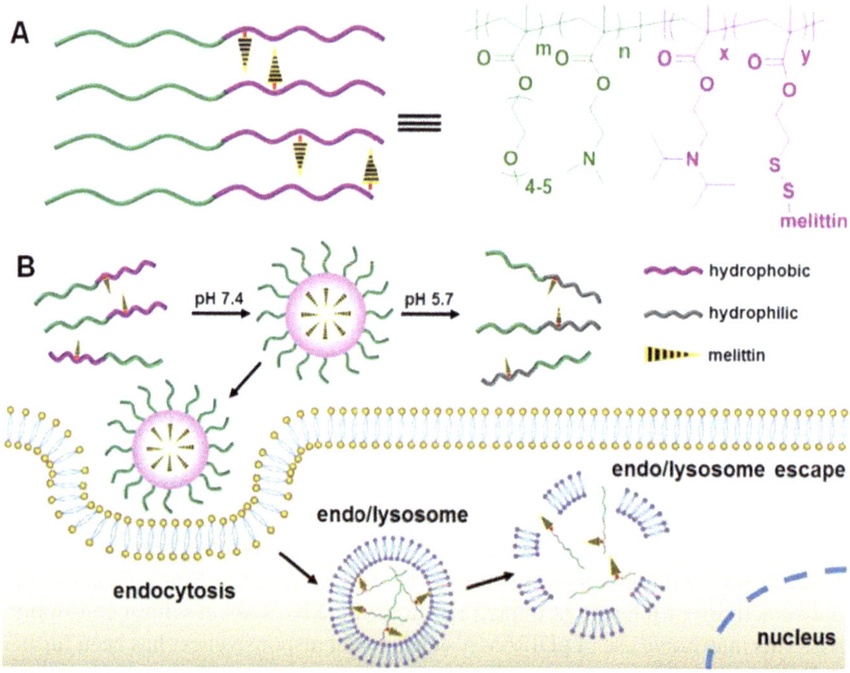


**Fig. 25.1** (**a**) Chemical Structure of VIPER: This panel displays the molecular composition of VIPER (virus-inspired polymer for endosomal release), highlighting its design elements that mimic viral mechanisms for efficient cellular entry and escape from endosomal entrapment. (**b**) Mechanism of VIPER-Induced Endo/Lysosomal Escape: This schematic illustrates the step-by-step process of how VIPER nanoparticles function within a cellular environment. Initially, at neutral pH, VIPER self-assembles into nanoparticles. These nanoparticles incorporate melittin molecules, which are sequestered within a pH-sensitive domain composed of the polymer pDIPAMA. Upon cellular uptake through endocytosis, these nanoparticles are trafficked to the acidic compartments of the endo/lysosomes. The acidic pH triggers a hydrophilic phase transition in pDIPAMA, leading to the exposure of melittin. This exposure is critical as melittin then interacts with the endo/lysosomal membrane, disrupting it and facilitating the release of the nanoparticle's cargo into the cytoplasm. This controlled disruption is pivotal for the therapeutic efficacy of the delivery system, allowing the delivered agents to bypass degradation pathways and enhance cellular bioavailability. (Figure is reproduced from [64] with permission from Angewandte Chemie)

exert its therapeutic effects, such as protein expression. By converting the cationic polymer into neutral components, the CART system minimizes the duration of exposure of cellular systems to potentially cytotoxic cationic materials. This reduction in cytotoxicity is crucial for enhancing the safety profile of nucleic acid delivery systems, ensuring that therapeutic efficacy is achieved without compromising cell viability. The versatility of the CART system was further demonstrated through the creation of libraries of CARTs with varying lipidic side groups. These variations allowed for the successful transfection of lymphocytes in vivo [70], and significant antitumor immune responses were observed following intratumoral injections [69]. In a more recent advancement, the original hydroxyethyl glycine (G) motif in the

CARTs was replaced with a hydroxyethyl lysine (K) motif [72]. This modification led to distinct CART carriers displaying specific organ targeting upon systemic administration in mice: K-CARTs showed over 90% selectivity for the lungs, whereas G-CARTs achieved more than 99% selectivity for the spleen.

## 25.9 Lipo-Xenopeptides

Lipo-xenopeptides refer to a class of bioactive compounds that combine features of lipids (liposomes or lipid-based carriers) and peptides derived from xenobiotic sources (organisms or substances foreign to the body, often from different species). This hybrid term suggests an innovative approach in drug delivery and therapeutic development. Xenopeptides are peptides derived from non-human sources, often from organisms such as bacteria, fungi, or other natural products. These peptides can exhibit diverse biological activities, including antimicrobial, anticancer, or immunomodulatory properties. Incorporating xenopeptides into drug delivery systems aims to harness their therapeutic potential while utilizing lipid-based carriers for effective transport and delivery. Lipo-xenopeptides leverage the targeting capabilities of liposomal carriers to deliver xenopeptides to specific tissues or cells. This targeted delivery minimizes off-target effects and enhances the therapeutic efficacy of the peptides. Liposomes provide a stable and biocompatible environment for xenopeptides, protecting them from enzymatic degradation and enhancing their circulation time in the body. Xenopeptides encompass a wide range of bioactivities that can be exploited for therapeutic purposes. Examples include peptides with antimicrobial properties for treating infections, anticancer peptides that selectively kill cancer cells, or peptides that modulate immune responses in autoimmune diseases. Combining liposomal technology with xenopeptides represents an innovative approach in drug development, potentially overcoming limitations of conventional drug therapies through improved efficacy, reduced toxicity, and novel mechanisms of action [73]. The advancement and precision of solid-phase peptide synthesis (SPPS) have facilitated the development of peptide [74–78] or xenopeptide [79–83] libraries. These libraries elucidate clear sequence–activity relationships, providing crucial insights into nucleic acid complexation and transfection efficacy. Mixson and colleagues conducted a comprehensive evaluation of peptide libraries comprising linear and branched structures. These peptides were designed with specific functionalities: lysines for nucleic acid binding and histidines for buffering endosomal pH. The study explored various ratios and configurations, including linear peptides and those with four or eight branches. The objective was to optimize delivery systems for efficient transfer of siRNA and mRNA. The incorporation of lysines in the peptides facilitates electrostatic interactions with nucleic acids, promoting complex formation that protects RNA molecules during circulation and cellular uptake. Histidines, known for their pH-buffering capacity, ensure that the peptides maintain stability in the acidic environment of endosomes. This buffering capability is crucial for preventing premature degradation of RNA and facilitating its release into the cytoplasm. By varying the ratio and branching configuration of lysines and histidines, Mixson and colleagues aimed to enhance the efficacy and specificity of RNA delivery systems. Different peptide structures may offer advantages in terms of stability, cellular uptake efficiency, and endosomal escape mechanisms, depending on the specific requirements for siRNA or mRNA delivery [74, 75].

In parallel, Lu and collaborators employed solid-phase peptide synthesis (SPPS), utilizing artificial oligoamine building blocks to assemble cationic lipopeptides designed as carriers for antitumoral siRNA. This innovative approach integrates peptide-like properties with functional groups optimized for cellular entry and endosomal escape, aiming to enhance the delivery efficiency and therapeutic efficacy of siRNA-based treatments for cancer. Solid-phase peptide synthesis allowed Lu et al. to systematically construct cationic lipopeptides with precise control over sequence and structure. The incorporation of oligoamine building blocks enhances the cationic nature of the peptides, promoting electrostatic interaction with negatively charged siRNA molecules and facilitating their complexation into stable polyplexes. Additionally, the lipopeptide structure provides hydrophobic domains that aid in membrane interaction and cellular uptake. Furthermore, these cationic lipopeptides are designed to overcome intracellular barriers such as endosomal entrapment. Functional groups within the peptides facilitate pH-responsive behaviors or membrane-disruptive activities upon encountering acidic endosomal environments, ensuring efficient release of siRNA into the cytoplasm where it can exert its therapeutic effects [83].

### *25.9.1 Double pH-Responsive Lipo-Xenopeptides*

Building on the principles of continuous chemical evolution, a novel class of highly potent, nonviral nanocarriers has been developed for the dynamic delivery of various nucleic acids including mRNA, pDNA, and siRNA. These carriers integrate the advantageous features of both polymer and lipid-based delivery systems and have demonstrated an efficacy that is several hundred-fold greater than that of previous generations [84]. The design of these innovative carriers involves incorporating at least two novel lipo amino fatty acids (LAFs) as cationizable motifs. These LAFs are complemented by polar cationizable succinoyl tetraethylene pentamine (Stp) aminoethylene units. The LAFs and Stp units are interconnected via lysines to form diverse structural topologies such as combs, T-shapes, bundles, and U-shapes, with varying ratios of Stp to LAF to optimize delivery characteristics for different types of nucleic acids. A key feature of these carriers is the switchable polarity of the LAFs, facilitated by a central tertiary amine that alters its hydrophobicity upon protonation. At physiological pH (7.4), the LAFs exhibit hydrophobic characteristics, as indicated by log $D$ values around +1. However, in the acidic pH environment typical of endosomes (pH 5.5), the tertiary amine becomes protonated, leading to a significant shift in log $D$ values to approximately −1, indicating hydrophilic behavior.

This "molecular chameleon character" enables pH-dependent structural and physical transformations within the carriers. It leverages lipid-mediated and cationic mechanisms to enhance the dynamic and rapid delivery capabilities of lipopolyplexes. Specifically, the hydrophobicity at neutral pH facilitates interactions with cell membranes and efficient cellular uptake, while the hydrophilicity at acidic pH promotes endosomal escape and nucleic acid release into the cytoplasm. The screening of different structural topologies, Stp/LAF ratios, and types of LAFs has resulted in the identification of tailor-made carriers optimized for specific types of nucleic acids. For pDNA and mRNA, carriers designed as bundles with short LAF hydrocarbon chains or as U-shapes with medium-length LAFs demonstrated the most effective delivery capabilities. In contrast, for siRNA, the U-shape topology exhibited superior potency, achieving significant effects even at doses approximately 170 times lower than those used in previous benchmarks. In vitro testing highlighted the exceptional potency of mRNA lipopolyplexes in tumor cells, even under challenging conditions with high serum presence (90%) and at ultralow dosages (as low as 3 pg of mRNA per cell). Their efficiency rivaled that of viral vectors traditionally used in gene therapy applications. Moreover, when evaluated in vivo, these carriers demonstrated robust activity across multiple organ systems, including the spleen, tumors, lungs, and liver, following systemic administration of 1–3 μg of mRNA in mice. This dual success in both in vitro efficacy and in vivo versatility underscores the significant potential of these custom-designed carriers in advancing gene therapy applications.

## 25.10 T-Shaped Lipo-OAA Carriers

The stabilization of polyplexes, particularly for the delivery of small nucleic acids like siRNA, benefits significantly from the incorporation of hydrophobic domains. To enhance this stabilization, lipo-oligoaminoamides (lipo-OAAs) have been designed with various structural topologies including i-shape, T-shape, and U-shape [80]. These designs involve the integration of two or four fatty acids or other lipidic residues, which help to stabilize the polyplexes by enhancing their interaction with the cellular membranes. Further structural stabilization is achieved through the addition of terminal tyrosine tripeptides and cysteines, which serve as additional stabilizing motifs [85]. Among these designs, the T-shaped lipo-OAA carriers have proven to be particularly effective for siRNA delivery. These carriers, depicted in Fig. 25.2, have shown a high suitability due to their structured design which facilitates efficient siRNA encapsulation and protection.

Enhancing these systems, surface modifications with ligand-conjugated polyethylene glycol (PEG) have been employed to target the polyplexes specifically to tumor cells. This targeted delivery system has been demonstrated in vivo, where tumor-targeted siRNA polyplexes successfully silenced the EG5 gene in mouse models of lymphoma and hepatoma [87, 88]. The addition of PEG not only improves the biocompatibility and circulation time of the polyplexes but also minimizes

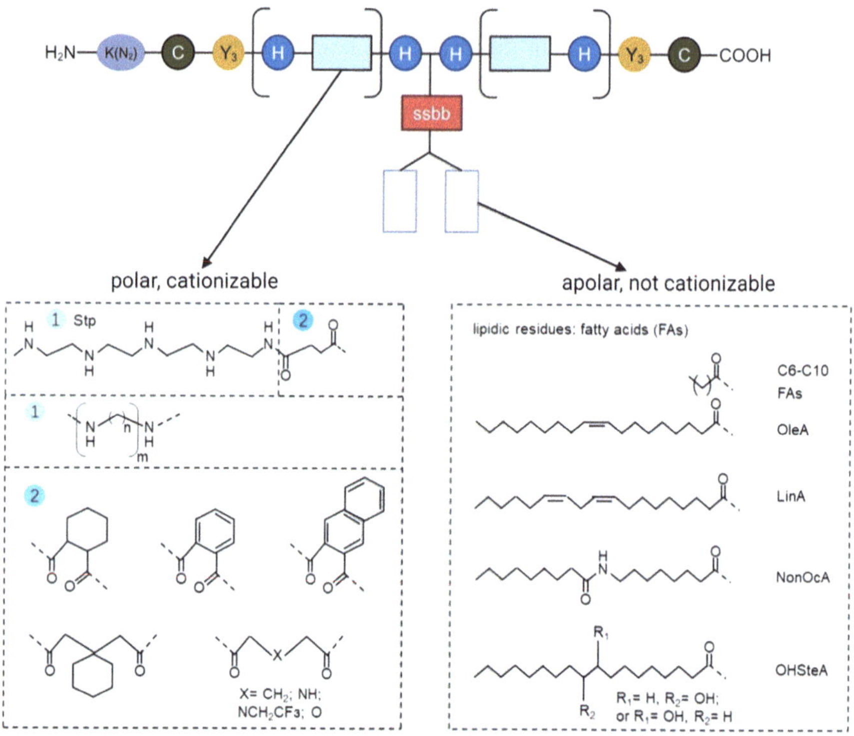


**Fig. 25.2** Structural design of lipo-oligoaminoamides (lipo-OAAs) for siRNA delivery. Top Panel: Schematic representation of lipo-OAA conjugate. The general structure of lipo-OAAs is shown, featuring various domains for enhanced stability and functionality. K($N_2$): Azide-functionalized lysine for further chemical modifications. C (Green): Cysteine residues providing cross-linking and additional stability. $Y_3$ (Yellow): Tyrosine tripeptides acting as stabilizing motifs. H (Blue): Histidine residues included for their buffering capacity in endosomal escape. ssBB (Red): Spacer-stabilized bifunctional block central to the structural integrity of the polymer. Bottom Left Panel: Polar, cationizable functional groups: Group 1: Polar cationizable groups such as spermidine (Stp) and various polyamine derivatives enhance solubility and cellular uptake. Group 2: Additional polar groups including cyclic and acyclic amines and esters, contributing to the interaction with cellular membranes. Bottom Right Panel: Apolar, non-cationizable lipidic residues: Fatty acids (FAs): Different lipidic residues such as: C6-C10 FAs: Short to medium-chain fatty acids for membrane interaction. OleA: Oleic acid for structural stability. LinA: Linoleic acid for flexibility and interaction. NonOcA: Nonanoic acid, enhancing lipid interaction. OHSteA: Hydroxystearic acid, with hydroxyl groups contributing to stabilization. This structural design integrates multiple hydrophobic domains, enhancing the stability of polyplexes and their interaction with cellular membranes. The inclusion of terminal tyrosine tripeptides and cysteine residues further stabilizes the structure, making T-shaped lipo-OAA carriers particularly effective for siRNA delivery by ensuring efficient encapsulation and protection of siRNA molecules. (Figure is adapted from Berger et al. published in PNAS [86] with permissions from the authors)

non-specific interactions, ensuring that the siRNA reaches and acts within the target tumor cells efficiently. This combination of hydrophobic stabilization, precise structural design, and targeted surface modification highlights the advanced capabilities of current gene delivery systems, showcasing significant potential for therapeutic applications, particularly in oncology.

The effective incorporation of hydrophobic fatty acid residues plays a crucial role in enhancing the efficiency of genome editing systems utilizing Cas9 protein/sgRNA RNP polyplexes. Lächelt and colleagues investigated different fatty acids within T-shaped structural designs and found that hydroxystearic acid (OHSteA) surpassed stearic acid in promoting genome editing by Cas9/sgRNA RNPs. OHSteA demonstrated superior performance, likely due to its specific chemical properties that enhance the stability and delivery efficiency of Cas9/sgRNA RNPs. The hydroxyl group in OHSteA may contribute to improved interactions with the Cas9 protein and sgRNA, facilitating the formation and stability of the polyplexes. This enhancement is critical for ensuring effective delivery of the genome editing components into target cells and subsequent cleavage of the target DNA sequences [89]. Building on this, Lin et al. refined the OHSteA-based Cas9 RNP carriers by adding folic acid (FolA) conjugated to PEG, which serves as a targeting ligand for receptor-mediated uptake, thereby enhancing gene editing capabilities [90]. The FolA-modified RNP formulations specifically targeted two immune checkpoint genes, PD-L1 and PVR. When injected into a CT26 colon cancer model in vivo, these targeted nanoparticles achieved significantly higher gene editing rates compared to their non-targeted counterparts. Moreover, the simultaneous disruption of PD-L1 and PVR facilitated the recruitment of $CD8^+$ T cells, leading to notable inhibition of CT26 tumor growth, highlighting the therapeutic potential of targeted gene editing. Further optimization of Cas9/sgRNA RNP polyplexes was achieved through systematic variations of the amphiphilic xenopeptide sequences, including the number and types of artificial oligoamino acids and the specific fatty acids used [91]. An extensive evaluation involving 78 different xenopeptide configurations established a correlation between the xenopeptides' log $D$ 7.4 (a measure of their hydrophobicity at physiological pH) and their genome editing efficacy. Among these, a particularly potent carrier named TFE-IDAtp1-LinA, which includes a trifluoroethyl-iminodiacetic acid analog of Stp, demonstrated remarkable efficiency in knocking out the green fluorescent protein, with an $ED_{50}$ of just 0.38 nM RNP. These mechanistic studies revealed that hydrophobic xenopeptides are more resistant to ionic stress and less prone to concentration-dependent dissociation. However, there exists an optimal range of log $D$ 7.4 for each xenopeptide architecture; deviation beyond this range (either too hydrophobic or too hydrophilic) dramatically reduces their activity. This insight is critical for designing future gene delivery systems, ensuring that they are finely tuned for maximum efficacy while minimizing potential off-target effects and toxicity. The effective delivery of larger nucleic acids like plasmid DNA (pDNA) and mRNA also hinges on finding an optimal balance between polyplex stability and efficient cargo release. Excessive stability of lipopolyplexes, while protective, can impede the transfection efficiency due to challenges in releasing the nucleic acid cargo within the cell [61, 92].

Research involving the screening of T-shaped lipo-OAA libraries, which included fatty acids varying in chain length from C2 to C18, revealed that carriers with shorter fatty acids (C6 to C10) offered less polyplex stability but enhanced endosomolytic capabilities and higher nucleic acid transfer activities compared to those with longer fatty acid chains [92]. This observation underscores the critical role that lipid chain length plays in balancing polyplex stability and functional delivery efficacy. For mRNA delivery, incorporating a bioreducible disulfide bond (ssbb) between the polymer backbone and the lipidic side chain is crucial for enhancing transfection effectiveness. This modification enables a responsive release mechanism in the reducing environment of the intracellular space, thereby facilitating more efficient release of mRNA. The bioreducible disulfide bond (ssbb) is designed to undergo cleavage in the presence of reducing agents, such as glutathione, which are abundant in the cytoplasm. This cleavage event triggers the detachment of the lipidic side chain from the polymer backbone, promoting the release of encapsulated mRNA into the cytosol. By mimicking natural processes of disulfide bond reduction, this mechanism ensures that mRNA can be released effectively once inside the target cells. In contrast, formulations lacking this dynamic tuning element may benefit from using less stabilizing shorter fatty acids. These shorter fatty acids are less likely to impede the release of mRNA after cellular uptake, thus ensuring that the therapeutic cargo can reach its intended site of action without hindrance [61]. Among the screened carriers, the top-performing lipo-OAA was distinguished by its T-shape configuration that incorporated histidines and a ssbb with oleic acid as the lipidic component. This specific carrier was notably effective in delivering mRNA to the lungs of mice when applied intratracheally. This success demonstrates the importance of meticulously designing carrier molecules that not only protect their nucleic acid cargo but also ensure timely and effective release at the target site, thereby maximizing therapeutic potential.

## 25.11 The Pathway to Clinical RNA Delivery

Regardless of their specific composition or intended targeting mechanisms, the development of nanoparticle-based RNA delivery systems for clinical applications is governed by a stringent preclinical testing framework known as "the nanoparticle discovery pipeline." This comprehensive pipeline initiates with an extensive high-throughput screening phase conducted within cell culture environments. At this early stage, researchers systematically evaluate a broad array of nanoparticle formulations to pinpoint those with the most favorable attributes for effective RNA delivery.

This initial screening process rigorously tests various key parameters, including nanoparticle size, surface charge, stability under physiological conditions, biocompatibility with human tissues, and the capacity to shield RNA payloads from enzymatic degradation. Such assessments are crucial as they directly impact the efficiency and safety of the nanoparticles when used in living organisms. The high-throughput nature of this screening allows for the simultaneous evaluation of multiple nanoparticle variants, streamlining the identification of promising candidates by assessing their performance in delivering RNA molecules into target cells effectively. This step helps identify formulations that demonstrate high transfection efficiency, low cytotoxicity, and specific targeting capabilities, among other desirable attributes. Insights gained from these in vitro tests are crucial for refining various nanoparticle characteristics. Based on the performance data gathered, a select few nanoparticles progress to testing in mice. Subsequently, an even smaller subset of these is chosen for further evaluation in rats, followed by a more restricted number tested in non-human primates (Fig. 25.3a). Only those nanoparticles that demonstrate a robust safety profile in non-human primates are then considered eligible for advancement into clinical trials. This structured approach ensures that only the most promising and safe RNA delivery systems are moved forward in the development pipeline (Fig. 25.3b).

## 25.12 Hallmarks of Clinically Relevant Delivery Systems

As the field of nanoparticle-based RNA delivery continues to evolve, there is significant potential to enhance next-generation delivery systems by adopting traits that are common among those currently approved for clinical use. For instance, detailed pharmacokinetic and clearance studies on Patisiran offer valuable insights [94, 95]. These studies not only highlighted the sustained gene silencing achievable over several months but also documented stable mean serum concentrations and pharmacodynamics across 18 months of repeated administration. Additionally, the FDA Center for Drug Evaluation and Research's report on Patisiran [96] provides detailed insights into the accepted animal models, toxicological assessments, and critical time points that were instrumental in its approval process, serving as a potential template for future RNA drug developments. More broadly, the characteristics of the three RNA drug delivery systems that have received FDA approval [97, 98] or an Emergency Use Authorization (EUA) [99] to date may offer a blueprint for successful delivery vehicles. These systems tend to exhibit six key features that, collectively, could be considered hallmarks of efficacious and safe RNA delivery vehicles (Fig. 25.4). By integrating such established attributes into the development of new nanoparticles, researchers can potentially streamline the approval process and improve the likelihood of clinical success.

## 25.13 Conclusion

Over the past decade, RNA therapies have emerged as highly promising strategies for treating a wide range of diseases, yet significant advancements are still required to fully harness their potential. The preclinical nanoparticle discovery pipeline is an

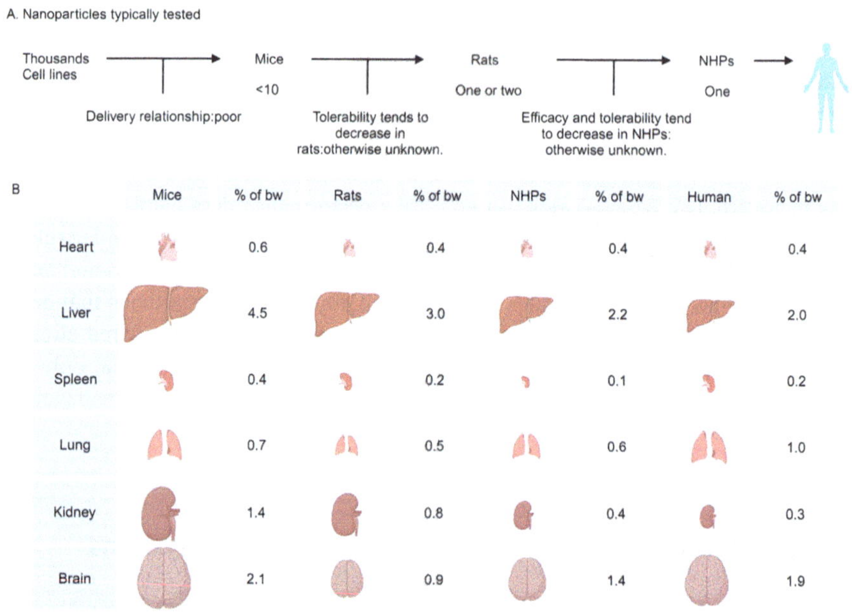


**Fig. 25.3** The preclinical nanoparticle discovery pipeline. Panel (**a**) depicts the initial phase where thousands of nanoparticles are synthesized and subjected to in vitro testing to evaluate their delivery capabilities. This stage serves to identify promising candidates for further evaluation. However, it is crucial to note that in vitro delivery outcomes often do not reliably predict in vivo efficacy. This discrepancy underscores the challenges in translating findings from cell culture settings to more complex biological environments, such as animal models and ultimately human trials. Following in vitro screening, a reduced number of nanoparticles are selected for evaluation in animal models, starting with mice. This step aims to assess biodistribution, pharmacokinetics, safety, and initial efficacy in vivo. Subsequently, a subset of these candidates may advance to testing in larger animal models such as rats or non-human primates (NHPs). Non-human primates are particularly valued in preclinical research due to their closer physiological resemblance to humans, offering insights into potential efficacy and safety in humans. However, optimizing the efficiency of the nanoparticle discovery pipeline requires a deeper understanding of interspecies drug delivery dynamics. Specifically, identifying which small-animal models best predict outcomes in NHPs and humans remains an underexplored area. (**b**) Tissue weight as a percentage of the total body weight in female animals across different species, highlighting that relative organ size can vary considerably. These variations may influence the on-target and off-target delivery of nanoparticles, affecting both efficacy and safety profiles. This underscores the importance of choosing appropriate animal models for preclinical testing to better predict human outcomes. (Figure is adapted from [93] with permission from Springer Nature (License number: 5902451025823))

essential systematic process that supports the development of nanoparticle-based RNA delivery systems, moving them closer to clinical application. This process begins with the synthesis of thousands of nanoparticle formulations followed by extensive in vitro screening to identify those with optimal delivery capabilities. Despite the successes achieved in vitro, translating these results to in vivo settings presents a substantial hurdle. This transition is crucial for demonstrating the efficacy and safety of RNA delivery systems in a living organism, thus emphasizing the

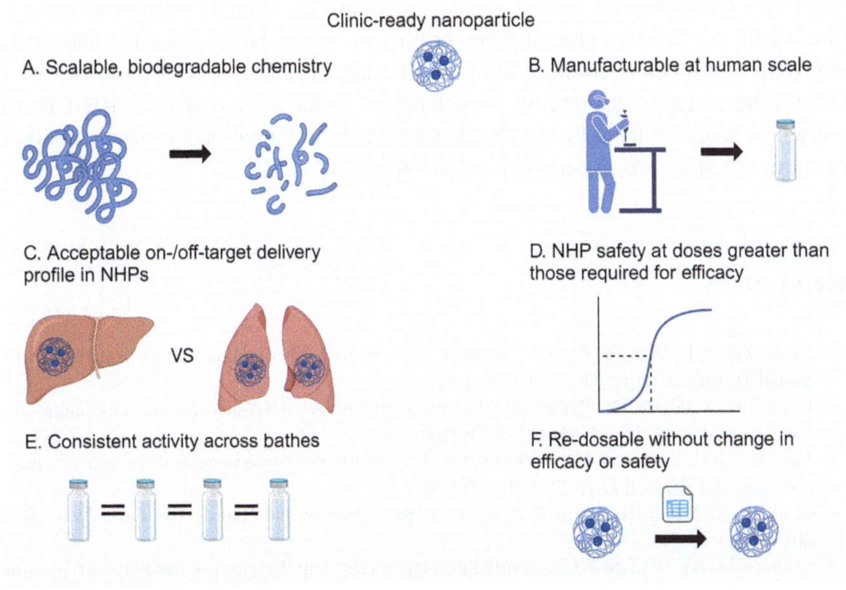


**Fig. 25.4** Hallmarks of a clinically relevant delivery system. The figure outlines six essential characteristics that, if assessed early in the development phase of a drug delivery system, can substantially improve the likelihood of receiving FDA approval. Criteria for clinic-ready nanoparticle development. (**a**) The nanoparticles must be synthesized from scalable, biodegradable chemistry to ensure efficient production and safe degradation. (**b**) They need to be manufacturable at a human scale, ensuring the production processes can meet clinical demand. (**c**) The delivery profile in non-human primates (NHPs) must be acceptable, demonstrating effective targeting with minimal off-target effects. (**d**) The nanoparticles should exhibit safety in NHPs at doses higher than those required for efficacy, ensuring a wide therapeutic window. (**e**) Consistency in activity across different batches is crucial for maintaining reliable therapeutic effects. (**f**) The nanoparticles should be redosable without a decrease in efficacy or safety, indicating their potential for repeated administration in clinical settings. These criteria collectively ensure that the nanoparticles are effective, safe, and practical for human therapeutic use. (Figure is adapted from [93] with permission from Springer Nature (License number: 5902450951344))

need for robust animal models. Mice, rats, and non-human primates (NHPs) are integral to this phase. NHPs are particularly valuable due to their close physiological and genetic similarities to humans, making them ideal for validating the effectiveness and safety of nanoparticles before progressing to human trials. However, optimizing the efficiency of the nanoparticle discovery pipeline requires a deep understanding of interspecies dynamics. This understanding ensures that findings from smaller animal models can predict responses in NHPs and, subsequently, in humans with higher accuracy. Addressing these interspecies differences is critical for refining the selection process of nanoparticle formulations and enhancing the predictability of how these formulations will perform in human subjects. Advancing this pipeline involves not only selecting the most promising nanoparticle candidates but also improving the methods by which these candidates are transitioned from preclinical findings to successful clinical outcomes. Continued research and innovation within this field are crucial. They promise to refine the stages of the nanoparticle discovery pipeline further, ultimately leading to more effective and precisely tailored treatments. This ongoing development is poised to transform RNA-based therapies, adapting them to specific diseases and individual patient needs, thereby opening new horizons in personalized medicine.

## References


1. Hu B, Zhong L, Weng Y, Peng L, Huang Y, Zhao Y, et al. Therapeutic siRNA: state of the art. Signal Transduct Target Ther. 2020;5(1):101.
2. Rhym LH, Anderson DG. Nanoscale delivery platforms for RNA therapeutics: challenges and the current state of the art. Med. 2022;3(3):167–87.
3. Vargason AM, Anselmo AC, Mitragotri S. The evolution of commercial drug delivery technologies. Nat Biomed Eng. 2021;5(9):951–67.
4. Couvreur P. Nanoparticles in drug delivery: past, present and future. Adv Drug Deliv Rev. 2013;65(1):21–3.
5. Chong ETJ, Ng JW, Lee P-C. Classification and medical applications of biomaterials—a mini review. BIO Integr. 2023;4(2):54–61.
6. Bulcha JT, Wang Y, Ma H, Tai PWL, Gao G. Viral vector platforms within the gene therapy landscape. Signal Transduct Target Ther. 2021;6(1):53.
7. Zhu H, Luo H, Chang R, Yang Y, Liu D, Ji Y, et al. Protein-based delivery systems for RNA delivery. J Control Release. 2023;363:253–74.
8. Saw PE, Kong N. SORTing the fate of nanodelivery systems. BIO Integr. 2022;3(1):41–2.
9. Sercombe L, Veerati T, Moheimani F, Wu SY, Sood AK, Hua S. Advances and challenges of liposome assisted drug delivery. Front Pharmacol. 2015;6:286.
10. Hojeong S, Se-Jin P, Yeajee Y, Jungho K, Chulwon C, Cheolhee W, et al. Recent advances in RNA therapeutics and RNA delivery systems based on nanoparticles. Adv Ther. 2018;1(7):1800065.
11. Barenholz Y. Doxil®—the first FDA-approved nano-drug: lessons learned. J Control Release. 2012;160(2):117–34.
12. Podesta JE, Kostarelos K. Chapter 17—Engineering cationic liposome siRNA complexes for in vitro and in vivo delivery. Methods Enzymol. 2009;464:343–54.
13. Hattori Y, Hara E, Shingu Y, Minamiguchi D, Nakamura A, Arai S, et al. siRNA delivery into tumor cells by cationic cholesterol derivative-based nanoparticles and liposomes. Biol Pharm Bull. 2015;38(1):30–8.
14. Kranz LM, Diken M, Haas H, Kreiter S, Loquai C, Reuter KC, et al. Systemic RNA delivery to dendritic cells exploits antiviral defence for cancer immunotherapy. Nature. 2016;534(7607):396–401.
15. Laouini A, Jaafar-Maalej C, Limayem-Blouza I, et al. Preparation, characterization and applications of liposomes: state of the art. J Colloid Sci Biotechnol. 2012;1(2):147–68.
16. Wang H, Zhang S, Lv J, Cheng Y. Design of polymers for siRNA delivery: recent progress and challenges. VIEW. 2021;2(3):20200026.
17. Elzahhar P, Belal ASF, Elamrawy F, Helal NA, Nounou MI. Bioconjugation in drug delivery: practical perspectives and future perceptions. Methods Mol Biol. 2019;2000:125–82.
18. Amini Y, Amel Jamehdar S, Sadri K, Zare S, Musavi D, Tafaghodi M. Different methods to determine the encapsulation efficiency of protein in PLGA nanoparticles. Biomed Mater Eng. 2017;28(6):613–20.

19. Devulapally R, Foygel K, Sekar TV, Willmann JK, Paulmurugan R. Gemcitabine and antisense-microRNA co-encapsulated PLGA-PEG polymer nanoparticles for hepatocellular carcinoma therapy. ACS Appl Mater Interfaces. 2016;8(49):33412–22.
20. Molina M, Asadian-Birjand M, Balach J, Bergueiro J, Miceli E, Calderón M. Stimuli-responsive nanogel composites and their application in nanomedicine. Chem Soc Rev. 2015;44(17):6161–86.
21. Kumari A, Yadav SK, Yadav SC. Biodegradable polymeric nanoparticles based drug delivery systems. Colloids Surf B Biointerfaces. 2010;75(1):1–18.
22. Zhang L, Chan JM, Gu FX, Rhee JW, Wang AZ, Radovic-Moreno AF, et al. Self-assembled lipid—polymer hybrid nanoparticles: a robust drug delivery platform. ACS Nano. 2008;2(8):1696–702.
23. Shi J, Xiao Z, Votruba AR, Vilos C, Farokhzad OC. Differentially charged hollow core/shell lipid-polymer-lipid hybrid nanoparticles for small interfering RNA delivery. Angew Chem Int Ed Engl. 2011;50(31):7027–31.
24. Xie X, Yue T, Gu W, Cheng W, He L, Ren W, et al. Recent advances in mesoporous silica nanoparticles delivering siRNA for cancer treatment. Pharmaceutics. 2023;15(10):2483.
25. Möller K, Müller K, Engelke H, Bräuchle C, Wagner E, Bein T. Highly efficient siRNA delivery from core-shell mesoporous silica nanoparticles with multifunctional polymer caps. Nanoscale. 2016;8(7):4007–19.
26. Paris JL, Vallet-Regí M. Mesoporous silica nanoparticles for co-delivery of drugs and nucleic acids in oncology: a review. Pharmaceutics. 2020;12(6):526.
27. Meka AK, Niu Y, Karmakar S, Hartono SB, Zhang J, Lin CXC, et al. Facile synthesis of large-pore bicontinuous cubic mesoporous silica nanoparticles for intracellular gene delivery. ChemNanoMat. 2016;2(3):220–5.
28. Lee JH, Kang S, Ahn M, Jang H, Min DH. Development of dual-pore coexisting branched silica nanoparticles for efficient gene-chemo cancer therapy. Small. 2018;14(7).
29. Kshatriya VV, Kumbhare MR, Jadhav SV, Thorat PJ, Bhambarge RG. Updated review on recent advances in silver nanoclusters in bioanalytical and biomedical applications. BIO Integr. 2024;5(1):e992.
30. Jiang Y, Huo S, Hardie J, Liang XJ, Rotello VM. Progress and perspective of inorganic nanoparticle-based siRNA delivery systems. Expert Opin Drug Deliv. 2016;13(4):547–59.
31. Varshosaz J, Taymouri S. Hollow inorganic nanoparticles as efficient carriers for siRNA delivery: a comprehensive review. Curr Pharm Des. 2015;21(29):4310–28.
32. Kshatriya VV, Kumbhare MR, Jadhav SV, Thorat PJ, Bhambarge RG. Properties and emerging applications of ruthenium nanoclusters. BIO Integr. 2024;5(1).
33. Zheng D, Giljohann DA, Chen DL, Massich MD, Wang XQ, Iordanov H, et al. Topical delivery of siRNA-based spherical nucleic acid nanoparticle conjugates for gene regulation. Proc Natl Acad Sci U S A. 2012;109(30):11975–80.
34. Li JM, Zhao MX, Su H, Wang YY, Tan CP, Ji LN, et al. Multifunctional quantum-dot-based siRNA delivery for HPV18 E6 gene silence and intracellular imaging. Biomaterials. 2011;32(31):7978–87.
35. Yang HW, Huang CY, Lin CW, Liu HL, Huang CW, Liao SS, et al. Gadolinium-functionalized nanographene oxide for combined drug and microRNA delivery and magnetic resonance imaging. Biomaterials. 2014;35(24):6534–42.
36. Kam NW, Liu Z, Dai H. Functionalization of carbon nanotubes via cleavable disulfide bonds for efficient intracellular delivery of siRNA and potent gene silencing. J Am Chem Soc. 2005;127(36):12492–3.
37. Molinaro R, Corbo C, Martinez JO, Taraballi F, Evangelopoulos M, Minardi S, et al. Biomimetic proteolipid vesicles for targeting inflamed tissues. Nat Mater. 2016;15(9):1037–46.
38. Molinaro R, Pastò A, Corbo C, Taraballi F, Giordano F, Martinez JO, et al. Macrophage-derived nanovesicles exert intrinsic anti-inflammatory properties and prolong survival in sepsis through a direct interaction with macrophages. Nanoscale. 2019;11(28):13576–86.

39. Corbo C, Cromer WE, Molinaro R, Toledano Furman NE, Hartman KA, De Rosa E, et al. Engineered biomimetic nanovesicles show intrinsic anti-inflammatory properties for the treatment of inflammatory bowel diseases. Nanoscale. 2017;9(38):14581–91.
40. Boada C, Zinger A, Tsao C, Zhao P, Martinez JO, Hartman K, et al. Rapamycin-loaded biomimetic nanoparticles reverse vascular inflammation. Circ Res. 2020;126(1):25–37.
41. Martinez JO, Molinaro R, Hartman KA, Boada C, Sukhovershin R, De Rosa E, et al. Biomimetic nanoparticles with enhanced affinity towards activated endothelium as versatile tools for theranostic drug delivery. Theranostics. 2018;8(4):1131–45.
42. Felgner PL, Barenholz Y, Behr JP, Cheng SH, Cullis P, Huang L, et al. Nomenclature for synthetic gene delivery systems. Hum Gene Ther. 1997;8(5):511–2.
43. Kumar R, Santa Chalarca CF, Bockman MR, Bruggen CV, Grimme CJ, Dalal RJ, et al. Polymeric delivery of therapeutic nucleic acids. Chem Rev. 2021;121(18):11527–652.
44. Lächelt U, Wagner E. Nucleic acid therapeutics using polyplexes: a journey of 50 years (and beyond). Chem Rev. 2015;115(19):11043–78.
45. Miyata K, Nishiyama N, Kataoka K. Rational design of smart supramolecular assemblies for gene delivery: chemical challenges in the creation of artificial viruses. Chem Soc Rev. 2012;41(7):2562–74.
46. Peng L, Wagner E. Polymeric carriers for nucleic acid delivery: current designs and future directions. Biomacromolecules. 2019;20(10):3613–26.
47. Uchida S, Kataoka K. Design concepts of polyplex micelles for in vivo therapeutic delivery of plasmid DNA and messenger RNA. J Biomed Mater Res A. 2019;107(5):978–90.
48. Yamauchi Y, Helenius A. Virus entry at a glance. J Cell Sci. 2013;126(Pt 6):1289–95.
49. Gonçalves C, Akhter S, Pichon C, Midoux P. Intracellular availability of pDNA and mRNA after transfection: a comparative study among polyplexes, lipoplexes, and lipopolyplexes. Mol Pharm. 2016;13(9):3153–63.
50. Scholz C, Wagner E. Therapeutic plasmid DNA versus siRNA delivery: common and different tasks for synthetic carriers. J Control Release. 2012;161(2):554–65.
51. Degors IMS, Wang C, Rehman ZU, Zuhorn IS. Carriers break barriers in drug delivery: endocytosis and endosomal escape of gene delivery vectors. Acc Chem Res. 2019;52(7):1750–60.
52. Xu Y, Szoka FC Jr. Mechanism of DNA release from cationic liposome/DNA complexes used in cell transfection. Biochemistry. 1996;35(18):5616–23.
53. Boussif O, Lezoualc'h F, Zanta MA, Mergny MD, Scherman D, Demeneix B, et al. A versatile vector for gene and oligonucleotide transfer into cells in culture and in vivo: polyethylenimine. Proc Natl Acad Sci U S A. 1995;92(16):7297–301.
54. Creusat G, Rinaldi AS, Weiss E, Elbaghdadi R, Remy JS, Mulherkar R, et al. Proton sponge trick for pH-sensitive disassembly of polyethylenimine-based siRNA delivery systems. Bioconjug Chem. 2010;21(5):994–1002.
55. Hall A, Lächelt U, Bartek J, Wagner E, Moghimi SM. Polyplex evolution: understanding biology, optimizing performance. Mol Ther. 2017;25(7):1476–90.
56. Kichler A, Leborgne C, Coeytaux E, Danos O. Polyethylenimine-mediated gene delivery: a mechanistic study. J Gene Med. 2001;3(2):135–44.
57. Sonawane ND, Szoka FC Jr, Verkman AS. Chloride accumulation and swelling in endosomes enhances DNA transfer by polyamine-DNA polyplexes. J Biol Chem. 2003;278(45):44826–31.
58. Winkeljann B, Keul DC, Merkel OM. Engineering poly- and micelleplexes for nucleic acid delivery—a reflection on their endosomal escape. J Control Release. 2023;353:518–34.
59. Itaka K, Harada A, Yamasaki Y, Nakamura K, Kawaguchi H, Kataoka K. In situ single cell observation by fluorescence resonance energy transfer reveals fast intra-cytoplasmic delivery and easy release of plasmid DNA complexed with linear polyethylenimine. J Gene Med. 2004;6(1):76–84.
60. Brunner S, Fürtbauer E, Sauer T, Kursa M, Wagner E. Overcoming the nuclear barrier: cell cycle independent nonviral gene transfer with linear polyethylenimine or electroporation. Mol Ther. 2002;5(1):80–6.

61. Krhač Levačić A, Berger S, Müller J, Wegner A, Lächelt U, Dohmen C, et al. Dynamic mRNA polyplexes benefit from bioreducible cleavage sites for in vitro and in vivo transfer. J Control Release. 2021;339:27–40.
62. Zintchenko A, Philipp A, Dehshahri A, Wagner E. Simple modifications of branched PEI lead to highly efficient siRNA carriers with low toxicity. Bioconjug Chem. 2008;19(7):1448–55.
63. Jarzębińska A, Pasewald T, Lambrecht J, Mykhaylyk O, Kümmerling L, Beck P, et al. A single methylene group in oligoalkylamine-based cationic polymers and lipids promotes enhanced mRNA delivery. Angew Chem Int Ed Engl. 2016;55(33):9591–5.
64. Cheng Y, Yumul RC, Pun SH. Virus-inspired polymer for efficient in vitro and in vivo gene delivery. Angew Chem Int Ed Engl. 2016;55(39):12013–7.
65. Feldmann DP, Cheng Y, Kandil R, Xie Y, Mohammadi M, Harz H, et al. In vitro and in vivo delivery of siRNA via VIPER polymer system to lung cells. J Control Release. 2018;276:50–8.
66. Meyer M, Dohmen C, Philipp A, Kiener D, Maiwald G, Scheu C, et al. Synthesis and biological evaluation of a bioresponsive and endosomolytic siRNA-polymer conjugate. Mol Pharm. 2009;6(3):752–62.
67. Peeler DJ, Thai SN, Cheng Y, Horner PJ, Sellers DL, Pun SH. pH-sensitive polymer micelles provide selective and potentiated lytic capacity to venom peptides for effective intracellular delivery. Biomaterials. 2019;192:235–44.
68. Wang T, Wang D, Yu H, Wang M, Liu J, Feng B, et al. Intracellularly acid-switchable multifunctional micelles for combinational photo/chemotherapy of the drug-resistant tumor. ACS Nano. 2016;10(3):3496–508.
69. Haabeth OAW, Blake TR, McKinlay CJ, Tveita AA, Sallets A, Waymouth RM, et al. Local delivery of Ox40l, Cd80, and Cd86 mRNA kindles global anticancer immunity. Cancer Res. 2019;79(7):1624–34.
70. McKinlay CJ, Benner NL, Haabeth OA, Waymouth RM, Wender PA. Enhanced mRNA delivery into lymphocytes enabled by lipid-varied libraries of charge-altering releasable transporters. Proc Natl Acad Sci U S A. 2018;115(26):E5859–66.
71. McKinlay CJ, Vargas JR, Blake TR, Hardy JW, Kanada M, Contag CH, et al. Charge-altering releasable transporters (CARTs) for the delivery and release of mRNA in living animals. Proc Natl Acad Sci U S A. 2017;114(4):E448–56.
72. Blake TR, Haabeth OAW, Sallets A, McClellan RL, Del Castillo TJ, Vilches-Moure JG, et al. Lysine-derived charge-altering releasable transporters: targeted delivery of mRNA and siRNA to the lungs. Bioconjug Chem. 2023.
73. Freitag F, Wagner E. Optimizing synthetic nucleic acid and protein nanocarriers: the chemical evolution approach. Adv Drug Deliv Rev. 2021;168:30–54.
74. He J, Xu S, Leng Q, Mixson AJ. Location of a single histidine within peptide carriers increases mRNA delivery. J Gene Med. 2021;23(2):e3295.
75. Leng Q, Scaria P, Zhu J, Ambulos N, Campbell P, Mixson AJ. Highly branched HK peptides are effective carriers of siRNA. J Gene Med. 2005;7(7):977–86.
76. Plank C, Tang MX, Wolfe AR, Szoka FC Jr. Branched cationic peptides for gene delivery: role of type and number of cationic residues in formation and in vitro activity of DNA polyplexes. Hum Gene Ther. 1999;10(2):319–32.
77. Stevenson M, Ramos-Perez V, Singh S, Soliman M, Preece JA, Briggs SS, et al. Delivery of siRNA mediated by histidine-containing reducible polycations. J Control Release. 2008;130(1):46–56.
78. Andaloussi SE, Lehto T, Mäger I, Rosenthal-Aizman K, Oprea II, Simonson OE, et al. Design of a peptide-based vector, PepFect6, for efficient delivery of siRNA in cell culture and systemically in vivo. Nucleic Acids Res. 2011;39(9):3972–87.
79. Hartmann L, Häfele S, Peschka-Süss R, Antonietti M, Börner HG. Tailor-made poly(amidoamine)s for controlled complexation and condensation of DNA. Chemistry. 2008;14(7):2025–33.

80. Schaffert D, Troiber C, Salcher EE, Fröhlich T, Martin I, Badgujar N, et al. Solid-phase synthesis of sequence-defined T-, i-, and U-shape polymers for pDNA and siRNA delivery. Angew Chem Int Ed Engl. 2011;50(38):8986–9.
81. Wolfe JM, Fadzen CM, Choo ZN, Holden RL, Yao M, Hanson GJ, et al. Machine learning to predict cell-penetrating peptides for antisense delivery. ACS Cent Sci. 2018;4(4):512–20.
82. Wolfe JM, Fadzen CM, Holden RL, Yao M, Hanson GJ, Pentelute BL. Perfluoroaryl bicyclic cell-penetrating peptides for delivery of antisense oligonucleotides. Angew Chem Int Ed Engl. 2018;57(17):4756–9.
83. Vaidya AM, Sun Z, Ayat N, Schilb A, Liu X, Jiang H, et al. Systemic delivery of tumor-targeting siRNA nanoparticles against an oncogenic LncRNA facilitates effective triple-negative breast cancer therapy. Bioconjug Chem. 2019;30(3):907–19.
84. Thalmayr S, Grau M, Peng L, Pöhmerer J, Wilk U, Folda P, et al. Molecular chameleon carriers for nucleic acid delivery: the sweet spot between lipoplexes and polyplexes. Adv Mater. 2023;35(25):e2211105.
85. Troiber C, Edinger D, Kos P, Schreiner L, Kläger R, Herrmann A, et al. Stabilizing effect of tyrosine trimers on pDNA and siRNA polyplexes. Biomaterials. 2013;34(5):1624–33.
86. Berger S, Lachelt U, Wagner E. Dynamic carriers for therapeutic RNA delivery. Proc Natl Acad Sci U S A. 2024;121(11):e2307799120.
87. Klein PM, Kern S, Lee DJ, Schmaus J, Höhn M, Gorges J, et al. Folate receptor-directed orthogonal click-functionalization of siRNA lipopolyplexes for tumor cell killing in vivo. Biomaterials. 2018;178:630–42.
88. Luo J, Schmaus J, Cui M, Hörterer E, Wilk U, Höhn M, et al. Hyaluronate siRNA nanoparticles with positive charge display rapid attachment to tumor endothelium and penetration into tumors. J Control Release. 2021;329:919–33.
89. Kuhn J, Lin Y, Krhac Levacic A, Al Danaf N, Peng L, Höhn M, et al. Delivery of Cas9/sgRNA ribonucleoprotein complexes via hydroxystearyl oligoamino amides. Bioconjug Chem. 2020;31(3):729–42.
90. Lin Y, Wilk U, Pöhmerer J, Hörterer E, Höhn M, Luo X, et al. Folate receptor-mediated delivery of Cas9 RNP for enhanced immune checkpoint disruption in cancer cells. Small. 2023;19(2):e2205318.
91. Lin Y, Luo X, Burghardt T, Dorrer S, Höhn M, Wagner E, et al. Chemical evolution of amphiphilic xenopeptides for potentiated Cas9 ribonucleoprotein delivery. J Am Chem Soc. 2023;145(28):15171–9.
92. Berger S, Krhač Levačić A, Hörterer E, Wilk U, Benli-Hoppe T, Wang Y, et al. Optimizing pDNA Lipo-polyplexes: a balancing act between stability and cargo release. Biomacromolecules. 2021;22(3):1282–96.
93. Paunovska K, Loughrey D, Dahlman JE. Drug delivery systems for RNA therapeutics. Nat Rev Genet. 2022;23(5):265–80.
94. Zhang X, Goel V, Attarwala H, Sweetser MT, Clausen VA, Robbie GJ. Patisiran pharmacokinetics, pharmacodynamics, and exposure-response analyses in the phase 3 APOLLO trial in patients with hereditary transthyretin-mediated (hATTR) amyloidosis. J Clin Pharmacol. 2020;60(1):37–49.
95. Zhang X, Goel V, Robbie GJ. Pharmacokinetics of patisiran, the first approved RNA interference therapy in patients with hereditary transthyretin-mediated amyloidosis. J Clin Pharmacol. 2020;60(5):573–85.
96. Center for Drug Evaluation and Research application number: 210922Orig1s000 FDA. 2018. https://www.accessdata.fda.gov/drugsatfda_docs/nda/2018/210922Orig1s000MultiRpdf.
97. Mannocci L, Zhang Y, Scheuermann J, Leimbacher M, De Bellis G, Rizzi E, et al. High-throughput sequencing allows the identification of binding molecules isolated from DNA-encoded chemical libraries. Proc Natl Acad Sci U S A. 2008;105(46):17670–5.
98. Shendure J, Ji H. Next-generation DNA sequencing. Nat Biotechnol. 2008;26(10):1135–45.
99. Saliba AE, Westermann AJ, Gorski SA, Vogel J. Single-cell RNA-seq: advances and future challenges. Nucleic Acids Res. 2014;42(14):8845–60.

# Chapter 26
# mRNA Vaccines

**Abstract** mRNA vaccines have garnered considerable interest as a modern alternative to traditional vaccine methods due to several compelling advantages. They offer robust efficacy, facilitated by their ability to stimulate potent immune responses against targeted antigens. This attribute is particularly advantageous in rapidly developing vaccines against emerging infectious diseases, where swift production and deployment are critical. Moreover, mRNA vaccine technology holds promise for economical large-scale production, as it bypasses the need for cell culture systems used in traditional vaccine manufacturing. Despite these benefits, the initial adoption of mRNA vaccines was hindered by significant challenges. One major hurdle was the inherent instability of mRNA molecules, which are prone to degradation before they can effectively produce the desired proteins within cells. Additionally, efficiently delivering mRNA vaccines into cells posed another obstacle, as they needed to evade degradation in the bloodstream and reach the target cells intact to elicit a robust immune response. Recent advancements in mRNA vaccine technology have largely addressed these obstacles. The success of these vaccines has accelerated interest and investment in mRNA platforms for other infectious diseases and even cancer immunotherapy. This chapter explores these advancements in depth, providing insights into the mechanisms behind mRNA vaccines, recent clinical successes, and the remaining challenges that must be overcome to fully realize their therapeutic potential across diverse medical fields.

## 26.1 Introduction

Vaccines have played a pivotal role in public health by preventing numerous illnesses and saving countless lives annually. Their widespread application has achieved remarkable milestones, such as the total eradication of smallpox and substantial reductions in diseases like polio, measles, and various childhood illnesses globally [1, 2]. Traditional vaccination approaches have demonstrated efficacy in conferring long-term immunity against severe diseases [3]. While traditional vaccines have been effective, the urgent demand for rapid development and mass



P. E. Saw, E. Song, *RNA Therapeutics in Human Diseases*,
https://doi.org/10.1007/978-981-96-3041-7_26

distribution during emerging viral outbreaks poses logistical hurdles [4]. Furthermore, while traditional vaccines are effective, the primary challenge for emerging viral vaccines lies not in their efficacy but in the necessity for swift development and mass distribution. Traditional methods may also be unsuitable for tackling non-infectious conditions like cancer, underscoring the critical need for more robust and adaptable vaccine platforms. The evolution of mRNA vaccines represents a promising advancement in addressing these challenges.

## 26.2 Advantages of mRNA Vaccine: Modulating Immunogenicity

Exogenous mRNA serves as a potent immunostimulant, engaging various innate immune receptors located on cell surfaces, within endosomes, and in the cytosol. This characteristic can present both advantages and challenges depending on its intended therapeutic application. In the context of vaccines, the immunostimulatory nature of mRNA can prove beneficial by augmenting adjuvant activity. This stimulation can lead to enhanced maturation of dendritic cells (DCs), crucial players in initiating adaptive immune responses, thereby promoting robust T and B cell activation. However, the innate immune recognition of mRNA can also pose challenges. In some cases, the activation of innate immune sensors can trigger pathways that hinder antigen expression. This interference may occur through mechanisms aimed at controlling viral infections but can inadvertently impair the effectiveness of mRNA vaccines by limiting protein translation and subsequent immune response activation [5–7]. It is now recognized that the immunogenic profile of mRNA can be modulated through several strategic approaches.

### *26.2.1 Purification of IVT mRNA*

In vitro transcribed (IVT) mRNA is usually synthesized from a DNA template using enzymes like T7 RNA polymerase. During this process, impurities such as double-stranded RNA (dsRNA), residual DNA templates, and incompletely synthesized mRNA strands can inadvertently trigger innate immune responses. Purification methods aim to remove these impurities, thereby reducing the potential for unwanted immune activation. Purification typically involves chromatography or precipitation methods tailored to remove contaminants specific to the IVT mRNA synthesis process. HPLC and column-based purification are commonly employed to achieve high purity [8].

### 26.2.2 Incorporation of Modified Nucleosides

Natural mRNA molecules synthesized in cells contain standard nucleosides (adenosine, guanosine, cytidine, uridine), which can be recognized by pattern recognition receptors (PRRs), especially by the innate immune system. To mitigate this issue, researchers have developed modified nucleosides, including pseudouridine and 5-methylcytidine, which are chemically altered versions of uridine and cytidine, respectively [9]. Modified nucleosides can increase mRNA stability by reducing susceptibility to degradation by nucleases present in the extracellular environment. By evading detection by innate immune sensors, modified nucleosides minimize immune activation and improve the overall translational efficiency of the mRNA into protein. This enhancement is crucial for achieving high levels of protein expression necessary for robust immune responses elicited by the vaccine. The details of modified nucleosides will also be discussed below.

### 26.2.3 Complexing mRNA with Carriers

Delivering mRNA into target cells poses several challenges, including susceptibility to degradation by extracellular nucleases and inefficient cellular uptake. To address these challenges, mRNA molecules are often complexed with delivery carriers, such as LNPs, polymers, or other encapsulation systems [10]. These carriers perform several critical functions as they shield mRNA from enzymatic degradation in the extracellular environment, enhancing its stability and preserving its integrity until it reaches the target cells. Carriers facilitate the cellular uptake of mRNA through mechanisms like endocytosis, overcoming the barriers that typically hinder naked mRNA entry into cells. Once inside cells, carriers assist in the controlled release of mRNA, ensuring optimal exposure to the cellular machinery responsible for translation into protein. The choice of carrier and its formulation can also influence the immunogenicity profile of the mRNA vaccine. For example, LNPs can trigger innate immune responses beneficial for vaccine efficacy while minimizing adverse reactions. Additionally, carriers can be engineered to modify the release kinetics of mRNA, influencing the duration and magnitude of protein expression, which is crucial for tailoring immune responses to specific vaccine targets [6, 7, 11, 12]. By employing these strategies, researchers aim to optimize mRNA vaccine formulations to achieve desired immune responses while minimizing potential inhibitory effects on antigen expression.

**Box 26.1 Unmodified, Unpurified mRNA Vs. Nucleoside-Modified, Purified mRNA in Vaccines: Stability, Immunogenicity, and Effectiveness**

*Unmodified, Unpurified mRNA*

Unmodified mRNA contains adenosine, guanosine, cytidine, and uridine and is synthesized without purification steps. This mRNA is inherently less stable because it is susceptible to degradation by extracellular nucleases and can form secondary structures that hinder efficient translation. Unmodified mRNA can be recognized by innate immune sensors, such as Toll-like receptors (TLRs) and RIG-I-like receptors (RLRs), due to the presence of unmethylated guanosine residues (cap structures) and double-stranded regions (dsRNA intermediates). This recognition can trigger robust innate immune responses, potentially leading to inflammation or reducing the mRNA's translational efficiency. The immunogenicity and instability of unmodified, unpurified mRNA can limit its effectiveness as a vaccine.

*Nucleoside-Modified, Purified mRNA*

Nucleoside-modified mRNA incorporates modified nucleosides (e.g., pseudouridine, 5-methylcytidine) during synthesis, which enhances stability by reducing susceptibility to degradation by nucleases. Purification processes further remove contaminants and residual DNA templates, ensuring high purity and integrity of the mRNA sequence. Modified nucleosides alter the mRNA's structure and chemical composition, reducing its recognition by innate immune sensors [9]. This modification decreases the likelihood of immune activation and inflammation, thereby allowing for more efficient translation of mRNA into protein [13]. The enhanced stability and reduced immunogenicity of nucleoside-modified, purified mRNA improve its effectiveness as a vaccine. These modifications enable higher and more sustained levels of protein expression within target cells, promoting robust immune responses against the encoded antigen. In summary, nucleoside modification and purification of mRNA represent critical advancements in vaccine technology (Fig. 26.1).

## 26.3 Innate and Adaptive Immune Stimulation

mRNA vaccines function by delivering synthetic mRNA into non-immune cells, prompting them to produce a specific antigen. mRNA is translated into protein within the cytosol, degraded by proteasomes and releasing antigenic epitopes. These epitopes are subsequently bound by molecules of the MHC class I, forming complexes on the surface of APCs, such as DCs. APCs, particularly those expressing the $CD8^+$ marker, interact with these antigen–MHC complexes. This interaction triggers the activation of cytotoxic T cells (i.e., $CD8^+$ T cells), enabling the immune system to recognize and mount a targeted response to cells expressing the antigen, aiding in the elimination of infected or abnormal cells. Moreover, mRNA vaccines

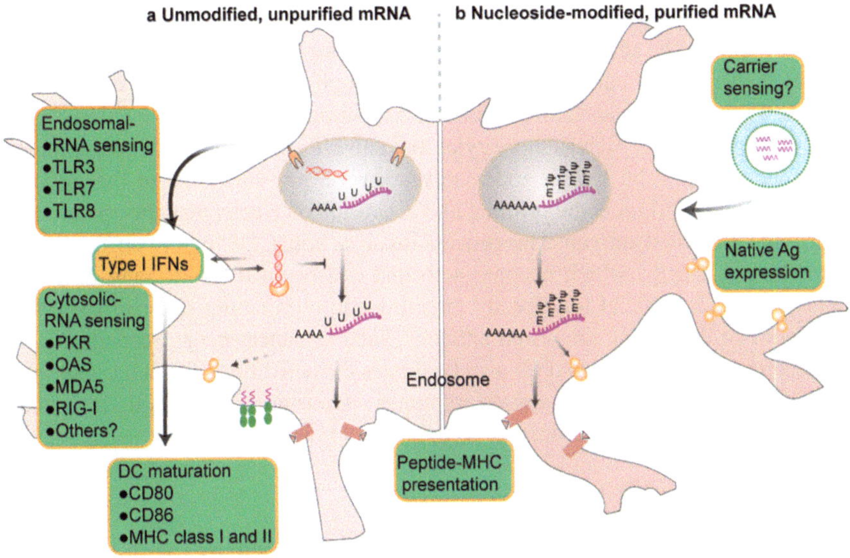


**Fig. 26.1** Innate immune recognition of mRNA vaccines by DCs. Key elements depicted include RNA sensors, antigens, DC maturation factors, and peptide–MHC complexes. RNA sensors: Highlighted in yellow, these sensors detect both ds- and unmodified ssRNAs present in mRNA vaccines. Antigens: Depicted in red, these represent the proteins encoded by the mRNA vaccines. DC maturation factors: Shown in green, these factors indicate the activation and maturation of DCs upon recognition of mRNA. Peptide–MHC complexes: Presented in light blue and red, these complexes are formed as a result of antigen processing and presentation by DCs. (Figure is adapted from [14] with permission from Springer Nature (License number 5902231383209))

can also indirectly activate bone marrow-derived DCs. These activated DCs are crucial in the initial priming of $CD8^+$ T cells [15]. mRNA vaccines additionally function by delivering their payload not only to non-immune cells but also to tissue-resident DCs and macrophages. These specialized immune cells are strategically positioned within tissues, including at the site of vaccine administration. Upon transfection with mRNA, DCs and macrophages in the local tissue uptake the synthetic genetic material. Inside these immune cells, the synthesized proteins are then processed and presented on the surface of these immune cells via MHC molecules. This local presentation of antigenic fragments by DCs and macrophages serves as a signal to nearby immune cells, initiating a targeted immune response right at the injection site. This response includes the activation of both $CD4^+$ helper T cells and $CD8^+$ cytotoxic T cells, which are crucial for coordinating and executing immune defenses against the specific antigen encoded by the mRNA [16].

Second, mRNA transfection of immune cells also leads to the presentation of antigens via the MHC class II pathway. In this pathway, antigens are processed and presented on the surface of APCs, including DCs, macrophages, and B cells, bound to MHC class II molecules to interact with T helper cells, specifically those expressing the CD4 marker. Activated CD4+ T cells can stimulate and help regulate the activity of other immune cells, including CD8+ T cells and B cells. They also contribute to the production of antibodies by B cells, enhancing the adaptive immune response against the antigen [17]. After administering mRNA vaccines, a portion of the injected material can migrate to nearby lymph nodes through the lymphatic system. Lymph nodes are critical hubs of the immune system, where they house various immune cells such as monocytes and naïve T and B cells. This presentation of antigens within the lymph nodes serves several key functions in the immune response. First, it triggers the priming and activation of T cells that specifically recognize the presented antigen. This process involves both $CD8^+$ cytotoxic T cells and $CD4^+$ helper T cells, which become activated and begin to proliferate. Activated $CD8^+$ T cells are crucial for directly targeting and eliminating cells infected with pathogens or displaying abnormal antigens, such as cancer cells. Second, mRNA transfection of lymph node APCs also facilitates the activation of B cells. This antibody production, known as humoral immunity, is essential for neutralizing pathogens and promoting their clearance from the body [18].

## 26.4 mRNA Vaccines Against Infectious Diseases

Developing prophylactic or therapeutic vaccines against infectious pathogens remains the cornerstone of controlling and preventing epidemics. However, traditional vaccine approaches have encountered challenges in generating effective vaccines for viruses that cause chronic or recurrent infections, such as HIV-1, herpes simplex virus, and respiratory syncytial virus (RSV). These viruses have complex immune evasion mechanisms, making it difficult to induce lasting protective immunity through conventional vaccine strategies. Additionally, the slow pace of commercial vaccine development and approval processes has proven inadequate in responding swiftly to the emergence of acute viral diseases. The Ebola and Zika virus outbreaks from 2014 to 2016 underscored the urgent need for more agile vaccine platforms capable of rapid deployment and adaptation to novel pathogens. To address these challenges, there is a critical push toward developing more dynamic and potent vaccine platforms. These next-generation platforms aim to harness advancements in vaccine technology, such as mRNA vaccines, viral vector vaccines, and nanoparticle-based vaccines [13]. These platforms offer advantages such as rapid production capabilities, flexibility in antigen design, and potential for enhanced immunogenicity. By leveraging these innovative approaches, researchers and pharmaceutical companies seek to improve vaccine efficacy and accelerate response times to emerging infectious threats.

Preclinical research on mRNA vaccines has shown promising attributes aligning with the criteria of an ideal clinical vaccine. These vaccines have demonstrated a favorable safety profile in animal studies, providing confidence in their potential for human use. One of their key advantages is the flexibility and rapidity with which they can be designed to target new infectious diseases. This agility stems from the straightforward process of synthesizing mRNA sequences based on the genetic information of the target pathogen. Moreover, mRNA vaccines can be manufactured at a scale suitable for good manufacturing practices (GMP), enabling mass production to meet global vaccination demands. Several companies are already advancing production capabilities, anticipating widespread deployment. Unlike traditional protein-based vaccines that rely on injected antigens, mRNA vaccines have shown the ability to induce robust CD8⁺ T cell responses. This occurs through the efficient presentation of antigens produced within cells on MHC class I molecules. Additionally, mRNA vaccines are effective in eliciting strong CD4⁺ T cell responses, which play crucial roles in coordinating immune responses and enhancing vaccine efficacy [19–21]. Furthermore, mRNA vaccines have demonstrated an advantage over DNA vaccines in generating potent neutralizing antibody responses in preclinical studies. Unlike DNA vaccines, which often require multiple doses or additional adjuvants to achieve sufficient antibody levels, mRNA vaccines have shown efficacy with just one or two low-dose immunizations in animal models.

### 26.4.1 Dendritic Cell mRNA (DC-mRNA) Vaccines

Ex vivo DC loading has emerged as a method widely used to stimulate cell-mediated immunity, primarily in cancer therapy, but it has also been explored for therapeutic vaccines against infectious diseases, albeit with limited application thus far. A notable example involves its use in developing a therapeutic vaccine for HIV-1. In this approach, individuals infected with HIV-1 and undergoing highly active antiretroviral therapy (HAART) received treatment with autologous DCs that had been electroporated with mRNA encoding various HIV-1 antigens. Although the treatment was found to be safe and successful in eliciting antigen-specific CD4⁺ and CD8⁺ T cell responses in patients, it did not result in significant clinical benefits. Despite the promising immunological responses observed, such as the activation of T cells specific to HIV-1 antigens, these responses were not sufficient to control viral replication or lead to measurable clinical improvements in the patients' conditions [22–27]. In a different study focusing on cytomegalovirus (CMV), researchers investigated a vaccine where DCs were loaded with mRNA encoding the CMV pp65 protein [28]. The results indicated that the vaccination was effective in inducing or expanding CMV-specific cellular immune responses, showcasing the potential of this strategy in enhancing protective immune responses against infectious diseases. These examples highlight the dual potential of ex vivo DC loading in both therapeutic and potentially prophylactic settings, although the clinical outcomes may vary.

### 26.4.2 Directly Injectable mRNA Vaccines

Unlike more complex vaccine platforms, such as viral vector or DNA-based vaccines, non-replicating mRNA vaccines offer a straightforward method of delivery that does not require viral vectors or live pathogens. However, it was only more recently that non-replicating mRNA vaccines have been shown to elicit robust and protective immune responses against various infectious diseases in preclinical and clinical studies. This includes vaccines targeting viruses like SARS-CoV-2, influenza, and Zika virus, among others. These advancements underscore the versatility and efficacy of mRNA vaccines in generating protective immunity, thereby bolstering their role as promising candidates for combating infectious diseases globally [29, 30]. This groundbreaking study showcased the effectiveness of intradermally administered mRNA vaccines using uncomplexed mRNA encoding influenza virus antigens, along with a protamine-complexed RNA adjuvant. The research demonstrated strong immunogenicity across multiple animal models and provided robust protection against lethal viral challenges. This approach highlights the versatility and potency of mRNA vaccines in eliciting protective immune responses. Similarly, the study also investigated the protamine-based RNActive platform, which encoded the rabies virus glycoprotein. This platform induced protective immunity against lethal virus challenges in mice and elicited potent neutralizing antibody responses in pigs. These findings underscore the adaptability of mRNA-based technologies in developing vaccines against various infectious pathogens, showcasing their potential to enhance vaccine efficacy and broaden protection across different species [19].

### *26.4.3 Nucleoside-Modified mRNA Vaccine*

Nucleoside-modified mRNA vaccines represent a significant advancement in vaccine technology, offering enhanced stability, reduced innate immune activation, and improved translational efficiency compared to unmodified mRNA. Building on these advantages, a recent study investigated the efficacy of a nucleoside-modified mRNA vaccine in a maternal vaccination model against Zika virus. In this study, pregnant mice were administered the nucleoside-modified mRNA vaccine. The results were striking, showing a profound reduction in Zika virus infection rates among fetal mice. Importantly, the vaccine not only protected against infection but also completely prevented the associated fetal viability issues observed in untreated groups. This outcome underscores the vaccine's ability to induce robust maternal immune responses that effectively cross the placental barrier and protect the developing fetuses from viral infection [31]. These outcomes underscore the potential of nucleoside-modified mRNA vaccines not only in protecting against infectious diseases but also in safeguarding vulnerable populations such as fetuses during outbreaks.

A recent study has significantly broadened the investigation into nucleoside-modified, LNP-complexed mRNA vaccines by evaluating their immunogenicity against H10N8 and H7N9 influenza viruses across multiple species, including mice, ferrets, non-human primates, and humans for the first time. This comprehensive study provided insights into the vaccine's effectiveness in eliciting protective immune responses across various animal models and humans. In mice, the study demonstrated that low doses of the mRNA vaccines administered via intradermal or intramuscular routes were capable of inducing robust immune responses that protected against homologous influenza virus challenges. These findings suggested the vaccine's potential efficacy and suitability for further development. Consistent results were observed in larger animal models, such as ferrets and cynomolgus monkeys. In these species, one or two doses of the vaccine, ranging from 50 to 400 μg, effectively induced immune responses that demonstrated protective efficacy against influenza virus strains. These outcomes underscored the vaccine's ability to generate immune responses in larger animals, including non-human primates, which are crucial for assessing translational relevance to humans [32].

## 26.5 mRNA Cancer Vaccines

mRNA cancer vaccines represent a pioneering approach in cancer immunotherapy, utilizing mRNA to encode tumor-specific antigens or proteins that stimulate immune responses against cancer cells. These vaccines are designed to harness the body's immune system to recognize and attack cancer cells selectively, offering several advantages. By delivering mRNA encoding tumor antigens directly into APCs or through direct injection, mRNA cancer vaccines can induce robust immune responses, including activation of CTLs and memory T cells. This immune activation targets cancer cells expressing the specific antigens encoded by the mRNA, potentially leading to tumor cell death and regression. Moreover, mRNA vaccines are versatile and can be rapidly adapted to target different types of cancers and tumor antigens, offering a personalized approach to cancer treatment that holds promise for enhancing overall patient outcomes in clinical trials and future therapeutic applications. mRNA-based cancer vaccines have recently undergone comprehensive review, detailing their development and potential in cancer treatment [33–37], and their advantages compared to conventional vaccines are detailed in Table 26.1 [38, 39].

Cancer vaccines are typically administered to patients who already have cancer. Unlike prophylactic vaccines, which prevent disease onset, therapeutic cancer vaccines aim to intervene in ongoing cancer progression. They seek to enhance the body's immune response against existing tumors, thereby potentially reducing tumor size, preventing recurrence, or achieving long-term disease control. The core strategy of cancer vaccines involves presenting tumor-specific antigens to the immune system. By boosting CTL activity, cancer vaccines aim to mount a targeted attack against malignant cells while minimizing damage to healthy tissues. Most cancer vaccines focus on TAAs or neoantigens that are specific to the patient's tumor cells, detailed in Box 26.2 [40]. The evolution of these vaccines reflects significant progress in understanding cancer immunology and the harnessing of mRNA technology to tailor highly specific treatments against individual tumors.

**Table 26.1** Detailed comparison between mRNA vaccines and conventional vaccines

| Feature | mRNA vaccine | Conventional vaccine |
|---|---|---|
| Type of vaccine | Utilizes mRNA to encode antigen(s) | Contains whole or partial forms of inactivated or attenuated pathogens, or proteins derived from them |
| Mechanism of action | mRNA instructs cells to produce antigen(s) | Pathogen or protein directly stimulates immune response |
| Antigen presentation | Antigen expressed by host cells | Antigen administered as part of the vaccine formulation |
| Production process | Synthesized in vitro using DNA templates and enzymes | Cultured and grown from pathogens or proteins in cell systems |
| Speed of development | Rapid development due to modular nature and synthetic production | Longer development timeline due to cultivation and testing |
| Flexibility | Easily modifiable for different antigens and variants | Limited by specific pathogen or protein target |
| Stability | Generally less stable, requiring lipid nanoparticle delivery | More stable in formulation, often refrigerated or frozen |
| Immunogenicity | Can induce robust immune responses, including T cell activation | Varies based on formulation and adjuvants used |
| Safety | Generally well-tolerated; short-lived mRNA in cells | Potential for adverse reactions due to pathogen components |
| Storage requirements | Often requires ultra-cold storage for lipid nanoparticles | Typically refrigerated or frozen, depending on formulation |
| Manufacturing scalability | Scalable due to synthetic production | Production scalability may vary based on manufacturing process |
| Clinical use | Novel technology with expanding applications in infectious diseases and cancer | Established history in preventing infectious diseases and some cancers |
| Examples | Pfizer-BioNTech, Moderna COVID-19 vaccines | Influenza, hepatitis B, polio vaccines, among others |

**Box 26.2 TAAs, TSAs, and Cryptic Antigens: Definition and Functions**

Tumor antigens can indeed emerge at various stages of protein synthesis and degradation, reflecting the dynamic changes and aberrations within cancer cells. Transcription: The process of transcription, where DNA is transcribed into mRNA, can lead to the production of aberrant mRNA transcripts in cancer cells. This may result from mutations, gene fusions, or epigenetic changes that alter the normal transcription patterns. During translation, mRNA is translated into proteins, and this process can be dysregulated in cancer. Errors in translation machinery or abnormal expression of translation factors may lead to the production of abnormal proteins or truncated protein products. These aberrant proteins can serve as tumor-specific antigens by virtue of their

(continued)

**Box 26.2** (continued)
altered structure or expression levels compared to normal cells. Cancer cells often exhibit dysregulated PTMs, such as phosphorylation, glycosylation, acetylation, and ubiquitination, among others. These modifications can alter protein structure, function, and antigenicity. In cancer, alterations in antigen processing and presentation pathways can lead to the presentation of tumor-specific peptides or altered peptide–MHC complexes that are recognized by T cells or other immune cells. Dysregulated protein degradation pathways in cancer cells, such as the ubiquitin–proteasome system or autophagy, can result in the accumulation of abnormal or truncated protein fragments. These fragments can serve as potential antigens if they are processed and presented effectively by APCs to initiate an immune response against cancer cells.

*Tumor-Associated Antigens (TAA)*

TAAs are antigens expressed by tumor cells that are recognized by the immune system, distinguishing them from normal cells. TAAs can be categorized into several types based on their expression patterns and roles in tumorigenesis. TAAs are normal cellular proteins that are expressed at higher levels in tumor cells compared to their normal counterparts. The immune system may recognize these overexpressed antigens as foreign or aberrant, triggering an immune response against the tumor. TAAs antigens are expressed at specific stages of cell differentiation and are often retained or re-expressed by tumor cells. Examples include melanoma-associated antigens (e.g., MART-1, gp100) and prostate-specific antigen (PSA) in prostate cancer; CTAs are proteins normally expressed in male germ cells but can also be aberrantly expressed in various cancers. They are attractive targets for immunotherapy because of their restricted expression in normal tissues outside the testes and their immunogenicity. Viral-associated antigens are also TAAs. Some tumors are associated with oncogenic viruses (e.g., HPV, EBV) that encode viral proteins expressed in infected cells. These viral antigens can be targeted by the immune system to eliminate virus-infected tumor cells.

*Tumor-Specific Antigens (TSAs)*

TSAs, as the name suggests, are antigens that are uniquely expressed by tumor cells and are not found in normal tissues. TSAs arise due to mutations, genomic rearrangements, or viral oncogene expression specific to tumor cells. TSAs are neoantigens derived from tumor-specific mutations (e.g., single nucleotide variants, insertions, deletions) in protein-coding genes. These mutations generate peptides that are presented on MHC molecules and recognized as foreign by the immune system. Neoantigens are highly specific to individual tumors and are considered ideal targets for personalized cancer immunotherapy. Some TSAs result from abnormal gene expression or alternative splicing events specific to tumor cells. These antigens may include truncated proteins, fusion proteins, or proteins normally expressed only during embryonic development.

(continued)

**Box 26.2** (continued)

*Cryptic Antigens*

Cryptic antigens originate from genomic regions that are traditionally considered non-coding or untranslated, known as the "dark matter" of the proteome. These regions do not correspond to the canonical exome-encoded proteome [41] but can still produce peptides that are presented by HLA class I molecules [42, 43]. This phenomenon suggests an additional reservoir of potential TSAs that have not been thoroughly explored. While research into cryptic antigens remains limited, emerging evidence supports their relevance in cancer immunity. Estimates suggest that cryptic peptides may constitute at least 10% of all peptides presented by HLA class I molecules, underscoring their potential significance in immune surveillance and therapeutic targeting of tumors [44]. Further investigation into these hidden proteomic elements promises to expand our understanding of TSAs and their implications for cancer immunotherapy. However, the exact quantity of these cryptic antigens remains uncertain due to the approximate nature of current estimates. Discovering such antigens poses significant challenges, yet advancements in MS/MS technology combined with cutting-edge sequencing software and techniques like RiboSeq offer promising technical foundations for improvement. Despite these advancements, the specificity of cryptic peptides to tumors remains unclear [42]. The categorization of cryptic antigens is dynamic, reflecting ongoing advancements in detection and analytical methods that enhance our ability to identify these antigens originating from introns, open reading frames, splice variants, and other sources. As these methods continue to improve, an increasing number of cryptic antigens will likely be annotated, providing insights into their origins and roles [43, 45]. This evolution promises to define the functions and relevance of cryptic antigens more precisely in the context of cancer immunology and potential therapeutic applications in the coming years. Once the mechanism becomes clear, one could then utilize and re-design these peptides as assemblies or clusters to be delivered systemically as tumor vaccines [46].

## 26.6 DC-mRNA Cancer Vaccines

DC functions can be significantly influenced by incorporating mRNA-encoded pro-inflammatory cytokines such as IL-12 or trafficking-associated molecules. A notable advancement in this area is the TriMix approach, which involves a combination of mRNA-encoded adjuvants—CD70, CD40L, and constitutively active TLR4—co-electroporated with antigen-encoding mRNAs. This innovative strategy has demonstrated efficacy in multiple preclinical studies by enhancing DC activation and driving a shift in $CD4^+$ T cell responses from regulatory to T helper 1 (Th1) type cells [47–49].

## 26.7 Administration Routes of mRNA Cancer Vaccines

The efficacy of mRNA cancer vaccines is heavily influenced by the route of administration and the delivery mechanism used. Researchers have explored a variety of administration routes to optimize the delivery of mRNA vaccines for cancer therapy. These include standard routes such as intradermal, intramuscular, subcutaneous, and intranasal injections, as well as more unconventional approaches such as intranodal (injection into lymph nodes), intravenous (directly into the bloodstream), intrasplenic (into the spleen), and intratumoral (directly into tumors). Each administration route offers distinct advantages and challenges. For instance, intradermal and intramuscular injections are commonly used due to their ease of access and ability to induce systemic immune responses. Subcutaneous administration is preferred for certain vaccine formulations that require slow release and prolonged exposure. Intranasal delivery targets mucosal immunity, potentially enhancing local and systemic immune responses against respiratory cancers.

*Intranodal Administration*, though less conventional, has proven to be an effective delivery method for mRNA vaccines. This technique involves direct injection of mRNA into secondary lymphoid tissues, which strategically targets antigen-presenting cells where T cell activation occurs, bypassing the need for DC migration to lymph nodes. Numerous studies have confirmed that naked mRNA, when injected intranodally, is preferentially taken up by DCs and capable of inducing strong prophylactic or therapeutic antitumor T cell responses [50, 51]; earlier research has shown similar efficacy with intrasplenic delivery [52]. Enhancing these responses, the coadministration of the dendritic cell-activating protein FMS-related tyrosine kinase 3 ligand (FLT3L) has been demonstrated to boost the effectiveness of intranodal mRNA vaccines in some cases [53, 54]. Expanding on this approach, the integration of the TriMix adjuvant—comprising CD70, CD40L, and an active form of TLR4—with intranodal injections of mRNAs encoding tumor-associated antigens has demonstrated potent antigen-specific CTL responses and effective control of tumor growth across various tumor models [55]. In recent research, this strategy was further validated using mRNA encoding the E7 protein of HPV combined with TriMix. The study showed significant enhancement in the infiltration of $CD8^+$ T cells into tumors and remarkable inhibition of tumor growth in a mouse model engineered to express HPV E7 antigen [56].

*Intranasal Vaccine Administration* offers a non-invasive, needle-free approach that facilitates rapid uptake of antigens by DCs, thereby enhancing immune responses. This delivery method has been effectively demonstrated in a study utilizing mRNA complexed with Stemfect lipid nanoparticles (LNPs), developed by Stemgent. In this study, intranasal administration of mRNA-LNP complexes resulted in delayed tumor onset and improved survival outcomes in mouse models. The models used the OVA-expressing E.G7-OVA T lymphoblastic cell line to simulate both prophylactic and therapeutic settings. The approach leveraged the ability of DCs in the nasal mucosa to capture and present antigens, thereby stimulating robust immune responses against the tumor cells expressing the OVA antigen [57].

This highlights the potential of intranasal administration for activating immune defenses directly at the mucosal surfaces, a primary entry point for many pathogens.

*Intratumoral mRNA Vaccination* offers a unique advantage by directly activating tumor-resident T cells in a targeted manner. Unlike conventional approaches that rely on tumor-associated antigens, this strategy utilizes mRNA encoding immune stimulatory molecules to stimulate tumor-specific immunity directly within the tumor microenvironment. In a glioblastoma mouse model, both forms of mRNA were able to inhibit tumor growth and confer protection, leveraging the innate immunogenic properties of mRNA itself [58]. Building on this foundation, recent research has explored more sophisticated approaches. For example, one investigation focused on intratumoral delivery of mRNA, encoding a fusion protein consisting of IFNβ and a TGFβ antagonist. This engineered cytokine configuration significantly enhanced the cytolytic activity of $CD8^+$ T cells and demonstrated a modest delay in tumor growth in mouse models of OVA-expressing lymphoma and lung carcinoma [59]. Another notable study highlighted the efficacy of intratumoral administration of TriMix mRNA, which includes mRNA encoding CD70, CD40L, and active TLR4, without encoding tumor-associated antigens. This formulation effectively activated $CD8\alpha^+$ DCs and promoted the generation of tumor-specific T cells, resulting in delayed tumor growth across various mouse models [60]. This method directly stimulates the immune system within the tumor microenvironment, potentially offering a potent therapeutic strategy to combat cancer by enhancing local immune responses against the tumor. These innovative approaches underscore the versatility and potential of mRNA-based vaccines administered via intranasal and intratumoral routes, highlighting their capability to elicit robust immune responses in a targeted and efficient manner.

*Intradermal Delivery of mRNA Vaccines* is a well-established method, particularly advantageous in cancer immunotherapy due to the abundance of APCs present in the skin. This approach has been validated in several key studies, demonstrating its potential in both preclinical and clinical settings [61]. An early pivotal study highlighted the efficacy of intradermal administration of total tumor RNA, which significantly delayed tumor growth in a fibrosarcoma mouse model. This approach capitalized on the skin's rich population of dendritic cells and other immune cells capable of processing and presenting tumor antigens [62]. The protamine-based RNActive platform represents a more refined approach, involving intradermal injection of mRNA encoding specific tumor antigens. This platform has shown effectiveness across various mouse cancer models and has transitioned into multiple clinical studies, both prophylactic and therapeutic [63].

## 26.8 Carrier-Based Delivery of mRNA Cancer Vaccine

Carrier-based delivery systems represent a pivotal advancement in the field of mRNA cancer vaccines, offering precise control over therapeutic payloads and enhancing the stability and delivery efficiency of mRNA molecules. By harnessing these carriers, researchers aim to optimize the immunogenicity and therapeutic efficacy of mRNA vaccines against cancer. This approach not only enhances antigen presentation to immune cells but also supports sustained and controlled release of mRNA, potentially overcoming challenges associated with rapid degradation and inefficient delivery.

**Cationic Polymeric/Lipid-Based Delivery** Cationic polymer/lipid-based delivery systems have revolutionized mRNA transfection in vitro, exemplified by commercially available reagents like TransIT-mRNA (Mirus Bio LLC) and Lipofectamine (Invitrogen). These systems efficiently deliver mRNA into various primary cells and cancer cell lines, yet their efficacy in vivo often falls short or introduces significant toxicity, as observed in recent studies. Overcoming these challenges has spurred advancements in developing safer and more efficient delivery methods suitable for in vivo applications, a topic extensively reviewed in recent literature [64–67]. Building upon a decade of experience with in vivo siRNA delivery, cationic lipids, polymers, and dendrimers now play pivotal roles in enhancing mRNA delivery efficiency and safety [68], as summarized in Table 26.2.

*Lipid nanoparticles (LNPs)* have emerged as a leading platform for mRNA delivery due to their ability to encapsulate and protect mRNA molecules, facilitate cellular uptake, and promote efficient endosomal release for cytoplasmic mRNA translation. LNPs typically consist of several key components, including ionizable cationic lipid (crucial for helping to facilitate endosomal escape by responding to the acidic pH within endosomes); PEGylation is used to coat LNPs, extending their circulation time by reducing clearance by the reticuloendothelial system (RES), incorporation of cholesterol to stabilize the lipid bilayer structure, other natural

**Table 26.2** Detailed analysis of various components and functions of nanoparticle-based mRNA vaccine carrier

| Component | Function | Advantages | Challenges |
|---|---|---|---|
| Cationic lipids | Formulate nanoparticles with mRNA to protect against degradation, facilitate cellular uptake | Efficient mRNA delivery into cells<br>Protection of mRNA from nucleases<br>Potentially targeting specific tissues or cells | Limited in vivo efficacy<br>Risk of toxicity and immunogenicity<br>Scaling is difficult |
| Cationic polymers | Complex with mRNA for form nanoparticles<br>Enhancing stability, cellular uptake, and endosomal escape | Protect mRNA<br>Sustained release<br>Able to modify surface properties for targeting specific cells | Cytotoxicity and immune response<br>Limited in vivo efficacy due to immune system clearance<br>Complexity in scale-up |
| Dendrimers | Highly branched polymer, complex with mRNA to form nanoparticles<br>Enhanced delivery, efficiency, and safety | Precise control over size and structure<br>High loading capacity | Clinical scale production |

phospholipids to contribute to the stability and biocompatibility of the nanoparticles, ensuring they can safely deliver mRNA payloads without causing undue immune responses or cytotoxicity. LNPs have shown significant promise in mRNA vaccine development and therapy, facilitating effective delivery of mRNA encoding various antigens or therapeutic proteins. Their design allows for precise modulation of mRNA release kinetics and immune responses, making them versatile tools for advancing mRNA-based treatments against infectious diseases and cancer. Efficient in vivo siRNA delivery via LNPs has been documented [69], and it has been more recently established that LNPs are highly effective for delivering both self-amplifying RNA [70] and conventional, non-replicating mRNA [71].

## 26.9 Combinatorial mRNA Vaccine + Immunotherapy for Cancer

While mRNA cancer vaccines have demonstrated immunogenicity in human trials, optimizing their clinical efficacy requires further refinement through advances in basic immunological research and improved vaccination strategies. Additionally, combining mRNA vaccines with other therapeutic modalities, such as chemotherapy, radiotherapy, and ICIs, has shown promise in enhancing treatment outcomes, as observed in preclinical studies. For example, in a notable study, the administration of cisplatin significantly potentiated the therapeutic effects of an mRNA vaccine that included HPV16 E7 oncoprotein and the TriMix adjuvant. This combination therapy led to complete regression of tumors in the female genital tract of a mouse model. This synergistic approach highlights the potential of integrating mRNA vaccines with conventional cancer treatments to achieve more robust and durable responses against tumors [56]. Furthermore, recent studies have indicated that combining mRNA vaccines with immunotherapy agents like antibodies against PD-1 could increase their effectiveness against diseases such as metastatic melanoma in humans [72] (Fig. 26.2).

## 26.10 Conclusion

The landscape of mRNA vaccines is currently thriving with extensive activity in both basic and clinical research. Over the past few years, numerous reports have emerged demonstrating the effectiveness of mRNA vaccines across a variety of applications. It is important to recognize that these results pertain to just a few iterations of mRNA vaccine platforms. Variations in the expression and immune-stimulating profiles of different mRNA vaccines could lead to different outcomes. There is a need for more research to understand how various animal models react to mRNA components and to decipher which immune signaling pathways are most effective in humans. With many promising mRNA platforms in development,

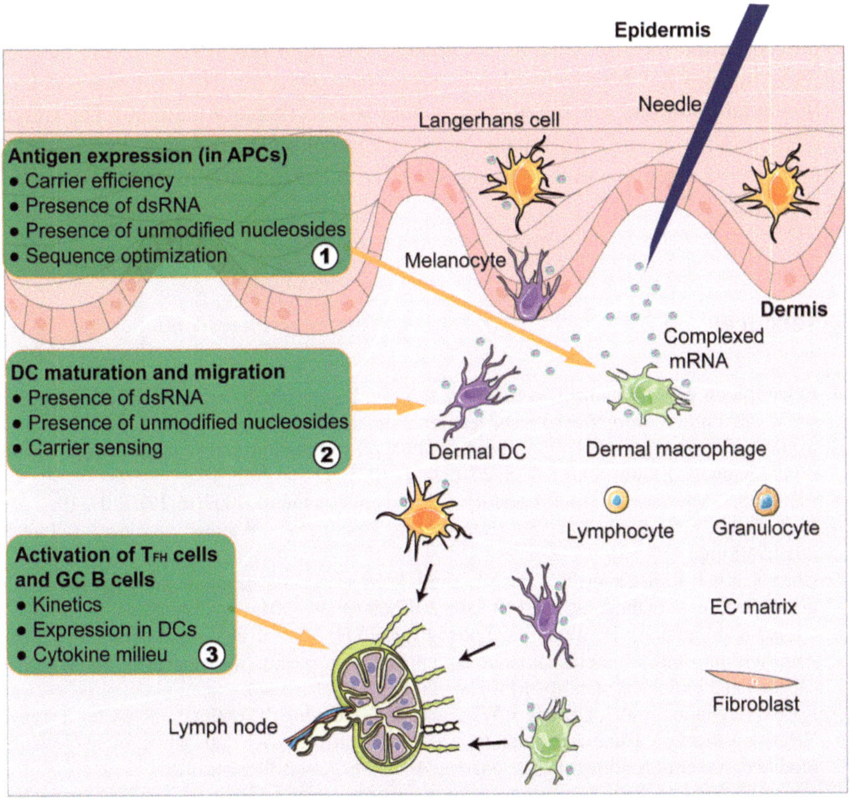


**Fig. 26.2** Key factors influencing the efficacy of directly injected mRNA vaccines. For mRNA vaccines administered via direct injection, several crucial factors determine their effectiveness in eliciting a robust immune response. These considerations can be visualized and understood through the example of an intradermal injection, which targets the delivery of mRNA directly into the skin where various immune cells are present. (i) Level of antigen expression in APCs. The delivery system or carrier used to encapsulate and transport the mRNA into cells plays a critical role in its success. Efficient carriers protect mRNA from degradation while facilitating its uptake by APCs, such as DCs. Components such as dsRNA or unmodified nucleosides can act as PAMPs. These molecules are recognized by PRRs on APCs, enhancing the immune response but also requiring careful modulation to avoid excessive inflammation. Adjustments in the mRNA sequence, such as optimizing codon usage for better translation efficiency, adjusting G:C content, and refining 5′ and 3′ UTRs, are crucial for maximizing protein synthesis and stability of the mRNA. (ii) DCs maturation and migration. The presence of PAMPs not only enhances initial antigen recognition but also stimulates the maturation of DCs. Mature DCs are pivotal in orchestrating an immune response as they migrate to secondary lymphoid tissues, where they prime T cells. Effective vaccines must also stimulate robust TFH cell responses. TFH cells are essential for helping B cells mature within germinal centers, a critical site for antibody production and affinity maturation. Although critical, the dynamics of TFH cell and GC B cell interactions and their optimization within the context of mRNA vaccines remain areas that are not fully understood and are currently under intense study. (Figure is adapted from [14] with permission from Springer Nature (License number: 590221268976))

detailed comparative studies are invaluable, helping concentrate efforts and resources on the most promising candidates. The rapid progression in the field of mRNA vaccines has been propelled by significant advancements in understanding RNA's innate immune sensing and improving in vivo delivery methods. The future of mRNA vaccines looks promising, with ongoing clinical trials and the increasing availability of resources poised to significantly enhance the field of mRNA-based therapeutics.

## References

1. Organization WH. Immunization coverage. Geneva: World Health Organization; 2023. https://www.who.int/news-room/fact-sheets/detail/immunization-coverage.
2. Younger DS, Younger AP, Guttmacher S. Childhood vaccination: implications for global and domestic public health. Neurol Clin. 2016;34(4):1035–47.
3. Plotkin SA. Vaccines: the fourth century. Clin Vaccine Immunol. 2009;16(12):1709–19.
4. Rodrigues CMC, Pinto MV, Sadarangani M, Plotkin SA. Whither vaccines? J Infect. 2017;74(Suppl 1):S2–s9.
5. Chen N, Xia P, Li S, Zhang T, Wang TT, Zhu J. RNA sensors of the innate immune system and their detection of pathogens. IUBMB Life. 2017;69(5):297–304.
6. Karikó K, Muramatsu H, Welsh FA, Ludwig J, Kato H, Akira S, et al. Incorporation of pseudouridine into mRNA yields superior nonimmunogenic vector with increased translational capacity and biological stability. Mol Ther. 2008;16(11):1833–40.
7. Karikó K, Muramatsu H, Ludwig J, Weissman D. Generating the optimal mRNA for therapy: HPLC purification eliminates immune activation and improves translation of nucleoside-modified, protein-encoding mRNA. Nucleic Acids Res. 2011;39(21):e142.
8. Kang DD, Li H, Dong Y. Advancements of in vitro transcribed mRNA (IVT mRNA) to enable translation into the clinics. Adv Drug Deliv Rev. 2023;199:114961.
9. Pardi N, Weissman D. Nucleoside modified mRNA vaccines for infectious diseases. Methods Mol Biol. 2017;1499:109–21.
10. Granados-Riveron JT, Aquino-Jarquin G. Engineering of the current nucleoside-modified mRNA-LNP vaccines against SARS-CoV-2. Biomed Pharmacother. 2021;142:111953.
11. Fotin-Mleczek M, Duchardt KM, Lorenz C, Pfeiffer R, Ojkić-Zrna S, Probst J, et al. Messenger RNA-based vaccines with dual activity induce balanced TLR-7 dependent adaptive immune responses and provide antitumor activity. J Immunother. 2011;34(1):1–15.
12. Rettig L, Haen SP, Bittermann AG, von Boehmer L, Curioni A, Krämer SD, et al. Particle size and activation threshold: a new dimension of danger signaling. Blood. 2010;115(22):4533–41.
13. Khurana A, Allawadhi P, Khurana I, Allwadhi S, Weiskirchen R, Banothu AK, et al. Role of nanotechnology behind the success of mRNA vaccines for COVID-19. Nano Today. 2021;38:101142.
14. Pardi N, Hogan MJ, Porter FW, Weissman D. mRNA vaccines—a new era in vaccinology. Nat Rev Drug Discov. 2018;17(4):261–79.
15. Lazzaro S, Giovani C, Mangiavacchi S, Magini D, Maione D, Baudner B, et al. CD8 T-cell priming upon mRNA vaccination is restricted to bone-marrow-derived antigen-presenting cells and may involve antigen transfer from myocytes. Immunology. 2015;146(2):312–26.
16. Alberer M, Gnad-Vogt U, Hong HS, Mehr KT, Backert L, Finak G, et al. Safety and immunogenicity of a mRNA rabies vaccine in healthy adults: an open label, non-randomised, prospective, first-in-human phase 1 clinical trial. Lancet. 2017;390(10101):1511–20.
17. Oberli MA, Reichmuth AM, Dorkin JR, Mitchell MJ, Fenton OS, Jaklenec A, et al. Lipid nanoparticle assisted mRNA delivery for potent cancer immunotherapy. Nano Lett. 2017;17(3):1326–35.

18. Firdessa-Fite R, Creusot RJ. Nanoparticles versus dendritic cells as vehicles to deliver mRNA encoding multiple epitopes for Immunotherapy. Mol Ther Methods Clin Dev. 2020;16:50–62.
19. Schnee M, Vogel AB, Voss D, Petsch B, Baumhof P, Kramps T, et al. An mRNA vaccine encoding rabies virus glycoprotein induces protection against lethal infection in mice and correlates of protection in adult and newborn pigs. PLoS Negl Trop Dis. 2016;10(6):e0004746.
20. Brito LA, Kommareddy S, Maione D, Uematsu Y, Giovani C, Berlanda Scorza F, et al. Self-amplifying mRNA vaccines. Adv Genet. 2015;89:179–233.
21. Chahal JS, Fang T, Woodham AW, Khan OF, Ling J, Anderson DG, et al. An RNA nanoparticle vaccine against Zika virus elicits antibody and CD8+ T cell responses in a mouse model. Sci Rep. 2017;7(1):252.
22. Van Gulck E, Vlieghe E, Vekemans M, Van Tendeloo VF, Van De Velde A, Smits E, et al. mRNA-based dendritic cell vaccination induces potent antiviral T-cell responses in HIV-1-infected patients. AIDS. 2012;26(4):F1–12.
23. Routy JP, Boulassel MR, Yassine-Diab B, Nicolette C, Healey D, Jain R, et al. Immunologic activity and safety of autologous HIV RNA-electroporated dendritic cells in HIV-1 infected patients receiving antiretroviral therapy. Clin Immunol. 2010;134(2):140–7.
24. Allard SD, De Keersmaecker B, de Goede AL, Verschuren EJ, Koetsveld J, Reedijk ML, et al. A phase I/IIa immunotherapy trial of HIV-1-infected patients with tat, rev and Nef expressing dendritic cells followed by treatment interruption. Clin Immunol. 2012;142(3):252–68.
25. Gandhi RT, Kwon DS, Macklin EA, Shopis JR, McLean AP, McBrine N, et al. Immunization of HIV-1-infected persons with autologous dendritic cells transfected with mRNA encoding HIV-1 gag and Nef: results of a randomized, placebo-controlled clinical trial. J Acquir Immune Defic Syndr. 2016;71(3):246–53.
26. Jacobson JM, Routy JP, Welles S, DeBenedette M, Tcherepanova I, Angel JB, et al. Dendritic cell Immunotherapy for HIV-1 infection using autologous HIV-1 RNA: a randomized, double-blind, placebo-controlled clinical trial. J Acquir Immune Defic Syndr. 2016;72(1):31–8.
27. Gay CL, DeBenedette MA, Tcherepanova IY, Gamble A, Lewis WE, Cope AB, et al. Immunogenicity of AGS-004 dendritic cell therapy in patients treated during acute HIV infection. AIDS Res Hum Retroviruses. 2018;34(1):111–22.
28. Van Craenenbroeck AH, Smits EL, Anguille S, Van de Velde A, Stein B, Braeckman T, et al. Induction of cytomegalovirus-specific T cell responses in healthy volunteers and allogeneic stem cell recipients using vaccination with messenger RNA-transfected dendritic cells. Transplantation. 2015;99(1):120–7.
29. Martinon F, Krishnan S, Lenzen G, Magné R, Gomard E, Guillet JG, et al. Induction of virus-specific cytotoxic T lymphocytes in vivo by liposome-entrapped mRNA. Eur J Immunol. 1993;23(7):1719–22.
30. Petsch B, Schnee M, Vogel AB, Lange E, Hoffmann B, Voss D, et al. Protective efficacy of in vitro synthesized, specific mRNA vaccines against influenza a virus infection. Nat Biotechnol. 2012;30(12):1210–6.
31. Richner JM, Jagger BW, Shan C, Fontes CR, Dowd KA, Cao B, et al. Vaccine mediated protection against Zika virus-induced congenital disease. Cell. 2017;170(2):273–83.e12.
32. Bahl K, Senn JJ, Yuzhakov O, Bulychev A, Brito LA, Hassett KJ, et al. Preclinical and clinical demonstration of immunogenicity by mRNA vaccines against H10N8 and H7N9 influenza viruses. Mol Ther. 2017;25(6):1316–27.
33. Diken M, Kranz LM, Kreiter S, Sahin U. mRNA: a versatile molecule for cancer vaccines. Curr Issues Mol Biol. 2017;22:113–28.
34. Fiedler K, Lazzaro S, Lutz J, Rauch S, Heidenreich R. mRNA cancer vaccines. Recent Results Cancer Res. 2016;209:61–85.
35. Grunwitz C, Kranz LM. mRNA cancer vaccines-messages that prevail. Curr Top Microbiol Immunol. 2017;405:145–64.
36. McNamara MA, Nair SK, Holl EK. RNA-based vaccines in cancer immunotherapy. J Immunol Res. 2015;2015:794528.
37. Sullenger BA, Nair S. From the RNA world to the clinic. Science. 2016;352(6292):1417–20.

38. Vigneron N. Human tumor antigens and cancer immunotherapy. Biomed Res Int. 2015;2015:948501.
39. Türeci Ö, Vormehr M, Diken M, Kreiter S, Huber C, Sahin U. Targeting the heterogeneity of cancer with individualized Neoepitope vaccines. Clin Cancer Res. 2016;22(8):1885–96.
40. Coulie PG, Van den Eynde BJ, van der Bruggen P, Boon T. Tumour antigens recognized by T lymphocytes: at the core of cancer immunotherapy. Nat Rev Cancer. 2014;14(2):135–46.
41. Diederichs S, Bartsch L, Berkmann JC, Fröse K, Heitmann J, Hoppe C, et al. The dark matter of the cancer genome: aberrations in regulatory elements, untranslated regions, splice sites, non-coding RNA and synonymous mutations. EMBO Mol Med. 2016;8(5):442–57.
42. Chong C, Müller M, Pak H, Harnett D, Huber F, Grun D, et al. Integrated proteogenomic deep sequencing and analytics accurately identify non-canonical peptides in tumor immunopeptidomes. Nat Commun. 2020;11(1):1293.
43. Andersen RS, Andersen SR, Hjortsø MD, Lyngaa R, Idorn M, Køllgård TM, et al. High frequency of T cells specific for cryptic epitopes in melanoma patients. Onco Targets Ther. 2013;2(7):e25374.
44. Laumont CM, Daouda T, Laverdure JP, Bonneil É, Caron-Lizotte O, Hardy MP, et al. Global proteogenomic analysis of human MHC class I-associated peptides derived from non-canonical reading frames. Nat Commun. 2016;7:10238.
45. Frankiw L, Baltimore D, Li G. Alternative mRNA splicing in cancer immunotherapy. Nat Rev Immunol. 2019;19(11):675–87.
46. Wu B, Wang H. Peptide assemblies as promising tumor vaccines: current platforms and progress. Bio Integr. 2023;4(2):45–51.
47. Bontkes HJ, Kramer D, Ruizendaal JJ, Meijer CJ, Hooijberg E. Tumor associated antigen and interleukin-12 mRNA transfected dendritic cells enhance effector function of natural killer cells and antigen specific T-cells. Clin Immunol. 2008;127(3):375–84.
48. Bontkes HJ, Kramer D, Ruizendaal JJ, Kueter EW, van Tendeloo VF, Meijer CJ, et al. Dendritic cells transfected with interleukin-12 and tumor-associated antigen messenger RNA induce high avidity cytotoxic T cells. Gene Ther. 2007;14(4):366–75.
49. Dörrie J, Schaft N, Müller I, Wellner V, Schunder T, Hänig J, et al. Introduction of functional chimeric E/L-selectin by RNA electroporation to target dendritic cells from blood to lymph nodes. Cancer Immunol Immunother. 2008;57(4):467–77.
50. Diken M, Kreiter S, Selmi A, Britten CM, Huber C, Türeci Ö, et al. Selective uptake of naked vaccine RNA by dendritic cells is driven by macropinocytosis and abrogated upon DC maturation. Gene Ther. 2011;18(7):702–8.
51. Cressina E, Chen L, Abell C, Leeper F, Smith A. Fragment screening against the thiamine pyrophosphate riboswitchthiM. Chem Sci. 2010;2:157–65.
52. Zhou WZ, Hoon DS, Huang SK, Fujii S, Hashimoto K, Morishita R, et al. RNA melanoma vaccine: induction of antitumor immunity by human glycoprotein 100 mRNA immunization. Hum Gene Ther. 1999;10(16):2719–24.
53. Kreiter S, Diken M, Selmi A, Petschenka J, Türeci Ö, Sahin U. FLT3 ligand as a molecular adjuvant for naked RNA vaccines. Methods Mol Biol. 2016;1428:163–75.
54. Kreiter S, Diken M, Selmi A, Diekmann J, Attig S, Hüsemann Y, et al. FLT3 ligand enhances the cancer therapeutic potency of naked RNA vaccines. Cancer Res. 2011;71(19):6132–42.
55. Van Lint S, Goyvaerts C, Maenhout S, Goethals L, Disy A, Benteyn D, et al. Preclinical evaluation of TriMix and antigen mRNA-based antitumor therapy. Cancer Res. 2012;72(7):1661–71.
56. Bialkowski L, van Weijnen A, Van der Jeught K, Renmans D, Daszkiewicz L, Heirman C, et al. Intralymphatic mRNA vaccine induces CD8 T-cell responses that inhibit the growth of mucosally located tumours. Sci Rep. 2016;6:22509.
57. Phua KK, Staats HF, Leong KW, Nair SK. Intranasal mRNA nanoparticle vaccination induces prophylactic and therapeutic anti-tumor immunity. Sci Rep. 2014;4:5128.
58. Scheel B, Aulwurm S, Probst J, Stitz L, Hoerr I, Rammensee HG, et al. Therapeutic anti-tumor immunity triggered by injections of immunostimulating single-stranded RNA. Eur J Immunol. 2006;36(10):2807–16.

59. Van der Jeught K, Joe PT, Bialkowski L, Heirman C, Daszkiewicz L, Liechtenstein T, et al. Intratumoral administration of mRNA encoding a fusokine consisting of IFN-β and the ectodomain of the TGF-β receptor II potentiates antitumor immunity. Oncotarget. 2014;5(20):10100–13.
60. Van Lint S, Renmans D, Broos K, Goethals L, Maenhout S, Benteyn D, et al. Intratumoral delivery of TriMix mRNA results in T-cell activation by cross-presenting dendritic cells. Cancer Immunol Res. 2016;4(2):146–56.
61. Clausen BE, Stoitzner P. Functional specialization of skin dendritic cell subsets in regulating T cell responses. Front Immunol. 2015;6:534.
62. Granstein RD, Ding W, Ozawa H. Induction of anti-tumor immunity with epidermal cells pulsed with tumor-derived RNA or intradermal administration of RNA. J Invest Dermatol. 2000;114(4):632–6.
63. Weide B, Pascolo S, Scheel B, Derhovanessian E, Pflugfelder A, Eigentler TK, et al. Direct injection of protamine-protected mRNA: results of a phase 1/2 vaccination trial in metastatic melanoma patients. J Immunother. 2009;32(5):498–507.
64. Kauffman KJ, Mir FF, Jhunjhunwala S, Kaczmarek JC, Hurtado JE, Yang JH, et al. Efficacy and immunogenicity of unmodified and pseudouridine-modified mRNA delivered systemically with lipid nanoparticles in vivo. Biomaterials. 2016;109:78–87.
65. Guan S, Rosenecker J. Nanotechnologies in delivery of mRNA therapeutics using nonviral vector-based delivery systems. Gene Ther. 2017;24(3):133–43.
66. Reichmuth AM, Oberli MA, Jaklenec A, Langer R, Blankschtein D. mRNA vaccine delivery using lipid nanoparticles. Ther Deliv. 2016;7(5):319–34.
67. Midoux P, Pichon C. Lipid-based mRNA vaccine delivery systems. Expert Rev Vaccines. 2015;14(2):221–34.
68. Hald Albertsen C, Kulkarni JA, Witzigmann D, Lind M, Petersson K, Simonsen JB. The role of lipid components in lipid nanoparticles for vaccines and gene therapy. Adv Drug Deliv Rev. 2022;188:114416.
69. Kanasty R, Dorkin JR, Vegas A, Anderson D. Delivery materials for siRNA therapeutics. Nat Mater. 2013;12(11):967–77.
70. Geall AJ, Verma A, Otten GR, Shaw CA, Hekele A, Banerjee K, et al. Nonviral delivery of self-amplifying RNA vaccines. Proc Natl Acad Sci U S A. 2012;109(36):14604–9.
71. Pardi N, Hogan MJ, Pelc RS, Muramatsu H, Andersen H, DeMaso CR, et al. Zika virus protection by a single low-dose nucleoside-modified mRNA vaccination. Nature. 2017;543(7644):248–51.
72. Sahin U, Derhovanessian E, Miller M, Kloke BP, Simon P, Löwer M, et al. Personalized RNA mutanome vaccines mobilize poly-specific therapeutic immunity against cancer. Nature. 2017;547(7662):222–6.

# Chapter 27
# circRNA Vaccines

**Abstract** Circular RNAs (circRNAs) are a unique class of single-stranded RNAs characterized by their covalently closed-loop structures. Unlike linear RNAs, circRNAs lack 3′ and 5′ ends, which makes them exceptionally stable and resistant to degradation by exonucleases. This intrinsic stability has garnered significant interest in circRNAs as potential candidates for novel therapeutic applications, including vaccine development. Remarkably, certain circRNAs possess specific structural features that enable them to be translated into proteins, further enhancing their potential utility in vaccine design. Vaccines are powerful tools in immunotherapy, playing a crucial role in the prevention of infectious diseases and the treatment of cancers. The advent of circRNA vaccines represents an exciting new frontier in this domain. Recent advancements have seen the successful synthesis and testing of several circRNA vaccines both in vitro and in vivo, showcasing their promise in eliciting robust immune responses. This chapter provides a comprehensive overview of the current understanding of translatable circRNAs, delving into their biology and functional mechanisms. We explore the molecular biology underlying circRNA formation, the various synthetic methods employed to generate circRNAs, and the strategies for their effective delivery. Additionally, we examine the current landscape of circRNA vaccines, highlighting recent developments and experimental results. Furthermore, we discuss the challenges that need to be addressed to advance the field of circRNA vaccines. These include optimizing circRNA synthesis and delivery, ensuring consistent and efficient translation, and understanding the long-term effects and safety of circRNA-based immunizations. By summarizing the progress made in recent years, we aim to outline the future directions for research and development in this promising area of vaccine technology.

## 27.1 Introduction

Vaccines are pivotal in maintaining public health by offering robust protection against infectious diseases and holding significant promise in the fight against cancer. The history of vaccination began in the late eighteenth century with Edward

 
P. E. Saw, E. Song, *RNA Therapeutics in Human Diseases*,
https://doi.org/10.1007/978-981-96-3041-7_27

Jenner, whose pioneering work marked the advent of immunization as a transformative medical practice. Vaccines serve as highly effective tools for preventing infectious diseases, reducing morbidity and mortality rates, and preventing outbreaks. By stimulating the immune system to recognize and remember specific pathogens, vaccines confer immunity without causing severe illness. This herd immunity not only protects vaccinated individuals but also shields vulnerable populations, including those unable to receive vaccines due to medical reasons. Edward Jenner, an English physician, is widely regarded as the father of vaccination. In 1796, Jenner conducted an experiment where he inoculated a young boy with cowpox, a related virus to smallpox, to protect against smallpox—a deadly and highly contagious disease that had ravaged populations for centuries [1]. This experiment proved successful when the boy later showed immunity to smallpox, demonstrating the concept of using a weakened or related pathogen to induce protective immunity against a more dangerous one. Vaccines have played a crucial role in eradicating diseases such as smallpox, which was officially declared eradicated in 1980 due to global vaccination efforts. Other diseases like polio and measles continue to be targeted for global eradication through vaccination programs [2].

## 27.2 Evolution of Vaccine Methodologies

Since Edward Jenner's breakthrough, vaccine methodologies have advanced significantly, moving beyond the initial reliance on inactivated or attenuated live pathogens. The evolution of vaccine methodologies spans centuries, marked by significant advancements in scientific understanding, technology, and manufacturing capabilities.

### *27.2.1 Live Attenuated Vaccines*

The concept of vaccination began in the eighteenth century with Edward Jenner's use of cowpox virus to protect against smallpox. His approach involved exposing individuals to a less virulent or attenuated form of the pathogen to induce immunity without causing severe disease. Live attenuated vaccines use pathogens that are weakened through serial passages in cell culture or by genetic modifications. These vaccines mimic natural (not weakened) infection, to stimulate a robust immune response that provides long-lasting protection in the host. Examples include the measles, mumps, rubella (MMR) vaccine and the oral polio vaccine (OPV). Live attenuated vaccines typically induce strong cellular and humoral immune responses, often requiring fewer doses to confer immunity. They closely mimic natural infection, providing broad immunity against various strains of pathogens. Concerns exist regarding the safety of live attenuated vaccines in immunocompromised individuals or pregnant women due to the potential for the attenuated pathogen to revert to a virulent form [3, 4].

### 27.2.2 Inactivated Vaccines

Inactivated vaccines were developed as an alternative to live attenuated vaccines to enhance safety and reduce the risk of disease transmission. These vaccines use pathogens that have been killed or inactivated through chemical or physical methods. Inactivated (or weakened) vaccines contain whole pathogens or purified antigens that cannot replicate in the body. Examples include the inactivated polio vaccine (IPV) and the hepatitis A vaccine. These vaccines are safer for immunocompromised individuals and pregnant women because they cannot cause infection. They can be produced using established manufacturing processes and often provide predictable immune responses. Inactivated vaccines may require multiple doses or adjuvants to enhance short-term and long-term immunogenicity. They typically induce weaker immune responses compared to live attenuated vaccines but are considered safer.

### 27.2.3 Subunit Vaccines

Subunit vaccines focus on using purified components of pathogens, such as proteins, polysaccharides, or conjugates, to elicit an immune response. In subunit vaccines, the virus as a whole will not be considered. Subunit vaccines aim to target specific antigenic components of pathogens to increase the height of immune response while minimizing reactogenicity. Subunit vaccines contain purified antigens or antigenic fragments that are selected to induce protective immune responses. Examples include the hepatitis B vaccine (HBV, recombinant surface antigen) and the human papillomavirus (HPV) vaccine (virus-like particles). These vaccines are highly purified and can be designed to include only the essential antigens, reducing the risk of adverse reactions. They are safer for widespread use and can induce targeted immune responses against specific pathogen components. Subunit vaccines may require adjuvants or multiple doses to enhance immune responses. Purification and manufacturing challenges may affect their scalability and cost-effectiveness.

### 27.2.4 Vector-Based Vaccines

Vector-based vaccines utilize harmless viruses or bacteria (vectors) as carriers to deliver genetic material encoding antigen proteins from the target pathogen into host cells. These vaccines aim to enhance immune responses by leveraging the replicative abilities of the vector. Vector-based vaccines deliver genetic instructions to host cells, instructing them to produce antigen proteins that stimulate immune responses. Examples include viral vector vaccines against Ebola virus and

COVID-19 [5, 6]. Vector-based vaccines can mimic natural infections, inducing robust cellular and humoral immune responses. They offer potential for single-dose efficacy and rapid vaccine development against emerging pathogens. Vector-based vaccines may face pre-existing immunity to the vector, which could reduce vaccine efficacy. Vector selection and manufacturing complexity can affect scalability and regulatory approval timelines.

### *27.2.5 Nucleic Acid Vaccines*

Nucleic acid vaccines represent the latest advancement in vaccine technology, utilizing DNA or mRNA encoding antigen proteins to induce immune responses within host cells. These vaccines offer rapid development capabilities and potential for personalized medicine. Nucleic acid vaccines directly deliver genetic material into host cells, where it is translated into antigen proteins that trigger immune responses (i.e., mRNA vaccines against COVID-19 developed by Moderna and Pfizer-BioNTech). Nucleic acid vaccines can be rapidly designed and manufactured, providing flexibility in responding to emerging pathogens. They do not require the production of viral proteins, enhancing safety and scalability. Stability of nucleic acids, delivery methods, and optimization of immune responses are ongoing challenges for nucleic acid vaccines. Regulatory approval and public acceptance are also critical factors [7–9].

## 27.3 Distinctive Features of circRNAs for Vaccine Development

circRNAs exhibit unique characteristics that render them promising candidates for vaccine development, particularly in terms of stability and efficiency. Its covalently closed structure makes it resistant to exonuclease degradation. circRNAs are formed by covalent bonds between the 3′ and 5′ ends, creating a closed-loop structure. This circularization eliminates free ends that are vulnerable to exonuclease-mediated degradation, which is a common mechanism of RNA degradation by cellular exonucleases. As a result, circRNAs exhibit enhanced stability compared to linear counterparts, including mRNA. This stability allows circRNAs to persist longer within cells and tissues, allowing for sustained expression of encoded antigens or therapeutic proteins [10–13]. Some circRNAs contain internal ribosome entry sites (IRES) or other translation initiation elements that enable cap-independent translation. This mechanism bypasses the need for a 5′ cap structure, which may be advantageous in environments where cap-dependent translation is inhibited, such as in certain immune cells or under cellular stress conditions. Finally, circRNAs can be engineered to encode specific antigenic proteins or peptides, which are then processed and presented on major MHC molecules on the cell surface, a feature that we will discuss in detail in the section below. This presentation triggers immune recognition and activation of T cells, contributing to adaptive immune responses against the encoded antigens [14–17]. In mammalian cells, circRNAs exhibit an intermediate half-life that is at least 2.5 times longer than that of linear mRNAs [18].

Indeed, circRNA-based vaccines offer distinct advantages due to their ability to translate antigens in the cytoplasm without integrating into the genome, along with their potential to activate innate immune responses through pattern recognition receptors. circRNAs are capable of translating antigens in the cytoplasm without integrating into the host genome. This characteristic ensures that circRNA-based vaccines do not alter the host cell's genetic information, addressing safety concerns associated with genomic integration. Synthetic circRNAs, particularly those derived from foreign introns, can activate innate immune responses through pattern recognition receptors (PRRs). One such receptor is retinoic-acid-inducible gene-I (RIG-I), which detects double-stranded RNA (dsRNA) patterns typically associated with viral infections. Activation of RIG-I triggers downstream pro-inflammatory signaling cascades that lead to the production of cytokines, interferons, and other immune mediators. This immune activation is crucial for initiating and amplifying adaptive immune responses against the encoded antigens.

circRNA-based vaccines exhibit innate adjuvant-like properties by activating the body's natural immune responses. This unique characteristic reduces reliance on external adjuvants typically used in vaccine formulations. By leveraging their inherent ability to stimulate innate immune pathways, circRNA-based vaccines streamline the design process and enhance safety profiles. This self-adjuvant effect not only simplifies vaccine development but also contributes to a more robust and targeted immune response, potentially improving vaccine efficacy against various pathogens [19–22]. Given their inherent stability, immunogenicity, potential for high translation efficiency through engineering, and the necessity for nucleotide modifications to improve mRNA stability without increasing unwanted immunogenicity [23], circRNAs possess greater potential for development as vaccines compared to mRNA.

## 27.4 circRNA Vaccine: Molecular Biology

The design of circRNA vaccines incorporates a variety of elements to optimize their stability, translational efficiency, and overall effectiveness as therapeutic agents. Each component of the circRNA structure plays a crucial role in ensuring that the vaccine is not only delivered effectively but also capable of eliciting a robust immune response [24].

### 27.4.1 Components of circRNA Vaccines [25–27]

**Coding Regions for Vaccine Antigens** Coding regions within circRNA vaccines contain sequences that encode specific protein or peptide antigens. These antigens are targeted to induce an immune response against particular pathogens or cancer cell markers. The selection of antigens is crucial and typically based on identifying proteins or peptides that are unique to the target pathogen or cancer cells. These coding sequences are optimized to ensure efficient translation and production of the antigenic proteins within the host cells [28, 29].

**Untranslated Regions (UTRs)** Untranslated regions flank the coding regions of circRNA at both the 5′ and 3′ ends. They play essential roles in regulating the stability and translational efficiency of the RNA molecule [30, 31]. UTRs contain regulatory elements that influence how ribosomes attach to the RNA and initiate the translation process. By modulating the accessibility of the ribosome-binding sites and interacting with RNA-binding proteins, UTRs can significantly impact the overall expression levels of the encoded antigenic proteins [32].

**Promoters** Promoters are DNA sequences that initiate the transcription of RNA molecules. In the context of circRNA vaccines, promoters are crucial for driving the transcription of the circular RNA precursor [24]. Although promoters themselves are not part of the circRNA structure, their selection and efficiency are critical for ensuring robust and controlled production of the circRNA transcript. Proper promoter usage can dictate the quantity of circRNA generated, influencing the potency and effectiveness of the vaccine.

**Internal Ribosome Entry Sites (IRES)** IRES are RNA elements positioned within circRNA vaccines upstream of the coding regions. They enable initiation through cap-independent translation, bypassing the need for a 5′ cap structure typically required for ribosome recruitment. Since IRES elements directly recruit ribosomes to the mRNA, facilitating protein synthesis, this is helpful under conditions where cap-dependent translation may be inhibited, such as during viral infections or in certain cellular stress responses. This feature enhances the translational efficiency of circRNA vaccines, ensuring robust antigen production within host cells [33].

**Permuted Intron–Exon (PIE) Systems** PIE systems involve specific rearrangements of intronic and exonic sequences during RNA splicing, which lead to the circularization of the RNA molecule [34]. These systems optimize the splicing efficiency, ensuring the consistent formation of circular RNA structures. Efficient circularization is crucial for stabilizing the RNA molecule and enhancing its resistance to degradation within host cells. PIE systems contribute to the structural integrity and functionality of circRNA vaccines, supporting their efficacy as antigen delivery platforms [35].

**RNA Spacers** RNA spacers are noncoding sequences strategically included within circRNA vaccines [16, 25]. They serve to separate functional elements such as the IRES, coding sequences, and UTRs. Spacers contribute to maintaining the structural integrity of circRNA molecules, preventing unwanted interactions between different RNA elements. By facilitating proper folding and processing of the RNA, spacers ensure the functional separation of regulatory regions and coding sequences, thereby optimizing circRNA stability and antigen expression.

**Homology Arms** Homology arms are sequences designed to facilitate the circularization process of circRNA vaccines [26]. These arms contain complementary sequences that promote the end-joining events necessary for forming the covalently closed-loop characteristic of circRNAs. Proper design of homology arms is critical for efficient circularization during RNA splicing. They enhance the yield and consistency of circRNA production, ensuring that each circRNA molecule is correctly structured to carry and express the desired antigenic coding sequences effectively [24].

The design and composition of circRNA vaccines are critical in leveraging the unique properties of circRNAs for therapeutic use. By combining these sophisticated molecular engineering strategies with effective delivery systems like LNPs, circRNA vaccines can be tailored to generate potent and durable immune responses against various diseases. This comprehensive approach in vaccine design ensures that circRNAs are not only stable and efficiently translated but also successfully engage the host immune system in a targeted and controlled manner [34].

**Box 27.1 Circular mRNA (circmRNA)**

Circular mRNA (circmRNA) is a unique molecular structure where the 3′-end of mRNA is covalently linked to the 5′-end through a process known as reverse splicing [36]. This circularization distinguishes circmRNA from traditional linear mRNA forms. One of the notable advantages of circmRNA is its ability to facilitate efficient, stable, and long-lasting protein expression within cells. This property makes circmRNA an appealing alternative to linear mRNAs for various therapeutic applications. In recent years, circmRNA research has advanced significantly, particularly in the fields of vaccines and cellular therapies. In vaccine development, circmRNA offers advantages such as enhanced antigen expression and prolonged antigen presentation, which are crucial for eliciting robust immune responses. Moreover, circmRNA-based strategies are being explored in cellular therapies to improve the delivery and expression of therapeutic proteins within target cells [36]. As ongoing

(continued)

**Box 27.1** (continued)
research continues to unveil the potential of circmRNA, its distinct properties are poised to contribute significantly to the evolving landscape of therapeutic reagents, offering novel opportunities for enhancing vaccine efficacy and advancing cellular therapies.

*Cytokine-encoding circmRNA for cancer therapy*

In this groundbreaking study, researchers introduced an innovative circular mRNA variant termed circmRNA, synthesized using the PIE system. This system incorporates a novel spacer1 sequence and E29 IRES, strategically designed to optimize protein translation efficiency. Unlike conventional linear mRNA, circmRNA is encapsulated within LNPs, which enhances its stability and delivery to target cells. The unique structural resilience of circmRNA enables it to outperform linear counterparts in terms of protein production capability. Both in vitro experiments with various tumor cell lines and in vivo studies using mouse models demonstrated circmRNA's efficacy in robustly expressing a diverse array of proteins. This versatility positions circmRNA as a promising candidate for applications requiring high-level and sustained protein expression.

A pivotal finding of this research was the potent intratumoral effect of circmRNA formulations encoding four cytokines. Upon intratumoral injection, these circmRNA constructs triggered a significant reduction in tumor size. This therapeutic effect was attributed to the activation of immune cells, particularly T cells, which were mobilized to infiltrate the tumor microenvironment. Remarkably, when combined with anti-PD-1 antibodies—a standard immunotherapy—circmRNA not only enhanced immune cell infiltration but also synergistically amplified the antitumor immune response. These findings underscore circmRNA's potential as a robust RNA-based platform for developing advanced intratumoral therapies. By leveraging its ability to induce potent immune responses and synergize with existing immunotherapies, circmRNA holds promise for treating various cancers and advancing personalized medicine approaches [37].

## 27.5 Advantages of circRNA Vaccine

circRNA vaccines represent a significant advancement over traditional vaccine platforms due to several key advantages. One major benefit is their capability to produce larger quantities of proteins over extended periods compared to linear RNAs. This enhanced protein production stems from the unique circular structure of circRNAs, which increases their stability and resistance to degradation within cells. The circular nature of circRNAs confers greater resilience against cellular enzymes that typically degrade linear RNA molecules. As a result, circRNAs can persist longer in the cytoplasm, facilitating prolonged translation of encoded proteins. This prolonged expression leads to higher translation efficiency, ensuring sustained levels of the desired protein over time. In contrast, linear RNAs are more susceptible to degradation, limiting their ability to maintain protein production for extended durations. Moreover, the stable structure of circRNAs enhances their capacity to function as efficient templates for protein synthesis. This characteristic not only supports higher protein yields per molecule but also contributes to more consistent and predictable expression profiles compared to linear RNA constructs [34].

Another compelling advantage of circRNA-based vaccines is their ability to generate highly immunogenic antigens even with relatively modest quantities of synthesized circRNAs. Research has demonstrated that circRNAs can efficiently produce antigens that are particularly effective in being presented by DCs, which play a crucial role in initiating and regulating immune responses by capturing, processing, and presenting antigens to T cells. The immunogenicity of antigens derived from circRNAs is attributed to several factors. The circular structure of circRNAs contributes to their stability within cells, allowing for prolonged antigen expression. This sustained presence of antigens enhances their uptake and processing by DCs, leading to robust activation of T cell responses. Additionally, circRNAs may possess intrinsic properties that facilitate their recognition by immune surveillance mechanisms, further enhancing their ability to stimulate immune responses. The efficient presentation of circRNA-derived antigens by DCs suggests that circRNA vaccines hold significant promise for developing highly effective vaccines. Even small quantities of circRNA can potentially induce strong and durable immune responses, making them suitable candidates for vaccines targeting infectious diseases, cancers, or other conditions requiring potent immune activation [38].

Another significant advantage of circRNA-loaded lipid nanoparticles (circRNA-LNPs) is their enhanced thermostability relative to linear mRNA-LNPs. Research studies have shown that circRNA vaccines encapsulated in LNPs exhibit remarkable resilience to temperature fluctuations, which is critical for vaccine storage and distribution logistics. The superior thermostability of circRNA-LNPs can be attributed to several factors. The circular structure of circRNAs inherently provides greater resistance to enzymatic degradation compared to linear RNAs, thereby contributing to their prolonged stability within lipid nanoparticles. This structural advantage allows circRNA-LNPs to maintain their integrity and functional efficacy over extended periods under varying storage conditions. Studies have specifically demonstrated that circRNA-LNPs remain viable for storage at 4 °C for at least 4 weeks and can endure room temperature conditions for approximately 30 days without significant degradation. This extended shelf life is essential for simplifying vaccine distribution logistics and ensuring vaccine efficacy in regions where cold chain storage may be challenging [39].

Despite their promising advantages, circRNA vaccines also present certain limitations and potential disadvantages that warrant consideration. One notable concern revolves around the immunogenicity of circRNAs, which represents a double-edged sword in vaccine development. On the one hand, the immunogenic properties of circRNAs enable the creation of highly efficient vaccines. CircRNAs can produce antigens that effectively stimulate immune responses, making them potentially valuable tools for developing vaccines against infectious diseases, cancers, and other conditions. Their ability to engage innate immune pathways and enhance antigen presentation by DCs contributes to robust immune activation. However, the heightened immunogenicity of circRNAs raises uncertainties and challenges. The strong immune response triggered by circRNA vaccines may lead to adverse reactions or immune-mediated side effects in some individuals. This could include inflammatory responses, autoimmune reactions, or hypersensitivity reactions, although the extent and severity of these reactions remain to be fully elucidated through rigorous clinical studies. Furthermore, the potent immunogenicity of circRNAs might complicate the design and optimization of vaccine formulations. Achieving a balance between eliciting a strong immune response for protective immunity and avoiding excessive inflammation or adverse effects poses a challenge. Strategies to modulate or control the immunogenicity of circRNA vaccines may be necessary to optimize their safety and efficacy profiles [40]. Additionally, safety concerns surrounding circRNA vaccines need further investigation. The techniques required for generating circRNA vaccines are also not yet mature enough for large-scale production (Fig. 27.1). Therefore, extensive research is required before circRNA vaccines can be advanced to clinical trials.

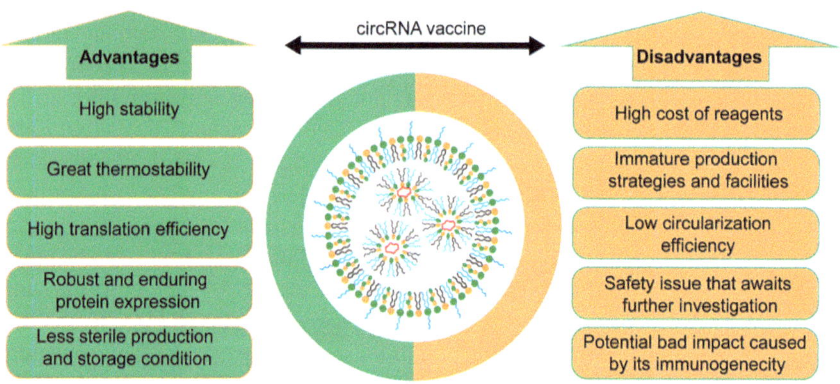


**Fig. 27.1** Depicting the advantages and disadvantages of circRNA vaccines. This figure presents a balanced overview of the advantages and disadvantages associated with circRNA vaccines. On the left, the advantages include high stability, great thermostability, high translation efficiency, robust and enduring protein expression, and the requirement for less sterile production and storage conditions. These benefits make circRNA vaccines a promising platform for vaccine development. On the right, the disadvantages are highlighted, which include the high cost of reagents, immature production strategies and facilities, low circularization efficiency, safety issues that require further investigation, and potential negative impacts caused by its immunogenicity. This comprehensive view underscores the potential of circRNA vaccines while also pointing out the challenges that need to be addressed for their successful implementation. (Figure is adapted from [26] under Creative Commons Attribution 4.0 International License)

## 27.6 Synthesis of circRNA Vaccine

The production of circRNAs typically involves several steps, starting with the synthesis of one or more precursor linear RNAs. These precursor RNAs are synthesized using standard molecular biology techniques, such as in vitro transcription (IVT) or polymerase chain reaction (PCR), to generate RNA sequences that contain the desired genetic information. After synthesis, the precursor linear RNAs undergo circularization to form covalently closed loops, which are characteristic of circRNAs. Circularization can be achieved through various methods, primarily involving chemical or enzymatic ligation techniques [41]. A major challenge in circRNA synthesis is their inability to utilize the standard cap-dependent mechanism to initiate translation. To address this, several strategies have been developed. Two primary mechanisms enable endogenous circRNA translation in a cap-independent manner: the methylation of the N6 position of adenine (m6A) near the start codon and the use of a viral IRES upstream of the ORF [39].

**m6A Methylation Near the Start Codon** One mechanism involves the presence of m6A modifications near the start codon of the circRNA. This modification is recognized by specific reader proteins that facilitate the recruitment of translation initiation factors. These factors then assemble at the m6A-modified site to initiate translation, bypassing the need for a 5′ cap structure. This process allows circRNAs to be translated into proteins efficiently despite lacking a traditional mRNA cap [42, 43]. This modification can enhance translation by modifying regions near the start codon.

**Viral IRES** Another mechanism involves the use of viral IRES elements located upstream of the ORF within the circRNA sequence. IRES elements are RNA sequences that directly recruit ribosomes to initiate translation internally, independent of the 5′ cap. This mechanism is often observed in viral circRNAs and certain endogenous circRNAs that have evolved to utilize IRES for efficient protein synthesis. The presence of an IRES allows ribosomes to directly bind to the circRNA and initiate translation at the ORF, facilitating protein production in a cap-independent manner [22].

Both mechanisms—m6A methylation near the start codon and the presence of viral IRES—enable endogenous circRNAs to undergo translation without relying on a 5′ cap structure. However, considering construct size, m6A modification is generally preferred over viral IRESs [44]. However, these two strategies can be combined to improve translation efficiency. For example, m6A can enhance the IRES-mediated translation of circZNF609 [45, 46]. Additionally, recent advancements have significantly improved circRNA translation. Chen et al. demonstrated that optimizing five elements—5′ and 3′ UTRs, vector topology, synthetic aptamers, and IRESs—not only increase protein yields from circRNAs by several 100-fold but also enable robust and sustained protein production in vivo [25].

### 27.6.1 *Precursor Linear RNA: The Synthesis Mechanism*

There are two primary methods for synthesizing precursor linear RNAs for circRNA production: chemical synthesis and enzymatic synthesis [47, 48]. Chemical synthesis techniques rely on synthesizer machines, utilizing phosphoramidites, which are derivatives of nucleotide triphosphates, as the fundamental building blocks of linear oligonucleotides [44]. Conversely, enzymatic strategies employ phage RNA polymerases for IVT. An IVT reaction typically involves a double-stranded DNA template, ribonucleotide triphosphates, and DNA-dependent RNA polymerase as its main components [44].

### 27.6.2 *RNA Circularization In Vitro*

The process of RNA circularization in vitro can be accomplished using three primary methodologies: chemical, enzymatic, and ribozymal circularization. Each approach leverages distinct mechanisms to create covalently closed loops in RNA molecules, which are pivotal for various biological and therapeutic applications [34].

#### 27.6.2.1 Chemical Strategies

Chemical circularization involves synthetic approaches where chemical reagents are used to join the ends of RNA molecules to form a covalent bond, thereby creating a closed-loop structure. This method is often preferred for its rapidity and the ability to process multiple RNA molecules simultaneously. Key reagents include bifunctional chemical linkers, which possess reactive groups capable of forming stable bonds with the phosphate groups at both ends of the RNA strand. The most common types of chemical bonds formed in this process are phosphodiester bonds, mirroring those found in natural RNA backbones. The process requires precise conditions—such as optimal pH, temperature, and linker concentration—to ensure efficient circularization without excessive degradation or side reactions. Chemical circularization is particularly useful in industrial settings where high throughput and scale are necessary. However, the method can sometimes introduce structural constraints or chemical modifications that might affect the RNA's biological activity or stability.

One of the widely used chemical approaches for RNA circularization involves the application of 1-ethyl-3-(3-dimethylaminopropyl) carbodiimide (EDC) or cyanogen bromide (CNBr). EDC is a zero-length crosslinking agent commonly used in bioconjugation and RNA chemistry. It facilitates RNA circularization by activating carboxyl groups (-COOH) on RNA molecules, promoting the formation of amide bonds between the RNA ends. EDC activates the carboxyl groups to form reactive intermediates that can undergo nucleophilic attack by adjacent amino groups on the RNA, such as amino-modified oligonucleotides or peptides. This process leads to the formation of stable amide bonds, thereby circularizing the RNA. CNBr is a chemical reagent used for cleaving peptide bonds at methionine residues in proteins. In RNA circularization, CNBr is used to activate specific sites on RNA molecules for subsequent ligation reactions. CNBr reacts with hydroxyl groups (-OH) on RNA, resulting in the formation of cyanate esters. These intermediates are highly reactive and can participate in nucleophilic substitution reactions with amino or hydroxyl groups present in nearby RNA molecules. Through this mechanism, CNBr facilitates the formation of covalent bonds between RNA segments, promoting circularization. Both EDC and CNBr are utilized to covalently link RNA ends, forming stable circular RNA molecules. Circular RNAs generated through these chemical methods exhibit enhanced stability compared to linear RNAs due to the absence of vulnerable linear ends susceptible to exonuclease degradation [49].

Additional chemical methods for RNA circularization focus on substituting native phosphate and hydroxyl groups of linear RNA with alternative functional groups. These methods aim to improve the efficiency, yield, and purity of circularized RNA molecules by enhancing the selectivity of bond formation. (1) Phosphate substitution strategies: In this approach, the non-bridging oxygen atoms in the phosphate backbone of RNA are substituted with sulfur atoms (phosphorothioate linkage). This modification alters the chemical reactivity of the phosphate groups, making them more susceptible to selective chemical reactions. Phosphorothioate-modified RNA segments can undergo sulfur-specific chemical reactions, such as reaction with disulfides or thiol groups, leading to covalent bond formation and subsequent circularization. This method enhances the efficiency and yield of RNA circularization by providing alternative reactive sites for specific bond formation, thereby reducing side reactions and improving product purity. (2) Hydroxyl substitution strategies: Modification of the 2′-hydroxyl group on ribose sugars with a methyl group (2′-O-methylation) alters the chemical reactivity of RNA molecules. This modification can be utilized to control the sites of RNA ligation reactions during circularization. 2′-O-methylated RNA segments exhibit altered hydrogen bonding properties and reduced reactivity compared to unmodified RNA, allowing for precise control over ligation sites and enhancing the fidelity of circularization reactions. This strategy improves the specificity and yield of RNA circularization by minimizing non-specific interactions and optimizing the conditions for RNA ligation reactions [50].

#### 27.6.2.2 Enzymatic Strategies

Enzymatic circularization utilizes natural or engineered enzymes to catalyze the covalent joining of RNA ends. RNA ligases, such as T4 RNA ligase, are commonly used for this purpose. These enzymes can specifically recognize the termini of the RNA and facilitate the formation of a covalent bond between the 3′ hydroxyl and 5′ phosphate ends. This method is highly specific and generally maintains the integrity and functionality of the RNA molecule, which is crucial for therapeutic applications where the biological activity of the RNA needs to be preserved. Enzymatic methods also allow for greater control over the circularization process, enabling the selective circularization of RNA molecules with specific sequences or structures. Despite these advantages, enzymatic circularization can be limited by the efficiency of the enzyme under various conditions and the necessity for precise experimental setups. Additionally, enzyme availability and cost can be limiting factors for some research or commercial applications. Enzymatic ligation of synthetic oligonucleotides is a fundamental technique in molecular biology, particularly for RNA circularization and other RNA engineering applications. Here is an elaboration on the three types of polynucleotide ligases commonly used for this purpose: T4 RNA ligase 1, T4 DNA ligase, and T4 RNA ligase 2.

#### T4 RNA Ligase 1

T4 RNA ligase 1 is an RNA ligase enzyme derived from bacteriophage T4. It catalyzes the formation of phosphodiester bonds between the 3′ hydroxyl group of one RNA molecule and the 5′ phosphate group of another RNA molecule [51]. T4 RNA ligase 1 is particularly useful for ligating RNA molecules with single-stranded nicks or overhangs, such as those generated during RNA circularization or RNA adapter ligation in RNA sequencing protocols (e.g., RNA-seq library preparation).

#### T4 DNA Ligase

Although primarily known for ligating DNA fragments, T4 DNA ligase can also ligate RNA molecules under certain conditions. It catalyzes the formation of phosphodiester bonds between the 5′ phosphate and 3′ hydroxyl ends of DNA or RNA molecules [52]. T4 DNA ligase is often employed in RNA–DNA hybrid molecule construction, such as in RNA–protein crosslinking studies or RNA modification mapping using chemically modified nucleotides.

#### T4 RNA Ligase 2

T4 RNA ligase 2, similar to T4 RNA ligase 1, catalyzes the ligation of RNA molecules but with different substrate specificity and reaction conditions. It can ligate RNA with 3′ hydroxyl and 5′ phosphate ends, as well as blunt-ended RNA molecules. T4 RNA ligase 2 is versatile and used in a variety of RNA manipulation techniques, including RNA circularization, RNA adapter ligation for small RNA sequencing, and RNA labeling experiments [53, 54].

#### 27.6.2.3 Ribozymatic Strategies

Ribozymatic strategies are frequently used when synthesizing larger sized circRNAs. Ribozymal circularization represents a sophisticated approach where ribozymes—RNA molecules with enzymatic activity—are engineered to catalyze their own end-to-end joining. This method utilizes the intrinsic properties of ribozymes to perform specific catalytic functions that include self-cleavage followed by self-ligation. The process can be finely tuned to achieve high specificity and efficiency, allowing for the circularization of RNA molecules under mild conditions that do not necessitate external enzymes or chemicals. One of the most compelling aspects of ribozymal circularization is the ability to design ribozymes that specifically recognize and circularize RNA sequences of interest. This method offers the potential for highly targeted applications, such as in gene therapy, where specific RNA sequences can be circularized to enhance their stability and function. However, the design and synthesis of functional ribozymes require a deep understanding of RNA structure and function and can be more technically challenging than chemical or enzymatic methods [50]. Group I and Group II introns are significant components in the generation of circRNAs through in vitro methods, although Group I introns are more frequently employed than Group II introns.

Group I Introns

Group I introns are RNA molecules capable of catalyzing their own removal from precursor RNA transcripts through a process called self-splicing [55]. They typically fold into complex secondary structures that include conserved domains necessary for splicing activity. Group I introns can facilitate the circularization of RNA molecules by catalyzing a reaction where the 5′ end of the intron ligates to the 3′ end, resulting in a covalently closed circular RNA structure. In vitro, Group I introns are harnessed to circularize exogenous RNA sequences, such as mRNA or synthetic RNA constructs, to produce stable circRNAs for various biotechnological applications, including RNA-based therapies and molecular biology research.

Group II Introns

Group II introns are also RNA molecules capable of self-splicing but involve a distinct splicing mechanism compared to Group I introns [56]. They typically fold into more complex secondary structures and exhibit greater sequence diversity across organisms. While Group II introns possess self-splicing capabilities that could potentially be adapted for in vitro circRNA production, they are less frequently utilized compared to Group I introns. Group II introns may pose challenges related to their structural complexity and the requirement for specific cofactors or conditions for efficient splicing and circularization [57–59].

Puttaraju et al. pioneered the use of the Group I intron system to synthesize circRNAs from a linear precursor containing introns. They utilized the Anabaena pre-tRNALeu gene to create a PIE construct. In this construct, the 5′ half of the Group I tRNA intron was relocated to the end of the tRNA exon, while the 3′ half of the intron was positioned at the start of the same exon, facilitating the formation of circRNAs through exon ligation [60]. Group I introns are favored in circRNA synthesis due to their well-characterized self-splicing mechanisms, high efficiency in circularizing RNA molecules, and adaptability to various RNA sequences and lengths. They can be engineered or modified to optimize their performance in generating circRNAs with specific structural and functional properties, enhancing their utility in biotechnological applications. Therefore, they are more commonly used. However, Group II introns can achieve in vitro RNA circularization more efficiently and does not require native exons [57].

## 27.7 Delivery of circRNA Vaccine

Effective vaccine delivery is pivotal for ensuring robust immunization outcomes. CircRNAs, which produce specific antigens within the host's cytoplasm, face challenges in freely diffusing across cell membranes due to their size. circRNAs are relatively large RNA molecules compared to small molecules like mRNA, which can freely diffuse through cell membranes. Their size can hinder efficient cellular uptake and distribution. To overcome diffusion barriers, circRNAs can utilize RNA delivery systems that facilitate their transport across cell membranes. This includes LNPs, liposomes, or other delivery vehicles designed to encapsulate and protect RNA molecules during transit [39, 61]. Although researchers have successfully delivered functional circRNAs in both naked and carrier-facilitated forms, the use of RNA delivery carriers, particularly nanoparticles, remains the most common method for vaccine delivery. Upon injection, circRNA-LNPs are internalized by cells through fusion with the endosomal membrane, facilitating the efficient delivery of circRNAs into the cytoplasm where the immune response can begin. Once circRNAs enter the cytosol, they are translated into proteins by ribosomes, leading to immune responses through several mechanisms [62].

### *27.7.1 Cellular Immune Response*

Intracellular antigens encoded by circRNA vaccines undergo a series of processes within host cells that lead to the activation of cellular immunity. Once translated within host cells, the intracellular antigens synthesized from circRNAs are subjected to degradation by the proteasome complex. The proteasome breaks down these proteins into smaller peptide fragments, typically around 8–10 amino acids in length. The peptides resulting from proteasomal degradation are then transported into the ER and loaded onto MHC I molecules. MHC I molecules are found on the surface of almost all nucleated cells and act as receptors that present these antigenic peptides to the immune system. $CD8^+$ cytotoxic T cells recognize and respond to antigens presented on MHC I molecules. Each $CD8^+$ T cell expresses a unique TCR that can specifically bind to a particular MHC I–peptide complex. Upon recognition of the antigenic peptide–MHC I complex, $CD8^+$ T cells become activated. This activation triggers a cascade of events, including the proliferation and differentiation of $CD8^+$ T cells into effector cytotoxic T cells. Activated cytotoxic T cells secrete cytokines such as TNF-α and IFN-γ. These cytokines play crucial roles in enhancing the immune response, stimulating other immune cells, and promoting inflammation at the site of infection. Cytotoxic T cells also release cytolytic molecules, including perforin and granzyme. Perforin forms pores in the membrane of target cells, facilitating the entry of granzyme and other proteases into the target cell. Granzyme activates pathways that lead to apoptosis of the infected or abnormal cell. The combined action of perforin and granzyme within target cells effectively disrupts cellular functions and induces apoptosis. This process is crucial for eliminating cells infected with intracellular pathogens, such as viruses, or cells displaying abnormal antigens, such as cancer cells [63, 64].

### 27.7.2 Humoral Immune Response

Secreted antigens produced from circRNAs are taken up by APCs such as DCs, macrophages, and B cells. Within these APCs, antigens are processed within endosomes, where they undergo proteolytic cleavage into smaller peptide fragments. Processed antigenic peptides are then loaded onto major MHC II molecules within the endosomal compartments of APCs. MHC II–peptide complexes are subsequently transported to the cell surface of APCs, where they are displayed to $CD4^+$ T helper cells. $CD4^+$ T helper cells interact with APCs presenting antigenic peptides on MHC II molecules. This interaction results in the activation of $CD4^+$ T cells, which play a central role in coordinating and enhancing the humoral immune response. Activated $CD4^+$ T cells secrete cytokines such as interleukins (e.g., IL-4, IL-5, IL-21) that promote various immune functions. Activated $CD4^+$ T cells stimulate B cell activation and differentiation, resulting in B cell differentiation into plasma cells, where some activated B cells differentiate into plasma cells, which are specialized cells that produce and secrete antibodies. Initially, IgM antibodies are produced rapidly in response to antigen exposure. Through interactions with CD4+ T cells and cytokine signaling, B cells undergo class switching, leading to the production of other antibody classes such as IgG. IgG antibodies are crucial for long-term immunity and provide enhanced specificity against the antigen. During B cell activation and differentiation, a subset of B cells develops into memory B cells. Memory B cells have a prolonged lifespan and retain the ability to recognize and respond rapidly to subsequent encounters with the same antigen. This establishment of immunological memory is essential for mounting a faster and more effective immune response upon re-exposure to the pathogen or antigen. APCs, including B cells and dendritic cells, can migrate to lymph nodes where they present antigen to CD4$^+$ T cells and further support B cell maturation. In lymph nodes, T cell-dependent interactions between APCs and B cells facilitate somatic hypermutation, affinity maturation, and the selection of B cells producing high-affinity antibodies. The coordinated activation of B cells, CD4$^+$ T cells, and APCs contributes to the production of high-affinity antibodies and the establishment of robust humoral immunity. This immune response is characterized by the production of specific antibodies that can neutralize pathogens, promote their clearance, and provide long-lasting protection against future infections [63, 64] (Fig. 27.2).

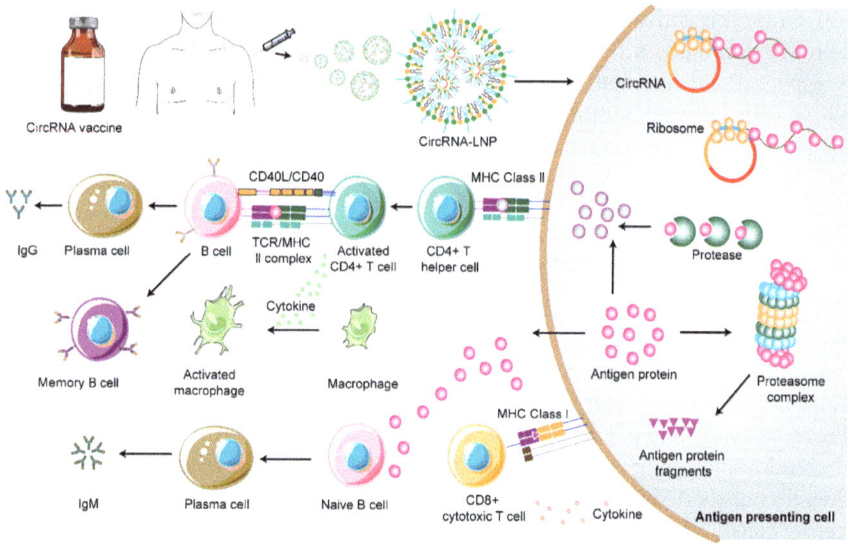


**Fig. 27.2** The circRNA vaccine is administered intramuscularly into the deltoid, where circRNA-LNPs are taken up by APCs through endosomal membrane fusion. Inside the cell, circRNAs are translated into antigen proteins by ribosomes. These intracellular antigens are broken down by the proteasome complex into smaller fragments and presented on the cell surface to CD8$^+$ T cells via MHC I molecules, leading to the activation of these T cells. Activated CD8$^+$ T cells then secrete cytokines and cytolytic molecules, initiating the cellular immune response. Secreted antigen proteins can activate B cells through their B cell receptors (BCR), transforming them into plasma cells that secrete IgM antibodies. These antigens can also be taken up by APCs, degraded, and presented on the cell surface by MHC II proteins, activating CD4$^+$ T cells. Activated CD4$^+$ T cells facilitate macrophage activation and somatic hypermutation in B cells, which can either become memory B cells or plasma cells secreting IgG antibodies, ensuring long-term immunity. (Figure is adapted from [26] under Creative Commons Attribution 4.0 International License)

## 27.8 Protein/Peptide-Coding Ability of circRNA

The discovery that certain circRNAs can encode proteins represents a significant breakthrough in molecular biology, significantly enhancing their potential for biomedical applications, particularly in vaccine development [31, 65]. Unlike traditional linear RNAs, which are prone to rapid degradation by cellular exonucleases, circRNAs possess a covalently closed loop structure that confers enhanced stability within cellular environments. This structural stability is crucial for vaccine applications, where the ability of RNA vaccines to persist in host cells can lead to more prolonged and effective immune responses. The closed-loop structure of circRNAs protects them from degradation, allowing for sustained expression of encoded proteins. This prolonged presence enhances the exposure of immune cells to antigenic proteins, thereby bolstering immune recognition and response. In the context of vaccine development, the stability of circRNAs facilitates their formulation into lipid nanoparticles or other delivery systems without significant loss of integrity or function. These delivery systems protect circRNAs from enzymatic degradation and facilitate their efficient uptake by target cells, further enhancing vaccine efficacy. Moreover, the ability of circRNAs to encode proteins opens new avenues for designing vaccines that can elicit specific immune responses against pathogens or cancer cells. By leveraging the stable expression of circRNA-encoded antigens, researchers can potentially develop vaccines that induce durable immunity and protection against infectious diseases or malignant tumors.

### *27.8.1 Protein-Encoding Capabilities of circRNAs*

The discovery of circRNAs initially challenged the conventional understanding that they are ncRNA molecules. Traditionally, circRNAs were thought to arise from splicing errors or as by-products of mRNA processing without protein-coding potential. However, subsequent research has revealed that certain circRNAs possess unique structural elements that enable them to function as templates for protein translation. Some circRNAs contain IRES or related elements. IRES elements allow ribosomes to initiate translation internally, bypassing the need for a 5′ cap structure that is typically required for conventional mRNA translation. This cap-independent translation mechanism enables circRNAs to be translated into proteins efficiently within host cells. CircRNAs that harbor IRES elements recruit specific translation initiation factors and ribosomal subunits directly to the RNA molecule [33]. This recruitment facilitates the assembly of the translation initiation complex, promoting the start of protein synthesis. By utilizing IRES-mediated translation initiation, circRNAs can overcome limitations associated with cap-dependent translation, particularly in environments where cap-dependent translation may be compromised, such as during viral infections or cellular stress responses. The protein-encoding capability of circRNAs opens avenues for their use in vaccine development [26, 29].

Researchers can design circRNAs to encode specific protein sequences that include epitopes—distinctive regions of antigens recognized by the immune system. By incorporating epitopes into circRNA-encoded proteins, researchers can precisely target immune responses against pathogens or cancer cells. circRNAs engineered to encode epitope-containing proteins can potentially enhance antigen expression and presentation within host cells [25]. This increased antigen exposure promotes robust immune recognition and response, leading to the activation of T cells and the production of specific antibodies by B cells. Such properties are particularly advantageous in vaccine design, where efficient antigen expression and immune activation are critical for eliciting protective immune responses.

### *27.8.2 Peptide-Encoding Capabilities of circRNAs*

As with protein, the translation of circRNAs into peptides typically occurs through IRES, which are RNA elements that allow ribosomes to bind and initiate translation directly at the start site of an ORF, bypassing the need for the 5′ cap structure that is essential in canonical translation processes. Additionally, some circRNAs may utilize a mechanism called m6A-driven translation, where m6A modification within the RNA sequence enhances the recruitment of initiation factors and ribosomes, facilitating translation. The peptides produced by circRNAs can vary significantly in size and function, depending on the length and sequence of the ORF. These peptides can serve various cellular roles, from regulatory functions, akin to those of small signaling molecules, to more structural or enzymatic roles [65]. The ability of circRNAs to encode peptides also suggests potential cross-talk between different cellular processes, where peptides derived from circRNAs could interact with or modulate the activity of other proteins. The discovery of peptide-coding circRNAs offers new avenues for research, particularly in understanding the complexity of gene regulation and protein networks within cells. Scientists can now investigate the specific functions of these peptides, their interaction with cellular machinery, and their impact on cellular physiology and pathology. In therapeutics, peptide-coding circRNAs present a novel resource for drug discovery and the development of new treatment strategies. In the context of vaccine development, circRNAs that encode immunogenic peptides could be engineered to enhance the immune response against infectious diseases or cancers. By presenting novel antigens to the immune system, these circRNA-based vaccines could potentially induce a stronger and more durable immune response compared to traditional vaccines.

Our recent study published in Nature delves into the intriguing concept of non-coding genomes encoding peptides that bind to HLA molecules as cryptic antigens, thereby triggering adaptive immune responses. We employed a comprehensive approach combining mass spectrometry analysis of the HLA class I peptidome with ribosome sequencing in human breast cancer samples. This innovative methodology allowed us to identify HLA-I-binding cryptic antigenic peptides originating from a tumor-specific circRNA known as circFAM53B. These cryptic peptides were found to efficiently prime naive $CD4^+$ and $CD8^+$ T cells in an antigen-specific manner, initiating robust antitumor immune responses. Importantly, our clinical investigations revealed that the expression levels of circFAM53B and its peptides correlated with increased infiltration of antigen-specific $CD8^+$ T cells and improved survival outcomes in patients with breast cancer and melanoma. Mechanistically, we demonstrated that circFAM53B-encoded peptides exhibit strong binding affinity to both HLA-I and HLA-II molecules, crucial for antigen presentation to cytotoxic T cells and helper T cells, respectively. In mouse models, vaccination using tumor-specific circRNA or its peptides effectively enhanced the infiltration of tumor-antigen-specific cytotoxic T cells, leading to potent tumor control. Overall, our findings underscore the potential of noncanonical translation of circRNAs as a pivotal driver of antitumor immunity. Targeting tumor-specific circRNAs through vaccination emerges as a promising immunotherapeutic strategy against various malignancies. This approach not only highlights the therapeutic potential of circRNA-encoded peptides in eliciting robust immune responses but also suggests novel avenues for developing personalized cancer vaccines tailored to individual tumor profiles [65].

## 27.9 Conclusion

Due to their notable advantages, such as enhanced stability and efficient protein translation, circRNAs have emerged as promising candidates for vaccine development. However, further research and advancements are required before circRNA vaccines can be widely adopted. Current studies suggest that circRNA vaccines do not induce clinical symptoms, yet comprehensive safety assessments are necessary. The immunogenicity of circRNAs and its implications for vaccine efficacy remain uncertain and could potentially hinder their progress. Moreover, current production capabilities are insufficient for large-scale manufacturing, limiting widespread application. Nevertheless, ongoing technological improvements are anticipated to broaden the clinical applications of circRNA vaccines, particularly in trials targeting infectious diseases and cancer suppression.

## References


1. Riedel S. Edward Jenner and the history of smallpox and vaccination. Proc (Baylor Univ Med Cent). 2005;18(1):21–5.
2. Meyer H, Ehmann R, Smith GL. Smallpox in the post-eradication era. Viruses. 2020;12:2.
3. Pollard AJ, Bijker EM. A guide to vaccinology: from basic principles to new developments. Nat Rev Immunol. 2021;21(2):83–100.
4. Minor PD. Live attenuated vaccines: historical successes and current challenges. Virology. 2015;479-480:379–92.

5. Heidary M, Kaviar VH, Shirani M, Ghanavati R, Motahar M, Sholeh M, et al. A comprehensive review of the protein subunit vaccines against COVID-19. Front Microbiol. 2022;13:927306.
6. Mellquist-Riemenschneider JL, Garrison AR, Geisbert JB, Saikh KU, Heidebrink KD, Jahrling PB, et al. Comparison of the protective efficacy of DNA and baculovirus-derived protein vaccines for EBOLA virus in Guinea pigs. Virus Res. 2003;92(2):187–93.
7. Francis MJ. Recent advances in vaccine technologies. Vet Clin North Am Small Anim Pract. 2018;48(2):231–41.
8. Plotkin SA, Plotkin SL. The development of vaccines: how the past led to the future. Nat Rev Microbiol. 2011;9(12):889–93.
9. Soni D, Van Haren SD, Idoko OT, Evans JT, Diray-Arce J, Dowling DJ, et al. Towards precision vaccines: lessons from the second international precision vaccines conference. Front Immunol. 2020;11:590373.
10. Pamudurti NR, Bartok O, Jens M, Ashwal-Fluss R, Stottmeister C, Ruhe L, et al. Translation of CircRNAs. Mol Cell. 2017;66(1):9–21.e7.
11. Kramer MC, Liang D, Tatomer DC, Gold B, March ZM, Cherry S, et al. Combinatorial control of drosophila circular RNA expression by intronic repeats, hnRNPs, and SR proteins. Genes Dev. 2015;29(20):2168–82.
12. Li X, Liu CX, Xue W, Zhang Y, Jiang S, Yin QF, et al. Coordinated circRNA biogenesis and function with NF90/NF110 in viral infection. Mol Cell. 2017;67(2):214–27.e7.
13. Wilusz JE. A 360° view of circular RNAs: from biogenesis to functions. Wiley Interdiscip Rev RNA. 2018;9(4):e1478.
14. Jeck WR, Sharpless NE. Detecting and characterizing circular RNAs. Nat Biotechnol. 2014;32(5):453–61.
15. Kristensen LS, Andersen MS, Stagsted LVW, Ebbesen KK, Hansen TB, Kjems J. The biogenesis, biology and characterization of circular RNAs. Nat Rev Genet. 2019;20(11):675–91.
16. Ju M, Kim D, Son G, Han J. Circular RNAs in and out of cells: therapeutic usages of circular RNAs. Mol Cells. 2023;46(1):33–40.
17. Fischer JW, Leung AK. CircRNAs: a regulator of cellular stress. Crit Rev Biochem Mol Biol. 2017;52(2):220–33.
18. Enuka Y, Lauriola M, Feldman ME, Sas-Chen A, Ulitsky I, Yarden Y. Circular RNAs are long-lived and display only minimal early alterations in response to a growth factor. Nucleic Acids Res. 2016;44(3):1370–83.
19. Chen YG, Kim MV, Chen X, Batista PJ, Aoyama S, Wilusz JE, et al. Sensing self and foreign circular RNAs by intron identity. Mol Cell. 2017;67(2):228–38.e5.
20. Perriman R, Ares M Jr. Circular mRNA can direct translation of extremely long repeating-sequence proteins in vivo. RNA. 1998;4(9):1047–54.
21. Abe N, Matsumoto K, Nishihara M, Nakano Y, Shibata A, Maruyama H, et al. Rolling circle translation of circular RNA in living human cells. Sci Rep. 2015;5:16435.
22. Chen CY, Sarnow P. Initiation of protein synthesis by the eukaryotic translational apparatus on circular RNAs. Science. 1995;268(5209):415–7.
23. Pardi N, Hogan MJ, Porter FW, Weissman D. mRNA vaccines—a new era in vaccinology. Nat Rev Drug Discov. 2018;17(4):261–79.
24. Choi S-W, Nam J-W. Optimal design of synthetic circular RNAs. Exp Mol Med. 2024;56(6):1281–92.
25. Chen R, Wang SK, Belk JA, Amaya L, Li Z, Cardenas A, et al. Engineering circular RNA for enhanced protein production. Nat Biotechnol. 2023;41(2):262–72.
26. Xie J, Ye F, Deng X, Tang Y, Liang JY, Huang X, et al. Circular RNA: a promising new star of vaccine. J Transl Int Med. 2023;11(4):372–81.
27. Niu D, Wu Y, Lian J. Circular RNA vaccine in disease prevention and treatment. Signal Transduct Target Ther. 2023;8(1):341.
28. Miao Q, Ni B, Tang J. Coding potential of circRNAs: new discoveries and challenges. PeerJ. 2021;9:e10718.

29. Zhao X, Zhong Y, Wang X, Shen J, An W. Advances in circular RNA and its applications. Int J Med Sci. 2022;19(6):975–85.
30. Hwang HJ, Kim YK. Molecular mechanisms of circular RNA translation. Exp Mol Med. 2024;56(6):1272–80.
31. Shi Y, Jia X, Xu J. The new function of circRNA: translation. Clin Transl Oncol. 2020;22(12):2162–9.
32. Leppek K, Das R, Barna M. Functional 5′ UTR mRNA structures in eukaryotic translation regulation and how to find them. Nat Rev Mol Cell Biol. 2018;19(3):158–74.
33. Chen C-K, Cheng R, Demeter J, Chen J, Weingarten-Gabbay S, Jiang L, et al. Structured elements drive extensive circular RNA translation. Mol Cell. 2021;81(20):4300–18.e13.
34. Wesselhoeft RA, Kowalski PS, Anderson DG. Engineering circular RNA for potent and stable translation in eukaryotic cells. Nat Commun. 2018;9(1):2629.
35. Umekage S, Kikuchi Y. In vivo circular RNA production using a constitutive promoter for high-level expression. J Biosci Bioeng. 2009;108(4):354–6.
36. Wang X, Dong J, Lu Y. Circular mRNA: a novel therapeutic agent. Biotechnol Notes. 2023;4:58.
37. Yang J, Zhu J, Sun J, Chen Y, Du Y, Tan Y, et al. Intratumoral delivered novel circular mRNA encoding cytokines for immune modulation and cancer therapy. Mol Ther Nucleic Acids. 2022;30:184–97.
38. Szabó GT, Mahiny AJ, Vlatkovic I. COVID-19 mRNA vaccines: platforms and current developments. Mol Ther. 2022;30(5):1850–68.
39. Liu X, Zhang Y, Zhou S, Dain L, Mei L, Zhu G. Circular RNA: An emerging frontier in RNA therapeutic targets, RNA therapeutics, and mRNA vaccines. J Control Release. 2022;348:84–94.
40. Ti D, Yan X, Wei J, Wu Z, Wang Y, Han W. Inducing immunogenic cell death in immuno-oncological therapies. Chin J Cancer Res. 2022;34(1):1–10.
41. Petkovic S, Müller S. Synthesis and engineering of circular RNAs. Methods Mol Biol. 2018;1724:167–80.
42. Yang Y, Muisha MB, Zhang J, Sun Y, Li Z. Research progress on N(6)-adenosylate methylation RNA modification in heart failure remodeling. J Transl Int Med. 2022;10(4):340–8.
43. Meyer KD, Saletore Y, Zumbo P, Elemento O, Mason CE, Jaffrey SR. Comprehensive analysis of mRNA methylation reveals enrichment in 3′ UTRs and near stop codons. Cell. 2012;149(7):1635–46.
44. Obi P, Chen YG. The design and synthesis of circular RNAs. Methods. 2021;196:85–103.
45. Legnini I, Di Timoteo G, Rossi F, Morlando M, Briganti F, Sthandier O, et al. Circ-ZNF609 is a circular RNA that can be translated and functions in Myogenesis. Mol Cell. 2017;66(1):22–37.e9.
46. Di Timoteo G, Dattilo D, Centrón-Broco A, Colantoni A, Guarnacci M, Rossi F, et al. Modulation of circRNA metabolism by m(6)a modification. Cell Rep. 2020;31(6):107641.
47. Usman N, Cedergren R. Exploiting the chemical synthesis of RNA. Trends Biochem Sci. 1992;17(9):334–9.
48. Rong M, He B, McAllister WT, Durbin RK. Promoter specificity determinants of T7 RNA polymerase. Proc Natl Acad Sci U S A. 1998;95(2):515–9.
49. Dolinnaya NG, Sokolova NI, Ashirbekova DT, Shabarova ZA. The use of BrCN for assembling modified DNA duplexes and DNA-RNA hybrids; comparison with water-soluble carbodiimide. Nucleic Acids Res. 1991;19(11):3067–72.
50. Müller S, Appel B. In vitro circularization of RNA. RNA Biol. 2017;14(8):1018–27.
51. Wang LK, Schwer B, Shuman S. Structure-guided mutational analysis of T4 RNA ligase 1. RNA. 2006;12(12):2126–34.
52. Moore MJ, Joining RNA. Molecules with T4 DNA ligase. In: Haynes SR, editor. RNA-protein interaction protocols. Totowa, NJ: Humana Press; 1999. p. 11–9.
53. Miller ES, Kutter E, Mosig G, Arisaka F, Kunisawa T, Rüger W. Bacteriophage T4 genome. Microbiol Mol Biol Rev. 2003;67(1):86–156, table of contents.

54. Remaut E, Tsao H, Fiers W. Improved plasmid vectors with a thermoinducible expression and temperature-regulated runaway replication. Gene. 1983;22(1):103–13.
55. Gomes R, Silva K, Theodoro RC. Group I introns: structure, splicing and their applications in medical mycology. Genet Mol Biol. 2024;47Suppl 1(Suppl 1):e20230228.
56. Xu L, Liu T, Chung K, Pyle AM. Structural insights into intron catalysis and dynamics during splicing. Nature. 2023;624(7992):682–8.
57. Mikheeva S, Hakim-Zargar M, Carlson D, Jarrell K. Use of an engineered ribozyme to produce a circular human exon. Nucleic Acids Res. 1997;25(24):5085–94.
58. Jarrell KA. Inverse splicing of a group II intron. Proc Natl Acad Sci U S A. 1993;90(18):8624–7.
59. Puttaraju M, Been MD. Circular ribozymes generated in Escherichia coli using group I self-splicing permuted intron-exon sequences. J Biol Chem. 1996;271(42):26081–7.
60. Puttaraju M, Been MD. Group I permuted intron-exon (PIE) sequences self-splice to produce circular exons. Nucleic Acids Res. 1992;20(20):5357–64.
61. Jeck WR, Sorrentino JA, Wang K, Slevin MK, Burd CE, Liu J, et al. Circular RNAs are abundant, conserved, and associated with ALU repeats. RNA. 2013;19(2):141–57.
62. Bartlett BL, Pellicane AJ, Tyring SK. Vaccine immunology. Dermatol Ther. 2009;22(2):104–9.
63. Chaudhary N, Weissman D, Whitehead KA. mRNA vaccines for infectious diseases: principles, delivery and clinical translation. Nat Rev Drug Discov. 2021;20(11):817–38.
64. Hill A, Beitelshees M, Pfeifer BA. Vaccine delivery and immune response basics. Methods Mol Biol. 2021;2183:1–8.
65. Huang D, Zhu X, Ye S, Zhang J, Liao J, Zhang N, et al. Tumour circular RNAs elicit antitumour immunity by encoding cryptic peptides. Nature. 2024;625(7995):593–602.

# Chapter 28
# Cancer Biomarker Based on ncRNA Liquid Biopsy

**Abstract** Historically dismissed as non-functional "junk" RNAs, non-coding RNAs have recently been acknowledged for their significant role in the pathogenesis of various cancers, overturning previous misconceptions. The past decade has not only highlighted their biological importance but also revealed frequent abnormalities in their expression levels across numerous cancers, sparking interest in their potential as cancer biomarkers. Among these, miRNAs can influence gene transcription epigenetically and are extensively researched as candidates for liquid biopsy biomarkers in cancer diagnosis. Advances in high-throughput sequencing technologies have allowed for detailed molecular profiling of non-coding RNA expression across different cancers. Moreover, innovations in technology that enable the detection of low-abundance RNAs in blood samples have led to the discovery of new, previously undetectable biomarkers for cancer. This paper provides an update on the recent advancements in using non-coding RNAs as non-invasive biomarkers for cancer, assessing various non-coding RNA types for their efficacy as blood-based markers and exploring how recent technological developments have enhanced the field of non-invasive molecular biomarkers in oncology.

## 28.1 Introduction

The concept of a "liquid biopsy" encompasses the collection and examination of non-solid biological materials such as blood, saliva, urine, and other body fluids, which serve multiple clinical purposes, including diagnosis of diseases and monitoring after treatment [1]. Distinct from traditional tissue biopsies, which are invasive, liquid biopsies are typically non-invasive (e.g., using saliva, urine, stool) or minimally invasive (e.g., blood), offering a way to evaluate the health or disease state of less accessible organs. In the past 20 years, circulating nucleic acids, particularly cell-free DNA (cfDNA), have been adopted in clinical settings for disease screening across various conditions, notably cancers [2–6]. While circulating tumor-derived DNA (ctDNA) fragments shed into the bloodstream by cancer cells are testable, significant challenges arise from the limited established mutations and



P. E. Saw, E. Song, *RNA Therapeutics in Human Diseases*,
https://doi.org/10.1007/978-981-96-3041-7_28

the variability of mutational loads detectable in early versus advanced cancer stages within cfDNA [7]. This variability underlines a key limitation in using ctDNA for the early, non-invasive detection of cancer. In addition to examining mutations in ctDNA, the field of cancer diagnostics is increasingly investigating other molecular markers such as transcriptional changes in cfRNA and epigenetic modifications, notably DNA methylation. These molecular alterations often occur early in the process of carcinogenesis and possess the unique characteristic of being dynamic, which significantly enhances their value as potential diagnostic indicators. Among these, ncRNAs, particularly miRNAs, have emerged as promising targets in the quest to develop non-invasive cancer biomarkers due to their stability in circulation and differential expression across various cancer types [18].

This chapter comprehensively collates the most recent studies focusing on a variety of ncRNAs, emphasizing their potential roles and effectiveness in the context of non-invasive cancer diagnostics. It also critically compares the diagnostic utility of ncRNAs with more traditional biomarkers such as cfDNA, within the framework of liquid biopsy platforms. A significant portion of the discussion is dedicated to the latest breakthroughs in ncRNA quantification methods, including advanced sequencing technologies and digital droplet PCR, which have dramatically improved the sensitivity and specificity of ncRNA detection. Furthermore, we explore the implications of these technological advancements for the future landscape of cancer diagnostics. The continuous refinement of ncRNA quantification techniques is likely to expand the clinical applicability of ncRNA-based biomarkers, facilitating earlier detection of malignancies, better monitoring of disease progression, and more tailored therapeutic interventions. Thus, this evolving area holds considerable promise for significantly advancing the field of non-invasive cancer diagnostics, ultimately contributing to more precise and personalized oncology care.

## 28.2 ncRNAs as Biomarkers

Once relegated to the category of "junk," ncRNAs have, over the past two decades, undergone a remarkable reevaluation in the scientific community. Emerging research has illuminated their pivotal roles in regulating various cellular processes and their implications in the pathogenesis of diseases, most notably cancer. This paradigm shift has underscored the significant potential of ncRNAs in clinical applications, particularly in the field of non-invasive diagnostics. ncRNAs, including miRNAs, lncRNAs, and circRNAs, have shown great promise as biomarkers for cancer detection and monitoring [1, 5, 6, 8–13]. Their high stability in body fluids, coupled with their disease-specific expression patterns, makes them ideal candidates for liquid biopsy applications. Liquid biopsies offer a less invasive alternative to traditional tissue biopsies, allowing for the frequent monitoring of disease progression and response to treatment through simple blood samples. The specific profiles of ncRNAs can provide critical insights into tumor type, stage, and potential therapeutic targets, enhancing personalized medicine approaches and improving patient outcomes. As research continues to reveal the intricate roles of ncRNAs in cellular biology and disease, their integration into clinical practice holds the potential to revolutionize cancer diagnostics and therapeutic strategies. Currently, the majority of research on ncRNA-based liquid biopsy biomarkers focuses on their application in diagnostics and cancer screening [5, 9]. Due to the common overexpression of many ncRNAs in cancers, these molecules are also considered promising for tracking cancer progression and detecting recurrences. Indeed, numerous studies have shown that circulating ncRNAs, including miRNAs and lncRNAs, can effectively monitor cancer [13, 14]. Additionally, the functional roles of various ncRNAs in cancer have been elucidated through tissue-based profiling, which has helped clarify their potential in oncology [15–17]. This growing body of evidence supports the exploration of these ncRNAs in blood samples for the discovery of new cancer biomarkers. This development suggests the potential of circulating ncRNAs to serve as companion diagnostics for these ncRNA-based therapies. While further research is needed to confirm the effectiveness of ncRNAs in these roles, the outlook is increasingly positive regarding their use in developing non-invasive cancer biomarkers. This optimism is bolstered by ongoing advancements in our understanding of ncRNA molecular profiles within the bloodstream, which could significantly impact future diagnostic and therapeutic approaches in oncology.

### 28.2.1 MicroRNAs (miRNAs)

MicroRNAs (miRNAs) are endogenous, single-stranded molecules containing 18–25 nucleotides that primarily attach to the 3′-UTR of target genes, potentially regulating various cellular processes by influencing RNA translation. miRNAs stand as some of the most extensively investigated ncRNAs in the field of oncology. Their prominence as candidates for liquid biopsy biomarkers is largely attributed to their substantial presence and exceptional stability in various bodily fluids, including blood, urine, stool, and saliva. These small, approximately 22 nucleotides long RNA molecules play crucial roles in gene regulation by binding to the messenger RNAs of protein-coding genes, thereby inhibiting their translation or leading to their degradation. The ability of miRNAs to circulate freely in body fluids, often encapsulated in extracellular vesicles such as exosomes or bound to protein complexes, protects them from RNAse-mediated degradation, thus enhancing their stability. This feature is particularly beneficial for diagnostic purposes as it allows for the reliable detection and quantification of miRNAs, even in small sample volumes [9, 19, 20].

Among the various miRNAs, miR-21 and miR-155 are particularly notable. Classified as "onco-miRs," these miRNAs function as oncogenes and are commonly upregulated across various cancers, making them well-recognized markers in circulating biofluids [21, 22]. They are consistently elevated in cancer patients and have been implicated in numerous cancer types [23–27]. For instance, microRNAs like miR-21 and miR-155 are elevated in various inflammatory diseases as well as in cancer. This dual presence complicates their utility as specific cancer biomarkers because elevated levels of these microRNAs might not necessarily indicate a malignancy. MiR-21 and miR-155 are known to be involved in key regulatory pathways that control cell proliferation, apoptosis, and immune response, which are pivotal not only in tumor biology but also in inflammatory responses. Their upregulation in inflammatory conditions suggests that they may be part of a general stress or injury response, rather than being specific to oncogenic processes [28]. The challenge lies in the specificity and selectivity of miRNAs; although abundant and stable, many miRNAs expressed in non-cancerous tissues are transcriptionally repressed in cancerous ones, which limits their effectiveness as diagnostic markers. This review will further assess the role of miRNAs in the context of liquid biopsies, considering their widespread tissue expression against the need for cancer-specific biomarkers. Despite these challenges, the unique attributes of miRNAs—particularly their abundance and stability in fluids—continue to make them attractive targets for developing non-invasive cancer diagnostics (Fig. 28.1).

### *28.2.2 Long Non-Coding RNAs (lncRNAs)*

Long non-coding RNAs (lncRNAs) are a category of ncRNAs exceeding 200 nucleotides in length, playing crucial roles in various biological functions [29]. Many lncRNAs are involved in the pathogenesis of different cancers, mainly by modulating significant cancer-related transcriptional regulators [30]. Notable among these are HOTAIR, MALAT1, and PVT1, which have been consistently found to be deregulated across various cancers, with their oncogenic roles confirmed in further research [31–33]. Although over 60,000 lncRNAs have been identified from human tumors and cancer cell lines, the vast majority remain uncharacterized and their biological implications unclear [34].

Recent advancements suggest that lncRNAs can be detected in the bloodstream as cell-free RNAs, positioning them as viable candidates for non-invasive cancer biomarkers [35]. Established cancer-associated lncRNAs such as urothelial carcinoma-associated 1 (UCA1), metastasis-associated lung adenocarcinoma transcript 1 (MALAT1), and H19 are garnering attention for their diagnostic potential due to their elevated levels in the serum and plasma of cancer patients. These lncRNAs, each with distinct roles in cellular processes, are frequently upregulated in various malignancies, correlating with tumor progression, metastasis, and poor prognosis. UCA1 has been particularly noted for its role in bladder cancer, where it influences cell proliferation and chemotherapy resistance. Similarly, MALAT1 is associated with the metastatic spread of lung cancer and several other cancer types, serving as a prognostic marker for poor survival. H19's involvement is noted in multiple cancer pathways, including epigenetic regulation and growth control [36–39]. However, additional research is needed to verify their cancer-specific origins and to establish their efficacy as reliable cancer biomarkers.

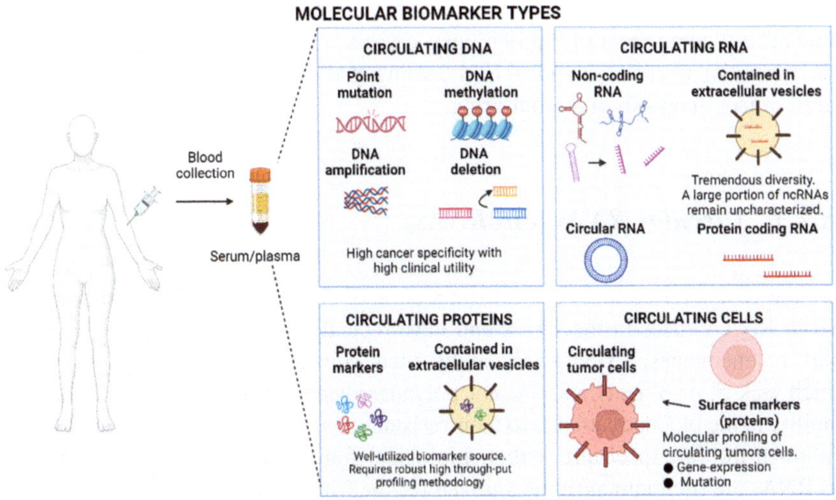


**Fig. 28.1** Overview of molecular biomarker types in blood samples for cancer diagnosis and monitoring. This infographic presents an integrated view of the types of molecular biomarkers that can be detected in blood samples, illustrating their potential applications in cancer diagnostics and monitoring. (i) Circulating DNA: Highlighted are various forms of DNA alterations detectable in blood, including point mutations, DNA methylation changes, and DNA deletions. These biomarkers offer high specificity and clinical utility in identifying and monitoring cancer. (ii) Circulating RNA: Focuses on different types of RNA found in the bloodstream, such as non-coding RNAs, circular RNAs, and protein-coding RNAs, often contained within extracellular vesicles. This panel emphasizes the diversity of RNA molecules and their potential roles in disease processes. (iii) Circulating proteins: Details on protein markers also contained in extracellular vesicles, which are widely used in clinical settings. The utility of these biomarkers is underscored by the need for robust, high-throughput profiling methods to effectively utilize them in clinical practice. (iv) Circulating cells: Describes circulating tumor cells characterized by specific surface proteins and other molecular features like gene expression profiles and mutations. These cells provide direct insights into the molecular landscape of the cancer, aiding in personalized treatment strategies. (Figure is drawn in BioRender.com)

Intriguingly, recent research has illuminated the presence of lncRNAs within extracellular vesicles (EVs) found in human serum and plasma. This discovery suggests that certain lncRNAs might not merely be byproducts of cellular breakdown but could be actively secreted into circulation encapsulated within tumor-derived EVs. Such a mechanism indicates a deliberate role for lncRNAs in intercellular communication and tumor microenvironment manipulation, which is pivotal for cancer progression and metastasis. The encapsulation of lncRNAs in EVs offers several diagnostic advantages. First, the protective environment within EVs shields lncRNAs from enzymatic degradation, enhancing their stability in the bloodstream. This stability is crucial for the reliability of liquid biopsies, as it ensures that the detected lncRNAs accurately reflect the current state of the tumor. Second, since EVs can be selectively taken up by specific cell types, lncRNAs within these vesicles might also provide targeted information about the tumor's biological state and microenvironment [40–42]. As evidence accumulates, the prospect of lncRNAs as a foundational component of liquid biopsy sources becomes increasingly promising. It is imperative to closely monitor forthcoming studies to fully ascertain their potential in cancer diagnosis and monitoring.

### 28.2.3 Circular RNAs (circRNAs)

Circular RNAs (circRNAs) are a distinct class of endogenous ncRNAs first identified in RNA viruses in the 1970s and have recently been acknowledged for their roles in oncogenesis. Characterized by their covalently closed loop structures, which lack 5′ or 3′ ends and are not polyadenylated, circRNAs exhibit enhanced stability in the bloodstream due to their resistance to exonuclease digestion [43, 44]. Although the full spectrum of their biological functions is yet to be determined, circRNAs are predominantly recognized for their role as miRNA sponges, thereby influencing gene regulation [45, 46]. circRNAs originate from a variety of genetic sources, including both protein-coding and non-coding genes, making them a diverse group of molecules with significant potential for cancer diagnostics. Interestingly, some circRNAs are associated with overexpressed oncogenes in various cancers, acting either as regulators of gene expression or as molecular sponges that bind and sequester microRNAs, thereby modulating the activity of oncogenic pathways. For example, circRNAs can enhance or inhibit the expression of oncogenes through their interaction with microRNAs. By sequestering microRNAs that would otherwise inhibit oncogene expression, circRNAs can indirectly promote oncogenic signaling pathways that contribute to tumor growth and progression. This regulatory capability has been observed in several types of cancers, where specific circRNAs are upregulated in concert with key oncogenes, contributing to the malignant phenotype [47, 48].

Recent studies have identified specific circRNAs that are differentially expressed in the context of cancer, including both upregulation and downregulation depending on the tumor type and its microenvironment. For instance, ciRS-7, known for its role as a sponge for miR-7, is upregulated in colorectal and gastric cancers, where it promotes cancer progression by influencing oncogenic pathways. Similarly, circHIPK3 has been found in elevated levels in the blood of patients with liver and lung cancers, correlating with tumor load and stage [49–52]. Notably, circRNAs have also been detected in EVs, such as exosomes, which are small lipid bilayer vesicles released by cells into the surrounding environment. RNA sequencing of these exosomal contents has revealed an intriguing aspect: most exosomal circRNAs originate from protein-coding exons. This finding is significant as it suggests that these circRNAs might play roles in cell-to-cell communication, potentially influencing cancer progression and metastasis by transferring genetic information between cells. The presence of circRNAs in exosomes offers a pathway for their potential function beyond the cell of origin. Since exosomes can travel through bodily fluids and interact with recipient cells, the circRNAs they carry can exert effects at distant sites, facilitating the creation of a microenvironment conducive to cancer growth and spread [8].

One significant challenge in harnessing circRNAs for liquid biopsies in cancer diagnostics lies in their detection and profiling. This difficulty stems from the limitations of conventional RNA-sequencing techniques, which are typically designed to target polyadenylated (polyA) transcripts. circRNAs, being non-polyadenylated, do not possess the polyA-tail that is commonly used to enrich RNA samples during sequencing preparation, making them less amenable to traditional RNA-seq methods. Standard short-read paired-end RNA sequencing does offer a way to identify the back-spliced junctions unique to circRNAs. These junctions are crucial markers because they confirm the circular nature of these RNA molecules. However, accurately profiling non-polyA transcripts like circRNAs requires specialized techniques that can bypass the limitations of polyA-targeted enrichment. This necessity for specialized methods complicates their use in routine diagnostics [45, 53]. To address the challenges associated with the detection and accurate measurement of circRNAs, several techniques have been refined. One prominent method involves the use of RNase R, an exoribonuclease that selectively degrades linear RNAs while sparing circular forms. This enzymatic treatment significantly reduces the background of linear mRNAs, thereby enriching the sample in circRNAs and enhancing the specificity and sensitivity of subsequent analyses [53]. The Ribo-Zero technique is another critical method in the toolkit for enhancing circRNA research, particularly by enriching circRNA samples while efficiently removing rRNAs, which often constitute a significant portion of RNA extracts and can interfere with RNA-sequencing workflows [54]. The adoption of exome capture probes in circRNA research represents a strategic enhancement in the specificity and efficiency of circRNA detection, especially in sequencing applications. Exome capture probes are traditionally used in genomics to isolate exons of the genome for targeted sequencing. Adapting this technology for circRNA detection leverages its high specificity to enrich for circRNA sequences, particularly those derived from exonic regions [55]. These advancements in RNA-sequencing technology have considerably boosted the detection and quantification of circRNAs in cancer, making them promising candidates for non-invasive cancer biomarkers.

### *28.2.4 Piwi-Interacting RNAs (piRNAs)*

Piwi-interacting RNAs (piRNAs) are a fascinating class of small non-coding RNAs that, while similar in size to miRNAs, ranging from 26 to 31 nucleotides, differ significantly in their biogenesis and function. Unlike miRNAs, piRNAs do not exhibit sequence conservation across different species, suggesting a highly specialized and context-dependent role in genome regulation. Their unique characteristics and modes of action set them apart within the small RNA landscape, particularly in their association with the Piwi subfamily of the Ago proteins [56]. piRNAs play crucial roles primarily in germ cells, where they are most abundantly expressed.

Their main function is the silencing of transposable elements, protecting the genome integrity from potentially harmful transposon activity that can lead to mutations and genomic instability. By forming complexes with Piwi proteins, piRNAs guide these complexes to complementary transposon transcripts, leading to the degradation of the transposon RNA or the repression of its transcription. Beyond transposon silencing, piRNAs are involved in the epigenetic regulation of genes and can influence post-transcriptional gene silencing, akin to the mechanisms observed with miRNAs.

In the context of oncology, numerous studies have identified piRNAs as significantly dysregulated in cancerous tissues, suggesting their potential roles as either tumor suppressors or oncogenes [57–61]. This dysregulation points to the utility of piRNAs in cancer diagnostics, particularly as biomarkers for liquid biopsies. Emerging research has consistently shown that changes in the levels of circulating piRNAs are associated with the presence of cancer. These changes may reflect alterations in gene expression and genomic stability typically observed in cancer cells, making piRNAs potential biomarkers for cancer detection and monitoring. Variations in the expression levels of specific piRNAs have been linked to various types of cancer, including breast, prostate, and pancreatic cancers. These piRNAs can potentially serve as early indicators of malignancy, even before clinical symptoms manifest, thereby facilitating early intervention [62–64]. piRNAs are promising for the development of liquid biomarkers not only for cancer but also for other diseases where genetic and epigenetic regulation play a critical role. Moving forward, it will be essential to delineate how cancers release piRNAs into the circulation and how this process compares with the release mechanisms for miRNAs. Understanding these mechanisms will enhance our ability to harness piRNAs effectively in the clinical setting, potentially leading to advanced diagnostic and therapeutic strategies in oncology.

### *28.2.5 Transfer RNAs (tRNAs)*

Transfer RNAs (tRNAs) are fundamental components of the cellular machinery, linking the genetic code in mRNA with the amino acid sequence of proteins. While their primary role is well-understood—transferring amino acids to ribosomes for protein synthesis—tRNAs also have intriguing derivatives that play roles beyond protein translation, particularly in the regulation of cellular processes and response to stress. In the bloodstream and other cellular contexts, tRNAs are often found in smaller, fragmented forms known as "transfer RNA-derived small RNAs" (tsRNAs). These fragments can be broadly categorized into two types: tRNA fragments (tRFs) and tRNA-derived stress-induced RNAs (tiRNAs). Both tRFs and tiRNAs are being studied for their potential as biomarkers for various diseases, including cancer, where changes in their levels might reflect tumor dynamics or the response to treatments. Since tsRNAs can be detected in blood and other bodily fluids, they offer a promising avenue for developing non-invasive diagnostic tests, which could be particularly valuable for monitoring disease progression or response to therapy [65].

There is speculation that tRNAs have higher turnover rates in cancerous tissues compared to normal tissues, particularly under stress conditions, leading to increased levels in cancers [66, 67]. This has been supported by studies showing elevated levels of various tRNAs not only in the serum and plasma but also in the urine of cancer patients [68, 69]. For instance, one study analyzing serum RNA-sequencing profiles found that the expression levels of serum tRFs are closely linked to clinicopathologic features in breast cancer patients [70]. Furthermore, recent research has identified numerous exosomal tRFs that are significantly enriched in the circulation of liver cancer patients relative to healthy individuals [71]. Although the precise role of tRNAs in cancer remains to be fully elucidated, the growing body of evidence suggests that they could be valuable as novel non-invasive biomarkers for cancer diagnosis. These findings highlight the potential of tsRNAs, particularly tRFs and tiRNAs, in oncological diagnostics, underscoring their relevance in reflecting pathological states such as cancer.

## 28.3 ncRNAs as Liquid Biopsies

A liquid biopsy provides a non-invasive alternative to tissue biopsies, utilizing bodily fluids to obtain diagnostic information. While blood is the most common fluid for such biopsies, the potential of cell-free ncRNAs in other fluids like saliva, urine, and stool is being actively explored, with each fluid type offering unique advantages depending on the cancer.

miRNAs are small non-coding RNAs that play crucial roles in regulating gene expression and have emerged as significant biomarkers in the field of cancer diagnostics. They are particularly appealing for use in alternative liquid biopsy mediums such as saliva, urine, and stool due to their high stability and abundance. These characteristics make miRNAs excellent candidates for non-invasive cancer screening and monitoring. miRNAs are notably stable in bodily fluids, protected from RNAse-mediated degradation by being enclosed in extracellular vesicles or associated with protein complexes or lipoproteins. miRNAs are abundant in various bodily fluids, which allows for their detection and quantification using current molecular techniques such as quantitative PCR, next-generation sequencing, and microarray analysis [72–74].

lncRNAs are emerging as pivotal players in the realm of non-coding RNA-based cancer diagnostics, second only to miRNAs in terms of research attention and potential clinical relevance. LncRNAs, typically longer than 200 nucleotides, do not code for proteins but are highly involved in regulating gene expression at various levels, from chromatin modification to mRNA stability and translation. Their detection in non-traditional liquid biopsy mediums like saliva, urine, and stool underscores their versatility and diagnostic potential. Several oncogenic lncRNAs have been identified in saliva, urine, and stool samples of cancer patients. For instance, the presence of specific lncRNAs in saliva has been linked to oral cancers, in urine to bladder and prostate cancers, and in stool to colorectal cancers. This broad spectrum of detectability across different body fluids highlights their potential as versatile markers for non-invasive testing [12, 75–77].

## 28.4 ncRNAs Vs. Other Non-Invasive Biomarkers

### *28.4.1 Circulating Tumor DNA*

Circulating tumor DNA (ctDNA) consists of DNA fragments that have originated from tumors and are present in the bloodstream. Typically, the primary components of ctDNA are genetic mutations linked to various cancers. This type of DNA is particularly specific to tumors, and research has indicated its efficacy in tracking cancer recurrence after pinpointing the specific mutations in the tumors [78, 79]. Moreover, numerous studies have confirmed ctDNA's effectiveness in detecting advanced-stage cancers [7]. Nonetheless, the use of ctDNA as an early detection biomarker for cancer is somewhat restricted due to the generally low occurrence of random mutations in precancerous conditions and initial stages of cancer, which challenges the sensitivity of ctDNA-based tests [80]. Thus, even with potential future enhancements in measuring ctDNA, the inherent biological constraints might still restrict its utility as a biomarker for diagnosing cancer. In contrast, RNA-based liquid biopsies are very dynamic and tend to be altered in the early phases of cancer, making them more promising as biomarkers for early cancer detection. Overall, it is crucial to analyze the molecular attributes of each biomarker and assess their appropriateness for specific clinical applications.

### *28.4.2 Circulating RNAs*

Among RNA-based liquid biopsy sources, ncRNAs and mRNAs have undergone considerable research. Yet, in the field of non-invasive cancer biomarkers, circulating miRNAs have received the most attention [5]. This predominance is mainly due to miRNAs being more abundant in the bloodstream than mRNAs. The presence of high levels of RNases and other RNA-degrading enzymes in blood initially fueled the belief that cfRNAs might not exist in serum and plasma. However, it was established in the 1970s that a fraction of cfRNAs avoid RNase degradation by being encapsulated within EVs [81]. Further research confirmed that EVs provide a stable environment for RNA, resisting degradation through freeze-thaw cycles and moderate heat [82, 83]. While EVs are a primary carrier of cfRNAs, specific RNA types, including miRNAs, are also stable outside EVs. These miRNAs can bind with the

Ago protein to form a stable complex independent of EVs [84, 85]. Recent findings have indicated a substantial presence of miRNAs in extravesicular locations, although mRNAs are seldom found outside EVs [86]. Overall, the high levels of circulating miRNAs have established them as the most extensively characterized sources of cfRNAs, providing a wealth of data that supports their use in every phase of cancer biomarker development, from target identification to biomarker validation [87].

### *28.4.3 Proteins*

Circulating protein-based biomarkers have become central to the field of oncology, representing some of the most extensively studied and clinically integrated non-invasive indicators of cancer [88]. These biomarkers are proteins found in bodily fluids, such as blood or urine, that can be indicative of cancerous activity within the body. Because they can be detected and measured with relatively straightforward laboratory techniques, protein-based biomarkers are widely used in clinical settings to aid in cancer diagnosis, monitor disease progression, and evaluate treatment effectiveness [89]. The value of protein-based biomarkers lies in their ability to serve as indirect markers for detecting and assessing malignancies. They may not pinpoint the exact location or type of tumor but can signal abnormalities or changes in the body that warrant further investigation. Common examples include prostate-specific antigen (PSA) for prostate cancer, cancer antigen 125 (CA-125) for ovarian cancer, and carcinoembryonic antigen (CEA) for colorectal and other types of cancer [90, 91]. The presence and levels of these proteins can help healthcare providers determine the likelihood of cancer and make decisions about the need for more definitive diagnostic procedures, such as biopsies or imaging studies [92].

Since proteins originate from mRNAs, their expression profiles are often similar to those of the corresponding mRNAs. Advances in DNA and RNA-sequencing technologies have significantly enhanced the capabilities for whole genome and whole transcriptome analysis. Nevertheless, a major hurdle in utilizing protein-based biomarkers is the challenge of accurately quantifying the entire proteome in the bloodstream. As methods for comprehensive genomic and transcriptomic profiling become more common, it is critical to develop similar technologies for proteome analysis to maintain parity in biomarker identification techniques. Despite the emergence of technologies capable of multiplexing the analysis of up to 5000 proteins at once, protein profiling still lags behind genomic and transcriptomic methods in terms of breadth and precision. This disparity highlights the need for more refined techniques that can match the thoroughness of nucleic acid sequencing in the evaluation of protein biomarkers.

### *28.4.4 DNA Methylation*

DNA methylation, which involves the addition of methyl groups to the cytosine or adenine nucleotides in DNA, is one of the most well-established and extensively studied epigenetic mechanisms. This biochemical process plays a crucial role in regulating gene expression by influencing the accessibility of the DNA to transcriptional machinery, thus controlling whether genes are turned on or off, without altering the underlying DNA sequence itself. This ability to modulate gene activity has profound implications for cellular differentiation, development, and the maintenance of cellular identity [10, 93, 94]. Over the past two decades, the potential of circulating methylated DNA as a source for cancer biomarkers has been vigorously explored and has shown considerable promise. As cancer cells often exhibit unique patterns of DNA methylation compared to normal cells, these differential patterns can be leveraged for diagnostic purposes. Circulating methylated DNA, which can be released into the bloodstream from tumor cells through cell death or secretion, provides a non-invasive means to detect and monitor various types of cancer. This approach not only facilitates early detection but also enables the monitoring of tumor dynamics and response to treatment in real-time. The clinical application of methylated DNA biomarkers could revolutionize cancer diagnostics by providing a sensitive, specific, and minimally invasive diagnostic tool that could greatly enhance patient management and outcomes [10, 95]. Extensive research into DNA methylation patterns has uncovered specific sites that are frequently associated with various cancers, offering significant potential for early detection and monitoring. A notable example of such research is the identification of methylation in the SEPT9 gene, which has been consistently observed in cases of colorectal cancer. This gene, when methylated, undergoes changes in its expression that are linked to tumorigenesis, making it a prime target for diagnostic purposes.

## 28.5 Technics in Quantifying Circulating ncRNAs

### *28.5.1 Molecular Profiling*

One of the most pivotal advancements in the realm of diagnostic cancer biomarkers has been the introduction and adoption of high-throughput technologies such as next-generation sequencing (NGS). This technology facilitates extensive analysis of the entire genome or transcriptome, offering two major benefits that have accelerated its use in non-invasive cancer biomarker research. First, the widespread availability of tissue-based sequencing databases, like TCGA [96–99], provides critical insights into the molecular signatures specific to various cancer types. These expansive databases allow for precise identification of genes consistently dysregulated across cancer types, enhancing the predictive accuracy of potential biomarkers found in circulation based on comparisons with tissue expression levels. Second, significant advancements in NGS technology have broadened its application in the field of liquid biopsies. Over the last decade, the costs associated with NGS have decreased dramatically, making it feasible to sequence a larger array of clinical samples and achieve deeper data coverage. This depth is crucial for identifying low-abundance molecular signals typical of cancer-derived elements in blood. Additionally, the advent of single-cell sequencing technologies has greatly refined the resolution at which these low-input samples can be analyzed, thereby improving the accuracy of transcriptomic profiling even from minimal amounts of starting material such as plasma or serum. Together, these technological improvements have not only reduced the barriers to entry for conducting detailed molecular assessments but have also enhanced the overall efficacy and reliability of using circulating biomarkers for cancer diagnosis. As a result, researchers now have the tools to perform more detailed evaluations of molecular profiles in the circulation, potentially leading to earlier and more accurate detection of cancer.

Systematic investigations into the molecular profiling of ncRNAs for cancer biomarker development are relatively scarce. Nevertheless, a number of studies have successfully leveraged tissue profiling data to pinpoint potential ncRNA biomarker targets, later validating these cancer-related ncRNA signatures in blood-based samples, such as plasma and serum. For instance, a recent study capitalized on multiple tissue sequencing datasets to pinpoint miRNAs that were significantly overexpressed in gastric cancer tissues. This research then tested the robustness of the identified miRNA signature for gastric cancer across various cohorts [100]. A similar methodology was employed in research on ovarian cancer; here, researchers utilized several ovarian cancer tissue and serum datasets to develop a miRNA-based diagnostic signature. This signature was subsequently validated using serum samples collected prospectively [101]. Furthermore, studies involving pancreatic cancer used multiple tissue datasets to identify miRNAs linked with poor prognosis among patients. The prognostic significance of these miRNA signatures was then confirmed through validation in a serum cohort [11]. These examples underscore the potential of using tissue-derived data for the identification and validation of ncRNA signatures as blood-based biomarkers for cancer. This approach not only enhances the predictive accuracy of these biomarkers but also broadens the scope of their applicability in clinical settings.

### *28.5.2 Bioinformatics Approaches*

One of the foremost challenges in cancer biomarker development is the intrinsic heterogeneity of cancers, which complicates the identification of consistent biomarkers across different cases. Molecular analyses of various cancer tissues have demonstrated that no single gene biomarker reliably differentiates cancerous from healthy tissue. To address the limited sensitivity and specificity offered by individual genes, researchers have turned to combining multiple biomarkers to create comprehensive gene signatures that enhance cancer detection. However, the process of analyzing extensive sequencing data to develop these signatures is inherently complex. A primary method for managing disease heterogeneity and the voluminous data from sequencing is the adoption of sophisticated bioinformatic algorithms, particularly those based on machine learning. Machine learning techniques help distill vast amounts of data into more manageable, interpretable formats [102]. These advanced computational methods offer substantial improvements over traditional single-marker or small-panel approaches by integrating large datasets to develop more accurate biomarkers. Machine learning's application in liquid biopsy biomarker development for various diseases, including cancer, is expanding rapidly [102]. The future will likely see more widespread adoption of these AI-driven techniques. Nonetheless, several fundamental challenges must be addressed to fully realize their potential. Among these, the quality of the datasets is paramount; the effectiveness of machine learning in developing biomarkers heavily depends on the resolution and integrity of the data used. Given the typically low concentration of cancer-derived molecular signals in the bloodstream, enhancing the resolution of these signals is crucial for improving the accuracy and utility of circulating biomarkers. This advancement is essential for pushing the boundaries of non-invasive cancer diagnostics and personalized medicine.

## 28.6 Future Outlook

### *28.6.1 Cancer Specificity of ncRNAs*

When developing circulating non-invasive biomarkers, achieving high specificity for a particular disease or organ is critical to enhancing the reliability and usefulness of these markers in clinical diagnosis. Biomarkers with high disease specificity can drastically reduce the incidence of false positives, making the diagnostic process more accurate and trustworthy. This specificity is particularly crucial in the context of cancer, where early and precise detection can significantly influence treatment decisions and outcomes. For example, biomarkers that are uniquely associated with a specific type of cancer not only provide clear diagnostic indicators but also aid in monitoring disease progression and response to therapy. This level of specificity ensures that the biomarker can reliably reflect the presence and state of a particular cancer, without interference from other conditions that might present similar symptoms or molecular profiles. Furthermore, certain genes known for their organ-specific expression provide valuable insights into the origin of cancer when they become dysregulated. These organ-specific genes act as precise indicators of where a malignancy may be developing within the body. If a gene typically expressed solely in the liver shows abnormal methylation patterns, it strongly suggests a pathological process within the liver, such as liver cancer. This approach to biomarker development not only aids in pinpointing the location of the cancer but also enhances the overall precision of the diagnostic process, guiding more targeted investigations and treatments.

Notably, mutations like those in KRAS, which are seldom found in healthy individuals, offer high cancer specificity. However, since such mutations can appear in various types of cancers, their presence in circulating DNA alone may not conclusively indicate the cancer's location. To date, many established miRNA biomarkers lack cancer specificity, but a potential solution lies in the study of isomiRs. IsomiRs, or isoforms of miRNAs, are produced in specific human tissues and their expression varies depending on the tissue and disease context [103, 104]. Analysis of isomiRs across multiple TCGA datasets has revealed that these ncRNAs display high tissue specificity [105], although their close sequence similarity to their parent miRNAs complicates quantification by qPCR. Fortunately, RNA sequencing, which is increasingly employed in biomarker research, can accurately quantify these sequences, thereby demonstrating the viability of isomiRs as non-invasive cancer biomarkers.

Moreover, studies have shown that some EVs, specifically exosomes released from particular organs, retain molecular signatures indicative of their tissue of origin. These tissue-specific exosomes have been identified via surface proteins unique to their originating tissue [106, 107]. Additionally, research has pinpointed several exosomal surface markers that are unique to particular cancers [108, 109], suggesting that analyzing the ncRNA content of these tissue-specific or cancer-specific exosomes could significantly enhance the specificity of cancer biomarkers. Although these specific exosomes might be rare in circulation, leveraging both the molecular characteristics of the exosomes and the ncRNA profiles within could unlock substantial potential in the realm of cancer diagnostics. This combined approach holds great promise for improving the accuracy and specificity of non-invasive cancer biomarkers, ultimately aiding in the early detection and precise localization of cancer.

### *28.6.2 Quantification Issues*

Quantifying circulating ncRNAs presents distinct challenges, particularly concerning the normalization of qPCR-based quantifications. This issue is critical when dealing with circulating miRNAs and necessitates reliable normalization methods to ensure accurate, reproducible results [9]. Currently, two principal normalization strategies are employed for miRNAs.

#### 28.6.2.1 Synthetic Spiked-In Controls

This method involves the addition of synthetic miRNA controls into the samples before the extraction process. Common synthetic controls include cel-miR-39 from *C. elegans* and ath-miR-159a from *Arabidopsis*. These controls are not present in human samples and thus serve as consistent standards to which miRNA expression levels can be normalized. The key advantage of using synthetic spiked-in controls is their consistency across different experimental setups and sample batches, making them invaluable for comparative studies across different time points and conditions. However, this method has limitations, particularly when dealing with archival samples that have been stored for long periods at −80 °C. RNA degradation under these conditions can affect the integrity of the RNA, leading to potential inaccuracies in quantification if the degradation affects the spiked-in controls differently than the endogenous miRNAs.

#### 28.6.2.2 Endogenous Controls

The alternative approach uses endogenous miRNAs as internal controls. Common choices include miR-16 and miR-451, which are generally abundant and stable across various types of samples. Another option is U6 small nuclear RNA, often used due to its consistent presence in cells. However, the major drawback of using endogenous controls lies in their potential regulation or dysregulation under different physiological or pathological conditions, such as in various cancers. This variability can lead to biased normalization and skewed results, as the supposed "constant" baseline may itself be altered by the disease state or experimental condition [110].

To address these limitations, leveraging NGS platforms for ncRNA quantification offers a promising alternative of RNA-sequencing normalization. RNA-Seq typically involves normalization strategies such as counts per million (CPM) or reads per million (RPM), which calculate gene expression levels as a relative abundance to the total gene counts detected. Additionally, transcripts per million (TPM) is used to adjust for gene length, ensuring that longer genes do not disproportionately influence the expression data. These RNA-Seq methodologies tend to be less susceptible to the variances seen with traditional qPCR normalization methods, offering a more stable and reliable framework for analyzing gene expression. However, these approaches do not fully account for variability in patients' overall levels of circulating RNAs, which can introduce another layer of complexity.

**Box 28.1 How Technology Has Shaped the Evolution of RNA as a Clinical Biomarker**

Historically, gene expression analysis in clinical settings has predominantly relied on single gene tests executed via quantitative reverse transcription PCR (qRT-PCR) [111], such as those used for detecting the influenza virus [112]. This technique boasts numerous advantages, including speed, precision, sensitivity, and cost-effectiveness, alongside its capacity for high-throughput analysis, which allows for the processing of numerous clinical samples with minimal sample input. Given these attributes and its longstanding use, qRT-PCR is often considered the "gold standard" for measuring RNA levels in

(continued)

**Box 28.1** (continued)
clinical diagnostics. Nevertheless, it does possess limitations, notably its restriction to a small number of markers per assay despite its ability to handle many samples simultaneously. Following the introduction of hybridization-based methods for evaluating multiple gene targets, which utilize arrayed probes on solid surfaces [113], microarray technology marked a significant advancement in clinical assay development. This technology enabled the simultaneous measurement of multiple RNA targets, known as "gene expression profiles," which are increasingly used as diagnostic or prognostic tools in research.

A prime example of microarray application in a clinical context is the MammaPrint test by Agendia [114], which assesses a 70-gene profile in breast tumors to predict recurrence risk. Despite their effectiveness, microarray assays can be limited by factors such as inter-laboratory variability in sample preparation that may impact reproducibility, and signal-to-noise issues that could influence detection limits. Interestingly, despite the potential of microarrays, a number of multi-gene profile tests for cancer such as OncoTypeDX (Genome Health) [115] for breast cancer and Prolaris (Myriad) [116] for prostate cancer have reverted to utilizing qRT-PCR. This preference is largely due to qRT-PCR's superior reliability, reproducibility, sensitivity, and accuracy, underscoring its continued relevance and predominance in contemporary clinical diagnostics. This ongoing reliance on qRT-PCR, especially for crucial multi-gene tests, highlights its foundational role in the evolution and operation of gene expression analysis within the medical field.

## 28.7 Conclusion

In this chapter, we have highlighted recent advancements in the use of ncRNAs for developing non-invasive cancer biomarkers. Initially, miRNAs dominated the landscape as the principal ncRNAs studied for their potential as blood-based indicators of cancer. However, a diverse array of ncRNAs is now being recognized as promising candidates for such biomarkers. Traditional methods relying on hypotheses have pinpointed several potential biomarkers for cancer diagnosis. Yet, the shift toward comprehensive molecular profiling facilitates a hypothesis-independent search, broadening the scope significantly. This unbiased methodology is poised to uncover novel biomarkers that could transform cancer diagnostics. Although the integration of high-throughput profiling techniques in non-invasive biomarker discovery is still in its nascent stages, the analysis of circulating ncRNA profiles in cancer patients is yielding valuable insights. These developments are not only enriching the research landscape but are also setting the stage for significant breakthroughs in clinical applications. The potential to develop game-changing non-invasive biomarkers through enhanced precision oncology is on the horizon, promising to revolutionize how cancer is detected and monitored.

## References

1. Crowley E, Di Nicolantonio F, Loupakis F, Bardelli A. Liquid biopsy: monitoring cancer-genetics in the blood. Nat Rev Clin Oncol. 2013;10(8):472–84.
2. Campos-Carrillo A, Weitzel JN, Sahoo P, Rockne R, Mokhnatkin JV, Murtaza M, et al. Circulating tumor DNA as an early cancer detection tool. Pharmacol Ther. 2020;207:107458.
3. Ignatiadis M, Lee M, Jeffrey SS. Circulating tumor cells and circulating tumor DNA: challenges and opportunities on the path to clinical utility. Clin Cancer Res. 2015;21(21):4786–800.
4. Schwarzenbach H, Hoon DS, Pantel K. Cell-free nucleic acids as biomarkers in cancer patients. Nat Rev Cancer. 2011;11(6):426–37.
5. Toiyama Y, Okugawa Y, Fleshman J, Richard Boland C, Goel A. MicroRNAs as potential liquid biopsy biomarkers in colorectal cancer: a systematic review. Biochim Biophys Acta Rev Cancer. 2018;1870(2):274–82.
6. Wada Y, Shimada M, Murano T, Takamaru H, Morine Y, Ikemoto T, et al. A liquid biopsy assay for noninvasive identification of lymph node metastases in T1 colorectal cancer. Gastroenterology. 2021;161(1):151–62.e1.
7. Bettegowda C, Sausen M, Leary RJ, Kinde I, Wang Y, Agrawal N, et al. Detection of circulating tumor DNA in early- and late-stage human malignancies. Sci Transl Med. 2014;6(224):224ra24.
8. Li Y, Zheng Q, Bao C, Li S, Guo W, Zhao J, et al. Circular RNA is enriched and stable in exosomes: a promising biomarker for cancer diagnosis. Cell Res. 2015;25(8):981–4.
9. Shigeyasu K, Toden S, Zumwalt TJ, Okugawa Y, Goel A. Emerging role of MicroRNAs as liquid biopsy biomarkers in gastrointestinal cancers. Clin Cancer Res. 2017;23(10):2391–9.
10. Okugawa Y, Grady WM, Goel A. Epigenetic alterations in colorectal cancer: emerging biomarkers. Gastroenterology. 2015;149(5):1204–25.e12.
11. Kandimalla R, Shimura T, Mallik S, Sonohara F, Tsai S, Evans DB, et al. Identification of serum miRNA signature and establishment of a nomogram for risk stratification in patients with pancreatic ductal adenocarcinoma. Ann Surg. 2022;275(1):e229–e37.
12. Bahn JH, Zhang Q, Li F, Chan TM, Lin X, Kim Y, et al. The landscape of microRNA, Piwi-interacting RNA, and circular RNA in human saliva. Clin Chem. 2015;61(1):221–30.
13. Volovat SR, Volovat C, Hordila I, Hordila DA, Mirestean CC, Miron OT, et al. MiRNA and LncRNA as potential biomarkers in triple-negative breast cancer: a review. Front Oncol. 2020;10:526850.
14. Zen K, Zhang CY. Circulating microRNAs: a novel class of biomarkers to diagnose and monitor human cancers. Med Res Rev. 2012;32(2):326–48.
15. Dvinge H, Git A, Gräf S, Salmon-Divon M, Curtis C, Sottoriva A, et al. The shaping and functional consequences of the microRNA landscape in breast cancer. Nature. 2013;497(7449):378–82.
16. Niknafs YS, Han S, Ma T, Speers C, Zhang C, Wilder-Romans K, et al. The lncRNA landscape of breast cancer reveals a role for DSCAM-AS1 in breast cancer progression. Nat Commun. 2016;7:12791.
17. Xu S, Kong D, Chen Q, Ping Y, Pang D. Oncogenic long noncoding RNA landscape in breast cancer. Mol Cancer. 2017;16(1):129.
18. Toden S, Zumwalt TJ, Goel A. Non-coding RNAs and potential therapeutic targeting in cancer. Biochim Biophys Acta Rev Cancer. 2021;1875(1):188491.
19. Link A, Balaguer F, Shen Y, Nagasaka T, Lozano JJ, Boland CR, et al. Fecal MicroRNAs as novel biomarkers for colon cancer screening. Cancer Epidemiol Biomarkers Prev. 2010;19(7):1766–74.
20. Okugawa Y, Toiyama Y, Goel A. An update on microRNAs as colorectal cancer biomarkers: where are we and what's next? Expert Rev Mol Diagn. 2014;14(8):999–1021.
21. Si ML, Zhu S, Wu H, Lu Z, Wu F, Mo YY. miR-21-mediated tumor growth. Oncogene. 2007;26(19):2799–803.

22. Tili E, Croce CM, Michaille JJ. miR-155: on the crosstalk between inflammation and cancer. Int Rev Immunol. 2009;28(5):264–84.
23. Abue M, Yokoyama M, Shibuya R, Tamai K, Yamaguchi K, Sato I, et al. Circulating miR-483-3p and miR-21 is highly expressed in plasma of pancreatic cancer. Int J Oncol. 2015;46(2):539–47.
24. Liu R, Liao J, Yang M, Shi Y, Peng Y, Wang Y, et al. Circulating miR-155 expression in plasma: a potential biomarker for early diagnosis of esophageal cancer in humans. J Toxicol Environ Health A. 2012;75(18):1154–62.
25. Shi C, Yang Y, Xia Y, Okugawa Y, Yang J, Liang Y, et al. Novel evidence for an oncogenic role of microRNA-21 in colitis-associated colorectal cancer. Gut. 2016;65(9):1470–81.
26. Sun Y, Wang M, Lin G, Sun S, Li X, Qi J, et al. Serum microRNA-155 as a potential biomarker to track disease in breast cancer. PLoS One. 2012;7(10):e47003.
27. Toiyama Y, Takahashi M, Hur K, Nagasaka T, Tanaka K, Inoue Y, et al. Serum miR-21 as a diagnostic and prognostic biomarker in colorectal cancer. J Natl Cancer Inst. 2013;105(12):849–59.
28. Urbich C, Kuehbacher A, Dimmeler S. Role of microRNAs in vascular diseases, inflammation, and angiogenesis. Cardiovasc Res. 2008;79(4):581–8.
29. Statello L, Guo CJ, Chen LL, Huarte M. Gene regulation by long non-coding RNAs and its biological functions. Nat Rev Mol Cell Biol. 2021;22(2):96–118.
30. Schmitt AM, Chang HY. Long noncoding RNAs in cancer pathways. Cancer Cell. 2016;29(4):452–63.
31. Gupta RA, Shah N, Wang KC, Kim J, Horlings HM, Wong DJ, et al. Long non-coding RNA HOTAIR reprograms chromatin state to promote cancer metastasis. Nature. 2010;464(7291):1071–6.
32. Gutschner T, Hämmerle M, Eissmann M, Hsu J, Kim Y, Hung G, et al. The noncoding RNA MALAT1 is a critical regulator of the metastasis phenotype of lung cancer cells. Cancer Res. 2013;73(3):1180–9.
33. Tseng YY, Moriarity BS, Gong W, Akiyama R, Tiwari A, Kawakami H, et al. PVT1 dependence in cancer with MYC copy-number increase. Nature. 2014;512(7512):82–6.
34. Iyer MK, Niknafs YS, Malik R, Singhal U, Sahu A, Hosono Y, et al. The landscape of long noncoding RNAs in the human transcriptome. Nat Genet. 2015;47(3):199–208.
35. Wang J, Gao Y, Wang X, Gao Y, Li L, Zhang J, et al. Circulating lncRNAs as noninvasive biomarkers in bladder cancer: a diagnostic meta-analysis based on 15 published articles. Int J Biol Markers. 2020;35(2):40–8.
36. Chen Q, Su Y, He X, Zhao W, Wu C, Zhang W, et al. Plasma long non-coding RNA MALAT1 is associated with distant metastasis in patients with epithelial ovarian cancer. Oncol Lett. 2016;12(2):1361–6.
37. Okugawa Y, Toiyama Y, Hur K, Toden S, Saigusa S, Tanaka K, et al. Metastasis-associated long non-coding RNA drives gastric cancer development and promotes peritoneal metastasis. Carcinogenesis. 2014;35(12):2731–9.
38. Zheng ZK, Pang C, Yang Y, Duan Q, Zhang J, Liu WC. Serum long noncoding RNA urothelial carcinoma-associated 1: a novel biomarker for diagnosis and prognosis of hepatocellular carcinoma. J Int Med Res. 2018;46(1):348–56.
39. Zhou X, Yin C, Dang Y, Ye F, Zhang G. Identification of the long non-coding RNA H19 in plasma as a novel biomarker for diagnosis of gastric cancer. Sci Rep. 2015;5:11516.
40. Lane JS, Hoff DV, Cridebring D, Goel A. Extracellular vesicles in diagnosis and treatment of pancreatic cancer: current state and future perspectives. Cancers (Basel). 2020;12(6):1530.
41. Yousefi H, Maheronnaghsh M, Molaei F, Mashouri L, Reza Aref A, Momeny M, et al. Long noncoding RNAs and exosomal lncRNAs: classification, and mechanisms in breast cancer metastasis and drug resistance. Oncogene. 2020;39(5):953–74.
42. Zang X, Gu J, Zhang J, Shi H, Hou S, Xu X, et al. Exosome-transmitted lncRNA UFC1 promotes non-small-cell lung cancer progression by EZH2-mediated epigenetic silencing of PTEN expression. Cell Death Dis. 2020;11(4):215.

43. Chen LL, Yang L. Regulation of circRNA biogenesis. RNA Biol. 2015;12(4):381–8.
44. Sanger HL, Klotz G, Riesner D, Gross HJ, Kleinschmidt AK. Viroids are single-stranded covalently closed circular RNA molecules existing as highly base-paired rod-like structures. Proc Natl Acad Sci U S A. 1976;73(11):3852–6.
45. Hansen TB, Jensen TI, Clausen BH, Bramsen JB, Finsen B, Damgaard CK, et al. Natural RNA circles function as efficient microRNA sponges. Nature. 2013;495(7441):384–8.
46. Qu S, Yang X, Li X, Wang J, Gao Y, Shang R, et al. Circular RNA: a new star of noncoding RNAs. Cancer Lett. 2015;365(2):141–8.
47. Chen J, Li Y, Zheng Q, Bao C, He J, Chen B, et al. Circular RNA profile identifies circPVT1 as a proliferative factor and prognostic marker in gastric cancer. Cancer Lett. 2017;388:208–19.
48. Hsiao KY, Lin YC, Gupta SK, Chang N, Yen L, Sun HS, et al. Noncoding effects of circular RNA CCDC66 promote colon cancer growth and metastasis. Cancer Res. 2017;77(9):2339–50.
49. Chen J, Cui L, Yuan J, Zhang Y, Sang H. Circular RNA WDR77 target FGF-2 to regulate vascular smooth muscle cells proliferation and migration by sponging miR-124. Biochem Biophys Res Commun. 2017;494(1–2):126–32.
50. Hang D, Zhou J, Qin N, Zhou W, Ma H, Jin G, et al. A novel plasma circular RNA circFARSA is a potential biomarker for non-small cell lung cancer. Cancer Med. 2018;7(6):2783–91.
51. Li Z, Zhou Y, Yang G, He S, Qiu X, Zhang L, et al. Using circular RNA SMARCA5 as a potential novel biomarker for hepatocellular carcinoma. Clin Chim Acta. 2019;492:37–44.
52. Lin J, Cai D, Li W, Yu T, Mao H, Jiang S, et al. Plasma circular RNA panel acts as a novel diagnostic biomarker for colorectal cancer. Clin Biochem. 2019;74:60–8.
53. Jeck WR, Sorrentino JA, Wang K, Slevin MK, Burd CE, Liu J, et al. Circular RNAs are abundant, conserved, and associated with ALU repeats. RNA. 2013;19(2):141–57.
54. Giannoukos G, Ciulla DM, Huang K, Haas BJ, Izard J, Levin JZ, et al. Efficient and robust RNA-seq process for cultured bacteria and complex community transcriptomes. Genome Biol. 2012;13(3):R23.
55. Vo JN, Cieslik M, Zhang Y, Shukla S, Xiao L, Zhang Y, et al. The landscape of circular RNA in cancer. Cell. 2019;176(4):869–81.e13.
56. Czech B, Hannon GJ. One loop to rule them all: the Ping-pong cycle and piRNA-guided silencing. Trends Biochem Sci. 2016;41(4):324–37.
57. Liu Y, Dou M, Song X, Dong Y, Liu S, Liu H, et al. The emerging role of the piRNA/piwi complex in cancer. Mol Cancer. 2019;18(1):123.
58. Ross RJ, Weiner MM, Lin H. PIWI proteins and PIWI-interacting RNAs in the soma. Nature. 2014;505(7483):353–9.
59. Watanabe T, Lin H. Posttranscriptional regulation of gene expression by Piwi proteins and piRNAs. Mol Cell. 2014;56(1):18–27.
60. Weng W, Li H, Goel A. Piwi-interacting RNAs (piRNAs) and cancer: emerging biological concepts and potential clinical implications. Biochim Biophys Acta Rev Cancer. 2019;1871(1):160–9.
61. Weng W, Liu N, Toiyama Y, Kusunoki M, Nagasaka T, Fujiwara T, et al. Novel evidence for a PIWI-interacting RNA (piRNA) as an oncogenic mediator of disease progression, and a potential prognostic biomarker in colorectal cancer. Mol Cancer. 2018;17(1):16.
62. Gu X, Wang C, Deng H, Qing C, Liu R, Liu S, et al. Exosomal piRNA profiling revealed unique circulating piRNA signatures of cholangiocarcinoma and gallbladder carcinoma. Acta Biochim Biophys Sin (Shanghai). 2020;52(5):475–84.
63. Mai D, Zheng Y, Guo H, Ding P, Bai R, Li M, et al. Serum piRNA-54265 is a new biomarker for early detection and clinical surveillance of human colorectal cancer. Theranostics. 2020;10(19):8468–78.
64. Qu A, Wang W, Yang Y, Zhang X, Dong Y, Zheng G, et al. A serum piRNA signature as promising non-invasive diagnostic and prognostic biomarkers for colorectal cancer. Cancer Manag Res. 2019;11:3703–20.
65. Jia Y, Tan W, Zhou Y. Transfer RNA-derived small RNAs: potential applications as novel biomarkers for disease diagnosis and prognosis. Ann Transl Med. 2020;8(17):1092.

66. Fu H, Feng J, Liu Q, Sun F, Tie Y, Zhu J, et al. Stress induces tRNA cleavage by angiogenin in mammalian cells. FEBS Lett. 2009;583(2):437–42.
67. Gebetsberger J, Polacek N. Slicing tRNAs to boost functional ncRNA diversity. RNA Biol. 2013;10(12):1798–806.
68. Gehrke CW, Kuo KC, Waalkes TP, Borek E. Patterns of urinary excretion of modified nucleosides. Cancer Res. 1979;39(4):1150–3.
69. Lakings DB, Waalkes TP, Borek E, Gehrke CW, Mrochek JE, Longmore J, et al. Composition, associated tissue methyltransferase activity, and catabolic end products of transfer RNA from carcinogen-induced hepatoma and normal monkey livers. Cancer Res. 1977;37(1):285–92.
70. Dhahbi JM, Spindler SR, Atamna H, Boffelli D, Martin DI. Deep sequencing of serum small RNAs identifies patterns of 5′ tRNA half and YRNA fragment expression associated with breast cancer. Biomark Cancer. 2014;6:37–47.
71. Zhu L, Li J, Gong Y, Wu Q, Tan S, Sun D, et al. Exosomal tRNA-derived small RNA as a promising biomarker for cancer diagnosis. Mol Cancer. 2019;18(1):74.
72. Duran-Sanchon S, Moreno L, Gómez-Matas J, Augé JM, Serra-Burriel M, Cuatrecasas M, et al. Fecal microRNA-based algorithm increases effectiveness of fecal immunochemical test-based screening for colorectal cancer. Clin Gastroenterol Hepatol. 2021;19(2):323–30.e1.
73. Hoshino I. The usefulness of microRNA in urine and saliva as a biomarker of gastroenterological cancer. Int J Clin Oncol. 2021;26(8):1431–40.
74. Li L, Wang A, Cai M, Tong M, Chen F, Huang L. Identification of stool miR-135b-5p as a non-invasive diagnostic biomarker in later tumor stage of colorectal cancer. Life Sci. 2020;260:118417.
75. Xie Z, Chen X, Li J, Guo Y, Li H, Pan X, et al. Salivary HOTAIR and PVT1 as novel biomarkers for early pancreatic cancer. Oncotarget. 2016;7(18):25408–19.
76. Zhan Y, Du L, Wang L, Jiang X, Zhang S, Li J, et al. Expression signatures of exosomal long non-coding RNAs in urine serve as novel non-invasive biomarkers for diagnosis and recurrence prediction of bladder cancer. Mol Cancer. 2018;17(1):142.
77. Gharib E, Nazemalhosseini-Mojarad E, Baghdar K, Nayeri Z, Sadeghi H, Rezasoltani S, et al. Identification of a stool long non-coding RNAs panel as a potential biomarker for early detection of colorectal cancer. J Clin Lab Anal. 2021;35(2):e23601.
78. Vidal J, Muinelo L, Dalmases A, Jones F, Edelstein D, Iglesias M, et al. Plasma ctDNA RAS mutation analysis for the diagnosis and treatment monitoring of metastatic colorectal cancer patients. Ann Oncol. 2017;28(6):1325–32.
79. Kruger S, Heinemann V, Ross C, Diehl F, Nagel D, Ormanns S, et al. Repeated mutKRAS ctDNA measurements represent a novel and promising tool for early response prediction and therapy monitoring in advanced pancreatic cancer. Ann Oncol. 2018;29(12):2348–55.
80. Ruiz-Bañobre J, Goel A. Genomic and epigenomic biomarkers in colorectal cancer: from diagnosis to therapy. Adv Cancer Res. 2021;151:231–304.
81. Stroun M, Anker P, Beljanski M, Henri J, Lederrey C, Ojha M, et al. Presence of RNA in the nucleoprotein complex spontaneously released by human lymphocytes and frog auricles in culture. Cancer Res. 1978;38(10):3546–54.
82. Kalra H, Adda CG, Liem M, Ang CS, Mechler A, Simpson RJ, et al. Comparative proteomics evaluation of plasma exosome isolation techniques and assessment of the stability of exosomes in normal human blood plasma. Proteomics. 2013;13(22):3354–64.
83. Sokolova V, Ludwig AK, Hornung S, Rotan O, Horn PA, Epple M, et al. Characterisation of exosomes derived from human cells by nanoparticle tracking analysis and scanning electron microscopy. Colloids Surf B Biointerfaces. 2011;87(1):146–50.
84. Fuji T, Umeda Y, Nyuya A, Taniguchi F, Kawai T, Yasui K, et al. Detection of circulating microRNAs with Ago2 complexes to monitor the tumor dynamics of colorectal cancer patients during chemotherapy. Int J Cancer. 2019;144(9):2169–80.
85. Winter J, Diederichs S. Argonaute proteins regulate microRNA stability: increased microRNA abundance by Argonaute proteins is due to microRNA stabilization. RNA Biol. 2011;8(6):1149–57.

86. Enderle D, Spiel A, Coticchia CM, Berghoff E, Mueller R, Schlumpberger M, et al. Characterization of RNA from exosomes and other extracellular vesicles isolated by a novel spin column-based method. PLoS One. 2015;10(8):e0136133.
87. Chen Y, Wu T, Zhu Z, Huang H, Zhang L, Goel A, et al. An integrated workflow for biomarker development using microRNAs in extracellular vesicles for cancer precision medicine. Semin Cancer Biol. 2021;74:134–55.
88. Veyssière H, Bidet Y, Penault-Llorca F, Radosevic-Robin N, Durando X. Circulating proteins as predictive and prognostic biomarkers in breast cancer. Clin Proteomics. 2022;19(1):25.
89. Xing X, Cai L, Ouyang J, Wang F, Li Z, Liu M, et al. Proteomics-driven noninvasive screening of circulating serum protein panels for the early diagnosis of hepatocellular carcinoma. Nat Commun. 2023;14(1):8392.
90. Desai S, Guddati AK. Carcinoembryonic antigen, carbohydrate antigen 19-9, cancer antigen 125, prostate-specific antigen and other cancer markers: a primer on commonly used cancer markers. World J Oncol. 2023;14(1):4–14.
91. Duffy MJ. Biomarkers for prostate cancer: prostate-specific antigen and beyond. Clin Chem Lab Med. 2020;58(3):326–39.
92. Chatterjee SK, Zetter BR. Cancer biomarkers: knowing the present and predicting the future. Future Oncol. 2005;1(1):37–50.
93. Goel A, Boland CR. Epigenetics of colorectal cancer. Gastroenterology. 2012;143(6):1442–60.e1.
94. Jung G, Hernández-Illán E, Moreira L, Balaguer F, Goel A. Epigenetics of colorectal cancer: biomarker and therapeutic potential. Nat Rev Gastroenterol Hepatol. 2020;17(2):111–30.
95. Lam K, Pan K, Linnekamp JF, Medema JP, Kandimalla R. DNA methylation based biomarkers in colorectal cancer: a systematic review. Biochim Biophys Acta. 2016;1866(1):106–20.
96. Cancer Genome Atlas Network. Comprehensive molecular portraits of human breast tumours. Nature. 2012;490(7418):61–70.
97. Cancer Genome Atlas Network. Comprehensive molecular characterization of human colon and rectal cancer. Nature. 2012;487(7407):330–7.
98. Cancer Genome Atlas Network. Comprehensive molecular profiling of lung adenocarcinoma. Nature. 2014;511(7511):543–50.
99. Cancer Genome Atlas Research Network. Integrated genomic characterization of oesophageal carcinoma. Nature. 2017;541(7636):169–75.
100. Izumi D, Zhu Z, Chen Y, Toden S, Huo X, Kanda M, et al. Assessment of the diagnostic efficiency of a liquid biopsy assay for early detection of gastric cancer. JAMA Netw Open. 2021;4(8):e2121129.
101. Kandimalla R, Wang W, Yu F, Zhou N, Gao F, Spillman M, et al. OCaMIR—a noninvasive, diagnostic signature for early-stage ovarian cancer: a multi-cohort retrospective and prospective study. Clin Cancer Res. 2021;27(15):4277–86.
102. Ko J, Baldassano SN, Loh PL, Kording K, Litt B, Issadore D. Machine learning to detect signatures of disease in liquid biopsies—a user's guide. Lab Chip. 2018;18(3):395–405.
103. Loher P, Londin ER, Rigoutsos I. IsomiR expression profiles in human lymphoblastoid cell lines exhibit population and gender dependencies. Oncotarget. 2014;5(18):8790–802.
104. Telonis AG, Loher P, Jing Y, Londin E, Rigoutsos I. Beyond the one-locus-one-miRNA paradigm: microRNA isoforms enable deeper insights into breast cancer heterogeneity. Nucleic Acids Res. 2015;43(19):9158–75.
105. Telonis AG, Magee R, Loher P, Chervoneva I, Londin E, Rigoutsos I. Knowledge about the presence or absence of miRNA isoforms (isomiRs) can successfully discriminate amongst 32 TCGA cancer types. Nucleic Acids Res. 2017;45(6):2973–85.
106. Mathivanan S, Lim JW, Tauro BJ, Ji H, Moritz RL, Simpson RJ. Proteomics analysis of A33 immunoaffinity-purified exosomes released from the human colon tumor cell line LIM1215 reveals a tissue-specific protein signature. Mol Cell Proteomics. 2010;9(2):197–208.

107. Vallabhajosyula P, Korutla L, Habertheuer A, Yu M, Rostami S, Yuan CX, et al. Tissue-specific exosome biomarkers for noninvasively monitoring immunologic rejection of transplanted tissue. J Clin Invest. 2017;127(4):1375–91.
108. Castillo J, Bernard V, San Lucas FA, Allenson K, Capello M, Kim DU, et al. Surfaceome profiling enables isolation of cancer-specific exosomal cargo in liquid biopsies from pancreatic cancer patients. Ann Oncol. 2018;29(1):223–9.
109. Melo SA, Sugimoto H, O'Connell JT, Kato N, Villanueva A, Vidal A, et al. Cancer exosomes perform cell-independent microRNA biogenesis and promote tumorigenesis. Cancer Cell. 2014;26(5):707–21.
110. Xiang M, Zeng Y, Yang R, Xu H, Chen Z, Zhong J, et al. U6 is not a suitable endogenous control for the quantification of circulating microRNAs. Biochem Biophys Res Commun. 2014;454(1):210–4.
111. Heid CA, Stevens J, Livak KJ, Williams PM. Real time quantitative PCR. Genome Res. 1996;6(10):986–94.
112. Wang W, Ren P, Mardi S, Hou L, Tsai C, Chan KH, et al. Design of multiplexed detection assays for identification of avian influenza a virus subtypes pathogenic to humans by SmartCycler real-time reverse transcription-PCR. J Clin Microbiol. 2009;47(1):86–92.
113. Lockhart DJ, Dong H, Byrne MC, Follettie MT, Gallo MV, Chee MS, et al. Expression monitoring by hybridization to high-density oligonucleotide arrays. Nat Biotechnol. 1996;14(13):1675–80.
114. Mook S, Van't Veer LJ, Rutgers EJ, Piccart-Gebhart MJ, Cardoso F. Individualization of therapy using MammaPrint: from development to the MINDACT trial. Cancer Genomics Proteomics. 2007;4(3):147–55.
115. Paik S, Shak S, Tang G, Kim C, Baker J, Cronin M, et al. A multigene assay to predict recurrence of tamoxifen-treated, node-negative breast cancer. N Engl J Med. 2004;351(27):2817–26.
116. Cooperberg MR, Simko JP, Cowan JE, Reid JE, Djalilvand A, Bhatnagar S, et al. Validation of a cell-cycle progression gene panel to improve risk stratification in a contemporary prostatectomy cohort. J Clin Oncol. 2013;31(11):1428–34.

# Chapter 29
# Clinical RNA Sequencing: From Molecular Basis to Clinics

**Abstract** The development of RNA sequencing (RNA-Seq) technologies has significantly expanded the capabilities of RNA-based biomolecules for use in diagnosis, prognosis, and therapeutic strategies across various medical fields, including oncology and infectious disease management. RNA-Seq technology excels in detecting gene fusions and variably expressed transcripts associated with diseases, providing immediate and valuable insights. More significantly, RNA-Seq's ability to identify a wide range of RNA species paves the way for innovative clinical applications. One of the most promising uses is employing extracellular RNAs as non-invasive biomarkers for disease diagnosis, offering a less intrusive option compared to traditional methods. Concerted efforts in several key areas are further enhancing RNA-Seq's clinical impact: developing benchmark standards, refining assays for compatibility with clinical workflows, and ensuring the consistency and reproducibility of these assays. These initiatives aim to establish a robust framework for the broader adoption of RNA-Seq in clinical settings. By addressing these technical and regulatory challenges, researchers and clinicians can leverage RNA-Seq to unlock a new era of precision medicine, where RNA-based diagnostics and treatments are tailored to individual patient profiles, enhancing the efficacy and efficiency of medical interventions.

## 29.1 Introduction

RNA, a multifaceted and dynamic biomolecule, is integral to numerous biological functions and holds significant potential within the realm of molecular diagnostics. The ability of RNA to act as a diagnostic, prognostic, and therapeutic tool is becoming increasingly vital in various aspects of healthcare, including disease detection, prognostic evaluations, and informing therapeutic decisions. The ongoing technological advancements, particularly in next-generation sequencing (NGS), have dramatically enhanced the applications of RNA measurements in clinical settings. The evolution of NGS technologies has particularly facilitated the growth of RNA-Seq methods, which now allow for an exhaustive exploration of the vast array of RNA

 
P. E. Saw, E. Song, *RNA Therapeutics in Human Diseases*,
https://doi.org/10.1007/978-981-96-3041-7_29

species. This capability introduces unprecedented opportunities for medical diagnostics and treatment strategies. RNA-Seq provides a comprehensive analysis of the transcriptome, capturing a complete picture of RNA transcript variations within a cell at any given moment. This level of detail not only aids in identifying the subtle genetic changes associated with different forms of diseases, such as novel gene fusions, alternative splicing events, and changes in gene expression levels, but also enhances our understanding of disease mechanisms at a molecular level [1]. RNA-Seq stands out as a highly versatile platform technology, offering numerous advantages over traditional gene expression microarrays. One of the most significant benefits is its broader dynamic range, which allows for the detection of a wide spectrum of gene expression levels, from very low to very high, with greater accuracy and sensitivity. This capability is crucial for identifying biologically significant changes that might be missed by technologies with narrower ranges. Further enhancing its utility, RNA-Seq is adept at identifying specific genetic alterations, including single nucleotide variants (SNVs), insertions, deletions, and a variety of transcript isoforms, such as alternative splicing events and chimeric gene fusions. This precision enables researchers to uncover intricate details about gene regulation and expression that are often critical for understanding complex diseases like cancer, where such mutations can drive disease progression or confer drug resistance. Moreover, RNA-Seq's ability to identify chimeric gene fusions is particularly valuable, including those that involve previously unknown genes or transcripts. This feature is vital for discovering new diagnostic markers and therapeutic targets, opening up possibilities for novel treatments and interventions. Despite the commercial availability of RNA-Seq assays [2, 3], these initial offerings only scratch the surface of what could be a much wider application of RNA-Seq in clinical diagnostics. This chapter evaluates some current and future clinical uses of RNA-Seq, concentrating on differential expression, the detection of rare or fusion transcripts, and allele-specific expression, while also touching on emerging fields like pathogen detection and ncRNA measurement [4–6]. Figure 29.1 depicts the overview of the RNA-Seq process.

## 29.2 RNA Sequencing: Step-By-Step Analysis

### *29.2.1 Step 1: RNA Isolation*

The first and foundational step in transcriptome sequencing is the extraction of RNA from a biological sample, which could include cells, tissues, or fluids. The success of an RNA-Seq experiment heavily relies on the quality of the RNA extracted, as this directly affects the subsequent steps of library preparation and the overall sequencing results. To ensure that the RNA extracted meets the necessary standards for high-quality sequencing, it is routinely assessed using sophisticated equipment such as an Agilent Bioanalyzer. This instrument evaluates RNA quality by assigning an RNA Integrity Number (RIN), which ranges from 1 to 10. A RIN of 10 represents RNA that is intact and largely undegraded, while lower values

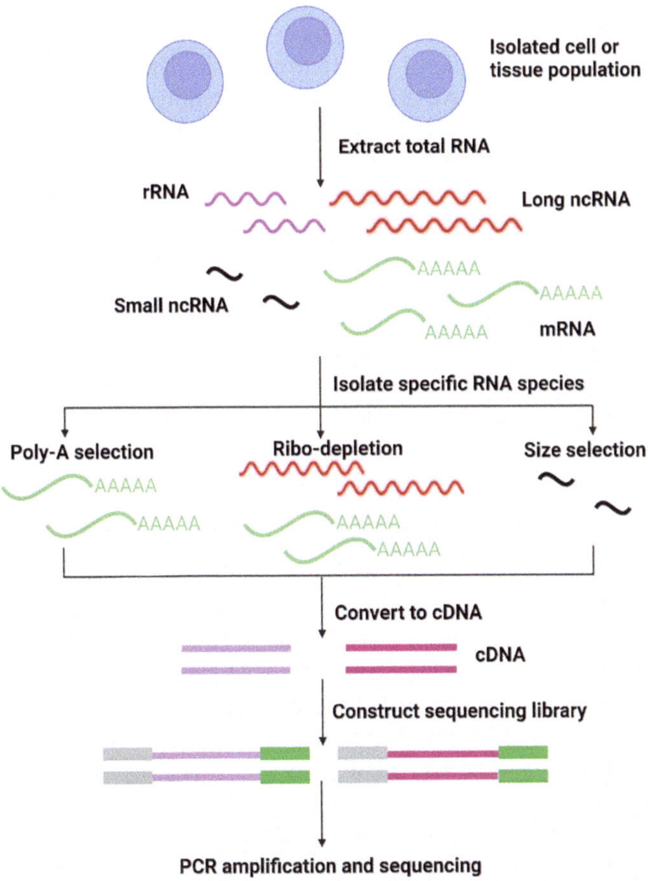


**Fig. 29.1** RNA-Seq process overview. The RNA-Seq process begins with the extraction of RNA from a biological sample, which could be derived from various sources such as cells or tissue specimens. This initial step is critical as the quality and integrity of the RNA extracted will significantly influence the subsequent analysis outcomes. Following extraction, the RNA undergoes a selection process to isolate specific subsets of RNA molecules, tailored to the research objectives. For instance, the poly-A selection method is employed to enrich mRNA populations by targeting polyadenylated transcripts, which are typically the focus in gene expression studies. Conversely, the ribo-depletion strategy is used to remove rRNAs, which constitute the majority of RNA in a cell but are generally not of interest in most RNA-Seq analyses. This step enhances the proportion of mRNA in the sample, thereby increasing the sensitivity and specificity of the subsequent sequencing. Once the desired RNA subset is isolated, it is converted into cDNA through a process called reverse transcription. This step is pivotal as DNA is more stable and easier to amplify than RNA, facilitating the next stages of the process. Next, sequencing adaptors are attached to the ends of the cDNA fragments. These adaptors are crucial for the sequencing process as they contain sequences that allow the cDNA fragments to bind to the sequencing instrument. The adaptors also contain barcodes that enable the identification of sequences from different samples after mixed sequencing runs. The final preparatory step involves PCR amplification of the cDNA, a process that increases the quantity of the cDNA to levels sufficient for high-throughput sequencing. This amplification step is essential to ensure that there is enough material for sequencing and that the data obtained are of high quality and depth. (Figure is drawn in BioRender.com)

indicate increasing levels of degradation, which could potentially compromise the accuracy and reliability of the RNA-Seq data. The RIN is determined through an analysis that resembles traditional gel electrophoresis. The Bioanalyzer uses this method to examine the relative abundances and integrity of the rRNA subunits, specifically the 28S and 18S rRNA bands. The ratio of these two bands serves as a critical indicator of RNA quality: a high 28S/18S ratio generally corresponds to high-quality RNA, which is crucial for constructing effective sequencing libraries. High-quality RNA ensures that the cDNA libraries used in sequencing are representative of the original RNA population, with minimal bias introduced by the degradation of specific RNA molecules. This is crucial for accurate transcript quantification and for the detection of low-abundance transcripts, which can be pivotal in understanding disease mechanisms, cellular responses, and developmental processes. Maintaining RNA integrity thus not only supports robust data but also enhances the reliability of interpretations drawn from the RNA-Seq experiment [7–9].

### 29.2.2 Step 2: Methods for RNA Library Preparation

Once RNA has been successfully isolated from a biological sample, the next crucial step in transcriptome sequencing is constructing the RNA-Seq library. This stage of the process is tailored according to the specific RNA species targeted for study and the NGS platform to be used. Each NGS platform may have slightly different requirements for library preparation, influencing the specific methods and reagents used. The RNA-Seq library construction process typically involves these four key steps (Fig. 29.2) [10, 11].

The meticulous construction of the RNA-Seq library is essential for the success of the sequencing experiment. It ensures that the library is compatible with the chosen NGS platform and that the data generated will be of high quality and suitable for answering the research questions at hand. Well-prepared libraries lead to efficient sequencing runs and reliable data for downstream analysis, impacting the overall effectiveness and accuracy of gene expression studies and other genomic research.

### 29.2.3 Step 3: Selection of RNA Species for RNA-Seq Libraries

Before constructing RNA-Seq libraries, one of the critical decisions involves choosing an appropriate library preparation method that best suits the study's goals. This choice is pivotal because it dictates which RNA species from the "total" RNA sample are either enriched or depleted, enhancing the focus on the RNA types of interest. The typical RNA types under consideration include rRNA, pre-mRNA, mRNA, and various classes of ncRNA. In most cellular samples, rRNA is the predominant component, making up over 95% of all cellular RNA. This abundance can pose a significant challenge for RNA-Seq because rRNA can saturate the sequencing reads

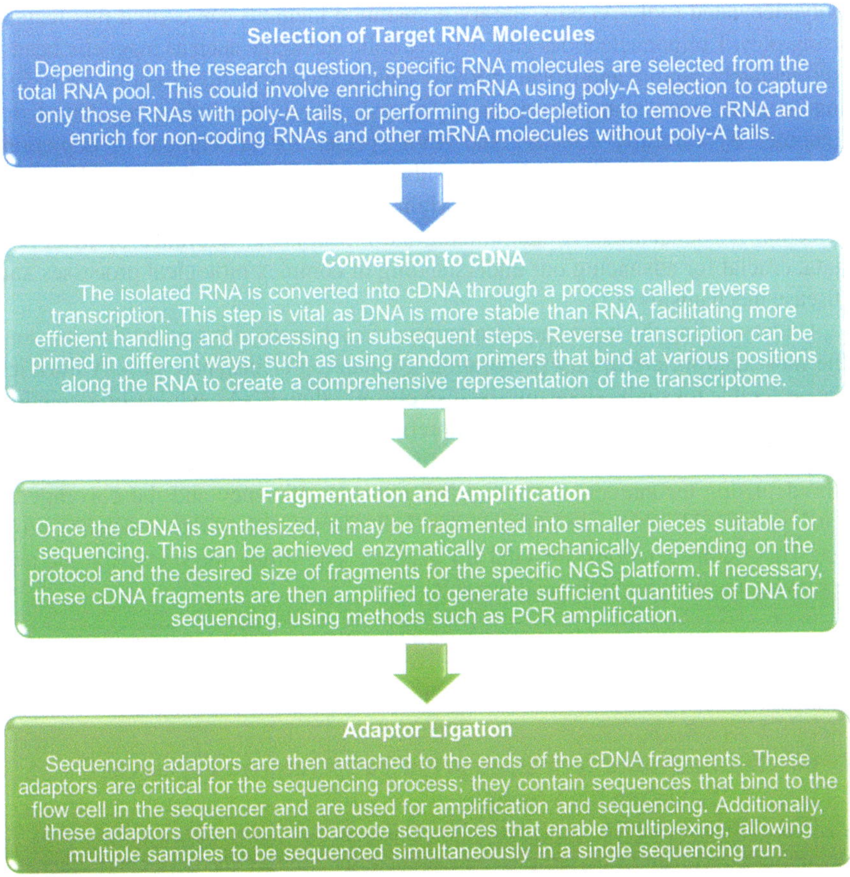


**Fig. 29.2** The four consecutive steps in the construction of RNA-Seq library

if not removed, severely reducing the sequence coverage of other RNA types that are present in much smaller quantities. Consequently, the effective profiling of the transcriptome requires strategies either to remove rRNA or to selectively enrich the RNA species of interest.

In order to gain high-quality data, there are some additional steps that one might need to carry out, for instance, the removal of rRNA from the sample. This is to prevent rRNA from overwhelming the sequencing data; many protocols are designed to selectively remove these molecules from the sample. This step is crucial as it enhances the visibility of other RNA species during sequencing, especially those that are less abundant but potentially of high biological significance. Techniques for rRNA depletion include using commercially available kits such as RiboMinus (Life Technologies) and RiboZero (Epicentre). These kits selectively bind and remove rRNA, thereby enriching the remaining RNA pool in the sample. Another common approach is to enrich mRNA in the library. Since most mRNA molecules have polyadenylated (poly-A) tails, library preparation methods can exploit this feature using poly-T oligonucleotides attached to a solid substrate, such as magnetic beads, to selectively capture and isolate mRNA. This method is highly effective for focusing on gene expression analysis as it primarily targets the coding regions of the genome. By carefully selecting the RNA species for enrichment or depletion, researchers can tailor their RNA-Seq libraries to their specific research needs, enabling more detailed and focused insights into the transcriptome. This strategic selection is foundational in maximizing the efficacy and relevance of RNA-Seq data, crucial for advancing our understanding of complex biological processes and disease mechanisms [12].

Understanding the biases and limitations of each method is critical for choosing the most appropriate library preparation technique. For instance, poly-A libraries are preferable for studies focusing exclusively on coding RNAs, while ribo-depletion libraries offer a better solution for comprehensive studies that include ncRNAs and unmodified pre-mRNAs. Additionally, it is important to consider the variations in ribo-depletion techniques, such as rRNA removal efficiency and the coverage of small genes [13], which can impact the overall effectiveness of the sequencing project. These considerations are vital for accurately profiling the transcriptome and achieving meaningful biological insights.

**Box 29.1 Innovative Techniques to Enrich for Targeted Regions of Interest Within the RNA Sample**

In the context of RNA sequencing, the ability to enrich for targeted regions of interest within an RNA sample can significantly enhance the depth and specificity of sequencing data. This enrichment is particularly useful for focusing on specific genes or transcript variants, enabling detailed studies of gene expression, splicing events, and post-transcriptional modifications. Several innovative techniques have been developed to achieve this targeted enrichment, each with unique advantages and applications, as summarized below.

*PCR-Based Strategies*

One of the most straightforward methods for enriching specific RNA sequences involves the use of PCR [14]. This technique can amplify desired RNA regions that have been converted into cDNA, making it possible to focus sequencing efforts on these targets. PCR-based enrichment is highly specific and can be used to amplify multiple regions simultaneously by designing specific primers for each region of interest. This method is ideal for applications where the regions of interest are well defined and limited in number.

*Hybrid Capture*

This technique involves the use of biotinylated RNA or DNA probes that hybridize specifically to the regions of interest in the RNA sample [15]. Once

(continued)

**Box 29.1** (continued)
hybridization occurs, the RNA–DNA hybrids are captured using streptavidin-coated magnetic beads. The non-hybridized RNA is washed away, leaving only the targeted regions bound to the probes. This method is highly scalable and can be used to enrich large numbers of target regions simultaneously, making it suitable for more complex transcriptomic studies.

*In-Solution Capture*

Similar to hybrid capture, in-solution capture uses oligonucleotide probes that bind to target RNA sequences in solution [16]. This method eliminates the need for solid-phase support (like magnetic beads) and can be performed entirely in a single tube, reducing sample loss and simplifying the workflow. In-solution capture is particularly useful for capturing large sets of targets and can be highly efficient and cost-effective for medium to high-throughput applications.

*Molecular Inversion Probes (MIPs)*

MIPs are a powerful tool for the targeted capture of RNA sequences. These probes are designed to have regions that are complementary to the target RNA sequence, with a gap between the two regions [17]. During the assay, the probes circularize around the target RNA, and the gap is filled in by a polymerase reaction. This circularization process selectively captures the target RNA, which can then be linearized, amplified, and sequenced. MIPs are highly specific and can be designed to capture multiple targets in a single experiment, making them efficient for both focused and broad-scale studies.

Each of these techniques offers distinct benefits and can be selected based on the specific requirements of the research project, such as the number of targets, the size of the regions of interest, and the complexity of the RNA sample. By employing these targeted enrichment strategies, researchers can enhance the resolution and accuracy of RNA-Seq data, providing deeper insights into RNA function and regulation in health and disease [18].

## 29.3 Selection of Small RNA Species for Library Preparation

In addition to the library preparation protocols previously detailed, specific methodologies have been tailored for isolating and profiling small RNA species, which are pivotal regulators of gene expression. These small RNAs, such as miRNA, siRNA, and piRNA, are typically short in length (ranging from 15 to 30 nucleotides), not polyadenylated, and often present in low abundance within cells. Their unique characteristics necessitate specialized strategies for effective profiling [19].

Due to their short length and lack of a poly-A tail, small RNAs cannot be isolated using the same poly-T oligo capture methods used for mRNA. Instead, library preparation protocols for small RNAs often involve the use of adapters that can ligate directly to the RNA molecules. These adapters are specially designed to attach to the 5′ and 3′ ends of the small RNAs, irrespective of their sequence. The efficiency of adapter ligation is crucial for the successful library preparation of small RNAs. This process must be highly efficient and accurate to ensure that even the scarcest small RNAs are captured and represented in the library. Special enzymes are used that recognize the RNA duplex formed by the small RNA and the adapter, facilitating a stable and effective ligation. Once adapters are ligated, the small RNAs are reverse transcribed into cDNA. This is followed by PCR amplification, which not only increases the quantity of the library but also introduces necessary sequencing adapters and barcode sequences. This step is critical as it prepares the small RNA library for high-throughput sequencing. Following amplification, size selection steps are often employed to ensure that only the cDNA corresponding to the original small RNAs, now flanked by adapters, is taken forward to sequencing. This step helps to eliminate adapter dimers or other non-specific products that could interfere with sequencing. Profiling small RNAs is essential due to their significant roles in gene regulation, developmental timing, pathogen response, and gene silencing mechanisms. Techniques that effectively capture the diversity and expression levels of small RNAs can provide insights into cellular processes at a regulatory level, revealing mechanisms of disease progression and potential therapeutic targets [20].

## 29.4 cDNA Synthesis in RNA-Seq Library Preparation

In the landscape of RNA-Seq library preparation, the transformation of RNA into cDNA stands as a fundamental step, critical for the success of the sequencing process. This necessity stems from the fact that most NGS technologies are optimized for DNA rather than RNA. The cDNA synthesis process is pivotal not only because it makes RNA sequences amenable to these technologies but also because it influences the quality and type of data obtained from sequencing [21–24].

### *29.4.1 Traditional cDNA Synthesis Protocols*

Traditional protocols for cDNA synthesis involve creating libraries where the strand orientation of the original RNA molecules is not preserved [21]. In these protocols, reverse transcription is followed by the synthesis of the second cDNA strand, thus forming double-stranded cDNA that represents both the mRNA and its complementary sequence. While this approach is efficient and maximizes the yield of cDNA, it results in sequencing reads that do not retain information about which strand of DNA the RNA was transcribed from. The loss of strand orientation can be a significant drawback, particularly when studying complex genomes or transcriptomes with high densities of overlapping genes. Without strand-specific information, it becomes challenging to determine whether a transcript originates from the sense or antisense strand, complicating the analysis of regulatory RNA elements, antisense transcripts, and overlapping genes located on opposite strands. This information is crucial for accurately mapping read data, especially for identifying overlapping transcripts and for de novo transcript discovery where annotation might not be available [25].

### 29.4.2 Strand-Specific cDNA Synthesis

To overcome these limitations, strand-specific or strand-preserving cDNA synthesis methods have been developed. These methods employ strategies such as marking the RNA with adapters that indicate the directionality of transcription or selectively degrading one strand of the cDNA after reverse transcription to retain only the strand that directly corresponds to the mRNA. Strand-specific protocols enhance the ability to accurately assign sequencing reads to their respective genomic locations and orientations, providing a more detailed and precise understanding of transcriptome complexity. They are particularly valuable in research focused on gene regulation, non-coding RNA functions, and transcriptome dynamics, where understanding the precise biogenesis and interplay of transcripts is essential [26].

### 29.4.3 Incorporation of Deoxy-UTP (dUTP)

A more widely adopted approach for maintaining strand orientation uses the incorporation of a chemical label, specifically deoxy-UTP (dUTP), during the synthesis of the second-strand cDNA. This labeled second-strand cDNA can then be selectively degraded using enzymes, effectively eliminating these strands post-library construction and ensuring that only the first-strand cDNA sequences are sequenced [21]. This method helps in accurately identifying which strand a given read originates from, preserving the strandedness of the RNA. While this dUTP method is preferred for its efficiency and accuracy in preserving strand information, it is not without its pitfalls. Specifically, antisense transcripts located near highly expressed genes must be interpreted with caution due to the potential for a small percentage of reads (approximately 1%) that may originate from the opposite strand as a result of sequencing errors or other biases [27]. This highlights the necessity for careful analysis and validation of RNA-Seq data, especially when using these advanced library preparation techniques.

## 29.5 Multiplexing in RNA-Seq Library Construction

Multiplexing is a crucial strategy in RNA-Seq library preparation that significantly enhances cost-efficiency and scalability. It involves indexing, or tagging, multiple samples with unique sequence identifiers (barcodes) that allow them to be pooled and sequenced simultaneously in a single sequencing run [28]. This approach maximizes the utility of sequencing resources by distributing the high-throughput capabilities of modern sequencing platforms across multiple samples. By combining multiple samples in one sequencing lane, multiplexing reduces the cost per sample and increases the throughput of sequencing projects. This is especially valuable in large-scale studies involving numerous samples, such as population-level genetic studies or comprehensive cancer genome projects. Advanced platforms like the Illumina HiSeq 2500 are capable of generating a vast number of reads, with up to 750 million paired-end reads per lane. This high capacity makes these platforms ideal for multiplexing, as they can accommodate the simultaneous sequencing of many differentially indexed samples without compromising the depth of coverage for individual samples. While the high sequencing depth achievable with modern platforms offers detailed genomic insights, it can result in over-sequencing, particularly for samples of lower complexity. In these cases, generating an excessive number of reads does not necessarily improve the quality of data but rather increases sequencing costs and data management burdens. Multiplexing addresses this issue by enabling the sequencing depth to be balanced and tailored to the complexity of each sample. Researchers can allocate more sequencing power to complex samples while reducing it for simpler ones, optimizing the use of sequencing capacity. Effective multiplexing strategies prevent the wasteful over-sequencing of samples. By judiciously assigning read capacity based on the needs of each sample, researchers can avoid unnecessary data accumulation, focusing resources on generating meaningful, interpretable data. This strategic allocation not only saves costs but also streamlines data analysis, making it more manageable and efficient [29].

To facilitate multiplexing, unique 6-base pair indices, or "barcodes," are added to each RNA-Seq library. These barcodes make it possible to mix multiple samples in a single sequencing batch and subsequently identify the source of each read during data analysis. Depending on the complexity of the samples and the goals of the study, a single lane can effectively accommodate between 2 and 20 samples, achieving sufficient coverage for robust transcriptome analysis [30, 31]. For samples where the goal is to quantify moderately to highly abundant transcripts, approximately 30–40 million reads are necessary. In contrast, to capture the full breadth of sequence diversity within complex transcript libraries, which includes rare and lowly expressed transcripts, as many as 500 million reads might be required [32]. Despite its advantages, multiplexing requires careful planning and execution. The choice of barcodes or adapters, the balancing of sample pool compositions, and the precise calculation of loading concentrations are critical to avoid issues such as barcode cross-talk, sample imbalance, or unequal sequencing depth across the pooled samples. Additionally, robust bioinformatics tools are needed to demultiplex the pooled data accurately and to ensure that the sequence reads are correctly assigned to their respective samples post-sequencing. Therefore, it is crucial to carefully consider the required depth of sequencing needed to confidently address the research questions posed, while making efficient use of NGS capabilities. This approach not only optimizes financial resources but also enhances the overall efficiency of the sequencing process.

## 29.6 Using Extracellular RNAs (exRNAs) to Monitor Diseases

The exploration of extracellular RNAs (exRNAs) in biofluids as a method for disease monitoring is an expanding field within diagnostic research. exRNAs, which are emitted by all cellular types throughout the body, are safeguarded from degradation in the extracellular environment by their encapsulation in extracellular vesicles or their association with RBPs and lipoproteins [33–37]. Utilizing exRNA as a diagnostic tool is attractive due to its non-invasive nature; biofluids like plasma are easier and less invasive to collect compared to tissue samples, allowing for more frequent and longitudinal monitoring. Transcripts from various tissues, including neuronal cells from the brain, can be detected in plasma, indicating the potential wide-reaching applications of exRNA testing [38]. One of the primary challenges with this method, however, is the lack of tissue specificity since biofluids contain exRNAs from multiple sources. The extent to which specific RNA profiles can be detected may be influenced by the size and proximity of the tissue or organ to the collection site. For instance, plasma is known to predominantly contain transcripts from the liver and heart, whereas saliva typically has higher levels of transcripts from the salivary glands and the esophagus.

To improve tissue-specific capture of exRNAs, recent innovations include techniques that isolate extracellular vesicles originating from particular tissues [39, 40]. This targeted approach helps in refining the detection and analysis of exRNAs, enhancing the diagnostic capabilities of this promising technology. The utilization of exRNAs for cancer detection shares similarities with extracellular DNA analysis, such as the ability to identify and quantify mutations in RNA transcripts, provided these mutations are transcribed. A key advantage of exRNAs is their abundance compared to DNA, which could potentially increase the sensitivity of tests. Additionally, variations in exRNA expression levels may indicate tissue or organ damage or disease presence, offering insights not possible through DNA analysis alone. The exRNA spectrum includes a vast array of mRNAs and various regulatory RNAs, all of which can be extensively profiled using RNA-Seq to assess their roles in disease.

The growing interest in exRNAs for non-invasive disease monitoring is evident in their use as markers for critical disease pathways and gauges of treatment effectiveness. Companies like Exosome Diagnostics are at the forefront, developing exRNA-based diagnostic tests under the Clinical Laboratory Improvement Amendments (CLIA) framework. One of their pioneering products, ExoDx Lung (ALK), detects EML4–ALK fusion transcripts in plasma-derived exosomes from patients with non-small cell lung cancer (NSCLC), showcasing the clinical potential of this technology [41]. Furthermore, the exploration of circulating RNA is not limited to cancer diagnostics but extends to prenatal health monitoring, where changes in fetal exRNA profiles provide valuable insights [42]. The US National Center for Advancing Translational Sciences (NCATS) has propelled this field forward with the inception of the Extracellular RNA Communication Consortium [43]. This initiative aims to refine exRNA diagnostic applications by supporting research to map exRNA profiles in healthy and diseased states [44, 45]. As support for exRNA research continues to grow, significant advancements in understanding and utilizing these biomolecules are expected, which will enhance their reliability and utility in clinical diagnostics. This progression promises to overcome current challenges such as variability in exRNA detection and to establish standardized methods for their analysis.

## 29.7 RNA-Seq for Infectious Disease Diagnosis

RNA-based diagnostics are pivotal in detecting a plethora of clinically significant RNA viruses, including HIV, Ebola, West Nile, dengue, hepatitis A, D, and E, coxsackie, influenza, as well as coronaviruses like SARS and MERS. The widespread clinical integration of qRT-PCR assays for viral identification and typing is well-documented [46–49]. These diagnostics are poised for evolution into RNA-based sequencing assays, which would enhance their breadth and precision. For instance, unbiased non-targeted metagenomic RNA-Seq has been effectively employed for the direct detection of influenza virus RNA within respiratory fluids and concurrently identifying other viral pathogens in certain instances [50]. This method's capacity for broad viral detection underscores its utility in comprehensive pathogen diagnostics. In public health, RNA-Seq proved instrumental during the 2014 Ebola outbreak in West Africa, tracing the virus's origin and transmission pathways [51].

Moreover, RNA-based amplicon sequencing is emerging as a crucial tool for assessing viral quasi-species—a mixed allele population seen in viruses like the hepatitis C virus (HCV) and HIV. This approach is essential in clinical settings to ascertain the presence and concentration of drug-resistance mutations, which may represent a minor fraction within a larger viral population, thus guiding patient treatment strategies [52, 53]. Despite these advancements, the regular use of RNA-based diagnostics for infectious diseases in clinical practice predominantly remains confined to qRT-PCR assays. The potential of RNA-Seq and other RNA-based methodologies lies in their ability to offer more detailed, sensitive, and comprehensive viral analysis, which is critical for effective disease management and control.

## 29.8 Microbial Exogenous Small RNA

The detection of a wide variety of exogenous RNAs from non-human sources in human plasma has unveiled complex interactions between hosts and various external entities such as microbiomes, dietary sources, and environmental factors [54, 55]. These microbial exogenous RNAs, whether or not they are enclosed in outer-membrane vesicles [56], are currently under intense investigation, especially in the context of infectious diseases [57]. They represent a promising frontier for the identification of novel diagnostic targets. Ongoing comprehensive studies—including in vitro, ex vivo, and in vivo analyses—are dedicated to cataloging small RNAs produced by pathogens. This effort aims to establish a detailed reference database of exogenous RNA signals, which could prove instrumental for future clinical studies related to infections [58–60]. For instance, ex vivo research on *Neisseria meningitidis* infections in human blood has identified numerous small RNAs linked with bacterial infections [61]. This discovery paves the way for these small RNAs to be considered as potential biomarkers for such conditions. Furthermore, extensive studies into the microRNAome of *Mycobacterium tuberculosis* have uncovered a variety of biomarkers that are now being examined for their diagnostic capabilities [62]. These biomarkers hold the potential not only for disease diagnosis but also for phenotype–genotype predictive diagnostics, such as determining the presence of multi-drug resistance [63]. The exploration of these exogenous RNAs could significantly enhance our diagnostic approaches, particularly in identifying and combating infectious diseases effectively.

### *29.8.1 Pathogen mRNA*

The use of microbial mRNA as a diagnostic marker offers promising enhancements in the detection of infections, particularly those at low levels such as bacteremia and infections in cerebrospinal fluid. This method focuses on the expression levels of mRNA, which can reveal active infection sites and provide a more accurate prediction of disease presence than genomic detection alone. To improve diagnostic accuracy and predict the likelihood of cervical cancer more reliably, RNA-based diagnostics have been developed to detect early oncoproteins E6/E7 mRNA of HPV. Unlike DNA detection, mRNA levels of E6/E7 serve as a surrogate for active infection, thereby offering a better predictive marker for determining the risk of developing cervical cancer [64]. This method exemplifies the shift from simple genomic detection to a more dynamic and informative RNA-based approach, enhancing the predictive accuracy and clinical relevance of infectious disease diagnostics. Through the measurement of microbial mRNA, healthcare providers can achieve earlier and more precise infection detection, which is critical for effective treatment and better patient outcomes.

### 29.8.2 Host RNA

The utilization of mRNA signatures from host responses offers promising potential for monitoring specific infections effectively. For example, the upregulation of host immune factors such as interferon beta 1 (IFNB1), interferon lambda 2 (IFNL2), and interferon lambda 3 (IFNL3) has been identified in patients with genotype 3 HCV infections [65, 66]. This genotype is notably associated with accelerated liver fibrosis and presents a higher risk for developing hepatocellular carcinoma compared to infections caused by other HCV genotypes. Moreover, the interactions between host microRNAs and genes during an infection response are emerging as valuable sources of diagnostic biomarkers. These markers not only help in identifying specific infections but also play a crucial role in differentiating between active and latent states of infections. A pertinent example is the use of microRNA-gene markers to distinguish between latent and active tuberculosis infections caused by *Mycobacterium tuberculosis* [67]. Despite the potential of these small RNA biomarkers, several challenges hinder their development and validation in infection assays. These include issues related to sensitivity and specificity, as well as the complex task of normalizing small RNA concentrations in clinical samples, which can vary significantly across different conditions, tissues, and individuals [68]. These variations make standardization a critical and challenging aspect of diagnostic development. Nevertheless, RNA-Seq stands out as a valuable orthogonal method to traditional genomics-based diagnostics in clinical microbiology. It not only complements DNA-based testing by providing additional layers of information through the assessment of RNA levels but also enhances the ability to pinpoint and monitor dynamic changes in pathogen and host responses over time. This capability makes RNA-Seq a potent tool in the ongoing enhancement of infectious disease diagnostics, allowing for more precise, real-time disease management.

## 29.9 Integrating DNA and RNA-Seq in Oncology

Recent developments in clinical sequencing have highlighted the considerable advantages of an integrated approach that combines RNA-Seq with DNA-seq. This comprehensive strategy amplifies the capabilities of genetic analysis, providing a more nuanced understanding of the genetic underpinnings of disease. By incorporating RNA-Seq with DNA-seq, clinicians and researchers can achieve a more detailed analysis of mutant alleles. DNA sequencing identifies variations at the DNA level, while RNA-Seq contributes additional layers of context by showing how these mutations influence gene expression. This combination allows for the assessment of the functional effects of genetic mutations, enhancing our understanding of their roles in disease processes. The integrated approach is particularly effective in identifying both known and novel gene fusions. DNA sequencing can pinpoint the location of potential fusion events, but RNA-Seq provides crucial evidence of the expression of these fusions, confirming their presence and biological relevance. This is vital for diseases like cancer, where gene fusions often play a critical role in tumor development and progression and may serve as therapeutic targets or diagnostic markers. While DNA sequencing can suggest the possibility of alternative splicing events, RNA-Seq directly observes these events by identifying the actual mRNA splice forms present in the cells. This capability makes RNA-Seq invaluable for recognizing splice variants that might be overlooked by DNA-seq alone, providing a more complete picture of gene expression and regulation. The integration of RNA-Seq with DNA sequencing provides a holistic view of the genome and transcriptome, offering insights into how genetic variations affect gene function and contribute to disease. This approach not only supports the identification and characterization of genetic alterations but also enhances the ability to link these changes to specific pathogenic mechanisms and potential therapeutic interventions [69–71]. All types of sequencing are summarized in Fig. 29.3.

## 29.10 Challenges and Future Directions

Despite these promising applications, several challenges remain in the widespread clinical implementation of integrated sequencing strategies. These include the complexity of data analysis, the need for robust bioinformatics tools to handle multimodal data, and the regulatory hurdles associated with validating these new methods for clinical use. However, the successful application of these techniques in research settings across various fields suggests substantial potential for their future integration into routine clinical practice. As these methods continue to evolve and improve, they are expected to offer unprecedented precision in the diagnosis, prognosis, and personalized treatment planning across a broad spectrum of diseases, heralding a new era of precision medicine based on a comprehensive genomic and transcriptomic understanding. RNA-Seq stands poised to transform clinical diagnostics with its comprehensive capacity to assess global gene transcript levels and various RNA species. This technique's broad applicability may significantly impact the diagnosis and management of numerous diseases. While recent developments have begun to standardize RNA-Seq through benchmark standards for technical and analytical practices, further proficiency and validation steps are essential to enhance its clinical utility.

To date, the push toward clinical application has involved the development of benchmark standards that address the need for consistent, reproducible, and precise results. However, achieving these standards in a clinical setting requires overcoming substantial hurdles, particularly concerning test proficiency and validation. Future paths toward clinical integration may likely involve transitioning from broad-spectrum discovery assays to targeted RNA-Seq applications. These targeted assays would be designed to detect specific RNA signatures indicative of particular conditions, much like how microarrays and qRT-PCR are currently used but with potentially greater sensitivity and comprehensiveness. The establishment of fixed RNA-Seq protocols, akin to the fixed nature of probe sets in microarrays or

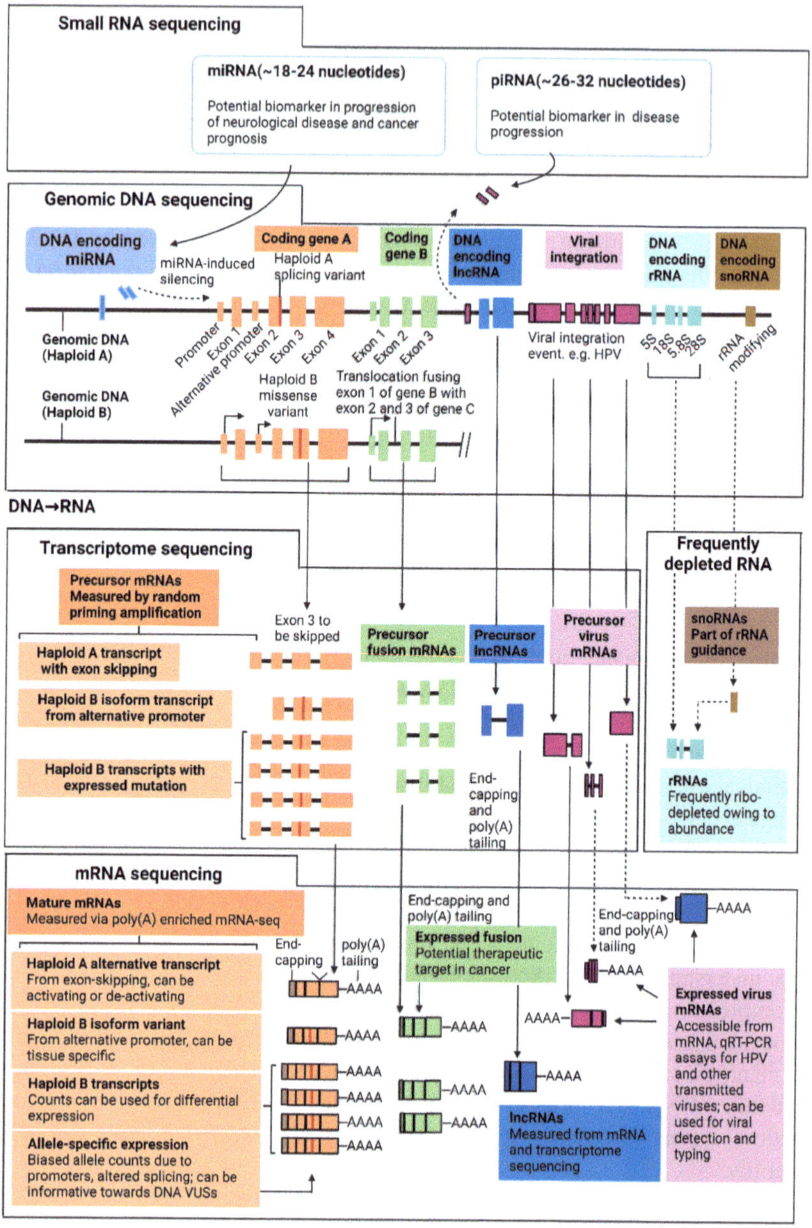


**Fig. 29.3** Scope of RNA species identification through RNA sequencing techniques. Various RNA-Seq methods facilitate the analysis of a wide range of clinically important RNA species. For instance, small RNA deep sequencing targets numerous small non-coding RNAs via size selection, including miRNAs and piRNAs. To measure precursor RNAs, random primer amplification is utilized, and oligo(dT) primers are employed for isolating polyadenylated transcripts. Furthermore, RNA-Seq enables the identification and quantification of alternative transcripts, chimeric gene fusion transcripts, and viral RNA transcripts, along with assessments for allele-specific expression. This comprehensive detection capability underscores the versatility of RNA-Seq in capturing a broad spectrum of RNA types that are pivotal in clinical diagnostics. (Figure is adapted from [72] PMC Open Access Subset license)

qRT-PCR, might offer a faster route to clinical application. Simplifying and standardizing RNA-Seq analysis to deliver reliable, reproducible data is crucial. As such, ongoing studies and trials are necessary to confirm RNA-Seq's analytical validity and operationalize its implementation in clinical labs.

## 29.11 Conclusion

In conclusion, RNA-Seq stands as a formidable platform for deciphering the complex and dynamic molecular landscape of human diseases. Its potential to revolutionize clinical practice hinges on rigorous efforts to establish and demonstrate its analytical validity. These efforts are essential to ensure that RNA-Seq can be reliably integrated into the framework of modern molecular diagnostics, where it is poised to significantly influence therapeutic decision-making and enhance patient outcomes. Before RNA-Seq can be broadly adopted in clinical settings, it is crucial to demonstrate that it can consistently produce accurate, reproducible, and reliable data. This involves rigorous testing to validate that RNA-Seq can detect, quantify, and differentiate various RNA species and their modifications across different biological and clinical conditions. To achieve reliable results that are comparable across different laboratories and studies, standardization of RNA-Seq protocols is imperative. This includes uniform procedures for sample collection, RNA extraction, library preparation, sequencing, and data analysis. Standardization ensures that RNA-Seq results are consistent and reproducible, key criteria for clinical diagnostics. RNA-Seq technologies must be validated not only in terms of laboratory performance but also regarding their clinical relevance. This means proving that RNA-Seq can provide actionable insights that improve clinical decision-making, such as identifying disease subtypes, predicting response to treatment, or monitoring disease progression and treatment efficacy. Navigating the regulatory landscape is another critical step toward the clinical adoption of RNA-Seq. Gaining regulatory approval requires demonstrating that the technology meets stringent criteria for safety, efficacy, and quality. Compliance with regulatory standards helps to assure healthcare providers and patients of the reliability and clinical utility of RNA-Seq-based diagnostics. As RNA-Seq is a complex and rapidly evolving technology, ongoing training and education for medical professionals are essential. Clinicians must understand the capabilities, limitations, and appropriate contexts for the use of RNA-Seq to make informed decisions about its application in patient care. Finally, integrating RNA-Seq into existing clinical workflows presents logistical and infrastructural challenges. This integration involves ensuring that the sequencing facilities are adequately equipped and that there are robust systems for data management and analysis, which are crucial for handling the large volumes of data generated by RNA-Seq. The journey toward the widespread clinical adoption of RNA-Seq will involve navigating these complexities with a focused commitment to standardization, validation, and regulation. By addressing these challenges, RNA-Seq can fulfill its potential as a cornerstone of precision medicine, transforming our approach to diagnosing and treating a wide range of diseases.

# References

1. Mortazavi A, Williams BA, McCue K, Schaeffer L, Wold B. Mapping and quantifying mammalian transcriptomes by RNA-Seq. Nat Methods. 2008;5(7):621–8.
2. Doebele RC, Davis LE, Vaishnavi A, Le AT, Estrada-Bernal A, Keysar S, et al. An oncogenic NTRK fusion in a patient with soft-tissue sarcoma with response to the tropomyosin-related kinase inhibitor LOXO-101. Cancer Discov. 2015;5(10):1049–57.
3. Sonu RJ, Jonas BA, Dwyre DM, Gregg JP, Rashidi HH. Optimal molecular methods in detecting p190 (BCR-ABL) fusion variants in hematologic malignancies: a case report and review of the literature. Case Rep Hematol. 2015;2015:458052.
4. Cech TR, Steitz JA. The noncoding RNA revolution-trashing old rules to forge new ones. Cell. 2014;157(1):77–94.
5. Wang Z, Gerstein M, Snyder M. RNA-Seq: a revolutionary tool for transcriptomics. Nat Rev Genet. 2009;10(1):57–63.
6. Luo S, Wang L, Xiao Y, Cao C, Liu Q, Zhou Y. Single-cell RNA-sequencing integration analysis revealed immune cell heterogeneity in five human autoimmune diseases. BIO Integr. 2023;4(4):145–59.
7. Tomita H, Vawter MP, Walsh DM, Evans SJ, Choudary PV, Li JZ, et al. Effect of agonal and postmortem factors on gene expression profile: quality control in microarray analyses of postmortem human brain. Biol Psychiatry. 2004;55:346–52.
8. Thompson KL, Pine PS, Rosenzweig BA, Turpaz Y, Retief J. Characterization of the effect of sample quality on high density oligonucleotide microarray data using progressively degraded rat liver RNA. BMC Biotechnol. 2007;7:57.
9. Rudloff U, Bhanot U, Gerald W, Klimstra DS, Jarnagin WR, Brennan MF, et al. Biobanking of human pancreas cancer tissue: impact of ex-vivo procurement times on RNA quality. Ann Surg Oncol. 2010;17(8):2229–36.
10. Head SR, Komori HK, LaMere SA, Whisenant T, Van Nieuwerburgh F, Salomon DR, et al. Library construction for next-generation sequencing: overviews and challenges. Biotechniques. 2014;56(2):61–4, 6, 8, passim.
11. Wu Y, Xiao N, Li Y, Gao Q, Ning Y, Yu L, et al. Identification and fine mapping of qPBR10-1, a novel locus controlling panicle blast resistance in Pigm-containing P/TGMS line. Mol Breed. 2021;41(12):75.
12. Christodoulou DC, Gorham JM, Herman DS, Seidman JG. Construction of normalized RNA-seq libraries for next-generation sequencing using the crab duplex-specific nuclease. Curr Protoc Mol Biol. 2011;Chapter 4:Unit4.12.
13. Huang R, Jaritz M, Guenzl P, Vlatkovic I, Sommer A, Tamir IM, et al. An RNA-Seq strategy to detect the complete coding and non-coding transcriptome including full-length imprinted macro ncRNAs. PLoS One. 2011;6(11):e27288.
14. Pellegrino M, Sciambi A, Yates JL, Mast JD, Silver C, Eastburn DJ. RNA-Seq following PCR-based sorting reveals rare cell transcriptional signatures. BMC Genomics. 2016;17(1):361.
15. Gaudin M, Desnues C. Hybrid capture-based next generation sequencing and its application to human infectious diseases. Front Microbiol. 2018;9:2924.
16. Singh RR. Target enrichment approaches for next-generation sequencing applications in oncology. Diagnostics (Basel). 2022;12(7):1539.
17. Lin S, Wang W, Palm C, Davis RW, Juneau K. A molecular inversion probe assay for detecting alternative splicing. BMC Genomics. 2010;11(1):712.
18. Querfurth R, Fischer A, Schweiger MR, Lehrach H, Mertes F. Creation and application of immortalized bait libraries for targeted enrichment and next-generation sequencing. Biotechniques. 2012;52(6):375–80.
19. Benesova S, Kubista M, Valihrach L. Small RNA-sequencing: approaches and considerations for miRNA analysis. Diagnostics (Basel). 2021;11(6):964.

20. Morin RD, Zhao Y, Prabhu A-L, Dhalla N, McDonald H, Pandoh P, et al. Preparation and analysis of microRNA libraries using the Illumina massively parallel sequencing technology. Methods Mol Biol. 2010;650:173–99.
21. Parkhomchuk D, Borodina T, Amstislavskiy V, Banaru M, Hallen L, Krobitsch S, et al. Transcriptome analysis by strand-specific sequencing of complementary DNA. Nucleic Acids Res. 2009;37(18):e123.
22. Vivancos AP, Güell M, Dohm JC, Serrano L, Himmelbauer H. Strand-specific deep sequencing of the transcriptome. Genome Res. 2010;20(7):989–99.
23. Mills JD, Kawahara Y, Janitz M. Strand-specific RNA-Seq provides greater resolution of transcriptome profiling. Curr Genomics. 2013;14(3):173–81.
24. Lister R, O'Malley RC, Tonti-Filippini J, Gregory BD, Berry CC, Millar AH, et al. Highly integrated single-base resolution maps of the epigenome in Arabidopsis. Cell. 2008;133(3):523–36.
25. Synthesis of complementary DNA. Nat Methods. 2005;2(2):151–2.
26. Levin JZ, Yassour M, Adiconis X, Nusbaum C, Thompson DA, Friedman N, et al. Comprehensive comparative analysis of strand-specific RNA sequencing methods. Nat Methods. 2010;7(9):709–15.
27. Zeng W, Mortazavi A. Technical considerations for functional sequencing assays. Nat Immunol. 2012;13(9):802–7.
28. Cheng J, Liao J, Shao X, Lu X, Fan X. Multiplexing methods for simultaneous large-scale transcriptomic profiling of samples at single-cell resolution. Adv Sci. 2021;8(17):2101229.
29. Smith AM, Heisler LE, St Onge RP, Farias-Hesson E, Wallace IM, Bodeau J, et al. Highly-multiplexed barcode sequencing: an efficient method for parallel analysis of pooled samples. Nucleic Acids Res. 2010;38(13):e142.
30. Birney E, Stamatoyannopoulos JA, Dutta A, Guigó R, Gingeras TR, Margulies EH, et al. Identification and analysis of functional elements in 1% of the human genome by the ENCODE pilot project. Nature. 2007;447(7146):799–816.
31. Blencowe BJ, Ahmad S, Lee LJ. Current-generation high-throughput sequencing: deepening insights into mammalian transcriptomes. Genes Dev. 2009;23(12):1379–86.
32. Fu GK, Xu W, Wilhelmy J, Mindrinos MN, Davis RW, Xiao W, et al. Molecular indexing enables quantitative targeted RNA sequencing and reveals poor efficiencies in standard library preparations. Proc Natl Acad Sci U S A. 2014;111(5):1891–6.
33. Valadi H, Ekström K, Bossios A, Sjöstrand M, Lee JJ, Lötvall JO. Exosome-mediated transfer of mRNAs and microRNAs is a novel mechanism of genetic exchange between cells. Nat Cell Biol. 2007;9(6):654–9.
34. Arroyo JD, Chevillet JR, Kroh EM, Ruf IK, Pritchard CC, Gibson DF, et al. Argonaute2 complexes carry a population of circulating microRNAs independent of vesicles in human plasma. Proc Natl Acad Sci U S A. 2011;108(12):5003–8.
35. Michael A, Bajracharya SD, Yuen PS, Zhou H, Star RA, Illei GG, et al. Exosomes from human saliva as a source of microRNA biomarkers. Oral Dis. 2010;16(1):34–8.
36. Skog J, Würdinger T, van Rijn S, Meijer DH, Gainche L, Sena-Esteves M, et al. Glioblastoma microvesicles transport RNA and proteins that promote tumour growth and provide diagnostic biomarkers. Nat Cell Biol. 2008;10(12):1470–6.
37. Patton JG, Franklin JL, Weaver AM, Vickers K, Zhang B, Coffey RJ, et al. Biogenesis, delivery, and function of extracellular RNA. J Extracell Vesicles. 2015;4:27494.
38. Koh W, Pan W, Gawad C, Fan HC, Kerchner GA, Wyss-Coray T, et al. Noninvasive in vivo monitoring of tissue-specific global gene expression in humans. Proc Natl Acad Sci U S A. 2014;111(20):7361–6.
39. Goetzl EJ, Boxer A, Schwartz JB, Abner EL, Petersen RC, Miller BL, et al. Altered lysosomal proteins in neural-derived plasma exosomes in preclinical Alzheimer disease. Neurology. 2015;85(1):40–7.
40. Shi M, Liu C, Cook TJ, Bullock KM, Zhao Y, Ginghina C, et al. Plasma exosomal α-synuclein is likely CNS-derived and increased in Parkinson's disease. Acta Neuropathol. 2014;128(5):639–50.

41. Brock G, Castellanos-Rizaldos E, Hu L, Coticchia C, Skog J. Liquid biopsy for cancer screening, patient stratification and monitoring. Transl Cancer Res. 2015;4(3):280–90.
42. Tsui NB, Jiang P, Wong YF, Leung TY, Chan KC, Chiu RW, et al. Maternal plasma RNA sequencing for genome-wide transcriptomic profiling and identification of pregnancy-associated transcripts. Clin Chem. 2014;60(7):954–62.
43. Ainsztein AM, Brooks PJ, Dugan VG, Ganguly A, Guo M, Howcroft TK, et al. The NIH extracellular RNA communication consortium. J Extracell Vesicles. 2015;4:27493.
44. Quinn JF, Patel T, Wong D, Das S, Freedman JE, Laurent LC, et al. Extracellular RNAs: development as biomarkers of human disease. J Extracell Vesicles. 2015;4:27495.
45. Laurent LC, Abdel-Mageed AB, Adelson PD, Arango J, Balaj L, Breakefield X, et al. Meeting report: discussions and preliminary findings on extracellular RNA measurement methods from laboratories in the NIH extracellular RNA communication consortium. J Extracell Vesicles. 2015;4:26533.
46. Sutter DE, Worthy SA, Hensley DM, Maranich AM, Dolan DM, Fischer GW, et al. Performance of five FDA-approved rapid antigen tests in the detection of 2009 H1N1 influenza a virus. J Med Virol. 2012;84(11):1699–702.
47. Santiago GA, Vergne E, Quiles Y, Cosme J, Vazquez J, Medina JF, et al. Analytical and clinical performance of the CDC real time RT-PCR assay for detection and typing of dengue virus. PLoS Negl Trop Dis. 2013;7(7):e2311.
48. Styer LM, Miller TT, Parker MM. Validation and clinical use of a sensitive HIV-2 viral load assay that uses a whole virus internal control. J Clin Virol. 2013;58(Suppl 1):e127–33.
49. Pinsky BA, Sahoo MK, Sandlund J, Kleman M, Kulkarni M, Grufman P, et al. Analytical performance characteristics of the Cepheid GeneXpert Ebola assay for the detection of Ebola virus. PLoS One. 2015;10(11):e0142216.
50. Fischer N, Indenbirken D, Meyer T, Lütgehetmann M, Lellek H, Spohn M, et al. Evaluation of unbiased next-generation sequencing of RNA (RNA-seq) as a diagnostic method in influenza virus-positive respiratory samples. J Clin Microbiol. 2015;53(7):2238–50.
51. Gire SK, Goba A, Andersen KG, Sealfon RS, Park DJ, Kanneh L, et al. Genomic surveillance elucidates Ebola virus origin and transmission during the 2014 outbreak. Science. 2014;345(6202):1369–72.
52. Gregori J, Esteban JI, Cubero M, Garcia-Cehic D, Perales C, Casillas R, et al. Ultra-deep pyrosequencing (UDPS) data treatment to study amplicon HCV minor variants. PLoS One. 2013;8(12):e83361.
53. Ekici H, Rao SD, Sönnerborg A, Ramprasad VL, Gupta R, Neogi U. Cost-efficient HIV-1 drug resistance surveillance using multiplexed high-throughput amplicon sequencing: implications for use in low- and middle-income countries. J Antimicrob Chemother. 2014;69(12):3349–55.
54. Wang K, Li H, Yuan Y, Etheridge A, Zhou Y, Huang D, et al. The complex exogenous RNA spectra in human plasma: an interface with human gut biota? PLoS One. 2012;7(12):e51009.
55. Beatty M, Guduric-Fuchs J, Brown E, Bridgett S, Chakravarthy U, Hogg RE, et al. Small RNAs from plants, bacteria and fungi within the order Hypocreales are ubiquitous in human plasma. BMC Genomics. 2014;15(1):933.
56. Ghosal A, Upadhyaya BB, Fritz JV, Heintz-Buschart A, Desai MS, Yusuf D, et al. The extracellular RNA complement of Escherichia coli. Microbiologyopen. 2015;4(2):252–66.
57. Westermann AJ, Förstner KU, Amman F, Barquist L, Chao Y, Schulte LN, et al. Dual RNA-seq unveils noncoding RNA functions in host-pathogen interactions. Nature. 2016;529(7587):496–501.
58. Chen SJ, Chen GH, Chen YH, Liu CY, Chang KP, Chang YS, et al. Characterization of Epstein-Barr virus miRNAome in nasopharyngeal carcinoma by deep sequencing. PLoS One. 2010;5(9):e12745.
59. Meshesha MK, Veksler-Lublinsky I, Isakov O, Reichenstein I, Shomron N, Kedem K, et al. The microRNA transcriptome of human cytomegalovirus (HCMV). Open Virol J. 2012;6:38–48.

60. Cucher M, Macchiaroli N, Kamenetzky L, Maldonado L, Brehm K, Rosenzvit MC. High-throughput characterization of Echinococcus spp. metacestode miRNomes. Int J Parasitol. 2015;45(4):253–67.
61. Fagnocchi L, Bottini S, Golfieri G, Fantappiè L, Ferlicca F, Antunes A, et al. Global transcriptome analysis reveals small RNAs affecting Neisseria meningitidis bacteremia. PLoS One. 2015;10(5):e0126325.
62. Latorre I, Leidinger P, Backes C, Domínguez J, de Souza-Galvão ML, Maldonado J, et al. A novel whole-blood miRNA signature for a rapid diagnosis of pulmonary tuberculosis. Eur Respir J. 2015;45(4):1173–6.
63. Ren N, Gao G, Sun Y, Zhang L, Wang H, Hua W, et al. MicroRNA signatures from multidrug-resistant mycobacterium tuberculosis. Mol Med Rep. 2015;12(5):6561–7.
64. Lie AK, Kristensen G. Human papillomavirus E6/E7 mRNA testing as a predictive marker for cervical carcinoma. Expert Rev Mol Diagn. 2008;8(4):405–15.
65. Kanwal F, Kramer JR, Ilyas J, Duan Z, El-Serag HB. HCV genotype 3 is associated with an increased risk of cirrhosis and hepatocellular cancer in a national sample of U.S. veterans with HCV. Hepatology. 2014;60(1):98–105.
66. Mitchell AM, Stone AE, Cheng L, Ballinger K, Edwards MG, Stoddard M, et al. Transmitted/founder hepatitis C viruses induce cell-type- and genotype-specific differences in innate signaling within the liver. MBio. 2015;6(2):e02510.
67. Wu LS, Lee SW, Huang KY, Lee TY, Hsu PW, Weng JT. Systematic expression profiling analysis identifies specific microRNA-gene interactions that may differentiate between active and latent tuberculosis infection. Biomed Res Int. 2014;2014:895179.
68. Barry SE, Chan B, Ellis M, Yang Y, Plit ML, Guan G, et al. Identification of miR-93 as a suitable miR for normalizing miRNA in plasma of tuberculosis patients. J Cell Mol Med. 2015;19(7):1606–13.
69. Andersson AK, Ma J, Wang J, Chen X, Gedman AL, Dang J, et al. The landscape of somatic mutations in infant MLL-rearranged acute lymphoblastic leukemias. Nat Genet. 2015;47(4):330–7.
70. Robinson D, Van Allen EM, Wu YM, Schultz N, Lonigro RJ, Mosquera JM, et al. Integrative clinical genomics of advanced prostate cancer. Cell. 2015;161(5):1215–28.
71. Mody RJ, Wu YM, Lonigro RJ, Cao X, Roychowdhury S, Vats P, et al. Integrative clinical sequencing in the management of refractory or relapsed cancer in youth. JAMA. 2015;314(9):913–25.
72. Byron SA, Van Keuren-Jensen KR, Engelthaler DM, Carpten JD, Craig DW. Translating RNA sequencing into clinical diagnostics: opportunities and challenges. Nat Rev Genet. 2016;17(5):257–71.

# Chapter 30
# Utilization of AI in Designing RNA Therapeutics

**Abstract** The early phases of disease modeling and target identification are essential to drug discovery, significantly impacting the overall success of drug development. Traditionally, the process of identifying drug targets has been exceptionally time-intensive, often spanning several years to decades, and typically originates within academic research. However, the advent of artificial intelligence (AI) is revolutionizing this domain by utilizing its capability to process extensive datasets and complex biological systems, thereby becoming a pivotal tool in contemporary drug target identification. This chapter explores the latest advancements in target discovery, with a particular emphasis on notable breakthroughs facilitated by AI in identifying therapeutic targets. Additionally, we emphasize the need to balance innovative methods with dependability in target selection. There is an increasing trend of AI-discovered targets undergoing experimental validation, with several AI-developed drugs progressing to clinical trial stages. We also discuss the current limitations in the field and suggest potential future directions.

## 30.1 Introduction

The drug discovery process is known for its lengthy, costly, and high-risk nature, often taking around a decade and costing approximately $2 billion to introduce a new drug to the market [1]. As of 2022, less than 500 drug targets had been successfully identified [2], representing a small fraction of the potential druggable targets in the human genome [3, 4]. Despite rigorous preclinical optimization, the average failure rate of drug candidates in clinical trials between 2009 and 2018 was 84.6%, primarily due to a lack of clinical efficacy in Phase 2 and Phase 3 trials [5]. This high failure rate results in substantial financial losses and wasted resources.

Target identification, the process of pinpointing biological molecules or cellular pathways that drugs can target for therapeutic effects, is increasingly critical in modern drug discovery. Despite significant advancements in experimental techniques and -omic technologies over recent decades, the challenge of identifying viable therapeutic targets remains substantial (Fig. 30.1). Recently, the integration



P. E. Saw, E. Song, *RNA Therapeutics in Human Diseases*,
https://doi.org/10.1007/978-981-96-3041-7_30

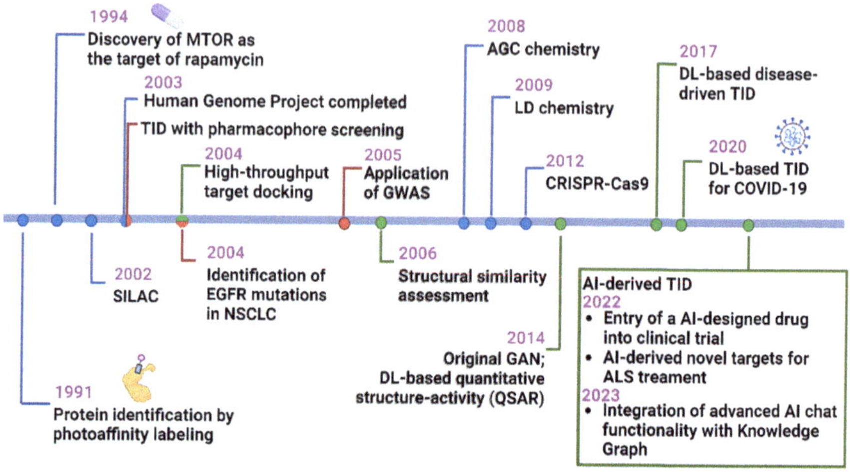


**Fig. 30.1** The progression of artificial intelligence (AI) in early drug development stages. Critical technological advancements in target identification are divided into three main categories: experimental methods (red), multi-omic approaches (blue), and computational techniques (green). Historically, experimental methods were the primary strategy for identifying therapeutic targets. The advent of big data has redirected attention toward the integrated and efficient analysis of multi-omic datasets for target discovery. Additionally, the latest developments in AI-driven biological analyses have enabled the identification of new targets, propelling AI-designed drugs into clinical trials. (Figure is adapted from [6] with permission from Elsevier (License number: 5902331232775))

of multi-omic data with AI algorithms has demonstrated considerable potential in overcoming these challenges. This review delves into traditional methods of target identification, emphasizing the transformative role of AI algorithms. This chapter aims to offer a forward-looking perspective on AI-driven drug discovery, advocating for the seamless integration of AI technologies into the drug discovery pipeline.

## 30.2 AI-Driven Target Identification

In recent years, there has been an extraordinary increase in the generation of biomedical data, ranging from fundamental research on disease mechanisms to extensive clinical studies involving patients. This deluge of data, while rich with potential insights, presents a significant challenge in terms of analysis and interpretation. Here, AI has emerged as an invaluable tool. AI's capabilities in processing and interpreting complex biomedical datasets enable the detection of patterns and relationships that might escape human researchers, thereby advancing our understanding and treatment of various diseases. AI has made significant strides in various domains, such as biomarker and target identification [7–9], prioritizing therapeutic indications [10], designing drug-like molecules [11, 12], predicting pharmacokinetics [13], analyzing drug–target interactions [14, 15], and designing clinical trials [16] (Fig. 30.2).

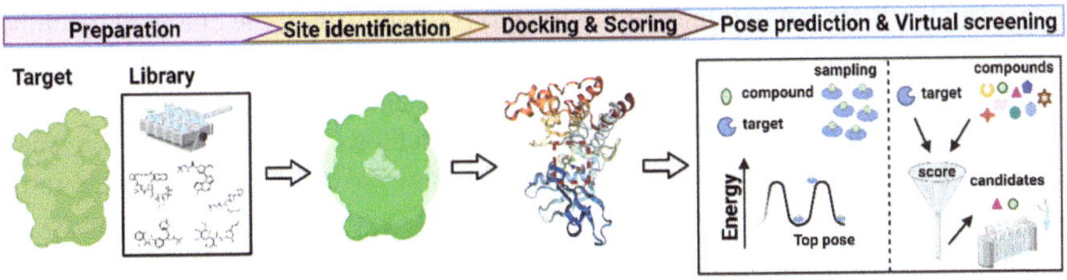


**Fig. 30.2** A representative workflow for structure-based RNA-targeted drug discovery. Step 1. Preparation: Begin by preparing the structure or structural ensemble of the disease-related target RNA and compiling the compound library. Ensure the RNA structure is accurately modeled and the compound library is comprehensive, including a diverse range of small molecules for screening. Part 2. Binding Site Identification: Identify potential binding sites or pockets within the target RNA. Highlight these sites using computational tools to detect regions on the RNA that could potentially interact with small molecules. Part 3. Docking and Scoring: Perform docking simulations and score the sampled RNA–small molecule complexes. Utilize docking algorithms to predict how small molecules might bind to the RNA and evaluate these interactions through scoring functions that estimate binding affinity. Part 4. Result Analysis: Analyze the docking results to meet specific project goals. This analysis can include: Predicting the native bound conformation from various sampled decoy conformations for accurate binding pose prediction. Identifying the most promising drug candidates for a particular RNA target from the compound library, enhancing the virtual screening process. By following these steps, researchers can systematically identify and optimize RNA-targeted small molecule therapeutics. (Figure is adapted from [17] with permission from Elsevier (License number: 5902300721180))

## 30.3 Application of Deep Learning Models in Target Discovery

In the past few years, machine learning algorithms, particularly deep learning methods, have attracted considerable interest and shown exceptional outcomes in pharmaceutical research. Machine learning (ML) is a broader category of AI that involves algorithms that can learn from and make predictions based on data [6]. It encompasses various methods, including supervised, unsupervised, and reinforcement learning, while deep learning (DL) is a subset of machine learning. DL uses neural networks with many layers (hence "deep") to model complex patterns in large datasets. These deep neural networks are capable of learning representations of data with multiple levels of abstraction [18]. This sophisticated approach allows for the intricate analysis and interpretation of vast and complex datasets, thereby enhancing the precision and efficiency of target discovery in drug development. The ability of deep learning models to uncover subtle patterns and relationships within biological data has propelled their adoption in identifying novel therapeutic targets and advancing our understanding of disease mechanisms. To enhance the identification of lead compounds that target specific RNA structures, researchers have developed various virtual screening (VS) methodologies. These techniques are broadly classified into two primary categories: ligand-based virtual screening (LBVS) and structure-based virtual screening (SBVS). We summarized this in Table 30.1.

Currently, there are two major classifications of algorithm: the classical machine learning (ML) and deep learning (DL). Classical ML approaches utilize traditional algorithms such as linear regression, logistic regression, decision trees, support vector machines, and random forests [32]. These methods are well-suited for small datasets and yield interpretable results, but they may have difficulty capturing intricate relationships within the data. Classical ML approaches in the context of

**Table 30.1** Comparisons between LBVS and SBVS in virtual screening process

| Screening process | Ligand-based virtual screening (LBVS) [19–22] | Structure-based virtual screening (SBVS) [23–31] |
|---|---|---|
| General attributes | LBVS approaches focus on identifying potential lead compounds without requiring detailed knowledge of the RNA target's 2D or 3D structure. These methods utilize predefined molecular descriptors, such as physicochemical properties, to screen large compound libraries through techniques like chemical similarity searches, pharmacophore searches, 3D shape matching, and quantitative structure–activity relationship (QSAR) analysis | In contrast, SBVS techniques depend on the structural information of the RNA target to address critical questions, such as identifying binding sites, determining which small molecules can bind, and predicting binding modes. Comprehensive reviews of SBVS strategies applied to RNA–small molecule interactions can be found in the literature |
| Methodology | LBVS relies on known ligands or small molecules that bind to the target RNA structure | SBVS involves using the three-dimensional structure of the RNA target to screen for potential binding compounds |
| Principle | It identifies new compounds based on their similarity to known active compounds | It predicts how well a candidate compound fits into the binding site of the RNA structure using docking simulations and scoring functions |
| Approach | LBVS uses computational techniques to compare the molecular properties, such as shape, electrostatic potential, and chemical features, of candidate molecules to those of known active ligands | SBVS requires accurate structural information of the RNA target, often obtained through experimental techniques like X-ray crystallography or NMR spectroscopy, or predicted through homology modeling |
| Advantages | LBVS can be effective when there are already known ligands for the target RNA, as it leverages existing structural and activity data | SBVS can explore a wider chemical space than LBVS since it does not rely on existing ligands. It can also accommodate various types of RNA structures, including those without known ligands |
| Limitations | It may struggle when there are no or few known ligands, limiting the scope of applicable targets | It heavily relies on the accuracy of the RNA structure prediction or experimental determination. Additionally, docking accuracy and scoring functions can impact the reliability of the results |

RNA–small molecule interactions involve a variety of well-established algorithms known for their simplicity and interpretability. These methods are typically advantageous for small datasets where DL techniques might be prone to overfitting [33]. Key classical ML and DL approaches and their characteristics are summarized in Table 30.2.

## 30.4 Limitations of Current ML-Based Models

In addition to the limited availability of experimental data for RNA–small molecule interactions, there are two other major limitations inherent in current ML-based models.

### 30.4.1 Accounting for RNA Dynamics

Current machine learning (ML) models in drug discovery typically focus on training with holo-form RNA structures, which already include bound ligands. This approach, however, often overlooks the importance of apo-form RNA structures, which do not contain bound ligands and can exhibit significant conformational changes when they interact with ligands. This neglect may lead to a significant drop in the accuracy and effectiveness of ML models when they are used to predict interactions with apo-form RNAs, since these models might not accurately capture the structural dynamics essential for ligand binding [46]. To address this limitation, one promising strategy in RNA-targeted virtual screening is to implement an ensemble-based approach. This involves generating a diverse array of RNA conformations to represent a spectrum of potential structural states that the RNA might adopt in reality. An ideal ensemble would encompass native-like conformations that feature well-defined binding pockets, enhancing the likelihood of identifying effective ligand interactions. For example, in their research, Xie and Frank utilized a tool called RNACavityMiner to carefully select 20 RNA conformers exhibiting prominently scored cavities from a substantial pool of 64,000 SimRNA-generated HIV-1 TAR structures. By employing this selective ensemble in virtual screening against a vast library of over 102,000 small molecules, they achieved significant enrichment factors. This indicates a strong predictive performance, demonstrating the potential of an ensemble-based approach to substantially improve the outcomes of RNA-targeted virtual screening by better mimicking the dynamic nature of RNA structures [47].

**Table 30.2** Key classical ML approaches and their characteristic

| Type | Overview | Application | Advantages |
|---|---|---|---|
| *ML approaches* [34–40] | | | |
| Linear regression | Linear regression models the relationship between a dependent variable and one or more independent variables by fitting a linear equation to observed data | In RNA–small molecule interactions, linear regression can predict binding affinities based on molecular descriptors | Simple to implement and interpret; provides clear insights into how changes in molecular features influence binding affinity |
| Logistic regression | Logistic regression is used for binary classification problems, estimating the probability that a given input belongs to a particular class | This method can classify molecules as either binders or non-binders to a specific RNA target | Effective for binary outcome prediction; provides probabilities that help in understanding classification confidence |
| Decision trees | Decision trees split the data into subsets based on feature values, creating a tree-like model of decisions | They can identify which molecular features are most important for predicting binding affinity or interaction strength | Easy to visualize and interpret; can handle both numerical and categorical data |
| Support vector machines (SVM) | SVMs find the hyperplane that best separates data into classes in a high-dimensional space | SVMs can classify RNA–small molecule interactions, distinguishing between different types of binding activities | Effective in high-dimensional spaces; robust to overfitting, especially in high-dimensional feature spaces |
| Random forests | Random forests are an ensemble method that combines multiple decision trees to improve classification or regression performance | They can be used to predict binding affinities and identify important molecular features influencing RNA–small molecule interactions | High accuracy due to ensemble learning; reduces overfitting by averaging multiple trees |
| *DL approaches* [41–45] | | | |
| Artificial neural networks (ANNs) | ANNs consist of interconnected layers of neurons that process input data and learn to map inputs to outputs through training | In RNA–small molecule interactions, ANNs can predict binding affinities, classify molecules, and identify key molecular features from raw data | Flexible and capable of learning complex patterns; can handle large datasets with diverse features |

(continued)

**Table 30.2** (continued)

| Type | Overview | Application | Advantages |
|---|---|---|---|
| Convolutional neural networks (CNNs) | CNNs are designed to automatically and adaptively learn spatial hierarchies of features through convolutional layers, making them suitable for grid-like data such as images | CNNs can analyze 3D molecular structures of RNA and small molecules, identifying interaction sites and predicting binding modes | Efficient at capturing local dependencies in data; effective for processing 3D molecular structures and images |
| Recurrent neural networks (RNNs) | RNNs are specialized for sequential data, maintaining a hidden state that captures information about previous inputs | They can model the dynamic interactions between RNA and small molecules over time, useful for simulating binding processes | Suitable for time-series data and sequences; can capture temporal dependencies in interaction data |
| Long short-term memory networks (LSTMs) | LSTMs are a type of RNN designed to overcome the vanishing gradient problem, allowing them to learn long-term dependencies | LSTMs can be used to predict the progression of RNA–small molecule binding interactions over time, providing insights into binding kinetics | Effective at capturing long-term dependencies; robust to the vanishing gradient problem |
| Graph neural networks (GNNs) | GNNs are designed to operate on graph-structured data, making them well-suited for molecular data where atoms and bonds form a natural graph | GNNs can model the intricate relationships within molecular structures, predicting binding affinities and interaction sites based on the graph representation of molecules | Capable of capturing complex molecular relationships; suitable for representing and learning from graph-structured data |

### 30.4.2 Consideration of Thermodynamic and Kinetic Properties

Current ML-based models in drug discovery often fail to consider crucial aspects such as the thermodynamic and kinetic properties of the binding process. Typically, these models are designed merely to assess whether a specific region on a molecule could potentially accommodate a ligand, ignoring critical physical properties such as binding affinity and the duration of the drug's residence at the target site. One of the primary challenges in integrating these properties into ML models is the lack of sufficient training data that encapsulate these complex characteristics. To overcome this issue, researchers like Liu and Frank have proposed and explored advanced strategies. They suggest leveraging more sophisticated computational techniques, such as molecular dynamics simulations, to generate detailed thermodynamic and kinetic profiles for individual RNA–ligand complexes or binding cavities. These simulations can provide a rich dataset that captures the dynamic interactions and stability of the binding events over time, offering insights that are typically not available from static data [48].

## 30.5 Case Studies Showcasing the Applications of ML-Based Models

Major examples of ML-based models and their applications in the RNA-targeted drug discovery workflow are highlighted here.

### *30.5.1 Identification of Potential Binding Sites Within the HIV-1 Trans-Activation Response (TAR) Element*

The HIV-1 TAR element, a critical RNA sequence, plays a pivotal role in trans-activating the viral promoter and thus facilitating the replication of the virus. This RNA segment is crucial because it directly influences the viral lifecycle by interacting with other molecules to enhance viral gene expression. Due to its central role in viral replication and its dynamic nature, the HIV-1 TAR element serves as an exemplary model for studying the interplay between RNA functionality and its structural dynamics. Additionally, this RNA sequence is highly relevant for RNA-targeted drug discovery efforts. Its essential function and structural accessibility make it an attractive target for designing drugs that can disrupt its interaction with viral proteins, thereby potentially inhibiting the viral replication process [49–52]. Structural investigations into the TAR RNA element from HIV-1, especially when it interacts with small molecules, peptides, or cations, reveal that this RNA structure is exceptionally dynamic. These studies indicate that the TAR RNA does not remain rigid but rather adopts various conformations that are crucial for effective ligand binding. This behavior exemplifies a mechanism known as "conformational selection," where the RNA dynamically shifts among multiple structures until the right conformation that optimally fits the ligand is achieved. This dynamic aspect of TAR RNA is significant because it suggests that the RNA can actively participate in the binding process, adjusting its shape and orientation to facilitate optimal interaction with various ligands [53–56]. Sequence-based RNA–ligand binding site prediction methods frequently encounter limitations due to their lack of spatial data integration. These methods primarily rely on the nucleotide sequence without considering the three-dimensional conformation of RNA, which is critical for understanding how RNA interacts with various molecules. Such approaches can miss crucial nuances of how RNA structures adapt to accommodate different ligands.

In an effort to evaluate whether the BiteNetN model could effectively account for these structural dynamics, researchers Kozlovskii and Popov undertook a detailed analysis of seven different TAR RNA structures complexed with small molecules. These structures exhibited substantial variability, with pair-wise root mean square deviation (RMSD) values spanning from 1.9 Ångström (Å) to 10.7 Å, indicating significant differences in their conformations. The performance of BiteNetN was impressive in capturing these dynamic structural changes. The results from their analysis showed that BiteNetN achieved a weighted receiver operating characteristic area under the curve (ROC AUC) of 0.872 and a Matthews correlation coefficient (MCC) of 0.510. These metrics signify a strong predictive ability of the model to identify RNA–ligand binding sites accurately, taking into account the conformational flexibility of RNA that is essential for effective drug targeting. This indicates a substantial advancement in RNA–ligand interaction prediction models, emphasizing the importance of incorporating structural dynamics for enhanced prediction accuracy [57].

### *30.5.2 Application of Nucleic Acid Binding Site (NABS) Model*

Jiang et al. developed a sophisticated model known as the nucleic acid binding site (NABS) model, employing an ensemble learning strategy specifically tailored to pinpoint cavities on RNA molecules where small molecules are likely to bind [58]. The core machine learning component of NABS incorporates a trio of advanced algorithms: Random Forest (RF), Extreme Gradient Boosting (XGB), and Light Gradient Boosting Machine (LGBM). These algorithms are well-regarded for their robustness and accuracy in handling complex predictive tasks. The NABS model is distinguished by its comprehensive use of diverse features to enhance its predictive accuracy. It integrates structural attributes such as solvent accessibility, depth and protrusion indices, and the Laplacian norm, which help in understanding the physical context of potential binding sites. Additionally, it considers preference features, which include the types of nucleotides, the functional groups present, and RNA secondary motifs that might influence binding affinity. It also takes into account sequence features, like the composition and distribution of nucleotides along the RNA strand, which can be critical for recognizing specific binding patterns. For training this model, the RNA–ligand (RL) dataset employed encompasses 53 structures, providing a robust foundation for the algorithms to learn from actual RNA–small molecule interactions. This diverse dataset ensures that the model can generalize well across various types of RNA structures and potential binding scenarios. By using such a multidimensional approach, the NABS model represents a significant step forward in the ability to accurately predict and understand how small molecules interact with RNA, which is crucial for advancing RNA-targeted drug discovery efforts.

### *30.5.3 Prediction of RNA–Small Molecule Binding Sites Using RLBind*

Wang et al. developed RLBind, a novel approach utilizing convolutional neural networks (CNNs) for predicting RNA–small molecule binding sites [59]. This model marks a significant departure from methods like BiteNetN, as it employs a unique strategy that involves a one-dimensional (1D) feature vector for each nucleotide within the RNA sequence. These feature vectors encapsulate both sequence-dependent properties, such as the type and order of nucleotides, and structure-dependent properties, which include the spatial arrangement and physical characteristics of the RNA strand. RLBind is designed to integrate both global and local information in its analysis. By considering global information, the model gains insights from the entire RNA sequence, which helps in understanding broad structural and functional themes. Simultaneously, it examines local details, focusing on specific segments of the sequence where interactions with small molecules are likely. This dual approach enables RLBind to capture a comprehensive picture of potential binding sites, significantly enhancing its ability to accurately predict where small molecules may interact with the RNA [59]. The model was trained using the same training, validation, and test sets as RNAsite [46], with a tenfold cross-validation procedure. RLBind outperformed Rsite [60], Rsite2 [61], RBind [62], and RNAsite [46] in terms of precision, recall, and MCC, albeit with a slightly lower AUC than RNAsite on one test set. These results emphasize the potential of CNN-based models in advancing RNA-targeted drug discovery.

### 30.5.4 Integration of Bioinformatics and DL Models

To pinpoint viable therapeutic targets for amyotrophic lateral sclerosis (ALS), Pun et al. employed a comprehensive approach that integrated several bioinformatics and deep learning models. These models were specifically trained using a broad array of disease-specific data, encompassing both multi-omic datasets (such as genomics, proteomics, and metabolomics) and text-based information from scientific literature and databases. This blend of diverse data types enabled the models to capture a holistic view of the molecular underpinnings of ALS. By applying these advanced computational techniques, the team was able to efficiently sift through vast amounts of biological and textual data to prioritize genes that are most likely to be effective targets for drug development. Their analysis successfully identified 18 promising targets for ALS treatment. These targets are not only potentially druggable but also relevant to the specific pathological processes of ALS, offering a focused direction for future therapeutic interventions [63].

### 30.5.5 Introduction of Sophisticated DL Methods

Additionally, Fabris et al. introduced a sophisticated deep learning method that features an innovative modular architecture designed to identify human genes associated with various age-related diseases. This approach represents a significant advancement in the field of biomedical research, particularly in gerontology and genomics. The novel method developed by Fabris et al. is engineered to extract and learn complex patterns from a range of biological data sources. Specifically, it analyzes gene or protein features that include gene ontology (GO) terms, which provide a standardized vocabulary to describe gene product attributes; protein–protein interactions (PPIs), which offer insights into the functional interactions and networks within the cell; and biological pathways, which detail the series of actions among molecules in a cell that lead to a certain product or change in the cell. By leveraging these rich datasets, the deep learning model is adept at detecting intricate correlations and interactions that might indicate a gene's involvement in the aging process and its related diseases. The modular architecture of the model allows it to be flexible and scalable, accommodating the integration of diverse types of biological data and making it a powerful tool for researchers exploring the genetic foundations of age-related conditions [64].

### *30.5.6 Creation of a DL Ensemble*

West et al. pioneered a breakthrough in developmental biology and regenerative medicine by creating a deep learning ensemble. This advanced computational model was trained on transcriptomic profiles derived from more than 12,000 embryonic and adult cells. The scale and diversity of this dataset enabled the model to capture a broad spectrum of cellular behaviors and developmental transitions. Through their analysis, this study uncovered a previously unrecognized gene, COX7A1, as a pivotal regulator of the embryonic-fetal transition. COX7A1 is known as an important subunit of cytochrome c oxidase and is vital in the assembly complex in the mitochondrial respiratory chain. In cancer research, COX7A1 regulates autophagy to suppress human lung cancer growth [65]. In recent years, some researchers have identified that COX7A1 is implicated in human cancer cell metabolism and therapy. This transition is a critical phase in early development where cells begin to differentiate and take on specialized functions. The identification of COX7A1 sheds light on the molecular mechanisms that drive this process, offering new perspectives on how cells evolve from their embryonic states to more mature forms. The implications of discovering COX7A1 extend beyond understanding normal development. It also has potential ramifications in the fields of epimorphic tissue regeneration and cancer research. In epimorphic regeneration, the body rebuilds large portions of tissue or even entire organs, and insights into embryonic transitions could inform strategies to enhance regenerative capabilities in humans. Similarly, in cancer, where cells often revert to less differentiated, more embryonic-like states, understanding the control mechanisms of early development could lead to novel approaches to cancer treatment [66].

## 30.6 Large Language Models (LLMs)

Large language models (LLMs) have become instrumental in advancing therapeutic target discovery by harnessing their capabilities in rapid biomedical text mining. These models are pretrained on extensive text datasets that include millions of scientific publications, allowing them to extract and synthesize vast amounts of complex information quickly and accurately. Models such as scGPT, Microsoft's BioGPT, and In Silico Medicine's ChatPandaGPT exemplify the application of LLMs in the biomedical field [67]. By leveraging these tools, researchers can access chat functionalities that intelligently parse and understand biomedical literature. These models are particularly adept at identifying and correlating critical pieces of information across disparate sources. They can effectively link diseases with potential genetic causes or associated biological processes, thereby illuminating paths to new therapeutic targets [68]. This capability facilitates the swift identification of biological mechanisms implicated in the development and progression of diseases, as well as potential drug targets and biomarkers [69]. LLMs excel in parsing natural language and interpreting complex scientific concepts, making them powerful accelerators in formulating disease-related hypotheses [70]. However, it is crucial to acknowledge that LLMs, often trained on human-generated texts, might not always accurately evaluate the validity and relevance of the input data. This can lead to the perpetuation of existing human biases and misconceptions. Additionally, because these models primarily rely on published data, their ability to uncover genuinely novel targets might be constrained.

## 30.7 Tools for Sequence Optimization

**LinearDesign** LinearDesign optimizes RNA sequences to enhance their stability, reduce their immunogenicity, and improve their translational efficiency. This is achieved by analyzing and modifying nucleotide sequences to avoid problematic motifs, such as unintended secondary structures or sequences that might trigger immune responses [71]. It can tailor RNA sequences to bind specifically to certain targets, which is crucial for applications like RNAi or CRISPR-based gene editing. This specificity ensures that the RNA molecules perform their intended function with minimal off-target effects. The tool also considers the environmental and experimental conditions under which the RNA will be used. For instance, RNA molecules designed for therapeutic use might be optimized for higher stability in the bloodstream, while those for laboratory experiments might be optimized for expression in a specific type of cell culture. LinearDesign integrates machine learning algorithms and bioinformatics tools to predict the outcomes of sequence changes. This integration allows for the simulation and evaluation of potential RNA designs before they are synthesized, saving time and resources in the development process. The tool is designed to be accessible to both biologists and computational scientists, offering a user-friendly interface that simplifies the complex process of RNA design. Users can input specific parameters and requirements, and LinearDesign handles the computational complexity behind the scenes [72].

**DNA Chisel** DNA Chisel allows users to define various biological constraints that the DNA sequence must meet. These can include avoiding specific sequences that might create unwanted restriction sites, maintaining or improving codon usage for better expression in a particular organism, or introducing mutations at specific locations. The software then optimizes the DNA sequence to meet these constraints without altering its original function. Users can specify objectives for their DNA design projects, such as maximizing the stability of mRNA, minimizing the immunogenicity of a therapeutic DNA, or altering GC content to meet specific experimental requirements. DNA Chisel evaluates possible sequences that satisfy both the constraints and the objectives, providing optimized solutions. DNA Chisel can refactor large DNA sequences, making it suitable for projects that involve the synthesis of entirely new genomes or the redesign of existing ones. This capability is particularly valuable in synthetic biology for creating synthetic organisms or for genome-scale engineering. The software can be integrated with other bioinformatics tools and software suites, enhancing its utility in larger workflow pipelines. This makes it easy to use DNA Chisel in conjunction with sequence analysis, gene synthesis, and other design tools, facilitating a comprehensive approach to genetic engineering. Because it is open-source, DNA Chisel allows advanced users to modify the code to fit their specific needs or to add new functionalities. This flexibility ensures that DNA Chisel can evolve with the advancing field of synthetic biology and continue to meet the changing needs of its users [73].

**OPTIMIZER** The primary function of OPTIMIZER is to align the codon usage of a target gene with that of the host organism's preferred codon usage. This alignment is crucial because different organisms tend to favor certain codons over others, which can significantly impact the efficiency of translation and protein synthesis. OPTIMIZER allows users to select from a wide range of host organisms whose codon usage tables are well-documented. This flexibility makes it applicable to a diverse array of projects, from bacterial and yeast expression systems to mammalian cells. Users can input their own codon usage tables if they are working with less common organisms or engineered strains with altered codon preferences. This customization is particularly valuable for advanced synthetic biology applications involving novel synthetic organisms. Beyond simple codon optimization, OPTIMIZER provides tools for analyzing the gene sequence. These tools can identify potential issues such as rare codons, repetitive sequences, or internal ribosome entry sites that might affect gene expression or stability [74].

**GeneOptimizer** Unlike simpler tools that may only adjust codon usage, GeneOptimizer considers a broader range of factors. These include codon usage adaptation, GC content adjustment, avoidance of unwanted sequences (such as RNA secondary structures, repeat sequences, cryptic splicing sites, and internal ribosome entry sites), and the removal of restriction enzyme sites that can hinder cloning and expression. GeneOptimizer analyzes the entire sequence for various potential issues that might affect transcription, translation, and the eventual stability of the mRNA and protein. By doing this, it ensures that the optimized gene is not only expressible but also robust in the chosen expression system. The tool is designed to work with a wide array of host organisms, from common ones like *E. coli* and yeast to mammalian and plant systems. This flexibility makes GeneOptimizer invaluable for projects that require expression across different biological systems. GeneOptimizer is often used in tandem with gene synthesis services. Once a sequence is optimized, it can be directly synthesized, streamlining the process from design to actual experimental application. This integration is particularly beneficial in industrial settings where speed and efficiency are crucial [75].

AI-driven target prioritization capitalizes on a variety of multi-model approaches, integrating diverse sources of publicly accessible omic and text data. Omic data encompass genomics, transcriptomics, proteomics, epigenomics, and metabolomics [76]. These datasets provide crucial insights into altered signaling pathways, molecular interactions, and protein–protein interactions, making them indispensable for target prioritization. Text-based data, sourced from funding reports, patents, publications, and clinical trials, complement these insights by offering context and current research directions. During the target prioritization process, multiple selection criteria are employed. These include the protein family class, the current development status of potential targets, their druggability, associated toxicity, and their novelty. By applying these criteria, the selection process is refined, ensuring that the AI-driven targets align with specific research objectives and are likely to contribute valuable new insights or therapeutic approaches. This methodical approach enables a strategic narrowing down of targets, optimizing resource allocation, and focusing efforts on the most promising candidates.

## 30.8 Developing RNA Therapeutics with TREAT Platform

Developing RNA therapeutics typically involves a systematic process that begins with targeted RNA screening, followed by the intricate tasks of drug design and optimization. Current tools for RNA screening, however, often fail to account for noncoding RNAs, despite their significant roles in regulating various biological processes and their implications in diseases [77]. Furthermore, the design of therapeutic RNAs is complicated by the need for multi-objective optimization, addressing various challenges such as reducing potential immunogenicity, increasing molecular stability, and enhancing the efficiency of translational production. To comprehensively explore the therapeutic possibilities of ncRNAs and improve the screening and design phases of RNA-based drugs, a new platform known as TREAT was introduced. This innovative platform is enriched with an extensive database comprising 43,087,953 regulatory relationships between coding and noncoding genes, gathered from 81 biological networks under diverse physiological conditions. TREAT utilizes advanced graph representation learning techniques, including Random Walk Diffusions, to effectively screen for disease-relevant targets. In addition, it employs well-established algorithms like Topological Degree and PageRank to further refine the screening process.

TREAT not only facilitates the screening of therapeutic targets but also supports the intricate design and optimization of large RNA molecules, including interfering RNAs, which play a critical role in gene silencing and regulation. To tackle the computational challenges associated with the multi-objective optimization of large RNAs, TREAT employs a novel stratification approach where features are divided into local and global categories. Local features are analyzed using bins of fixed or dynamic lengths, drawing inspiration from AI language models used for protein sequences. This methodology allows for a more nuanced understanding of the RNA's local structural and functional properties. Following the local analysis, a global assessment is conducted on the refined set of candidate RNAs. This step is crucial as it effectively reduces the computational search space, focusing resources on the most promising candidates for further development. Such a structured approach to optimization ensures that the designed RNAs not only meet specific therapeutic goals but also adhere to critical pharmacological properties like stability and efficacy.

## 30.9 Innovative Platforms for RNA Target Identification in TREAT

**Open Targets** Open Targets is a pioneering public–private initiative designed to transform drug discovery through the strategic use of human genetics and genomics data. This collaboration brings together expertise from academic institutions, biotech firms, and pharmaceutical companies to enhance the process of identifying and prioritizing potential drug targets. By integrating a wide range of data sources, including genetic associations, functional genomics, and phenotypic data, Open Targets aims to create a more structured and data-driven approach to drug development. The core objective of Open Targets is to leverage the vast amount of available genetic information to systematically pinpoint targets that are most likely to be therapeutically effective. This approach not only streamlines the early stages of drug discovery but also increases the likelihood of success in later phases of drug development. By focusing on targets with a strong genetic justification, Open Targets reduces the risks associated with drug development, potentially leading to more effective and safer treatments. Furthermore, Open Targets emphasizes transparency and accessibility, making its findings and methodologies available to the wider scientific community. This openness fosters a collaborative environment where researchers from various sectors can contribute to and benefit from the project's resources, accelerating the pace of innovation and discovery in drug development [78].

**Network Analyst** Network Analyst is an advanced molecular network analysis and visualization platform that offers comprehensive tools for integrating and exploring protein–protein interaction (PPI) networks and gene regulatory networks. This platform is designed to aid researchers in assessing the relevance and significance of specific biological targets within the complex web of molecular interactions. By synthesizing data from both PPI networks and gene regulatory networks, Network Analyst provides a multifaceted view of how proteins and genes interact within cellular processes. These interactions are crucial for understanding the pathways that lead to disease states and identifying potential intervention points. The platform's visualization capabilities allow users to clearly see these interactions and relationships, making it easier to pinpoint influential nodes and connections that could be effective drug targets. Network Analyst's powerful analytical tools enable users to delve deep into network dynamics, offering insights into the structural and functional properties of molecular systems. This can include identifying key regulators within networks or discovering sub-networks that are critical to disease progression. By providing a detailed and accessible way to analyze these networks, Network Analyst supports the identification of novel therapeutic targets and helps advance our understanding of complex biological systems [79].

## 30.10 Innovative Tools for Sequence Optimization in TREAT

The TREAT platform incorporates three sophisticated ranking algorithms—Degree Centrality, PageRank, and Random Walk—to enhance its capability in screening for disease-relevant targets. These algorithms are part of graph theory and are particularly suited to analyze complex networks such as biological systems where various genes and proteins interact. Each algorithm offers a unique approach to determining the importance or influence of nodes within a network, aiding in the identification of key targets that could be pivotal in disease mechanisms or therapeutic interventions.

### *30.10.1 Degree Centrality*

Degree Centrality is a fundamental measure used in network analysis. It quantifies the importance of a node based on the number of connections (or edges) it has to other nodes. In the context of disease-relevant target screening, a node (such as a gene or protein) with a high degree centrality is considered crucial because it interacts with many other nodes. This could indicate a hub of activity essential for the normal functioning of a biological pathway or potentially a critical point whose disruption could lead to disease.

### 30.10.2 *PageRank*

Originally developed for ranking web pages, PageRank is a sophisticated algorithm that measures the importance of nodes in a network based on the quality, not just the quantity, of connections. Each node's importance is enhanced by the importance of the nodes that link to it. In biological networks, PageRank can identify influential genes or proteins that, although may not be highly connected, hold strategic positions due to their connections with other highly influential nodes. This can be particularly useful in identifying targets that, if modulated, could have widespread effects on a disease state [79, 80].

### 30.10.3 *Random Walk*

The Random Walk algorithm is a probabilistic model that simulates a random path through a network, visiting nodes according to certain probabilities. In TREAT, the Random Walk algorithm helps identify nodes that are frequently accessed pathways in the network, suggesting their role in key functional pathways. This approach is beneficial for discovering targets that may not be immediately obvious through direct connections alone but are crucial in connecting various important pathways or functions within the biological system [81]. Beyond its capabilities for optimizing large RNA molecules directly, TREAT also provides support for designing interfering RNAs (iRNAs). These are crucial tools in gene regulation and therapeutic interventions, where they can selectively silence or modulate gene expression. By including this functionality, TREAT enhances its utility in therapeutic contexts where precise gene regulation is necessary. TREAT is also equipped with advanced algorithms that not only assist in target identification and RNA optimization but also in evaluating and predicting the efficacy of RNA designs. The platform's user-friendly interface ensures that these complex functionalities are accessible to researchers with varying levels of computational expertise, promoting broader use and experimentation. Furthermore, TREAT offers exploratory visualization tools that allow users to visually assess and explore the interactions and properties of RNAs within biological networks. These tools are invaluable for understanding the contextual implications of potential therapeutic targets and for presenting complex data in a comprehensible manner (Fig. 30.3).

## 30.11 The Use of AI-Generated Synthetic Data for Target Identification

"Synthetic data" refers to data that is artificially generated rather than obtained by direct measurement from real-world events. This type of data is specifically designed to mimic the patterns, characteristics, and behaviors observed in real datasets [6].

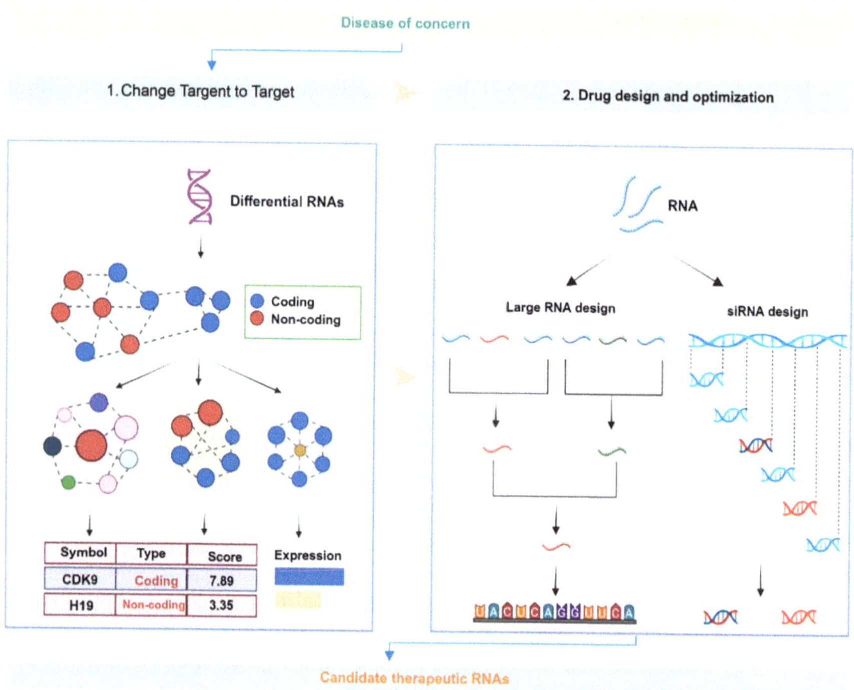


**Fig. 30.3** Framework of TREAT. TREAT consists of two computational components: "target screening" and "drug design and optimization." The "target screening" component is dedicated to identifying and ranking disease-relevant targets. The "drug design and optimization" component focuses on the design and optimization of large RNAs and siRNAs. The gray bar represents an RNA sequence, the blue circles on the left signify local bins, which are used for evaluating local features, and the red lines below the RNA sequence represent modified bases. The red and blue rectangles on the right illustrate candidate siRNAs, each with different scores, indicating their potential efficacy. (Figure is adapted from [77] with permission from Elsevier (License number: 5904631375749))

The generation of synthetic data utilizes sophisticated AI algorithms, which can model and replicate complex biological phenomena and interactions. The use of synthetic data is particularly valuable in the field of biomedical research, where real data can be scarce, difficult to obtain, or ethically sensitive. For example, in drug development, synthetic data can simulate how new drugs might interact with various biological systems, providing an initial assessment of efficacy and safety before actual clinical trials. This can accelerate research timelines and reduce costs significantly [82–84].

This methodology is especially beneficial in therapeutic fields where experimental data is characterized by limited patient data availability; traditional research approaches may face significant hurdles due to the small sample sizes that cannot effectively support robust statistical analysis. In such scenarios, AI can play a critical role by generating synthetic data that extrapolates from existing knowledge and observed patterns in similar or related conditions. By employing synthetic data, researchers can effectively expand their datasets without needing additional physical experiments or patient recruitment. This data serves as a surrogate, enabling the training of AI models with a broader range of data inputs that might not otherwise be available. These AI models can then analyze complex interactions and dependencies within the data, potentially uncovering therapeutic targets and pathways that would be difficult to identify using limited real-world data alone [85].

Furthermore, synthetic data plays a crucial role in validating predictions made by AI algorithms, thereby adding an additional layer of reliability to the target discovery process. This validation is vital, as it helps confirm the robustness and accuracy of AI-driven insights into potential therapeutic targets, ensuring that subsequent research and development efforts are based on credible predictions. Moreover, synthetic data is instrumental in addressing issues of data imbalance and bias, which are common challenges in medical research. Often, in therapeutic areas, certain patient demographics are underrepresented in available datasets. This lack of representation can skew AI models and lead to biased or incomplete analyses, potentially overlooking viable targets for specific populations.

To counter this, AI can generate synthetic data that mirrors these underrepresented groups, effectively filling the gaps in real data. This inclusion allows for a more comprehensive and inclusive analysis, ensuring that therapeutic developments benefit a broader range of patient groups. For instance, in conditions predominantly affecting specific ethnicities or age groups with insufficient real-world data, synthetic data can be tailored to reflect the genetic backgrounds and health profiles of these groups. This broader representation aids in discovering more targeted and effective therapeutic interventions that are inclusive of all affected populations [86].

## 30.12 Validation and Quality Control of Synthetic -Omic Data

**Comparative Analyses** This involves comparing the distributions of variables in both synthetic and real datasets. Techniques like Kolmogorov-Smirnov tests, histograms, and density plots can be employed to visually and quantitatively assess if the synthetic data captures the central tendencies, variability, and skewness of real data distributions [87]. Analyzing the correlation patterns within the datasets helps determine if the synthetic data maintains the relationships observed between different variables in the real data. Correlation matrices and scatter plots can be useful tools for this analysis, providing insights into the strength and direction of the relationships that synthetic data is supposed to emulate [88]. This detailed analysis looks at specific features or attributes within the data to ensure that important characteristics are accurately reflected in the synthetic version. This might involve examining summary statistics, such as mean, median, range, and standard deviations for each feature across both datasets. Where possible, benchmarking against known ground-truth data is highly valuable. This process involves comparing synthetic data against a dataset whose accuracy and characteristics are well-established. This benchmarking provides a concrete measure of the synthetic data's accuracy and performance, offering a direct indication of its reliability for further research or application [89].

**Functional Analyses** Functional analyses are crucial for evaluating whether synthetic data accurately encapsulates the underlying biological knowledge and maintains coherent functional relationships as seen in real data [90]. This process involves assessing how well the synthetic data can replicate biological functions and interactions, ensuring that it is not just statistically similar to real data but also biologically meaningful [91]. In functional analysis, there are four main techniques used.

(a) *Biological Fidelity Check*: This involves verifying that the synthetic data retains essential biological characteristics and behaviors observed in real data. For instance, if the dataset pertains to gene expression, the analysis would confirm whether gene regulation and signaling pathways are realistically represented.
(b) *Representation of Specific Features*: In contexts like single-cell research, it is important to check whether specific cell types or states are accurately represented within the synthetic dataset. This might include evaluating whether the proportions and distributions of cell types match those in authentic biological samples, and if the synthetic data can reproduce the typical cellular responses and interactions.
(c) *Consistency of Functional Relationships*: Functional analyses also examine whether relationships and dependencies between variables (such as genes or proteins) are maintained. Techniques such as pathway analysis or network modeling can be used to see if synthetic data reflects the complex web of biological interactions with fidelity.
(d) *Simulation of Biological Processes*: Another aspect of functional analysis is simulating biological processes using synthetic data to observe if outcomes are consistent with those driven by real data. For example, using synthetic data to model disease progression or response to treatment and comparing the results with known clinical outcomes.

**Expert Involvement and Peer Review** Incorporating domain experts and conducting rigorous peer reviews are critical steps in the process of creating and validating synthetic data, especially when it is used for critical applications like target identification in biomedical research. This practice involves engaging with seasoned professionals who have extensive knowledge and experience in the relevant fields to scrutinize and critique the synthetic data, ensuring it meets high standards of accuracy and applicability. Engaging domain experts and peer reviewers is essential to ensure that synthetic data is not only scientifically sound but also relevant and robust enough for high-stakes applications like pharmaceutical development and personalized medicine. This scrutiny is particularly important in fields where the cost of errors can be high, both in terms of financial resources and human health. Moreover, expert involvement and peer review lend an added layer of transparency to the process, reinforcing the scientific integrity of the research. It helps build trust among the scientific community and the general public, especially when the outcomes influence health policies or clinical practices [84].

## 30.13 Experimental Validation of AI-Identified Targets in Clinical Diseases

Target validation using cell lines, organoids, or animal models in preclinical tests is crucial in reducing the project attrition rate and the cost of drug development. An increasing number of AI-identified targets are being successfully validated in vitro and in vivo.

### *30.13.1 ALS Treatment Research*

In recent studies focused on finding new treatments for amyotrophic lateral sclerosis (ALS), 28 potential therapeutic targets were identified using AI algorithms. These AI-proposed targets were subsequently tested in a Drosophila (fruit fly) model that mimics the neurodegenerative aspects of ALS, specifically to assess their impact on neurodegeneration in the fly's eye—a commonly used model for studying neural health and disease. The validation process in this model organism revealed eight novel targets that had not been previously reported in the context of ALS. When these targets were suppressed, there was a significant reduction in the neurodegeneration of the eye tissues in these flies. This outcome suggests that these targets play a crucial role in the progression of neurodegenerative processes and that modulating their activity could offer a new avenue for ALS treatment [63].

### *30.13.2 KANK1 in ALS*

In groundbreaking research, Zhang et al. utilized a machine learning-based approach to discover a novel gene, KANK1, associated with ALS. This identification marks a significant advancement in the understanding of the genetic factors contributing to ALS. Following the computational identification, the research team employed the CRISPR-Cas9 gene-editing technology to introduce mutations in KANK1 within human neuronal cells, thereby validating the neurotoxic effects of these mutations. The identification of KANK1 as a contributor to ALS invites further investigation into its biological functions and interactions with other genes and proteins. Researchers can explore how KANK1 mutations influence neuronal health and connectivity, potentially linking these effects to the symptoms experienced by ALS patients [92].

### 30.13.3 HDAC6 in Cardioprotection

Recent research has spotlighted the inhibition of HDAC6 (histone deacetylase 6) as a promising strategy for cardioprotection. This innovative approach was initially identified through deep learning analysis, which utilized complex algorithms to sift through vast amounts of data and pinpoint potential therapeutic targets for heart disease. Following this computational discovery, the efficacy of HDAC6 inhibition was empirically validated in a specialized mouse model, specifically a BAG3 cardiomyocyte-knockout model that mimics human dilated cardiomyopathy. Inhibiting HDAC6 could potentially enhance cardiac function by several mechanisms, including the upregulation of protective stress responses in heart muscle cells or improvement in cellular structural integrity. The study likely delved into these mechanisms, providing insights into how modifying HDAC6 activity could lead to improved outcomes in heart disease [93].

### 30.13.4 CDK20 in HCC

Recent advancements in the field of oncology research have highlighted CDK20 as a novel therapeutic target for hepatocellular carcinoma (HCC), the most common form of liver cancer. This discovery was facilitated by the application of deep learning-based methods, which analyzed vast datasets to identify potential targets that could be pivotal in the progression of HCC. Following the identification of CDK20 as a promising target, researchers used generative AI to design a potent small molecule inhibitor. This inhibitor was subsequently tested and demonstrated significant selective anti-proliferative activity against an HCC cell line [94].

### 30.13.5 ROR-γt in Multiple Sclerosis

In their pioneering work, Zeng et al. introduced deepDTnet, an advanced computational platform designed to identify molecular targets for existing drugs. This platform integrates data from 15 heterogeneous types, encompassing chemical properties, genomic data, phenotypic characteristics, and cellular networks. This multidimensional data approach allows for a comprehensive in silico exploration of potential drug–target interactions, enhancing the possibility of repurposing existing drugs for new therapeutic uses. The use of deepDTnet to identify new applications for existing drugs is a cost-effective strategy that can significantly shorten the drug development timeline. Drug repurposing not only saves time and resources but also offers a quicker path to treatment for diseases that currently have limited options [95].

# 30.14 Future Perspectives

## *30.14.1 AI in Target Discovery for Drug Development*

Target discovery stands as a foundational phase in the modern drug development process. Despite the vast potential within the human genome for druggable targets, only a small fraction has been effectively identified and tapped into for therapeutic development. This gap highlights the urgent need for more sophisticated and efficient methods for discovering these targets. AI is anticipated to play a pivotal role in addressing some of the most challenging diseases, including cancer, neurodegenerative disorders, and autoimmune conditions. These diseases are characterized by intricate molecular mechanisms that are often difficult to decipher using traditional research methods. Diseases like cancer, Alzheimer's, Parkinson's, and autoimmune disorders involve a myriad of molecular interactions and pathways that contribute to their progression. These complex biological systems require sophisticated tools to be thoroughly understood. AI excels in managing and analyzing large-scale biological data, enabling researchers to uncover patterns and interactions that might be missed through conventional methods.

## *30.14.2 AI-Driven Target Discovery*

By leveraging machine learning and other AI technologies, scientists can systematically analyze genetic, proteomic, and other omics data to identify novel therapeutic targets. AI can predict how different genes, proteins, and other molecules interact within the cellular environment, highlighting potential targets for intervention. This capability is particularly valuable for diseases with multifactorial etiologies, where pinpointing a single target can be challenging. AI's ability to integrate and interpret vast datasets from diverse sources allows it to identify new pathways involved in disease processes. For instance, in cancer research, AI can analyze tumor genetics and patient data to discover previously unknown mechanisms of tumor growth and resistance to therapy. Similarly, in neurodegenerative diseases, AI can help identify molecular changes that occur long before symptoms appear, offering insights into early intervention strategies. The identification of novel targets and pathways through AI-driven methods paves the way for developing more precise and effective treatments. These targeted therapies can potentially improve patient outcomes by focusing on the underlying causes of the disease rather than just alleviating symptoms. For example, targeted cancer therapies that inhibit specific proteins driving tumor growth have shown significant success and improved survival rates. Traditional research methods often involve labor-intensive and time-consuming processes. AI reduces these barriers by automating data analysis and generating hypotheses at a much faster rate. This efficiency allows researchers to explore a broader range of possibilities in a shorter time frame, leading to quicker advancements in understanding and treating complex diseases [96].

### 30.14.3 AI's Role in Identifying Synergistic Combinations

AI excels in integrating and analyzing vast and diverse datasets. This includes genomic data that provides insights into genetic variations and mutations, patient records that offer clinical histories and treatment responses, and synthetic lethality data that highlights potential vulnerabilities in cancer cells. By bringing together these data sources, AI can uncover complex interactions and dependencies that might be missed by traditional methods. Many diseases, particularly those that are chronic or multifactorial like cancer, diabetes, and neurodegenerative disorders, involve multiple molecular pathways. AI can help map these pathways and identify points where they intersect or interact. Understanding these intersections can reveal potential combinations of therapeutic targets that, when addressed together, can produce a synergistic effect, enhancing the overall efficacy of the treatment. Traditional monotherapies often fall short in treating complex diseases due to their inability to address the multifaceted nature of these conditions. AI can predict which combinations of drugs or targets will work synergistically to overcome these limitations. By modeling and simulating various therapeutic combinations, AI can identify pairs or groups of drugs that, when used together, enhance each other's effectiveness, reduce resistance, and minimize side effects. AI-driven analyses can also contribute to the development of personalized medicine. By taking into account an individual's genetic makeup, disease profile, and treatment history, AI can suggest tailored combinations of therapies that are most likely to be effective for that particular patient. This personalization can lead to better outcomes and a higher quality of life for patients. The traditional drug discovery process is often lengthy and expensive. AI can streamline this process by rapidly identifying and validating novel drug targets and combinations, reducing the time and cost associated with bringing new treatments to market. This acceleration is particularly critical in addressing emergent health challenges where time is of the essence. Complex and emergent health challenges, such as antibiotic-resistant infections, cancer, and emerging viral diseases, require innovative treatment strategies. AI can help by identifying not only new drug targets but also effective combinations of existing therapies that can be repurposed to tackle these challenges. This flexibility in finding solutions can be crucial in responding to public health crises.

## 30.15 Outstanding Challenges in AI-Driven Target Discovery

Despite the significant advancements in AI-driven target discovery, several critical challenges and questions remain to be addressed to ensure the responsible and effective deployment of AI in drug development.

### 30.15.1 Ethical Considerations

Ensuring the responsible and ethical use of AI in drug development is paramount. This involves addressing concerns related to the ethical implications of AI-driven decisions and ensuring that these decisions are made transparently and fairly. It is essential to consider the broader societal impacts of AI applications and to establish guidelines that prevent misuse and ensure equitable access to AI-driven healthcare advancements.

### 30.15.2 Data Privacy

Protecting sensitive patient data is crucial for maintaining public trust and complying with data protection regulations such as GDPR and HIPAA. AI systems often require access to large datasets that include personal health information. Ensuring that these data are anonymized, securely stored, and only used with informed consent is vital for ethical AI deployment. Implementing robust data security measures and maintaining transparency about data usage are key to gaining public and regulatory confidence.

### 30.15.3 Regulatory Frameworks

Developing and adhering to comprehensive regulatory frameworks that govern the use of AI in drug discovery is essential. These frameworks should ensure that AI-generated insights meet the stringent standards required for clinical application. Regulatory bodies need to establish clear guidelines for the validation, approval, and monitoring of AI tools used in drug development. This includes setting standards for data quality, algorithm transparency, and the reproducibility of AI-driven findings.

### *30.15.4 Integration with Existing Research Paradigms*

Integrating AI-driven insights with traditional research methodologies can be challenging. Ensuring that AI tools complement and enhance existing scientific processes, rather than replacing them, is important for the seamless integration of AI into drug discovery. Collaborative efforts between AI specialists and domain experts are necessary to bridge the gap between computational predictions and biological validation.

## 30.16 Conclusion

In conclusion, AI has emerged as a revolutionary force in the realm of target discovery and drug development, significantly altering the landscape of how we identify new drug targets and repurpose existing ones. The integration of AI technologies has streamlined and enhanced the drug discovery process, making it more efficient and precise. As AI technology continues to evolve, we can expect even greater contributions to this field. The collaborative efforts of researchers, combining expertise from AI, biology, chemistry, and medicine, will further drive these advancements. This synergy will not only speed up the development of safe and effective therapeutics but also broaden the scope of treatable conditions, addressing diseases that have so far been intractable. Looking ahead, AI is poised to become indispensable in the pharmaceutical industry, from early-stage research to clinical trials and beyond. Its ability to process vast amounts of data, predict outcomes, and uncover hidden patterns ensures that AI will play a crucial role in the future of medicine. By accelerating the pace of drug discovery and development, AI promises to significantly enhance human health and well-being, bringing new hope and improved quality of life to patients worldwide.

## References

1. Hinkson IV, Madej B, Stahlberg EA. Accelerating therapeutics for opportunities in medicine: a paradigm shift in drug discovery. Front Pharmacol. 2020;11:770.
2. Zhou Y, Zhang Y, Lian X, Li F, Wang C, Zhu F, et al. Therapeutic target database update 2022: facilitating drug discovery with enriched comparative data of targeted agents. Nucleic Acids Res. 2022;50(D1):D1398–D407.
3. Kana O, Brylinski M. Elucidating the druggability of the human proteome with eFindSite. J Comput Aided Mol Des. 2019;33(5):509–19.
4. Finan C, Gaulton A, Kruger FA, Lumbers RT, Shah T, Engmann J, et al. The druggable genome and support for target identification and validation in drug development. Sci Transl Med. 2017;9(383):eaag1166.
5. Sun D, Gao W, Hu H, Zhou S. Why 90% of clinical drug development fails and how to improve it? Acta Pharm Sin B. 2022;12(7):3049–62.

6. Pun FW, Ozerov IV, Zhavoronkov A. AI-powered therapeutic target discovery. Trends Pharmacol Sci. 2023;44(9):561–72.
7. Mamoshina P, Volosnikova M, Ozerov IV, Putin E, Skibina E, Cortese F, et al. Machine learning on human muscle transcriptomic data for biomarker discovery and tissue-specific drug target identification. Front Genet. 2018;9:242.
8. Zhavoronkov A, Li R, Ma C, Mamoshina P. Deep biomarkers of aging and longevity: from research to applications. Aging (Albany NY). 2019;11(22):10771–80.
9. Muslu O, Hoyt CT, Lacerda M, Hofmann-Apitius M, Frohlich H. GuiltyTargets: prioritization of novel therapeutic targets with network representation learning. IEEE/ACM Trans Comput Biol Bioinform. 2022;19(1):491–500.
10. Liu R, Wei L, Zhang P. A deep learning framework for drug repurposing via emulating clinical trials on real-world patient data. Nat Mach Intell. 2021;3(1):68–75.
11. Zhavoronkov A, Ivanenkov YA, Aliper A, Veselov MS, Aladinskiy VA, Aladinskaya AV, et al. Deep learning enables rapid identification of potent DDR1 kinase inhibitors. Nat Biotechnol. 2019;37(9):1038–40.
12. Ivanenkov YA, Polykovskiy D, Bezrukov D, Zagribelnyy B, Aladinskiy V, Kamya P, et al. Chemistry42: an AI-driven platform for molecular design and optimization. J Chem Inf Model. 2023;63(3):695–701.
13. Obrezanova O. Artificial intelligence for compound pharmacokinetics prediction. Curr Opin Struct Biol. 2023;79:102546.
14. Chen R, Liu X, Jin S, Lin J, Liu J. Machine learning for drug-target interaction prediction. Molecules. 2018;23(9):2208.
15. Ye Q, Hsieh CY, Yang Z, Kang Y, Chen J, Cao D, et al. A unified drug-target interaction prediction framework based on knowledge graph and recommendation system. Nat Commun. 2021;12(1):6775.
16. Kavalci E, Hartshorn A. Improving clinical trial design using interpretable machine learning based prediction of early trial termination. Sci Rep. 2023;13(1):121.
17. Zhou Y, Chen SJ. Advances in machine-learning approaches to RNA-targeted drug design. Artif Intell Chem. 2024;2(1):100053.
18. McCulloch WS, Pitts W. A logical calculus of the ideas immanent in nervous activity. 1943. Bull Math Biol. 1990;52(1–2):99–115. discussion 73-97
19. Stokes JM, Yang K, Swanson K, Jin W, Cubillos-Ruiz A, Donghia NM, et al. A deep learning approach to antibiotic discovery. Cell. 2020;180(4):688–702.e13.
20. Bagnolini G, Luu TB, Hargrove AE. Recognizing the power of machine learning and other computational methods to accelerate progress in small molecule targeting of RNA. RNA. 2023;29(4):473–88.
21. Cai Z, Zafferani M, Akande OM, Hargrove AE. Quantitative structure-activity relationship (QSAR) study predicts small-molecule binding to RNA structure. J Med Chem. 2022;65(10):7262–77.
22. Yazdani K, Jordan D, Yang M, Fullenkamp CR, Calabrese DR, Boer R, et al. Machine learning informs RNA-binding chemical space. Angew Chem Int Ed Engl. 2023;62(11):e202211358.
23. Tessaro F, Scapozza L. How 'Protein-Docking' translates into the new emerging field of docking small molecules to nucleic acids? Molecules. 2020;25(12):2749.
24. Chen L, Calin GA, Zhang S. Novel insights of structure-based modeling for RNA-targeted drug discovery. J Chem Inf Model. 2012;52(10):2741–53.
25. Philips A, Łach G, Bujnicki JM. Computational methods for prediction of RNA interactions with metal ions and small organic ligands. Methods Enzymol. 2015;553:261–85.
26. Stefaniak F, Chudyk EI, Bodkin M, Dawson WK, Bujnicki JM. Modeling of ribonucleic acid–ligand interactions. Wiley Interdiscip Rev Comput Mol Sci. 2015;5(6):425–39.
27. Luo J, Wei W, Waldispühl J, Moitessier N. Challenges and current status of computational methods for docking small molecules to nucleic acids. Eur J Med Chem. 2019;168:414–25.
28. Manigrasso J, Marcia M, De Vivo M. Computer-aided design of RNA-targeted small molecules: a growing need in drug discovery. Chem. 2021;7(11):2965–88.

29. Zhou Y, Jiang Y, Chen SJ. RNA-ligand molecular docking: advances and challenges. Wiley Interdiscip Rev Comput Mol Sci. 2022;12(3):e1571.
30. Feng Y, Yan Y, He J, Tao H, Wu Q, Huang SY. Docking and scoring for nucleic acid-ligand interactions: principles and current status. Drug Discov Today. 2022;27(3):838–47.
31. Bernetti M, Aguti R, Bosio S, Recanatini M, Masetti M, Cavalli A. Computational drug discovery under RNA times. QRB Discov. 2022;3:e22.
32. Eloranta S, Boman M. Predictive models for clinical decision making: deep dives in practical machine learning. J Intern Med. 2022;292(2):278–95.
33. Xu L, Yan Y, Huang X. Classical deep learning models. Deep learning in solar astronomy. Singapore: Springer Nature; 2022. p. 7–17.
34. Shen C, Ding J, Wang Z, Cao D, Ding X, Hou T. From machine learning to deep learning: advances in scoring functions for protein–ligand docking. Wiley Interdiscip Rev Comput Mol Sci. 2020;10(1):e1429.
35. Li H, Sze KH, Lu G, Ballester PJ. Machine-learning scoring functions for structure-based drug lead optimization. Wiley Interdiscip Rev Comput Mol Sci. 2020;10(5):e1465.
36. Li H, Sze KH, Lu G, Ballester PJ. Machine-learning scoring functions for structure-based virtual screening. Wiley Interdiscip Rev Comput Mol Sci. 2021;11(1):e1478.
37. Dhakal A, McKay C, Tanner JJ, Cheng J. Artificial intelligence in the prediction of protein–ligand interactions: recent advances and future directions. Brief Bioinform. 2022;23(1):bbab476.
38. Staszak M, Staszak K, Wieszczycka K, Bajek A, Roszkowski K, Tylkowski B. Machine learning in drug design: use of artificial intelligence to explore the chemical structure–biological activity relationship. Wiley Interdiscip Rev Comput Mol Sci. 2022;12(2):e1568.
39. Crampon K, Giorkallos A, Deldossi M, Baud S, Steffenel LA. Machine-learning methods for ligand–protein molecular docking. Drug Discov Today. 2022;27(1):151–64.
40. Yang C, Chen EA, Zhang Y. Protein–ligand docking in the machine-learning era. Molecules. 2022;27(14):4568.
41. Li Y, Liu Z, Li J, Han L, Liu J, Zhao Z, et al. Comparative assessment of scoring functions on an updated benchmark: 1. Compilation of the test set. J Chem Inf Model. 2014;54(6):1700–16.
42. Li Y, Han L, Liu Z, Wang R. Comparative assessment of scoring functions on an updated benchmark: 2. Evaluation methods and general results. J Chem Inf Model. 2014;54(6):1717–36.
43. Li Y, Su M, Liu Z, Li J, Liu J, Han L, et al. Assessing protein–ligand interaction scoring functions with the CASF-2013 benchmark. Nat Protoc. 2018;13(4):666–80.
44. Huang N, Shoichet BK, Irwin JJ. Benchmarking sets for molecular docking. J Med Chem. 2006;49(23):6789–801.
45. Mysinger MM, Carchia M, Irwin JJ, Shoichet BK. Directory of useful decoys, enhanced (DUD-E): better ligands and decoys for better benchmarking. J Med Chem. 2012;55(14):6582–94.
46. Su H, Peng Z, Yang J. Recognition of small molecule-RNA binding sites using RNA sequence and structure. Bioinformatics. 2021;37(1):36–42.
47. Boniecki MJ, Lach G, Dawson WK, Tomala K, Lukasz P, Soltysinski T, et al. SimRNA: a coarse-grained method for RNA folding simulations and 3D structure prediction. Nucleic Acids Res. 2016;44(7):e63.
48. Liu Y, Frank AT. Using selectively scaled molecular dynamics simulations to assess ligand poses in RNA aptamers. J Chem Theory Comput. 2022;18(9):5703–9.
49. Ganser LR, Kelly ML, Herschlag D, Al-Hashimi HM. The roles of structural dynamics in the cellular functions of RNAs. Nat Rev Mol Cell Biol. 2019;20(8):474–89.
50. Ganser LR, Chu CC, Bogerd HP, Kelly ML, Cullen BR, Al-Hashimi HM. Probing RNA conformational equilibria within the functional cellular context. Cell Rep. 2020;30(8):2472–80.e4.
51. Borkar AN, Bardaro MF Jr, Camilloni C, Aprile FA, Varani G, Vendruscolo M. Structure of a low-population binding intermediate in protein-RNA recognition. Proc Natl Acad Sci U S A. 2016;113(26):7171–6.

52. Stelzer AC, Frank AT, Kratz JD, Swanson MD, Gonzalez-Hernandez MJ, Lee J, et al. Discovery of selective bioactive small molecules by targeting an RNA dynamic ensemble. Nat Chem Biol. 2011;7(8):553–9.
53. Bannwarth S, Gatignol A. HIV-1 TAR RNA: the target of molecular interactions between the virus and its host. Curr HIV Res. 2005;3(1):61–71.
54. Abulwerdi FA, Le Grice SFJ. Recent advances in targeting the HIV-1 tat/TAR complex. Curr Pharm Des. 2017;23(28):4112–21.
55. Frank AT, Stelzer AC, Al-Hashimi HM, Andricioaei I. Constructing RNA dynamical ensembles by combining MD and motionally decoupled NMR RDCs: new insights into RNA dynamics and adaptive ligand recognition. Nucleic Acids Res. 2009;37(11):3670–9.
56. Lu J, Kadakkuzha BM, Zhao L, Fan M, Qi X, Xia T. Dynamic ensemble view of the conformational landscape of HIV-1 TAR RNA and allosteric recognition. Biochemistry. 2011;50(22):5042–57.
57. Kozlovskii I, Popov P. Structure-based deep learning for binding site detection in nucleic acid macromolecules. NAR Genom Bioinform. 2021;3(4):lqab111.
58. Jiang Z, Xiao SR, Liu R. Dissecting and predicting different types of binding sites in nucleic acids based on structural information. Brief Bioinform. 2022;23(1):bbab411.
59. Wang K, Zhou R, Wu Y, Li M. RLBind: a deep learning method to predict RNA-ligand binding sites. Brief Bioinform. 2023;24(1):bbac486.
60. Zeng P, Li J, Ma W, Cui Q. Rsite: a computational method to identify the functional sites of noncoding RNAs. Sci Rep. 2015;5:9179.
61. Zeng P, Cui Q. Rsite2: an efficient computational method to predict the functional sites of noncoding RNAs. Sci Rep. 2016;6:19016.
62. Wang K, Jian Y, Wang H, Zeng C, Zhao Y. RBind: computational network method to predict RNA binding sites. Bioinformatics. 2018;34(18):3131–6.
63. Pun FW, Liu BHM, Long X, Leung HW, Leung GHD, Mewborne QT, et al. Identification of therapeutic targets for amyotrophic lateral sclerosis using PandaOmics—an AI-enabled biological target discovery platform. Front Aging Neurosci. 2022;14:914017.
64. Fabris F, Palmer D, Salama KM, de Magalhães JP, Freitas AA. Using deep learning to associate human genes with age-related diseases. Bioinformatics. 2020;36(7):2202–8.
65. Zhao L, Chen X, Feng Y, Wang G, Nawaz I, Hu L, et al. COX7A1 suppresses the viability of human non-small cell lung cancer cells via regulating autophagy. Cancer Med. 2019;8(18):7762–73.
66. West MD, Labat I, Sternberg H, Larocca D, Nasonkin I, Chapman KB, et al. Use of deep neural network ensembles to identify embryonic-fetal transition markers: repression of COX7A1 in embryonic and cancer cells. Oncotarget. 2018;9(8):7796–811.
67. Cui H, Wang C, Maan H, Pang K, Luo F, Duan N, et al. scGPT: toward building a foundation model for single-cell multi-omics using generative AI. Nat Methods. 2024;21:1470.
68. Luo R, Sun L, Xia Y, Qin T, Zhang S, Poon H, et al. BioGPT: generative pre-trained transformer for biomedical text generation and mining. Brief Bioinform. 2022;23(6):bbac409.
69. Yang R, Tan TF, Lu W, Thirunavukarasu AJ, Ting DSW, Liu N. Large language models in health care: development, applications, and challenges. Health Care Sci. 2023;2(4):255–63.
70. Sahoo SS, Plasek JM, Xu H, Uzuner Ö, Cohen T, Yetisgen M, et al. Large language models for biomedicine: foundations, opportunities, challenges, and best practices. J Am Med Inform Assoc. 2024;31:2114.
71. Blakney AK. A tool for optimizing messenger RNA sequence. Nature. 2023;621(7978):262–4.
72. Samelson AJ, Tran QD, Robinot R, Carrau L, Rezelj VV, Kain AM, et al. BRD2 inhibition blocks SARS-CoV-2 infection by reducing transcription of the host cell receptor ACE2. Nat Cell Biol. 2022;24(1):24–34.
73. Ramkumar P, Abarientos AB, Tian R, Seyler M, Leong JT, Chen M, et al. CRISPR-based screens uncover determinants of immunotherapy response in multiple myeloma. Blood Adv. 2020;4(13):2899–911.
74. Chakraborty S, Hosen MI, Ahmed M, Shekhar HU. Onco-multi-OMICS approach: a new frontier in cancer research. Biomed Res Int. 2018;2018:9836256.

75. Nurk S, Koren S, Rhie A, Rautiainen M, Bzikadze AV, Mikheenko A, et al. The complete sequence of a human genome. Science. 2022;376(6588):44–53.
76. Kamya P, Ozerov IV, Pun FW, Tretina K, Fokina T, Chen S, et al. PandaOmics: an AI-driven platform for therapeutic target and biomarker discovery. J Chem Inf Model. 2024;64(10):3961–9.
77. Luo Y, Liu L, He Z, Zhang S, Huo P, Wang Z, et al. TREAT: therapeutic RNAs exploration inspired by artificial intelligence technology. Comput Struct Biotechnol J. 2022;20:5680–9.
78. Ochoa D, Hercules A, Carmona M, Suveges D, Gonzalez-Uriarte A, Malangone C, et al. Open targets platform: supporting systematic drug-target identification and prioritisation. Nucleic Acids Res. 2021;49(D1):D1302–D10.
79. Zhou G, Soufan O, Ewald J, Hancock REW, Basu N, Xia J. NetworkAnalyst 3.0: a visual analytics platform for comprehensive gene expression profiling and meta-analysis. Nucleic Acids Res. 2019;47(W1):W234–W41.
80. Bu D, Xia Y, Zhang J, Cao W, Huo P, Wang Z, et al. FangNet: mining herb hidden knowledge from TCM clinical effective formulas using structure network algorithm. Comput Struct Biotechnol J. 2021;19:62–71.
81. Ruiz C, Zitnik M, Leskovec J. Identification of disease treatment mechanisms through the multiscale interactome. Nat Commun. 2021;12(1):1796.
82. Shayakhmetov R, Kuznetsov M, Zhebrak A, Kadurin A, Nikolenko S, Aliper A, et al. Molecular generation for desired transcriptome changes with adversarial autoencoders. Front Pharmacol. 2020;11:269.
83. Viñas R, Andrés-Terré H, Liò P, Bryson K. Adversarial generation of gene expression data. Bioinformatics. 2022;38(3):730–7.
84. Beaulieu-Jones BK, Wu ZS, Williams C, Lee R, Bhavnani SP, Byrd JB, et al. Privacy-preserving generative deep neural networks support clinical data sharing. Circ Cardiovasc Qual Outcomes. 2019;12(7):e005122.
85. Song J, Xu Z, Cao L, Wang M, Hou Y, Li K. The discovery of new drug-target interactions for breast cancer treatment. Molecules. 2021;26(24):7474.
86. Chawla NV, Bowyer KW, Hall LO, Kegelmeyer WP. SMOTE: synthetic minority over-sampling technique. J Artif Intell Res. 2002;16:321–57.
87. Ghasemi A, Zahediasl S. Normality tests for statistical analysis: a guide for non-statisticians. Int J Endocrinol Metab. 2012;10(2):486–9.
88. Mishra P, Pandey CM, Singh U, Gupta A, Sahu C, Keshri A. Descriptive statistics and normality tests for statistical data. Ann Card Anaesth. 2019;22(1):67–72.
89. El Emam K, Mosquera L, Fang X, El-Hussuna A. Utility metrics for evaluating synthetic health data generation methods: validation study. JMIR Med Inform. 2022;10(4):e35734.
90. Bhat BA, Singh G, Sharma R, Yaseen M, Ganai NA. Biological networks: tools, methods, and analysis. In: Shaik NA, Hakeem KR, Banaganapalli B, Elango R, editors. Essentials of bioinformatics. Understanding bioinformatics: genes to proteins, vol. 1. Cham: Springer International; 2019. p. 255–86.
91. Treppner M, Salas-Bastos A, Hess M, Lenz S, Vogel T, Binder H. Synthetic single cell RNA sequencing data from small pilot studies using deep generative models. Sci Rep. 2021;11(1):9403.
92. Zhang S, Cooper-Knock J, Weimer AK, Shi M, Moll T, Marshall JNG, et al. Genome-wide identification of the genetic basis of amyotrophic lateral sclerosis. Neuron. 2022;110(6):992–1008.e11.
93. Yang J, Grafton F, Ranjbarvaziri S, Budan A, Farshidfar F, Cho M, et al. Phenotypic screening with deep learning identifies HDAC6 inhibitors as cardioprotective in a BAG3 mouse model of dilated cardiomyopathy. Sci Transl Med. 2022;14(652):eabl5654.
94. Ren F, Ding X, Zheng M, Korzinkin M, Cai X, Zhu W, et al. AlphaFold accelerates artificial intelligence powered drug discovery: efficient discovery of a novel CDK20 small molecule inhibitor. Chem Sci. 2023;14(6):1443–52.
95. Zeng X, Zhu S, Lu W, Liu Z, Huang J, Zhou Y, et al. Target identification among known drugs by deep learning from heterogeneous networks. Chem Sci. 2020;11(7):1775–97.
96. Ong E, Wong MU, Huffman A, He Y. COVID-19 coronavirus vaccine design using reverse vaccinology and machine learning. Front Immunol. 2020;11:1581.

MIX
Papier aus verantwortungsvollen Quellen
Paper from responsible sources
FSC® C105338

If you have any concerns about our products,
you can contact us on
**ProductSafety@springernature.com**

In case Publisher is established outside the EU,
the EU authorized representative is:
**Springer Nature Customer Service Center GmbH**
**Europaplatz 3, 69115 Heidelberg, Germany**

Printed by Libri Plureos GmbH
in Hamburg, Germany